Springer Undergraduate Texts in Philosophy

The Springer Undergraduate Texts in Philosophy offers a series of self-contained textbooks aimed towards the undergraduate level that covers all areas of philosophy ranging from classical philosophy to contemporary topics in the field. The texts will include teaching aids (such as exercises and summaries) and will be aimed mainly towards more advanced undergraduate students of philosophy.

The series publishes:

- All of the philosophical traditions
- Introduction books with a focus on including introduction books for specific topics such as logic, epistemology, German philosophy etc.
- Interdisciplinary introductions – where philosophy overlaps with other scientific or practical areas

This series covers textbooks for all undergraduate levels in philosophy particularly those interested in introductions to specific philosophy topics.

We aim to make a first decision within 1 month of submission. In case of a positive first decision the work will be provisionally contracted: the final decision about publication will depend upon the result of the anonymous peer review of the complete manuscript. We aim to have the complete work peer-reviewed within 3 months of submission.

Proposals should include:

- A short synopsis of the work or the introduction chapter
- The proposed Table of Contents
- CV of the lead author(s)
- List of courses for possible course adoption

The series discourages the submission of manuscripts that are below 65,000 words in length.

For inquiries and submissions of proposals, authors can contact Ties.Nijssen@Springer.com

More information about this series at http://www.springer.com/series/13798

Sven Ove Hansson • Vincent F. Hendricks
Editors

Esther Michelsen Kjeldahl
Assistant Editor

Introduction to Formal Philosophy

Springer

Editors
Sven Ove Hansson
Division of Philosophy
Royal Institute of Technology (KTH)
Stockholm, Sweden

Vincent F. Hendricks
Center for Information and Bubble Studies
University of Copenhagen
Copenhagen, Denmark

ISSN 2569-8737 ISSN 2569-8753 (electronic)
Springer Undergraduate Texts in Philosophy
ISBN 978-3-030-08454-7 ISBN 978-3-319-77434-3 (eBook)
https://doi.org/10.1007/978-3-319-77434-3

Printed on acid-free paper

This Springer imprint is published by the registered company Springer International Publishing AG part of Springer Nature.
The registered company address is: Gewerbestrasse 11, 6330 Cham, Switzerland

Preface

In 1974, a wonderful little book came out entitled *Formal Philosophy: Selected Papers of Richard Montague*, edited by Richmond H. Thomason. The book was a beautiful testimony to the fact that formal methods may indeed clarify, sharpen and solve philosophical problems, defusing airy philosophical intuitions in clear, crisp and concise ways while at the same time turning philosophical wonder into scientific inquiry.

Today, formal philosophy is a thoroughly interdisciplinary package. Methods from logic, mathematics, computer science, linguistics, physics, biology, economics, game theory, political theory, psychology, etc. all chip in and have their place in the methodological toolbox of formal philosophy. Thus, formal philosophy is not yet another puristic philosophical province but rather a discipline gaining its momentum and content from its close shaves with the methods of science in general.

Introduction to Formal Philosophy intends to present the formal philosophy landscape in all its splendour. In self-contained entries written by experts in the field, the book introduces the methods of formal philosophy and provides an overview over the major areas of philosophy in which formal methods play crucial roles. The presentations are comparatively non-technical in the sense that definitions and theorems are stated with standard formal rigour, but much emphasis is placed on clarifying the relationships between formal constructions and the informal notions that they represent. Proofs and derivations are normally not presented. The main focus is on showing how formal treatments of philosophical problems may help us understand them better, solve some of them and even present new philosophical problems that would never have seen the light of day without the use of a formal apparatus.

Introduction to Formal Philosophy has a pedagogical but also an unabashed propagandistic purpose. While in no way denigrating other methodologies, we hope to show the versatility, forcefulness and efficiency of treating philosophical problems with formal methods. Hopefully, this will serve to increase the self-consciousness of formal philosophy for the benefit of scientific inquiry in general.

We express our gratitude to Ties Nijssen and his colleagues at Springer for taking on this project. Henrik Boensvang, Rasmus K. Rendsvig and Esther Michelsen Kjeldahl have rendered efficient editorial assistance for which we are most grateful. Last but not least, we thank the contributing authors for all the time and efforts they have spent on this project.

Stockholm, Sweden
Copenhagen, Denmark
January 2018

Sven Ove Hansson
Vincent F. Hendricks

Contents

Part I The Scope and Methods of Formal Philosophy

1 Formalization 3
Sven Ove Hansson

Part II Reasoning and Inference

2 Argument 63
Henry Prakken

3 Formal Methods and the History of Philosophy 81
Catarina Dutilh Novaes

4 Nonmonotonic Reasoning 93
Alexander Bochman

5 Induction 105
Rafal Urbaniak and Diderik Batens

6 Conditionals 131
John Cantwell

7 Neural Network Models of Conditionals 147
Hannes Leitgeb

8 Proof Theory 177
Jeremy Avigad

9 Logics of (Formal and Informal) Provability 191
Rafal Urbaniak and Pawel Pawlowski

Part III Metaphysics and Philosophy of Language

10 Theory of Concepts 241
Erich Rast

11 **Categories** 251
Jean-Pierre Marquis

12 **Can Natural Language Be Captured in a Formal System?** 273
Martin Stokhof

13 **Reference and Denotation** 289
Robert van Rooij

14 **Indexicals** 297
Philippe Schlenker

15 **Necessity and Possibility** 323
Melvin Fitting

16 **Bivalence and Future Contingency** 333
Gabriel Sandu, Carlo Proietti, and François Rivenc

Part IV Epistemology

17 **Epistemic Logic and Epistemology** 351
Wesley H. Holliday

18 **Knowledge Representation for Philosophers** 371
Richmond H. Thomason

19 **Representing Uncertainty** 387
Sven Ove Hansson

20 **Belief Change** 401
Sven Ove Hansson

21 **Probability Theory** 417
Darrell P. Rowbottom

22 **Bayesian Epistemology** 431
Erik J. Olsson

23 **Coherence** 443
Sven Ove Hansson

Part V Philosophy of Science

24 **Computational Models in Science and Philosophy** 457
Paul Thagard

25 **Models of the Development of Scientific Theories** 469
Gerhard Schurz

26 **Space and Time** 487
John Byron Manchak

Part VI Value Theory and Moral Philosophy

27 **Formal Investigations of Value** 499
Sven Ove Hansson

28 **Value Theory (Axiology)** 523
Erik Carlson

29 **Preference and Choice** 535
Sven Ove Hansson

30 **Preference Change** 549
Fenrong Liu

31 **Money-Pumps** 567
Sven Ove Hansson

32 **Deontic Logic** 577
Sven Ove Hansson

33 **Action Theories** 591
Andreas Herzig, Emiliano Lorini, and Nicolas Troquard

Part VII Decision Theory and Social Philosophy

34 **Decision Theory: A Formal Philosophical Introduction** 611
Richard Bradley

35 **Dynamic Decision Theory** 657
Katie Steele

36 **Causal Decision Theory** 669
Brad Armendt

37 **Social Choice and Voting** 693
Prasanta K. Pattanaik

38 **Judgment Aggregation** 705
Philippe Mongin

39 **Logical Approaches to Law** 721
John Woods

About the Authors

Brad Armendt is Associate Professor of Philosophy at Arizona State University. His research centers on rational decision-making and the epistemology of rational belief. It includes work on decision theory, subjective probability, and belief updating. He is also interested in causal models and causal inference, models of social interaction such as evolutionary games, and the philosophical work of Frank Ramsey. E-mail: armendt@asu.edu.

Jeremy Avigad is Professor in the Department of Philosophy and the Department of Mathematical Sciences at Carnegie Mellon University. He has research interests in mathematical logic, proof theory, the philosophy of mathematics, formal verification, automated reasoning, and the history of mathematics. His work has been supported by the National Science Foundation, the Air Force Office of Scientific Research, the Andrew W. Mellon Foundation, and the John Templeton Foundation. He has held visiting research positions at Microsoft Research Redmond and the Microsoft Research/Inria Joint Centre in Saclay, France. He is currently involved in the development of Lean, a new system for interactive theorem proving. E-mail: avigad@cmu.edu.

Diderik Batens is Professor Emeritus at the Centre for Logic and Philosophy of Science, University of Ghent. He is a logician and epistemologist, and has a particular interest in adaptive and paraconsistent logics. In the philosophy of science, he promotes a fallabilist view that is informed by insights from paraconsistent logic. E-mail: Diderik.Batens@UGent.be.

Alexander Bochman received his PhD in philosophy from Tel-Aviv University. He is now an Associate Professor in Computer Science at the Holon Institute of Technology (HIT), Israel. His research focuses on nonmonotonic reasoning and its applications in belief revision, defeasible reasoning, logic programming, theories of action and change, formal argumentation theory, and causal reasoning. He has published numerous papers and two books, *A Logical Theory of Nonmonotonic Inference and Belief Change* (2001) and *Explanatory Nonmonotonic Reasoning* (2005). E-mail: bochmana@hit.ac.il.

Richard Bradley is Professor of Philosophy in the Department of Philosophy, Logic, and Scientific Method at the London School of Economics and Political Science, and an editor of *Economics and Philosophy*. His research is concentrated in decision theory, formal epistemology and ethics, and the theory of social choice, but he also works on the semantics of conditionals and hypothetical reasoning. His book *Decision Theory with a Human Face*, recently published with Cambridge University Press, gives an account of decision-making under conditions of severe uncertainty, suitable for rational but bounded agents. Email: r.bradley@lse.ac.uk.

John Cantwell is Professor in Philosophy at the Royal Institute of Technology, Stockholm. He is a member of the editorial board of *Theoria.* His research focuses on foundational issues in formal epistemology, semantics, and logic, with a particular emphasis on their interaction. He has well over 30 papers in refereed journals on conditionals, expressivism, epistemic modals, and belief revision. E-mail: cantwell@kth.se.

Erik Carlson is Professor in Practical Philosophy at Uppsala University. His areas of research include axiology, normative ethics, measurement theory, and the problems of free will and determinism. He has published one book and more than fifty papers in journals and anthologies. E-mail: Erik.Carlson@filosofi.uu.se.

Catarina Dutilh Novaes is Professor at the Department of Philosophy of the Vrije Universiteit, Amsterdam, as well as Editor-in-Chief of the journal *Synthese*. Her research over the years has focused on a number of different topics, in particular history of logic (especially Latin medieval logic) and philosophy of logic and mathematics. She is the author of *Formalizing Medieval Logical Theories* (Springer 2007), *Formal Languages in Logic* (CUP 2012), and co-editor (with Stephen Read) of the *Cambridge Companion to Medieval Logic* (CUP 2016). She has published a number of articles in journals such as *Philosophical Studies*, *Erkenntnis*, the *Journal of Philosophical Logic*, and the *Journal of the History of Philosophy*, among others. She led the NWO-funded project "The Roots of Deduction" (2011–2016) and will be leading the ERC-funded project "The Social Epistemology of Argumentation" (2018–2023). E-mail: c.dutilhnovaes@vu.nl.

Melvin Fitting was born in Troy, New York, in 1942. His undergraduate degree was from Rensselaer Polytechnic Institute in mathematics, and his 1968 PhD was supervised by Raymond Smullyan at Yeshiva University. His dissertation became his first book, *Intuitionistic Logic, Model Theory, and Forcing* (1969). He has worked in many areas, including intensional logic, semantics for logic programming, and theory of truth. Much of his work has involved developing tableau systems for nonclassical logics, thus generalizing the classical systems of his mentor Smullyan. In 2012, he received the Herbrand Award from the Conference on Automated

Deduction, largely for this work. He was at Lehman College of the City University of New York from 1969 to his retirement in 2013. He was also on the faculty of the City University Graduate Center, in the Departments of Mathematics, Computer Science, and Philosophy. He has authored or co-authored nine books as well as numerous research papers, covering philosophical logic, computability, automated theorem proving, and, with Raymond Smullyan, set theory. He is currently an emeritus professor and very much active. E-mail: melvin.fitting@gmail.com.

Sven Ove Hansson is Professor in Philosophy at the Royal Institute of Technology, Stockholm. He is Editor-in-Chief of *Theoria* and the book series *Outstanding Contributions to Logic*. He is also member of the editorial boards of the *Journal of Philosophical Logic*, *Studia Logica*, *Synthese*, and several other journals. His logical research focuses on belief revision, preference logic, and deontic logic. His other philosophical research includes contributions to decision theory, philosophy of science and technology, the philosophy of risk, and moral and political philosophy. He has published over 350 papers in refereed journals and books. His books include *A Textbook of Belief Dynamics: Theory Change and Database Updating* (1999), *The Structure of Values and Norms* (2001), *David Makinson on Classical Methods for Non-Classical Problems* (2014, edited), *Descriptor Revision* (2017), *Technology and Mathematics: Philosophical and Historical Investigations* (2018, edited), and *Belief Change: Introduction and Overview* (with Eduardo Fermé, 2018). E-mail: soh@kth.se.

Vincent F. Hendricks is Professor of Formal Philosophy at the University of Copenhagen. He is Director of the Center for Information and Bubble Studies (CIBS) sponsored by the Carlsberg Foundation and was awarded the Elite Research Prize by the Danish Ministry of Science, Technology, and Innovation and the Roskilde Festival Elite Research Prize both in 2008. He was Editor-in-Chief of *Synthese: An International Journal for Epistemology, Methodology and Philosophy of Science* between 2005 and 2015. He is the author of numerous books and articles on logic, epistemology, information theory, and democratic processes. Two of his most recent books are *Readings in Formal Epistemology* (edited with Horacio Arló-Costa and Johan van Benthem, 2016) and *Infostorms: Why Do We "Like"? Explaining Individual Behavior in the Social Net* (2nd edition, with Pelle G. Hansen, 2016). E-mail: vincent@hum.ku.dk.

Andreas Herzig is a CNRS (Centre National de la Recherche Scientifique) researcher at the Institut de Recherche en Informatique de Toulouse (IRIT) of the University of Toulouse. His main research topic is the investigation of logical models of interaction, with a focus on logics for reasoning about knowledge, belief, time, action, intention and obligation, and the development of theorem-proving methods for them. He investigates applications in belief-desire-intention logics, multi-agent planning, and argumentation theory. He co-authored an introductory book on modal logics and tableaux methods (2014, Springer-Birkhäuser), and co-edited a book on conditional logics (1995, Oxford University Press). He has supervised or co-supervised 20 PhD theses. He is Editor-in-Chief of the *Journal Applied Non-Classical Logics* since 2015. He is member of the editorial boards of *Artificial Intelligence Journal* and the *Journal of Philosophical Logic*. E-mail: Andreas.Herzig@irit.fr.

Wesley H. Holliday is Associate Professor of Philosophy and a faculty member of the Group in Logic and the Methodology of Science at the University of California, Berkeley. He currently serves as an editor of the *Review of Symbolic Logic.* His research in formal philosophy has focused on modal and intuitionistic logic, epistemic logic and epistemology, logic and natural language, and logic and probability. His dissertation on epistemic logic won the E. W. Beth Dissertation Prize from the Association for Logic, Language, and Information. E-mail: wesholliday@berkeley.edu.

Hannes Leitgeb is Chair of Logic and Philosophy of Language and Founder at the Munich Center for Mathematical Philosophy at Ludwig-Maximilians-University, Munich. He is Editor-in-Chief of *Erkenntnis* and member of the editorial boards of several other journals. His research interests are in logic and philosophy of language (truth, paradox, modality, conditionals, nonclassical logic), epistemology and general philosophy of science (belief, inference, Bayesianism, induction, empirical content), philosophy of mathematics (structuralism, informal provability, abstraction), philosophy of cognitive science (representation, metacognition), and the history of logical empiricism. In 2010, he received an Alexander von Humboldt Professorship Award from the Alexander von Humboldt Foundation. He has published numerous articles and books, including his recent monograph *The Stability of Belief: How Rational Belief Coheres with Probability* (Oxford University Press, 2017). E-mail: Hannes.Leitgeb@lmu.de.

Fenrong Liu is Changjiang Distinguished Professor of Logic at Tsinghua University, Beijing. She is the holder of the Amsterdam-China logic chair at the University of Amsterdam, and co-director of the Tsinghua–UvA Joint Research Centre for Logic. She works mainly in the field of logics for rational agency. Her research work includes studies of formally structured models of preference dynamics and logical modeling of different types of agents. She has published a number of papers and books on these topics, notably, *Reasoning About Preference Dynamics* (Springer 2011). Her recent interest lies in understanding the features of information flow and decision-making in social settings, and establishing more realistic models. In addition, she maintains active interests in ancient Chinese logic. She is currently editing a *Handbook of Logical Thought in China*. She is Editor-in-Chief of the new book series of the Studia Logica Library: *Logic in Asia*, editor of the *Australasian Journal of Logic*, associate editor of *Studia Logica* and *Studies in Logic*, and member of the editorial boards of *Synthese* and *Topoi*. E-mail: fenrong@tsinghua.edu.cn.

Emiliano Lorini is CNRS (Centre National de la Recherche Scientifique) researcher and co-head of the LILaC (Logique, Interaction, Langue, et Calcul) group at the Institut de Recherche en Informatique de Toulouse. His main interest is in the application of logic and game theory to modeling sociocognitive concepts and phenomena. He is member of the Institute for Advanced Study in Toulouse. He was awarded the CNRS Bronze Medal in 2014 for his early achievements in artificial intelligence. He has authored more than 130 articles in journals and international conferences and workshops in the fields of logic and AI. He was programme chair of 11th European Workshop on Multi-agent Systems (EUMAS 2013); the First European Conference on Social Intelligence (ECSI 2014); and the Second International Workshop on Norms, Actions, and Games (NAG 2016). He is member of the editorial board of *Topoi*. E-mail: Emiliano.Lorini@irit.fr.

John Byron Manchak is Professor of Logic and Philosophy of Science at the University of California, Irvine. For the most part, he thinks about space and time within the context of the general theory of relativity. E-mail: jmanchak@uci.edu.

Jean-Pierre Marquis is Professor of Philosophy at the University of Montreal, in Montreal, Canada. His research focuses on logic, philosophy, and the foundations of mathematics. His book *From a Geometrical Point of View: a Study in the History and Philosophy of Category Theory* was published by Springer in 2009. He has published numerous papers on category theory, categorical logic, algebraic topology, algebraic geometry, homotopy theory, and the foundations of mathematics. His article "Stairway to Heaven: The Abstract Method and Levels of Abstraction in Mathematics" was included in *The Best Writing in Mathematics 2017* (Princeton University Press, 2018). E-mail: Jean-Pierre.Marquis@umontreal.ca.

Philippe Mongin is a Research Professor at CNRS (Centre National de la Recherche Scientifique) and a Professor at the HEC School of Management (École des hautes études commerciales), Paris. Starting from philosophy of economics at large, he moved to the mathematical disciplines of economics, and more specifically, decision theory and social choice theory. His work in formal philosophy has dealt with epistemic logics for game theory, the decision-theoretic foundations of probability, and above all, the theories of collective preferences and judgments. His other philosophical work concerns methodological issues in the social sciences, especially the role that rationality assumptions and value judgments play there. Besides co-editing *Epistemic Logic and the Theory of Games and Decisions* (1997) and co-authoring *Utility Theory and Ethics* (1998), he has published in many international journals, e.g., *Journal of Economic Theory*, *American Economic Journal*, *Games and Economic Behavior*, *Management Science*, *Synthese*, *Erkenntnis*, *Journal of Philosophical Logic*, *Artificial Intelligence*, and *Economics and Philosophy*. E-mail: mongin@greg-hec.com.

Erik J. Olsson is Professor and Chair in Theoretical Philosophy at Lund University, Sweden. His areas of research include epistemology, philosophical logic, pragmatism, and, more recently, philosophy of the Internet. He is best known for his work on coherence and probability, and for his defense of the reliabilist theory of knowledge. He is associate editor and board member of several book series and journals, including *Theoria*. Olsson has published over 100 articles in refereed journals and collections. His books include *Belief Revision Meets Philosophy of Science* (ed. with S. Enqvist, Springer, 2011), *Knowledge and Inquiry: Essays on the Pragmatism of Isaac Levi* (ed., Cambridge University Press, 2006), *Against Coherence: Truth, Probability, and Justification* (Oxford University Press, 2005), *Logik in der Philosophie* (ed. with W. Spohn and P. Schroeder-Heister, Synchron Publishers, 2005), *Pragmatisch denken* (ed. with A. Fuhrmann, Ontos Verlag, 2004), and *The Epistemology of Keith Lehrer* (ed., Kluwer, 2003). E-mail: erik_j.olsson@fil.lu.se.

Prasanta K. Pattanaik is Professor Emeritus and Professor of the Graduate Division at the Department of Economics, University of California, Riverside. The main areas of his current research interest are welfare economics and the theory of social choice, decision theory, and the measurement of multidimensional well-being and deprivation of societies. He has published several books and papers in professional journals. His books include *Voting and Collective Choice* (1971), *Strategy and Group Choice* (1978), and *Essays on Individual Decision-Making and Social Welfare* (2009). He is a Fellow of the Econometric Society. E-mail: prasanta.pattanaik@ucr.edu.

Pawel Pawlowski is pursuing his PhD degree in philosophy at Ghent University, working with Rafal Urbaniak and Joke Meheus. His dissertation is focused on formal explications of the notion of informal mathematical provability. He is interested in formal representations of philosophical notions within the framework of first-order arithmetical theories. E-mail: pawel.pawlowski@ugent.be.

Henry Prakken is Lecturer in Artificial Intelligence at the Department of Information and Computing Sciences of Utrecht University, and Professor of legal informatics and legal argumentation at the Law Faculty of the University of Groningen. His main research interests concern computational models of argumentation and their application in multi-agent systems, legal reasoning, and other areas. He has published over 200 papers in journals and books. Prakken is a past president of the International Association for AI and Law (IAAIL), of the JURIX Foundation for Legal Knowledge-Based Systems, and of the steering committee of the COMMA conferences on Computational Models of Argument. He is on the editorial board of several journals, including *Artificial Intelligence* (from 2017 as an associate editor). E-mail: h.prakken@uu.nl.

Carlo Proietti is Researcher at the Department of Philosophy of the University of Lund (Sweden). His main areas of research are epistemology and philosophical logic. The specific focus of his current research is the application of techniques from abstract argumentation and multi-agent logical modelling to the analysis of group polarization. He has published several articles on international philosophy journals, including *Synthese*, *Journal of Philosophical Logic, Erkenntnis* and *History and*

Philosophy of Logic. He defended his PhD thesis in Philosophy in 2008 at the University of Paris I – Sorbonne (*The future contingents problem and the Fitch's paradox: A unified approach to two problems in modal logic*). E-mail: Carlo.proietti@fil.lu.se.

Erich Rast works as Senior Researcher at the IFILNOVA Institute of Philosophy of the New University of Lisbon under a postdoctoral research fellowship by the Portuguese Foundation for Science and Technology. He has published a book on indexicality and written articles on linguistic context-dependence, type-driven semantics for quantifier domain restriction, and abductive reasoning in higher-order logics. He has also worked on philosophical and formal aspects of value disagreement, value structure, philosophical aspects of de se attitudes, and the role of computationalism in the philosophy of mind. Email: erast@fcsh.unl.pt.

François Rivenc is Professor Emeritus of Philosophy at the University Paris1 Pantheon-Sorbonne. His logical research focuses on the history and philosophy of logic. His books include *Introduction à la logique* (1989), *Deux recherches sur l'universalisme logique* (1993), *Introduction à la logique pertinente* (2005), and *Entre logique et langage* (2009, with Gabriel Sandu). E-mail: francois.rivenc@orange.fr.

Darrell P. Rowbottom is currently Head and Professor of philosophy at Lingnan University, Hong Kong. He is an associate editor of the *Australasian Journal of Philosophy* and a coordinating editor of *Theory and Decision*. He has published over 70 items, including a textbook on the philosophy of probability (which is presently being translated into simplified Chinese and Japanese); a monograph on the contemporary relevance of Popper's philosophy; and numerous articles on social epistemology, scientific method, and scientific realism. He has also published on a more occasional basis in other areas, such as metaphysics, philosophy of mind, and logic. His most recent articles are "What Is (Dis)Agreement?" in *Philosophy and Phenomenological Research* and "Scientific Realism: What It Is, The Contemporary Debate, and New Directions" in *Synthese.* He has also recently completed the chapter on instrumentalism for *The Routledge Handbook of Scientific Realism.* E-mail: darrellrowbottom@ln.edu.hk.

Gabriel Sandu is Professor of Theoretical Philosophy at the University of Helsinki. He is a former research director at the Centre National de la Recherche Scientifique, Paris, and professor at Paris 1, Pantheon-Sorbonne. His research focuses on game-theoretical semantics, independence-friendly logic (IF logic), and, more broadly, philosophical logic. E-mail: gabriel.sandu@helsinki.fi.

Philippe Schlenker is Senior Researcher at the Institut Jean-Nicod (Centre National de la Recherche Scientifique, Paris) and Global Distinguished Professor at New York University. He was the managing editor of *Journal of Semantics* between 2009 and 2012 and currently heads its advisory board. His research has been devoted to formal semantics and pragmatics (indexicality, intensionality, anaphora, presupposition, supplements), philosophical logic (self-reference, liars, strengthened liars), and sign language semantics (with special reference to anaphora and iconicity). Some of his recent research pertains to the formal analysis of alarm calls in some non-human primates, gesture semantics, and music semantics. His research has been supported by grants from the NSF, the European Science Foundation (Euryi Award, 2007), and the European Research Council (Advanced Grant, 2013–2018). E-mail: philippe.schlenker@gmail.com.

Gerhard Schurz is Professor of Theoretical Philosophy at the Department of Philosophy, Heinrich Heine University, Düsseldorf, Germany, and Director of the Duesseldorf Center for Logic and Philosophy of Science. He is President of the German Association for Philosophy of Science since 2016, and member of the editorial boards of *Synthese*, *Erkenntnis*, *Episteme*, *Grazer Philosophische Studien*, and the *Journal for General Philosophy of Science*. He has published 7 books, edited 25 anthologies, and published more than 220 papers in refereed journals and books. His selected books are *Philosophy of Science: A Unified Account* (Routledge 2013) and *The Is-Ought Problem* (Kluwer 1997). His recent papers are "Probability, Approximate Truth, and Truthlikeness" (with G. Cevolani, *Australasian Journal of Philosophy* 2016), "Impossibility Theorems for Rational Belief" (*Noûs* 2017), and "No Free Lunch Theorem, Inductive Skepticism, and the Optimality of Meta-Induction" (*Philosophy of Science* 2017). E-mail: schurz@phil-fak.uni-duesseldorf.de.

Katie Steele is Associate Professor in Philosophy at the Australian National University. She is an associate editor of *Philosophy of Science*. Her research spans a number of topics concerning justified choice and inference under uncertainty. She is also interested in the science-policy interface, with a focus on climate policy. She has published a number of papers in top philosophy journals on these topics. E-mail: katie.steele@anu.edu.au.

Martin Stokhof is Professor in Philosophy of Language at the Institute for Logic, Language and Computation (ILLC) and the Department of Philosophy of the University of Amsterdam, and at the Department of Philosophy of Tsinghua University in Beijing. Together with Jeroen Groenendijk, he has published extensively on the semantics of questions, dynamic semantics, and other topics in formal semantics. He is co-author of the two-volume Gamut textbook *Logic, Language and Meaning* that has appeared in Dutch, English, Spanish, and Chinese. He has also published a textbook in Dutch on the philosophy of language, and written a monograph on ethics and ontology in Wittgenstein's early work, *World and Life as One* (Stanford University Press, 2002). His current research focuses on methodological issues in semantics and linguistic theory in general, and on topics related to Wittgenstein's philosophy. E-mail: M.J.B.Stokhof@uva.nl.

Paul Thagard is Distinguished Professor Emeritus of Philosophy at the University of Waterloo. He is a Fellow of the Royal Society of Canada, the Cognitive Science Society, and the Association for Psychological Science. His most recent books are *The Brain and the Meaning of Life* (Princeton University Press, 2010) and *The Cognitive Science of Science* (MIT Press, 2012). He is now completing a three book *Treatise on Mind and Society* to be published by Oxford University Press. E-mail: pthagard@uwaterloo.ca.

Richmond H. Thomason is Professor of Philosophy, Linguistics, and Computer Science at the University of Michigan. Thomason's central interests are in logic. He is particularly concerned with adapting logical theories for applications beyond the purely mathematical sciences, and especially in linguistics and artificial intelligence. E-mail: rthomaso@umich.edu.

Nicolas Troquard is a researcher at the Free University of Bozen-Bolzano, Italy. He is a computer scientist whose research is concerned with logic, artificial intelligence, multiagent systems, ontologies, games, social choice, and formal philosophy. He received a joint PhD in 2007, in information technologies from the University of Trento and in artificial intelligence from the University of Toulouse. He was the principal investigator of the Marie Curie FP7-Trentino project "Logical Analysis of Socio-Technical Systems" that was carried out between 2011 and 2014 at the Laboratory for Applied Ontology (ISTC-CNR), Trento. He has also been a research associate at the Department of Computer Science at the University of Liverpool (2007–2010), and an assistant professor in computer science at the Université Paris-Est Créteil (2014–2016). E-mail: nicolas.troquard@unibz.it.

Rafal Urbaniak completed his PhD in the logic and philosophy of mathematics at the University of Calgary (working with Prof. R. Zach) in 2008, focusing on the development of Lesniewski's foundations of mathematics. Currently, he is a Postdoctoral Fellow of the Research Foundation Flanders at the Centre for Logic and Philosophy of Science at Ghent University in Belgium, and an Associate Professor at the Department of Philosophy, Sociology, and Journalism at the University of Gdansk, Poland. His major interest is the application of formal methods to philosophical problems, such as theories of rationality, belief revision, philosophy of mathematics, philosophy of mind, theories of conditionals, and philosophy of thought experiments. Currently, his main research project pertains to the use of probabilistic methods in juridical fact-finding. More details and papers are available at http://ugent.academia.edu/RafalUrbaniak. E-mail: rafal.urbaniak@ugent.be.

Robert van Rooij is Professor of Logic and Cognition at the Institute of Logic, Language, and Computation (ILLC) at the University of Amsterdam. He has published numerous articles on the semantics and pragmatics of natural language, philosophy of language, and philosophical logic in international journals and books. Robert van Rooij is well known for his work on, among other topics, propositional attitudes, dynamic semantics, vagueness, conversational implicatures, and game theoretical semantics. The tools he uses include non-monotonic and many-valued logic, decision theory, and game theory. He has been an associate editor of the *Journal of Semantics* and is now associate editor of the *Review of Symbolic Logic*. He is also member of the editorial boards of various other journals. E-mail: R.A.M.vanRooij@uva.nl.

John Woods has done foundational work on fallacies, fiction, and Aristotle's logic. He has explored prospects for naturalizing the logics of human inference and applying them to legal reasoning in criminal trials. Recent books include *Errors of Reasoning: Naturalizing the Logic of Inference* (2014); *Aristotle's Earlier Logic,* 2nd edition (2014); and *Is Legal Reasoning Irrational?* (2015). With Dov Gabbay, he is editor of the eleven-volume *Handbook of the History of Logic.* With Gabbay and Paul Thagard, he is general editor of the sixteen-volume *Handbook of the Philosophy of Science.* Formerly president of the Academy of Humanities and Social Sciences of the Royal Society of Canada, and president of the Canadian Federation of Humanities, he is currently director of the Group on Abductive Systems at the University of British Columbia. Woods is a life member of the Society of Fellows of the Netherlands Institute for Advanced Study. His *Truth in Fiction: Rethinking its Logic* is currently in production for the *Synthese Library.* Email: john.woods@ubc.ca.

Part I
The Scope and Methods of Formal Philosophy

Chapter 1
Formalization

Sven Ove Hansson

Abstract This introduction to formal philosophy has its focus on the basic methodology of formalization: the selection of concepts for formalization, appropriate splittings and merges of concepts to be formalized, the idealization that is necessary prior to formalization, the identification of variables and their domains, and the construction of a formal language. Other topics covered in this chapter are the advantages and pitfalls of formal philosophy, the relationships between formal models and that which they represent, and the use of non-logical models in philosophy.

1.1 Introduction

Few issues in philosophical style and methodology are as controversial among philosophers as formalization. Some philosophers are anti-formalists who consider texts making use of logical or mathematical notation as non-philosophical and not worth reading. Others are pan-formalists who consider non-formal treatments as – at best – useful preparations for the real work to be done in a formal language. But discussions on the pros and cons of formalization are more common at the coffee tables of philosophy departments than in scholarly books and journal articles. That is unfortunate since formalization has important methodological issues in need of systematic treatment.

This chapter is devoted to the use of formal methods in philosophy. It has a (non-exclusive) emphasis on logic which is the most commonly used formal language in philosophical investigations. We will have a close look at what formal logic is (Sect. 1.2) and how it can contribute to philosophical clarification (Sect. 1.3), the process that takes us from natural to logical language (Sect. 1.4), the construction of a logical language (Sect. 1.5), some philosophical uses of logical inference

S. O. Hansson (✉)
Division of Philosophy, Royal Institute of Technology (KTH), Stockholm, Sweden
e-mail: soh@kth.se

S. O. Hansson, V. F. Hendricks (eds.), *Introduction to Formal Philosophy*, Springer Undergraduate Texts in Philosophy, https://doi.org/10.1007/978-3-319-77434-3_1

(Sect. 1.6), and the philosophical use of non-logical formal models (Sect. 1.7). Finally, we will summarize some of the dangers and difficulties in the philosophical use of formal methods (Sect. 1.8).

1.2 Formal Logic as a Tool for Philosophy

Formal philosophy began with logic, and logic is still its dominating formal language. A good case can be made for increased use of non-logical formal methods, but in a general exposition of formal philosophy, logic is still the best starting-point.

1.2.1 The Origins of Logic

Logic is concerned with how we draw conclusions. Its systematic study begins with the observation that some inferences fall into general patterns. These patterns are characterized by being insensitive to the meaning of certain elements of that which we say or think, and even unaffected by the uniform substitution of these elements. Following Gottfried Wilhelm Leibniz (1646–1716), we can use the term "formal arguments" for arguments in which "the form of reasoning has been demonstrated in advance so that one is sure of not going wrong with it" [49, p. 479].[1] Consider the following argument:

Rich men are condescending.
Therefore: Non-condescending men are not rich.

The changeable elements here are of course "rich men" and "condescending men". We will call them variables. The example exhibits three important features of variables in a logical argument. First, the validity of an argument is unaffected by vagueness in its variables. In most other contexts, the use of vague terms makes it difficult to determine whether that which is said is valid or not. Thus, the sentence "He is rich" is vague because the term "rich" is vague, and for a similar reason so is the sentence "He has condescending manners". But the inclusion of both these vague terms into the above argument does not affect its validity.

Secondly, the validity of an argument does not depend on whether that which is said about the variables is true or false. Suppose that you meet the richest man in the world and he turns out to be a friendly and respectful person. Then the premise of the argument is not true, but the argument is still valid, i.e. it is still true that the conclusion follows from the premise.

---

[1] "... des argumens en forme; parce que leur forme de raisonner a esté prédemontrée, en sorte qu'on est seur de ne s'y point tromper" [48, 478–479].

Thirdly, we can freely substitute the variables for something else, if we do so uniformly. By uniformity is meant that all instances of a variable are substituted by the same new element. We can for instance make the following substitution in the above argument:

Baroque music is beautiful.
Therefore: Non-beautiful music is not Baroque music.

We know that this argument is valid since the previous one is valid. They are instances of the same argument form. When analyzing an inference, it is useful to express it in such an argument form.

The above examples represent an argument form with one premise and one conclusion. In Aristotle's (384–322 BCE) logic, such arguments are called conversion rules. Aristotle referred for instance to the argument form exemplified by the conclusion from "No pleasure is good" to "No good is a pleasure" ([2], I:ii, 25a). However, the major focus in Aristotelian logic was on arguments with two premises and one conclusion, called syllogisms. The following is an example of a syllogism:

All logicians are philosophers.
Some logicians are cacographers.
Therefore: Some philosophers are cacographers.

The validity of this syllogism is not disturbed by the vagueness of the terms "logician" and "philosopher". Even more importantly, to confirm the validity of this argument one need not know what a cacographer is – or for that matter what a philosopher or a logician is.

Archimedes (c.287-c.212 BCE) is reported to have said: "Give me a place to stand on, and I will move the Earth" [16]. For a lever to work properly, we need a rigid and reliable pivot. Similarly in logic, in order for some terms, namely the variables, to be flexible in meaning and indeed exchangeable, we need other terms that provide a rigid and immutable platform on which the movements and exchanges of variables can take place. The terms that have this function are called *logical constants*. In the above examples, "all", "some", and "not" have the role of logical constants. Syllogistic logic, which held sway from Aristotle's time until the late nineteenth century, was devoted to these three logical constants and the argument forms that could be constructed with them. But there were also three parallel traditions in logic that employed other logical constants.

One of these was sentential logic, the logic of sentences, first developed by Chrysippus (c.279-c.206 BCE) and other Stoics. In sentential logic, the variables are sentences or propositions, rather than parts of sentences as in syllogistic logic. Chrysippus accurately identified a proposition as "that which is capable of being denied or affirmed as it is in itself" [21, pp. 69–70]. The logical constants are words like "and", "or", "if", and "not". An argument in sentential logic can be as follows:

Either I laugh or you cry.
I do not laugh.
Therefore: You cry.

Here, the variables are the sentences “I laugh” and “You cry”, and the logical constants are “not” and “and”.[2] Sentential logic lived a marginal existence in the shadow of syllogistic logic but gained in importance through the work of George Boole (1815–1864) and others in the nineteenth century.

The second of these traditions was modal logic, the logic of necessity, possibility, and related concepts. Its two most important logical constants are “necessarily” and “possibly”. The oldest texts on modal logic are by Aristotle himself. Just like sentential logic, modal logic was overshadowed by standard syllogistic logic. It was revived in the early twentieth century by C.I. Lewis (1883–1964).

The third tradition is somewhat more difficult to pinpoint. It has its origin in what Aristotle called the *topoi*, or topics. These were valid arguments in which the role of logical constants was played by a wider range of concepts. These include “good”, “better”, and “child”. Studies of the topics continued through the ages, although usually with somewhat less precision than in the dominant logical pursuit, namely syllogistic logic [23]. The importance of such argumentation was emphasized by Leibniz when he wrote:

> “It should also be realized that there are *valid non-syllogistic inferences* which cannot be rigorously demonstrated in any syllogism unless the terms are changed a little, and this altering of the terms is the non-syllogistic inference. There are several of these, including arguments from the direct to the oblique – e.g. ‘If Jesus Christ is God, then the mother of Jesus Christ is the mother of God’. And again, the argument-form which some good logicians have called relation-conversion, as illustrated by the inference: ‘If David is the father of Solomon, then certainly Solomon is the son of David.” ([48], p. 479; translation from [49], p. 479)

1.2.2 The “Newtonian” Revolution in Logic

These traditions in logic – studies of syllogisms as well as the other, subsidiary subject areas – had one important limitation in common: They were devoted to single argumentative steps. Actual argumentation usually proceeds by a whole series of steps. This restriction to single steps, taken one at time, turns out to be a serious limitation since some arguments cannot be fully understood unless one takes a more comprehensive approach. Clear examples of this can be found in mathematical reasoning. In his *Elements*, Euclid (fl.300 BCE) often introduced an assumption only in order to refute it. After making the assumption he presented a multi-step argument based on it. Many steps later he arrived at an inconsistent conclusion, based on which he inferred that the assumption was false (“reductio ad absurdum”, reduction to absurdity) This is a type of argumentation that logicians had great difficulties in accounting for since they dealt with each step separately [45, p. 597].

In the middle of the nineteenth century, logic was still a particularistic discipline, dealing with small argumentative steps in isolation, and lacking a unifying theory

[2]This is the argument form later known as Modus tollendo ponens or the disjunctive syllogism [5].

for the various types of argumentative steps. We can compare its status to that of mechanics two hundred years earlier. Before Isaac Newton's (1642–1727) *Principia* (1687), there were two branches of mechanics: terrestrial mechanics that dealt with the movements of objects on earth and celestial mechanics that dealt with the movements of heavenly bodies. Newton managed to unite the two disciplines by providing a mathematical model that was sufficiently general to cover the movements of both earthly and heavenly bodies. His new framework covered not only single events, but also complex interactions among a large number of objects, such as the bodies of the solar system.

In 1879 Gottlob Frege (1848–1925) published his *Begriffsschrift* which did to logic what the *Principia* had done to mechanics [19]. Frege's major invention was a notation (quantifiers) that could express the logical constants "all" and "some" in a much more versatile manner, and made them easily combinable with sentential constants such as "and" and "or". His new framework was a general logical calculus lacking the limitation to small steps that was inherent in the Aristotelian system of syllogisms. Instead of considering just two premises it was now possible to consider any set of premises, however large. This made it possible to ask questions that did not even arise in the logic of syllogisms. For any given a set of premises, one could ask whether a particular conclusion follows from it. Sometimes that question could be answered affirmatively by providing a step-by-step proof. In other cases it could be answered negatively by showing that no combination of valid argumentative steps can lead to the conclusion. With Frege, logic took the giant leap from an atomistic study of the smallest parts of arguments to a holistic analysis of what can and cannot be inferred from given premises.

Frege's system was limited to the logical constants that had been studied for more than two millennia in syllogistic and sentential logic: "all", "some", "not", "and", "or", "if", and "if and only if". Including them all in one and the same system was a major achievement, not least since arguments using these logical constants cover a large part of mathematical reasoning. But for philosophy this was still not enough. In philosophical argumentation the structural properties of other terms than these have crucial roles. For instance, if we wish to scrutinize Kant's views on whether ought implies can, then we do not have much use for the logical principles governing words like "all" or "and". Instead, our focus will have to be on properties of the concepts expressed by the words "ought" and "can" [75]. In the twentieth and twenty-first centuries, philosophical logicians fully realized this, and developed logical systems in which the role of logical constants is played by terms representing a wide variety of notions such as "necessary", "possible", "know", "believe", "do", "try", "after", "permit", "decide", "will", "right", "good", "blameworthy", "duty", "better", "cause", "freedom", "vague", and a wealth of others. Many of these had been studied by logicians in previous centuries, as part of the modal or the topics tradition. However, after Frege they could be included in holistic systems of argumentation, rather than being used in rules referring to a single, isolated step of reasoning. Through all these extensions, formal logic has expanded its territory most substantially, and this expansion is still an on-going process. We can see it as the second step of the "Newtonian" revolution in logic, after the first step for which Gottlob Frege was himself responsible.

1.2.3 The Actual Truth or a Model of the Truth?

The remarkable achievements of Frege's system of logic inspired many philosophers, and some believed that logical analysis could now replace other, more uncertain methods used by philosophers. Bertrand Russell (1872–1970) maintained that "every philosophical problem, when it is subjected to the necessary analysis and purification, is found either to be not really philosophical at all, or else to be, in the sense in which we are using the word, logical" [71, p. 14]. He and many others believed that logic would make it possible to reach a more fundamental level of philosophical insight, thereby resolving philosophical problems that could not be solved in natural language due to its lack of precision.

It was soon discovered, however, that philosophers can disagree about a problem expressed in logical terms just as they can disagree about one expressed in natural language. Russell's own analysis of definite descriptions provides a clear example of this. By a definite description is meant one that applies to exactly one object. In English, definite descriptions are often expressed with the definite article "the" followed by a singular: "the teapot on the lowest shelf", "the current president of South Africa", etc. The problematic cases are those in which there is either no object or more than one object answering to the description. If I ask you to take out the teapot on the lowest shelf, you will have problems in following the instruction if there is either no teapot or two or more teapots on that shelf. The following standard example has been used in the discussion:

The king of France is wise.

According to Russell [69], this should be interpreted as follows in predicate logic, with K standing for "is the king of France" and W for "is wise":

$$(\exists x)(Kx \;\&\; (\forall y)(Ky \rightarrow x = y) \;\&\; Wx)$$

This can be paraphrased as follows: "There is ($\exists$) someone (x) who is king of France (K). Everyone ($\forall y$) who is king of France is identical to him. He is wise (W)." It follows directly from this analysis that (as long as France remains a republic) the quoted sentence is false.

In a criticism of Russell's account, P.F. Strawson (1919–2006) contended that if someone uttered the sentence "The king of France is wise", then the question whether that sentence was true or false "simply didn't arise, because there was no such person as the king of France" [76, p. 330]. In Strawson's view, our sentence can be formalized in the simple way

$$Wk$$

where W denotes "is wise" and k denotes "the king of France". According to Strawson, this sentence expressed a true statement when uttered in the reign of Louis XIV, and a false statement when pronounced in the reign of Louis XV. But when asserted during the time of a French republic it expresses no statement at all, and consequently the question whether it expresses a true or a false statement does not even arise. Russell [70] disagreed and defended his original standpoint. The debate has continued since then [17].

This and many other examples show that merely translating a philosophical problem into logical language cannot be expected to solve it. Philosophical dispute can continue, now referring to the logical formulation. What logic can do, however, is to provide more precise statements of the problem and of alternative standpoints pertaining to its solution or dissolution. This, as we will see, can be an important enough achievement.

1.2.4 A Guarded Defence of Formalization

In a larger perspective, the rise of modern symbolic logic can be seen as part of a more general, long-term, trend: More and more scientific and scholarly disciplines have become dependant on mathematical modelling. Astronomy is the only empirical branch of learning that has been thoroughly mathematized ever since antiquity. Physics became gradually more and more mathematized from the late Middle Ages onwards, and chemistry since the late eighteenth century. But the great rush came in the twentieth century, when discipline after discipline adopted mathematical methods. One of the best examples is economics, which has gone from almost no use of mathematics to being dominated by theories expressed in mathematical language [14]. In the last few decades, formal models, in particular game theory, have had a strong and increasing influence throughout the social sciences. At the same time, the mathematization of the natural sciences has accelerated. Today, large parts of biology and the earth sciences, such as ecology, population genetics, and climatology, are thoroughly mathematized.

The reason why mathematical tools were adopted in these and many other areas is of course that they have proven efficient; they have improved the predictive and explanatory capacities of the disciplines in question. The increased role of formal methods in philosophy has a similar explanation: we have introduced formal tools in order to express problems more precisely and obtain solutions in new ways. But there is a caveat: The usefulness of formal tools is not quite as overwhelming in philosophy as in the empirical disciplines. The difference can be seen from a comparison between philosophy and early physics.

We usually think of mathematical physics as beginning with Galilei Galileo (1564–1642), but mathematical methods were used in physics already in the fourteenth century. When medieval physicists (the so-called calculatores) developed mathematical models of physical phenomena, they proceeded in much the same way as Euclidean geometers. A geometer used "pure thought" to determine the laws that govern lines, surfaces, and three-dimensional bodies. In much the same way, physicists used their intuition when attempting to find the laws that govern the movement of bodies. And importantly, intuition had a double role: Not only was the development of these mathematical models guided by intuition, it was also against intuition that they were tested. This was before the great revolution in physics led by Galileo. Although Galileo used his intuition as a starting-point when developing mathematical models of physical phenomena, he went on to test these models against experiments and exact observations. Since our mechanical intuitions are rather consistently wrong, this reality check was necessary to correct errors in the previous models [54, 74, 83].

Today, this is the standard approach to mathematical models in the empirical sciences. Mathematical models are tested against measurements whose values are expected to correspond to the variables of these models. Obviously, this can only be done if accurate measurement methods are in place. Before the thermometer was invented (in the seventeenth century), physicists had no better means to assess theories about heat than to compare them with everyday experiences of heat and cold. Exact measurement of temperature was a necessary condition for developing accurate mathematical theories of heat (thermodynamics). Today, no physicist would argue in favour of a thermodynamic principle by referring to our vague everyday experiences of heat and cold.

This is a general pattern in science. Measurement is our bridge between theories and observations. Mathematics is the medium in which we can transport information across that bridge, a medium unsurpassed in its information-carrying capacity. Today the bridge of measurement is quite crowded, carrying loads of information back and forth that are used on one side for the improvement of theories, and on the other side for the construction of new experiments and observations.

As we saw, physics had access to the mathematical medium long before it learnt how to avail itself of the bridge. Unfortunately, philosophy is in a situation comparable to that of pre-Galilean physics: we have the mathematical medium, but we do not have the bridge of measurement. And this is not a deficiency that can easily be mended within the confines of philosophy as we conceive it today. Philosophers studying concepts such as knowledge, truth, goodness, and permission are operating with constructs of the human mind that do not necessarily have exact empirical correlates. Our situation can to some extent be compared to that of mathematicians, who have all of their foundations on the theoretical side. Their research can improve the theories that are used in empirical work, but the information received back from empirical investigations does not normally lead to corrections of the mathematics. Similarly, philosophy can sometimes be used to improve theories in other disciplines, and the exactness of formal philosophy is often needed to match the precision required in these disciplines. But at least in most philosophical subject areas, empirical observations cannot support or disprove a theoretical statement in the same clear-cut way as in the empirical sciences.

Therefore, the claims that can be made for formalization are weaker in philosophy than in the natural and social sciences. In philosophy, the major virtue of formalization is the same as that of idealization in informal languages: By isolating important aspects it helps to bring them to light. In philosophical discussions we usually deviate from the general-language meanings of key terms such as "knowledge" or "value", giving them meanings that are more streamlined and more accessible to exact definition. This does not necessarily mean that we have access to a true philosophical meaning that these concepts should be adjusted to. A much more credible justification is that such simplifications are necessary in order to obtain the precision needed for philosophical analysis. However, this is a sail between Scylla and Charybdis (on the bridgeless waters just referred to). We have to deviate from general language in order to make a sufficiently precise analysis. But if we deviate so far as to lose contact with general-language meanings, then the rationale for the whole undertaking may well be lost. This precarious situation applies, of course, to formal and informal philosophy alike.

All this boils down to a rather guarded defence of formalization in philosophy. It is a language in which we can build more precise models of philosophical subject matter, and as we will see, there are philosophical topics for which this increased precision is indispensable. However, formalization is no panacea. Mistaken ideas can be as easily formalized as valid ones. But although formalization is no safe road to philosophical truth, it is one of the best tools that we have for expressing, criticizing, and improving philosophical standpoints. It is an obvious but important corollary of this line of defence that we should not expect to find a uniquely "correct" formal analysis of philosophical subject matter. Different formalizations may capture different properties of our concepts [33, 38, 39].

1.3 Formalization as Clarification

The use of formalization in philosophy is part of our general strivings for clarity and precision in philosophical discussions. In this sense, formalization is continuous with the development of specialized (non-formal) philosophical language. Since antiquity, philosophers have spent much effort on clarifying the central concepts of the discussions they have taken part in, and almost invariably such clarifications have led to new distinctions and opened up for the formulation of new standpoints and new questions. We find such linguistic analysis in Plato's Socratic dialogues, for instance the discussions on virtue in *Meno* and knowledge in *Theaetetus*. We also find it in ancient texts from other civilizations, for instance in writings in the Mohist tradition in China that in many ways anticipated modern developments in the philosophy of language [52, 53].

1.3.1 The Need for Clarity

Clarity is still a major criterion of philosophical quality. We need precise concepts in order to develop and criticize philosophical arguments, and therefore careful analysis and development of our own terminology is an essential part of modern philosophy. This type of work is also an important part of philosophy's contributions to other disciplines. In interdisciplinary co-operations, it is often the role of philosophers to work out precise definitions and distinctions [34]. The importance of precision has been pointed out by many of the great philosophers, for instance by

Aristotle and (arguably with some amount of rhetorical exaggeration) by Ludwig Wittgenstein (1889–1951):

> "Our discussion will be adequate if it has as much clearness as the subject-matter admits of, for precision is not to be sought for alike in all discussions, any more than in all the products of the crafts." (Aristotle, Nicomachean Ethics I:iii, 1094b [3])

> "Everything that can be said can be said clearly." (Wittgenstein, Tractatus logico-philosophicus 4.116 [85])

So why should we strive for clarity and exactness? To begin with, we do so in order to facilitate communication. In everyday life we appreciate exactness whenever information is important for us. When listening to my stories about what I have seen in the streets of Berlin you probably do not worry much about how accurately I describe the geographical relations between the different streets, but if I give you directions to your hotel you will expect me to be quite precise about such details. If someone tells you about the medicine her aunt took against arthritis you may even prefer not to hear all the details about dosage and the like, but if your doctor recommends you to take a drug you want her to be very clear about doses and timing. As philosophers we are professionally interested in issues and details that most people seldom worry about, and therefore we often strive for exactness and clarity in respects that are usually disregarded in other contexts.

In addition to facilitating communication, exactness also facilitates investigation. If it is unclear to you exactly what I have said, how can you verify or repudiate my statement? As noted by Karl Popper (1902–1994), a statement has to be precise in order to be accessible to falsification or corroboration [47, 66]. This applies, of course, not only to philosophy but to science in general. One of the major virtues of mathematical theories in the natural and social sciences is that they provide us with predictions that are precise enough for testing.

In philosophy, as well as other disciplines, we often have to extend our language in order to express new thoughts and talk about that which we have not spoken of before. This is taken for self-evident in most academic disciplines. No one would expect a natural language to contain beforehand all the terms and distinctions needed to express new developments in chemistry, mathematics, or economics. In philosophy as well, new terms have been introduced along with new ideas and concepts. "Supervene", "induction", "modality", "consequentialism", and "prioritarianism" are examples of this.

Unfortunately, though, some philosophers seem to have believed that philosophical insights are in some way hidden in the language (mostly their own mother tongue). They have attempted to do philosophy by looking for meanings or connotations that only a person with an accurate feeling for the finest nuances of the language can pick up. But very few insights of lasting or general philosophical interest have been obtained in that way. The so-called ordinary language philosophy was a cul-de-sac. In order to develop philosophical terminology, we need to carefully construct and delimit new distinctions that have no obvious counterparts in non-philosophical language, and assign terms to them.

Since antiquity onwards, philosophy and poetry have been each other's antithesis in terms of their approaches to language. This may seem paradoxical since philosophy and poetry are closely related in another important respect: They both deal with "big issues" such as existence, meaning, knowability, and morality. But the two pursuits deal with these issues in different ways – ways that are complementary rather than competing. These differences have large effects on their respective linguistic ideals. In poetry, elegance usually has precedence over precision. In philosophy the reverse is usually the case, as keenly pointed out by C.S. Peirce when advocating

> ". . . a suitable technical nomenclature, whose every term has a single definite meaning universally accepted among students of the subject, and whose vocables have no such sweetness or charms as might tempt loose writers to abuse them – which is a virtue of scientific nomenclature too little appreciated." [65, pp. 163–164]"

In poetry, and in belles lettres in general, disambiguation is no goal. To the contrary, ambiguity and imprecision are often necessary means to achieve the desired literary effect [46]. Philosophy does the very contrary: It tries to achieve as much precision as possible, even though its subject matter often makes this particularly difficult [77].

1.3.2 *What is Exactness?*

Clarity is a wider concept that exactness. In order for a statement to be clear it is not sufficient for it to be exact. It also has to be expressed in a way that makes it reasonably easy to understand. Something that is clear should, in Descartes' words, be "open to the attending mind"[3] ([15, p. 22], [20]). For our present purposes we can focus on the somewhat narrower concept of exactness. ("Exact" can be taken to be synonymous with "precise".) This is a concept with two clearly distinguishable meanings. The following examples serve to show the difference:

(a) The colour of that laser beam is green.
(b) The colour of that laser beam is yellowish green.
(c) The colour of that laser beam lies somewhere in the wavelength interval 495–570 nanometres.
(d) She is in the centre of Paris.
(e) She is close to Notre Dame.
(f) She is in one of the first six arrondissements of Paris.

When going from (a) to (b) we restrict the scope of colours. Fewer colours answer to the latter than the former description. If we instead go from (a) to (c), we do not reduce the number of possible colours, or at least we are not sure to do so since "green" corresponds approximately to the stated wavelength interval. However,

[3] "Claram voco illam, quae menti attenditi praesens et aperta est."

(c) is considerably less vague than (a) since we have in practice eliminated the borderline cases that might be classified as either green or not green. Both the move from (a) to (b) and that from (a) to (c) can be described as moves in the direction of exactness, but these are different types of exactness. We can describe (b) as more restricted than (a), and (c) as more definite than (a). Similarly, (e) is more restricted than (d) and (f) more definite than (d). Restrictedness and definiteness are the two major forms of exactness.

It turns out that philosophically speaking, not even exactness itself is a sufficiently exact concept! [47, 77, 84] This can be seen clearly if we ask the simple question which of (b) and (c) is the more exact statement. The best answer to that question is to refuse answering it, and instead distinguish between the two notions of exactness, restrictedness and definiteness.

In philosophy, both types of exactness are important, but lack of definiteness tends to be more detrimental than lack of restrictedness. We can for instance use a wide concept of "action" that includes omissions (refraining from acting) and various non-intentional behaviour. Such a wide concept may be impractical for some purposes, but if its boundaries are sharp enough it will not create communicative hurdles that we cannot deal with. A concept of action that lacks in definiteness will be much more problematic, in particular if the undetermined borderline cases are among those that we need to attend to. Needless to say, the importation of such indefiniteness into a formal model will make the latter just as loose and ill-defined as its informal counterpart, and perhaps even more dangerously so if its vagueness is obscured by the seemingly exact paraphernalia of a mathematical language.

1.3.3 Can Inexactness Be Described Exactly?

We have to be realistic. Using the tools of philosophical analysis, we can make our concepts more specific and, in particular, more definite. But this is one of the many human activities in which perfection is in practice unattainable. Even after considerable efforts, many of our concepts will remain imprecise. Furthermore, some of the concepts that we wish to include in our analysis may be "essentially inexact", i.e. inexactness is part of what they express, and therefore their meaning cannot be mirrored by a definition from which the vagueness has been removed [34]. The relational concept "near" may be a case in point. Any precise definition of that concept, for instance as "within a distance smaller than 5.3 km" can be accused of missing essential features of nearness, namely that it comes in degrees and that it is judged differently in different contexts. (For instance, 5.3 km is near if you are driving on the motorway, but not if you are travelling by foot on an arduous mountain trail.) The same applies to concepts such as "bald" and "tall".

In such cases, instead of a vagueness-resolving definition we may opt for a vagueness-preserving one. The question then arises: Can we provide a formal representation that preserves the vagueness? The most obvious way to do so would be to construct a model in which the concept in question comes in degrees. We can for instance construct a model in which I am tall to the degree 0.45 and the late basketball player Manute Bol (who was 46 cm taller than me) tall to the degree 0.99. But if the notion of tallness is essentially imprecise, as we have supposed, then it cannot be captured by such exact numbers. Perhaps we should make the numbers less precise, and assign to me tallness to the degree 0.40–0.55? But then both the lower and the upper limit appear to be artificially precise. Perhaps we should replace each of them by something less precise, such as an interval? In this way we are caught in an infinite regress of dissolving boundaries that seems very difficult to stop. Arguments like these have led some philosophers to question whether vague concepts can at all be adequately represented in a formal language [72, 78].

But there is a way out, for which we have already prepared the ground. A model should not be expected to correspond exactly to that which it is a model of. All that we can expect is that some features of the model should be structurally interrelated in the same way as some important features of the original. The grass mats used in a model railway may consist of plastic, but they represent lawns and meadows, not plastic mats. Similarly, the exact numbers in our model of degrees of tallness do not represent precise degrees. Instead, they represent the vagueness of our intuitive concept of tallness. They do this remarkably better than a model with all-or-nothing tallness, but they do not correspond perfectly to the intuitive concept. This should not be a problem, once we have realized that our formal construction is a model of our intuitive notion, not the "real truth" behind it. "[W]e can have mathematical precision in the semantics without attributing it to the natural language being studied by making use of the logic as modelling picture" [12, p. 246].[4]

1.4 From Natural Language to Logical Representation

Any representation of a concept in logic or some other formal language is the outcome of a streamlining of the concept, a simplification for the sake of clarity, in other words an idealization. In this section we are first going to have a close look at philosophical idealization, and in particular the relationship between formal and informal idealizations. After that we will turn to some of the major problems that have to be solved in the process of formalization, or idealization into a formal language as it can also be called. Throughout this section, the examples will concern logical formalization although most of the principles discussed are also relevant for formalization into other formal languages.

[4]Cf. Williamson's [84, pp. 270–275] notion of a "variable margin model".

1.4.1 The Nature of Idealization

Formal models are ideals in the sense of "[s]omething existing only as a mental conception". (OED) To idealize in this sense means to perform a "deliberate simplifying of something complicated (a situation, a concept, etc.) with a view to achieving at least a partial understanding of that thing. It may involve a distortion of the original or it can simply mean a leaving aside of some components in a complex in order to focus the better on the remaining ones" [61, p. 248].

This sense of idealization must be distinguished from the more common sense of expressing a (too) high opinion of something. Formal models may or may not represent something as "perfect or supremely excellent in its kind". (OED) In (formal and informal) philosophy, both types of idealization are common. In particular, the concepts that we use when philosophizing on human behaviour tend to be both (1) idealizing–simplifying, i.e. they leave out many of the complexities of real life, and (2) idealizing–perfecting, usually by representing patterns that satisfy higher standards of rationality than what most humans live up to [37]. Since formal philosophy has its starting-points in informal philosophy, it tends to inherit both types of idealization.

The reason why we idealize–simplify is that philosophical subject-matter is typically so complex that an attempt to cover all aspects will entangle the model to the point of making it useless. A reasonably simple model has to leave out some philosophically relevant features. For a simple example of this we can consider philosophical usage of the term "better". In ordinary language, "A is better than B" and "B is worse than A" are not always exchangeable. It would for instance be strange to say: "Zubin Mehta and Daniel Nazareth are two excellent conductors. Nazareth is worse than Mehta." Given the first sentence, the second should be: "Mehta is better than Nazareth." Generally speaking, we only use "worse" when emphasizing the badness of the lower-ranked alternative ([25, p. 13]; [80, p. 10]; [11, p. 244]). There may also be other psychological or linguistic asymmetries between betterness and worseness [79, p. 1060]. However, a long-standing philosophical tradition persists in not making this distinction in regimented philosophical language [7, p. 97]. The reason for this is that the distinction does not seem to have enough philosophical significance to be worth the complications that it would give rise to. The logic of preference adheres to this tradition from informal philosophy, and $A > B$ is taken to represent "B is worse than A" as well as "A is better than B".

Idealization–simplifying – be it formal or informal – always involves deviations from that which we model. Therefore, counter-arguments can always be made against an idealized account of philosophical subject matter. It is for instance easy to find examples in which betterness and worseness are not interdefinable in the way described above. Such deviations will always have to be judged in relation to the purpose of the model and how it is used. Does the deviation show that an important aspect of the subject matter has been "lost in idealization"? If it does, then we have to consider how much we would lose in simplicity by including it. Sometimes, the best strategy is to replace the idealization by a richer account. On other occasions, it may be better to continue its use while keeping in mind how it deviates from that which it is intended to capture. In this respect idealizations are like maps: They always require a compromise between overview and detail, and it is often advisable to use different maps for different purposes.

The reason why we idealize–perfect is that as philosophers we are at least as interested in what should be as in what is. Throughout the long history of our discipline, philosophers have tried to answer questions about how to think and how to behave. Requirements of rationality are usually important parts of the answers to such questions, and therefore idealization–perfection is commonly concerned with the ideal of rationality. Philosophical investigations of inferences, beliefs, decisions, and moral behaviour usually expound on the behaviour of rational thinkers, believers, decision makers and moral agents. We idealize–perfect in order to get a grip on what rationality demands of us, and sometimes also in order to gain insights on other normative demands such as those of morality.

It is important to keep track of one's idealizations and the reasons for them. Unfortunately, that is often not done. A particularly problematic confusion is that between the two forms of idealization. As one example of this, most accounts of human preferences depict them as transitive, i.e. someone who prefers a to b and b to c is assumed to also prefer a to c. That is not always the case for real-life preferences.[5] The reason why transitivity is assumed may be that the concept has been idealized–simplified, idealized–perfected, or both. In a discussion of divergences between the model and actual human behaviour it is important to know why the model assumes transitivity. Our analysis of such divergences may differ depending on whether transitivity was assumed for perfecting or simplifying reasons.

1.4.2 An Idealization in Two Steps

Formalization in philosophy typically results from an idealization in two steps, first from common language to a regimented philosophical language, and then from regimented into mathematical or logical language. For example, consider the derivation of the permission predicate (P) of deontic logic from the non-philosophical concept of a permission. We can use the following example from non-regimented language:

(1) Li-Hua is permitted to drive the forklift.

Here, the permission refers to an action. In regimented philosophical language, it is common to represent each action by a sentence denoting the state of affairs consisting in that action taking place. Hence:

---

[5] See Chap. 29.

(2) It is $\text{permitted}_{\text{phil}}$ that Li-Hua drives the forklift.

where "$\text{permitted}_{\text{phil}}$" is the philosophical idealization of the "permitted" of ordinary language. "$\text{Permitted}_{\text{phil}}$" differs from "permitted" in referring exclusively to what conscious agents do. It also differs in other ways. In non-philosophical usage, "when saying that an action is permitted we mean that one is at liberty to perform it, that one may either perform the action or refrain from performing it." In regimented philosophical language, however, "being permitted to perform an action is compatible with having to perform it" [67, p. 161].

The difference is perhaps best illustrated by the fact that in ordinary language we do not call something "permitted" that is in fact obligatory. Suppose that someone pays you in advance for cleaning their house. It would seem strange to say that you are then "permitted" to clean the house, since that would give the impression that you have a choice to do otherwise. However, according to philosophical usage of the term, it would be correct to say that you are permitted to do the cleaning. More generally, in philosophical language a permission is assumed to hold whenever the corresponding obligation holds ([67, p. 161]; [1, p. 55]; [10, p. 76]).

The second step of idealization takes us from "$\text{permitted}_{\text{phil}}$" to the deontic predicate P. This means that we go from (2) to

(3) Pa,

where P is a predicate expressing permission and a the sentence (or the proposition represented by the sentence) "Li-Hua drives the forklift". There are major differences in meaning between "permitted" and P. It should be noted, though, that in terms of most of the more philosophically significant differences, "$\text{permitted}_{\text{phil}}$" is closer to P than to "permitted". This applies for instance to the property of "$\text{permitted}_{\text{phil}}$" that we focused on above, namely that it holds for whatever is obligatory. This corresponds rather exactly the property of P that for all actions a, Oa implies Pa, where Oa is the corresponding predicate of obligation.

Intuitively speaking, most of the idealization in this example took place in the first step (from ordinary language to regimented philosophical language) rather than in the second (from regimented to formal language). And this is not untypical. Informal idealizations can sometimes be quite far-reaching. For instance, the concept of a person used in some philosophical discussions on personal identity is remarkably remote from the concept of a person in everyday language.

As all this should make clear, the difference between logical treatments of philosophical subject matter and treatments of the same matter in regimented natural language is *not* their distance to everyday concepts. The major differences are instead the mathematical skills that the formal models require and the characteristic types of questions that can be asked and answered with their help. Some philosophers who complain about the lacking realism of formal representations may to some extent confuse unfamiliarity in appearance with dissimilarity in meaning.

1.4.3 Selecting Concepts for Formal Representation

Logic is concerned with reasoning, but not all types of reasoning are included in the subject matter of logic. When discussing the pros and cons of different cars, we use arguments couched in terms such as “safe”, “comfortable”, “easy to drive”, etc. These terms and their interrelations are not part of logic. Similarly, the terms used in wine tasting, such as “earthy” and “fruity” do not seem to have been subject to logical (or other formal) analysis. The same applies, of course, to the vast majority of terms that we use in different types of arguments. Logic is only concerned with a small fraction of the concepts and thought patterns employed in argumentation and reasoning. Whereas virtually every concept with some role in philosophy has been subject to some degree of informal idealization, only few of them have been formalized. Those that have been formalized are characterized by having wide usage and a role in inferences that is largely independent of context.

The core concepts of logic are the truth-functional concepts “and”, “or”, “not”, “if... then”, “if and only if”, “some”, and “all”. These are concepts that we assume, for good reason, to have the same role in inferences in widely different contexts. They are often called the logical constants. (However, the definition of a logical constant is controversial [56].) But very few of the issues about valid argumentation that arise in philosophy (or outside of philosophy) concern the properties of words like these. It is more common for such issues to be concerned with the rules governing our usage of terms such as “know”, “believe”, “try”, “do”, “good”, “better”, “ought”, “forbidden”, and “permitted”. These are also concepts with interesting structural interrelations that are fairly constant across contexts, and they have all been subject to logical formalization.

The choice of concepts for formalization should ultimately depend on whether the resulting models will be useful for philosophical and other worthwhile purposes. Since the shaping of new formal models is a creative rather than a rule-bound process, the following five critera for what to formalize should be read as tentative suggestions and nothing more.

First, the promising candidates usually have a meaning that is reasonably *constant across contexts*. This applies for instance to the words “good” and “bad”. The meaning of these terms is presumably the same if we isolate them from a discussion on good and bad teachers as if we isolate them from a discussion on good and bad refrigerators [29]. This makes “good” and “bad” more promising candidates than, say, “earthy” or “sweet”.

Secondly, the promising candidates usually provide a *structure* into which other, more context-specific, concepts can be inserted. This applies pre-eminently to the truth-functional concepts, but also to many others, such as our examples “good” and “bad”. We can for, instance, talk about a collection containing both good and bad books, or an organ having both good and bad registers. Other examples are the action-theoretical concepts “do”, “try”, “refrain from”, and “see to it that”, to which we can affix more context-specific expressions denoting various types of actions.

Thirdly, it is usually a positive sign if we can combine the concept with some kind of *logical or other mathematical operations*. The most common examples are truth-functional and set-theoretical operations. As an example of the former, the concept "see to it that" can be combined with sentences describing states of affairs, and such sentences can be negated, combined into conjunctions and disjunctions, etc. As an example of the latter, in discussions about collective action we can talk about different groups of people, and one such group may for instance be a subset of another.

Fourthly, promising candidates tend to come with interesting issues about *potential structural properties* that seem to be generalizable across contexts. We can for instance ask whether something can be at the same time both good and bad. We can also ask whether someone who sees to it that a thereby also sees to that a-or-b.

Fifthly and finally, it is also a good sign if *connections with previous formalizations* are in sight. For instance, a logic of "good" and "bad" has obvious connections with the logic of "better". (If a is good and b bad, can we conclude that a is better than b?) Similarly, a logic of collective action can be connected with previously developed logics of individual action.

1.4.4 Structuralizing

After we have chosen a concept for formalization, we have to idealize it to make it suitable for formal treatment. As noted above, much of that idealization has often already been performed in informal philosophy. But for the purpose of formalization we may need to streamline the structural properties of the concept somewhat further. We can call this form of idealization *structuralizing*. In practice it often consists of the unification or splitting of concepts and the search for definability relations.

The *unification* of concepts is usually advisable when we are dealing with conceptually closely related terms in the informal language that have important structural properties in common. Such terms often differ in fine details that we cannot capture in the formal language without losing too much in simplicity. We have already seen one example that answers to this description, namely the unification of betterness and (converse) worseness. Another example is the collection of words used in ordinary language to denote obligatoriness: "must", "should", "ought", "have to", etc. These words are not exact synonyms. Typically, "obligations" originate from promises or agreements, whereas "duties" are associated with roles and offices in organizations and institutions [6, 18, 62]. Already in informal moral philosophy it is nevertheless common to regard "Yasmin ought to ...", "It is a duty for Yasmin to ...", and "Yasmin has an obligation to ..." as synonymous. The reason for this is that the differences in meaning between these expressions have little or no relevance in most philosophical discussions. In deontic logic this simplification is even more useful. Therefore deontic logic standardly contains only a single prescriptive predicate (denoted O) rather than several predicates corresponding to different prescriptive natural-language terms. The prescriptive predicate of the formal language can be seen as representing the common core of the various prescriptive expressions in natural language. Arguably, this core is more streamlined and more suitable for formal treatment than each of the natural-language predicates that were the starting-points of the formalization.

The opposite operation of *splitting* concepts is useful, sometimes necessary, when the concept we wish to formalize has meanings that differ in their structural properties. The splitting of concepts is, of course, quite common also in informal philosophy, but in preparing for formalization we have to pay particular attention to structural properties when deciding whether or not to split a concept.

Again, we can use prescriptive terms from moral philosophy as examples. Consider the following two sentences:

(a) "You must help her."
(b) "You must be wrong."

(a) expresses an obligation. (b) does not. Instead it expresses necessity. This is reason enough for the informal philosopher to distinguish between the two meanings of the word. For the formal philosopher there is an additional reason, namely that the two senses have different structural (logical) properties. To see this, consider the following property:

If $Must(X)$ then X.

This property holds for the "must" of our second example. If I am right in saying that you must be wrong, then surely you are wrong. We can easily verify that the property also holds in other cases where "must" is used in the same sense. But it does not hold in the first example. Even if I am right in saying that you must help the person referred to, it certainly does not follow that you actually do so. Again, we can verify that the same applies to other sentences where "must" has the same meaning. Such a consistent difference in terms of (logical) structure is a sure sign that for the purposes of formalization, "must" has to be split into two concepts. It is only obligation−must that can be unified with the other prescriptive predicates into the deontic operator O. Necessity−must can instead be unified with "necessary", "unavoidable" and the like.

Next, consider the following two uses of the word "ought":

(c) "You ought to help your destitute brother."
(d) "There ought to be no suffering in the world."

(c) expresses a prescription, something that someone should do. Alternatively, we could express the same statement with some other such term, saying for instance "You have a duty to help your destitute brother". In this respect, (d) is quite different. It expresses a wish about the state of the world, or an evaluation of such a state. It does not directly prescribe or recommend any action. This is a well-known distinction. The "ought" of (c) is called ought-to-do (Tunsollen) and that of (d) ought-to-be (Seinsollen or ideal ought) ([68, p. 195]; [13]).

This double usage is specific for "ought", and does not apply to prescriptive predicates in general. It would not make much sense to say that there is a duty for the world not to contain any suffering. Since deontic logic is concerned with prescriptions in general, not only those expressed by the English word "ought", ought-to-be and ought-to-do have to be split. Only the latter should be unified with the other prescriptive predicates into the deontic O operator. Just like necessity–must, ought-to-be should be treated as a separate concept, not to be merged or confused with the prescriptive ones.

Unfortunately, this has not always been realized. (Perhaps one of the reasons for this is that the O operator is usually read "ought", and we are not sufficiently often reminded that in spite of this, it represents the common core of several natural language expressions.) A considerable amount of confusion has been created by attempts to unify ought-to-do with ought-to-be. This is usually done by reconstructing ought-to-do as ought-to-be referring to actions, in the way shown in the following two examples:

Person i ought to do x. = It ought to be the case that person i does x.
Person i ought to do x. = The world ought to be such that person i does x.

But this does not work. "You ought to sing in tune" means something quite different from "The world ought to be such that you sing in tune." And more generally speaking, that which we ought to do does not coincide with that which the world ought to be such that we do. The world ought to be free of racism, and in such a world no one would help victims of racism (since there would be none). Recently, a newly wed woman was killed by a robber. It certainly ought not to be the case that her husband went to her funeral less than a month after they married. But of course he ought to go to the funeral. The distinction between ought-to-do and ought-to-be is fundamental, and the two notions should be kept apart in both formal or informal moral discourse [35].

If one concept is *definable* in terms of another, then we can focus on the latter, and treat the defined concept as a mere abbreviation in the formal language. It is not uncommon for philosophically important concepts to be definable in terms of each other. One example is the interdefinability among the three modal concepts of necessity, possibility and impossibility. To be impossible means not to be possible, and something is necessary if and only if it is not possible that it is not the case. Letting $\Box$ stand for necessity, $\Diamond$ for possibility, and $\not\Diamond$ for impossibility, we can express these relationships as follows:

$$\Box a \leftrightarrow \not\Diamond \neg a$$

$$\Box a \leftrightarrow \neg \Diamond \neg a$$

$$\not\Diamond a \leftrightarrow \neg \Diamond a$$

It follows, of course, that in a modal logic we can take any of these three concepts as primitive (undefined), and define the others in terms of it. It is unimportant which of them we select to be the primitive notion.

In other cases, definability comes only in one direction. We can for instance define "best" in terms of "better" in the following way:

x is (uniquely) best if and only if for all y other than x: x is better than y.

However, there is no corresponding way to define "better" directly in terms of "best".[6] Therefore, "best" is in practice always treated as a defined concept in formal languages.

In general, logical languages with fewer primitive (undefined) concepts tend to be more manageable. The aim to have as few primitives as possible is called *definitional economy*. In order to achieve it we have to investigate carefully if some of the concepts on our agenda for formalization can be defined in terms of some of the others.

1.4.5 Introducing Formulas

As we have already seen, the concepts that are subject to formalization tend to owe much of their usefulness to the ways in which they can be connected to various more specific expressions. The common truth-functional connectives can be combined with any sentences carrying truth-values. To the deontic operators P and O we attach action-describing sentences. To a "stit" (see-to-it-that) operator we connect a name representing a person and a sentence describing a potential outcome of an action by that person, for instance:

$stit_i a$

where i is a person and a the outcome of that person's action. These attachments are called "variables".

Variables are essential components of formal languages; without them non-trivial formalization would not be possible. Historically, they are an important invention. In medieval times, names (such as "Socrates") were used to denote arbitrary persons. That practice is still frequent in philosophical texts, but it is also common to use single letters to denote persons. ("If A borrows money from B and then gives it to C,. . . .") Informal philosophical discourse also contains symbols representing objects that do not have proper names in other contexts. ("If the state of affairs a obtains at time t,. . . .") In logic, we do more of the same. The following series of synonymous statements illustrates the different degrees of compactness of notation:

---

[6] We can do so if we manipulate the sets of alternatives, see Chap. 27.

Ordinary language:
The first cause took place either before or at the same time as the second cause, and the second cause took place before the effect.

Informal philosophical language:
c_1 preceded or was simultaneous with c_2, and c_2 preceded e.

Logical language:
$c_1 \leq c_2 \;\&\; c_2 < e$.

The following series of restatements of a definition illustrates the pros and cons of the more compact notation that formalization makes available.

Ordinary language:
A cousin is a person with whom one has at least one grandparent in common but no parent in common.

Semi-formal language 1:
Person i is a cousin of person j if and only if (1) there is a person who is a grandparent of both i and j, but (2) there is no person who is a parent of both i and j.

Semi-formal language 2:
Person i is a cousin of person j if and only if (1) there is a person x who is a grandparent of both i and j, but (2) there is no person v who is a parent of both i and j.

Semi-formal language 3:
Person i is a cousin of person j if and only if (1) there are persons x, y, and z such that x is a parent of y who is a parent of i and x is also a parent of z who is a parent of j, but (2) there is no person v who is a parent of both i and j.

Logical language:
iCj if and only if :
$(\exists x)(\exists y)(\exists z)(xPyPi \;\&\; xPzPj) \;\&\; \neg(\exists v)(vPi \;\&\; vPj)$

The first of these statements is clearly the most easily readable one, and the last is the most precise and compact one. For many purposes, some compromise between readability and precision may be desirable; then one of the intermediate, semi-formal options may be optimal. The cases when logical notation is most useful are those in which we want to prove some property of the concepts we are working with. Box 1.1 on page 25 shows how the compactness of formal notation makes a proof easier to follow.

Box 1.1 Two versions of the same argument
Consider the two relations on points in time: "precedes or is equal to" ($\leq$) and "precedes" ($<$). We are going to show that if the former of these is transitive, then so is the latter.

In natural language

Consider three points in time such that the first precedes the second and the second precedes the third. Then clearly the first precedes or is equal to the second, and the second precedes or is equal to the third. Since the relation "precedes or is equal to" is transitive, we can conclude that the first precedes or is equal to the third. Now suppose that the first does not precede the third. Since the first precedes or is equal to third, we can conclude that the third is equal to the first. Thus the third precedes or is equal to the second. But this is impossible since the second precedes the third. We have derived a contradiction from the assumption that the first does not precede the third. Thus the first precedes the third. This shows that the relation "precedes" is transitive.

In formal language

Let t_1, t_2, and t_3 be three points in time such that $t_1 < t_2$ and $t_2 < t_3$. Then $t_1 \leq t_2$ and $t_2 \leq t_3$, and transitivity yields $t_1 \leq t_3$. Now suppose that $t_1 < t_3$ is not the case. It then follows from $t_1 \leq t_3$ that $t_1 = t_3$. We can then substitute t_3 for t_1 in $t_1 \leq t_2$, and obtain $t_3 \leq t_2$. But that is impossible since $t_2 < t_3$. It follows from this contradiction that $t_1 < t_3$. [36]

1.4.6 Determining the Number of Variables

In ordinary language, one and the same concept can be associated with different numbers of variables:

Cynthia is a mother.
Cynthia is Peter's mother.

It would be tempting to follow the same pattern in formal language, and (with the predicate M denoting motherhood) translate the sentences as follows:

Mc
Mcp

This would require that we allow one and the same predicate to appear with different numbers of variables. However, the introduction of such flexible predicates would

Table 1.1 Four usages of the term "free" that differ with respect to the variables

Example	Schema
She is free now.	i is free.
She is free from all those debts now.	i is free from the obstacle x.
Finally she was free to take up her studies again.	i is free to perform the action y.
She is now free from any legal obstacles to leave the country.	i is free from the obstacle x to perform the action y.

leave the exact relationship between formulas such as Mc and Mcp unclear. There is a much better way to deal with this, namely to introduce M as a two-place predicate, which means that an expression containing M can only be well-defined if each instance of M has two variables. The single-variable expression "Cynthia is a mother" is synonymous with "Cynthia is someone's mother", which we can express with the existential quantifier $\exists$ as follows:

$$(\exists x)Mcx$$

When introducing a predicate or a relation into the formal language, it is important to choose the right number and type(s) of variables. It is often preferable to include representations of all the variables that can be attached to the corresponding informal expressions, and then define uses with a reduced number of variables in the way we just did for motherhood.

The term "free" as used in political philosophy is an interesting example of this. If we classify uses of "free" in informal language according to the variables, then we will find at least four variants. Table 1.1 gives examples of these, and it also provides general schemata for each of the variants. These variants represent different notions of freedom, notions that are controversial in political philosophy. Some political thinkers have claimed that all true freedoms can be fully expressed by statements of the second type, "freedom from" (negative freedom). Others have put much emphasis on freedoms representable by the third type of expressions, "freedom to" (positive freedom). They see freedom largely as ability to make and implement one's own choices [4]. The fourth variant is less common, but it is quite useful since all the others can be defined in terms of it [55]. In formal analysis it would take the form of a three-place predicate

$$F(i, x, y)$$

where i is an individual, x an obstacle, and y some action that the individual can potentially perform. In this case it is much more difficult than for motherhood to determine how the three-place predicate should be used to define the two-place and one-place ones. As a first attempt we could define "freedom to" as $(\forall x)F(i, x, y)$, i.e. one is free to y if and only if one is free from all obstacles that might prevent the attainment of y. However, that may seem somewhat extreme. Arguably I am free

Table 1.2 Two usages of the term "duty" that differ in terms of the variables

Example	Schema
It is his duty to answer the phone on all times of the day.	Person i has a duty to do x.
The lawyer has a duty towards the client to defend her interests.	Person i has a duty towards person j to do x.

to read the morning newspaper even if a snowdrift makes it impossible for me to get hold of it. A distinction between different classes of obstacles may have to be introduced. Similar problems arise for the reduction of the three-place predicate to the two-place "freedom from". However, these difficulties should not be counted against the three-place predicate. To the contrary, these are real philosophical difficulties in the analysis of political freedom. The three-place predicate is a tool to present these difficulties more clearly, thereby making them more amenable to precise analysis.

But this is a controversial area. Traditionally, the negative notion of freedom is associated with right-leaning and the positive notion with left-oriented political ideas. Not surprisingly, the three-place predicate has been accused of both a left-wing and a right-wing bias ([22]; [64, p. 253]). Nevertheless, it has the advantage of allowing us to represent "freedom from" and "freedom to" in one and the same format, rather than just treating them as mutually incompatible notions.

In doubtful cases it is usually better to include than to exclude a variable when introducing a formal predicate. But of course, there are cases when one or other of the variables has such a small role that it can for most purposes be excluded. For a possible example, let us consider the notion of a duty, as shown in the examples and schemas of Table 1.2. Common usage of the term "duty" is dominated by the first variant mentioned in the table, two-place duty. The second variant, the three-place notion of a duty, is more uncommon. The two-place notion has the advantage that it can be unified with other prescriptive notions in the way discussed above. (Some of these, such as "morally required", do not have a three-place variant.) It is indeed common practice in philosophy to treat duty as a two-place concept. There are good reasons for this practice, but it has a price: We lose the ability to express that someone owes something to a specific person. Such relationships will then have to be treated in separate investigations, using a different formal representation [30, 58]. As noted above, there is nothing wrong with using different formal representations of a concept for different purposes.

But something more can be learned from this example. Even the two-place format "Person i has a duty to do x" does not correspond to the standard deontic operator for obligations, namely Ox which only has place for one variable. How is that possible? Obligations are normally tied to persons, and surely it makes a difference who is subject to an obligation? The explanation is that x in Ox is normally taken to refer to an action by a specified agent. If x represents the action consisting in me paying my rent, then we can take it for granted that I am the duty-holder in Ox. However, this is a rather precarious principle since information that is not stated explicitly runs a risk of being forgotten or misunderstood. The suppression of the person variable can make us forget about its existence, so that we treat moral prescriptions as impersonal although they are not. This may be one of the sources of the confusion about ought-to-be that was referred to above.

1.4.7 Specifying the Domains of the Variables

For each variable-place attached to a predicate we need a well-defined domain (source), i.e. a set whose elements represent the objects that variables in that place can stand for. In some cases the same domain can be used for more than one variable-place. This applies to the two-place predicate of motherhood. Here we can use the same domain, namely the set of all human beings, for both variable-places. For the three-place predicate of freedom the situation is quite different. We need three sets of variables, representing persons, obstacles, and actions.

In a formal treatment it is important to assign well-defined domains to all variable-places, and to be careful not to transgress them. There are two ways to deal with the complication that different variable-places refer to different groups of objects. To exemplify this, consider a simple logic of parenthood relationships with the predicates F and M, such that Fxy means that x is father of y and Mxy that x is mother of y. We can assume that fathers are men, mothers are women, and their children can be either. One way to express this is to use two sets of variables, $\mathbb{W}$ representing women and $\mathbb{M}$ representing men, and then introduce the distinction in the requirements for formulas to be well-formed, as follows:

Mxy is a well-formed formula if and only if $x \in \mathbb{W}$ and $y \in \mathbb{M} \cup \mathbb{W}$.
Fxy is a well-formed formula if and only if $x \in \mathbb{M}$ and $y \in \mathbb{M} \cup \mathbb{W}$.

The other alternative is to have only one domain, namely the domain $\mathbb{H}$ consisting of human beings, and include the restrictions in the logic rather than in the formation rules for the language. This can be done with one-place predicates denoting "is male" and "is female":

Each of Mxy and Fxy is a well-formed formula if and only if $x \in \mathbb{H}$ and $y \in \mathbb{H}$.
From Mxy it follows logically that Lx, where L denotes "is female".
From Fxy it follows logically that Gx, where G denotes "is male".

The two approaches are equivalent, and the choice between them is a matter of taste and convenience. The latter approach places the restrictions in the logic rather than in the language. This can be seen as an advantage since it makes the restrictions somewhat more accessible to modifications and adjustments. For instance, if $(\forall x)(Lx \vee Gx)$ holds in our original statement of the logic, then we can easily remove this principle in order to include people who are neither female or male. It may be an advantage to be able to do this without changing the language, which is considered to be a more drastic change of the framework.

1.5 Building a Logical Language

The distinguishing feature of logical language as compared to other formal languages is its focus on the representation of propositions (statements), by which we usually mean something that can be either true or false. In natural language we express propositions with sentences. One and the same proposition can be expressed by different sentences. Thus, "Dana is married to Lou" expresses the same proposition as "Lou is married to Dana" (and of course the same proposition can also be expressed by sentences in other natural languages).

Other formal languages than logic also contain sentences expressing propositions. In the appropriate contexts, $x^2 = y^2 + z^2$ represents a proposition about the relationships between the lengths of the hypotenuse and the legs of a right-angled triangle, $E = mc^2$ one about the mass–energy equivalence in relativity theory, etc. However, logic is distinguished by the generality of its treatment of sentences and by its suitability for formal work related to the conclusions that can be drawn from sets of sentences.

Sometimes, logical expressions are used to represent sentences that are not to be classified as true or false, but rather according to some other dichotomy, such as that between morally approved and morally unapproved actions or states of affairs. There are also logical systems, called many-valued logics, in which the traditional true/false dichotomy is replaced by a classification containing more than two alternatives, such as true/false/unknown. These distinctions have little impact on the construction of logical languages, and they will therefore not be considered in this section.

The simplest tools for building a logical language are those that treat sentences as wholes and do not contain separate representations of their parts. These constructions will be the topic of Sect. 1.5.1. In Sect. 1.5.2 we turn to the construction of sentences from their parts, and in Sect. 1.5.3 to formal elements that refer to the parts of the sentences thus formed. Section 1.5.4 shows how the formation rules for a formal language are usually expressed.

1.5.1 From Atomic to Composite Sentences

Let us start with a set of (proposition-representing) sentences. We can call them $a, b, \ldots$. To begin with, we will treat them as "atoms" ("atomic sentences"), i.e. we disregard their internal structure. This is of course a choice of a level of abstraction.

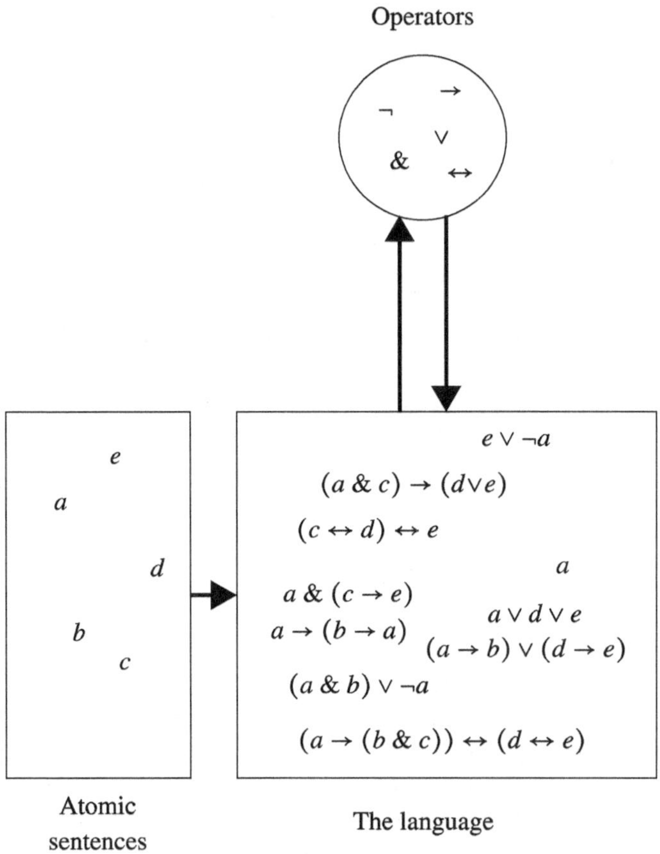

Fig. 1.1 A language formation diagram for sentential (propositional) logic

It provides us with a sort of bird's-eye view that has turned out to be quite useful for the study of phenomena such as conclusions and assumptions.

In order to get things going we need means to combine atomic sentences to form composite, or as we usually say, molecular sentences. The construction elements used for this purpose are called sentential operators, since they are operators that take us from a sentence (or several sentences) to a new sentence. The simplest sentential operator is negation, often denoted $\neg$. It takes us from a sentence a to its negation $\neg a$. If a represents the same proposition as "I am tired", then $\neg a$ represents the same proposition as "I am not tired". Other such operators are conjunction ("and", &), disjunction ("or", $\vee$), material implication ("if ... then", $\rightarrow$ or $\supset$), and equivalence ("if and only if", $\leftrightarrow$ or $\equiv$). (All these are truth-functional operators, but that is not a property of the language but one of the logic.)

Figure 1.1 shows how these operators can be used to form the full language of propositional logic. Two important features should be noted in that diagram. First, the atomic sentences are themselves directly introduced into the language, as shown

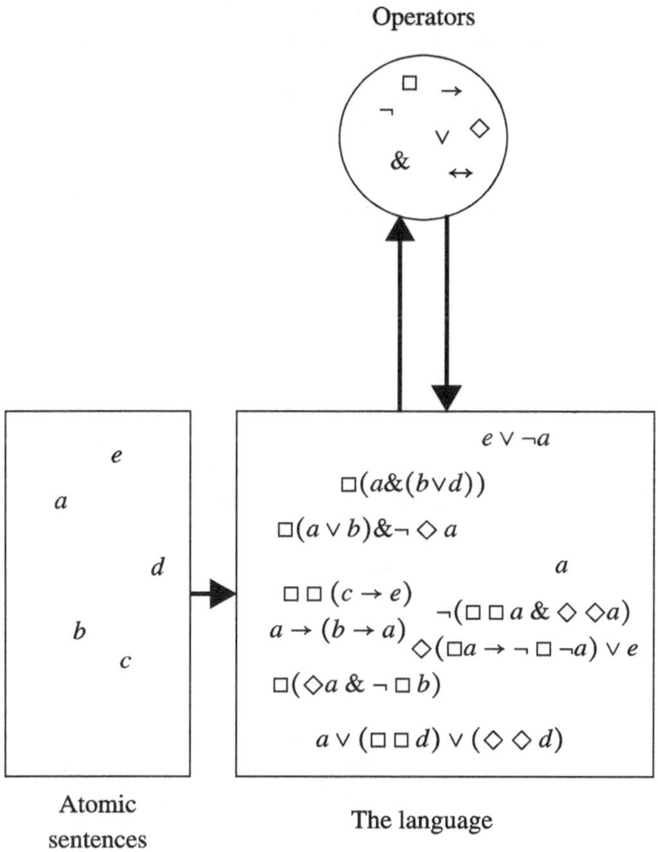

Fig. 1.2 A language formation diagram for modal logic. □ stands for necessity and ◇ for possibility

with the horizontal arrow. Secondly, the operators ¬, & etc. can be applied not only to atomic but also to molecular formulas. This means that unlimitedly complex formulas can be formed, such as $\neg a \vee \neg(b \vee c)$, etc.

Other operators can be added to the language in the same way. In a discussion about necessity and possibility we will need the unary (single input) operators □ ("it is necessary that ...") and ◇ ("it is possible that ..."), and often also the binary (two input) operator of strict implication ⇒ ("if ... then necessarily ..."). These are inserted into the logical language in Fig. 1.2. An important feature of this language is that □ and ◇ can take as inputs sentences in which they are themselves already present. We can therefore form sentences such as $\Box\Box b$ and $\Box(a \rightarrow \Diamond(a \vee b))$. From an interpretational point of view this is not quite uncontroversial. It can for instance be questioned whether a sentence such as $\Box\Box b$ ("it is necessary that it is necessary that *b*") is at all meaningful. Is necessity iterable?

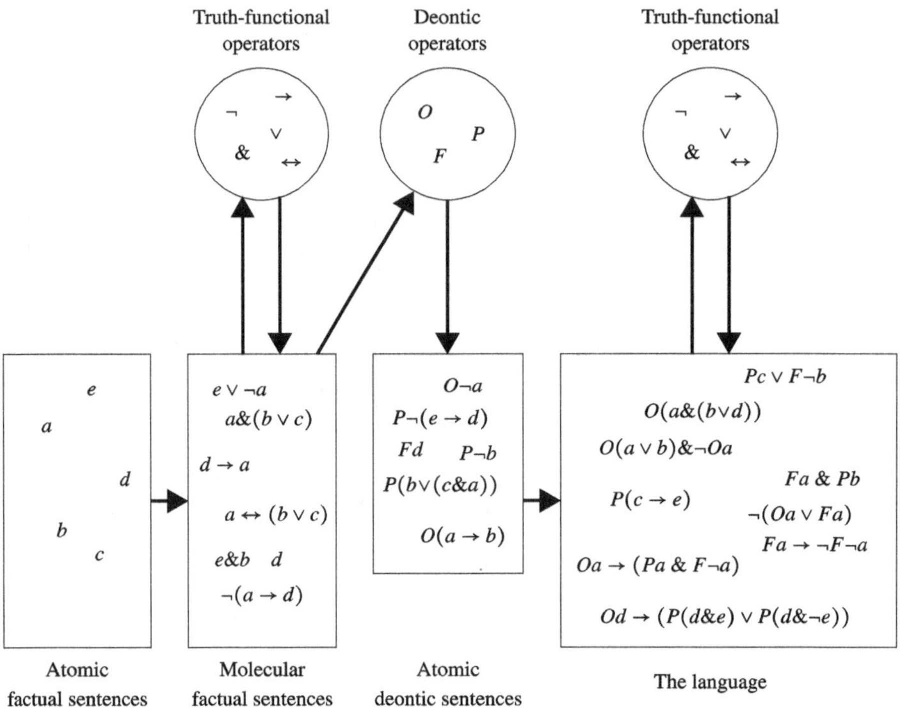

Fig. 1.3 A language formation diagram for a deontic logic that does not allow the iteration of deontic operators. O stands for obligation, P for permission, and F for prohibition

One operator whose repeated use in a formula has often been questioned is the deontic operator O that stands for moral requirement. From a factual statement a representing some human action we can form the sentence Oa saying that a is morally required. But how meaningful is the sentence OOa? Does it say that it is morally required that it is morally required that a? Then, exactly what does that mean? There are reasonable interpretations of O that make this sentence meaningless. In order to block the formation of such sentences we need to construct a somewhat more complex language formation diagram, as shown in Fig. 1.3. Here we are not allowed to affix O to sentences already containing O. Therefore neither OOa nor $O(Oa \vee O\neg a)$ are well-formed formulas, which means that although they consist of parts of the language, they are not themselves parts of the language. However, we can apply truth-functional operators to sentences containing O, forming sentences such as $\neg O(a \rightarrow b)$ and $Oa \vee Ob$.

Figure 1.4 shows an alternative language formation diagram for a deontic language that does not allow "repeated" application of the deontic operators. The difference between Figs. 1.3 and 1.4 is that in the latter, atomic and molecular factual statements such as a, $a \vee b$ etc. are directly included in the deontic language. Here

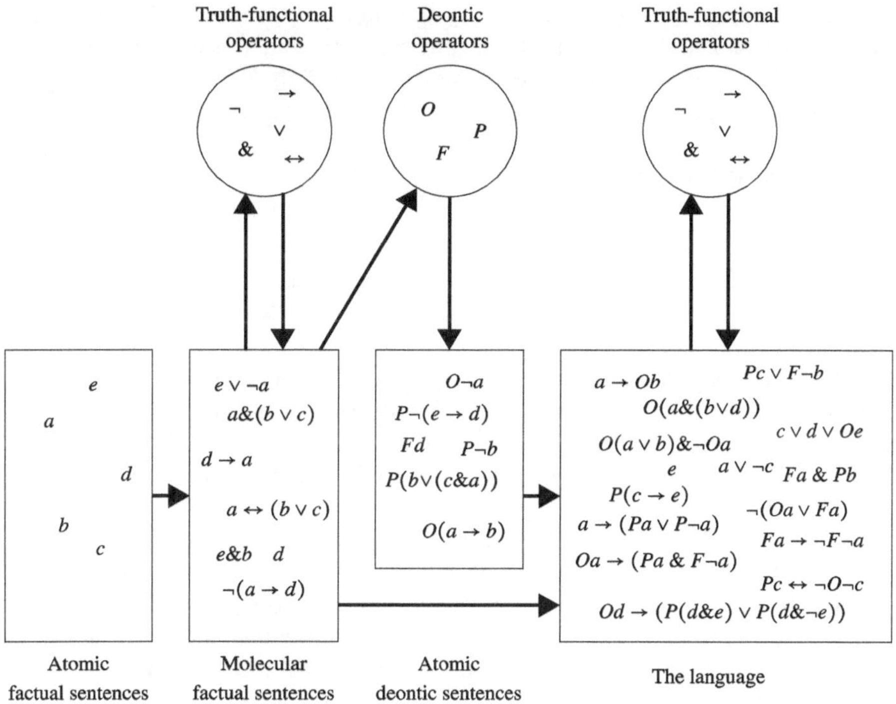

Fig. 1.4 Another language formation diagram for a deontic logic that does not allow the iteration of deontic operators. It differs from the diagram of Fig. 1.3 in allowing for the direct introduction of factual sentences into the language

it is also possible to form sentences such as $Op\ \&\ \neg p$ and other sentences with "mixed" deontic and factual contents.

1.5.2 Decomposing the Atoms

Factual sentences in many natural languages tend to have a standard grammatical form containing two main parts, a subject and a predicate. The subject represents that which we say something about, and the predicate that which we say about it:

Socrates	wrote no book.	Nevertheless,	his thoughts	changed the world.
subject	*predicate*	*connective*	*subject*	*predicate*
sentence 1			*sentence 2*	

As we have already seen, formal logic has taken over this structure from natural language. Predicate language that is based on the subject/predicate distinction is by far the most common formal representation used to decompose the logical atoms and scrutinize their components. When we translate the sentence "The author is a bore" into predicate logic, we identify the subject ("The author") and assign a symbol such as i to it. Similarly, we identify the predicate ("is a bore") and assign to it a symbol such as B. The sentence is then denoted Bi.

Many natural language predicates refer to some specific individual or other object of thought. In grammars of natural language there are two competing ways to analyze such sentences:

First analysis:	*Second analysis:*
Angelina is in love with Barbara.	Angelina is in love with Barbara.
subject (Angelina) — *predicate* (is in love with Barbara.)	*subject* (Angelina) — *predicate* (is in love with) — *object* (Barbara.)
sentence	*sentence*

For the purposes of logic the second analysis is preferred, since it allows for more detailed investigations. It makes use of the predicate "is in love with" which takes two variables, one of which corresponds to the subject and the other to the object of the natural language sentence. The sentence can then be rendered by the formula Lab, where L represents "is in love with".

As mentioned in Sect. 1.4.6, each predicate always takes the same number of variables (often called arguments). The number of variables is often called the arity of the predicate. A predicate is called unary (monadic, 1-ary, one-place) if it takes one variable, binary (dyadic, 2-ary, two-place) if it takes two, ternary (3-ary, three-place) if it takes three, and for any natural number n it is called n-ary (n-place) if it takes n variables. A 0-ary (nullary, zero-place) predicate, i.e. a predicate without variables, functions in the same way as an atomic sentence. That a predicate is nullary does not mean that there is nothing that it says something about. Instead, this means that we have chosen not to decompose it and introduce variables representing one or more of its components.

When formalizing natural language, it is a good general rule to use predicates with the lowest arity that is compatible with an adequate representation of the subject matter. In particular, if predicates with high arity can be replaced by truth-functional combinations of predicates with lower arity, then that should be done. "The author and the bookseller are bores" should be translated as $Ba \mathbin{\&} Bb$, not as $\overline{B}ab$ with a dyadic predicate $\overline{B}$. Similarly, "Ivan and Joanna are Kelly's parents" should be translated as $Pik \mathbin{\&} Pjk$, not as $\overline{P}ijk$.

Two warnings are warranted. First, use nothing else than the existential quantifier ($\exists$) to represent "exists". Do *not* introduce a predicate to represent "exists", since doing so gives rise to complications that you would like to avoid [8]. Secondly, use nothing else than the equality sign ($=$) to represent "is equal to" or "is the same as" as a binary predicate. It is important to follow the standard rules for predicate logic with identity, which can be found in most textbooks on elementary logic.

Figure 1.5 shows a formation diagram for a simple predicate language with monadic and dyadic predicates. Note that the components of sentences (variables and non-nullary predicates) are not themselves included in the language, contrary to the atomic sentences in Figs. 1.1, 1.2, and 1.4.

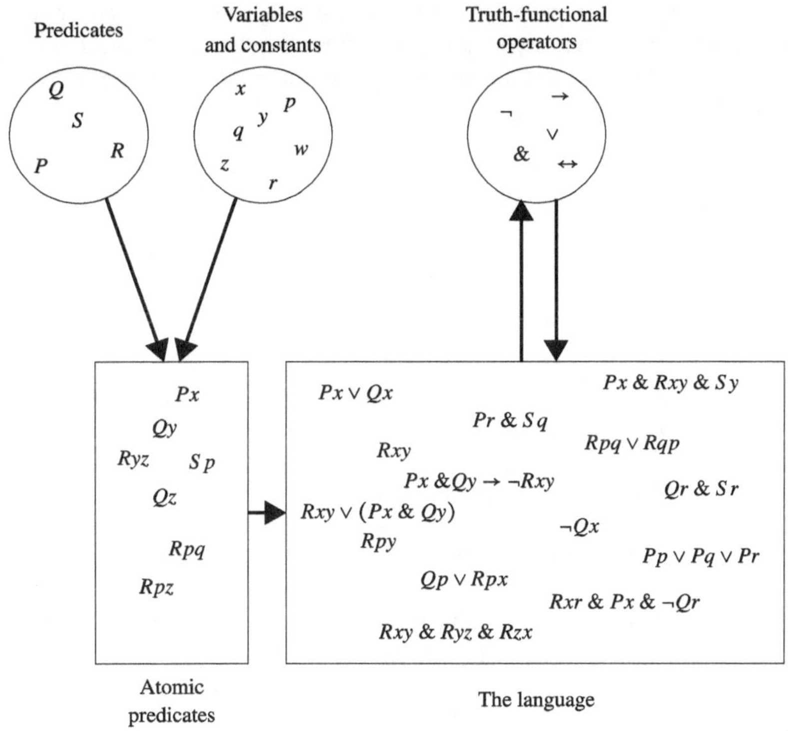

Fig. 1.5 A language formation diagram for standard predicate logic without quantifiers

1.5.3 Quantifiers

In order to make efficient use of the decomposition of atomic sentences into predicates and variables, we need to employ Frege's great invention, quantifiers. The quantifiers ∀ ("all") and ∃ ("some") are a type of sentential operators. Just like the monadic operators referred to in Sect. 1.5.1, they take us from a sentence to another sentence, hence if Fxy denotes "x has y as a friend" and i denotes the author, then $(\exists y)Fiy$ says that the author has some friend. Similarly, the sentence

$$(\forall x)(\forall y)(Fxy \rightarrow Fyx)$$

says that friendship is always mutual, whereas the sentence

$$(\exists x)(\exists y)(Lxy \,\&\, \neg Lyx)$$

where L denotes "loves" expresses the most unfortunate fact that the same does not apply to love.

The translation of sentences from natural language into predicate logic is not always straightforward, and sometimes it requires considerable changes in structure. Often, sentences with one and the same structure in natural language require quite different translations:

The dog is a Mastiff.
The giraffe is a mammal.

The first sentence is preferably translated into

$$Md$$

where M is the predicate "is a Mastiff" and d the particular dog referred to. The second sentence is best translated as

$$(\forall x)(Gx \rightarrow \overline{M}x)$$

where G is the predicate "is a giraffe" and $\overline{M}$ the predicate "is a mammal".

In a language with quantifiers we need to distinguish between constants and variables. A constant, such as d in our formula Md, refers invariably to a particular object, and it is not affected by quantifiers. It can be compared to a unique name such as "Louis XIV" in natural language. A variable, such as x in our formula $(\forall x)(Gx \rightarrow \overline{M}x)$, has no meaning in itself but acquires meaning in the context, just like pronouns such as "that", "this", and "it" in natural language.

The use of variables makes predicate logic well suited to keep track of complex relationships. The resources of natural language are much less suited for that purpose. We can distinguish between "this" and "that", but we do not use them repeatedly with persistent reference. We can introduce phrases like "the first person" and "the second person", but talk using such expressions is usually difficult to follow. (See Box 1.1 on p. 25.)

The language formation diagram in Fig. 1.6 (an extension of Fig. 1.5) summarizes the construction of predicate logic with quantifiers. Note that in this language, quantifiers cannot be applied to predicates. For instance, a formula such as $(\exists P)(\forall x)Px$ is not well-formed. Due to this limitation, the logic based on this language is called *first-order* predicate logic. In second-order predicate logic, $(\exists P)(\forall x)Px$ is a well-formed formula. (It can be interpreted as "There is a property that everything has.")

Ordinary language contains many expressions that have similar functions in sentences as "all" and "some":

Most Icelanders understand Norwegian.
Very few Germans understand Chinese.
At most three Government members have experience of blue-collar work.
There are *infinitely many* prime numbers.
The committee has an *an odd number of* members.

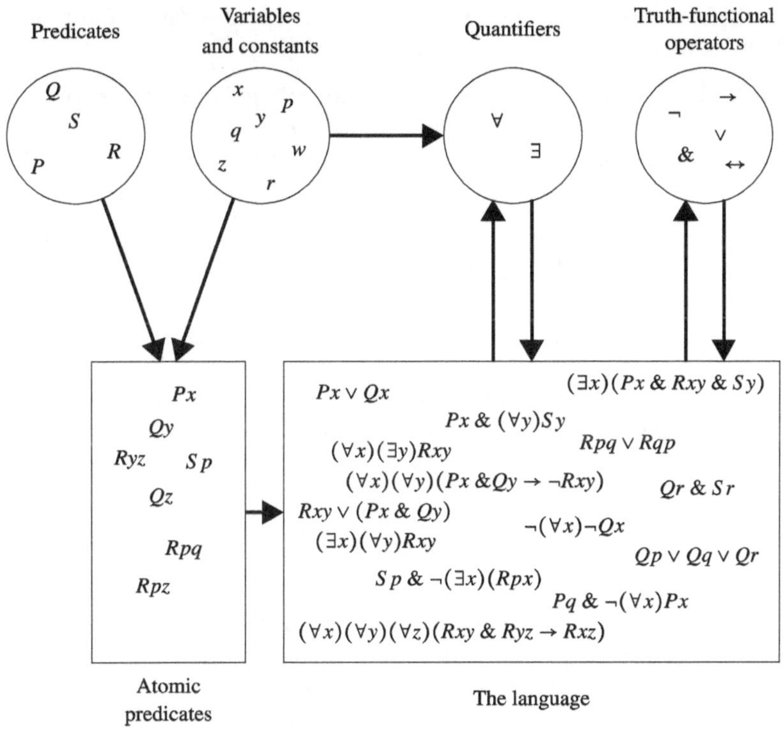

Fig. 1.6 A language formation diagram for standard predicate logic with quantifiers

The formal representation of such *generalized quantifiers* is of considerable philosophical interest. The same applies to second- and higher-order logics, in which the predicates themselves are treated as variables of quantifiers.

1.5.4 Specifying the Language

Formal languages are usually defined recursively, i.e. the definition identifies their smallest elements and then proceeds to specify how these elements can gradually be combined into larger and larger linguistic expressions. Our language formation diagrams show how this recursive process proceeds with repeatable steps of concatenation. In the specialized literature, this is expressed in more compact fashion. There are two common ways to specify a logical language. One is to list a set of language formation rules. In the following example this is done for the modal language presented in Fig. 1.2:

$\mathcal{L}$ is the language consisting exactly of the sentences obtainable through the following rules:

1. $T \subseteq \mathcal{L}$, where $T = \{a_1, a_2, \ldots\}$ is a countably infinite set of sentences.
2. If $\alpha \in \mathcal{L}$, then $\neg\alpha \in \mathcal{L}$, $\Box\alpha \in \mathcal{L}$, and $\Diamond\alpha \in \mathcal{L}$.
3. If $\alpha \in \mathcal{L}$ and $\beta \in \mathcal{L}$, then $\alpha \mathbin{\&} \beta \in \mathcal{L}$, $\alpha \vee \beta \in \mathcal{L}$, $\alpha \rightarrow \beta \in \mathcal{L}$, and $\alpha \leftrightarrow \beta \in \mathcal{L}$.

The other method, particularly common in computer science, is an abbreviation of the former as a so-called Backus-Naur Grammar clause:

$$\phi ::= a_1, a_2, \ldots \mid \neg\phi \mid \phi \mathbin{\&} \psi \mid \phi \vee \psi \mid \phi \rightarrow \psi \mid \phi \leftrightarrow \psi \mid \Box\phi \mid \Diamond\phi$$

Here, ::= denotes that the symbol to the left should be replaced by one of those on the right, and | denotes a choice among different such substitutions.

1.6 The Uses of Logical Inference

Translations into logical language can to some extent be clarifying in themselves. This applies for instance to translations from the rather erratic quantifiers of natural language to the more regular ones of predicate logic. But the most important advantages of formalization are only obtainable when we go beyond mere translation, and investigate, with logical tools, the properties of the models that we have built. It is a major advantage of formal models that they are so precisely described that such properties can be determined with certainty. Their major disadvantage, of course, is that these properties may be different from those of that which they are a model of. Efficient use of formal models requires both that we investigate the formal properties of our models *and* that we critically evaluate how these properties relate to those of the phenomena that led us to develop the models.

This is not the place to delve into the methodology of logico-mathematical work, how to construct axioms and prove lemmas and theorems. Instead, this section is devoted to the connections between the construction of a system of logical inferences and the process of formalization. Section 1.6.1 discusses the choice between extensional and non-extensional logic for sentential operators. Section 1.6.2 shows how logical analysis can reveal distinctions that are less obvious in natural language, thereby contributing to the development of new philosophical concepts. This is followed by a discussion of how logical analysis can lead to improvements of the formal framework itself. Sometimes minor adjustments are sufficient (Sect. 1.6.3). On other occasions, logical analysis forces us back to the drawing-board in search for a better formal model (Sect. 1.6.4).

Box 1.2 Two ways to use a name
"Charles-Édouard Jeanneret" was the legal full name and "Le Corbusier" the pseudonym of a famous architect. Consider the following sentences:

(1) Le Corbusier was born in 1887.
(2) Charles-Édouard Jeanneret was born in 1887.
(3) Le Corbusier is a pseudonym.
(4) Charles-Édouard Jeanneret is a pseudonym.

In the first two sentences, the names refer to the person. These two sentences have the same truth conditions, and they are indeed both true. In the last two sentences, the names refer to themselves, and these two sentences do not have the same truth conditions. (3) is true and (4) false.

In sentence (1), "Le Corbusier" is used *extensionally*, by which is meant that the truth-value of this sentence is not changed if "Le Corbusier" is replaced by another expression with the same extension. (The extension of an expression is the collection of objects to which it refers, in this case a collection consisting of one person.) In sentence (3), "Le Corbusier" is used non-extensionally.

1.6.1 Intersubstitutivity of Logical Equivalents

It is important in philosophy to distinguish between extensional and non-extensional uses of an expression. (See Box 1.2 for a reminder.) Therefore, when constructing the logic of a sentential operator, we have to decide whether to give it an extensional or a non-extensional logic.[7] We can illustrate this with the sentences about the Dodo (*Raphus cucullatus*) that can be formed with the following notation:

d The Dodo is extinct.
r *Raphus cucullatus* is extinct.
E There is sufficient scientific evidence that ....
K Alix knows that ....

Since r and d are equivalent, so are Er and Ed. More generally speaking, if we can replace a sentence attached to the operator E by an equivalent sentence, then the truth-value is not changed. An operator with this property is said to be *extensional* or satisfy *intersubstitutivity of logical equivalents*.

The operator K does not have this property, since Kr and Kd are not logically equivalent. It is both possible and quite common to know that the Dodo is extinct

[7]Rudolf Carnap [9, pp. 57–63] claimed that all non-extensional concepts can be reconstructed as extensional, but his mode of reconstruction has not caught on and does not seem to be practicable.

without knowing that *Raphus cucullatus* is extinct. This is a feature that K shares with most other operators representing attitudes such as believing, doubting, wishing, preferring etc.

When constructing a logical system that contains sentential operators, it is important to specify which of these operators satisfy intersubstitutivity and which do not. (Note that this is a property of the inference pattern applied to the language, not a property of the language itself.) It might seem obvious that a logical operator that represents a non-extensional concept in natural language, such as a knowledge or belief operator, should also be non-extensional. In practice, however, it is quite common to use extensional operators to represent non-extensional concepts. The reason for this is that non-extensionality usually comes with a high price: it makes the logic of the operator so weak that very little can be proved. Intersubstitutivity of logical equivalents is an idealization that allows us to have a much richer logic to work with. There are two major ways to justify that idealization.

The most common justification is that the "non-extensional" uses of the concept can in most cases easily be identified. We can therefore use an operator with an extensional logic and just bear in mind that it is inadequate to deal with problems where non-extensional properties of the underlying concept have a role. We can for instance develop a logic of belief with an extensional belief operator B and assign properties to it such as $Ba \;\&\; Bb \rightarrow B(a\&b)$ and $Ba \rightarrow \neg B\neg a$. A disadvantage with this approach is that the outer limits of the logic's area of application cannot be specified in precise terms.

The other, somewhat more sophisticated, approach is to change the interpretation of the operator so that it does not refer to the original ordinary-language concept but to some variant of it that can be expected to allow for the substitution of logical equivalents. For the belief operator such a reinterpretation has been proposed by Isaac Levi [50, 51]. His solution is to interpret B as referring to what the agent is committed to believe rather than what she actually believes.[8] A person who believes in the above statement d ("The Dodo is extinct"), is also committed to believe in r ("*Raphus cucullatus* is extinct") upon understanding its meaning. This approach has the advantage over the previous one that the delimitation is more precise and therefore more accessible to criticism and improvement.

For another example, consider again the predicate O of moral requirement. In deontic logic, O is usually taken to be extensional. But examples are not difficult to find in which this assumption gives rise to strange results. Let a_1 signify that John kills his wife's murderer, a_2 that he kills only other persons than his wife's murderer, and b that he does not kill anybody at all. Then $\neg a_1$ is logically equivalent with

[8]Arguably, this interpretation deviates from the common understanding of what it means to be committed to something. In ordinary parlance, commitment seems to be subject to a "committed implies can" restriction that parallels the "ought implies can" restriction. If I am committed to believe in all true mathematical statements, then this is a commitment in an entirely different sense from that in which I am committed to keep my promises and repay my loans. In a more exact analysis, such a commitment would have to be conditional on knowledge or knowability.

$a_2 \vee b$. If O is extensional, then it follows from this that $O\neg a_1$ and $O(a_2 \vee b)$ are equivalent, and in particular that

$$O\neg a_1 \rightarrow O(a_2 \vee b)$$

In words: If John ought not to kill his wife's murderer, then he ought to kill either only other persons than his wife's murderer, or no one at all. This is the revenger's paradox [31]. It can be avoided by giving up intersubstitutivity. However, that would be a far-reaching weakening of deontic logic. Therefore, just as in epistemic logic, it is customary in deontic logic to retain intersubstitutivity in spite of the problems that it can give rise to. In the case of deontic logic, one way to justify this is that the sentences that cause trouble tend to be expressed in misleading ways so that we easily overlook that they are synonymous. For instance, although the two sentences "John is obliged not to kill his wife's murderer" and "John is obliged to either kill only other persons than his wife's murderer, or no one at all" mean exactly the same thing, only the second makes it explicit that no prohibition to kill other persons than his wife's murderer is pronounced. In deontic logic, just as epistemic logic, the tradition is to accept extensionality and avoid the "intensional contexts" that give rise to trouble.

1.6.2 Logical Inference as a Means to Discover New Concepts

We can use the treatment of moral dilemmas in deontic logic as an example of how logic can be used to analyze philosophical concepts in a precise way that also gives rise to new philosophical concepts. Suppose that there is some action representable by the sentence a, such that both Oa and $O\neg a$ hold, in other words both a and not-a are morally required. This means that the dictates of the O operator cannot be completely complied with. This is the most obvious case of a moral dilemma. Indeed, moral dilemmas are often defined as situations with two conflicting obligations.

But need they be two? Suppose that someone needs to be able to reach me urgently, so that I am morally required to keep my mobile phone on. At the same time I am, for quite different reasons, obliged to be in the audience when my child performs in a school play. But members of the audience are required to keep their mobile phones turned off during the performance. Letting a denote that I have my phone on and b that I attend the performance, I am then under the three obligations Oa, Ob, and $O\neg(a\&b)$. It is easy to check that each combination of two of these three obligations is fully compatible, so there is no dilemma according to the standard definition that refers to two conflicting moral requirements. Still, the situation seems dilemmatic enough. The reason for this is of course that the combined contents of all three obligations is inconsistent. This should lead us to define moral dilemmas in terms of such combined inconsistency rather than in terms of two conflicting obligations.

For another example, suppose that I am morally required both to be in Stockholm at 10.00 a.m. (c) and to be in the neighbouring town Uppsala at 10.30 a.m. the same day (d). This is by no means logically impossible; it would indeed be practically possible if I had access to a helicopter. But I don't. The set consisting of the two sentences c and d is not logically inconsistent. Still, this appears to express a true moral dilemma. Although the set in question is not inconsistent, it is impossible to satisfy the contents of both its elements. If we want examples like this to be regarded as moral dilemmas, then we will have to revise our definition of dilemmas so that it refers to impossibility rather than inconsistency. Realizing all elements of a finite set of sentences means the same as realizing their conjunction. We can therefore express this condition with a possibility operator $\Diamond$: A set of obligations $\{Oa_1, Oa_2, \ldots, Oa_n\}$ gives rise to a moral dilemma if and only if $\neg \Diamond (a_1 \& a_2 \& \ldots \& a_n)$, where $\Diamond$ denotes possibility.

In this way, we have generalized our original notion of a moral dilemma to the more general notion of (lack of) joint possibility (compossibility) of a set of moral obligations. This opens up for further distinctions since there are different notions of possibility. We can now speak of moral dilemmas of different types, depending on how we interpret $\Diamond$. If we interpret it as logical possibility, then we are concerned with "moral dilemmas with respect to logical possibility" which are of course much fewer than the "moral dilemmas with respect to practical possibility" that we obtain with a weaker interpretation of $\Diamond$.

Once we have formulated the issue of joint possibility of a set of moral obligations, we can generalize it further, and discuss the joint possibility of sets of norms that may contain permissions. Should we treat permissions in the same way as obligations? In other words, must the contents of a set of permissions be jointly possible in order for the set of permissions to be consistent? It is easy to show that such a requirement would be unreasonable. Just consider the set $\{Pa, P\neg a\}$, where P denotes permission and a that you take part in the weekly ceremonies of a local religious establishment. Since $a \& \neg a$ is inconsistent, such a requirement would render this set of permissions inconsistent. This is unconvincing since the very idea of religious freedom is to let us choose between such, mutually incompatible alternatives. For a set of permissions to be consistent, it seems to be sufficient that each of them, taken alone, is consistent (or possible).

Next, let us consider sets of norms that contain both obligations and permissions. There is a rather obvious way to combine the above two criteria into a single criterion that covers this more general case: Each combination of the contents of all the obligations with that of any single one of the permissions should be jointly possible. This criterion was proposed by Georg Henrik von Wright (1916–2003) ([82]; cf. [26]). It seems to work fairly well when put to test in various examples. However, calling all situations in which some permission cannot be used a "dilemma" would seem to stretch the term too far. Therefore, it may be better to use a different terminology for these cases, such as the following:

> A set of norms (obligations and permissions) is *compossible* if and only if the set consisting of the contents of all its elements is jointly possible.
>
> A set of norms (obligations and permissions) is *realizable* if and only if each subset containing all the obligations and at most one of the permissions is compossible.

> A set of norms (obligations and permissions) gives rise to a *moral dilemma* if and only if the subset consisting of all its obligations is not compossible.

As we have already noted, these definitions come in different variants, depending on the standard of possibility that we apply.

We have just expressed these distinctions in natural, rather than formal, language. So what was the point of formalization in this case? The point is that it is no coincidence that these distinctions were developed by deontic logicians, rather than by moral philosophers working without the aid of a formal language. In this and many other cases, the formal language directs our attention to inference-related considerations that turn out to be helpful for the development of philosophical terminology. The usefulness of formal models is confirmed, not disconfirmed, when they give rise to distinctions that can also be expressed and used in informal philosophy.

1.6.3 Reconsidering the Formalization: Splitting Concepts

As emphasized in Sect. 1.4.4, the process of formalization should include careful consideration of whether or not terms from natural language can be treated uniformly in the formal language. However, in spite of the formalizer's best efforts, it is not uncommon that once rules of logical inference have been introduced, new problems are discovered that reveal a need to modify the original formalization. On occasions, a need for additional splitting of concepts is discovered. Consider the following two sentences, said to someone who beats a cat:

(1) "You must stop beating Mei-Yin."
(2) "You are not allowed to be cruel to animals."

(1) differs from (2) in offering a norm for only one situation, namely the present one. In contrast, (2) exemplifies the most common type of norms referring to several situations, namely normative rules.[9] This distinction is not easily extracted from deontic discourse in natural language, since most languages use the same linguistic forms for both purposes. This applies to conditional as well as non-conditional norms. Consider the following examples:

(3) "If a president from the left is elected, then rich people will have to pay more taxes."
(4) "If you bribe the head of department, then you will be permitted to take part in the extra retake."
(5) "If you borrow money, then you must pay it back."

[9]This distinction was made in [28]. Similarly, Carlos Alchourrón [1] distinguished between "a norm for a single possible circumstance (which may be the actual circumstance)" and a norm for "all possible circumstances", and David Makinson [59] distinguished between norms "in all circumstances" and norms "in present circumstances".

(6) "If you pay the exam fee at least one week in advance, then you will be permitted to take part in the extra retake."

(3) and (4) are conditional statements saying that *if* the situation satisfies (will satisfy) a certain characteristic, *then* certain actions are (will be) obligatory, respectively permitted. These statements do not report any normative rules; they only tell us what will be the case (normatively) under certain conditions. Contrastingly, (5) and (6) express rules stating that in situations satisfying the given criteria, a particular norm holds. The similarity between (4) and (6) illustrates that linguistic form does not help us to distinguish between the two types of statements. In fact, natural language provides no cue about the different types of conditionality in (4) and (6). It is our knowledge of what legal and administrative rules usually look like that makes us infer that (6) reports a permissive rule and (4) a statement about what will in fact be permitted under certain circumstances.

In order to explore the logical significance of this distinction, we can use the standard notation for conditional obligation and permission: We write $O(a \mid b)$ for "a is obligatory, given b", and similarly $P(a \mid b)$ for "a is permitted, given b".[10] Now consider the following two logical principles:

If b is true and Oa holds, then so does $O(a \mid b)$.
If b is true and Pa holds, then so does $P(a \mid b)$.

Let us first try them out on statements expressing situation-specific norms. Suppose that before the presidential election I made the statement denoted (3) above. A left-wing president is elected, and after the election it turns out that rich people are indeed required to pay more taxes. It would then be strange to claim that what I said was wrong. In particular, a rebuttal could not be based on the claim that (3) does not hold in general – the statement only referred to the specific situation. The same analysis applies, perhaps even more clearly, to statement (4). In fact, these principles apply, although perhaps less obviously, if the sentences a and b are completely unrelated. This is due to properties of "if... then ..." that are unrelated to the normative component of the sentences. In a non-normative context, we would admit the following inference as valid (albeit somewhat awkward):

Xiu-xiu has a blue shirt.
Xiu-xiu knows the ancient Greek language.
______
If Xiu-xiu has a blue shirt, then she knows the ancient Greek language.

For the same reason we should accept the following inference:

Xiu-xiu has a blue shirt.
Xiu-xiu is permitted to read classified government documents.
______
If Xiu-xiu has a blue shirt, then she is permitted to read classified government documents.

[10] This notation was introduced by Bengt Hansson [27].

But this can only be true provided that we do not read the conditional statement as expressing a normative rule. From the facts that Xiu-xiu has a blue shirt and that she is permitted to read classified government documents we certainly cannot conclude that there is a rule to the effect that if she has a blue shirt, then she is permitted to read these documents. And it is easily checked that the two inference rules do not hold for rule-reporting normative statements such as (6). From the two facts that I paid the exam fee more than one week in advance and that I was permitted to take part in the exam, it does not necessarily follow that there is a rule to the effect that if one pays the fee within this time then one is allowed to take part in the exam.

Since situation-specific and rule-expressing norms are expressed in the same way in natural language, the distinction between them has often gone unnoticed. It received attention when the formal structure was put to test in logical investigations, and it turned out that they differ in what logical rules they obey. The logical differences between situation-specific and rule-expressing norms is nevertheless a good reason to make this distinction in both formal and informal philosophy, despite the fact that ordinary language does not distinguish between them.

1.6.4 Reconsidering the Formalization: Radical Reform

To illustrate how logical investigations can reveal the need for a radical reform of a formalization, we can consider the problem of so-called free-choice permissions [81, pp. 21–22]. These are permissions for someone to make a choice, for instance:

> You are allowed to marry either a man or a woman.
>
> The surgeon is permitted to take out either the patient's left or his right kidney, and transplant it to the patient's daughter.

An obvious first attempt to formalize free choice permission is to represent "or" with ordinary truth-functional disjunction ($\vee$), and this is indeed the formalization that was the starting-point of the discussion. It would then seem rather obvious that the following postulate should hold:

$$P(a \vee b) \rightarrow Pa \,\&\, Pb$$

This postulate looks innocuous when presented in connection with an example of permitted choice. However, if we also require intersubstitutivity for logically equivalent sentences, then we can make derivations with highly implausible outcomes, such as the following [44, pp. 176–177]:

$P((a\&b)\vee(a\&\neg b)) \rightarrow P(a\&b)\ \&\ P(a\&\neg b)$

(a substitution instance of the postulate)

$Pa \rightarrow P(a\&b)\ \&\ P(a\&\neg b)$

(follows from the intersubstitutivity of logically equivalent sentences)

$Pa \rightarrow P(a\&b)$

The endpoint of this derivation is obviously absurd, and it gives us reason to either give up the formalization of free choice permission as $P(a \vee b)$, or else modify the framework in which the derivation took place. Following the first line, some authors have tried to solve the problem by replacing the standard permission operator P by some other operator, but such alternative operators have invariably been shown to have implausible properties [41]. The underlying problem is that all these constructions are based on the assumption that free choice permission to p or q can be represented as a property of the sentence $p \vee q$. However, if intersubstitutivity holds, then this *single sentence assumption* is not at all plausible. The reason for this is that it has the following rather obvious consequence:

If $a \vee b$ is equivalent with $c \vee d$, then there is a free choice permission to a or b if and only if there is a free choice permission to c or d.

It is not difficult to find examples showing that this leads to absurd conclusions:

The vegetarian's free lunch [41]
In this restaurant I may have a meal with meat or a meal without meat. Therefore I may either have a meal and pay for it or have a meal and not pay for it.

Proof
Let m denote that you have a meal with meat, v that you have a meal without meat, and p that you pay. $((m\vee v)\&p) \vee ((m\vee v)\&\neg p)$ is equivalent with $m \vee v$. Therefore, it follows from the single sentence assumption that $((m\vee v)\&p) \vee ((m\vee v)\&\neg p)$ is (free choice) permitted if and only if $m\vee v$ is (free choice) permitted.

To sum up, in a framework with intersubstitutivity of logical equivalents, (free choice) permission to perform either a or b cannot be represented as a function of the single sentence $a\vee b$. Instead, we can treat it as a function of the two sentences a and b, i.e. as a function of two variables, not one. Similarly, (free choice) permission to perform either a, b, or c can be treated as a function of three variables, etc. Alternatively, we can treat free choice permission as a property of a set of action-describing sentences ($\{a, b\}$ respectively $\{a, b, c\}$) in these examples) [41]. In this case, logical investigations of what initially seemed to be a quite straightforward formalization revealed the need for a rather drastic reform of that representation. In the process, we also learned something about the underlying informal notion that would otherwise not have been easy to discover.

1.7 Going Beyond Logic

The previous sections have been devoted to the use of logic in philosophy, and this for good reasons. Much philosophical subject matter is well represented by sentences, and logic provides us with powerful tools to investigate how sentences connect with each other.

But in spite of these advantages, there are no a priori grounds why logical languages should be better suited than other symbolic languages for modelling each and every subject matter studied by philosophers. In some cases, other formal approaches can capture features of the subject matter that are difficult to express in logic. It is also important to note that there is no clear demarcation between logical and "non-logical" formal methods. Arguably, much if not most of mathematics can be reconstructed in a logical framework, and conversely, logic can be seen as a branch of mathematics. But for practical purposes we can distinguish between those symbolic languages that are taught in courses and textbooks on logic and those that one has to learn elsewhere. The following subsections will briefly introduce three formal approaches of the latter category that have fairly widespread use in philosophy, namely numerical models, decision matrices, and choice functions.

1.7.1 Numbers

Numbers are ubiquitous in most of the sciences. Physicists, economists, ecologists, demographers, and scientists of almost any other discipline make frequent use of models whose variables take numerical values. Philosophy is an exception, and this for a reason that we discussed in Sect. 1.2.4: The variables that are relevant in philosophy usually cannot be correlated with empirical measurements, and therefore the most important advantage that numerical models have in other disciplines does not apply in philosophy. But nevertheless, there are cases when models involving numbers are useful in philosophy.

In value theory, it is often assumed that *value*, for instance moral value, can be expressed numerically. Moral value can then be represented by a function u, such that for each object a of evaluation, $u(a)$ is a number that represents its value. Since there is no measurement-based unit for moral value, a fictive unit is employed, often called "utile" or "util". But although the unit is elusive, the use of a numerical value function imposes a structure with considerable impact on how value is conceived. In particular, it allows us to add and multiply values. If some event has the consequences a, b, and c, we feel free to speak of their total value and calculate it as $u(a)+u(b)+u(c)$. This, of course, is the basic structure of utilitarian moral philosophy. In contrast, deontological ethics is usually conceived in terms of the binary distinction between duties and non-duties, and therefore it has much less use for quantitative measures of value.

Standard logical models are not good at representing *time*. The reason for this is that we usually assume that for any two (non-identical) points in time, there is some third point in time that is posited between them. Such a structure is beyond elementary logic; the best way to introduce it is to employ rational (or real) numbers.

A third important area for numerical representation is *probability*. It can be introduced through a function p on event-representing propositions, such that if a is an event, then $p(a)$ is the probability of that event. Probabilities are used in epistemology, decision theory, and many other areas of philosophy. They can be given either an objective or a subjective interpretation. "Objective" probabilities represent frequencies or tendencies pertaining to events in nature. "Subjective probabilities" represent an agent's degree of belief in statements. Notably, the term "probability" should only be used about measures that have the same mathematical properties as the objective probabilities that we know from examples with coins, dice and other randomizing devices. Mathematically, this means that probabilities have to satisfy the Kolmogorov axioms.[11] It can plausibly be argued that our subjective degrees of belief should be represented by degree-of-belief functions that do not satisfy these axioms, but then they should not be called "probabilities".

In epistemology, probabilistic and logical models have complementary strengths and weaknesses. Logical models can provide us with a reasonable account of the inferential relationships among beliefs, in other words how acceptance of one belief can lead us to accept or reject some other belief. However, logical models have difficulties in representing the relations of strength among beliefs, i.e. how one belief can be stronger or weaker than another. For probabilistic models it is the other way around. They can provide good accounts of the differences in strength among beliefs, but not of the inferential connections among them [57]. Neither type of model is well suited to represent both these aspects of belief systems. That is why we need them both.

1.7.2 Decision Matrices

In a formal model of decision problems, several prominent components need to be represented. There is a set of *alternatives* that the decision-maker can choose among. In many real-life problems, the set of alternatives is open in the sense that new alternatives can be invented or discovered [32, 43]. A typical example is your decision how to spend tomorrow evening. In other decision problems, the set of alternatives is closed, so that no new alternatives can be added. Your decision how to vote in the upcoming elections will probably be an example of this. There will be

---

[11] See Chap. 19.

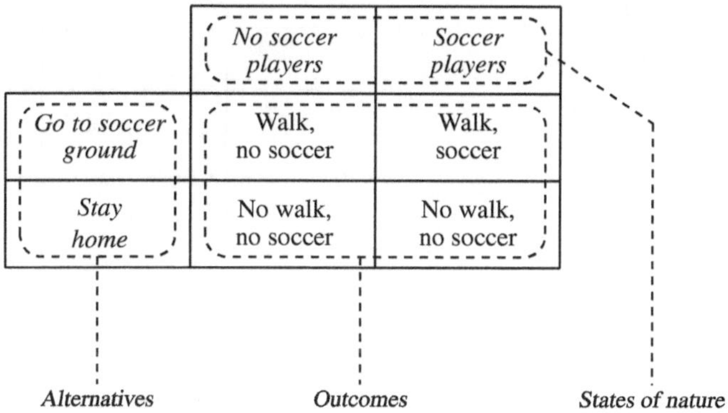

Fig. 1.7 The basic construction of a decision matrix

a limited number of alternatives (candidates or parties) that you can choose among. In decision theory, the alternative set is almost invariably assumed to be closed. The major reason for this is that formal treatment is much easier if the alternative set is closed. For the same reason, it is also commonly assumed that the alternatives are mutually exclusive, i.e., it is not possible to choose more than one of them.

The effects of a decision depend not only on the decision-maker's choice but also on various factors beyond her control. In decision theory, these extraneous factors are usually summarized into a number of cases, called *states of nature*. The states of nature include natural events but also decisions by other persons. As an example, consider a young boy, Peter, who makes up his mind whether or not to go to the local soccer ground to see if there is any soccer going on that he can join. The effect of that decision depends on whether there are any soccer players present. In decision theory, this situation can be described in terms of two states of nature, "players present" and "no players present".

The possible *outcomes* of a decision are determined by the combined effects of the chosen alternative and the state of nature that turns out to prevail. Hence, if Peter goes to the soccer ground and there are no players present, then the outcome can be summarized as "walk and no soccer". If he goes and there are players present, then the outcome is "walk and soccer". If he does not go, then the outcome is "no walk and no soccer".

The basic idea of a *decision matrix* is to tabulate alternatives against states of nature in order to show which outcome results from each combination. The decision matrix for Peter's decision is shown in Fig. 1.7. Such a matrix provides a clear presentation of the decision, but it does not contain all the information that the decision-maker needs in order to make the decision. The most important missing information concerns how the outcomes are valued and how plausible the states of affairs are.

The *values of outcomes* are usually expressed with numbers. Sometimes an empirical value measure is available, such as economic costs or gains, or the number of persons killed in an accident. But often fictitious numbers have to be used. In our

Fig. 1.8 A utility matrix

	No soccer players	*Soccer players*
Go to soccer ground	0	10
Stay home	3	3

Fig. 1.9 A probabilistic utility matrix

	.7	*.3*
Go to soccer ground	0	10
Stay home	3	3

example, we may for instance assume that Peter assigns the value 0 to walking to the soccer ground but finding no opportunity to play soccer, the value 3 to staying at home, and the value 10 to walking to the soccer ground and playing soccer there. We can then replace the basic decision matrix of Fig. 1.7 by a *utility matrix* (*payoff matrix*) in which these values take the place of the outcome descriptions, see Fig. 1.8.

Peter's decision will be influenced by how probable he believes it to be that there are any players at the soccer ground. Suppose that he takes this probability to be 0.3. Then he can replace the states of nature by probabilities, as in Fig. 1.9. This type of matrix is the starting-point in much of decision theory.

Game theory differs from decision theory in that there are two or more agents, each of whom has a set of alternatives to choose among. In game theory it is usually assumed that the outcome depends only on the decisions of the agents, so that no distinction is made between different states of nature. (This is an idealization that may of course sometimes be problematic.) In the basic game theoretical matrix for two agents, the decisions of the agents are tabulated against each other, and the outcome is determined by the combinations of their decisions. Figure 1.10 shows an example of this. Two agents, Rosa and Carmen, are going to meet for a meal. Rosa will make the food and Carmen will bring a bottle of wine. Rosa prefers red wine for meat and white wine for fish, whereas Carmen prefers white wine for all kinds of food.

Just as in decision matrices, the outcome descriptions of game matrices are often replaced by numerical values representing the values of the outcomes. In games it is important to distinguish between values for the different players. Therefore, outcome values are represented by vectors. When there are two agents, the vector $\langle x, y\rangle$ represents a situation in which the agent choosing among the rows in the

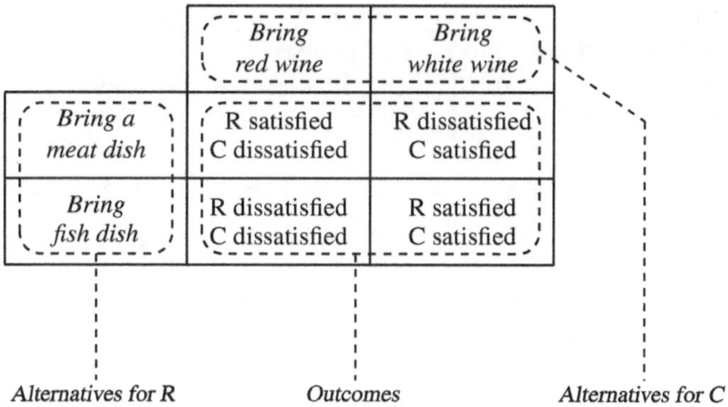

Fig. 1.10 The basic construction of a game matrix for two agents (players)

Fig. 1.11 A utility matrix for the same game as in Fig. 1.10

	Bring red wine	*Bring white wine*
Bring a meat dish	$\langle 1, 0\rangle$	$\langle 0, 1\rangle$
Bring fish dish	$\langle 0, 0\rangle$	$\langle 1, 1\rangle$

matrix assigns the value x to the outcome, whereas the other agent assigns the value y. Figure 1.11 is a utility version of the matrix in Fig. 1.10.

Both decision matrices and game matrices have turned out to be quite useful in moral and political philosophy. In particular, game matrices put focus on coordination problems that are not so easily treated in traditional logic-based models. Game matrices are also increasingly used in (social) epistemology in order to capture collective epistemic processes.

1.7.3 Choice Functions

A *choice function* is a representation of an agent's choice tendencies. If C represents your choice tendencies, then $C(\{x, y, z\}) = \{x\}$ means that if you have to choose among x, y, and z, then you will choose x. Choice functions are usually applied to sets whose elements are mutually exclusive, and they allow for ties; thus $C(\{x, y, z\}) = \{x, y\}$ means that in the choice among x, y, and z you have a tendency to choose either x or y, but you have no inclination to choose one of these rather than the other. Choice functions have an important role in decision theory, in particular in social decision theory where a central issue is how the choices of individuals can be combined into a social choice.[12]

In addition, choice functions have turned out to be useful in several other applications, such as belief revision, non-monotonic reasoning, and the logic of conditionals. Consider an agent who initially believes in the two statements a and b, but then receives information showing that they cannot both be true. In the terminology of belief revision, the agent has to contract by the sentence $a\&b$. She must then give up either her belief in a or that in b, or both. A choice function can be used to model such choices. Virtually all belief revision models make use of choice functions, but they differ in what the choice functions are applied to: beliefs to remove, beliefs to retain, belief states that can be the outcome of the operation, etc. The application of choice functions to different types of objects in a model of a human belief system can give rise to operations of belief change with different properties [42].

1.7.4 Combinations

The use of non-logical formal tools does not mean that logic is discarded. To the contrary, the different non-logical tools are often combined with components from logic. It is for instance convenient to apply the utility function u to sentences that represent the states of affairs under evaluation, and the same applies to the probability function p. Increasingly, logicians are working with "hybrid systems" that combine logic's unsurpassed ability to represent statements and their interrelations with various non-logical formal tools that enable us to treat and rearrange these statements in ways that logic alone does not have resources for: choose among them, assign values and probabilities to them, arrange them in temporal order, etc. Such hybrid systems can sometimes make it possible to combine the advantages of two or several formal representations.

1.8 Aberrations in Formal Models

In spite of all its advantages, formalization is not always useful. In some cases it has given rise to more confusion than clarity. And even when it is useful, it is seldom if ever without problems. As pointed out in Sect. 1.2.4, a formal model is always the outcome of a trade-off between simplicity and faithfulness to the object of study. If the subject-matter is complex, then a reasonably simple model will usually have to leave out some of its philosophically relevant features.

[12] See Chap. 37.

Due to this trade-off, an uncriticizable formal model of philosophical subject matter is in practice unachievable. It will always be possible to develop a criticism that puts focus on the simplifications that are inherent in the model. However, even if such a counter-argument convincingly discloses an imperfection in the model, it does not necessarily follow that the model is unfit for use. If a problem in the model cannot be solved without substantial losses of simplicity, then it may be appropriate to continue using the model, bearing in mind its weaknesses (and perhaps supplementing it with other models that have other strengths and weaknesses). The same applies, of course, to inaccuracies in informal models and approaches in philosophy. The "adversary method" [63] in philosophy which takes any flaw in a philosophical theory as proof that the theory should be rejected in toto, is equally misguided in formal as in informal philosophy.

This is the reason why this section is called "Aberrations in formal models", rather than for instance "Faults in formal models". Depending on what we use the model for, some aberrations may be acceptable whereas others are not.

We can divide aberrations in formal models into two major types: those concerning what can be expressed in the model's language and those concerning what can be inferred in the model. Each of these types can be further subdivided depending on whether the aberration concerns an unjustified addition to what can be expressed respectively inferred, or an unjustified subtraction from it.

1.8.1 Aberrations of Expression

Almost all formal models have a conspicuous deficit in what they can express. This is mainly because in order to construct a workable formal model, the number of primitive notions has to be kept to a minimum. A few examples will be sufficient to show the rather drastic limitations in the expressive power of most formal languages. In formal value theory, only a few value concepts are represented, primarily "better" and "at least as good as", and those that can be defined from these, such as "best" and perhaps "good". In contrast, ordinary language is rich in value terms, most of which are seldom if ever included in formal accounts: "acceptable", "fairly good", "worthless", "invaluable" etc. As discussed in Sect. 1.4.4, deontic logic usually has only one concept of moral requirement (O), whereas natural language has a whole collection of prescriptive terms that differ in strengths and connotations, such as "must", "should", "ought", "has to", "duty", "obligation", etc. Epistemic logic has its focus on representations of the two terms "know" and "believe", mostly leaving out other epistemic terms such as "assume", "guess", "be convinced", "doubt", "be aware that", "have a hunch that", "suspect", etc.

Some formal languages contain superfluous expressions that do not correspond to anything meaningful that can be said about their subject matter. One way in which this comes about is through the formation rules of logical languages. If the language contains a sentential operator G for "good", then the formation rules allow us, for any sentence a, to form a statement Ga meaning "a is good". Then this will also apply to tautologies and contradictions, and we can form sentences such as $G(b \vee \neg b)$ and $G(c\&\neg c)$ that do not seem to have a meaningful interpretation. For similar reasons, the language of deontic logic contains the sentence $O(a\&\neg a)$. It seems to express some form of moral impossibility, but it does not correspond to how we normally think about moral conundrums. Although one can say "I am obliged both to be here and not to be here", this refers to two separate obligations. ("I am obliged to be here and I am also obliged not to be here", Oa & $O\neg a$.) It does not refer to a single obligation to do something impossible ($O(a\&\neg a)$).

Such superfluous expressions can, if we so wish, be excluded from the language. Technically, this requires somewhat more complicated language formation rules than the conventional ones. In our examples, we can postulate that G and O can only be affixed to sentences that are neither logically true nor logically false. However, most logicians would be reluctant to employ language formation rules that refer to what can be logically inferred. There are good reasons to construct the language prior to, and independently of, the rules of inference. Therefore, it is much more common to retain these artefacts in the logical language, and either treat them as uninterpreted anomalies or, if possible, provide them with an interpretation that corresponds to the conditions under which they can be inferred. We can for instance treat $O(p\&\neg p)$ as an indicator of the presence of a moral conflict or dilemma.

The choice between these different ways to deal with artefacts in the logical language is largely a matter of convenience, and not very important. What is important, however, is that we do not take it for granted that all expressions in a formal language are meaningful just because they are constructed from meaningful language elements.

1.8.2 Aberrations of Inference

In some cases, the formal language does not support inferences that are reasonable and can be drawn in ordinary language. One example of this is the inference from "a is permitted" to "not-a is permitted" that we can draw in ordinary language with its bilateral notion of permission, but not in deontic logic with its unilateral notion.

But the major problem with inferences is usually the opposite one: formal models tend to support excessive inferences, i.e. inferences that are allowed by the formal system but do not correspond to any properties of that which is modelled. Arguably, most of the more problematic aberrations in formal models consist in such superfluous inference patterns. In Sect. 1.6.1 we noted that the intersubstitutivity of logically equivalent sentences produces superfluous inferences, but we also noted that for many purposes this may be an aberration that is worth its price.

A somewhat related idealization is the use of logically closed sets for various purposes in formal models, perhaps most conspicuously to represent an epistemic agent's set of beliefs. A set of sentences is logically closed if and only if everything that follows logically from it is among its elements. Hence if both a and $a \rightarrow b$

are elements of a logically closed set, then so is b. In the logic of belief revision, a logically closed set (called a "belief set") is the standard representation of an agent's beliefs. Since all mathematical truths that are expressible in a language are logical truths in that language, this means that she believes in all mathematical truths that can be expressed in the language. Such logico-mathematical omniscience is of course far beyond human capabilities. The best justification for this aberration from our actual doxastic behaviour seems to be the reinterpretation of belief sets proposed by Isaac Levi: They do not represent what an agent actually believes but what she is committed to believe. (Cf. Sect. 1.6.1 where this solution was applied to the belief operator.)

Another interesting example of excessive inferences can be taken from deontic logic. Consider the following three properties of a deontic logic:

Existence of moral dilemmas:
There are action-describing sentences a and b such that Oa & Ob, although a&b is logically inconsistent.

Agglomeration:
If Oa and Ob then $O(a\&b)$.

Necessitation:
If Oa, and a logically implies b, then Ob.

Each of these principles has immediate intuitive appeal, as can easily be confirmed with examples. But in combination they lead to an absurd conclusion. According to Existence of moral dilemmas, there are sentences a and b such that Oa & Ob and a&b is logically inconsistent. According to Agglomeration, $O(a\&b)$. Since a&b is inconsistent, it holds for any sentence c that a&b implies c. Necessitation yields Oc, and we have proved the following remarkably undesirable property:

Universal obligatoriness:
Oc

Obviously, the formal inference from Oa, Ob and the inconsistency of a&b to Oc does not correspond to how we normally reason or argue about our moral obligations. From "I ought to be at home with my children this evening" and "I ought to work all night at the office", we do not conclude "I ought to spend this evening boozing in a nightclub". Therefore, the derivation of Universal obligatoriness from three seemingly quite plausible postulates is a logical artefact that has nothing to do with the subject matter of deontic logic. Universal obligatoriness is so damaging that any system implying it will have to be rejected. Consequently, a workable system of deontic logic cannot contain all three of the principles Existence of moral dilemmas, Agglomeration, and Necessitation. The most common solution is to give up Existence of moral dilemmas. But for some purposes, such as the study of moral dilemmas, one of the other two principles will have to go instead.

If we replace necessitation by the weaker assumption of intersubstitutivity,

Intersubstitutivity of logical equivalents:
If Ox and x is logically equivalent with y, then Oy.

then the derivation of Oc (universal obligatoriness) will be blocked, but we can instead derive $O(c\&\neg c)$ (obligatory inconsistency) from Oa and Ob, given that $a\&b$ is logically inconsistent. This is also an artefact of the formal model, but as argued in Sect. 1.8.1, $O(c\&\neg c)$ does not do much damage. Arguably, it can be tolerated, and treated as an innocuous artefact of the formal system. Possibly, it can even be given a meaningful interpretation, as an indicator of the presence of an inconsistency.

1.8.3 Conclusion

In philosophy, like other disciplines, formal models are useful tools that allow us to express ideas more precisely and to probe their implications. As in other disciplines, we can only use formal models efficiently if we keep track of their strengths and weaknesses. Since all formal models are idealizations, they all have imperfections, and we should never expect to find the uniquely best formal model that will tell us the whole truth and nothing but the truth about some philosophical subject matter. But there can be no doubt that formal models are indispensable tools in philosophical investigations. Today, no philosopher can afford to be ignorant of how they can contribute to new philosophical insights.

References and Proposed Readings

Asterisks (∗) indicate recommended readings.

1. Alchourrón, C. (1993). Philosophical foundations of deontic logic and the logic of defeasible conditionals. In J.-J. C. Meyer & R. J. Wieringa (Eds.), *Deontic logic in computer science* (pp. 43–84). Chichester: John Wiley & Sons.
2. Aristotle. (1928). Prior analytics (A. J. Jenkinson, Trans.). In W. D. Ross (Ed.), *The works of Aristotle* (Vol. 1). London: Oxford University Press.
3. Aristotle. (1942). *The Nicomachean ethics* (W. D. Ross, Trans.). London: Oxford University Press.
4. Berlin, I. (1969). *Four essays on liberty*. Oxford: Oxford University Press.
5. Bobzien, S. (2002). The development of modus ponens in antiquity: From Aristotle to the 2nd century AD. *Phronesis, 47*, 359–394.
6. Brandt, R. B. (1965). The concepts of obligation and duty. *Mind, 73*, 374–393.
7. Brogan, A. P. (1919). The fundamental value universal. *Journal of Philosophy, Psychology, and Scientific Methods, 16*, 96–104.
8. Campbell, R. (1974). Real predicates and 'exists'. *Mind, 83*, 95–99.
9. Carnap, R. ([1928] 1961). *Der logische Aufbau der Welt*. Berlin: Weltkreis-Verlag.

10. Castañeda, H. N. (1981). The paradoxes of deontic logic: The simplest solution to all of them in one fell swoop. In R. Hilpinen (Ed.), *New studies in deontic logic* (pp. 37–95). Dordrecht: Reidel.
11. Chisholm, R. M., & Sosa, E. (1966). On the logic of intrinsically better. *American Philosophical Quarterly, 3*, 244–249.
12. Cook, R. T. (2002). Vagueness and mathematical precision. *Mind, 111*, 225–247.
13. Dayton, E. (1981). Two approaches to deontic logic. *Journal of Value Inquiry, 15*, 137–147.
14. Debreu, G. (1986). Theoretic models: Mathematical form and economic content. *Econometrica, 54*, 1259–1270.
15. Descartes, R. ([1644] 1905). Principia Philosophiae. In C. Adam & P. Tannery (Eds.), *Oeuvres de Descartes* (Vol. VIII). Paris: Léopold Cerf.
16. Drachmann, A. G. (1958). How Archimedes expected to move the Earth. *Centaurus, 5*, 278–282.
17. Elbourne, P. (2013). *Definite descriptions*. Oxford: Oxford University Press.
18. Forrester, M. (1975). Some remarks on obligation, permission, and supererogation. *Ethics, 85*, 219–226.
19. Frege, G. (1879). *Begriffsschrift, eine der aritmetischen nachgebildete Formelsprache des reinen Denkens*. Halle.
20. Gewirth, A. (1943). Clearness and distinctness in Descartes. *Philosophy, 18*, 17–36.
21. Gould, J. (1970). *The philosophy of Chrysippus*. Leiden: E.J. Brill.
22. Gould, J. (1980). Freedom: Triadic or tripartite? *Modern Schoolman, 58*, 47–52.
23. Green-Pedersen, N. J. (1984). *The tradition of the topics in the middle ages: The commentaries on Aristotle's and Boethius' 'topics'*. München, Wien: Philosophia Verlag.
24. ∗ Haack, S. (1978). *Philosophy of logics*. Cambridge: Cambridge University Press. [An accessible introduction to the major philosophical issues in logic.].
25. Halldén, S. (1957). *On the logic of 'better'*. Lund: Library of Theoria.
26. Hansen, J. (2014). Reasoning about permission and obligation. In S. O. Hansson (ed.) *David Makinson on classical Methods for Non-Classical Problems* (pp. 287–333) Dordrecht: Springer.
27. Hansson, B. (1969). An analysis of some deontic logics. *Noûs, 3*, 373–398.
28. Hansson, S. O. (1988). Deontic logic without misleading alethic analogies. *Logique et Analyse, 31*, 337–353, 355–370.
29. Hansson, S. O. (1990). Defining 'good' and 'bad' in terms of 'better'. *Notre Dame Journal of Formal Logic, 31*, 136–149.
30. Hansson, S. O. (1990). A formal representation of declaration-related legal relations. *Law and Philosophy, 9*, 399–416.
31. Hansson, S. O. (1991). The revenger's paradox. *Philosophical Studies, 61*, 301–305.
32. Hansson, S. O. (1996). Decision-making under great uncertainty. *Philosophy of the Social Sciences, 26*, 369–386.
33. ∗ Hansson, S. O. (2000). Formalization in philosophy. *Bulletin of Symbolic Logic, 6*, 162–175. [Short introduction to formalization, with an emphasis on its advantages and disadvantages.]
34. ∗ Hansson, S. O. (2006). How to define – A tutorial. *Princípios, Revista de Filosofia, 13*(19–20), 5–30. [Advice for the informal definitional work that should precede the construction of a formal definition of a philosophical concept.]
35. Hansson, S. O. (2006). Ideal worlds – Wishful thinking in deontic logic. *Studia Logica, 82*, 329–336.
36. Hansson, S. O. (2006). Editorial: A dialogue on logic. *Theoria, 72*, 263–268.
37. Hansson, S. O. (2006). Levi's ideals. In E. J. Olsson (Ed.), *Knowledge and inquiry. Essays on the pragmatism of Isaac Levi* (pp. 241–247). Cambridge: Cambridge University Press.
38. Hansson, S. O. (2009). The demise of modern logic? In F. Stoutland (Ed.), *Philosophical probings: von Wrights later work* (pp. 169–176). Copenhagen: Automatic Press.
39. Hansson, S. O. (2010). Editorial: Methodological pluralism in philosophy. *Theoria, 76*, 189–191.

40. Hansson, S. O. (2011). Twelve theses on the use of logic in moral philosophy. In A. Gupta & J. van Benthem (Eds.), *Logic and philosophy today* (Vol. 2, pp. 65–84). London: College Publications.
41. Hansson, S. O. (2013). Varieties of permission. In D. Gabbay, J. Horty, X. Parent, R. van der Meyden, & L. van der Torre (Eds.), *Handbook of deontic logic and normative systems* (Vol. 1, pp. 195–240). London: College Publications.
42. Hansson, S. O. (2017). *Descriptor revision. Belief change through direct choice*. Cham: Springer.
43. Hansson, S. O. (2018). Risk analysis under structural uncertainty. In T. Aven & E. Zio (Eds.), *Knowledge in risk assessment and management*. Hoboken, NJ: Wiley.
44. Hilpinen, R. (1982). Disjunctive permissions and conditionals with disjunctive antecedent. *Acta Philosophica Fennica, 35*, 175–194.
45. Hodges, W. (2009). Traditional logic, modern logic and natural language. *Journal of Philosophical Logic, 38*, 589–606.
46. Kaplan, A., & Kris, E. (1948). Esthetic ambiguity. *Philosophy and Phenomenological Research, 8*, 415–435.
47. Koj, L. (1991). Exactness and philosophy. In G. Schurz & G. J. W. Dorn (Eds.), *Advances in scientific philosophy, essays in honour of Paul Weingartner on the occasion of the 60th anniversary of his birthday* (pp. 599–609). Amsterdam: Rodopi.
48. Leibniz, G. W. ([1704] 1962). *Nouveaux Essais sur L'entendement Humain*. In *Philosophische Schriften* (Vol. 6). Berlin: Akademie-Verlag.
49. Leibniz, G. W. ([1704] 1996). *New essays on human understanding* (P. Remnant & J. Bennett, Trans.). Cambridge: Cambridge University Press.
50. Levi, I. (1977). Subjunctives, dispositions and chances. *Synthese, 34*, 423–455.
51. Levi, I. (1991). *The fixation of belief and its undoing*. Cambridge, MA: Cambridge University Press.
52. Liu, F., & Yang, W. (2010). A brief history of Chinese logic. *Journal of Indian Council of Philosophical Research, 27*, 101–123.
53. Liu, F., & Zhang, J. (2010). New perspectives on moist logic. *Journal of Chinese Philosophy, 37*, 605–611.
54. Livesey, S. J. (1986). The Oxford calculatores, quantification of qualities, and Aristotles prohibition of metabasis. *Vivarium, 24*, 50–69.
55. MacCallum, G. (1967). Negative and positive freedom. *Philosophical Review, 76*, 312–334.
56. * MacFarlane, J. (2015). Logical constants. *Stanford encyclopedia of philosophy*. http://plato.stanford.edu/entries/logical-constants. [Clarifying summary of the philosophical discussions on the notion of a logical constant.]
57. Makinson, D. (1965). The paradox of the preface. *Analysis, 25*, 205–207.
58. Makinson, D. (1986). On the formal representation of rights relations. *Journal of Philosophical Logic, 15*, 403–425.
59. Makinson, D. (1999). On a fundamental problem of deontic logic. In P. McNamara & H. Prakken (Eds.), *Norms and information systems. New studies on deontic logic and computer science* (pp. 29–53). Amsterdam: IOS Press.
60. * Makinson, D. (2012). *Sets, logic and maths for computing*. Dordrecht: Springer. [Although written primarily for computer scientists, this is an excellent introduction for philosophers who wish to acquaint themselves with the major formal tools used in philosophy. Exercises are included.]
61. McMullin, E. (1985). Galilean idealization. *Studies in history and philosophy of science, 16*, 247–273.
62. Mish'alani, J. K. (1969). 'Duty', 'obligation' and 'ought'. *Analysis, 30*, 33–40.
63. Moulton, J. (1983). A paradigm of philosophy: The adversary method. In S. G. Harding & M. B. Hintikka (Eds.), *Discovering reality: Feminist perspectives on epistemology, metaphysics, methodology, and philosophy of science* (pp. 149–164). Dordrecht: D. Reidel.
64. Parent, W. (1983). Recent work on the concept of liberty. In K. G. Lucey & T. R. Machan (Eds.), *Recent work in philosophy* (pp. 247–275). Totowa, N.J.: Rowman and Allanheld.

65. Peirce, C. S. (1905). What pragmatism is. *Monist, 15*, 161–181. Reprinted in Peirce, C. S. (1935) Pragmatism and pragmaticism. In C. Hartshorne & P. Weiss (Eds.), *Collected papers* (Vol. 5). Cambridge, MA: Harvard University Press.
66. Popper, K. (1935). *Logik der Forschung: zur Erkenntnistheorie der modernen Naturwissenschaft*. Wien: Julius Springer. [Translated into English as *The logic of scientific discovery*.]
67. Raz, J. (1975). Permissions and supererogation. *American Philosophical Quarterly, 12*, 161–168.
68. Robinson, R. (1971). Ought and ought not. *Philosophy, 46*, 193–202.
69. Russell, B. (1905). On denoting. *Mind, 14*, 479–493.
70. Russell, B. (1957). Mr. Strawson on referring. *Mind, 66*, 385–389.
71. Russell, B. ([1914] 1969). *Our knowledge of the external world, as a field for scientific method in philosophy*. London: Allen & Unwin.
72. Sainsbury, R. M. (1996). Concepts without boundaries. In R. Keefe & P. Smith (Eds.), *Vagueness: A reader* (pp. 251–264). Cambridge, MA: MIT Press.
73. * Sainsbury, R. M. (2001). *Logical forms: An introduction to philosophical logic*. Oxford: Blackwell. [An introductory textbook in logic that pays much attention to the philosophical issues. Exercises are included.]
74. Schemmel, M. (2014). Medieval representations of change and their early modern application. *Foundations of Science, 19*, 11–34.
75. Stern, R. (2004). Does ought imply can? And did Kant think it does? *Utilitas, 16*, 42–61.
76. Strawson, P. F. (1950). On referring. *Mind, 59*, 320–334.
77. Trapp, R. W. (1978). Exaktheit in der Philosophie. *Zeitschrift für allgemeine Wissenschaftstheorie, 9*, 307–336
78. Tye, M. (1994). Vagueness: Welcome to the quicksand. *Southern Journal of Philosophy, 33*(Supplement), 1–22.
79. Tyson, P. D. (1986). Do your standards make any difference? Asymmetry in preference judgments. *Perceptual and motor skills, 63*, 1059–1066.
80. von Wright, G. H. (1963). *The logic of preference*. Edinburgh: Edinburgh University Press.
81. von Wright, G. H. (1968). An essay in deontic logic and the general theory of action. *Acta Philosophica Fennica, 21*, 1–110.
82. von Wright, G. H. (1999). Ought to be – Ought to do. In G. Meggle (Ed.), *Actions, norms, values. Discussions with Georg Henrik von Wright* (pp. 3–9). Berlin: De Gruyter.
83. Wallace, W. A. (1969). The 'Calculatores' in early sixteenth-century physics. *British Journal for the History of Science, 4*, 221–232.
84. Williamson, T. (1994). *Vagueness*. London: Routledge.
85. Wittgenstein, L. (1922). *Tractatus logico-philosophicus*, with an introduction by Bertrand Russell. New York: Humanities Press.

Part II
Reasoning and Inference

Chapter 2
Argument

Henry Prakken

Abstract This chapter discusses how formal models of argumentation can clarify philosophical problems and issues. Some of these arise in the field of epistemology, where it has been argued that the principles by which knowledge can be acquired are defeasible. Other problems and issues originate from the fields of informal logic and argumentation theory, where it has been argued that outside mathematics the standards for the validity of arguments are context-dependent and procedural, and that what matters is not the syntactic form but the persuasive force of an argument.

Formal models of argumentation are of two kinds. Argumentation logics formalise the idea that an argument only warrants its conclusion if it can be defended against counterarguments. Dialogue systems for argumentation regulate how dialogue participants can resolve a conflict of opinion. This chapter discusses how argumentation logics can define non-deductive consequence notions and how their embedding in dialogue systems for argumentation can account for the context-dependent and procedural nature of argument evaluation and for the dependence of an argument's persuasive force on the audience in an argumentation dialogue.

2.1 Introduction

Introductions to logic often portray logically valid inference as 'foolproof' reasoning: an argument is valid if the truth of its premises guarantees the truth of its conclusion. However, we all construct arguments from time to time that are not foolproof in this sense but that merely make their conclusion plausible when their premises are true. For example, if we are told that Peter, a professor in economics, says that reducing taxes increases productivity, we conclude that reducing taxes increases productivity since we know that experts are usually right within their domain of expertise. Sometimes such arguments are defeated by counterarguments.

H. Prakken (✉)
Department of Information and Computing Sciences, Utrecht University & Faculty of Law, University of Groningen, Groningen, Netherlands
e-mail: h.prakken@uu.nl

S. O. Hansson, V. F. Hendricks (eds.), *Introduction to Formal Philosophy*, Springer Undergraduate Texts in Philosophy, https://doi.org/10.1007/978-3-319-77434-3_2

For example, if we are also told that Peter has political ambitions, we have to retract our previous conclusion that he is right about the effect of taxes if we also believe that people with political ambitions are often unreliable when it comes to taxes. Or, to use an example of practical instead of epistemic reasoning, if we accept that reducing taxes increases productivity and that increasing productivity is good, then we conclude that the taxes should be reduced, unless we also accept that reducing taxes increases inequality, that this is bad and that equality is more important than productivity. However, as long as such counterarguments are not available, we are happy to live with the conclusions of our fallible arguments. The question is: are we then reasoning fallaciously or is there still logic in our reasoning?

An answer to this question has been given in the development of argumentation logics. In a nutshell, the answer is that there is such logic but that it is inherently dialectic: an argument only warrants its conclusion if it is acceptable, and an argument is acceptable if, firstly, it is properly constructed and, secondly, it can be defended against counterarguments. Thus argumentation logics must define three things: how arguments can be constructed, how they can be attacked by counterarguments and how they can be defended against such attacks.

Argumentation logics are a form of nonmonotonic logic, since their notion of warrant is nonmonotonic: new information may give rise to new counterarguments defeating arguments that were originally acceptable. Besides a logical side, argumentation also has a dialogical side: notions like argument, attack and defence naturally apply when (human or artificial) agents try to persuade each other to adopt or give up a certain point of view.

This chapter[1] aims to show how formal models of argumentation can clarify philosophical problems and issues. Some of these arise in the field of epistemology. Pollock [10] argued that the principles by which knowledge can be acquired are defeasible. Later he made this precise in a formal system [11], which inspired the development of argumentation logics in artificial intelligence (AI). Rescher [20] also stressed the dialectical nature of theories of knowledge and presented a disputational model of scientific inquiry.

Other issues and problems originate from the fields of informal logic and argumentation theory. In 1958, Stephen Toulmin launched his influential attack on the logic research of those days, accusing it of only studying mathematical reasoning while ignoring other forms of reasoning, such as commonsense reasoning and legal reasoning [21]. He argued that outside mathematics the standards for the validity of arguments are context-dependent and procedural: according to him an argument is valid if it has been properly defended in a dispute, and different fields can have different rules for when this is the case. Moreover, in his famous argument scheme he drew attention to the fact that different premises can have different roles in an argument (data, warrant or backing) and he noted the possibility of exceptions to rules (rebuttals). Perelman argued that arguments in ordinary discourse should not be evaluated in terms of their syntactic form but on their rhetorical potential to persuade an audience [9]. These criticisms gave rise to the fields of informal logic

[1] An earlier version of this chapter has appeared as [14].

and argumentation theory, which developed notions like argument schemes with critical questions and dialogue systems for argumentation. Many scholars in these fields distrusted or even rejected formal methods, but one point of this chapter is that formal methods can also clarify these aspects of reasoning. Another claim often made in these fields is that arguments can only be evaluated in the context of a dialogue or procedure. A second point of this paper is that this can be respected by embedding logical in dialogical accounts of argumentation.

The philosophical problems to be discussed in this chapter then are:

- Can argumentation-based standards for non-deductive inference be defined?
- To what extent are these standards procedural?
- To what extent are they context-dependent?
- What is the nature of argument schemes?
- Can the use of arguments to persuade be formalised?

2.2 Dung's Abstract Argumentation Frameworks

In 1995 Phan Minh Dung introduced a now standard abstract formalism for argumentation-based inference, which assumes as input nothing but a set (of arguments) ordered by a binary relation (by Dung called 'attack' but in this chapter the term 'defeat' will be used).

Definition 2.1 An *abstract argumentation framework* (AF) is a pair $\langle \mathcal{A}, Def \rangle$, where $\mathcal{A}$ is a set arguments and $Def \subseteq \mathcal{A} \times \mathcal{A}$ is a binary relation of defeat. We say that an argument A *defeats* an argument B iff $(A, B) \in Def$, and that A *strictly defeats* B if A defeats B while B does not defeat A. A set S of arguments is said to defeat an argument A iff some argument in S defeats A.

Dung [4] defined four alternative semantics for AFs (over the years further semantics have been proposed; cf. Baroni et al. [1]). A semantics for AFs characterises so-called argument extensions of AF's, that is, subsets of $\mathcal{A}$ that are in some sense coherent. One way to define extensions is with *labellings* of AFs, which assign to zero or more members of $Args$ either the label *in* or *out* (but not both) satisfying the following constraints:

1. an argument is *in* iff all arguments defeating it are *out*.
2. an argument is *out* iff it is defeated by an argument that is *in*.

Stable semantics labels all arguments, while *grounded semantics* minimises and *preferred semantics* maximises the set of arguments that are labelled *in*, and *complete semantics* allows all labellings satisfying the two constraints. Let $S \in$ {stable, preferred, grounded, complete} and (*In*, *Out*) an S-status assignment. Then *In* is defined to be an S-extension.[2]

[2]This definition is different from but equivalent to Dung's [4] definition of extensions.

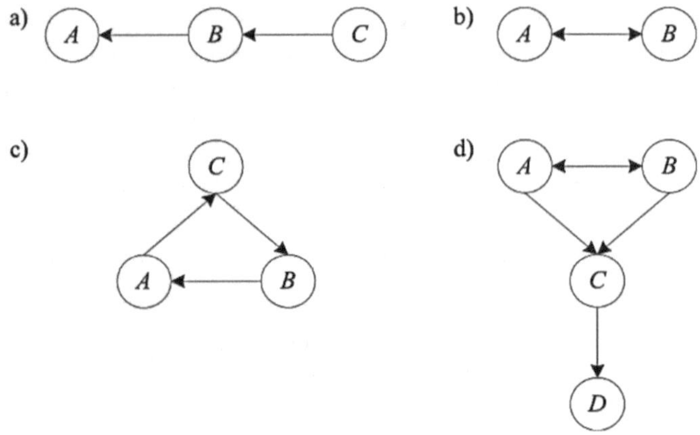

Fig. 2.1 Four argumentation frameworks

Some known facts (also holding for the corresponding extensions) are that each grounded, preferred or stable labelling of an AF is also a complete labelling of that AF; the grounded labelling is unique but all other semantics allow for multiple labellings of an AF; each AF has a grounded and at least one preferred and complete labelling, but there are AFs without stable labellings; and the grounded labelling of an AF is contained in all other labellings of that AF.

Then the acceptability status of arguments can be defined as follows:

Definition 2.2 For grounded semantics an argument A is *justified* iff A is in the grounded extension; *overruled* iff A is not in the grounded extension but defeated by a member of the grounded extension; *defensible* otherwise. For stable and preferred semantics an argument A is *justified* iff A is in all stable/preferred extensions; *overruled* iff A is in no stable/preferred extension; *defensible* otherwise.

Figure 2.1 illustrates the definitions with some example argumentation frameworks, where defeat relations are graphically depicted as arrows.

In AF (a) all semantics produce the same unique labelling. Argument C is *in* by constraint (1) since it has no defeaters, so B is *out* by constraint (2) since it is defeated by C, so A is *in* by constraint (1) since C defeats B. So all semantics produce the same, unique extension, namely, $\{A, C\}$. Hence in all semantics A and C are justified while B is overruled. It is sometimes said that C *reinstates*, or *defends* A by defeating its defeater B.

In AF (b) grounded semantics does not label any of the arguments while preferred and stable semantics produce two alternative labellings: one in which A is *in* and B is *out* and one in which B is *in* and A is *out*. Hence the grounded extension is empty while the preferred-and-stable extensions are $\{A\}$ and $\{B\}$. All these extensions are also complete. Hence in all semantics both A and B are defensible.

AF (c) has no stable extensions since no argument can be labelled both *in* and *out* while there is a unique grounded, preferred and complete extension, which is empty, generated by a labelling which does not label any argument. Note that if a fourth argument D is added with no defeat relations with the other three arguments, there is still no stable extension while the unique grounded, preferred and complete extension is $\{D\}$.

Finally, AF (d) shows a difference between grounded and preferred semantics. The grounded extension is empty, since A and B can be left unlabelled so that C and D are also unlabelled, while the two preferred (and stable) extensions are $\{A, D\}$ and $\{B, D\}$. Thus while in grounded semantics all arguments are defensible, in preferred and stable semantics A and B are defensible, D is justified and C is overruled.

The above definitions characterise *sets* of arguments that are in some sense acceptable. In addition, procedures have been studied for determining whether a given argument is a member of such a set. Some take the form of an *argument game* between two players, a proponent and an opponent of an argument. The precise rules of the game depend on the semantics the game is meant to capture. The rules should be chosen such that the existence of a winning strategy (in the usual game-theoretic sense) for the proponent of an argument corresponds to the investigated semantic status of the argument, for example, 'justified in grounded semantics' or 'defensible in preferred semantics'.

Because of space limitations we can give only briefly one example game. The following game is sound and complete for grounded semantics in that the proponent of argument A has a winning strategy just in case A is in the grounded extension. The proponent starts a game with an argument and then the players take turns, trying to defeat the previous move of the other player. In doing so, the proponent must strictly defeat the opponent's arguments while he is not allowed to repeat his own arguments. A game is terminated if it cannot be extended with further moves. The player who moves last in a terminated game wins the game. Thus the proponent has a winning strategy if he has a way to make the opponent run out of moves (from the implicitly assumed AF) whatever choice the opponent makes.

As remarked in the introduction, argumentation logics must define three things: how arguments can be constructed, how they can be attacked and how they can be defended against attacks. Dung's abstract formalism only answers the third question. To answer the first two questions, accounts are needed of argument construction and the nature of attack and defeat. We next discuss a general framework for formulating such accounts.

2.3 An Abstract Framework for Structured Argumentation

The *ASPIC*$^+$ framework [7, 8, 13] aims to integrate and further develop the main current formal models of structured argumentation. While some of its design choices can perhaps be debated, the framework is still representative of work in the field, for which reason we present it here. *ASPIC*$^+$ gives structure to Dung's arguments and defeat relation. It defines arguments as inference trees formed by applying strict ($\rightarrow$)

or defeasible ($\Rightarrow$) inference rules to premises formulated in some logical language. Informally, if an inference rule's antecedents are accepted, then if the rule is strict, its consequent must be accepted *no matter what*, while if the rule is defeasible, its consequent must be accepted *if there are no good reasons not to accept it*. Arguments can be attacked on their 'ordinary' premises and on their applications of defeasible inference rules. Some attacks succeed as *defeats*; whether this is so is partly determined by preferences. The acceptability status of arguments is then defined by applying any of [4] semantics for abstract argumentation frameworks to the resulting set of arguments with its defeat relation.

ASPIC$^+$ is not a system but a framework for specifying systems. To start with, it defines the notion of an abstract *argumentation system* as a structure consisting of a logical language $\mathcal{L}$ with a negation symbol $\neg$,[3] a set $\mathcal{R}$ consisting of two subsets $\mathcal{R}_s$ and $\mathcal{R}_d$ of strict and defeasible inference rules, and a naming convention n in $\mathcal{L}$ for defeasible rules in order to talk about the applicability of defeasible rules in $\mathcal{L}$. Thus, informally, $n(r)$ is a wff in $\mathcal{L}$ which says that rule $r \in \mathcal{R}$ is applicable. (As is usual, the inference rules in $\mathcal{R}$ are defined *over* the language $\mathcal{L}$ and are not elements *in* the language.)

Definition 3.1 An *argumentation system* is a triple $AS = (\mathcal{L}, \mathcal{R}, n)$ where:

- $\mathcal{L}$ is a logical language with a negation symbol $\neg$.
- $\mathcal{R} = \mathcal{R}_s \cup \mathcal{R}_d$ is a set of strict ($\mathcal{R}_s$) and defeasible ($\mathcal{R}_d$) inference rules of the form $\varphi_1, \ldots, \varphi_n \rightarrow \varphi$ and $\varphi_1, \ldots, \varphi_n \Rightarrow \varphi$ respectively (where φ_i, φ are meta-variables ranging over wff in $\mathcal{L}$), and $\mathcal{R}_s \cap \mathcal{R}_d = \emptyset$.
- $n : \mathcal{R}_d \longrightarrow \mathcal{L}$ is a naming convention for defeasible rules.

We write $\psi = -\varphi$ just in case $\psi = \neg\varphi$ or $\varphi = \neg\psi$ (we will sometimes informally say that formulas φ and $-\varphi$ are each other's negation).

Henceforth, a set $S \subseteq \mathcal{L}$ is said to be *directly consistent* iff $\nexists\, \psi, \varphi \in S$ such that $\psi = -\varphi$, otherwise S is *directly inconsistent*. And S is said to be *indirectly (in)consistent* if its closure under application of strict inference rules is directly (in)consistent.

Definition 3.2 A *knowledge base* in an $AS = (\mathcal{L}, \mathcal{R}, n)$ is a set $\mathcal{K} \subseteq \mathcal{L}$ consisting of two disjoint subsets $\mathcal{K}_n$ (the *axioms*) and $\mathcal{K}_p$ (the *ordinary premises*).

Intuitively, the axioms are certain knowledge and thus cannot be attacked, whereas the ordinary premises are uncertain and thus can be attacked.

Definition 3.3 An *argumentation theory* is a tuple $AT = (AS, \mathcal{K})$ where AS is an argumentation system and $\mathcal{K}$ is a knowledge base in AS.

ASPIC$^+$ arguments are now defined relative to an argumentation theory $AT = (AS, \mathcal{K})$, and chain applications of the inference rules from AS into inference graphs (which are trees if no premise is used more than once), starting with elements

[3] In most papers on *ASPIC*$^+$ negation can be non-symmetric. In this paper we present the special case with symmetric negation.

from the knowlege base $\mathcal{K}$. Arguments thus contain subarguments, which are the structures that support intermediate conclusions (plus the argument itself and its premises as limiting cases). In what follows, for a given argument the function Prem returns all its premises, Conc returns its conclusion, Sub returns all its subarguments, DefRules returns all defeasible rules of an argument and TopRule returns the final rule applied in the argument.

Definition 3.4 An *argument* A on the basis of an argumentation theory with a knowledge base $\mathcal{K}$ and an argumentation system $(\mathcal{L}, \mathcal{R}, n)$ is any structure obtainable by applying one or more of the following steps finitely many times:

1. φ if $\varphi \in \mathcal{K}$ with $\texttt{Prem}(A) = \{\varphi\}$; $\texttt{Conc}(A) = \varphi$; $\texttt{Sub}(A) = \{\varphi\}$; $\texttt{DefRules}(A) = \emptyset$; $\texttt{TopRule}(A)$ = undefined.
2. $A_1, \ldots A_n \rightarrow/\Rightarrow \psi$[4] if $A_1, \ldots, A_n$ are arguments such that there exists a strict/defeasible rule $\texttt{Conc}(A_1), \ldots, \texttt{Conc}(A_n) \rightarrow/\Rightarrow \psi$ in $\mathcal{R}_s/\mathcal{R}_d$.
   $\texttt{Prem}(A) = \texttt{Prem}(A_1) \cup \ldots \cup \texttt{Prem}(A_n)$,
   $\texttt{Conc}(A) = \psi$,
   $\texttt{Sub}(A) = \texttt{Sub}(A_1) \cup \ldots \cup \texttt{Sub}(A_n) \cup \{A\}$.
   $\texttt{DefRules}(A) = \texttt{DefRules}(A_1) \cup \ldots \cup \texttt{DefRules}(A_n)$;
   $\texttt{TopRule}(A) = \texttt{Conc}(A_1), \ldots, \texttt{Conc}(A_n) \rightarrow/\Rightarrow \psi$.

Then A is: *strict* if $\texttt{DefRules}(A) = \emptyset$; *defeasible* if $\texttt{DefRules}(A) \neq \emptyset$; *firm* if $\texttt{Prem}(A) \subseteq \mathcal{K}_n$; *plausible* if $\texttt{Prem}(A) \subseteq \mathcal{K}_p$.

Example 3.5 Consider a knowledge base in an argumentation system with $\mathcal{R}_s = \{p, q \rightarrow s;\ u, v \rightarrow w\}$; $\mathcal{R}_d = \{p \Rightarrow t;\ s, r, t \Rightarrow v\}$; $\mathcal{K}_n = \{q\}$; $\mathcal{K}_p = \{p, r, u\}$. An argument for w is displayed in Fig. 2.2. The type of a premise is indicated with a superscript and defeasible inferences and attackable premises and conclusions are displayed with dotted lines. Formally the argument and its subarguments are written as follows:

A_1: p	A_5: $A_1 \Rightarrow t$
A_2: q	A_6: $A_1, A_2 \rightarrow s$
A_3: r	A_7: $A_5, A_3, A_6 \Rightarrow v$
A_4: u	A_8: $A_7, A_4 \rightarrow w$

We have that

$\texttt{Prem}(A_8) =$	$\{p, q, r, u\}$
$\texttt{Conc}(A_8) =$	w
$\texttt{Sub}(A_8) =$	$\{A_1, A_2, A_3, A_4, A_5, A_6, A_7, A_8\}$
$\texttt{DefRules}(A_8) =$	$\{p \Rightarrow t;\ s, r, t \Rightarrow v\}$
$\texttt{TopRule}(A_8) =$	$u, v \rightarrow w$

[4] $\rightarrow/\Rightarrow$ means that the rule is a strict, respectively, defeasible rule.

Fig. 2.2 An argument

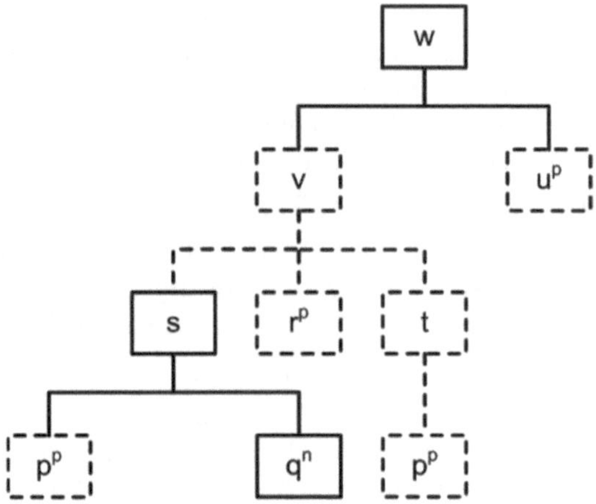

Arguments can be attacked in three ways: on their premises (undermining attack), on their conclusion (rebutting attack) or on an inference step (undercutting attack). The latter two are only possible on applications of defeasible inference rules.

Definition 3.6 *A attacks B* iff *A undercuts*, *rebuts* or *undermines B*, where:

- *A undercuts* argument B (on B') iff $\texttt{Conc}(A) = -n(r)$ for some $B' \in \texttt{Sub}(B)$ such that B''s top rule r is defeasible.
- *A rebuts* argument B (on B') iff $\texttt{Conc}(A) = -\varphi$ for some $B' \in \texttt{Sub}(B)$ of the form $B_1'', \ldots, B_n'' \Rightarrow \varphi$.
- Argument *A undermines B* (on B') iff $\texttt{Conc}(A) = -\varphi$ for some $B' = \varphi, \varphi \in \mathcal{K}_p$.

In Example 3.5 argument A_8 can be undercut on two of its subarguments, namely, A_5 and A_7. An undercutter of A_5 must have a conclusion $-\varphi$ where $n(p \Rightarrow t) = \varphi$ while an undercutter of A_5 must have a conclusion $-\varphi$ where $n(s,\ r,\ t \Rightarrow w) = \varphi$. Argument A_8 can be rebutted on A_5 with an argument for $-t$ and on A_7 with an argument for $-v$. Moreover, if the rebuttal of A_5 has a defeasible top rule, then A_5 in turn rebuts the argument for $-t$. However, A_8 itself does not rebut that argument, except in the special case where $w = --t$. Finally, argument A_8 can be undermined with an argument that has conclusion $-p$, $-r$ or $-u$.

Attack relations between arguments can be resolved with an ordering on arguments. To formalise this, the notion of a structured argumentation framework is introduced.

Definition 3.7 Let AT be an *argumentation theory* (AS, KB). A *structured argumentation framework* (SAF) defined by AT is a triple $\langle \mathcal{A}, Att, \preceq \rangle$ where

- $\mathcal{A}$ is the set of all arguments on the basis of AT;
- $\preceq$ is an ordering on $\mathcal{A}$;
- $(X, Y) \in Att$ iff X attacks Y.

Modgil and Prakken [7] also study a variant of this definition in which arguments are required to have indirectly consistent premises.

Now attacks combined with the argument ordering yield three kinds of defeat. For undercutting attack no preferences are needed to make it succeed, since undercutters are explicit exceptions to the rule they undercut. Rebutting and undermining attacks succeed only if the attacked argument is not stronger than the attacking argument.

Definition 3.8 A *defeats* B iff: A undercuts B, or; A rebuts/undermines B on B' and $A \not\prec B'$.[5] A strictly *defeats* B iff A *defeats* B and B does not *defeat* A

The success of rebutting and undermining attacks thus involves comparing the conflicting arguments at the points where they conflict. The definition of successful undermining exploits the fact that an argument premise is also a subargument.

The *ASPIC*$^+$ framework assumes the argument ordering as given. It may depend on all sorts of standards, such as statistical strength of generalisations, reliability of information sources, preferences over outcomes of actions, or norm hierarchies. In many contexts such standards can themselves be argued about. One way to formalise this is by using Modgil's [6] idea to decompose the defeat relation of Dung's [4] abstract argumentation frameworks into a more basic *attack* relation and to allow *attacks on attacks* in addition to attacks on arguments. Combined with *ASPIC*$^+$, the idea is that if argument C claims that argument B is preferred to argument A, and A attacks B, then C undermines the success of A's attack on B (i.e., A does not *defeat* B) by pref-attacking A's attack on B.

Recall that argumentation logics must define three things: how arguments can be constructed, how they can be defeated and how they can be defended against defeating counterarguments. While Dung's abstract argumentation semantics addresses the last issue, we can now combine it with the *ASPIC*$^+$ framework to address the first two issues.

Definition 3.9 An *abstract argumentation framework (AF) corresponding to a* SAF $= \langle \mathcal{A}, Att, \preceq \rangle$ is a pair $(\mathcal{A}, Def)$ such that Def is the defeat relation on $\mathcal{A}$ determined by $\langle \mathcal{A}, Att, \preceq \rangle$.

The justified arguments of the above defined AF are then defined under various semantics, as in Definition 2.2. We now see that an argument can be defended against attacks in two ways: by showing that the attacker is inferior to it or by defeating the attacker with a counterattack that reinstates the original argument.

We can now finally define an argumentation-based consequence notion for well-formed formulas (relative to an AT and with respect to any given semantics):

Definition 3.10 A wff $\varphi \in \mathcal{L}$ is *justified* if φ is the conclusion of a justified argument, and *defensible* if φ is not justified and is the conclusion of a defensible argument.

An alternative definition of a justified wff is to say that every extension contains an argument with the wff as its conclusion. Unlike the above definition, this alternative

[5] $X \prec Y$ means as ususal that $X \preceq Y$ and $Y \not\preceq X$.

definition allows different extensions containing different arguments for a justified conclusion. This is similar to the different treatments that semantics for abstract argumentation give to Fig. 1.1d.

One possible analysis of this difference is that some semantics, or some definitions of justification, are better than others, but an alternative analysis is that different definitions capture different senses or strengths of justification, which each may have their use in certain contexts. For example, in the law, criminal cases require higher proof standards than civil cases. And while in domains like the law and medicine defeasible arguments are acceptable, in the field of mathematics all arguments must, of course, be deductive. Thus we see how our formal framework for argumentation can make sense of Toulmin's claim that the standards for the validity of arguments are context-dependent.

In addition, the kind of reasoning can be relevant, such as the distinction between epistemic and practical reasoning. If, for instance, two incompatible actions (say reducing and increasing taxes) have two different good consequences (say increasing productivity and increasing equality in society) and there is no reason to prefer one consequence over the other, then an arbitrary choice is (all other things being equal) rational. If, on the other hand, two experts disagree about whether reducing taxes increases productivity, then an arbitrary choice for one of them seems irrational. So it might be argued that in practical reasoning a defensible conclusion can be good enough while in epistemic reasoning we should aim for justified conclusions.

2.4 The Nature of Inference Rules

While we now have a general framework for the definition of argumentation logics, much more can be said. To start with, the framework can be instantiated in many ways, so there is a need for principles that can be used in assessing the quality of instantiations. Caminada and Amgoud [3] formulated several so-called rationality postulates, namely, that each extension should be closed under subarguments and under strict rule application, and be directly and indirectly consistent. $ASPIC^+$ unconditionally satisfies the two closure postulates while Prakken [13] and Modgil and Prakken [7] identify conditions under which some broad classes of instantiations satisfy the two consistency postulates.

The next question is, what are 'good' collections of strict and defeasible inference rules? In AI there is a tradition to let inference rules express domain-specific information, such as *all penguins are birds* or *birds typically fly*. This runs counter to the usual practice in logic, in which inference rules express general patterns of reasoning, such as modus ponens, universal instantiation and so on. This practice is also followed in systems for so-called classical argumentation [2], in which arguments from a possibly inconsistent knowledge base are classical proofs from consistent subsets of the knowledge base. These systems are in fact a special case of the $ASPIC^+$ framework with $\mathcal{L}$ being the language of standard propositional or first-order logic, the strict rules being all valid propositional or first-order inferences, with no defeasible rules and no axiom premises, and with the premises of all arguments required to be indirectly consistent. In this approach (which can be generalised to other deductive logics) arguments can thus only be sensibly attacked on their premises.

While this approach has some merits, it is doubtful whether all argumentation can be reduced to inconsistency handling in some deductive logic. In particular John Pollock strongly emphasized the importance of *defeasible* reasons in argumentation. He was quite insistent that defeasible reasoning is not just some exotic, exceptional, add-on to deductive reasoning but is, instead, an essential ingredient of our cognitive life:

> ... we cannot get around in the world just reasoning deductively from our prior beliefs together with new perceptual input. This is obvious when we look at the varieties of reasoning we actually employ. We tend to trust perception, assuming that things are the way they appear to us, even though we know that sometimes they are not. And we tend to assume that facts we have learned perceptually will remain true, as least for a while, when we are no longer perceiving them, but of course, they might not. And, importantly, we combine our individual observations inductively to form beliefs about both statistical and exceptionless generalizations. None of this reasoning is deductively valid. [12, p. 173]

Here the philosophical distinction between *plausible* and *defeasible* reasoning is relevant; see Rescher [19, 20] and Vreeswijk [23, Ch. 8]. Plausible reasoning is valid deductive reasoning from an uncertain basis while defeasible reasoning is deductively invalid (but still rational) reasoning from a solid basis. In these terms, models of deductive argumentation formalize plausible reasoning, while Pollock modeled defeasible reasoning and the $ASPIC^+$ framework gives a unified account of these two kinds of reasoning.

There is also semantic support for the idea of defeasible inference rules. Consider, for example, the statistical generalisation *men usually have no beard.* Concluding from this that *people with a beard are usually not men* is a so-called 'base rate fallacy' [22]. If (epistemic) defeasible reasoning is reduced to inconsistency handling in deductive logic, such fallacies are easily committed. Likewise, it has been argued that reasons of practical and normative reasoning are inherently defeasible; cf. e.g. [18].

While the case for defeasible inference rules thus seems convincing, the question remains what are 'good' defeasible inference rules, especially if they are to express general patterns of inference. Here two bodies of philosophical work are relevant, namely, Pollock's [10, 11] notion of *defeasible reasons* and argumentation-theory's notion of *argument schemes* [26]. Pollock's defeasible reasons are general patterns of epistemic defeasible reasoning. He formalised reasons for perception, memory, induction, temporal persistence and the statistical syllogism, as well as undercutters for these reasons. In the $ASPIC^+$ framework Pollock's defeasible reasons can be expressed as schemes (in the logical sense, with metavariables ranging over $\mathcal{L}$) for defeasible inference rules. The same analysis applies to argument schemes, which are stereotypical non-deductive patterns of reasoning. Uses of argument schemes are evaluated in terms of critical questions specific to the scheme. In the literature on argumentation theory many collections of argument schemes have been proposed, both for epistemic, practical and evaluative reasoning. An example of an epistemic argument scheme is the scheme from expert opinion [26, p. 310]:

> E is an expert in domain D, E asserts that P is true, P is within D, therefore presumably P is true

Walton [26] give this scheme six critical questions: (1) Is E credible as an expert source? (2) Is E an expert in domain D? (3) What did E assert that implies P? (4) Is E personally reliable as a source? (5) Is P consistent with what other experts assert? (6) Is E's assertion of P based on evidence?

A practical argument scheme is the scheme from good (bad) consequences (here in a formulation that deviates from Walton [26] to stress its abductive nature):

> Action A results in P, P is good (bad), therefore all other things being equal A should (not) be done.

This scheme is usually given two critical questions: (1) Does A result in P? (2) Does A also result in something which is bad (good)? (3) (When P is concluded to be good) Is there another way to realise P?

In *ASPIC*$^+$, argument schemes can also be formalised as schemes for defeasible inference rules; then critical questions are pointers to counterarguments. In the scheme from expert opinion questions (2) and (3) point to underminers (of, respectively, the first and second premise), questions (4), (1) and (6) point to undercutters (the exceptions that the expert is biased or incredible for other reasons and that he makes scientifically unfounded statements) while question (5) points to rebutting applications of the expert opinion scheme. In the scheme from good (bad) consequences question (1) points to underminers of the first premise, question (2) points to rebuttals using the opposite version of the scheme while question (3) points to undercutters.

This account of argument schemes can also clarify Toulmin's [21] distinction between *warrants* (rule-like premises) and *backings* of warrants. For example, a warrant can be that smoking causes cancer while its backing can be an expert opinion: then the defeasible inference rule expressing the scheme from expert opinion allows to infer the warrant from the backing.

Let us illustrate the just-proposed modelling of defeasible reasons and argument schemes with an example. The logical language $\mathcal{L}$ is informally assumed to be a first-order language augmented with a conditional for defeasible generalisations, $\mathcal{R}_s$ consists of all deductively valid inferences over $\mathcal{L}$ and $\mathcal{R}_d$ consists of the above schemes from expert opinion (e) and from good (gc) and bad (bc) consequences, plus a modus ponens scheme (dmp) for defeasible generalisations. Consider then the following arguments (where premise arguments are assumed to be in $\mathcal{K}_p$ and defeasible inferences are labelled with the inference rule they apply).

A_1: *P says "lowering taxes increases productivity"*
A_2: *P is an expert in economics*
A_3: *"lowering ... productivity" is about economics*
A_4: $A_1, A_2, A_3 \Rightarrow_e$ *lowering ... productivity*
A_5: *Increased productivity is good*
A_6: $A_4, A_5 \Rightarrow_{gc}$ *taxes should be lowered*

B_1: *lowering taxes increases inequality*
B_2: *Increased inequality is bad*
B_3: $B_1, B_2 \Rightarrow_{bc}$ *taxes should not be lowered*

C_1: *P has political ambitions*
C_2: *people with political ambitions are usually not reliable about taxes*
C_3: $C_1, C_2 \Rightarrow_{dmp}$ *P is not reliable about taxes*

C_4: *Rule e does not apply to unreliable people*
C_5: $C_3, C_4 \rightarrow$ *Rule e does not apply to P*

D_1: *P is never on TV*
D_2: *people who are never on TV usually have no political ambitions*
D_3: $D_1, D_2 \Rightarrow_{dmp}$ *P has no political ambitions*

Arguments A_6 and B_3 rebut each other. Assume $B_3 \prec A_6$ so A_6 strictly defeats B_3. Assuming the obvious naming convention, argument C_5 undercuts A_6 on A_4 and so defeats both, while D_3 undermines C_5 on C_1 and C_1 in turn rebuts D_3. At this point we know that all unattacked premise arguments are justified in any semantics, since they have no defeaters. For the remaining arguments, suppose first $D_3 \prec C_1$. Then C_1 strictly defeats D_3, so in any semantics D_3, A_4 and A_6 are overruled, while all C_i and B_3 are justified. Suppose next $C_1 \prec D_3$. Then D_3 strictly defeats C_3 and C_5 by strictly defeating C_1, so in any semantics D_3 and all A_i are justified, while C_1, C_3, C_5 and B_3 are overruled. Suppose finally that neither $C_1 \prec D_3$ nor $D_3 \prec C_1$. Then C_1 and D_3 defeat each other so, even though D_3 still strictly defeats C_3 and C_5, in any semantics all non-premise arguments plus C_1 are defensible.

2.5 Argumentation as a Form of Dialogue

As stated in the introduction, argumentation theorists often claim that arguments can only be evaluated in the context of a dialogue or procedure. More specifically, Walton [24] regards argument schemes as dialogical devices, determining dialectical obligations and burdens of proof. An argument is a move in a dialogue and the scheme that it instantiates determines the allowed and required responses to that move. At first sight, our account of argument schemes as defeasible inference rules would seem to be incompatible with Walton's dialogical account. However, these two accounts can be reconciled by embedding argumentation logics in dialogue systems for argumentation.

While argumentation logics define notions of consequence from a given body of information, dialogue systems for argumentation [25] regulate disputes between real agents, who each have their own body of information, and who may be willing to learn from each other so that their information state may change. Moreover, during the dialogue they may construct a joint theory on the issue in dispute, which also evolves over time. Essentially, dialogue systems define a communication language (the well-formed utterances) and a protocol (when a well-formed utterance may be made and when the dialogue terminates).

Consider the following simple example, with a dialogue system that allows players to move arguments and to challenge, concede or retract premises and conclusions of these arguments. Each challenge must be answered with a ground for the challenged statement or else the statement must be retracted. The two agents have their own knowledge base but a shared $ASPIC^+$ argumentation system with a propositional language and three defeasible inference rules: $p \Rightarrow q$, $r \Rightarrow p$ and $s \Rightarrow \neg r$. Paul's and Olga's knowledge bases contain, respectively, single ordinary premises p and r. Let us assume that all arguments are of equal preference. Paul wants to persuade Olga that q is the case. He can internally construct the following argument for q: A_1: r, A_2: $A_1 \Rightarrow p$, A_3: $A_2 \Rightarrow q$. However, a well-known argumentation heuristic says that arguments in dialogue should be made as sparse as possible in order to avoid attacks. Therefore, Paul only utters the last step in the argument, hoping that Olga will accept p so that Paul does not have to defend r. This leads to the following dialogue.

P_1:	*q since p*	O_1:	*why p*
P_2:	*p since r*	O_2:	*¬r since s*
P_3:	*retract r, retract q*		

What has happened here? If Olga had been a trusting person who concedes a statement if she cannot construct an argument for the opposite, then she would have conceded p and q after P_1. But q is not a justified conclusion from the joint knowledge bases, so this outcome is undesirable. In fact, Olga was less trusting and first asked Paul for his reasons for p. Since Paul was honest, he gave his true reasons, which allowed Olga to discover that she could attack Paul with an undermining counterargument. Paul could not defend himself against this attack, so he realised that he cannot persuade Olga that q is true; he therefore retracted r and q.

Argumentation logic applies here in several ways. It can model the agents' internal reasoning but it can also be applied at each dialogue stage to the joint theory that the agents have created at that stage. For example, after O_2 the logic says that q is overruled on the basis of $\mathcal{K}_n = \varnothing, \mathcal{K}_p = \{p, r, s\}$ while after P_4 the logic says that no argument for q can be constructed on the basis of $\mathcal{K}_n = \varnothing, \mathcal{K}_p = \{p, s\}$. Argumentation logic can also be used as a component of notions of soundness and completeness of protocols, such as:

- A protocol is *sound* if whenever at termination p is accepted, p is justified by the participants' joint knowledge bases.
- A protocol is *weakly* complete if whenever p is justified by the participants' joint knowledge bases, there is a legal dialogue at which at termination p is accepted.

– A protocol is *strongly* complete if whenever p is justified by the participants' joint knowledge bases, all legal dialogues terminate with acceptance of p.

These notions can also be defined relative to the joint theory constructed during a dialogue, or made conditional on particular agent strategies and heuristics (for example, a protocol could be sound and complete on the condition that all agents are honest but not trusting).

We can now without giving up the idea of an argumentation logic make sense of the claim that arguments should be evaluated in the context of a dialogue or procedure. The dialogue provides the relevant statements and arguments at each stage of the dialogue. The logic then determines the justified arguments at that stage. The logic also points at the importance of investigation. Since arguments can be defeated by counterarguments, the search for information that gives rise to counterarguments is an essential part of testing an argument's viability: the more thorough this search has been, the more confident we can be that an argument is justified if we cannot find defeaters. The ultimate justification of an argument is then determined by applying the logic to the final information state. Thus the ultimate justification of an argument depends on both logic and dialogue, or more generally on both logic and investigation.

On this account the critical questions of argument schemes have a dual role. On the one hand they define possible counterarguments to arguments constructed with the scheme (logic) while on the other hand they point at investigations that could be done to find such counterarguments (dialogue and procedure). This account also gives a further explanation why argument evaluation is context dependent, since different contexts may require different protocols for dialogue: when a decision has to be reached in reasonable time (as in a business meeting), a protocol may be more restrictive than in settings like academic debate. For example, the right to give alternative replies to a move may be restricted so that agents are forced to think what is their best reply.

Finally, on this account persuasiveness of arguments can be modelled as follows. Each dialogical agent has an internal argumentation theory and evaluates incoming arguments in terms of how they fit with the AF that it can internally generate. Given an *acceptance attitude* the agent will either accept the argument's premises and/or conclusion, or attack it with a counterargument, or ask for further grounds for a premise. Personality models can help modelling which types of arguments an agent of a certain type tends to accept. This gives a third way in which argument evaluation is context-dependent: the persuasive force of an argument depends on the listener. Current work of this kind is still preliminary but fascinating and promising (see e.g., the proceedings of the annual *ArgMas* workshops on argumentation in multi-agent systems). In fact this work provides a formal or even computational account of Perelman's New Rhetoric [9].

2.6 Conclusion

In this chapter we discussed five philosophical problems concerning argumentation. We first showed how argumentation-based standards for non-deductive inference can be defined, by presenting an abstract framework for argument evaluation given a set of arguments and their attack and defeat relations, and by supplementing it with accounts of argument construction and the nature of attack and defeat. We then clarified how a dialogical account of argument evaluation can be given in formal terms, by discussing the embedding of argumentation logics in dialogue systems for argumentation. This embedding also clarified the nature of argument schemes: argument schemes can be seen as defeasible inference rules and their critical questions as pointers to counterarguments. We also clarified how the use of arguments to persuade can be formalised, by adding the notions of argumentation strategies and heuristics and suggesting the use of personality models of argumentative agents. Finally, we gave several reasons why argument evaluation is context-dependent: different domains may have different sets of argument schemes, different contexts may require more or less strict semantics and/or protocols for dialogue and the persuasive force of arguments may depend on the listener.

References and Proposed Reading

1. *Baroni, P., Caminada, M., & Giacomin, M. (2011). An introduction to argumentation semantics, *The Knowledge Engineering Review, 26*, 365–410. [A comprehensive survey of semantics for abstract argumentation.].
2. Besnard, P., & Hunter, A. (2008). *Elements of argumentation*. Cambridge, MA: MIT Press.
3. Caminada, M., & Amgoud, L. (2007). On the evaluation of argumentation formalisms. *Artificial Intelligence, 171*, 286–310.
4. Dung, P. (1995). On the acceptability of arguments and its fundamental role in nonmonotonic reasoning, logic programming, and *n*–person games. *Artificial Intelligence, 77*, 321–357.
5. *Hunter, A. (Ed.). (2014). *Argument and Computation* (Vol. 5). Special issue with Tutorials on Structured Argumentation. [Contains tutorial introductions to four alternative accounts of structured argumentation.]
6. Modgil, S. (2009). Reasoning about preferences in argumentation frameworks. *Artificial Intelligence, 173*, 901–934.
7. Modgil, S., & Prakken, H. (2013). A general account of argumentation with preferences. *Artificial Intelligence, 195*, 361–397.
8. Modgil, S., & Prakken, H. (2014). The ASPIC+ framework for structured argumentation: a tutorial. *Argument and Computation 5*, 31–62.
9. Perelman, C., & Olbrechts-Tyteca, L. (1969). *The new rhetoric. A treatise on argumentation*. Notre Dame: University of Notre Dame Press.
10. Pollock, J. (1974). *Knowledge and justification*. Princeton: Princeton University Press.
11. *Pollock, J. (1995). *Cognitive carpentry. A blueprint for how to build a person*. Cambridge, MA: MIT Press. [A classic philosophical account of defeasible argumentation.].
12. Pollock, J. (2009). A recursive semantics for defeasible reasoning. In I. Rahwan & G. Simari (Eds.), *Argumentation in artificial intelligence* (pp. 173–197). Berlin: Springer.
13. Prakken, H. (2010). An abstract framework for argumentation with structured arguments. *Argument and Computation, 1*, 93–124.

14. Prakken, H. (2011). An overview of formal models of argumentation and their application in philosophy. *Studies in Logic, 4*, 65–86.
15. *Prakken, H. (2018). Historical overview of formal argumentation. In P. Baroni, D. Gabbay, M. Giacomin, & L. van der Torre (Eds.), *Handbook of formal argumentation* (Vol. 1). London: College Publications.
16. *Prakken, H., & Vreeswijk, G. (2002). Logics for defeasible argumentation. In D. Gabbay & F. Günthner (Eds.), *Handbook of philosophical logic* (Vol. 4, 2nd Ed., pp. 219–318). Dordrecht/Boston/London: Kluwer Academic Publishers. [A systematic, although somewhat outdated introduction to argumentation logics.].
17. *Rahwan, I., & Simari, G. (Eds.) (2009). *Argumentation in artificial intelligence*. Berlin: Springer. [A collection of survey papers on all aspects of formal and computational argumentation.].
18. Raz, J. (1975). *Practical reason and norms*. Princeton: Princeton University Press.
19. Rescher, N. (1976). *Plausible reasoning*. Assen: Van Gorcum.
20. Rescher, N. (1977). *Dialectics: A controversy-oriented approach to the theory of knowledge*. Albany: State University of New York Press.
21. Toulmin, S. (1958). *The uses of argument*. Cambridge: Cambridge University Press.
22. Tversky, A., & Kahneman, D. (1974). Judgement under uncertainty: heuristics and biases. *Science, 185*, 1124–1131.
23. Vreeswijk, G. (1993). *Studies in defeasible argumentation*. Doctoral dissertation, Free University Amsterdam.
24. Walton, D. (1996). *Argumentation schemes for presumptive reasoning*. Mahwah: Lawrence Erlbaum Associates.
25. Walton, D., & Krabbe, E. (1995). *Commitment in dialogue. Basic concepts of interpersonal reasoning*. Albany: State University of New York Press.
26. Walton, D., Reed, C., & Macagno, F. (2008). *Argumentation schemes*. Cambridge: Cambridge University Press.

Chapter 3
Formal Methods and the History of Philosophy

Catarina Dutilh Novaes

Although not (yet) entirely mainstream, uses of formal methods for the study of the history of philosophy, the history of logic in particular, represent an important trend in recent philosophical historiography. In this chapter, I discuss what can (and cannot) be achieved by the application of formal methods to the history of philosophy, addressing both motivations and potential pitfalls. The first section focuses on methodological aspects, and the second section presents three case studies of historical theories which have been investigated with formal tools: Aristotle's syllogistic, Anselm's ontological argument, and medieval theories of supposition.

3.1 Methodological Considerations

3.1.1 Why (Not) Apply Formal Methods to the History of Philosophy?

Let us begin by discussing motivations and potential objections to the use of formal methods in the study of the history of philosophy. A recurring concern pertains to the risk of *anachronism*: formal methods are for the most part recent inventions, and applying these modern frameworks to theories of the past is bound to bring along a range of presuppositions and assumptions that have no counterpart in the historical framework in question.

However, while this issue may be more acute in the case of formal methods, it in fact pertains to philosophical historiography in general. Indeed, a certain amount of

C. Dutilh Novaes (✉)
Department of Philosophy, Vrije Universiteit, Amsterdam, The Netherlands
e-mail: c.dutilhnovaes@vu.nl

S. O. Hansson, V. F. Hendricks (eds.), *Introduction to Formal Philosophy*, Springer Undergraduate Texts in Philosophy, https://doi.org/10.1007/978-3-319-77434-3_3

anachronism is inherent to any historical analysis, and it is not immediately obvious why the anachronism brought in by formal methods would be substantially different from the anachronism brought in by other recent methodologies and frameworks. Thus, even acknowledging that philosophical theories bear a mark of historicity, formal methods can still be seen as legitimate interpretive tools for historical investigations.

Nonetheless, the risk of excessive anachronism when employing formal methods is real, and perhaps more acute than with other methodologies. Thus, the historian of philosophy who employs formal methods must remain particularly alert so as to minimize or in any case take into account the inevitable traces of anachronism in her investigations. The choice of the formalism to be used must be judicious, as for a given historical analysis some formalisms will bring in a lesser degree of anachronism and inadequacy than others.

This being said, formal methods can in fact be valuable tools in the context of textual exegesis. Much of what the historian of philosophy does consists in working with *texts*, and formalization may help elucidate particularly thorny passages or arguments.[1] (However, it must be stressed that a formalization of a historical theory usually does not consist in taking the very linguistic expression of the theory in the original text as its object.[2]) In other words, formal methods can serve as a hermeneutical tool among others; by engaging in the formalization of a given historical theory, the interpreter may obtain a deeper understanding of the theory, possibly an understanding that other interpretive methods could not provide.

Indeed, formal methods seem particularly well-placed to unveil certain aspects of the target theory. A formalization presupposes a deconstruction of the historical theory so that some of its key elements are isolated from the others, thus outlining its logical scaffolding. Furthermore, formal methods may disclose 'hidden' aspects of a historical theory, which are not visible to the 'naked eye' (to pursue Frege's metaphor of a formalism as a microscope, in the preface of the *Begriffsschrift*).

Hence, provided they are used with caution and that their inherent anachronism is taken into account, formal methods can be irreplaceable items in a historian's toolkit. But their use is only justified if they truly shed new light on the object of analysis; unless new insight is obtained, fancy formalization may simply be overkill.

[1]More recently, computational methods have been gaining quite a lot of traction for research in history of philosophy, under the umbrella of 'digital humanities'. These are exciting developments that may well change substantially how historians of philosophy approach their topics, but for now they are still at early stages. While these can be broadly understood as formal methods, in this piece I do not discuss them any further for reasons of space.

[2]In fact, I have argued elsewhere ([11], chap. 3) that it is a mistake to think about formalizations in general merely as taking portions of 'natural language' as their starting point and translating them into a formal language.

3.1.2 How (Not) to Apply Formal Methods in History of Philosophy

How does a historian of philosophy who applies formal methods proceed? It cannot be sufficiently emphasized that the formal historian remains above all a *historian*: it is solely on the basis of solid conceptual knowledge of her object of study that she can successfully apply formal methods in her investigations. While formal tools can be instrumental even in the interpretive process of textual analysis, ultimately the formal historian must be thoroughly familiar with the historical framework in question before formalization begins.

Next, an important step is the choice of an adequate formalism. The first uses of formal methods to study the history of philosophy, in the second half of the twentieth century, tended to adopt uncritically the 'standard' logical systems of the time, in particular classical predicate logic. But as we will see with the case studies below, uncritically adopting an inadequate framework is likely to lead to poor results. An inadequate formalism will bring along unwarranted assumptions and presuppositions, and/or fail to capture some key components of the historical theory if they have no counterpart in the formalism.

The point is not that there will be at most one adequate formal framework for each historical theory; there may well be different, equally adequate frameworks, or frameworks adequate for different aspects of the theory. In other words, conceptual as well as semi-pragmatic considerations will play a role, but some formalisms are hopelessly unsuitable for a given historical theory. The choice of a formalism is already an *interpretive* choice; there is no such thing as a theoretically neutral formalization.

A formalization is always a process of abstraction, but one which promises to offer further insight precisely because it separates what is crucial from what is secondary in a given theory (relative to a given purpose), allowing for a more uniform analysis. In any formalization, some elements of the target phenomenon are represented by certain features of the model – what Shapiro[3] refers to as the *representors* – while other features of the model are *artifacts* (again in Shapiro's terminology), introduced for convenience. So a good formalization is not one where every aspect of what is being formalized is represented, but rather one where the tradeoff between simplification and accuracy of representation is favorable.

In particular, the chosen formalism must have the right level of *granularity* with respect both to the target historical theory and the purpose of the formalization: it must abstract the right amount of information away – not too much, not too little. The formalization is too coarse if it fails to capture important aspects of the historical theory, and it is too fine-grained if it projects distinctions and concepts into the theory that are not there to start with (naturally, it can be both too coarse and too fine-grained).

[3] Shapiro [30].

Note that these general considerations must be viewed as no more than schematic guidelines for what is to count as an adequate formalization. Actual criteria must be discussed on a case-by-case basis, as will be illustrated by the case studies below.

3.1.3 Interpreting the Results

Assume that the historian has undertaken a formalization of an episode in the history of philosophy, and is now looking at the end-product. What does the formalization mean? Has it succeeded in outlining aspects of the historical theory that alternative methodologies had failed to identify?

There is a sense in which the goal of a formalization (of a historical theory or otherwise) is precisely to reveal novel, hidden aspects of its object of study. In some sense, the goal is to obtain a situation of *mismatch* between one's initial beliefs about a given historical theory and the results of the formalization.[4] But if a formalization makes a prediction that is not explicitly to be found in the informal theory being formalized (or vice-versa) – i.e. if there is a mismatch between formalization and what is formalized – then this may mean two things: either the formalization is not sufficiently faithful to the informal theory – in which case it is a 'bad' formalization; or the formalization in fact 'sees' something in the original theory that was not immediately apparent – in which case it is a 'good' formalization in that it is illuminating.

If, however, the historian's prior views on the historical theory and the results of the formalization match completely, then on the one hand one may say that the formalization is entirely accurate and adequate, but on the other hand one may also say that it is uninformative in that it produced no new insights. So there is a sense in which precisely the cases of mismatch are the interesting ones; when mismatch occurs, further analysis is required in order to establish whether it is indeed a novel result revealed by the formalization or rather a sign that it is inadequate.[5]

Again, there is no one-fits-all answer here; in each case, further analysis is required to establish whether a mismatch between initial expectations and the results of the formalization signals inadequacy, or alternatively, novelty and informativeness. This may also be done with a critical stance, i.e. the formalization may be able to outline shortcomings and flaws in the historical theory itself (e.g. the potential invalidity of Anselm's ontological argument). But often, what may appear to be a shortcoming in the historical theory is, on further scrutiny, an unwarranted projection of presuppositions (e.g. some of Łukasiewicz's criticisms of Aristotelian syllogistic). Thus, although a certain amount of critical stance is to be commended, the principle of charity remains an important guideline for the formal historian of philosophy.

[4]See [13].

[5]For an example of formal analysis actually revealing something new about a historical theory, see [8] on Bradwardine's solution to the Liar paradox.

3.2 Case Studies

To appreciate the (initially) innovative character of applying modern mathematical logic to the analysis of so-called 'traditional logic', it is important to bear in mind that much (though not all) of modern mathematical logic emerged as a *rejection* of traditional logic. But since then (first half of twentieth century), much has changed, and formal methods have been regularly used for the analysis of philosophical theories of the past. In what follows, I discuss three case studies: Aristotle's syllogistic; Anselm's ontological argument; and medieval theories of supposition. By its very nature, the history of *logic* is particularly amenable to formal analysis, but Anselm's ontological argument illustrates a fruitful application of formal methods outside the history of logic.

3.2.1 Syllogistic

The founder of 'formal history of philosophy' is the Polish logician Jan Łukasiewicz, well known for his work on mathematical logic; the historical theory he set out to formalize was Aristotle's syllogistic. In the *Prior Analytics*, Aristotle presents the logical system which became known as syllogistic, whose language contains only four kinds of sentences (*a* and *b* are arbitrary terms):

A: All *a* is *b*
I: Some *a* is *b*
E: No *a* is *b*
O: Some *a* is not *b*

Aristotle develops a theory of the pairs of such sentences yielding conclusions that 'follow of necessity' – the famous syllogistic arguments. Of the 256 possible combinations, 24 are said by Aristotle to constitute valid arguments. Łukasiewicz became interested in Aristotle's syllogistic already in the 1920s, but his major work on the topic was published only in 1951: his monograph *Aristotle's Syllogistic from the Standpoint of Modern Formal Logic* [20]. Łukasiewicz's account of Aristotelian syllogistic can be thus summarized:

> The logic of Aristotle is a theory of the relations A, E, I, and O (in their mediaeval senses) in the field of universal terms. [. . .] As a logic of terms, it presupposes a more fundamental logic of propositions, which, however, was unknown to Aristotle and was discovered by the Stoics in the century after him. Aristotle's theory is an axiomatized deductive system, in which the reduction of the other syllogistic moods to those of the first figure is to be understood as the proof of these moods as theorems by means of the axioms of the system. ([23], 134)

Crucially, Łukasiewicz formulates syllogistic as an *axiomatic theory* embedded in a *propositional logic*, thus disregarding its original term-based nature. He arrives at the same results as Aristotle (at least in terms of which arguments are deemed valid or invalid), but his derivations are nothing like Aristotle's own. In particular, he criticizes Aristotle's *per impossibile* proofs of the syllogisms *Baroco* and *Bocardo* (in the medieval terminology) as incorrect, simply because they are not deemed correct within his axiomatic approach. He himself acknowledges that, taking valid syllogisms to be *rules of inference* rather than axioms, Aristotle's proofs are correct, but rather than viewing this as a sign that his axiomatic interpretation might be inadequate, he prefers to attribute the error to Aristotle.[6] Łukasiewicz's formalization in fact imposes "an order on Aristotle's syllogistic, rather than discovering the order within it" ([33], 192).

In the early 1970s, John Corcoran [5, 6] and Timothy Smiley [31] independently presented alternative formalizations of Aristotle's syllogistic; *contra* Łukasiewicz's axiomatic approach, they emphasized the role of rules of inference in the system. Corcoran, for instance, views syllogistic as a *term-based natural deduction system*. Thus, a valid syllogism such as "All *a* is *b*, all *b* is *c*, thus all *a* is *c*", which is rendered as an axiom by Łukasiewicz (in Polish notation):

$$CKAbcAabAac$$ [7]

is formalized by Corcoran as a rule of inference:

$$Azy + Axz \ |{=}Axy$$

In this way, "Corcoran succeeds, as Lukasiewicz did, in reproducing Aristotle's results, and he succeeds, as Lukasiewicz did not, in reproducing Aristotle's method step by step, so that the annotated deductions of his system D are faithful translations of Aristotle's exposition." ([23], 134) Undoubtedly, Corcoran's formalization (as Smiley's) is a great improvement over Łukasiewicz's from the point of view of historical accuracy.

Alongside a presentation of Aristotle's syllogistic as a natural deduction system, Corcoran also introduces a formal semantics for the system, on the basis of families of non-empty sets. He proves that his deductive system is sound and complete with respect to this semantics, and then goes on to argue that this establishes the adequacy of his deductive system. But why is it that *this* particular semantics should serve as yardstick for the adequacy of the deductive system? Corcoran does not offer much motivation for the choice of this semantics, and indeed other semantics for syllogistic have been proposed in the literature [2].

There is no doubt that formal analysis has greatly improved our understanding of syllogistic as a logical system.[8] But the divergences between Łukasiewicz's formalization and Corcoran's also outline the extent to which conceptual, historical analysis of the texts remains crucial, and illustrate the open-ended nature of formalization in history of philosophy.

---

[6]See ([29], 37–39).

[7]Polish notation is based on prefixing operators. 'C' stands for implication and 'K' for conjunction, so this expression roughly means *'Abc & Aab → Aac'*.

[8]See for example [1] for some interesting meta-theoretical results, and [18] for a formal analysis of Buridan's modal syllogism.

3.2.2 Anselm's Ontological Argument

Anselm's so-called ontological argument (most famously presented in chapter II of the *Proslogion*, written c. 1077-78) purports to demonstrate the existence of God on the basis of a seemingly plausible definition of God as 'that than which nothing greater can be thought'. More precisely, it purports to show from this definition alone that a contradiction can be derived from the assumption that God does not exist.[9]

Anselm's argument is one of the most discussed arguments in the history of philosophy, and continues to puzzle commentators. Structurally, it is prima facie a plausible argument, but there is something highly unsettling about deriving such a strong conclusion (God exists) from apparently modest premises, by an apparently valid reasoning. Commentators widely disagree on where the problem lies; as summarized by Uckelman ([35], section 5),

> The verdict on the premises range from "obviously true" to "obviously false", and similarly for the validity of the argument(s). The difficulty of determining the soundness and validity of the argument is also located in different places, with some of the various possibilities put forward including the problem of counterfactual reasoning, the role played by the term 'God', the analysis of definite descriptions, substitution into opaque contexts, the definition of perfection, and the nature of possibility. Others believe that the real error of the proof is still to be found, while some believe that the error is as simple as begging the question or the fallacy of equivocation.

(Uckelman provides extensive references to the different commentators holding these views.) Given this interpretive conundrum, it seems that the application of modern logical apparatuses could be of great use to the interpreter. In effect, an adequate formalization might be able to unveil the logical structure of the argument, making hidden assumptions explicit, and bringing to the fore each of the inferential steps in the argument. However, the different formalizations of Anselm's argument proposed in the literature disagree significantly on how best to interpret and analyze it, which again illustrates the fluidity of formalization in research on the history of philosophy: even a single argument, originally expressed in what amounts to half a page of text, is susceptible to receiving highly diverging formal analyses.

Two notable applications of formal tools to Anselm's argument were proposed by Jacquette [16] and Oppenheimer and Zalta [24, 25]. Jacquette argues that the argument has a strong modal component, more precisely an intensional/epistemic component, introduced by the notion of 'that than which nothing greater *can be thought*'. (In fact, arguably there are two intensional layers: one introduced by 'can be' and the other introduced by 'thought'.) On his reconstruction, the argument commits the fallacy of substitution in opaque contexts, as two co-referential terms

[9]See [34] for a concise presentation of the argument.

(the definiens and the definiendum in the proposed definition of God) cannot be used interchangeably in opaque contexts. Thus, according to Jacquette, the argument is not valid.

While Jacquette focuses on the intensional/epistemic component of the argument, Oppenheimer and Zalta highlight the fact that the definiens in the proposed definition of God corresponds to a *definite description*.[10] Rather than eliminating the definite description, they maintain that, in an analysis/reconstruction of the argument, the phrase should be explicitly represented as such. For this end, they resort to the framework of free logic, which allows for terms or expressions having no denotation. On their reconstruction, the argument comes out as valid, once the proposed logical behavior of definite descriptions is properly spelled out.

Arguably, each of these two formal analyses of Anselm's argument has illuminated a particular central aspect thereof: the intensional/epistemic component for Jacquette, and the definite description component for Oppenheimer and Zalta. In itself, this is not particularly remarkable; as argued in Sect. 3.1.2, a formalization always entails a decision to focus on certain aspects of its object at the expense of others. Thus, it is perfectly conceivable that there might be more than one adequate formalization for the same object. Nonetheless, the fact that these two analyses disagree on their verdict regarding the validity of Anselm's argument does suggest that they cannot both be equally 'right'. Perhaps a unified analysis taking both elements into account would be required to adjudicate the issue.

In any case, formalizations of Anselm's argument illustrate applications of formal methods in history of philosophy going beyond the history of logic strictly speaking. They also illustrate the fact that formalizations always entail theoretical choices, but suggest as well that, while there is typically room for more than one adequate formalization, at times two formalizations turn out to be true competitors that cannot both be adequate.

3.2.3 *Medieval Theories of Supposition*

Supposition is a key concept in Latin medieval semantics, but the phrase 'medieval theories of supposition' covers a rather heterogeneous group of theories, ranging from the twelfth to the fifteenth century [9]. The fragments of theories of supposition having attracted the attention of contemporary philosophers and logicians are primarily those (seemingly) related to the modern concept of *quantification*, especially the so-called modes of personal supposition [10]. This is in itself quite revealing: in first instance, modern philosophers were mostly interested in the *similarities*, rather than in the differences, between the historical theories in question and modern frameworks. Indeed, from early on, the 'quantificational fragment' of supposition

[10]([24], 509).

theories was viewed from the perspective of modern conceptions of quantification: "The theory of supposition is, to a very large extent, one with the modern theory of quantification ..." ([3], 28).[11]

The different modes of personal supposition offer a semantic account of a wide range of what the medieval authors referred to as *syncategorematic terms* ('every', 'not', 'no', 'some', 'only' etc.).[12] This is spelled out by means of inferential relations between sentences where such terms occur, and sentences of the form 'This *a* is *b*', where '*a*' and '*b*' are terms occurring in the original sentences; the latter, the *categorematic terms*, are those said to have such-and-such supposition. (There are also rules specifying in which syntactic contexts, defined by the syncategorematic terms and word order, a term would have such-and-such supposition)

The main modes of personal supposition can be defined as follows. Let (S) and (Q) stand for any syncategorematic term (or combination thereof), and the general form of a sentence P be '(Q) *a* is (S) *b*'. The generic definitions of the modes of personal supposition in terms of inferential relations are[13]:

- A term *a* has *determinate* supposition in P **if and only if**: A disjunction of sentences of the form 'This *a* is (S) *b*' can be inferred from P, but a conjunction of sentences of the form 'This *a* is (S) *b*' cannot be inferred from P.
- A term *a* has *confused and distributive* supposition in P **if and only if**: A conjunction of sentences of the form 'This *a* is (S) *b*' can be inferred from P.
- A term *a* has *merely confused* supposition in P **if and only if**: A sentence with a disjunctive subject term of the form 'This *a*, or that *a* etc ... is (S) *b*' can be inferred from P, but neither a disjunction nor a conjunction of propositions of the form 'This *a* is (S) *b'* can be inferred from P.

The same applies *mutatis mutandis* to predicate terms, so that P can be fully analyzed in terms of disjunctions and conjunctions of simpler sentences (possibly including disjunctive terms). For example, in 'Every *a* is *b*', '*a*' has confused and distributive supposition and 'b' has merely confused supposition; in 'No *a* is *b*' both terms have confused and distributive supposition; in 'Some *a* is *b*' both terms have determinate supposition; and in 'Some *a* is not *b*' '*a*' has determinate supposition and '*b*' has confused and distributive supposition.

Earlier interpreters noted that, while modern theories of quantification are expressed in the formal language of predicate calculus, medieval theories were expressed in the regimented form of Latin used at the time. But if this is merely a superficial difference in modes of expression – that is, if theories of supposition are indeed "one" with modern theories of quantification – then the translation into the language of predicate calculus should be a straightforward affair. Matthews ([21], 99) was the first to challenge this assumption, noting that "Ockham [and medieval

[11] See also [22], and [4] for an overview focusing specifically on the scholarship on Ockham.

[12] See [26] for an overview from a contemporary perspective.

[13] See (Ockham [36], chap. 70) and ([17], chaps. 4.3.5 and 4.3.6) for some of the original formulations of these definitions.

authors in general] quantifies over terms whereas modern logicians quantify over variables"; thus, "Ockham and the moderns are not free to agree on the interpretation of any categorical propositions". In a similar vein, Henry [14] suggested that, rather than variable-based theories of quantification, an alternative system, namely Leśniewiski's Ontology, would be the right modern system to formalize medieval theories of supposition. Indeed, Ontology is term-based, and the basic sentential form is the traditional subject-copula-predicate form, thus being closer in spirit to the medieval framework. But it brings along yet other presuppositions alien to the supposition framework, and at any rate it never became widely adopted by historians of philosophy. For the most part, formal treatments of supposition theory continued to rely on standard predicate logic [19, 32], with mixed results.

Another challenge for any formalization of supposition theory with modern predicate logic is the definition of merely confused supposition. As seen above, merely confused supposition relies on term-disjunction: "This *a* or that *a* or that other *a* etc. is *b*". Now, in its standard versions, modern predicate logic does not contain the device of term-disjunction (or of term-conjunction, for that matter). It is not an insurmountable problem, and Priest and Read [28] adapted standard predicate logic so as to accommodate term-disjunction. Nevertheless, the need for such adaptations suggests once again that the equation between medieval theories of modes of supposition and modern standard approaches to quantification is by no means straightforward.

Does this mean that medieval theories of supposition are not amenable to investigations with modern logical tools? This conclusion would be unwarranted. Given the striking similarities between portions of Latin medieval semantics and the modern enterprise of formal semantics, it would seem that formal tools can indeed be fruitfully applied here.[14] Nevertheless, as stressed in section "How (not) to apply formal methods in history of philosophy", formalization requires prior and extensive *conceptual analysis*: one must first grasp the historical theory in its own terms so as to determine which modern formalism, if any, might be adequate for a formalization. With respect to theories of supposition, rather than hastily concluding that they are "one" with modern quantification theory, some of the questions to be asked are: what did theories of supposition represent for the medieval authors themselves? What were the purposes assigned to them by these authors? ([7], chap. 1; [4], 11–15)

There is no doubt that the modes of personal supposition deal with 'quantificational phenomena' broadly construed, but a formalization must also do justice to the profound differences between how medieval authors conceptualized these phenomena and the presuppositions underlying modern systems such as predicate logic. Generally, it would seem that the latter is not a particularly suitable system to formalize the former, especially given the centrality of the concept of variable in

[14][27] is a particularly ambitious and impressive recent example of applications of modern formal tools borrowed from logic and linguistic to medieval logical theories.

the latter and its complete absence in the former. Indeed, it would seem that tailor-made formalisms are more likely to offer informative analyses of these (and other) medieval theories.

3.3 Conclusion

I have here attempted to offer a nuanced picture of the role of formal methods in the study of the history of philosophy. Views on the matter tend to be extreme, split between those who maintain that the application of formal methods for historical analysis is hopelessly anachronistic and thus unwarranted; and those who deem it entirely unproblematic. I have suggested that formalization can be an illuminating approach for the historian of philosophy, but also that it requires careful reflection and conceptual analysis. I have also suggested that, while generally there is not one unique correct formalization of a historical theory, some formalizations are definitely more adequate than others. Ultimately, a formalization must strive to balance the orthogonal desiderata of faithfulness and informativeness; not an easy task, but one with potentially fruitful results.

References[15]

1. Andrade-Lotero, E. J., & Becerra, E. (2008). Establishing connections between Aristotle's natural deduction and first-order logic. *History and Philosophy of Logic, 29*(4), 309–325.
2. Andrade-Lotero, E. J., & Dutilh Novae, C. (2012). Validity, the squeezing argument and alternative semantic systems: The case of Aristotelian syllogistic. *Journal of Philosophical Logic, 41*, 387–418.
3. Boehner, P. (1952). *Medieval logic: An outline of its development from 1250 – c. 1400*. Manchester: Manchester University Press.
4. Cameron, M. (2011). Methods and methodologies: An introduction. In M. Cameron & J. Marenbon (Eds.), *Methods and methodologies* (pp. 1–26). Leiden: Brill.
5. Corcoran, J. (1972). Completeness of an ancient logic. *The Journal of Symbolic Logic, 37*(4), 696–702.
6. Corcoran, J. (1974). Aristotle's natural deduction system. In J. Corcoran (Ed.), *Ancient logic and its modern interpretations* (pp. 85–131). Dordrecht: D. Reidel Publishing Company.
7. Dutilh Novaes, C. (2007). *Formalizing medieval logical theories: Supposition, obligationes and consequentia*. Berlin: Springer.
8. Dutilh Novaes, C. (2011a). Lessons on truth from medieval solutions to the Liar paradox. *The Philosophical Quarterly, 61*, 58–78.
9. Dutilh Novaes, C. (2011b). Medieval theories of supposition. In H. Lagerlund (Ed.), *Encyclopedia of medieval philosophy* (pp. 1229–1236). Berlin: Springer.
10. Dutilh Novaes, C. (2011c). Medieval theories of quantification. In H. Lagerlund (Ed.), *Encyclopedia of medieval philosophy* (pp. 1093–1096). Berlin: Springer.

[15]There is not much literature specifically on the application of formal methods for the study of the history of philosophy, but the interested reader can consult in particular [12, 15, 33].

11. Dutilh Novaes, C. (2012). *Formal languages in logic – A philosophical and cognitive analysis*. Cambridge: Cambridge University Press.
12. Dutilh Novaes, C. (2015). The formal and the formalized: The cases of syllogistic and supposition theory. *Kriterion, 131*, 253–270.
13. Dutilh Novaes, C., & Reck, E. (2017). Carnapian explication, formalisms as cognitive tools, and the paradox of adequate formalization. *Synthese, 194*(1), 195–215.
14. Henry, D. P. (1964). Ockham, *suppositio*, and modern logic. *Notre Dame Journal of Formal Logic, 5*, 290–292.
15. Hodges, W., & Johnston, S. (2017). Medieval modalities and modern methods: Avicenna and Buridan. *IfCoLog Journal of Logics and their Applications, 4*(4), 1029–1073.
16. Jacquette, D. (1997). Conceivability, intensionality, and the logic of Anselm's modal argument for the existence of god. *International Journal for Philosophy of Religion, 42*(3), 163–173.
17. Buridan, J. (2001). *Summulae de Dialectica* (G. Klima, Trans.). New Haven: Yale University Press.
18. Johnston, S. (2015). A formal reconstruction of Buridan's modal syllogism. *History and Philosophy of Logic, 36*(1), 2–17.
19. Karger, E. (1976). *A study in William of Ockham's modal logic*, PhD dissertation, University of California, Berkeley.
20. Łukasiewicz, J. (1957). *Aristotle's syllogistic from the standpoint of modern formal logic* (2nd edn.). Oxford: Oxford University Press.
21. Matthews, G. (1964). Ockham's supposition theory and modern logic. *Philosophical Review, 73*, 91–99.
22. Moody, E. (1953). *Truth and consequence in medieval logic*. Amsterdam.
23. Mulhern, M. (1974). Corcoran on Aristotle's logical theory. In J. Corcoran (Ed.), *Ancient logic and its modern interpretations* (pp. 133–150). Dordrecht: Reidel.
24. Oppenheimer, P. E., & Zalta, E. N. (1991). On the logic of the ontological argument. *Philosophical Perspectives, 5*, 509–529.
25. Oppenheimer, P. E., & Zalta, E. N. (2007). Reflections on the logic of the ontological argument. *Studia Neoaristotelica, 4*(1), 28–35.
26. Parsons, T. (2008). The development of supposition theory in later 12th through 14th centuries. In D. Gabbay & J. Woods (Eds.), *Handbook of the history of logic* (pp. 157–281). Amsterdam: Elsevier.
27. Parsons, T. (2014). *Articulating medieval logic*. Oxford: Oxford University Press.
28. Priest, G., & Read, S. (1977). The formalization of Ockham's theory of supposition. *Mind, 86*, 109–113.
29. Van Rijen, J. (1989). *Aspects of Aristotle's logic of modalities*. Dordrecht: Kluwer.
30. Shapiro, S. (1998). Logical consequence: Models and modality. In M. Schirn (Ed.), *Philosophy of mathematics today* (pp. 131–156). Oxford: Oxford University Press.
31. Smiley, T. (1973). What is a syllogism? *Journal of Philosophical Logic, 2*(1), 136–154.
32. Spade, P. V. (1988). The logic of the categorical: The medieval theory of ascent and descent. In N. Kretzman (Ed.), *Meaning and inference in medieval philosophy: Studies in memory of Jan Pinborg* (pp. 187–224). Dordrecht: Kluwer.
33. Thom, P. (2011). On formalizing the logics of the past. In M. Cameron & J. Marenbon (Eds.), *Methods and methodologies* (pp. 191–206). Leiden: Brill.
34. Uckelman, S. (2011). The ontological argument. In M. Bruce, & S. Barbone (Eds.), *Just the argument: 100 of the most important arguments in western philosophy* (pp. 25–27). New York: Wiley-Blackwell.
35. Uckelman, S. (2012). The reception of St. Anselm's logic in the 20th and 21st centuries. In G. Gasper & I. Logan (Eds.), *Saint Anselm of Canterbury and his legacy* (pp. 405–426). Toronto: Pontifical Institute of Mediaeval Studies.
36. William of Ockham. (1998). *Summa Logicae Part I* (M. Loux, Trans.). St. Augustine's Press: South Bend.

Chapter 4
Nonmonotonic Reasoning

Alexander Bochman

Abstract Nonmonotonic reasoning is a theory of the rational use of assumptions. We describe the relations between NMR and Logic, and two main paradigms of NMR, preferential and explanatory one.

4.1 Nonmonotonic Reasoning Versus Logic

Nonmonotonic reasoning (NMR) is an essential part of the logical approach to Artificial Intelligence. Its birth is due to the research methodology suggested in McCarthy [16] whose objective was a logical formalization of *common sense reasoning* for dealing with AI problems. NMR itself was born, however, as a result of dissatisfaction with traditional logical methods. Reasoning necessary for an intelligent behavior and decision making has appeared to be impossible to represent as deductive inferences in a logical system. The essence of the problem was formulated in Minsky [21] that questioned the suitability of representing commonsense knowledge in a form of a deductive system. Minsky also pointed to monotonicity of logical systems as a source of the problem:

> *Monotonicity*: ... In any logistic system, all the axioms are necessarily "permissive" - they all help to permit new inferences to be drawn. Each added axiom means more theorems, none can disappear. There simply is no direct way to add information to tell such the system about kinds of conclusions that should not be drawn!

Long before the first nonmonotonic formalisms, there have been problems and applications in AI that required some forms of nonmonotonic reasoning. Initial solutions to these problems worked, and this was an incentive for trying to provide them with a more systematic logical basis [15].

A. Bochman (✉)
Computer Science Department, Holon Institute of Technology, Holon, Israel
e-mail: bochmana@hit.ac.il

S. O. Hansson, V. F. Hendricks (eds.), *Introduction to Formal Philosophy*, Springer Undergraduate Texts in Philosophy, https://doi.org/10.1007/978-3-319-77434-3_4

NMR is intimately related to traditional philosophical problems of natural kinds and ceteris paribus laws. These notions resist precise logical definition, but involve description of normal cases. Reasoning with such concepts is inherently *defeasible*, so it fails to 'preserve truth' under all circumstances, which has always been considered a standard for logical reasoning.

Natural kinds have reappeared in AI as a practical problem of building taxonomic hierarchies for large knowledge bases that are allowed to have exceptions. The theory of reasoning in such taxonomies has been called *nonmonotonic inheritance* (see [10]). The guiding principle in resolving conflicts in such hierarchies was a *specificity principle*: more specific information should override more generic information in cases of conflict. Thus, a knowledge base may contain both *Birds fly* and *Penguins don't fly*, but then, given that Tweety is a penguin, we univocally infer that it does not fly, since *Birds fly* is a less specific claim. Though nonmonotonic inheritance relied more on graph-based representations than on traditional logical tools, it has managed to provide a plausible analysis of reasoning in this restricted context.

Nonmonotonicity of a different kind occurs in databases, logic programming and planning algorithms. A common assumption in such systems is that positive assertions that are not explicitly stated or derivable should be considered false. Thus, a database of students enrolled in a particular course implicitly presupposes that students that do not appear in the list are not enrolled in the course. Databases embody such negative information by appealing to the *closed word assumption*, which states that if a positive fact is not derivable from the database, its negation is assumed to hold. A similar principle is employed in programming languages for AI such as Prolog and Planner. Thus, in Prolog, the goal **not** *G* succeeds if the attempt to find a proof of G fails. Prolog's negation **not** is a nonmonotonic operator: if G is not provable from some axioms, it needn't remain nonprovable from an enlarged axiom set. This negation-as-failure has been used to implement important forms of commonsense reasoning, which eventually has led to developing modern declarative logic programming as a general representation formalism for AI (see [1]).

But first and foremost, nonmonotonicity has appeared in reasoning about actions. The main problem here was the *frame problem*: how efficiently determine which things remain the same in a changing world (e.g., a red block remains red after we have put it on top of another block). The frame problem arises in the context of predictive reasoning that is essential for planning and formalizing intelligent behavior, though neglected in traditional logic. Prediction involves the inference of later states from earlier ones. Changes in this setting do not merely occur, but occur for a reason. Furthermore, we usually assume that most things will be unchanged by the performance of an action. It is this *inertia assumption* that connects reasoning about action and change with NMR. What complicates the problem, however, is a *ramification problem*, the necessity of taking into account derived effects (ramifications) of actions. Suppose we have a suitcase with two locks, and it is opened if both locks are open. Then the action of opening one lock produces an indirect effect of opening the suitcase if the other lock is open. Such derived effects override the inertia assumption. The ramification problem has raised general questions about the role of causation in dynamic reasoning, and has led, eventually, to the so-called causal approach to the frame problem (see [8]).

Last but not least, there was the *qualification problem*, the problem of specifying conditions for a given action to have its intended effect. If I turn the ignition key in my car, I expect the car to start. However, many conditions have to be true for this: the battery must be alive, there must be gas in the tank, there is no potato in the tailpipe, etc. – an open-ended list of qualifications. Still, we normally *assume* that turning the key will start the car. This is obviously a special instance of a general philosophical problem of ceteris paribus laws, laws or generalizations that are valid under 'normal' circumstances which are usually impossible to specify exactly. It has become, however, an urgent practical problem for the representation of action and change in AI.

The above problems and their first solutions provided the starting point and basic objectives for the first nonmonotonic theories. These origins explain, in particular, an eventual discrepancy that has developed between NMR and commonsense reasoning. Though the latter has often appeared to be a promising way of solving AI problems, the study of 'artificial reasoning' need not be committed to it. Still, in trying to cope with principal commonsense reasoning tasks, the suggested formalisms have succeeded in capturing important features of the latter and thereby have broken new territory for logical reasoning. Today, nonmonotonic reasoning is not yet another application of logic, but a relatively independent field of logical research that has a great potential in informing, in turn, general logical theory and many areas of philosophical inquiry.

4.2 What Is Nonmonotonic Reasoning?

In everyday reasoning, we usually have incomplete information about a given situation, and we use a lot of assumptions about how things normally are in order to carry out further reasoning. Without such assumptions, it would be impossible to accomplish the simplest human reasoning tasks. Speaking generally, human reasoning is not reducible to collecting facts and deriving their consequences, but involves also making assumptions (and wholesale theories) about the world and acting in accordance with them. In this sense, commonsense reasoning is a simplified form of a general scientific methodology.

NMR is a theory of the rational use of assumptions. Now, assumptions are just beliefs, so they are abandoned when we learn new facts that contradict them. However, NMR assigns a special status to assumptions; it makes them *default* assumptions. Default assumptions are seen as always acceptable unless they conflict with current evidence. This presumptive reading has a semantic counterpart in the notion of *normality*; defaults are considered as holding for normal circumstances, and the nonmonotonic reasoning always assumes that the world is as normal as is compatible with known facts. This kind of belief commitment is a novel contribution of NMR to a general theory of reasoning.

This form of reasoning is distinct from deductive inference already because the latter is monotonic: if C is provable from a set a, it will be provable from a larger set $a \cup \{A\}$. Assumption-based reasoning is not monotonic, however, because adding new facts may invalidate some of the assumptions.

The default *Birds fly* is not a statement that is true or not of the world; some birds fly, some do not. Rather, it is an assumption used in building our theory of the world. NMR does not make any claims about the objective status of the assumptions it uses, so it does not depend on the objective confirmation of the latter. What it cares about, however, is the internal coherence of the choice of assumptions in particular situations. Of course, if we make an entirely inappropriate claim a default assumption, it will either be useless (inapplicable in most situations) or, worse, it may produce wrong conclusions. This makes nonmonotonic reasoning a risky business. Still, in most cases assumptions we make are useful and give desired results, and hence they are worth the risk of making an error. But what is even more important, more often than not we simply have no 'safe' replacement for such a reasoning strategy. That is why it is worth to teach robots and computers to reason in this way.

4.3 Two Problems of Default Assumptions

The primary problem of NMR is how we can make and consistently use default assumptions. Three initial nonmonotonic formalisms, namely circumscription [17], default logic [23] and modal nonmonotonic logic [20] have provided rigorous answers to this problem. The formalisms used three different languages – the classical language in circumscription,[1] a set of inference rules in default logic, and a modal language in modal nonmonotonic logic. Still, a common idea was to represent commonsense conditionals as ordinary conditionals with additional assumptions that could readily be accepted in the absence of contrary information. The differences between the three theories amounted, however, to different mechanisms of making default assumptions. In fact, default logic and modal nonmonotonic logics embodied the same nonmonotonic mechanism. However, the differences between both of them and circumscription were more profound. In order to articulate them, we should consider yet another important problem of default assumptions.

In order to preserve consistency of the resulting solutions, default assumptions should not be used when they contradict known facts and other defaults. Clearly, if a default plainly contradicts the facts, it should be 'canceled'. But if a number of defaults are jointly inconsistent with the facts, although each of them taken alone is consistent with them, then we have a *selection problem*: which of the defaults should be retained, and which abandoned in each particular case? An apparent solution is to choose all maximal consistent subsets of defaults; this solution was implicitly

[1] Circumscription amounts to using only *minimal* models satisfying a first-order description.

used in the circumscription approach of [17]. Unfortunately, it has turned out to be inadequate as a general solution to the selection problem. The main reason is that commonsense defaults are not born equal, and in most cases there is an additional structure of dependence and priority among the defaults themselves. As a result, not all consistent combinations of defaults turn out to be adequate as options for choice. We mentioned, for instance, that the choice of defeasible rules in nonmonotonic inheritance is constrained by the specificity principle: the two rules *Birds fly* and *Penguins don't fly* are jointly incompatible with a fact that Tweety is a penguin, but we univocally drop only the first rule in this situation, since it is a less specific claim than *Penguins don't fly*. Speaking generally, commonsense defaults involve much more structure than just a set of assumptions. That is why a solution to the primary problem of NMR, how to make default assumptions, does not necessary provide a solution to the selection problem. The latter requires a deeper understanding of the use of assumptions in commonsense reasoning.

A general way of handling the selection problem in the framework of circumscription, called *prioritized* circumscription, has been suggested by Lifschitz and endorsed in McCarthy [18]. The solution amounted to imposing priorities among minimized predicates. In fact, it was one of the origins of a general *preferential approach* to NMR (see below).

Default and modal nonmonotonic logics suggested a different, *explanatory* approach to the selection problem. In fact, this approach has 'borrowed' a much larger piece of commonsense methodology than circumscription. In both scientific and commonsense discourse, a particular law may fail to explain the actual outcome due to interference with other mechanisms and laws that contribute to the combined result. In other words, violations of laws are always *explainable* (at least in principle) by other laws that are active. It is this justificational aspect of reasoning that has been formalized in the notion of extension in default logic and corresponding models of modal nonmonotonic logic. An extension is a model generated by a set of defaults that is not only consistent, but also, and most importantly, explains away, or refutes, all other defaults that are left out. The latter requirement constitutes a very strong constraint on the coherence of potential choices, which goes far beyond plain consistency. Using this requirement, an explanatory theory can be 'tuned' to intended combinations of defaults by supplying the underlying logic with appropriate refutation rules for default assumptions. In a hindsight, this might be seen as one of the reasons why these formalisms have been relatively slow in realizing the complexity of the selection problem. In fact, the problem has 'survived' initial attempts of formalization, and has reappeared in a most dangerous form as a *Yale Shooting Anomaly* in Hanks and McDermott [9], where it was demonstrated that apparently plausible representations of defaults in default logic and other formalisms still do not provide an intended choice of assumptions for the solution of the frame problem. Nevertheless, despite initial, radically anti-logicist, reactions (cf. [19]), subsequent studies have shown that the Yale Shooting problem can be resolved, after all, in the framework of these formalisms.

4.4 Logic in Nonmonotonic Reasoning

The first nonmonotonic systems have re-shaped the initial contrast between NMR and logic. Namely, it has been shown that a nonmonotonic formalism can be defined by supplying some logical formalism with a *nonmonotonic semantics*, which forms a distinguished subset of the corresponding *logical* semantics determined by the logical formalism itself. Thus, for circumscription, the underlying logical formalism is just the classical logic (and its semantics), while the nonmonotonic semantics is given be the set of minimal models.

Unfortunately, this latter description has also brought to life a problematic 'shortcut' notion of *nonmonotonic logic* as a formalism determined directly by syntax and associated nonmonotonic semantics. On this view, a nonmonotonic logic has become just yet another logic determined by an unusual (nonmonotonic) semantics. However, this view has actually hindered in a number of ways an adequate understanding of nonmonotonic reasoning.

In ordinary logical systems, the semantics determines the set of logical consequences of a given theory, but also, and most importantly, it provides an interpretation for the syntax itself. Namely, it provides propositions and rules of a formalism with *meaning*, and its theories with *informational content*. By its very design, however, the nonmonotonic semantics is defined as a certain subset of logically possible models, and consequently it does not determine, in turn, the meaning of the propositions and rules of the syntax. Two radically different theories may (accidentally) have the same nonmonotonic semantics. Furthermore, such a difference cannot be viewed as apparent, since it may well be that by adding further rules or facts to both these theories, we obtain new theories that already have different nonmonotonic models (see [3] for further discussion).

The above situation is remarkably similar to the distinction between meaning (intension) and extension of logical concepts, a distinction that is fundamental for modern logic. Nonmonotonic semantics provides, in a sense, the extensional content of a theory in a particular context of its use. In order to determine the meaning, or informational content, of a theory, we have to consider all potential contexts of its use, and hence 'retreat' to the underlying logic. This distinction suggests the following more adequate understanding of nonmonotonic reasoning:

$$Nonmonotonic\ Reasoning = Logic + Nonmonotonic\ Semantics$$

Logic and its associated logical semantics are responsible for providing the meaning of the rules of the formalism, while the nonmonotonic semantics provides us with nonmonotonic consequences of a theory in particular situations.

In addition to a better understanding of the structure of nonmonotonic formalisms, the above two-layered structure has important benefits in comparing different formalisms. In particular, it allows us to see many of them as instantiations of the same nonmonotonic mechanisms in different underlying logics.

4.5 Preferential Nonmonotonic Reasoning

In solving the selection problem of default assumptions, preferential approach follows the slogan *"Choice presupposes preference"*, which makes it an instance of a general methodology that is at least as old as decision theory and the theory of social choice. According to this approach, the choice of assumptions should be made by establishing preference relations among them.

Generalizing prioritized circumscription, [25] defined a model preference logic based on an arbitrary preference ordering on interpretations.

Definition An interpretation i is a *preferred model* of A if it satisfies A and there is no better interpretation $j > i$ satisfying A. A *preferentially entails* B (written $A \mid\!\sim B$) if all preferred models of A satisfy B.

Shoham's approach was very appealing, and apparently suggested a unifying perspective on NMR. Kraus et al. [11] provided an axiomatization of such inference relations. This has established logical foundations for a research program that attracted many researchers both in AI and in logic. A detailed description of the preferential approach can be found in [15].

A representation of preferential entailment more suitable for real NMR can be based on the following model, where belief states correspond to admissible combinations of default assumptions (see [2]):

Definition An *epistemic state* is a triple $(\mathcal{S}, l, \prec)$, where $\mathcal{S}$ is a set of *belief states*, $\prec$ a preference relation on $\mathcal{S}$, while l is a labeling function assigning a deductively closed *belief set* to every belief state from $\mathcal{S}$.

Epistemic states can determine what to believe in particular situations. Changes in facts do not automatically lead to changes in epistemic states: the actual assumptions made in particular situations are obtained by choosing preferred belief states that are consistent with the facts.

A preferentially entails B in an epistemic state if $A \supset B$ holds in all preferred belief states consistent with A. Though apparently different from the original definition of Shoham, it is actually equivalent to the latter.

It is tempting to conclude from the above that preferential approach has assimilated nonmonotonic reasoning to plain deductive reasoning in a certain 'nonmonotonic' logic. This conclusion would be premature, however.

Preferential entailment is called nonmonotonic for the obvious reason that its rules do not admit Strengthening: $A \mid\!\sim B$ does not imply $A \wedge C \mid\!\sim B$. However, it is a monotonic, logical system in the more important sense that addition of new rules preserves previous derivations. Furthermore, the above semantics determines the *meaning* of conditionals, and hence preferential entailment describes precisely their *logic*. This inevitably implies, however, that it cannot capture the associated nonmonotonic reasoning with such defaults.

Preferential inference is severely sub-classical and does not allow us, for example, to infer *Red birds fly* from *Birds fly*. Clearly, there are good reasons for not accepting such a derivation as a *logical* rule; otherwise *Birds fly* would imply also

Penguins fly. Still, we could accept *Red birds fly* as a reasonable *default* conclusion from *Birds fly* in the absence of contrary information. By doing this, we would follow the general strategy of NMR of making reasonable assumptions on the basis of available information. This kind of reasoning will be defeasible, or *globally* nonmonotonic, since addition of new rules can block some of the conclusions made earlier. We can follow the idea of NMR also on the semantic side, namely by choosing 'most normal' epistemic states that satisfy a given set of conditionals. By doing this, we will accept rules that would not be derivable by preferential inference alone.

Summing up, the logic of preferential entailment should be extended to a nonmonotonic formalism by defining the associated nonmonotonic semantics. In fact, the literature is abundant with attempts to define such a theory.

Lehmann and Magidor [12] described a semantic construction, called *rational closure*, that allows us to make default conclusions from a set of conditionals.[2] This was a starting point in the quest for an adequate theory of defeasible entailment. A large number of modifications have been suggested, but a consensus has not been achieved. A general approach to this problem can be found in Geffner [6]. Finally, nonmonotonic inheritance (see [10]) can be viewed as a syntactic approach to defeasible entailment. Though it deals with conditionals restricted to literals, it has achieved a remarkable correspondence between what is derived and what is expected intuitively.

Most systems of defeasible entailment assume that classical implications corresponding to conditionals should serve as defaults in the associated nonmonotonic reasoning. Already this choice allows us to derive *Red birds fly* from *Birds fly* in the absence of conflicting information about redness. It is still insufficient, however, for capturing some further reasoning patterns. Suppressing details[3], what needs to be added here is a principled way of constructing a preference order on default sets. Recall, however, that establishing preferences among defaults is the main tool used by the preferential approach for resolving the selection problem of NMR. Accordingly the problem of defeasible entailment boils down again to the general selection problem for defaults. Unfortunately, this problem has turned out to be far from being trivial, or even univocal. Geffner's conditional entailment and nonmonotonic inheritance still remain the most plausible solutions suggested in the literature on preferential reasoning.

The preferential approach to NMR has suggested a powerful research program that significantly advanced our understanding of nonmonotonic reasoning and even of commonsense reasoning in general. Its most important achievement consists in formalizing a plausible logic of default conditionals that could serve as a logical basis for a full, nonmonotonic theory of defeasible reasoning. Unfortunately, it has not succeeded in achieving this latter goal.

---

[2] An equivalent construction, called system Z, has been suggested in Pearl [22].

[3] See [2].

4.6 Explanatory Nonmonotonic Reasoning

The explanatory approach encompasses almost all nonmonotonic formalisms that are actively investigated in AI today, including logic programming, argumentation and causal reasoning. Explanation can be seen as its basic ingredient. Propositions may not only hold in a model, but some of them are explainable (or caused) by other facts and rules. Furthermore, explanatory NMR is based on principles of *Explanation Closure* or *Causal Completeness* (see [24]), according to which any fact holding in a model should be explained.

By the above description, abduction and causation are integral parts of explanatory NMR. In some domains, explanatory reasoning adopts simplifying assumptions that exempt certain facts from the burden of explanation. Thus, the *Closed World Assumption* stipulates that negative assertions do not require explanation. In fact, minimization of models employed in McCarthy's circumscription can be seen as a by-product of this stipulation.

Simple default theories. Recall that a Tarski consequence relation is a set of rules $a \vdash A$ (where A is a conclusion, and a a set of premises) that satisfies the usual postulates. Its associated provability operator is $\mathrm{Cn}(u) = \{A \mid u \vdash A\}$. A consequence relation is *supraclassical* if it subsumes classical entailment.

For a set Δ of rules, let Cn_Δ denote the provability operator of the least supraclassical consequence relation containing Δ. Then $A \in \mathrm{Cn}_\Delta(u)$ precisely when A is derivable from u using the rules from Δ and classical entailment.

Now, a simple way of defining a nonmonotonic theory consists in combining a logical theory, given by a set of (Tarski) rules, and a set of default assumptions:

Definition A *simple default theory* is a pair $(\Delta, \mathcal{A})$, where Δ is a set of rules, and $\mathcal{A}$ a distinguished set of propositions called *defaults*.

Reasoning in this setting amounts to deriving plausible conclusions using rules and defaults. Explanatory reasoning requires here that a reasonable set of defaults explains why the rest of the defaults should be rejected.

Definition

- A set $\mathcal{A}_0$ of defaults is *stable* if and only if it is consistent and refutes any other default: $(\neg A) \in \mathrm{Cn}_\Delta(\mathcal{A}_0)$, for any $A \in \mathcal{A} \backslash \mathcal{A}_0$.
- A set s of propositions is an *extension* of a simple default theory iff $s = \mathrm{Cn}_\Delta(\mathcal{A}_0)$, for some stable set of defaults $\mathcal{A}_0$. Extensions determine the *nonmonotonic semantics* of a default theory.

Simple default theories provide a transparent description of explanatory NMR. Despite its simplicity, however, this formalism is equivalent to Reiter's default logic (see [4]). It is also closely related to the general argumentation (or assumption-based) framework of [5].

Generalizing the logic. For actual reasoning tasks of AI, we have to generalize the logical basis from Tarski rules to disjunctive rules $a \vdash b$, where b is a set of propositions. Informally, such a rule says that if all a's hold, then at least one of b's should hold. The theory of disjunctive inference is actually a well-developed part of general logical theory. A set of such rules forms a *Scott consequence relation* if and only if it satisfies the following postulates:

(Reflexivity) $A \vdash A$.
(Monotonicity) If $a \vdash b$ and $a \subseteq a', b \subseteq b'$, then $a' \vdash b'$;
(Cut) If $a \vdash b, A$ and $a, A \vdash b$, then $a \vdash b$.

Let $\overline{u}$ denote the complement of a set u of propositions. Then u is a *theory* of a Scott consequence relation if $u \nvdash \overline{u}$.[4] A Scott consequence relation in a classical language is *supraclassical*, if it satisfies:

Supraclassicality If $a \vDash A$, then $a \vdash A$.
Falsity $\mathbf{f} \vdash$.

The Falsity postulate excludes, in effect, classically inconsistent models.

Simple default theories can be naturally extended to disjunctive rules. The resulting formalism will be equivalent to a disjunctive generalization of default logic [7], and even to powerful formalisms suggested in Lin and Shoham [14] and Lifschitz [13] as unified formalisms for nonmonotonic reasoning and logic programming.

Biconsequence Relations. For a detailed analysis of explanatory NMR, we can employ reasoning with respect to a *pair* of contexts. On the interpretation suitable for NMR, one of these contexts is the main (objective) one, while the other context provides assumptions that justify inferences in the main context.

A *bisequent* is an inference rule of the form $a : b \Vdash c : d$, where a, b, c, d are sets of propositions. On the explanatory interpretation, it says 'If a's hold then one of c's holds *provided* no b is assumed, and all d's are assumed'.

A *biconsequence relation* is a set of bisequents satisfying the rules:

Monotonicity $\dfrac{a : b \Vdash c : d}{a' : b' \Vdash c' : d'}$, if $a \subseteq a', b \subseteq b', c \subseteq c', d \subseteq d'$;

Reflexivity $A : \Vdash A :$ and $: A \Vdash : A$;

Cut $\dfrac{a : b \Vdash A, c : d \quad A, a : b \Vdash c : d}{a : b \Vdash c : d} \qquad \dfrac{a : b \Vdash c : A, d \quad a : A, b \Vdash c : d}{a : b \Vdash c : d}$.

A biconsequence relation can be seen as a product of two Scott consequence relations. A pair (u, v) of sets of propositions is a *bitheory* of a biconsequence relation if $u : \overline{v} \nVdash \overline{u} : v$. A set u is a *theory* if (u, u) is a bitheory. A bitheory (u, v) is *positively minimal*, if there is no bitheory (u', v) such that $u' \subset u$. Finally, a biconsequence relation is *supraclassical* if both its component contexts respect the classical entailment.

Nonmonotonic semantics of a biconsequence relation is a set of theories that are explanatory closed in the sense that all their propositions are explained (i.e., derived) when the theory itself is taken as the assumption context.

[4] Or, equivalently, if $a \vdash b$ and $a \subseteq u$, then $u \cap b \neq \emptyset$.

Definition A set u is an *extension* of a biconsequence relation, if (u, u) is a positively minimal bitheory. A *default nonmonotonic semantics* of a biconsequence relation is the set of its extensions.

A direct correspondence between default logic and biconsequence relations can be established by representing Reiter's default rules $a : b/A$ as bisequents $a{:}\neg b \Vdash A{:}$. Then the above nonmonotonic semantics will correspond precisely to the semantics of extensions in default logic. Moreover, many other nonmonotonic formalisms, such as logic programming, modal and autoepistemic logics, and the causal calculus can be expressed in this framework by varying the underlying logic (see [3] for details).

References and Recommended Readings

1. Baral, C. (2003). *Knowledge representation, reasoning and declarative problem solving*. Cambridge/New York: Cambridge University Press.
2. Bochman, A. (2001). *A logical theory of nonomonotonic inference and belief change*. New York: Springer.
3. Bochman, A. (2005). *Explanatory nonmonotonic reasoning*. Hackensack: World Scientific.
4. Bochman, A. (2008). Default logic generalized and simplified. *Annals of Mathematics and Artificial Intelligence, 53*, 21–49.
5. Bondarenko, A., Dung, P. M., Kowalski, R. A., & Toni, F. (1997). An abstract, argumentation-theoretic framework for default reasoning. *Artificial Intelligence, 93*, 63–101.
6. Geffner, H. (1992). *Default reasoning. Causal and conditional theories*. Cambridge: MIT Press.
7. Gelfond, M., Lifschitz, V., Przymusińska, H., & Truszczyński, M. (1991). Disjunctive defaults. In *Proceedings of Second International Conference on Principles of Knowledge Representation and Reasoning, KR'91*, Cambridge, MA (pp. 230–237).
8. Giunchiglia, E., Lee, J., Lifschitz, V., McCain, N., & Turner, H. (2004). Nonmonotonic causal theories. *Artificial Intelligence, 153*, 49–104.
9. Hanks, S., & McDermott, D. (1987). Non-monotonic logics and temporal projection. *Artificial Intelligence, 33*, 379–412.
10. Horty, J. F. (1994). Some direct theories of nonmonotonic inheritance. In D. M. Gabbay, C. J. Hogger, & J. A. Robinson (Eds.), *Handbook of logic in artificial intelligence and logic programming 3: Nonmonotonic reasoning and uncertain reasoning*. Oxford: Oxford University Press.
11. Kraus, S., Lehmann, D., & Magidor, M. (1990). Nonmonotonic reasoning, preferential models and cumulative logics. *Artificial Intelligence, 44*, 167–207.
12. Lehmann, D., & Magidor, M. (1992). What does a conditional knowledge base entail? *Artificial Intelligence, 55*, 1–60.
13. Lifschitz, V. (1994). Minimal belief and negation as failure. *Artificial Intelligence, 70*, 53–72.
14. Lin, F., & Shoham, Y. (1992). A logic of knowledge and justified assumptions. *Artificial Intelligence, 57*, 271–289.
15. Makinson, D. (2005). *Bridges from classical to nonmonotonic logic*. London: King's College Publications.
16. McCarthy, J. (1959). Programs with common sense. In *Proceedings of the Teddington Conference on the Mechanization of Thought Processes* (pp. 75–91). London: Her Majesty's Stationary Office.

17. McCarthy, J. (1980). Circumscription – a form of non-monotonic reasoning. *Artificial Intelligence, 13*, 27–39.
18. McCarthy, J. (1986). Applications of circumscription to formalizing common sense knowledge. *Artificial Intelligence, 13*, 27–39.
19. McDermott, D. (1987). Critique of pure reason. *Computational Intelligence, 3*(3), 149–160.
20. McDermott, D., & Doyle, J. (1980). Nonmonotonic logic. *Artificial Intelligence, 13*, 41–72.
21. Minsky, M. (1974). A framework for representing knowledge (Technical report 306). Artificial Intelligence Laboratory, MIT.
22. Pearl, J. (1990). System Z: A natural ordering of defaults with tractable applications to default reasoning. In *Proceedings of the Third Conference on Theoretical Aspects of Reasoning About Knowledge (TARK'90)* (pp. 121–135), San Mateo: Morgan Kaufmann.
23. Reiter, R. (1980). A logic for default reasoning. *Artificial Intelligence, 13*, 81–132.
24. Reiter, R. (2001). *Knowledge in action: Logical foundations for specifying and implementing dynamic systems*. Cambridge/London: MIT Press.
25. Shoham, Y. (1988). *Reasoning about change*. Cambridge, MA: Cambridge University Press.

Chapter 5
Induction

Rafal Urbaniak and Diderik Batens

5.1 Introductory Remarks

Inductive reasoning, initially identified with enumerative induction (inferring a universal claim from an incomplete list of particular cases) is nowadays commonly understood more widely as any reasoning based on only partial support that the premises give to the conclusion. This is a tad too sweeping, for this includes any inconclusive reasoning. A more moderate and perhaps more adequate characterization requires that inductive reasoning not only includes generalizations, but also any (ideally, rational) predictions or explanations obtained in absence of suitable deductive premises. Inductive logic is meant to provide guidance in choosing the most supported from a given assembly of conjectures. (Some authors think that this has to be done by capturing the notion of partial support, but this conviction is by no means universally accepted.)

The authors would like to express gratitude to Mathieu Beirlaen and Frederik Van De Putte for reading and commenting on an earlier draft of this paper. Work on this paper was supported by the Special Research Fund of Ghent University through project [BOF07/GOA/019] and by Polish National Science Centre grant 2016/22/E/HS1/00304.

R. Urbaniak (✉)
Centre for Logic and Philosophy of Science, Ghent University, Ghent, Belgium

Institute of Philosophy, Sociology and Journalism, University of Gdańsk, Gdańsk, Poland
e-mail: rafal.urbaniak@ugent.be

D. Batens
Centre for Logic and Philosophy of Science, Ghent University, Ghent, Belgium
e-mail: Diderik.Batens@UGent.be

S. O. Hansson, V. F. Hendricks (eds.), *Introduction to Formal Philosophy*, Springer Undergraduate Texts in Philosophy, https://doi.org/10.1007/978-3-319-77434-3_5

Approaches to inductive reasoning are so varied that it is difficult to find a more specific characterization of all of them. In an attempt to draw at least a partial connection, let us observe that among requirements which such a logic is often expected to satisfy [20] are:

Connection with deduction: Deductive consequence and logical contradiction should fit into an inductive logic as extreme cases of support that a conclusion can obtain from premises.

Objectivity: If premises support the conclusion, this fact depends only on the meaning of the premises and the conclusion.

Connection with probability: Some notion of probability should play an important role in the development of inductive logic.

As we will see later on, the last two requirements are not universally accepted.

To fix the ideas, recall the standard axiomatization of probability theory, as given in 1933 by Kolmogorov [44]:

$$\Pr(p) \geq 0 \qquad \text{for any proposition } p \tag{5.1}$$

$$\Pr(p) = 1 \qquad \text{if } p \text{ is necessary} \tag{5.2}$$

$$\Pr(p \vee q) = \Pr(p) + \Pr(q) \qquad \text{if } p \text{ and } q \text{ exclude each other} \tag{5.3}$$

The first stab at capturing the notion of the support that a piece of evidence E gives to a hypothesis H might be to identify it with the probability of the material conditional $E \rightarrow H$. Alas, this approach does not work. For the probability of $E \rightarrow H$ is the same as the probability of $\neg E \vee H$, which means that even if there is no connection between E and H whatsoever, if the probability of H is high enough or the probability of E is low enough, the probability of $E \rightarrow H$ is still high (at least as high as the probability of H or the probability of $\neg E$). (In fact, *mutatis mutandis*, on this approach you can run any of the paradoxes usually associated with material implication.) Thus, if there is a connection between inductive support and probability, it has to be more sophisticated.

The *received view* is that the degree of confirmation is to be identified with the *conditional probability* of the hypothesis given the evidence, defined by:

$$\Pr(H \mid E) = \frac{\Pr(H \wedge E)}{\Pr(E)} \qquad \text{if } \Pr(E) \neq 0 \tag{5.4}$$

$\Pr(p)$ is usually caled the *absolute probability of* p, as opposed to the *conditional probability* of p given q, noted as $\Pr(p \mid q)$. Probability theory which tells one how probabilities are related is not a full confirmation theory, though. To complete the story we also have to explain and justify the basic assignment of probabilities to propositions involved – their probability measure.

The first mathematically developed proposal following this path was put forward by Carnap [11], and we start with presenting his approach (meant to satisfy all three above-mentioned requirements) in Sect. 5.2. In Sect. 5.3 we will briefly survey

Reichenbach's attempt to satisfy all three requirements. In Sect. 5.4 we discuss one of the main theories on today's market, Bayesianism, which drops the second requirement. Next, in Sect. 5.5, we discuss Popper's approach (which is a serious attempt to drop the third requirement). Finally, in Sect. 5.6 we discuss the adaptive approach to inductive generalization, which proceeds qualitatively and drops the third requirement, not taking any degrees of confirmation as necessary for inductive inferences.

A very important issue which we will not discuss in detail is the philosophical problem of finding a general justification of inductive methods. The problem, raised by Hume [33], has received enough attention in the literature (see for instance the survey by Vickers [75]) and we could not do it justice in this short essay meant to focus on formal methods (one exception is Sect. 5.3, where we look at an attempt of justifying induction by means of certain results about a formal method). Another thing which we won't mention are causal and abductive inferences. They do fall under our general notion of induction, but we decided to focus on more crucial phenomena in the development of formal methods of induction instead.

5.2 Carnap and Induction

5.2.1 Preliminaries

The main notion which Carnap's approach to induction [11] is meant to explicate is the logical notion of *the degree of confirmation of a hypothesis* H *by a given body of evidence* E : $c(H, E)$. If E is the conjunction of the available observational data, $c(H, E)$ expresses the degree of confidence or belief that one should assign to H.[1]

Consider a first-order language containing a finite number of logically independent monadic predicates, a finite number of individual constants and standard Boolean connectives. A *literal* in such a language is either an atomic formula or its negation.

A *state description* in such a language is a conjunction which for any predicate and any constant contains exactly one literal composed of them (e.g. either Ga or $\neg Ga$ but not both). Thus, a state description for any property and any object says whether this object has this property. Every sentence is logically equivalent to the disjunction of the state descriptions which entail it. If every object in the domain is named by a different individual constant, then the set of all state descriptions exhausts the possible states of the domain as describable in the language.

[1] Early defenders of the logical approach include Keynes [43] and Johnson [36, 37].

A *structure description* generated from a given state description ϕ is the set of all state descriptions that result from ϕ by a permutation of individual constants (sometimes, it's identified with the disjunction of its elements). Structure descriptions can be interpreted as encoding information about the numerical distribution of properties among objects.

For instance, take a language with only one predicate G and only two constants a and b. There are four state descriptions:

$$\text{(a) } Ga \wedge Gb, \text{ (b) } Ga \wedge \neg Gb, \text{ (c) } \neg Ga \wedge Gb, \text{ (d) } \neg Ga \wedge \neg Gb.$$

and three structure descriptions:

(A)	$\{Ga \wedge Gb\}$	('all objects have property G'),
(BC)	$\{Ga \wedge \neg Gb, \neg Ga \wedge Gb\}$	('exactly one object has property G'),
(D)	$\{\neg Ga \wedge \neg Gb\}$	('no object has property G').

A *probability measure* assigns probabilities to state descriptions, so that the sum of the probability measures of all state descriptions is 1. As state descriptions are mutually exclusive, the probability measure of a disjunction of state descriptions is the sum of the probability measures of all disjuncts. Each sentence is equivalent to a disjunction of state descriptions, so the probability measure covers all sentences. Given a probability measure m, $c(H, E)$ (the degree of confirmation of H by E) can be defined by:

$$c(H, E) = \frac{m(H \wedge E)}{m(E)} \tag{5.5}$$

That is, the degree to which evidence E confirms hypothesis H is the proportion of the probability of the hypothesis and the evidence to the probability of the evidence.[2] Thus, various confirmation functions arise from various probability measures.

5.2.2 Probability Measures $m^\dagger$ and m^*

One way to define a probability measure, introduced by Carnap, is to divide the probabilities equally among the state descriptions. If there are k available (up to logical equivalence) state descriptions, and exactly n of those state descriptions logically imply sentence H, the probability of H is defined by $m^\dagger(H) = n/k$.

Each state description in our example is assigned the $m^\dagger$-value of $1/4$. So, $m^\dagger(\neg Ga) = m^\dagger(c) + m^\dagger(d) = 1/4 + 1/4 = 1/2$. As it turns out, the degree of confirmation of $\neg Ga$ by Gb is also $1/2$:

[2] In case no evidence is available, hypothesis H is evaluated against any logical theorem $\top$, so that $c(H, \emptyset) = c(H, \top) = m(H)$.

$$c^{\dagger}(\neg Ga, Gb) = \frac{m^{\dagger}(\neg Ga \wedge Gb)}{m^{\dagger}(Gb)} = \frac{m^{\dagger}(c)}{m^{\dagger}(a) + m^{\dagger}(c)} = \frac{1/4}{1/2} = 1/2.$$

Analogous calculations show that $m^{\dagger}(Ga) = c^{\dagger}(Ga, Gb) = 1/2$. But this shows that observing Gb has no impact on the degree of belief one should assign to Ga.

The problem with this independence generalizes. Even for a thousand objects $a_1, a_2, \ldots, a_{1000}$ the following will hold:

$$c(P(a_1), P(a_2)\wedge\cdots\wedge P(a_{1000})) = c(\neg P(a_1), P(a_2)\wedge\cdots\wedge P(a_{1000})) = m(P(a_1)).$$

But this means that no amount of evidence will have any impact on the level of confirmation of $P(a_1)$.

This led Carnap to consider a different probability measure, m^*. The method of assigning m^* is quite simple: first divide probability 1 equally among the available (up to logical equivalence) structure descriptions, thus building in the assumption that each structure description is equally probable. Then, divide the probability of each structure description equally among its members.

In our example, each of three structure descriptions is assigned probability measure $1/3$. Since (A) and (D) contain exactly one state description, each of those state descriptions is assigned probability measure $1/3$. On the other hand, each element of (BC) obtains the value $1/6$.

To see how this probability measure favors homogenous descriptions and deals with the independence issue, compare the probability measure of $\neg Ga$ with the confirmation of the hypothesis that $\neg Ga$ on the evidence that Gb (intuitively, the latter should be lower). $\neg Ga$ holds in (c) and (d) and hence $m^*(\neg Ga) = m^*(c) + m^*(d) = 1/3 + 1/6 = 1/2$. On the other hand:

$$c^*(\neg Ga, Gb) = \frac{m^*(\neg Ga \wedge Gb)}{m^*(Gb)} = \frac{m^*(c)}{m^*(a) + m^*(c)} = \frac{1/6}{3/6} = 1/3.$$

As expected, $c^*(\neg Ga, Gb) < m^*(\neg Ga)$. Similarly, $c^*(Ga, Gb) = 2/3 > m^*(Ga) = 1/2$, so Gb (partially) confirms Ga and (partially) disconfirms $\neg Ga$.

5.2.3 The λ-Continuum of Confirmation Functions

As it turns out, there is a wide variety of confirmation functions [12]. To see how such a variety arises, consider the following. If $F_1, F_2, \ldots, F_k$ are all the monadic predicates of a given language, we say that a Q-formula predicated of a constant a is of the form:

$$\pm F_1 a \wedge \pm F_2 a \wedge \cdots \wedge \pm F_k a$$

where each $\pm$ stands either for a negation or for nothing. Q-formulas of such a language can be enumerated, let's pick the i-th one and call it Q_i. One of the key confirmation assignments that we would like to calculate is that of $c(H_{Q_i}, E_Q)$ where H_{Q_i} is the Q_i-formula predicated of a certain constant a and E_Q is a conjunction of certain Q-formulas predicated of some constants different from a. (That is, we would like to be able to measure how complete information about certain objects observed so far confirms a given complete description of a new object.)

As Carnap suggests, there are at least two important factors in our assessment of $c(H_{Q_i}, E_Q)$. One is the empirical factor of the relative frequency of Q_is in E_Q: s_i/s (where s_i is the number of occurrences (modulo logical equivalence) of Q_i in E and s is the number of non-equivalent Q-formulas in E). The other factor is the logical one: the logical factor of Q_i equals $1/K$, where K is the number of all Q-predicates of the language. Following Carnap, $c(H_{Q_i}, E_Q)$ should be somewhere between these two values. A convenient way of representing this is to take it to be their weighted mean defined by:

$$c(H_{Q_i}, E_Q) = \frac{\frac{w_1 s_i}{s} + \frac{w_2}{K}}{w_1 + w_2} \tag{5.6}$$

where w_1 and w_2 are weights. Actually, since what matters is the ratio of the weights, one of them can be parametrized. Carnap suggested parametrizing w_1 and taking it to be s, thus making sure that the empirical factor gains weight as more observations are being made. The other weight is usually represented as λ:

$$c(H_{Q_i}, E_Q) = \frac{s_i + \lambda/K}{s + \lambda} \tag{5.7}$$

Any choice of λ in (5.7) gives a new confirmation function in the sense of (5.5). Consider what happens when we take $\lambda = 0$. In this case

$$c(H_{Q_i}, E_Q) = \frac{s_i + 0/K}{s + 0} = s_i/s$$

For instance, suppose there are only three constants a, b, c and only one predicate F and that we so far observed only two of them, which turned out to be F. What are the confirmation values of the hypothesis that the last object will also be F and of the opposite hypothesis, if $\lambda = 0$?

$$c(Fc, Fa \wedge Fb) = s_F/s = 2/2 = 1$$
$$c(\neg Fc, Fa \wedge Fb) = S_{\neg F}/s = 0/2 = 0$$

If however, one observed object is F and another one isn't, we get:

$$c(Fc, Fa \wedge \neg Fb) = s_F/s = 1/2$$

In this sense, (5.7) for $\lambda = 0$ assigns maximal role to the evidence and no role whatsoever to the logical possibilities (it corresponds to Reichenbach's straight rule — see Sect. 5.3).

For comparison, consider what happens as λ approaches ∞: in the limit (5.7) yields $1/K$. Thus, in our example, no matter whether we observed any other objects which are F, the confirmation of the hypothesis that the next object will be F is just the prior logical probability of that hypothesis[3]:

$$c(Fc, Fa \wedge Fb) = c(Fc, \neg Fa \wedge \neg Fb) = m(Fc) = 1/2.$$

So taking $\lambda = \infty$ assigns maximal importance to the logical factor and no role to the evidence and does not allow for learning from experience. In fact, the confirmation function thus defined is $c^{\dagger}$, which we already discussed.

The above choices of λ are two extremes of a continuum of confirmation functions (the lower λ, the more important the impact of the evidence on the confirmation value of the hypothesis). Where is the c^* in this continuum? It is obtained by equating λ to K, in which case (5.7) yields the following:

$$c(H_{Q_i}, E_Q) = \frac{s_i + K/K}{s + K} = \frac{s_1 + 1}{s + K}. \tag{5.8}$$

5.2.4 Challenges and Tweaks

One difficulty is that the above framework provides a variety of probability measures without indicating why we should prefer any of them over the others. Hájek [28] and Glaister ([22]: 569) see this as a serious challenge. Vickers [75] is more moderate: given certain basic restrictions,[4] even if the confirmation function is not unique, quite a few useful claims hold no matter which non-extreme function we pick. Initially, Carnap felt quite strongly about m^*, but eventually this embarrassment of riches motivated Carnap to accept a somewhat subjectivist attitude consisting in saying that there is a wide variety of options which remain open, even after all methodological considerations have been brought it.[5] Some others, like Fitelson [20], see nothing wrong in relativizing confirmation to probability measures and using the logically objective 'given such-and-such probability measure, the confirmation degree in this case is…' (Fitelson compares this to special relativity theory in which it is not velocity but rather velocity with respect to a frame of reference that is objective.)

[3]This holds as long as the evidence does not contain any constant occurring in the hypothesis.

[4]Most notably, regularity (every state description has non-zero probability) and symmetry (complete permutations of individual constants and predicates of the same type do not change the value of the function).

[5]See [79] for historical remarks.

Another problem with Carnap's inductive logic is that it is not very successful at handling reasoning by analogy. Intuitively speaking, the more primitive properties two objects share, the more likely it should be that they would agree on other properties. Yet, c^* fails to capture this intuition.[6]

For instance, suppose we have a language with two predicates F and G and two constants a and b. If reasoning by analogy worked, then the fact that $Fa \wedge Fb \wedge Ga$ should give more support to the hypothesis that Gb than just the evidence Ga:

$$c^*(Gb, Fa \wedge Fb \wedge Ga) > c^*(Gb, Ga) \tag{5.9}$$

And yet, (5.9) fails, because in this case both degrees of confirmation are equal:

$$c^*(Gb, Ga) = \frac{m^*(Ga \wedge Gb)}{m^*(Ga)} = \frac{1/3}{1/2} = 2/3$$

$$c^*(Gb, Fa \wedge Fb \wedge Ga) = \frac{m^*(Fa \wedge Fb \wedge Ga \wedge Gb)}{m^*(Fa \wedge Fb \wedge Ga)} = \frac{1/9}{3/18} = 2/3$$

Carnap attempted to deal with such issues [13] (he introduced yet another parameter apart from λ, usually called η), but the success is quite limited. Some attempts to deal with analogical reasoning within a (widely) Carnapian framework are [15, 44, 53, 70] and [50].

Once we generalize the notions to infinite domains, Carnap's inductive methods a priori assign zero probabilities to universal generalizations. This is considered a problem [2] because usually laws of nature are taken to be universal, and if it were true that no finite evidence can provide support for any universal statement, this would go against our intuitions that certain scientific hypotheses are better confirmed than others. The requirements put on confirmation functions can be modified to allow for non-zero probabilities of universal generalizations [77], and some attempts to give a systematic account of non-zero probabilities of universal claims have been put forward. Most notable are those by Hintikka [30], who introduced yet another parameter α dependent on the number of constants available in the language to contribute to the non-zero confirmation of universal claims (the theory has been extended in Hintikka and Niiniluoto [31]) and Kemeny [39], who even with almost-zero confirmation degrees of universal hypotheses allowed to compare their support in model-theoretic terms.[7] Hintikka's approach only enables one to assign non-zero probabilities to really general hypotheses, such as 'all G are F', but not to objective probabilistic sentences like 'the ratio of F within the set of Gs is r'.

[6]The problem was noticed already by Kemeny [40]. See however ([3]: 92–96) and [51] for more details.

[7]See also [3, 5, 78] and [57] for more detailed accounts.

Fitelson [20] worries that on a Carnapian account nothing warrants the relevance of the evidence to the hypothesis, and irrelevant evidence may highly confirm a hypothesis just because the hypothesis is highly likely or the evidence highly unlikely. He suggests [19] that the only historically proposed definition of confirmation that obeys certain basic relevance requirements is that of Kemeny and Oppenheim [42], which identifies it with

$$\frac{\Pr(E|H) - \Pr(E|\neg H)}{\Pr(E|H) + \Pr(E|\neg H)}.$$

Thus, he suggests, relevance requirements help to deal with the initial embarrassment of riches.[8]

A challenge to a purely syntactic approach to confirmation has been posed by Goodman [26]. Say we have drawn a marble from a certain bowl on each of the past ninety days and they all have been red. Thus, it seems, the evidence that the first ninety marbles were red increases the confirmation of the hypothesis that the next one will be red as well. But take another predicate, S, defined as 'drawn up to today and red, or drawn after today and blue.' Our evidence tells us that the ninety marbles observed so far were S, and so, if Carnapian theory was straightforwardly adequate, that the next one will be S too. But this is clearly not the case: our evidence does not confirm the hypothesis that the next marble will be blue.[9] The main lesson to be drawn is that which predicates can be sensibly used in inductive reasonings is an extralogical issue.

Carnap set out to solve the problem of confirmation in terms of logical probability, apparently expecting that there would be a single adequate probability measure. After 1952 it turned out that he had to justify the choice of a probability measure. The only sensible way of achieving this which he saw was in terms of empirically motivated methodological considerations. In a sense this turned his program upside down. For instance, choosing different w_1 in (5.6) leads to a new variety of measures and parameterizing on s in (5.7) already presupposes that induction is justified (for example, weighing it with $1/s$ leads to an anti-inductive measure).

Despite the difficulties, Carnap's contributions were among the first technically elaborate attempts to explicate the notions involved. The Carnapian program encountered its difficulties, but their very appearance motivated researchers to follow many different paths and led to a variety of ongoing research projects.

[8] In fact, many other attempts of redefining c have been observed. See [18, 32] and [1] for a variety of options.

[9] The paradox is slightly better known in the version from 1953, where Goodman speaks of 'blue' and 'grue' [27].

5.3 Reichenbach's Straight Rule and Pragmatic Justification of Induction

Reichenbach identifies the probability of an event with the limit of the relative frequency of events of the same kind.[10] Given that we normally observe only finite sequences of events, the question arises as to how we are to assess the relative frequency at the limit and how our general strategy of achieving this is to be justified. Reichenbach's response to the first question is that we should apply what he calls the *straight rule* (SR), which roughly speaking, says that one should take observed relative frequencies to be the limiting relative frequencies (and adjust as new observations are made) [4].

Reichenbach [4, 65] attempted to motivate the acceptance of SR, and hence induction by the following considerations. Either there is an inductive method which succeeds, or there is none. If there is none, we do not lose anything by using SR. If there is one, then SR will succeed as well. This justification turns out to be problematic ([3]: 152–153), for there are many inductive methods which agree with SR on past success ratio, vary from it in the predictions about the future which they legitimize at any finite point [67], and converge to the same value. Reichenbach provides no way of picking SR from among all its rivals. Even if it is SR which in fact makes the right predictions, when assessed in terms of past successes it does not stand out from a crowd of so far equally successful methods (although, for a defense of SR against this qualm see [38]).

A related difficulty is that the type of convergence involved in SR is somewhat weak because if one wants to obtain knowledge about infinitely many probabilistic relations there might be no single upper limit on the number of observations that have to be made even if for each such relation an upper limit exists ([17]: 375).[11]

5.4 Bayesian Approaches to Induction

5.4.1 Bayesianism and Subjective Probability

The embarrassment of riches which haunts the Carnapian objectivist program is embraced by Bayesians. While the logical approach faces the difficulty of finding a justification for a specific choice of initial probabilities, the personalistic Bayesians take the choice of initial probabilities to be an extralogical (and personal) issue. For them, an important task of a formal theory of inductive reasoning is to explain how,

[10] Reichenbach developed a slightly unorthodox probability calculus, see [17] for details.

[11] Reichenbach's approach also has to face all the challenges which haunt any frequentist approaches to probability (like the need for a sensible account of the probabilities of singular events). For a discussion, see [3, 28].

given certain initial degrees of beliefs, one has to revise their commitment when faced with new evidence.

Bayesians take personal probabilities (degrees of beliefs, also called subjective probabilities or credences) to be strongly connected with bets [16, 64]. Suppose you bet an amount n on a certain outcome S and I bet $3n$ against S. If S takes place, you win $4n$ (gaining $3n$) and I lose $3n$. If S does not take place, I win $4n$ (gaining n) and you lose n. In such a case we say that the stake is $4n$ (the sum of all bets), your betting rate is $1/4$ and my betting rate against S is $3/4$. In general, a betting rate is just the bet divided by the stake. (A conditional bet is just like that, with the difference that if the condition is not satisfied, the bet is off.) A bet on S at rate k is called fair if there is no advantage in betting on S at rate k rather than against S at rate $1 - k$. The degree of your belief in S is within the Bayesian framework identified with what you consider the fair betting rate on S.[12]

An important role in updating beliefs in face of new evidence is played by a theorem of probability theory called Bayes' Theorem. Before we describe how Bayesian updating works, let us introduce the theorem.

5.4.2 Understanding and Applying Bayes' Theorem

Bayes' Theorem in its simple formulation states:

$$\Pr(H \mid E) = \frac{\Pr(E \mid H)\Pr(H)}{\Pr(E)} \tag{5.10}$$

The denominator can be rewritten in terms of conditional probabilities. By the law of total probability, if $A_1, \ldots A_n$ are mutually disjoint hypotheses such that the sum of their probabilities is 1,

$$\Pr(E) = \Pr(E|A_1)\Pr(A_1) + \cdots + \Pr(E|A_n)\Pr(A_n).$$

Applied to (5.10), this yields:

$$\Pr(H \mid E) = \frac{\Pr(E \mid H)\Pr(H)}{\Pr(E|A_1)\Pr(A_1) + \cdots + \Pr(E|A_n)\Pr(A_n)}. \tag{5.11}$$

In particular, we can use H, $\neg H$ as elements of the partition, in which case we have:

$$\Pr(H \mid E) = \frac{\Pr(E \mid H)\Pr(H)}{\Pr(E|H)\Pr(H) + \Pr(E|\neg H)\Pr(\neg H)}.$$

[12]As almost always in philosophy, the devil is in the details, and various worries arise when one really wants to measure degrees of belief in terms of bets, but those issues lie beyond the scope of our survey.

The most interesting feature of Bayes' Theorem is that it determines the conditional probability of a hypothesis given a body of evidence in terms of other probabilities (which is quite helpful if those other probabilities are easier to ascertain). In general, determining $\Pr(E|H_i)$ is often much easier than determining $\Pr(H_i|E)$ (and there may be good reasons for assigning equal probabilities to all H_i).

An important role in Bayesianism is played by a procedure called *conditionalization*. It consists in changing our belief in a hypothesis H once new evidence E is obtained in the following way. Take your initial probabilities involved in the right-hand side of (5.11) at time t. If you already have the right-hand side probabilities, Bayes' Theorem allows you to calculate the probability of H conditional on E at time t: $\Pr_t(H \mid E)$. Now, if new evidence E is provided at some later time t', your $\Pr_{t'}(H)$ should be identical to $\Pr_t(H \mid E)$. That is, if at a certain time you believe that the probability of a certain hypothesis given E is k, this is the probability you should assign to that hypothesis once you find out that E (and you don't find out anything else that might have impact on the relevant probabilities).[13]

The Bayesian framework allows for a number of ways of making sense of the confirmation that a piece of evidence gives to a hypothesis. A piece of evidence E (incrementally) confirms hypothesis H if $\Pr(H \mid E) > \Pr(H)$ and the confirmation level of H by E is often identified either with the *difference measure* $\Pr(H \mid E) - \Pr(H)$ or the *ratio measure* $\Pr(H \mid E)/\Pr(H)$.

5.4.3 Arguments for Bayesianism

Why would a rational agent's degrees of belief satisfy the axioms of probability? The claim is supported by considerations meaning to show that the acceptance of the axioms of probability theory is required to avoid being susceptible to sure loss. A Dutch Book against an agent is a bet (or a series thereof) which, collectively taken, the agent has to lose. Agents are called *coherent* if they are not susceptible to a Dutch Book. De Finetti [16] proved that if one's degrees of belief do not comply to the axioms of probability theory, one is not coherent. Kemeny [41], Shimony [69] and Lehman [49] proved that the implication in the opposite direction also holds.

One might be worried that grounding an epistemic standard in pragmatic considerations is inappropriate. For people with such concerns, another class of arguments developed from the perspective of *epistemic utility theory*, is available [5]. Think of truth as 1 and falsehood as 0. Pick a measure of distance between a given degree of belief and the given sentence's truth-value (for instance, one can use squared difference). The lower the score, the greater the accuracy of your belief. Define some sensible way of aggregating inaccuracies of one's beliefs into

[13] Jeffrey [35] provides a more general formulation which applies also to cases where one only finds out that E is probable to a certain degree. A Dutch Book argument (see Sect. 5.4.3) for this general formulation has been given by Armendt [1] (see also [70]).

one global measure of inaccuracy. Now, *Accuracy theorem* is available to the effect that if a set of degrees of beliefs violates the axioms of probability, there is a set of probabilistic degrees of belief which are more accurate, no matter what truth-values the beliefs have, and the *Converse accuracy theorem* says that no probabilistic set of beliefs is so dominated by a non-probabilistic one. From this perspective, not being Bayesian is irrational, because it entails being further from the truth, no matter what the truth is.

5.4.4 Challenges to Bayesianism

Let's briefly list the main concerns that the Bayesians have to deal with (some of them apply also to Carnap's approach):

▷ Bayesianism does not say anything about the choice of initial probabilities of E, of H and of $E \mid H$, so the same evidence might legitimately motivate two researchers to assign quite different probabilities to a hypothesis, if their initial probabilities are sufficiently different.

The Bayesian response to this difficulty is that one can prove that as the amount of evidence increases, probabilities assigned to relevant hypotheses will converge (almost) independently of what the initial subjective probabilities are [68]. The problem is that (i) this works only if the initial subjective probabilities are not 0 or 1, and (ii) extreme initial probabilities (close to 1 or close to 0) prevent rapid convergence and make the further search for evidence practically useless.

▷ In actual reasoning, rational agents rarely can assign (or even decently approximate) subjective probabilities to the relevant factors, and it is unclear whether betting preferences are a sufficient and correct way of discovering the priors [35].

▷ If $\Pr(E) = 1$, then $\Pr(H \mid E) = \Pr(H)$, so old evidence cannot confirm any hypothesis even if one realizes now that the evidence is relevant for the hypothesis, for example because it is implied by it (this is called the *problem of old evidence* [18, 23–25]).[14] Some (like [21]) try to avoid this by weakening the assumption that agents are logically omniscient,[15] but it is not clear what modifications of the Bayesian formal apparatus this move entails. Some try to apply pre-formal philosophical discussion to massage the phenomenon into the Bayesian framework [14, 48].

▷ Bayesianism in a sense disregards the structure of explanation and does not take into account such factors as its simplicity or the unity of the underlying theory. That is, all that is considered when we evaluate a given theory is our prior probabilities and available evidence in its favor: factors like simplicity or unity,

[14]For a discussion, see [1].

[15]That is, the assumption that they know all logical consequences of what they know.

intuitively important for the evaluation of a theory, are not explicitly considered in the evaluation procedure. Sure, one can take these factors to be incorporated among prior probabilities, but if that is the case, bayesianism does not really explain how these factors are to be assessed and sweeps them under the carpet of unexplained prior belief degrees.

▷ Bayesianism tells a story about rationality and its relation to betting behavior. Yet, it does not say much about why being rational in this sense should put one in an epistemologically privileged position. Why does the fact that I obey rules which would help me to avoid a Dutch Book result in Bayesian updating being the best way to go about scientific reasoning? It is not immediately clear why scientific success and winning bets should be related [10].

▷ As already mentioned, Bayes' Theorem establishes a connection between certain probabilities. The connection is useful if the probabilities on the right-hand side are easier to ascertain than the one we attempt to assess. Perhaps, $\Pr(E \mid H)$ often can be easily assessed, but the other probabilities on the right-hand side of (5.11) may be more problematic. For example, $\Pr(E \mid \neg H)$ seems at least as mysterious as $\Pr(H \mid E)$ if H is a general hypothesis. For instance, it is not really clear how to establish the probability of observing a black raven if not all ravens are black or the probability of Eddington's observation if relativity theory is false.

▷ Conditionalization does not follow from Bayes' Theorem and is not justified as an a priori rule of rationality. It does not follow from Bayes' Theorem, because one can obey Bayes' Theorem at each moment while completely changing one's degrees of beliefs between moments. Nor does it seem a priori, because it is diachronic, which means that it incorporates a prediction about what will happen at a later time based on what has happened so far (and such moves are usually not considered a priori since Hume). Some diachronic Dutch Book arguments have been given by David Lewis (as reported by [73]), but they rely on stronger assumptions which themselves do not seem a priori.

▷ The claim that rational agents should obey the laws of probability implies their logical omniscience (insofar as deductive logic is involved). This difficulty Bayesianism shares with many formal approaches to epistemology.[16]

Given a variety of troubles that a fully subjectivist approach to priors encounters, various unorthodox versions of Bayesianism are being put forward [34, 66, 76] which try to put some additional constraints on priors without running into the problems that fully objectivist and syntactical accounts run into [for a survey of early papers of Bayesianism see [46], and for a survey and further refences see 3].[17]

---

[16]A twist to this problem is that once classical logic becomes the underlying logic, Bayesianism is unable to account for the possibility of the underlying logic being revised and to explain how evidence might motivate a change of underlying logic [72].

[17]It is also worth mentioning that one of the strength of Bayesianism lies in various applications of the framework to classical philosophical problems. For instance, the framework is used to describe and assess more precisely various arguments in the philosophy of religion (see e.g., [29]).

5.5 Popper

As is well-known, Popper rejected the early Vienna Circle's verificationism from his (1935) on. For him, the central mechanism of scientific methodology is falsification. Good scientists try to falsify theories, and our best theories are the outcome of such attempts. Popper also rejects the idea of confirmation in the sense in which it was used before in this paper. No finite set of observational data can justify one to raise one's degree of belief in a theory[18]; a single falsification to the contrary justifies one's rejection of the theory. The very idea of inductive logic is rejected. For Popper, all logic is deductive (and coincides with classical logic).

Popper's disagreement with the Vienna Circle, not to mention personalists, lies in his different conception of science. Scientific theories are not mere generalizations of observations, but express lawlike connections. They are not justified by (passive) observations, but by actions: attempts to falsify the theories. This requires that one looks for specific observations or, even more typically, performs specific experiments. Finding 'confirming instances' is too easy.[19] But so is the duplication of experiments that are likely to succeed. This is why Popper requires severe tests, tests that are most likely to lead to falsification. The stress on theories is Popper's. Separate generalizations cannot be tested because their falsification can always be reasoned away by modifying another generalization.[20] Popper pushed the idea of falsification to its extreme consequences—we shall see only part of that here.

Popper invoked formal methods to make all this precise. These methods invoke classical logic. They also invoke logical probabilities. This, however, did not cause any embarrassment of riches for Popper. As he explained in appendix $*vii$ of [61], he considered Carnap's $m^\dagger$ as the only methodologically acceptable measure function for logical probability. All other measure functions can only be justified by non-logical considerations. So any occurrence of Pr in this section should be interpreted in terms of $m^\dagger$.

Testing a theory means trying to bring about an observable fact that falsifies the theory. So the first question for Popper's methodology is which theories one should test first.[21] A theory is *falsifiable* if a possible observable fact contradicts it—non-

---

[18]Compare this to the fact that if the number of constants is infinite, then every measure function m from Carnap's λ-continuum gives $m(h) = 0$ whenever h is a universally quantified formula, and gives $c(h, e) = 0$ whenever h is a universally quantified formula and e is the conjunction of finitely many singular formulas.

[19]In the appendix of (1979) Popper moreover rejects the common sense 'bucket theory' of knowledge.

[20]Compare this to Quine's arguments in "Two dogmas of empiricism" [63], which led Quine to a holistic position.

[21]Many of Popper's ideas stem from (what since Kuhn is called) revolutionary science and this requires conceptual change. Yet Popper's formal criteria (like all approaches discussed in the previous sections) presuppose a given language.

falsifiable theories are deemed unscientific.[22] A theory is more falsifiable (has a higher degree of testability) to the extent that more logically possible facts contradict it. This brings Popper to two criteria: generality and specificity. A hypothesis is more general to the extent that it concerns a logically larger set of objects; it is more precise to the extent that it specifies more about those objects. To get the flavor: where P, Q, and R are logically independent predicates, $\forall x(Px \supset Qx)$ is more general than $\forall x((Px \wedge Rx) \supset Qx)$: the former is contradicted by every sentence of the form $P\alpha \wedge \neg Q\alpha$ whereas the latter is only contradicted by sentences of the form $P\alpha \wedge R\alpha \wedge \neg Q\alpha$; $\forall x(Px \supset (Qx \wedge Rx))$ is more specific than $\forall x(Px \supset Qx)$: the former is contradicted by sentences of the form $P\alpha \wedge \neg Q\alpha$ as well as by sentences of the form $P\alpha \wedge \neg R\alpha$, whereas the latter is only falsified by the former sentences.[23] Popper identifies the *content* of a sentence A with the falsifiability of A and measures it, for example, by $1 - \Pr(A)$, which is $\Pr(\neg A)$. (By the way, despite using probability to define the content of a sentence, Popper did not use probability to explicate the notion of confirmation.) Note that, where A and B are logically independent,[24] $A \wedge B$ has an intuitively higher content than A and indeed $\Pr(\neg(A \wedge B)) > \Pr(\neg A)$.

Needless to say, $m^{\dagger}$ is unable to capture the differences between the general sentences from the previous paragraph if the domain is infinite. All those sentences have probability zero. These probabilities are defined by a limit for the number of elements of the domain going to infinity. In appendix $^{*}vii$ of (1935), Popper introduces a "fine-structure of probability". Even if $\Pr(A) = \Pr(B) = 0$, it is possible that $\Pr(A \mid B) > \Pr(B \mid A)$, and this indicates that B has a higher content than A. A ready example is obtained by letting A be $\forall x(Px \supset Qx)$ and letting B be $\forall x(Px \supset (Qx \wedge Rx))$. In this case $1 = \Pr(A \mid B) > \Pr(B \mid A)$. A different way to look at the criterion is by noting that the limits of the probabilities of both A and B converge to zero as the domain increases, but that the ratio of these probabilities is always larger than 1 (and goes to infinity).

So the objective is clear: formulate and test *bold hypotheses*. If the hypothesis survives the tests, one obtains a corroborated informative hypothesis—see below. If it fails, one may still move to a non-falsified hypothesis that has the next highest content (degree of falsifiability). Of course, no single (non-falsified) hypothesis has the highest content. In the propagandistic style that was usual for those days, Popper does not stress this. Here (as elsewhere), he is a free-market pal: pick yours and go for it. The market (sorry, the facts) will decide.

To compare theories, Popper [59, 60] introduced (and in [62] elaborated on) the notion of *verismilitude* or *truthlikeness*. Its qualitative version (as opposed to the quantitative formulation, mentioned below) is as follows: take an interpreted theory

[22] But Popper hastens to relativize this 'demarcation criterion'. 'Metaphysical' ideas play a central role in generating scientific theories.

[23] Compare also "all heavenly bodies move in circles" to "all planets of the sun move in ellipses", remembering that all circles are ellipses.

[24] Note this entails they're contingent.

T and let T^1 (T^0) be the set of its true (false) sentences. T is more truthlike than a theory S iff both $S^1 \subseteq T^1$ and $T^0 \subseteq S^0$, and either $S^1 \neq T^1$ or $T^0 \neq S^0$ (That is, a theory to be more truthlike has to surpass the other in its truth content without surpassing it in its falsity content, or to have smaller falsity content without being ahead in its truth content.) Miller [52] and Tichý [74] provide a compelling criticism of Popper's definitions.[25]

Suppose then some theories survived the imposed tests. How good are they? Here too, Popper formulates a measure, which he calls the degree of corroboration of a hypothesis. Here is a definition from appendix $^*ix$ of [61]:

$$C(H, E) = \frac{\Pr(E \mid H) - \Pr(E)}{\Pr(E \mid H) - \Pr(E \wedge H) + \Pr(E)}$$

So, where E is the conjunction of the available empirical evidence, the degree of corroboration of the hypothesis H is proportional to the difference between the probability of the evidence given the hypothesis and the absolute probability of the evidence. The denominator is a normalizing factor, which keeps the values between -1 and $+1$. If E contradicts H, $\Pr(E \mid H) = \Pr(E \wedge H) = 0$. So the degree of corroboration of H is -1. This indicates that H is falsified. The maximal value to which H may be corroborated is obtained if E is identical to H—this will apply if H is a singular statement or if, being God, you see that H obtains. In this case $\Pr(E/H) = 1$ and $\Pr(E \wedge H) = \Pr(H)$. The degree of corroboration of H then reduces to $1 - \Pr(H)/1$, in other words $\Pr(\neg H)$. So the maximal degree to which a hypothesis H may be corroborated is the content of H (the falsifiability degree of H). The higher the content of a hypothesis, the higher its potential degree of corroboration.

It is amusing to see that falsifiability turns up again here. Yet, putting the formal machinery in perspective, Popper stresses that the corroboration of H is only significant if H was subjected to the severest possible tests. We have seen before that these are the tests that are most likely to falsify the hypothesis. That Popper never offered a formal criterion for this, is presumably related to a weak spot in his formalisms. Intuitively, repeating an experiment that did not lead to falsification is not a severe test. But why is that? Apparently because, in view of previous instances of the test, the next instance is likely not to lead to falsification. But why is that so? Apparently this conclusion can only be drawn if we presume that the outcome of the next instance of the test is likely similar to the outcome of previous instances.

[25] Popper introduced also two quantitative notions of verisimilitude, which employed the notion of probability. Tichý [74] argues that both attempts have highly counterintuitive consequences. This is not to say that the project of defining truthlikeness is doomed. There are various interesting attempts to define the concept after Popper's initial failure (see e.g., [54, 58]). Even though no particular account is currently agreed on by everyone, certain progress has been made, and the issue is a lively topic (for a survey, see [55, 56]).

To presume so, however, is to presume a measure function different from Carnap's $m^\dagger$, viz. one that assigns a non-zero weight to the empirical factor. And this Popper does not want.

Indeed, Popper always stressed that a (non-falsifying) degree of corroboration of H should not affect our degree of rational belief in H. He nevertheless advised one to use the best tested theory as basis for action, and some take this to mean that the degree of rational belief of those theories is raised. In Section 9 of [62], Popper tried to remove this confusion. He distinguished preferring the best tested theory as basis for action from relying on that theory. Preferring such theory is justified, because of the merits the theory proved to have in the past. But this says nothing about the future. So we cannot rely on the theory; no theory was shown true or can be shown true. Our present most corroborated theories embody the best knowledge available today. Only fools take alternative theories as better. But even our best theories may be falsified tomorrow.

5.6 Inductive Generalization in Terms of a Logic

The approaches discussed before have clearly sensible application contexts. More problematic is their explicit or implicit claim on universality. Why, for example, should the decision to act on a certain scientific theory be arrived at by the same method as the decision to participate in a certain lottery?

Once a plurality of methods is accepted, there can hardly be any objection against phrasing some of these as logics: functions that assign a consequence set to every premise set. In the present section, we shall present such an approach, the one we are most familiar with: adaptive logics. The discussion will be restricted to logics of inductive generalization. These are logics that enable one to infer hypotheses of the form "all A are B" from sets of data, and which (most importantly) provide such consequence operation with a proof theory.[26]

An advantage of this approach to the 'acceptance' of scientific hypotheses is that it is more realistic than approaches in terms of degrees of (rational) belief. While data, hypotheses, and theories may be rejected in view of new observations or in view of a new systematization, they are provisionally considered as 'given'. Next, it is easily possible to consider scientific research as problem solving in the presence of provisionally accepted background knowledge. Moreover, there is no need to assign specific degrees of certainty to data, background knowledge, and inferred generalizations. Still, as we shall see, it is possible to express that some background theories are more 'reliable' than others.

[26]To complete the picture, we would need adaptive logics that enable one to derive hypotheses of the form $\Pr(A \mid B) = r$, in which Pr is an objective probability, and we would need adaptive logics that enable one to derive predictions that do not follow from derived general hypotheses.

5.6.1 An Example

A logic that enables one to derive general hypotheses from sets of data is obviously ampliative with respect to **CL** (Classical Logic); the derived general hypotheses are not derivable from the premises by **CL**. Moreover, such derivations are risky in several senses. For one thing, new evidence may become available and it may falsify some of the formerly derived hypotheses. Moreover, in the presence of background knowledge (formerly 'accepted' theories), it may be impossible to show, within a finite period of time, that a certain generalization is incompatible with the data and background knowledge.[27]

This situation has several consequences for the proofs of logics of inductive generalization. All ampliative conclusions drawn at some point in a proof, may later have to be revoked for one of the two reasons mentioned in the previous paragraph. This means that such proofs are dynamic and hence that one needs a device to control this dynamics. In adaptive logics, the control is exerted by, on the one hand, introducing ampliative conclusions on a (non-empty) condition and, on the other hand, providing a marking definition. At every stage of a proof, the marking definition settles which lines are marked and which unmarked. Marked lines are considered as OUT: the formula of a line that is marked at a stage of the proof is considered as not derived on that line at the stage. A stage of a proof is a sequence of lines that are correct according to the rules of the logic. A stage s' extends another stage s if all lines that occur in s occur in the same order in s'.

Marks may come and go with every new stage of the proof; a line may unmarked at a certain stage, marked at a later stage, unmarked again at a still later stage, and so on. So, apart from *derivability at a stage*, we need a stable notion of derivability, which is called *final derivability*. The premises may not enable one to show by means of a finite proof that a generalization is finally derived. If this is the case and we applied the right heuristic means, the proof provides us with a reason to *prefer* the derived generalizations as basis for action, but it does not allow us to *rely* on them—we borrow this distinction from Popper, as the reader will remember from Sect. 5.5.

Introducing adaptive logics in general would take too much space; we refer for example to [4, 7] for that. Here we shall start with a toy example proof and next introduce the machinery in as far as we need it. The idea behind the proof will be falsification. One may introduce any generalization in the proof on the condition that the negation of the generalization can be considered as false in view of the

[27] There is a mechanical procedure which for any particular inconsistent premise set will show that it is inconsistent. But there is no mechanical procedure which for any particular premise set will decide whether it is inconsistent. So, technically speaking, the set of inconsistent sets of formulas is semi-recursive (or semi-decidable) but not recursive (not decidable).

premises.[28] Let the premises be those of lines 1–4 of the proof, in which we also introduce some consequences.

1	$Pa \wedge Qa \wedge Sa$	premise	$\emptyset$
2	$Pb \wedge \neg Qb \wedge \neg Rb \wedge Sb$	premise	$\emptyset$
3	$\neg Sc$	premise	$\emptyset$
4	$\neg Pd$	premise	$\emptyset$
5	$\forall x(Px \supset Sx)$	RC	$\{\neg\forall x(Px \supset Sx)\}$
6	$\forall x(Px \supset Qx)$	RC	$\{\neg\forall x(Px \supset Qx)\}$
7	$\forall x(Px \supset (Qx \wedge Sx))$	5, 6; RU	$\{\neg\forall x(Px \supset Sx), \neg\forall x(Px \supset Qx)\}$
8	$\forall x Px$	RC	$\{\neg\forall x Px\}$
9	$\forall x \neg Rx$	RC	$\{\neg\forall x \neg Rx\}$
10	$\forall x(Qx \supset Rx)$	RC	$\{\neg\forall x(Qx \supset Rx)\}$
11	$\neg Pc$	3, 5; RU	$\{\neg\forall x(Px \supset Sx)\}$

Apart from the premise rule, two generic rules are used in the proof. The conditional rule RC may be read provisionally as: introduce any generalization A on the condition $\{\neg A\}$. The unconditional rule RU may be read as: if B is **CL**-derivable from $A_1, \ldots, A_n$ and $A_1, \ldots, A_n$ occur in the proof (on some conditions), one may derive B on the condition that is the union of the conditions of $A_1, \ldots, A_n$. Formulas introduced by the premise rule receive the empty set as their condition: premises need never be revoked. Note that 11 is a prediction derived from a derived generalization in view of the premises (the data).

The reader will have noted that some of the generalizations are falsified by the premises. In the subsequent extension of the proof, we show how this is handled. We do not repeat the premises.

5	$\forall x(Px \supset Sx)$	RC	$\{\neg\forall x(Px \supset Sx)\}$	
6	$\forall x(Px \supset Qx)$	RC	$\{\neg\forall x(Px \supset Qx)\}$	$\checkmark^{12}$
7	$\forall x(Px \supset (Qx \wedge Sx))$	5, 6; RU	$\{\neg\forall x(Px \supset Sx), \neg\forall x(Px \supset Qx)\}$	$\checkmark^{12}$
8	$\forall x Px$	RC	$\{\neg\forall x Px\}$	$\checkmark^{13}$
9	$\forall x \neg Rx$	RC	$\{\neg\forall x \neg Rx\}$	
10	$\forall x(Qx \supset Rx)$	RC	$\{\neg\forall x(Qx \supset Rx)\}$	
11	$\neg Pc$	3, 5; RU	$\{\neg\forall x(Px \supset Sx)\}$	
12	$\neg\forall x(Px \supset Qx)$	2; RU		
13	$\neg\forall x Px$	4; RU	$\emptyset$	

By deriving 12 from 2, we obtain a member of the conditions of lines 6 and 7. So the conditions of these lines cannot be considered as false, because a member of them has to be true if the premises are true. Similarly, line 8 is marked in view of line 13. These are plain cases of falsification.

There are also some unexpected features, which we illustrate in the following extension of the proof, in which we leave out the marked lines for reasons of pagination.

[28] We do not say "on the condition that the generalization is not falsified by the premises". Soon, the reason will become clear.

5	$\forall x(Px \supset Sx)$	RC	$\{\neg\forall x(Px \supset Sx)\}$	
...				
9	$\forall x\neg Rx$	RC	$\{\neg\forall x\neg Rx\}$	$\checkmark^{16}$
10	$\forall x(Qx \supset Rx)$	RC	$\{\neg\forall x(Qx \supset Rx)\}$	$\checkmark^{16}$
11	$\neg Pc$	3, 5; RU	$\{\neg\forall x(Px \supset Sx)\}$	
12	$\neg\forall x(Px \supset Qx)$	2; RU		
13	$\neg\forall x Px$	4; RU	$\emptyset$	
14	$Ra \vee \neg Ra$	RU	$\emptyset$	
15	$Ra \vee (Qa \wedge \neg Ra)$	1, 14; RU	$\emptyset$	
16	$\neg\forall x\neg Rx \vee \neg\forall x(Qx \supset Rx)$	15; RU	$\emptyset$	
17	$Pc \vee \neg Pc$	RU	$\emptyset$	
18	$(Pc \wedge \neg Sc) \vee \neg Pc$	3, 17; RU	$\emptyset$	
19	$\neg\forall x(Px \supset Sx) \vee \neg\forall x Px$	18; RU	$\emptyset$	

Note that 14 and 17 are theorems of **CL** and that RU allows one to introduce these anywhere in a proof. The first interesting case is line 16, at which a disjunction of negations of generalizations is derived. What this tells us is that either $\neg\forall x\neg Rx$ or $\neg\forall x(Qx \supset Rx)$ is true, but is does not tell us which of them is true. So should we mark lines 9 and 10 or not? We cannot have both unmarked because they jointly contradict the premises (as line 16 shows). And no logical consideration allows us to mark one rather than the other. So we had better mark both lines.

Why does line 19 not lead to marking line 5 (together with line 8)? Line 19 tells us that either $\neg\forall x(Px \supset Sx)$ or $\neg\forall x Px$ is true and it also does not tell us which of them is true. But line 13 tells us that: $\neg\forall x Px$ is true (on these premises). So $\neg\forall x(Px \supset Sx)$ is off the hook: if we know that A as well as $A \vee B$ are true and start to consider as many members of $\{A, B\}$ as false as is possible, we can safely consider B as false.

For the present logic, negations of generalizations are the *abnormalities*; the formulas of which as many as possible are considered as false. Let the set $U_s(\Gamma)$ comprise all disjuncts of minimal disjunctions of abnormalities that are derived on the condition $\emptyset$ at stage s of the proof. The members of $U_s(\Gamma)$ are the abnormalities that are *unreliable* at stage s. The marking definition goes as follows: a line is marked at a stage s if and only if its condition contains a member of $U_s(\Gamma)$.[29]

In general, an adaptive logic in standard format is defined as a triple: a lower limit logic (required to have certain properties), a set of abnormalities characterized by a logical form, and an adaptive strategy. The logic informally introduced before is called $\mathbf{LI}^r$—see [5] and [8]. Its lower limit logic is **CL**, its set of abnormalities is the set comprising all negations of generalizations, and its strategy is Reliability.

We promised to introduce final derivability. Let A be the formula of line l of a stage s of a proof from Γ. A is finally derived from Γ at line l if and only if l is

[29] This is the so-called Reliability strategy. The Minimal Abnormality strategy is slightly different from Reliability and offers a few more consequences than Reliability (and never less consequences). We shall not introduce it here.

unmarked at stage s and every stage s' that extends s and in which l is marked, can be further extended in such a way that l is unmarked. It is handy to see this in game-theoretic terms: the proponent writes s, the opponent extends it to s', and then the proponent is allowed to further extend s'. For Reliability both extensions are finite.

Adaptive logics also have a semantics. Consider the minimal disjunctions of abnormalities that are **CL**-derivable from Γ and let $U(\Gamma)$ be the set of their disjuncts. A reliable model of Γ is a **CL**-model of Γ that verifies no other abnormalities than members of $U(\Gamma)$. For logics that have Reliability as their strategy, semantic consequences of Γ are the formulas verified by all Reliable models of Γ. An interesting feature of the (occasionally mentioned) standard format is that it provides every adaptive logic in standard format with its proof theory, its semantics, soundness and completeness proofs, and proofs for lots of other metatheoretic properties. The standard format provides also certain criteria for final derivability: if one follows a certain (proof or tableau) procedure, final derivability will, for some premise sets and consequences, be established after finitely many steps.

Back to inductive generalization. Basic moves underlying $\mathbf{LI}^r$ can arguably provide guidance in actual research. The logic suggests one to obtain certain observations, possibly by experimental means. These observations might falsify certain generalizations. If that happens, shorter disjunctions of abnormalities become derivable, and by the same token, other, often more specific generalizations may become derivable.

5.6.2 Some Alternatives

Logics should not make methodological decisions, but should offer means to express methods in a precise way. So there should be many adaptive logics of inductive generalization, from which a scientist may choose on methodological grounds. This is indeed the case.

A first series of them is obtained by varying the elements of the adaptive logic. The Minimal Abnormality strategy was mentioned in Footnote 29; no other strategies seem sensible in the present context. Variants of the set of abnormalities have been studied. For some logics, this set comprises the formulas of the form $\exists x A(x) \wedge \exists x \neg A(x)$ in which $A(x)$ is a disjunction of literals. The effect of this change is a logic richer than $\mathbf{LI}^r$ in which $\forall x(Px \supset Qx)$ can only be introduced in a proof if a formula of the form $\neg P\alpha \vee Q\alpha$ (an instance of the generalization in the logician's sense) is derivable from the premises. A still richer logic is obtained if the set of abnormalities comprises the formulas of the form $\exists x(A(x) \wedge \pi x) \wedge \exists x(A(x) \wedge \neg \pi x)$, In which πx is a literal and $A(x)$ is a conjunction of literals. In this case $\forall x(Px \supset Qx)$ can only be introduced in a proof if a 'positive instance' of it is available, a formula of the form $P\alpha \wedge Q\alpha$. One might of course also vary the lower limit logic.

A different series of variants is obtained by combining adaptive logics (not necessarily those mentioned so far). There are several ways to do so. One of them goes as follows: first apply an adaptive logic that has one of the above sets of abnormalities restricted to the case where only one predicate occurs in the abnormality; to the resulting consequence set, apply an adaptive logic that has abnormalities in which at most two predicates occur; and so on. This leads to a serious enrichment and agrees with Poppers requirement that one should first test hypotheses that have the highest content.

An interesting extension is where the person applying the logic is allowed to introduce certain preferences. A scientist may have several reasons to do so, going from relying on 'established' science to personal preferences and mere guesses. These preferences may be expressed by defeasibly denying certain abnormalities in a prioritized way: this abnormality is almost certainly false, that one is probably false, etc. The priorities are expressed in the language by operators, which basically have a comparative effect. Several adaptive logics to handle such prioritized rejections are available in the literature [4, 6, 10] and each of them can be combined with an adaptive logic for inductive inference.

Most realistic applications require that handling background knowledge is combined with inductive inference. Background knowledge will drastically extend the data, but it may be falsified by them. Sometimes this is a reason to reject the whole theory, sometimes this is a reason to reject only falsified consequences of the theory and provisionally go with the others, hoping for a new systematization in the future. Again, adaptive logics to handle both cases (even jointly) are available. For a more complete overview of materials in this section and for further references we refer to Batens [6].

References

1. Armendt, B. (1980). Is there a Dutch Book argument for probability kinematics? *Philosophy of Science, 47*, 583–589.
2. Ayer, A. (1972). *Probability and evidence*. London: Macmillan.
3. Batens, D. (1975). *Studies in the logic of induction and in the logic of explanation containing a new theory of meaning relations*. Brugge: De Tempel.
4. Batens, D. (2005). On a logic of induction. In R. Festa, A. Aliseda, & J. Peijnenburg (Eds.), *Confirmation, empirical progress, and truth approximation* (Essays in debate with Theo Kuipers, Vol. 1, pp. 221–242). Amsterdam/New York: Rodopi.
5. Batens, D. (2006). On a logic of induction. *L&PS – Logic & Philosophy of Science, IV*(1), 3–32.
6. Batens, D. (2011). Logics for qualitative inductive generalization. *Studia Logica, 97*(1), 61–80.
7. Batens, D. (2018). *Adaptive logics and dynamic proofs. Mastering the dynamics of reasoning*. [Manuscript in progress, available online at http://logica.ugent.be/adlog/book.html].
8. Batens, D., & Haesaert, L. (2001). On classical adaptive logics of induction. *Logique et Analyse, 173–175*, 255–290. Appeared 2003.

9. Batens, D., & Haesaert, L. (2003). On classical adaptive logics of induction. *Logique et Analyse, 46*, 225–290.
10. Bird, A. (1998). *Philosophy of science*. New York: Routledge.
11. Carnap, R. (1950). *Logical foundations of probability*. London: Routledge.
12. Carnap, R. (1952). *The continuum of inductive methods*. Chicago: University of Chicago Press.
13. Carnap, R. (1959). *Induktive Logik und Wahrscheinlichkeit*. Wolfgang Stegmüller.
14. Christensen, D. (2004). *Putting logic in its place: Formal constraints on rational belief*. Oxford: Clarendon Press.
15. Constantini, D. (1983). Analogy by similarity. *Erkenntnis, 20*, 103–114.
16. De Finetti, B. (1937). La prévision: Ses lois logiques, ses sources subjectives. *Annales de l'Institut Henri Poincaré, 7*, 1–68. (Translated as "Foresight: Its Logical Laws, Its Subjective Sources" in Kyburg & Smokler (1964)).
17. Eberhardt, F., & Glymour, C. (2009). Hans Reichenbach's probability logic. In J. W. Stephan Hartmann & D. Gabbay (Eds.), *Handbook of the history of logic. Volume 10: Inductive logic* (pp. 357–389). Amsterdam/Boston: Elsevier.
18. Eells, E. (2005). Confirmation theory. In J. Pfeifer & S. Sarkar (Eds.), *Philosophy of science: An encyclopedia* (pp. 144–150). New York: Routledge.
19. Fitelson, B. (2001). *Studies in Bayesian confirmation theory*. Ph.D. thesis, University of Wisconsin-Madison.
20. Fitelson, B. (2005). Inductive logic. In J. Pfeifer & S. Sarkar (Eds.), *Philosophy of science: An encyclopedia* (pp. 384–394). New York: Routledge.
21. Garber, D. (1983). Old evidence and logical omniscience in Bayesian confirmation theory. In J. Earman (Ed.), *Testing scientific theories. Midwest studies in the philosophy of science* (Vol. X, pp. 99–131). Minneapolis: University of Minnesota Press.
22. Glaister, S. (2002). Inductive logic. In D. Jacquette (Ed.), *A companion to philosophical logic* (pp. 565–581). Malden: Blackwell.
23. Glymour, C. (1980). *Theory and evidence*. Princeton: Princeton University Press.
24. Good, I. (1968). Corroboration, explanation, evolving probability, simplicity, and a sharpened razor. *British Journal of Philosophy of Science, 19*, 123–143.
25. Good, I. (1985). A historical comment concerning novel confirmation. *British Journal of Philosophy of Science, 36*, 184–186.
26. Goodman, N. (1946). A query on confirmation. *Journal of Philosophy, 43*, 383–385.
27. Goodman, N. (1978). *Fact, fiction and forecast*. New Yor: The Bobbs-Merrill Company, Inc.
28. Hájek, A. (2010). Interpretations of probability. In E. N. Zalta (Ed.), *The Stanford encyclopedia of philosophy*. Spring 2010 edition. Stanford: The Metaphysics Research Lab, Stanford University.
29. Harrison, V., & Chandler, J. (Eds.). (2012). *Probability in the philosophy of religion*. Oxford: Oxford University Press.
30. Hintikka, J. (1966). A two-dimensional continuum of inductive methods. In J. Hick (Ed.), *Aspects of inductive logic* (pp. 113–132). Amsterdam: North Holland Publishing Company.
31. Hintikka, J., & Niiniluoto, I. (1980). An axiomatic foundation for the logic of inductive generalization. In R. Jeffrey (Ed.), *Studies in inductive logic and probability* (pp. 157–181). Berkeley: University of California Press.
32. Huber, F. (2007). Confirmation and induction. In J. Fieser & B. Dowden (Eds.), *Internet encyclopedia of philosophy*.
33. Hume, D. (1739). *A treatise of human nature* (2nd ed.). Clarendon Press; 1985 edition, edited by L.A. Selby-Bigge.
34. Jaynes, E. (2003). *Probability theory: The logic of science*. Cambridge: Cambridge University Press.
35. Jeffrey, R. (1965). *The logic of decision*. Chicago: University of Chicago.
36. Johnson, W. E. (1924). *Logic, part III. The logical foundations of science*. Cambridge: Cambridge University Press.
37. Johnson, W. E. (1932). Probability: The deductive and inductive problems. *Mind, 41*, 409–423.

38. Juhl, C. (1994). The speed-optimality of Reichenbach's straight rule of induction. *British Journal of Philosophy of Science, 45*, 857–863.
39. Kemeny, J. (1953). A logical measure function. *Journal of Symbolic Logic, 18*, 289–308.
40. Kemeny, J. G. (1953). Review of Carnap (1952). *Journal of Symbolic Logic, 18*, 168–169.
41. Kemeny, J. (1955). Fair bets and inductive probabilities. *Journal of Symbolic Logic, 20*, 263–273.
42. Kemeny, J., & Oppenheim, P. (1952). Degrees of factual support. *Philosophy of Science, 19*, 307–324.
43. Keynes, J. (1921). *Treatise on probability*. London: Macmillan. (Reprinted in 1962).
44. Kolmogorov, A. N. (1956). *Foundations of the theory of probability*. New York: Chelsea Publishing Company. (Translated by Nathan Morrison; first edition published in 1950; the original published in 1933 in the Ergebnisse Der Mathematik).
45. Kuipers, T. (1984). Two types of analogy by similarity. *Erkenntnis, 21*, 63–87.
46. Kyburg, H. E. (1970). *Probability and inductive logic*. London: Macmillan.
47. Kyburg, H., & Smokler, H. (Eds.). (1964). *Studies in subjective probability*. New York: Wiley.
48. Lange, M. (1996). Calibration and the epistemological role of Bayesian conditionalization. *Journal of Philosophy, 96*, 294–324.
49. Lehman, R. S. (1955). On confirmation and rational betting. *Journal of Symbolic Logic, 20*(3), 251–262.
50. Maher, P. (2000). Probabilities for two properties. *Erkenntnis, 52*, 63–91.
51. Maher, P. (2001). Probabilities for multiple properties: The models of Hesse and Carnap and Kemeny. *Erkenntnis, 55*, 183–216.
52. Miller, D. (1974). Popper's qualitative theory of verisimilitude. *British Journal for the Philosophy of Science, 25*(2), 166–177.
53. Niiniluoto, I. (1981). Analogy and inductive logic. *Erkenntnis, 16*, 1–34.
54. Niiniluoto, I. (1987). *Truthlikeness*. Dordrecht: Reidel.
55. Niiniluoto, I. (1998). Verisimilitude: The third period. *British Journal for the Philosophy of Science, 49*(1), 1–29.
56. Niiniluoto, I. (2005). Verisimilitude. In J. Pfeifer & S. Sarkar (Eds.), *Philosophy of science: An encyclopedia* (pp. 854–857). New York: Routledge.
57. Niiniluoto, I. (2009). The development of the Hintikka program. In Dov M. Gabbay, Stephan Hartmann, & John Woods (Eds.), *Handbook of the history and philosophy of logic* (Inductive logic, Vol. 10 pp. 265–309). Elsevier: Amsterdam.
58. Oddie, G. (1986). *Likeness to truth*. Dordrecht: Reidel.
59. Popper, K. (1962). Some comments on truth and the growth of knowledge. In E. Nagel, P. Suppes, & A. Tarski (Eds.), *Logic, Methodology and Philosophy of Science. Proceedings of the 1960 International Congress* (pp. 285–292). Stanford: Stanford University Press.
60. Popper, K. (1963). *Conjectures and refutations*. London: Routledge.
61. Popper, K. R. (1935). *Logik der Forschung*. Wien: Verlag von Julius Springer.
62. Popper, K. R. (1979). *Objective knowledge: An evolutionary approach*. Oxford: Oxford University Press.
63. Quine, W. V. O. (1953). *From a logical point of view*. Cambridge, MA: Harvard University Press.
64. Ramsey, F. (1978). Truth and probability. In D. H. Mellor (Ed.), *Foundations: Essays in philosophy, logic, mathematics and economics* (pp. 58–100). London: Routledge. [Originally published in 1926].
65. Reichenbach, H. (1938). *Experience and prediction*. Chicago: University of Chicago Press.
66. Rosenkrantz, R. (1981). *Foundations and applications of inductive probability*. Atascadero: Ridgeview Publishing.
67. Salmon, W. (1966). *The foundations of scientific inference*. Pittsburgh: University of Pittsburgh Press.
68. Savage, L. (1954). *The foundations of statistics*. New York: Wiley.
69. Shimony, A. (1955). Coherence and the axioms of confirmation. *Journal of Symbolic Logic, 20*(1), 1–28.

70. Skyrms, B. (1984). *Pragmatics and empiricism*. New Haven: Yale University Press.
71. Skyrms, B. (1993). Analogy by similarity in hyperCarnapian inductive logic. In J. Earman, A. I. Janis, G. J. Massey, & N. Rescher (Eds.), *Philosophical problems of the internal and external worlds* (pp. 273–282). Pittsburgh: University of Pittsburgh Press.
72. Talbott, W. (2008). Bayesian epistemology. In E. N. Zalta (Ed.), *The Stanford encyclopedia of philosophy*. Fall 2008 edition. Stanford: The Metaphysics Research Lab, Stanford University.
73. Teller, P. (1976). Conditionalization, observation, and change of preference. In *Foundations of probability theory, statistical inference, and statistical theories of science*. Dordrecht: Reidel.
74. Tichý, P. (1974). On Popper's definitions of verisimilitude. *British Journal for the Philosophy of Science, 25*(2), 155–160.
75. Vickers, J. (2010). The problem of induction. In E. N. Zalta (Ed.), *The Stanford encyclopedia of philosophy*. Fall 2010 edition. Stanford: The Metaphysics Research Lab, Stanford University.
76. Williamson, J. (2010). *In defence of objective Bayesianism*. Oxford: Oxford University Press.
77. Zabell, S. (1996). Confirming universal generalizations. *Erkenntnis, 45*, 267–283.
78. Zabell, S. (1997). The continuum of inductive methods revisited. In J. Earman & J. D. Norton (Eds.), *The cosmos of science* (pp. 351–385). Pittsburgh: University of Pittsburgh Press.
79. Zabell, S. L. (2009). Carnap and the logic of inductive inference. In J. W. Stephan Hartmann & D. Gabbay (Eds.), *Handbook of the history and philosophy of logic* (Inductive logic, Vol. 10, pp. 265–309). Elsevier: Amsterdam.

Recommended Readings

1. Earman, J. (1992). *Bayes or bust? A critical examination of Bayesian confirmation theory*. Cambridge, MA: MIT Press.
2. Hacking, I. (2001). *An introduction to probability and inductive logic*. Cambridge: Cambridge University Press.
3. Hájek, A., & Hall, N. (2002). Induction and probability. In P. K. Machamer & M. Silberstein (Eds.), *The Blackwell guide to the philosophy of science* (pp. 149–172). Malden: Blackwell.
4. Kuipers, T. (1978). *Studies in inductive probability and rational experimentation*. Dordrecht: Reidel.
5. Pettigrew, R. (2016). Epistemic utility arguments for probabilism. In E. N. Zalta (Ed.), *The Stanford encyclopedia of philosophy*, Spring 2016 edition. Stanford: Metaphysics Research Lab, Stanford University.
6. Reichenbach, H. (1949). *The theory of probability*. Berkeley: University of California Press.

Chapter 6
Conditionals

John Cantwell

Abstract Conditional constructions – constructs of the form If A, then B – have for over a century been subject to intense study in a wide variety of philosophical areas, as well as outside of philosophy. One important reason is that such constructs allow one to encode *connections* and *dependencies*, be they causal, epistemic, conceptual, or metaphysical. This chapter briefly outlines some of the main formal models that have been employed to analyze such constructs, as well as their philosophical motivation.

6.1 Background

The conditional construction – here I include such constructions as *If A then B*, *B even if A*, *B only if A*, and so on – is a small unassuming construction that for decades (in some cases centuries) has attracted massive interest from philosophers, logicians, linguists, computer scientists and cognitive psychologists. The huge interest in this small construction can be traced to two circumstances. First, the conditional is one of our primary vehicles for talking about, describing and representing *connections* and *dependencies*, be the connections and dependencies causal, conceptual, metaphysical, epistemic, or logico-semantic. Second, the problems one has in accounting for the meaning of the conditional to a large extent overlaps with the problems one has in accounting for the nature of these connections and dependencies. So the study of the conditional is one pathway into a large body of issues with ramifications far beyond the seemingly minor issue of the semantics of a small unassuming construction. Importantly, the formal methods that have been introduced in order to deal with the problems posed by conditionals have proved to be useful and illuminating in a range of other areas. Not surprisingly, the area – with its connections to many of the deepest problems of philosophy – is rife with controversy.

J. Cantwell (✉)
Division of Philosophy, Royal Institute of Technology (KTH), Stockholm, Sweden
e-mail: cantwell@kth.se

S. O. Hansson, V. F. Hendricks (eds.), *Introduction to Formal Philosophy*, Springer Undergraduate Texts in Philosophy, https://doi.org/10.1007/978-3-319-77434-3_6

Consider the following conditionals:

(1) If x is even and greater than 2, then x is not prime.
(2) If the conditions of the Versaille treaty had not been so severe, there would have been no WWII.
(3) If Shakespeare didn't write Hamlet someone else did.

The first of these can be used to express that *even* and *prime* are conceptually related properties; the second conditional can be used to express that there is a causal connection between the conditions set up in the Versaille treaty and the breakout of WWII; the third conditional can be be used to express that there is some degree of epistemic independence between one's conviction that *Shakespeare* wrote Hamlet and one's conviction that *someone* wrote Hamlet (e.g. some of the evidence that one has for the latter, is not also evidence that one has for the former). Three conditionals that can be used to express three very different kinds of connections and dependencies.

A substantial portion of the literature on conditionals deals with the semantics of conditionals in natural language. Various fundamental semantic questions have been addressed. Is there at some deeper semantic level only one type of conditional or are there different semantic kinds of conditionals? Compare the difference between (3) above, which is in the grammatically *indicative* mood, and the grammatically *subjunctive*:

(4) If Shakespeare hadn't written Hamlet someone else would have.

Clearly they don't have the same meaning, and this has convinced many that that there are different semantic kinds of conditionals. Further questions: Does the semantic value of a conditional depend on the (conceptual/causal/epistemic) connections and dependencies it is used to express, or does the conditional express the connections and dependencies that it does through the pragmatics of assertion? Can conditionals have truth values and, if so, do they always have truth values (or do they express 'gappy' propositions, propositions that can lack truth value)? Is the truth value of a conditional (if it has one) a function of the truth values of its constituent sentences or is the conditional semantically intensional? Note that some of these questions may have different answers for different kinds of conditionals.

Many, however, have approached the study of conditionals from another direction. Seeing that conditionals are linguistic vehicles for expressing connections and dependencies one may ask what kind of connections and dependencies – perhaps encoded in some underlying structure – they can be used to express. In such studies felicity to natural language usage is of secondary importance (although the link and appeal to linguistic intuitions is seldom abandoned altogether), instead more general structural phenomena are investigated.

One important phenomenon is *defeasibility* (sometimes referred to as *failure of antecedent strengthening* or *nonmonotonicity*). From the fact that one accepts $A \to C$ (I here use $\to$ as a generic conditional) it does not follow that one thereby should accept $(A \wedge B) \to C$. For instance, from the fact that one accepts:

(5) If the match is struck it will light.

it does not follow that one should accept:

(6) If the match is *submerged in water* and struck it will light.

The phenomenon of defeasibility reflects the fact that connections and dependencies often do not hold unconditionally or by necessity; often they depend on things being as they *normally* are, or on contingencies that just happen– as far as we know – to be the case or on other things *being equal* (*ceteris paribus*). A fundamental problem is that typically it is *impossible* to spell out in full detail how things normally are, what contingencies are necessary and sufficient or what the 'other things' are that are to be considered 'equal'. The expressive power of conditionals to a large extent derives from the fact that they implicitly invoke dependencies that are impossible to spell out. As nearly all discourse outside the realm of mathematics (and other aprioristic disciplines) deals with such defeasible connections and dependencies, it is of the first importance to understand how conditionals do the magic of depending on what cannot be spelled out. Note that we will make no progress here by merely *mentioning* that a conditional depends on that things are as they normally or on other things being equal. For from the fact that one accepts:

(7) If the match is struck it will, *other things being equal*, light.

it still does not follow that one should accept:

(8) If the match is submerged in water and struck it will, *other things being equal*, light.

The phenomenon of defeasibility is a core feature of the conditional itself, and the development of formal methods for analysing this phenomenon has been of central interest in the literature on conditionals.

In this paper I shall briefly discuss some of the main approaches to the analysis of conditionals, outline the formal structures that have made the analyses possible, and briefly indicate their philosophical underpinnings, with a particular eye towards how they account for how conditionals can be used to express connections and dependencies. I will consider two basic kinds of analyses: those that assign truth conditions to the conditional, and those that instead assign acceptance conditions (using the *Ramsey Test*).[1]

[1]The field is far too wide to enable an exhaustive survey in these few pages and so the present overview largely reflects the author's own interests and prejudices. Many important issues are ignored or treated only in passing and countless important contributions will not be credited. References given reflect (but by no means exhaust) works of seminal importance, works that give a more thorough overview of the issues, as well as work that may point towards interesting new developments.

6.2 Conditionals with Truth Values

6.2.1 *Formal Preliminaries*

Assume a language with atoms $p, q, r, \ldots$ closed under $\neg$ (negation), $\wedge$ (conjunction) and $\vee$ (disjunction). The language will subsequently be extended with the generic conditional $\rightarrow$. Non-atomic sentences will be denoted A, B, C, ...

The semantics for this language will be given relative a set U of *states*; these can be thought of as assignments of truth values to the atoms or as *possible worlds*, or as pairs of possible worlds and moments in time, what is crucial is that each state determines the truth values of the atomic sentences.

1. $u \models p$ if and only if p is true at u.
2. $u \models \neg A$ if and only if it is not the case that $u \models A$.
3. $u \models A \wedge B$ if and only if $u \models A$ and $u \models B$.
4. $u \models A \vee B$ if and only if $u \models A$ or $u \models B$.

In the form of truth-tables, at any given state u:

A	$\neg A$
t	f
f	t

A	B	$A \wedge B$
t	t	t
t	f	f
f	t	f
f	f	*f*

A	B	$A \vee B$
t	t	t
t	f	t
f	t	t
f	f	f

A sentence B is a *consequence* of $A_1, \ldots, A_n$, in symbols $A_1, \ldots, A_n \models B$ if and only if $u \models B$ whenever $u \models A_1, \ldots, u \models A_n$ (that is, if and only if B is true whenever each A_i is true).

6.2.2 *The Material Conditional*

The *material conditional*, here written $A \supset B$, is typically the first fully analysed conditional that students of logic encounter. In its classical form it is a truth-functional connective with interpreted according to the truth-table:

A	B	$A \supset B$
t	t	t
t	f	f
f	t	t
f	f	t

The interpretation has the virtue of simplicity (truth-tables belong to the simplest of formal structures), furthermore it can be derived from seemingly compelling logical principles; for instance, one can show that it is the interpretation that $\supset$ must have if it is to be truth-functional and satisfy (given the classical interpretation of negation):

$$A, A \supset B \models B. \text{ (Modus ponens)}$$

$$\neg B, A \supset B \models \neg A. \text{ (Modus tollens)}$$

$$\text{If } A \models B, \text{ then } \models A \supset B.$$

It is clear from the semantics of the material conditional that its truth value does not depend in any interesting way on connections and dependencies between antecedents and consequents: the material conditional is only sensitive to their truth value. As a result, if we take, say, the indicative conditional of natural language to have the semantics of the material conditional, semantics alone does very little to explain how and when such conditionals are asserted and denied. On the basis of semantics alone one would predict that both of the following conditionals would be generally accepted:

(9) If Shakespeare didn't write Hamlet, then his grandmother did.
(10) Even if someone other than Shakespeare wrote Hamlet, Shakespeare wrote Hamlet.

For if Shakespeare wrote Hamlet, then both these conditionals are true (as their antecedent is false). Yet those who believe that Shakespeare wrote Hamlet are not in general inclined to accept the conditionals. Indeed most would be inclined to *reject* these conditionals; so there is a deep and disturbing mismatch between the truth values of the conditionals and speakers' inclinations to use them. Such systematic mismatch is sometimes labeled 'paradoxes' of the material conditional and can be traced back to various logical properties such as:

$$\neg A \models A \supset B.$$

$$B \models A \supset B.$$

If the semantics of the material conditional is to have any credibility as a semantics of the conditional in natural language (specifically: the indicative conditional) then most of the explanatory work – including how such conditionals are used to express connections and dependencies – must be relegated elsewhere: to pragmatics. For instance, one can hold that the reason why one is not inclined to accept (9) is the same as the reason why one is not inclined to assert

(11) Either Shakespeare wrote Hamlet or his grandmother did.

even though one believes that it is true (due to one of the disjuncts being true): it would be misleading and would convey less information than the simple "Shakespeare wrote Hamlet". This was the strategy proposed by Grice [12] and has subsequently been defended by Lewis [22] and elaborated by Jackson [13]. However, many have come to the conclusion that the discrepancy between the semantics of the material conditional and the way in which indicative conditionals are used is too great: the semantics lacks credibility.

Some have argued (e.g. [5, 7]) that the most blatant collisions between semantics and pragmatics (e.g. cases where one is inclined to reject a conditional that one, allegedly, believes is true) can be avoided by allowing conditionals to have 'gappy' truth conditions:

A	B	$A \to B$
t	t	t
t	f	f
f	t	–
f	f	–

That is, the conditional is taken to lack truth value when the antecedent is false. This still leaves much of the explanatory work to pragmatics (e.g. the way the conditional is used to express connections and dependencies) but at least does not force pragmatics to explain why true conditionals are rejected. The semantics also has support from the psychological literature on how people assess conditionals. Nevertheless, the account is still viewed with scepticism, mainly regarding the intelligibility of truth-value gaps and how such gaps are to be accommodated in a wider story where conditionals embed in more complex sentences (see [23, 24]).

6.2.3 The Strict Conditional

An example of a conditional that semantically reflects a stronger connection between the antecedent and the consequent is the *strict* conditional, with the truth-conditions:

$u \models A \to B$ if and only if $v \models B$ for every state v such that $v \models A$.

The strict conditional is not supposed to reflect any particular construction in natural language, but it is a nice simple example of how a language can contain conditionals that reflect interesting connections between their antecedents and consequents, in this case: the connection of semantic consequence (for with the present truth-conditions, $A \to B$ is true if and only if B is a semantic consequence of A). If the English indicative conditional had these truth conditions then

(12) If you are a bachelor then you are not married,

would be true, while

(13) If Jim was run over by a truck, he died,

would be false (even if Jim was run over by a truck and died).

Notably, as semantic consequence involves the preservation of a property (the property of being true at a state), the strict conditional is not defeasible, that is, we have the property:

$$A \to C \models (A \wedge B) \to C.$$

6.2.4 Ontic Selection Functions: Counterfactuals

An explosion of interest in the formal structures used to represent conditionals came with the work of Stalnaker [29, 31] and Lewis [19]. In different ways they introduced the idea of a selection function γ that for each state u and each sentence A picks out a sub-set of the states in which A is true. Given such a selection function one can give truth conditions for a conditional as follows:

$$u \models A \to B \text{ if and only if } v \models B \text{ for every } v \text{ in } \gamma(u, A).$$

A selection function gives rise to many degrees of freedom both in terms of abstract structural properties and in terms of substantive interpretations. Common examples of structural constraints are[2]:

If v is in $\gamma(u, A)$, then $v \models A$.
If $u \models A$, then $\gamma(u, A)$ is the unit set $\{u\}$. (Centering)
If $v \models B$ for each v in $\gamma(u, A)$, then $\gamma(u, A) = \gamma(u, A \wedge B)$.

In these cases each structural constraint gives rise to a logical property:

$$\models A \to A.$$

$$A, A \to B \models B.$$

$$A \to B, A \to C \models (A \wedge B) \to C.$$

Notably, the semantics allows for *defeasibility*, that is, we do not in general have:

$$A \to C \models (A \wedge B) \to C.$$

For instance, consider the graphical representation in Fig. 6.1 of the states in which A, B and C are true and the areas selected by the selection function:
From this picture it is clear that we can have $u \models A \to C$ although we do not have $u \models (A \wedge B) \to C$.

Relationships between structural properties of the selection function and logical properties of the conditional are, of course, interesting in their own right. But it

[2] Arló-Costa [3] presents a thorough overview of the logic of conditionals and how they relate to structural conditions.

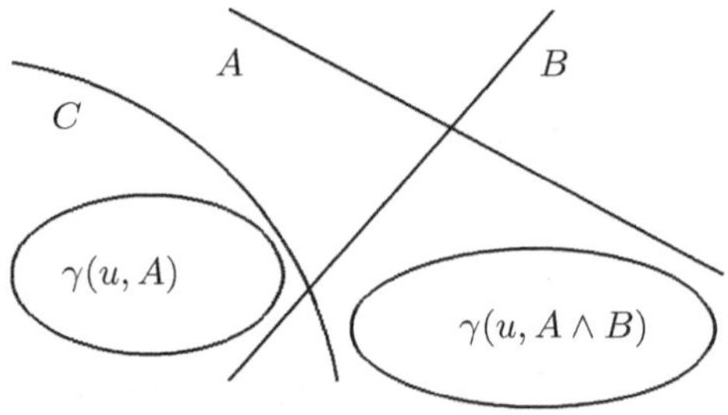

Fig. 6.1 $\gamma(u, A)$ is a subset of the C-states while $\gamma(u, A \wedge B)$ is not

has been the prospect that selection functions can be used to represent fundamental structures – the kind of structures that underlie the connections and dependencies that we express by natural language conditionals – that has been the main philosophical driving force for investigating selection-function semantics for conditionals.

By far the most influential interpretation is due to David Lewis (e.g. [19, 21]) who suggested that the selection function $\gamma(u, A)$ selects those A-worlds (on his account states are possible worlds) that are *most similar* to u; that is, the selection function is based on a similarity relation $v \leq_u w$ between possible worlds ($v \leq_u w$ holds if v is no less similar to u than is w). The similarity relation is standardly taken to be *reflexive*, *transitive* and *complete*[3] and to satisfy some proviso – *the limit assumption* – to ensure that in a given set of worlds there is at least one world that is *most similar* to the target world, that is one needs to guarantee that there are no infinitely descending chains of similarity.[4]

Similarity relations between worlds can be thought of as providing a substantive interpretation of the *ceteris paribus*-intuition in the evaluation of conditionals: a conditional is true if it is the case that if the antecedent were true, and *as much as possible* (this is where the similarity relation kicks in) remained the same, then the consequent would be true.

The introduction of a similarity relation shifts the focus to the question: *Similar in what way?* There are various ways of approaching this question. Lewis, who followed a tradition broadly deriving from David Hume, took special interest in similarity criteria that would allow for a reductive analysis of causal relations in terms of the similarity relation (and so for a reductive analysis of causal relations in terms of counterfactuals).

Counterfactual analyses of causality have been criticized (e.g. [32]), but this is not the place to evaluate Lewis' theory of causality. What is clear however is that a Lewis-style similarity semantics provides a robust and convincing model (even if one may be sceptical about the underlying metaphysical assumptions) for the

[3]**Reflexivity:** $v \leq_u v$. **Transitivity:** If $v \leq_u w$ and $w \leq_u z$, then $v \leq_u z$. **Completeness:** Either $v \leq_u w$ or $w \leq_u v$.

[4]The limit assumption is not, strictly speaking, necessary, but if one omits this constraint the semantic clause becomes more complex.

analysis of *counterfactual* conditionals, a class of conditionals that largely coincides with conditionals in the *subjunctive* mood:

(14) If the match had been struck, it would have lit.
(15) The Vietnam war would have escalated even if Kennedy had not been murdered.

As we typically take the truth values of such conditionals to depend on underlying causal connections and dependencies, this provides strong support for the idea that similarity relations can be used to encode important aspects of the causal structure of the world. Accordingly, the Lewis-Stalnaker analysis remains the dominant paradigm in the semantics of counterfactual conditionals, one where the *truth-value* of a conditional is sensitive to underlying causal connections and dependencies between antecedents and consequents.

Within the more linguistically oriented study of the semantics of conditionals, the dominant tradition is to take conditionals to have an underlying Lewis-Stalnaker-style semantic structure. Some of the most influential work here is due to Angelika Kratzer (see e.g. [14, 15], see collection in [16]). Kratzer combines a Lewis-style semantics with a 'restrictor' analysis of conditionals: the antecedent of a conditional is taken to restrict the space of possibilities relative to which the consequent is evaluated. This becomes particularly important when the consequent of a conditional contains a modality of some sort, as in:

(16) If the die is thrown lands on an even number, the probability that it will show a six is $1/3$.
(17) If you kill him, you should kill him gently.

In the first case the antecedent constrains the possible outcomes that is relevant for the probability modality (for note that absent the antecedent, the probability that the result of a throw of a fair die will show a six is $1/6$). In the second case the antecedent constrains the space of possible actions to be considered in deliberating what one *should* do (and note that it is *not* the case that you should kill him gently, but *if* you kill him, then a gentle killing seems to be the most humane course of action).

6.3 Epistemic Interpretations

Epistemic interpretations of conditionals[5] cannot be discussed without mention of Frank Ramsey's [28] famous footnote:

> If two people are arguing 'if p will q?' and both are in doubt as to p, they are adding p hypothetically to their stock of knowledge and arguing on that basis

[5]Some key references here are Stalnaker [30], Adams [1], Edgington [9], Levi [18] and Bennett [4].

> about q. We can say that they are fixing their degrees of belief in q given p. If p turns out false, these degrees of belief are rendered void. If either party believes $\neg p$ for certain, the question ceases to mean anything to him except as a question about what follows from certain laws and hypotheses.

In the footnote Ramsey identifies the main feature of what has become known as the *Ramsey Test*: conditionals are evaluated on the basis of what one would take to hold on the hypothetical assumption that the antecedent is true. As hypothetical reasoning is a key way of exploring epistemic connections such as evidential relations and other structures of justification, the Ramsey Test points the way to explaining how conditionals can be used to express such connections.

For instance, say that you have quite convincing testimony from the butler that suggests that either the maid or the gardener commited the murder. So on supposing, hypothetically, that it wasn't the maid, you conclude (hypothetically) that it was the gardener; that is, you accept

(18) If the maid didn't do it, the gardener did.

On the other hand, on supposing that neither the maid nor the gardener did it, you come to the conclusion that there must be something wrong with the butler's testimony, indeed that would suggest that *he* is the culprit; so you accept:

(19) If neither the maid nor the gardener did it, the butler did it.

The fact that you accept both conditionals reveal something important about how you evaluate the situation and what counts as evidence for what (note, in particular, that the examples show that the epistemic interpretation allows for defeasible conditionals).

Epistemic interpretations of conditionals typically do not take the connections and dependencies expressed by the use of a conditional to reside in its truth-conditions. The epistemic connections and dependencies expressed by a conditional are not taken to be objective features of the world, but are rather features of one's current epistemic state.

There is wide agreement[6] that the epistemic intepretation provides a good analysis of stand-alone (non-embedded) *indicative* conditionals (conditionals in the indicative mood) like (3), (18), and (19) above.[7] From the point of view of meaning theoretical orthodoxy the big problem with the epistemic interpretation is that it doesn't provide truth-conditions for the conditional. This creates a problem both in accounting for its logic, as the semantic consequence relation is based on sentences taking a truth value, and in accounting for its interaction with other connectives like negation ($\neg$) and disjunction ($\vee$), as these are truth-functional. Some have thus sought to combine epistemic interpretations with truth-functional

[6]See [4] and [10] for extensive discussions and references, but compare [26] and [25] for putative counterexamples (e.g. [8] argues that McGee fails to establish a counterexample).

[7]Some have argued that counterfactual conditionals can be given an epistemic interpretation, but this is a matter of considerable controversy.

accounts and take the epistemic interpretation to spell out, in a systematic way, the pragmatics of indicative conditionals, thus maintaining the traditional semantic-pragmatics distinction (see the discussion in Sect. 6.2.2). Others have taken this to show that the indicative conditional is inherently different from other truth-conditional constructions and has no semantics proper.

6.3.1 The Ramsey Test and Logic

In its most abstract form, the epistemic interpretation relies on the notion of an epistemic state $\mathcal{E}$ and on a function $*$ that takes a sentence A and returns the epistemic state $\mathcal{E} * A$ that corresponds to the epistemic state in which it is hypothetically assumed that A. As neither the acceptance conditions of conditionals nor their logic are derivable from their truth-conditions, one also needs a separate account of acceptance conditions and of *epistemic* consequence.

Let $\mathcal{E} \Vdash\!\!= A$ stand for *in the epistemic state $\mathcal{E}$ one is committed to accepting A* (some models, such as the probabilistic model, allow also for a notion of *degree* of acceptability, but this notion will not be covered here). With this in place one can state the Ramsey Test as follows:

$$\mathcal{E} | \!\models A \to B \text{if and only if } \mathcal{E} * A | \!\models B.$$

The Ramsey Test by itself gives no indication of the acceptance conditions for other kinds of sentences or of their logic. So let $A_1, \ldots, A_n | \!\models B$ stand for *B is an epistemic consequence of $A_1, \ldots, A_n$*. A reasonable minimal requirement is that as long as the sentences are not conditionals (nor contain conditionals) we have:

Semantic Closure If $A_1, \ldots, A_n \models B$, then $A_1, \ldots, A_n | \!\models B$.

A further reasonable requirement is:

Epistemic Closure If $\mathcal{E} | \!\models A_1, \ldots, \mathcal{E} | \!\models A_n$ implies $\mathcal{E} | \!\models B$ for every epistemic state $\mathcal{E}$, then $A_1, \ldots, A_n | \!\models B$.

If we allow for the converse direction then the epistemic consequence relation can be fully analyzed by acceptance conditions:

Reverse Epistemic Closure If $A_1, \ldots, A_n | \!\models B$, then $\mathcal{E} | \!\models A_1, \ldots, \mathcal{E} | \!\models A_n$ implies $\mathcal{E} | \!\models B$.

The logical properties of the conditional will to a large extent depend on properties of the $*$-operator. Here are two candidate properties (again, see Arló-Costa [3] for thorough overview):

Success $\mathcal{E} * A | \!\models A$.
Vacuity If $\mathcal{E} | \!\models A$, then $\mathcal{E} * A = E$.

These give rise to the properties:

$$\Vert\!\models A \to A.$$

$$A, A \to B \Vert\!\models B.$$

Another possible requirement is:

Iteration $\mathcal{E} * A * B = E * (A \wedge B)$.

This gives rise to the logical export-import-properties:

$$A \to (B \to C) \Vert\!\models (A \wedge B) \to C.$$

$$(A \wedge B) \to C \Vert\!\models A \to (B \to C).$$

Importantly, it is typically not assumed that $*$ satisfies *monotonicity*:

Monotonicity If $\mathcal{E} \Vert\!\models B$, then $\mathcal{E} * A \Vert\!\models B$.

Accordingly, the epistemic conditional can be defeasible: it is not in general the case that $A \to C \Vert\!\models (A \wedge B) \to C$.

Sometimes weaker properties than monotonicity are assumed, such as:

Preservation If $\mathcal{E} \Vert\!\models B$ and $\mathcal{E} \not\Vert\!\models \neg A$, then $\mathcal{E} * A \Vert\!\models B$.

Together with Logical Closure this entails:

If $\mathcal{E} \Vert\!\models A \supset B$ and $\mathcal{E} \not\Vert\!\models \neg A$, then $\mathcal{E} \Vert\!\models A \to B$.

This can be combined with the requirement:

If $\mathcal{E} * A \Vert\!\models B$ and $\mathcal{E} \not\Vert\!\models \neg A$, then $\mathcal{E} \Vert\!\models A \supset B$.

Together with Preservation and Logical Closure this entails that as long as one doesn't reject the antecedent of a conditional the acceptance conditions of the epistemic conditional coincides with acceptance conditions of the material conditional. This is important as the counterintuitive properties of material implication (as a model for the indicative conditional) mainly emerge in cases when the antecedent is believed to be false; the epistemic conditional thus keeps the 'good' parts of the material analysis of the conditional but avoids its problematic parts. For instance, in the absence of the monotonicity requirement we do not have:

$$\neg A \Vert\!\models A \to B.$$

$$B \Vert\!\models A \to B.$$

6.3.2 Formal Models

Two main types of formal structures are often used to model such epistemic states: models that assign *probabilities* to sentences and a broad range of models – closely related or identical to the models applied in the area of *belief revision* – that rank sentences according to their *entrenchment* (*plausibility*, etc.). Often (but not always) the latter kind of models use qualitative relations rather than numerical measures.

Probabilistic models (see in particular Adams [1]) follow a Bayesian tradition in which epistemic states are taken to be probability measures.[8] The revision operator $*$ in such a model takes a probability measure $\mathcal{E}_P$ and a sentence A and returns a new probability measure $\mathcal{E}_P * A$ by *conditionalisation*[9]:

$$(\mathcal{E}_P * A)(B) = \mathcal{E}_P(B \mid A) = \mathcal{E}_P(A \wedge B)/\mathcal{E}_P(A).$$

In a simple model of probabilistic acceptance one accepts all and only those non-conditional sentences that exceed some threshhold α ($.5 \leq \alpha \leq 1$), so that, when A is not a conditional:

$$\mathcal{E}_P \Vdash\!\!= A \text{ if and only if } \mathcal{E}_P(A) > \alpha.$$

For the epistemic consequence relation there are different options; Adams [2] makes a case for the p-consequence relation (the uncertainty of the conclusion cannot be greater than the sum of the uncertainty of the premises):

$A_1, \ldots, A_n \Vdash\!\!= B$ if and only if $(1-\mathcal{E}_P(B)) \leq (1-\mathcal{E}_P(A_1)) + \cdots + (1-\mathcal{E}_P(A_n))$, for all $\mathcal{E}_P$.

These jointly ensure that Semantic Closure and Epistemic Closure are satisfied. The model ensures that we have a defeasible conditional that satisfies important logical properties like modus ponens and export-import.

As an example of a qualitative representation of a epistemic state one can consider an epistemic state to be represented as an epistemic selection function $\mathcal{E}_\gamma$ that for any sentence B that has truth conditions picks out the set $\mathcal{E}_\gamma(B)$ of *most plausible* B-states (the B-states, recall, are the states in U where B is true). So here again we have a selection function, but notice that while selection functions in the ontic models were relativized to the states (worlds) themselves, here they are relativized to *epistemic states*, making the evaluation procedure speaker-dependent. The epistemic state that results from making the hypothetical assumption that A,

---

[8] A probability measure is, as is standard, here taken to be a real-valued function P that take sentences as their arguments and satisfies (a) $0 \leq \mathrm{P}(A) \leq 1$, (b) $\mathrm{P}(\neg A) = 1 - \mathrm{P}(A)$, and (c) $\mathrm{P}(A \vee B) = \mathrm{P}(A) + \mathrm{P}(B) - \mathrm{P}(A \wedge B)$.

[9] Conditionalisation only covers the case when $\mathrm{P}(A) > 0$; to deal with the case when $\mathrm{P}(A) = 0$ one can use Popper-measures or let $\mathrm{P} * A$ be a non-standard measure that assigns probability 1 (or 0) to all sentences.

$\mathcal{E}_\gamma * A$, can then be defined to be the epistemic selection function that for any sentence B picks out the set of *most plausible* B-states that are also A-states (i.e. the most plausible $A \wedge B$-states):

$$(\mathcal{E}_\gamma * A)(B) = \mathcal{E}_\gamma(A \wedge B).$$

A non-conditional sentence is accepted if it is true in all the most plausible models, i.e. letting $\top$ be an arbitrary tautology ($\mathcal{E}_\gamma(\top)$ will pick out the most plausible of *all* states) and A a non-conditional sentence:

$$\mathcal{E}_\gamma \Vdash\!\!= A \text{ if and only if } v \models A, \text{ for all } v \in \mathcal{E}_\gamma(\top).$$

The epistemic consequence relation can be defined as:

$A_1, \ldots, A_n \Vdash\!\!= B$ if and only if $\mathcal{E}_\gamma \Vdash\!\!= A_1, \ldots, \mathcal{E}_\gamma \Vdash\!\!= A_n$ implies $\mathcal{E}_\gamma \Vdash\!\!= B$, for all $\mathcal{E}_\gamma$.

These jointly ensure that Semantic Closure, Epistemic Closure and Reverse Epistemic Closure are satisfied. The model also ensures that we have a defeasible conditional that satisfies important logical properties like modus ponens and export-import.

This is not the place to discuss the relative merits of probabilistic versus qualitatitve models. Both kinds of models allow for a rich (but by no means exhaustive) representation of evidential relations and justificational structures and are able to explain intuitive judgements about indicative conditionals quite well.

6.3.3 Impossibility Results

As noted, epistemic interpretations of conditionals often treat them as *exceptional*, particularly, by not assigning truth-conditions to them. However, could it not be the case that the conditional has truth-conditions that are such that they just happen to also satisfy the Ramsey Test? For instance, couldn't the conditional have truth-conditions that, given the axioms of probability, forced the equality:

$$\mathrm{P}(A \to B) = \mathrm{P}(B \mid A)?$$

No, Lewis [20] showed (and this result has since been strenghtened in a number of ways) that a language with a conditional that embeds just like other connectives (which we would expect if the conditional had truth-conditions) cannot, on pain of triviality, satisfy this equality. As long as the standard axioms of probability are satisfied, the equality forces the probability of A to be either 0 or 1. Gärdenfors [11] has established a similar impossibility result for the qualitative case.

So to satisfy the epistemic interpretation the conditional must truly be exceptional. This conclusion is not undermined by the fact that the above equality can be non-trivially satisfied if one allows for *gappy* truth-conditions (see Sect. 6.2.2) as gappy truth-conditions are exceptional in their own right and probability measures on gappy propositions will not in general satisfy the standard axioms of probability (see, e.g. [6] for a discussion).

6.4 Concluding Remarks

Over the past century the conditional has been the locus of a rich and diverse range of philosophical debates. Considerable progress has been made in the understanding of how conditionals can be used to express connections and dependencies, as well as in the understanding of their defeasible character. To a large extent the progress can be traced back to the careful study of the formal apparatus used to represent the underlying structures. Furthermore, the insights have shed new light on issues of independent interest, a consequence of the fact that conditionals feed on structures that are of core interest in epistemology and metaphysics in general. The area of conditionals is thus one of the success stories of formal philosophy.

Many issues remain, however. Some hold that causal relations cannot be fully analyzed by counterfactuals, and so cannot be fully analyzed by similarity relations between worlds. Given that many counterfactuals are parasitic on the underlying causal structure of the world, this suggests that there may be alternative representations of this structure that could serve as the semantic basis of counterfactuals (e.g. Pearl's [27] work on causal models and Leitgeb's [17] work on probabilistic models to name just two such alternatives). Here there is a lot of work to be done. Likewise, the status of the indicative conditional under the epistemic interpretation is far from settled: Can one find representations that elegantly account for their special nature? Representations that also account for their semantic connections to counterfactuals?

References

* Indicates recommended reading.

1. Adams, E. W. (1975). *The logic of conditionals*. Dordrecht: Reidel.
2. Adams, E. W. (1998). *A primer of probability logic*. Stanford: CSLI.
3. *Arló-Costa, H. (2009). The logic of conditionals. In E. N. Zalta (Ed.), *The stanford encyclopedia of philosophy* (Spring 2009 ed.).
4. *Bennett, J. F. (2003). *A philosophical guide to conditionals*. Oxford: Oxford University press.
5. *Bradley, R. (2002). Indicative conditionals. *Erkenntnis, 56*, 345–378.
6. Bradley, R. (2007). A defence of the Ramsey test. *Mind, 116*(461), 1–21.
7. Cantwell, J. (2008). Indicative conditionals: Factual or epistemic? *Studia Logica, 88*, 157–194.
8. Cantwell, J., Lindström, S., & Rabinowicz W. (2017). McGee's counterexample to the Ramsey test. *Theoria, 83*(2), 154–168.
9. *Edgington, D. (1995). On conditionals. *Mind, 104*, 235–329.
10. Evans, J., & Over, D. (2004). *If*. Oxford: Oxford University Press.

11. Gärdenfors, P. (1986). Belief revision and the Ramsey test for conditionals. *Philosophical Review, 95*, 81–93.
12. *Grice, H. P. (1989). Indicative conditionals. In *Studies in the way of words* (pp. 58–87). Cambridge, MA: Harvard University Press.
13. Jackson, F. (1979). On assertion and indicative conditionals. *Philosophical Review, 88*, 565–589.
14. Kratzer, A. (1977). What 'must' and 'can' must and can mean. *Linguistics and Philosophy, 1*(3), 337–356.
15. Kratzer, A. (1979). Conditional necessity and possibility. In R. Bäuerle, U. Egli & A. von Stechow (Eds.), *Semantics from different points of view* (pp. 117–147). Berlin: Springer.
16. *Kratzer, A. (2012). *Modals and conditionals: New and revised perspectives*. Oxford: Oxford University Press.
17. Leitgeb, H. (2012). A probabilistic semantics for counterfactuals. *The Review of Symbolic Logic, 5*, 26–84.
18. Levi, I. (1996). *For the sake of the argument*. Cambridge: Cambridge University Press.
19. *Lewis, D. (1973). *Counterfactuals*. Cambridge: Harvard University Press.
20. *Lewis, D. (1976). Probabilities of conditionals and conditional probabilities. *Philosophical Review, 85*, 297–315.
21. Lewis, D. (1979). Counterfactual dependence and Time's arrow. *Nous, 13*, 455–476.
22. Lewis, D. (1986). Probabilities of conditionals and conditional probabilities II. *Philosophical Review, 95*, 581–589.
23. McDermott, M. (1996). On the truth conditions of certain 'If'-sentences. *Philosophical Review, 105*, 1–37.
24. *McGee, V. (1989). Conditional probabilities and compounds of conditionals. *Philosophical Review, 98*, 485–542.
25. McGee, V. (2000). To tell the truth about conditionals. *Analysis, 60*, 107–111.
26. Morton, A. (2004). Against the Ramsey test. *Analysis, 64*(4), 294–299.
27. Pearl, J. (2000). *Causality: Models, reasoning, and inference*. Cambridge: Cambridge University Press.
28. Ramsey, F. P. (1931). *Foundations of mathematics and other essays*. New York: Routledge and Kegan Paul.
29. Stalnaker, R. (1968). A theory of conditionals. In *Studies in logical theory* (No. 2 in American philosophical quarterly monograph series). Oxford: Blackwell.
30. Stalnaker, R. (1975). Indicative conditionals. *Philosophia, 5*, 269–286. Reprinted in *Ifs*, ed. W. L. Harper, R. Stalnaker & G. Pearce 1976 by Reidel.
31. Stalnaker, R. (1984). *Inquiry*. Cambridge, MA: The MIT press.
32. Strevens, M. (2003). Against Lewis's new theory of causation: A story with three morals. *Pacific Philosophical Quarterly, 84*(4), 398–412.

Chapter 7
Neural Network Models of Conditionals

Hannes Leitgeb

Abstract This chapter explains how artificial neural networks may be used as models for reasoning, conditionals, and conditional logic. It starts with the historical overlap between neural network research and logic, it discusses connectionism as a paradigm in cognitive science that opposes the traditional paradigm of symbolic computationalism, it mentions some recent accounts of how logic and neural networks may be combined, and it ends with a couple of open questions concerning the future of this area of research.

7.1 Introduction

Neural networks are abstract models of brain structures capable of adapting to new information. The learning abilities of artificial neural networks have given rise to successful computer implementations of various cognitive tasks, from the recognition of facial images to the prediction of currency movement. Under the heading 'deep learning', neural networks have become prominent again lately as major tools in the field of machine learning.

Logic deals with formal systems of reasoning; in particular, inductive logic studies formal systems of reasoning towards plausible but uncertain conclusions. As evidence accumulates, the degree to which it supports a hypothesis, as measured by the logic, should tend to indicate that the hypothesis is likely to be true.

Although sharing a joint focus on information and reasoning, until recently these two areas developed in opposition to each other: neural networks are quantitative dynamic systems, while logical reasoners must be symbolic systems; networks are described by mathematical equations, whereas logic is subject to normative statements about how we ought to reason; neural networks have been studied by

H. Leitgeb (✉)
Ludwig-Maximilians-University, Munich, Germany
e-mail: Hannes.Leitgeb@lmu.de

S. O. Hansson, V. F. Hendricks (eds.), *Introduction to Formal Philosophy*, Springer Undergraduate Texts in Philosophy, https://doi.org/10.1007/978-3-319-77434-3_7

scientists, whilst the classical "problem of induction" is regarded as belonging to philosophy. And so forth.

In recent years, however, this assessment has been changing: the emergence of logical formalisms for uncertain reasoning and the discovery that these formalisms apply to neural net processes on the representational level give rise to the expectation that the dynamics of artificial neural networks can be understood in terms of logically valid, and thus rational, rules of inference. As neural networks, commonsense reasoning, and maybe even scientific induction seem to conform to similar logical systems, a joint theoretical framework might be in the offing which might lead to new insights into the logical and cognitive basis of everyday reasoning, language, and science.

In this article we will focus on one outcome of these new developments: neural network semantics for conditionals. We will start with McCulloch's and Pitts' original interpretation of neural network components in terms of formulas of classical propositional logic, we will summarize the main features of connectionism which emerged as an alternative paradigm of cognitive science that was thought to be in opposition to logical takes on reasoning, and we will sketch how recent theories nevertheless attempt to describe states and processes in neural networks by means of logical terms. Finally, we will deal with one of these theories in more detail. Along the way we will also present very brief recaps of nonmonotonic reasoning and of the logical and philosophical literature on conditionals, as far as this serves the purpose of illuminating the neural networks models of conditionals that are the topic of this paper. We end with some tentative philosophical conclusions and with a list of interesting open questions. Clearly, this new field of research is relying heavily on the application of formal methods, mostly from logic and the mathematical theory of dynamical systems.[1]

7.2 Neural Networks as Models of Reasoning

In their famous article "A Logical Calculus of the Ideas Immanent in Nervous Activity", McCulloch and Pitts [45] first introduced artificial neural networks as mathematical abstractions from neural circuits in the brain. A McCulloch-Pitts network consists of a set of nodes and a set of connections between these nodes. Each node can be in one of two possible states: it fires (1), or it does not (0). Each connection is of one of two possible kinds: along inhibitory connections, nodes receive inhibitory signals by which they get deactivated at the next point of time (on a discrete time scale). Via excitatory connections, signals are transferred from

[1]This article is a revised and extended version of: Leitgeb [41]. Some material contained in Leitgeb [37, 38], Ortner and Leitgeb [46], and in the popular and non-technical exposition of logic and neural networks in Leitgeb [39] was used, too.

We are grateful for generous support received from the Alexander von Humboldt Foundation.

one node to another which have a stimulating effect on the target node: if the node does not get inhibited, and if the number of all incoming excitatory signals exceed or are identical to some fixed threshold value that is associated with the node, then the node fires at the next point of time. Although these appear to be quite simple devices, McCulloch and Pitts [45] effectively established that in principle every finite automaton can be realized by such a McCulloch-Pitts network (a formal result which was later made perfectly precise by the logician S. Kleene in his [31]). Furthermore, the state transitions which take place in such networks allow for a description in logical terms: if the activity of a node is considered as a truth value, then the node itself may be regarded as an entity which *has* a truth value, i.e., as a formula or proposition. If the "truth value" of a node does not depend on the "truth values" of other nodes (but, say, only on some given input), then it is indeed natural to regard such nodes as *atomic* formulas or propositions. Accordingly, if nodes are put together in a network, such that connections between nodes can cause the "truth values" of other nodes to be altered, then the latter nodes may be taken to correspond to *complex* formulas; the semantic dependency of the truth value of a complex formula on the truth values of its component formulas is thus represented by the network topology and the choice of thresholds.

As an example, consider two very elementary McCulloch-Pitts networks: In the first network, excitatory connections lead from nodes p and q to a third node. If this latter node has a threshold value of 2, then the node is going to fire if and only if both p and q were active at the previous point of time. So we can associate the formula $p \wedge q$ with this node:

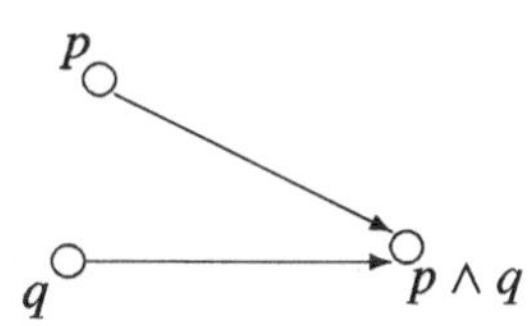

In the second network, two excitatory lines lead from p to the output node, whereas q is connected to the latter by an inhibitory edge. If e.g. the output node has a threshold of 2, it will be activated at the next point of time if and only if p is set to 1 and q is set to 0 (and therefore does not have any inhibitory influence). Hence, the third node in the network corresponds to the formula $p \wedge \neg q$:

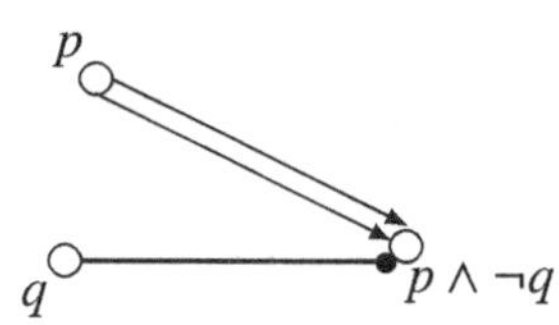

This way of associating nodes in networks with formulas in the language of classical propositional logic extends to more interesting networks with multiple layers of nodes and with more complex patterns of excitatory and inhibitory connections. E.g., it would be easy to extend the second network by a node that represents $\neg(p \wedge \neg q)$, i.e., a formula which is logically equivalent to the material conditional $p \supset q$. If our brains were, at least on some level, similar to neural networks of the McCulloch-Pitts kind, they could thus be understood as collections of simple logical units put together in order to calculate binary truth values from external or internal input. The calculation of the truth values of material conditionals would be a special case of this form of computational processing.

Of course, the McCulloch-Pitts networks are, in several respects, much too simple to be plausible models of actual neural networks in animal or human brains. In particular, they are not yet able to learn. The next decisive step in the development of artificial neural networks was to introduce variable weights that are attached to connections and which encode the degree of influence that nodes can exert on their target nodes via these connections. By sophisticated learning algorithms, these weights can be adjusted in order to map inputs to their intended outputs, e.g., facial images of persons to the names of these persons, or verbs to their correct past tenses. Despite some initial success in the 1950s and 1960s – mainly associated with F. Rosenblatt's *Perceptrons* which famously came under attack by M. Minsky's and S. Papert's monograph with the same title – it was only in the 1980s that artificial neural network models of cognition became serious contenders to the dominant symbolic computation paradigm in artificial intelligence. These new approaches to cognition are usually subsumed under the term 'connectionism'.[2] As we will explain below, the more recent neural network models do not only differ from the original McCulloch-Pitts networks in terms of complexity and learning abilities, they also differ in terms of the interpretation of their components: instead of assigning meaning – expressed by formulas – to *single* nodes, the modern approach emphasizes that it is rather *patterns* or *sets* of nodes which receive an interpretation.

How does 'cognition by neural networks' relate to the traditional 'cognition by symbolic computation' paradigm of cognitive science (exemplified by classical Artificial Intelligence)? According to the latter, (i) intelligent cognition demands structurally complex mental representations, such that (ii) cognitive processing is only sensitive to the form of these representations, (iii) cognitive processing conforms to rules, statable over the representations themselves and articulable in the format of a computer program, (iv) (standard) mental representations have syntactic structure with a compositional semantics, and (v) cognitive transitions conform to a computable cognitive-transition function (we adopt this characterization essentially from [30], with slight deviations). Intelligent cognition is supposed to be "systematic" and "productive" (see [21]), i.e., the representational capacities of intelligent agents are supposed to be necessarily closed under various representation-transforming and representation-generating operations (e.g., if an

[2]Rumelhart et al. [51] is still something like the "bible" of connectionism; Rojas [50] is a nice introduction to neural networks, and at ⟨http://plato.stanford.edu/entries/connectionism/⟩ the entry on connectionism in the *Stanford Encyclopedia of Philosophy* can be found – have a look at these for more background information.

agent is able to represent that aRb, it is also able to represent that bRa, etc.). This capacity is hypothesized to be due to the combinatorial properties of languages of mental symbols based on a recursive grammar. A cognitive agent that conforms to the symbolic computation paradigm has the belief that φ if and only if a corresponding sentence φ is stored in the agent's symbolic knowledge base. The rules that govern cognitive processes according to the symbolic computation paradigm are either represented within the cognitive agent as symbolic entities themselves, or they are hard-wired. Inference processes are taken to be internalizations of derivation steps within some logical system, and the alleged "systematicity" of inferences (see again [21]) is explained by the internal representation or hard-wiring of rules which are only sensitive to the syntactic form of sentential representations.

Cognition by artificial neural networks, on the other hand, belongs to the so-called dynamical systems paradigm of cognitive science which can be summarized by what van Gelder [59] calls the "dynamical hypothesis": "for every kind of cognitive performance exhibited by a natural cognitive agent, there is some quantitative [dynamical] system instantiated by the agent at the highest relevant level of causal organization [i.e., at the level of representations], so that performances of that kind are behaviors of that system" [59, p. 622]. A dynamical system may be regarded as a pair of a state space and a set of trajectories, such that each point of the space corresponds to a total cognitive state of the system, and every point of the space lies precisely on one trajectory. If a certain point corresponds to the system's total cognitive state at some time, the further evolution of the system follows the trajectory emanating at this point. Usually, such systems are either defined by differential equations, or by difference equations, defined over the points of the state space: in the first case one speaks of continuous dynamical systems with continuous time, while in the latter case one speaks of discrete dynamical systems with discrete time. In the discrete case, the set of trajectories may be replaced by a state-transition mapping, such that each trajectory is generated by the iterated application of the mapping. A *cognitive* dynamical system is a dynamical system with representations, i.e., where states and state transitions can be ascribed content or interpretation. The dynamic systems paradigm assumes that intelligent cognition takes place in the form of state-transitions in quantitative systems, i.e., systems in which a metric structure is associated with the points of the state space, and where the dynamics of the system is systematically related to the distances measured by the metric function. The distances between points may be regarded as a measure of their similarity *qua* total cognitive states. Moreover, the typical dynamical systems that are studied within the dynamical systems paradigm also have a vector space structure, and thus they "support a geometric perspective on system behaviour" [59, p. 619].

Connectionism is the most important movement within the dynamical systems paradigm: Artificial neural networks are the dynamical systems that the connectionists are interested in. Smolensky [54] characterizes connectionism by the following hypotheses: (i) "The connectionist dynamical system hypothesis: The state of the intuitive processor at any moment is precisely defined by a vector of numerical values (one for each unit). The dynamics of the intuitive processor are governed by a differential equation. The numerical parameters in this equation constitute the processor's program or knowledge. In learning systems, these parameters change according to another differential equation." (ii) "The subconceptual unit hypothesis: The entities in the intuitive processor with the semantics of conscious concepts of the task domain are complex patterns of activity over many units. Each unit participates in many such patterns." (iii) "The subconceptual level hypothesis: Complete, formal, and precise descriptions of the intuitive processor are generally tractable not at the conceptual level, but only at the subconceptual level." The subconceptual level is the level of analysis that is preferred by the connectionist paradigm, or, as Smolensky expresses it, by the *subsymbolic* paradigm; it lies "below" the conceptual level that is preferred by the symbolic computation paradigm, but "above" the neural level preferred by neuroscience.

Claim (i) proves connectionism to belong to the dynamical systems paradigm. The subconceptual unit hypothesis (ii) and the subconceptual level hypothesis (iii) highlight the main differences between the old McCulloch & Pitts approach presented above and modern day connectionism: by (ii), single nodes or single connections in a neural network are normally not supposed to carry any meaning at all; the representing units are distributed patterns of activation that involve a great number of nodes or even the network topology as a whole (see van Gelder [60] on "Distributed versus local representation"). In more metaphorical terms: there is not generally anything like a "grandmother cell", i.e., a single neuron that would correspond to a very complex formula which describes your grandmother and which would fire if and only if your grandmother were perceived. Rather, your grandmother's being perceived is represented by some complex pattern of activation which spreads throughout parts of the network at the time of perception. Furthermore, by (iii), if symbols can be attached to the activation patterns of nodes or to some other "global" aspects of neural networks at all, the transitions from one representing item – one pattern – to another will no longer be effected on the level of these representing items themselves but rather on the sub-symbolic level of nodes and edges. Therefore, for connectionists in the sense described, it seems impossible to translate the computations on the sub-symbolic level into sequences of rules on the symbolic level, let alone into logical rules which apply to complex symbolic expressions. Thus, McCulloch and Pitts' original *logical* approach to neural networks became something like the paradigmatic antagonist of the movement, and hence it had to be given up, or so it seemed. Instead of analyzing cognition in terms of localized representations of formulas – "hard constraints" – Smolensky [54, p. 18], suggests that connectionist cognition proceeds by means of "soft constraints": "Formalizing knowledge in soft constraints rather than hard rules has important consequences. Hard constraints have consequences singly; they are rules that can be applied separately and sequentially – the operation of each proceeding independently of whatever other rules may exist. But soft constraints have no implications singly; any one can be overridden by the others. It is only the entire set of soft constraints that has any implications. Inference must be a cooperative process [. . .] Furthermore, adding additional soft constraints can repeal conclusions that were formerly valid: Subsymbolic inference is fundamentally nonmonotonic." If human reasoning is as connectionists describe it, then McCulloch and Pitts' account of reasoning in terms of neural network implementations of truth functions in classical logic can hardly be adequate.

Even if this very last statement about McCulloch and Pitts' theory is true, this does not yet entail that the symbolic computation paradigm and the dynamical systems paradigm themselves have to be completely mutually exclusive, i.e.: significant aspects of the two paradigms could actually turn out to be compatible with each other. As Gärdenfors [23, p. 67f], suggests, the two paradigms might even be complementing each other: "they are best viewed as two different perspectives that can be adopted when describing the activities of various computational devices." Results on symbol manipulation in networks (see e.g. [12, 13, 55]), neural networks approaches to grammar representation (see e.g. [33, 56]), and hybrid systems that involve both neural network and symbolic components (see e.g. [10, 49]) indicate that there might be continuous paths of transition from the one paradigm to the other. In particular, the analysis of neural networks in terms of *logical laws and rules* has become a topic of research again in recent years, and on it we are going to focus now.

Here are some relevant references on logical accounts of neural network cognition (they can also be found in the bibliography – note that this is a *very* incomplete list though!):

- A.S. d'Avila Garcez, K. Broda, and D.M. Gabbay [15].
- A.S. d'Avila Garcez, K.B. Broda, and D.M. Gabbay [16].
- A.S. d'Avila Garcez, L.C. Lamb, and D.M. Gabbay [17].
- A.S. d'Avila Garcez et al. [18].
- S. Bader and P. Hitzler [3].
- C. Balkenius and P. Gärdenfors [4].
- R. Blutner [9].
- E.-A. Dietz, S. Hölldobler, and L. Palacios [19].
- P. Hitzler, S. Hölldobler, and A.K. Seda [26].
- S. Hölldobler [27].
- S. Hölldobler [29].
- S. Hölldobler and Y. Kalinke [28].
- H. Leitgeb [35].
- H. Leitgeb [37].
- H. Leitgeb [38].
- R. Ortner and H. Leitgeb [46].
- K. Stenning and M. van Lambalgen [57]

The main idea behind all of these theories is that if classical logic is replaced by a different logical calculus – in particular, by a system of nonmonotonic reasoning that is closer to the commonsense reasoning that our brains are usually involved in – then a logical description or characterization of neural network states and processes might be possible in a way, such that: (i) "The connectionist dynamical system hypothesis" is satisfied, maybe even in combination with (ii) "The subconceptual unit hypothesis", yet (iii) "The subconceptual level hypothesis" turns out to be false (for the precise statements of these theses see above). In other words: Logical descriptions of reasoning might become tractable again at the conceptual level, even when reasoning is realized in terms of the dynamics of an artificial neural network.[3]

Here is a brief, and very sketchy, guide to the literature as cited above:

A.S. d'Avila Garcez et al. [18], Bader and Hitzler [3], and Hölldobler [29] are very useful survey papers. Many authors and papers in this area of research can be found by checking the websites of the "NeSy" events in the workshop series on Neural-Symbolic Learning and Reasoning, which has been an ongoing endeavour since 2005.

The Hölldobler et al. group in Dresden has done pioneering work on how to generate neural networks from *logic programs*.[4] (See also Stenning and van Lambalgen, Chapter 7.) A logic program consists of rules which may look like this:

$$CanFly(Tweety) \Leftarrow Bird(Tweety), \neg Penguin(Tweety)$$

This is to be read as: if one has the information that Tweety is bird *but one lacks the information that Tweety is a penguin*, then one may infer that Tweety can fly. '$\Leftarrow$' here is much like the (non-material) if-then symbol in the sequent calculus of classical logic which connects the two sides of a sequent. Rational inferences that are based on such rules are *nonmonotonic*: given additional information, such as that Tweety is in fact a penguin, the inference would not longer be supported. Note that negation here is what is called *default negation*: e.g., $\neg Penguin(Tweety)$ expresses the *absence* of the positive information $Penguin(Tweety)$. What Hölldobler et al. managed to show was that it is possible to transform such logic programs into artificial neural networks, so that: the atomic formulas used in a given logic program correspond to the input nodes and to the output nodes in a feed-forward network; the rules in the logic program correspond to the nodes in the hidden layers; positive and negative information in the bodies of rule clauses correspond to excitatory and inhibitory connections, respectively; and additional feedback connections from the output nodes to the input nodes enable the network to converge on a model for the rules of the logic program, such that the model corresponds to a stable network state.

The group around d'Avila Garcez et al. has built on, and added to, this work, amongst others (i) by suggesting extraction methods that reverse the process just described by generating logic programs from (trained) neural networks, and (ii) by extending the results to logic programs that involve modal operators or that are based on intuitionistic logic. What the theories of these two groups have in common, too, is that they lie on the *consistency-based fixed point operator side*

---

[3]See Brewka et al. [11] for a very nice overview of nonmonotonic reasoning, Makinson [44] for a comprehensive logical treatment of the subject, and Schurz and Leitgeb [53] for a compendium of articles on cognitive aspects of nonmonotonic reasoning. Ginsberg [24] is an outdated collection of articles but it is still very useful if one wants to see what nonmonotonic reasoning derives from.

[4]Brewka et al. [11] includes a very clear and accessible introduction to logic programming.

of nonmonotonic reasoning: explained in terms of the rule above, as long as it is *consistent* to assume that Tweety is not a penguin, one may infer that Tweety is bird; what one is ultimately supposed to believe given evidence is computed by generating a fixed point of an immediate-consequence operator that is determined by the logic program.[5]

However, there is also the more recent *preference-based nonmonotonic inference relation* side of nonmonotonic reasoning, which became prominent through the now classical articles by Kraus, Lehmann, and Magidor [32] and Lehmann and Magidor [34]. In these approaches, metalinguistic statements such as

$$Bird(Tweety) \mid\!\sim CanFly(Tweety)$$

are considered which are now interpreted as saying: *in the most preferred (most normal, most plausible) worlds* in which Tweety is a bird, Tweety is also able to fly. Here, '$\mid\!\sim$' is a binary metalinguistic predicate which is syntactically like the symbol '$\models$' for classical logical consequence. If one replaces such metalinguistic statements by conditionals in the object language, such as by

$$Bird(Tweety) \Rightarrow CanFly(Tweety),$$

and one makes the preference-based semantics for these conditionals precise, the resulting semantics ends up being very close indeed to standard semantics for conditional logic as developed by philosophical logicians since the 1960s and 1970s (about which more in the next section). This preference-based approach is characterized by having much nicer logical properties than its consistency-based fixed-point operator counterpart. For instance, in all of the preference-based calculi, the following two rules (now spelled out in terms of conditionals)

$$\frac{\varphi \Rightarrow \psi,\ \varphi \wedge \psi \Rightarrow \rho}{\varphi \Rightarrow \rho} \quad \text{(Cautious Cut)}$$

$$\frac{\varphi \Rightarrow \psi,\ \varphi \Rightarrow \rho}{\varphi \wedge \psi \Rightarrow \rho} \quad \text{(Cautious Monotonicity)}$$

are logically valid. The combination of these two rules is usually referred to by the term 'cumulativity' (see [32]). Cumulativity expresses that adding inferred formulas to the evidence neither increases nor decreases the inferential strength of the evidence. However, the rule

$$\frac{\varphi \Rightarrow \rho}{\varphi \wedge \psi \Rightarrow \rho} \quad \text{(Monotonicity)}$$

---

5 Almost all of the classical approaches to nonmonotonic reasoning from the 1980s, such as default logic, inheritance networks, truth maintenance systems, circumscription, and autoepistemic logic belong to this class of nonmonotonic reasoning mechanisms.

is not logically valid anymore, which is why one cannot simply infer from

$$Bird(Tweety) \Rightarrow CanFly(Tweety)$$

that also

$$Bird(Tweety) \wedge Penguin(Tweety) \Rightarrow CanFly(Tweety)$$

holds. In contrast with the former approach to nonmonotonic reasoning, exceptions do not have to be stated explicitly anymore in the relevant rules or conditionals.[6]

It is nonmonotonic inference in this latter preferential sense that Balkenius and Gärdenfors represented in terms of state transitions within artificial neural networks, and which they studied by means of concrete experiments in computer simulations. Leitgeb's work builds on Balkenius and Gärdenfors' approach but adds soundness and completeness proofs for systems of nonmonotonic reasoning or conditional logic based on a corresponding neural network semantics. Blutner also starts from Balkenius and Gärdenfors but represents nonmonotonic inferences in so-called weight-annotated Poole systems by means of state-transitions in Hopfield networks, relating the results so obtained to Harmony or Optimality Theory in the sense of Smolensky and Legendre [56]. One point of difference between Blutner's and Leitgeb's theories – and one of agreement between Blutner's theory and the theories by Hölldobler et al. and d'Avila Garcez et al. – is that while Blutner represents atomic formulas in neural networks in terms of *nodes*, Leitgeb represents atomic formulas as *distributed patterns of activity*. Accordingly, generally, connections between nodes cannot be assigned any local symbolic interpretation anymore in Leitgeb's account. Since distributed representation was supposed to be one of the hallmarks of connectionism – in correspondence with (ii) "The subconceptual unit hypothesis" from above – we will concentrate on Leitgeb's theory in Sect. 7.4, where we will present the theory as a neural networks semantics for conditionals.

If any of these logical accounts of neural network cognition were to prove successful in the long run (logically, philosophically, and in applications), the gap between the dynamic systems paradigm and the symbolic computation paradigm in cognitive science would be bridged, or, at the very least, diminished. This would also constitute an important step in understanding what neural networks actually do; otherwise, we might be stuck with an ingenious technical machinery that maps an input to its desired output, but where the process that leads from the one to the other remains uninterpreted, unexplained, and unjustified. While it is certainly true that current implementations of machine learning do not themselves rely on the application of logical methods (see Wheeler [61]), logic might still play a role in the *rational reconstruction* and *assessment* of machine learning: in checking whether

[6]For more on the differences between the two sides of nonmonotonic reasoning, see Brewka et al. [11].

the "black box" conforms to norms of rationality and, perhaps, morality. Progress on the logic of neural networks might also lead to new insights in uncertain reasoning, induction, and even the philosophy of science – we will return to this in the final section of this article, which will include a list of open questions.[7]

7.3 A Brief Recap on Conditionals

Before I turn to a concrete example of a neural network semantics for conditionals, let me say a bit more about the conditionals that will be involved. Conditionals are sentences of an 'if... then...' form; so, the logical form of a conditional is an expression of the form

$$\text{If } \varphi, \text{ then } \psi$$

or, more formalized,

$$\varphi \Rightarrow \psi$$

where φ is called the 'antecedent' of the conditional and ψ its 'consequent'; both the antecedent and the consequent of a conditional are sentences. (See also John Cantwell's chapter 6 on conditionals in this handbook.)

Conditionals are crucial in everyday communication, especially when we want to convey information that goes beyond the currently present perceptual situation. Conditionals also play a major role in philosophical theories about dispositions, causality, laws, time, conditional norms, probability, belief, belief revision, and so forth. Finally, conditionals are closely related closely to quantifiers, such as 'All φ are ψ', 'There are φ which are ψ', 'Most φ are ψ', etc.[8] But note that in these latter cases, 'φ' and 'ψ' are place holders for *open formulas* – formulas with a free variable – rather than sentences.

Amongst conditionals in natural language, usually the following distinction is made[9]:

1. If Oswald had not killed Kennedy, then someone else would have.
2. If Oswald did not kill Kennedy, then someone else did.

[7]We should add that there are also results concerning the description of neural network states and processes by means of *classical* logic, over and above the traditional McCulloch and Pitts approach: see Pinkas [48] and Bechtel [5] for examples.

[8]See van Benthem [58] for a nice discussion of this relationship between conditionals and quantifiers; more can be found by consulting the theory of *generalized quantifiers* – see e.g. Peters and Westerstahl [47].

[9]The following famous example is due to Ernest Adams.

2 is accepted by almost everyone, whilst we do not seem to know whether 1 is true. This invites the following classification: a conditional such as 2 is called *indicative*, whereas a conditional like 1 is called *subjunctive*. In conversation, the antecedents of subjunctive conditionals are often assumed or presupposed to be false: in such cases, one speaks of these subjunctive conditionals as *counterfactuals*. Roughly, indicative conditionals represent the denoted act or state as an objective fact, while subjunctive conditionals represent a denoted act or state not as fact but as contingent or possible. Subjunctive and indicative conditionals may have precisely the same antecedents and consequents (as in the example above) while differing only in their conditional connectives, i.e., their 'if'-'then' occurrences having different meanings.

When logic developed into a serious philosophical and mathematical discipline in the late nineteenth and the early twentieth century, logicians quickly came up with two suggestions of how to formalize conditionals, whether indicative or subjunctive:

- $\varphi \supset \psi$: Formalization by means of material conditionals (material implications).
- $\varphi \dashv\!\!\! \exists \psi$: Formalization by means of strict conditionals (strict implications).

From an axiomatic point of view, the meaning of the former is given by any of the typical deductive systems for classical propositional logic. The logical systems for the latter were investigated intensively by C.I. Lewis, however it was only after the axiomatic systems of normal modal logic had been developed by S. Kripke that the analysis of $\varphi \dashv\!\!\! \exists \psi$ in terms of $\Box(\varphi \supset \psi)$ emerged as a standard (where $\Box$ is the necessity operator studied by modal logicians). On the semantic side, the meaning of $\supset$ is given by its well-known truth table, whereas the semantics of $\dashv\!\!\! \exists$ can be stated on the basis of the usual Kripkean possible worlds semantics of $\Box$.

These formalizations of the 'if... then...' in classical logic proved to be enormously successful, especially in the formalization of mathematical theories and of fragments of empirical theories. However, there was still a problem: both $\supset$ and $\dashv\!\!\! \exists$ are *monotonic*, i.e., the rule $\frac{\varphi \Rightarrow \psi}{\varphi \wedge \rho \Rightarrow \psi}$ is logically valid if '$\Rightarrow$' is replaced by either of the two connectives. On the other hand, there seem to be many instances of indicative and subjunctive conditionals in natural language which are *nonmonotonic*, i.e., for which the rule $\frac{\varphi \Rightarrow \psi}{\varphi \wedge \rho \Rightarrow \psi}$ should not assumed to be valid. E.g., 'If it rains, I will give you an umbrella' does not seem to logically imply "If it rains and I am in prison, I will give you an umbrella", nor does 'If it rained, I would give you an umbrella' seem to logically imply "If it rained and I were in prison, I would give you an umbrella". Accordingly, add e.g. '…and Kennedy in fact survived all attacks on his life' to the antecedent of 'If Oswald did not kill Kennedy, then someone else did' and the resulting conditional does not seem acceptable anymore. Therefore, philosophical logicans started to investigate new logical systems in which monotonicity (or *strengthening of the antecedent*) would not turn out to be logically valid. Lewis [43] is the classic treatise on counterfactuals as nonmonotonic conditionals, in which subjunctive conditionals are evaluated based on similarity orderings of possible worlds (which are similar to the preference orderings of possible worlds used in nonmonotonic reasoning). Since the nonmonotonicity phenomenon had already been well known in probability theory – a conditional probability $P(Y|X)$ being high does not entail the conditional probability $P(Y|X \cap Z)$ being high – it is not surprising that some of the modern accounts of conditionals instead relied on a probabilistic semantics: indeed, Adams [1] famously developed a probabilistic theory of indicative conditionals that does not support monotonicity (see Adams [2] for a more general overview of probability logic).[10]

It is these axiomatic and semantic systems of conditionals which got rediscovered a bit later (note: philosophy was *first*!) by theoretical computer scientists who initiated the field of nonmonotonic reasoning. For assume you want to represent in a computer system what happens to your car when you turn the ignition key: well, you might say, the car starts, so 'if the ignition key is turned in my car, then the car starts' seems to describe the situation properly. But what if the gas tank is empty? You better improve your description by saying 'if the ignition key is turned in my car and the gas tank is not empty, then the car starts'. However, this could still be contradicted by a potato that is clogging the tail pipe, or by a failure of the battery, or by an extra-terrestrial blocking your engine, or... The possible exceptions to 'if the ignition key is turned in my car, then the car starts' are countless, heterogeneous, and unclear. Nevertheless, we seem to be able to communicate information with such simple conditionals, and, equally importantly, we are able to reason with them in a rational manner. In order to do so we make use of a little logical "artifice": we do not really understand 'if the ignition key is turned in my car, then the car starts' as expressing that it is not the case that the ignition key is turned and the car does not start – after all, what is negated here might indeed be the case in exceptional circumstances – but rather that *normally*, or with a *high probability*, given the ignition key is turned, the car starts. Instead of trying to enumerate the indefinite class of exceptions in the if-part of a material or strict conditional, we tacitly or explicitly qualify 'if the ignition key is turned in my car, then the car starts' as holding only in normal or likely circumstances, whatever these circumstances may look like. As a consequence, the logic of such normality claims again differs from the logic of material or strict conditionals: 'if Tweety is a bird, then [normally] Tweety is able to fly' is, presumably, true, but 'if Tweety is a penguin bird, then [normally] Tweety is able to fly' is not, and neither is 'if Tweety is a dead bird, then [normally] Tweety is able to fly' or 'if Tweety is a bird with his feet set in concrete, then [normally] Tweety is able to fly'. So computer scientists found themselves in need of describing reality in terms of nonmonotonic normality conditionals on the basis of which computers should be able to draw justified inferences about the everyday world while being unaffected by the omnipresence of exceptions. And this need eventually led to conclusions very similar to those drawn by philosophers who cared about the logic and semantics of conditionals in natural language.

In the next section we will suggest that so-called *interpreted dynamical systems* may be used to yield a semantics for nonmonotonic conditionals. The logical

[10] For a textbook-like overview of the philosophical literature on indicative and subjunctive conditionals, see Bennett [7].

systems which turn out to be sound and complete with respect to such semantics are standard systems of conditional logic which have been studied both in philosophical logic and nonmonotonic reasoning. Interpreted artificial neural networks will be shown to be the paradigm case examples of such interpreted dynamical systems. Although the conditionals that are satisfied by such interpreted artificial neural networks are represented distributedly by these networks, the logical rules they obey are precisely the rules of systems which had been developed in order to make computers cope with the real world by means of symbolic computation, and which had been investigated even before by philosophers who intended to give a proper logical analysis of indicative and subjunctive conditionals. Since the dynamics of state changes in interpreted neural networks can be described correctly and completely by sets of conditionals that are closed under the rules of such logical systems, neural networks may be understood as nonmonotonic reasoners who, when they evolve under an input towards a state of "minimal energy", draw conclusions that follow from premises in all minimally abnormal cases.

7.4 From Dynamical Systems to Conditionals: Interpreted Dynamical Systems

Following Gärdenfors' proposal mentioned above, we will study cognitive dynamical systems from two complementary perspectives. On the one hand, cognitive dynamical systems such as neural networks can be described in terms of differential or difference equations, i.e., as *dynamical systems*. On the other hand, they exemplify cognitive states and processes that can be ascribed propositional contents which may in turn be expressed by sentences or formulas; so they are also *cognitive agents* or *reasoners*.

Here is an example. For the sake of simplicity, let us forget about the weights again that are attached to the edges of a typical neural network, and let us also assume that the activation functions that are defined for the nodes in such a network are as straight-forward and simple as in the case of the McCulloch-Pitts networks. Then we might, e.g., end up with a simple qualitative neural network that looks like this[11]:

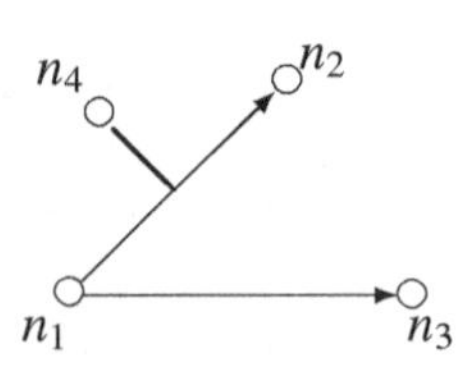

[11]Such networks are called 'inhibition networks' in Leitgeb [35].

This is a network with four nodes. n_1 is connected both to n_2 and n_3 by excitatory connections. In contrast with traditional McCulloch-Pitts networks, there is also an inhibitory connection that leads from n_4 to the *excitatory connection* from n_1 to n_2. So, if n_4 is active, this is not going to directly inhibit the activation of some other node at the next point of time, but instead any activity by n_4 will have the effect that no excitatory stimulus will be able to pass the edge from n_1 to n_2 at the next point of time.

Now, say, node n_1 gets activated by some external stimulus, e.g., by some sensory signal coming from outside. We will assume that such inputs always remain constant for sufficiently long, hence, in the present example, one should think of n_1 as being activated from the outside until the computational process that we are interested in has delivered its final output. Formally, we can describe what is going on in the following way: the network is in an initial state s_0; e.g., the state in which no node fires. This state s_0 may be regarded as a mapping from the set of nodes into the set $\{0, 1\}$, such that each node is mapped to 0 or "inactivity". Furthermore, the network is fed an input s^* that makes n_0 fire but which activates no other node: it is useful to identify such an input with the network state that the input would generate just by means of external influences on the network. Thus, in our case, s^* will be the state in which the node n_0 is mapped to 1 and in which all other nodes are mapped to 0. The resulting dynamics of the network can be described by means of a state transition mapping F_{s^*} that is given relative to the (constant) incoming input – s^* – and which is applied to the initial state s_0 in order to determine the next state s_1 of the network. Since no node is active in s_0, the only nodes which will be active in s_1 will be those activated by the input itself, i.e., n_1. So we have:

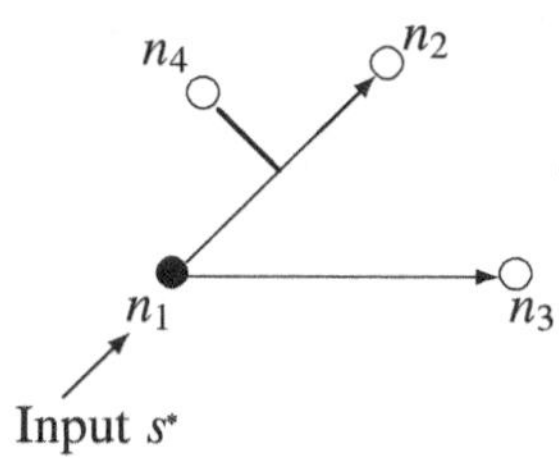

$$s_1 = F_{s^*}(s_0)$$

Accordingly, in order to determine the next state s_2 of the network, the state transition mapping F_{s^*} is applied again. The state transition will be such that the activity of n_1 in s_1 spreads to n_2 and n_3, which yields:

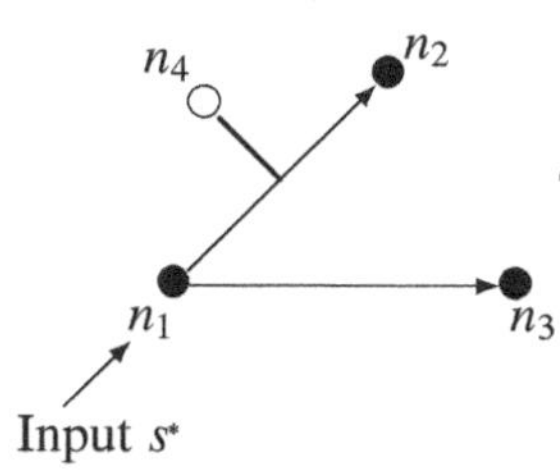

$$s_2 = F_{s^*}(s_1)$$

If the state transition mapping is applied again, then nothing is going to happen anymore (until the input to the network changes): hence, $s_3 = F_{s^*}(s_2) = s_2$. Connectionists regard such a *stable* or *equilibrium* state as a network's "answer" to the "question" posed by the input. So, s_2 – the state in which only n_1, n_2, n_3 fire – is the output that belongs to the input s^*. As we will also say, s_2 is an s^*-stable state.

What would happen if we applied a different input to the same initial state? Let s^{**} be the state in which both n_1 and n_4 fire, i.e., where the external input now causes these two nodes to become active. Then we have, by the same token as before:

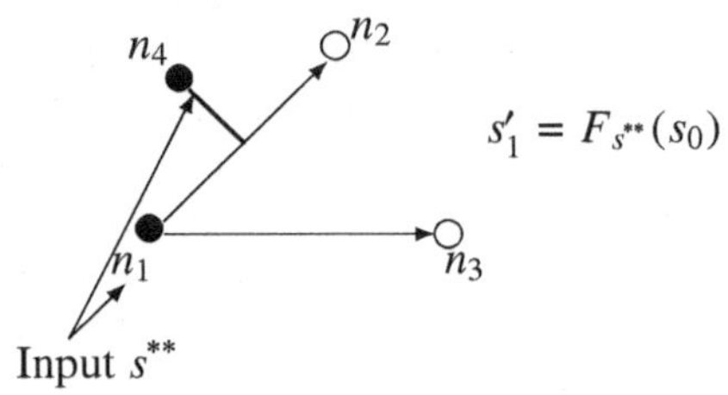

But now the state transition will be such that the activity of n_4 in s_1 blocks the excitation of n_2 by n_1. In other words:

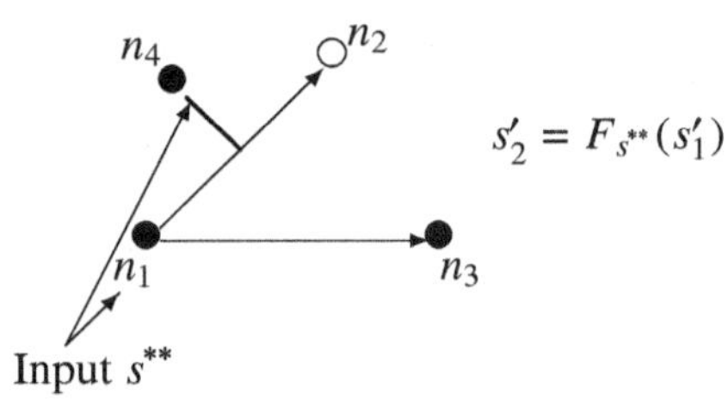

Once again, a stable state is reached after two computation cycles, and this time the output to the input state s^{**} is the state in which n_1, n_3, n_4 fire, i.e., s'_2 is an s^{**}-stable state.

What we have said so far constitutes a typical description of (simplified) network processes in the language of the theory of dynamical systems. Out goal is now to complement this description by one according to which cognitive dynamical systems have beliefs, draw inferences, and so forth. So if x is a neural network, we want to say things like

- x believes that $\neg\varphi$
- x infers that $\varphi \vee \psi$ from φ

  ⋮

where φ and ψ are *sentences*. Our task is thus to associate *states* of cognitive dynamical systems with *sentences* or *propositions*: the states of such dynamic systems ought to carry information that can be expressed linguistically.

Let us make this idea more precise. In order to do so, we first have to abstract from the overly simplified dynamical systems that were given by the qualitative neural networks sketched above. Indeed, we want to leave open at this point what our dynamical systems will be like – whether artificial neural networks or not – as long as they satisfy a few abstract requirements.

Here is what we will presuppose: we are dealing with a discrete dynamical systems with a set S of states. On S, a partial order[12] $\leqslant$ is defined, which we will interpret as an ordering of the amount of information that is carried by the states in question; so, $s \leqslant s'$ will be read as: s' carries at least as much information as s does. We will also assume that $\leqslant$ is "well-behaved" in so far as for every two states s and s' there is a uniquely determined state $sup(s, s')$ (i) that carries at least as much information as s, (ii) that carries at least as much information as s', and (iii) which is the state with the least amount of information among all those states for which (i) and (ii) hold. Formally, such a state $sup(s, s')$ is the *supremum* of s and s' with respect to the partial order $\leqslant$. Finally, an internal next-state function is defined for the dynamical system, such that this next-state function is like the state transition mapping described above except that – for the moment – we will disregard possible inputs to the system. Hence, in the examples above, an application of the corresponding next-state mapping would lead to the transmission of the activity of n_1 to n_3 once n_1 gets activated, but it will never lead to any activation of n_1 itself since n_1 can only be activated by external input.

Summing up, we determine what is called an 'ordered discrete dynamical system' by Leitgeb [38]:

Definition 1 $\mathcal{S} = \langle S, ns, \leqslant \rangle$ is an ordered discrete dynamical system if and only if

1. S is a non-empty set (the set of states).
2. $ns : S \to S$ (the internal next-state function).
3. $\leqslant \subseteq S \times S$ is a partial order (the information ordering) on S, such that for all s, $s' \in S$ there is a supremum $sup(s, s') \in S$ with respect to $\leqslant$.

In the example networks above, we had $S = \{s \mid s : N \to \{0, 1\}\}$ with $N = \{n_1, n_2, n_3, n_4\}$ being the set of nodes. In order to define a suitable information ordering $\leqslant$ on S, we can, e.g., use the following idea: the more nodes are activated in a state, the more information the state carries. Thus we would have, e.g.:

[12] A partial order $\leqslant$ (on S) is a reflexive, antisymmetric, and transitive binary relation, i.e.: for all $s \in S$: $s \leqslant s$; for all $s, s' \in S$: if $s \leqslant s'$ and $s' \leqslant s$ then $s = s'$; for all $s_1, s_2, s_3 \in S$: if $s_1 \leqslant s_2$ and $s_2 \leqslant s_3$ then $s_1 \leqslant s_3$.

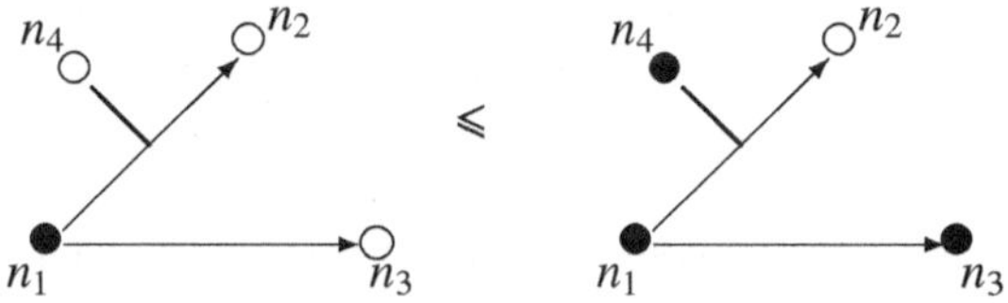

If $\leqslant$ is defined in this way, then $sup(s, s')$ turns out to be the union of the activation patterns that correspond to s and s'; in such a case one may also speak of $sup(s, s')$ as the "superposition of the states s and s'". The internal dynamics of the network is captured by the next-state mapping ns that is determined by the pattern of excitatory and inhibitory edges in the network.

Just as in the examples above, we are now ready to also consider an input, which is regarded to be represented by a state $s^* \in S$, and which is supposed to be held fixed for a sufficiently long duration of time. The state transition mapping F_{s^*} can then be defined by taking both the internal next-state mapping and the input s^* into account: the next state of the system is given by the superposition of s^* with the next internal state $ns(s)$, i.e.,

$$F_{s^*}(s) := sup(s^*, ns(s)).$$

The dynamics of our dynamical systems is thus determined by applying F_{s^*} iteratively to the initial state. Fixed points s_{stab} of F_{s^*}, i.e., where $F_{s^*}(s_{stab}) = s_{stab}$, are again regarded to be the "answers" the system gives to s^*; any such state s_{stab} is called s^*-stable (relative to the given ordered discrete dynamical system). Note that in general there may be *more than just one stable state* for the state transition mapping F_{s^*} that is determined by the input s^* (and by the given dynamical system), and there may also be *no stable state* at all for F_{s^*}: in the former case, there is more than just one "answer" to the input, in the latter case there is no "answer" at all. The different stable states may be reached by starting the computation in different initial states of the system.

Finally, we are ready to assign formulas to the states of ordered discrete dynamical system. These formulas are supposed to express the content of the information that is represented by these states. For this purpose, we fix a propositional language $\mathcal{L}$ which (i) includes finitely many propositional variables $p, q, r, \ldots$, and (ii) is closed under the application of the standard classical propositional connectives, i.e., $\neg, \wedge, \vee, \supset, \top, \bot$, where $\top$ is the *logical verum* (a tautology) and $\bot$ is the *logical falsum* (a contradiction). The formulas of $\mathcal{L}$ do not yet include any of the nonmonotonic conditional signs such as $\Rightarrow$ that we are interested in. The assignment of formulas to states is achieved by an interpretation mapping $\Im$. If φ is a formula in $\mathcal{L}$, then $\Im(\varphi)$ is the state that carries exactly the information that is expressed by φ, i.e., not less or more than what is expressed by φ. So we presuppose that for every formula in $\mathcal{L}$ there is a uniquely determined state the total information of which is expressed by that formula. If expressed in terms of belief, we can say that in the state $\Im(\varphi)$ *all the system believes is that* φ, i.e., the system only believes φ and all the propositions which are contained in φ from the viewpoint of the system (compare [42] on the modal logic of the 'all I know' operator). We will not demand that every state necessarily receives an interpretation but just that every formula in $\mathcal{L}$ will be the interpretation of some state. Furthermore, not just any assignment of states to formulas will do, but we will additionally assume certain postulates to be satisfied which will guarantee that $\Im$ is compatible with the information ordering that was imposed on the states of the system beforehand. An ordered discrete dynamical system together with such an interpretation mapping is called an 'interpreted ordered system' (cf. [38]). This is the definition stated in detail:

Definition 2 $\mathcal{S}_\Im = \langle S, ns, \leqslant, \Im \rangle$ is an interpreted ordered system if and only if

1. $\langle S, ns, \leqslant \rangle$ is an ordered discrete dynamical system.
2. $\Im : \mathcal{L} \to S$ (the interpretation mapping) is such that the following postulates are satisfied:
   (a) Let $\mathcal{TH}_\Im = \{\varphi \in \mathcal{L} \,|\, \text{for all } \psi \in \mathcal{L}\text{: } \Im(\varphi) \leqslant \Im(\psi)\}$:
   then it is assumed that for all $\varphi, \psi \in \mathcal{L}$: if $\mathcal{TH}_\Im \models \varphi \supset \psi$, then $\Im(\psi) \leqslant \Im(\varphi)$.
   (b) For all $\varphi, \psi \in \mathcal{L}$: $\Im(\varphi \wedge \psi) = sup(\Im(\varphi), \Im(\psi))$.
   (c) For every $\varphi \in \mathcal{L}$: there is an $\Im(\varphi)$-stable state.
   (d) There is an $\Im(\top)$-stable state s_{stab}, such that $\Im(\bot) \nleqslant s_{stab}$.

$\mathcal{S}_\Im$ satisfies the uniqueness condition if and only if for every $\varphi \in \mathcal{L}$ there is precisely one $\Im(\varphi)$-stable state.

How can these postulates be justified? First of all, $\mathcal{TH}_\Im$ is the set of formulas that are the interpretations of states which carry less information than, or an equal amount of information as, *any* other state with an interpretation. Hence, if $\varphi \in \mathcal{TH}_\Im$, then the information expressed by φ is contained in every interpreted state of the system. If spelled out in terms of belief, we may say: φ is believed by the system in every state that has an interpretation. For the same reason, such a belief cannot be revised by the system – it is "built" into the interpreted ordered system independently of its current input or state, as long as the state it is in has an interpretation at all. In more traditional philosophical terms, we might say that every such formula is believed a priori by the system. So if a material conditional $\varphi \supset \psi$ follows logically from $\mathcal{TH}_\Im$, then – since (rational) belief is closed under logical deduction – $\varphi \supset \psi$ must also be (rationally) believed by the system in every interpreted state whatsoever; indeed we may think of such a conditional as a strict a priori conditional: it is a material conditional which is epistemically necessary in the sense of being entailed by $\mathcal{TH}_\Im$, hence, if $\Box$ expresses entailment by $\mathcal{TH}_\Im$, then for every conditional $\varphi \supset \psi$ that is derivable from $\mathcal{TH}_\Im$ it holds that $\Box(\varphi \supset \psi)$. But if this is so, then the system must regard the propositional information that is expressed by ψ to be included in the propositional information that is expressed by φ – from the viewpoint of the system, φ must express a stronger proposition than ψ. In this case, with respect to the information ordering of the system, the state that belongs to ψ should be "below" the state that is associated with φ, or at worst the two states should be equal in the information ordering. In other words, $\Im(\psi) \leqslant \Im(\varphi)$

ought to be the case. That is exactly what is expressed by postulate 2a. $\mathcal{TH}_{\Im}$ may be interpreted as the set of "hard laws" or "strict laws" represented by the interpreted system.

Postulate 2b is more easily explained and justified: the state that belongs to a conjunctive formula $\varphi \wedge \psi$ should be the supremum of the two states that are associated with the two conjuncts φ and ψ, just as the proposition expressed by a conjunctive sentence is the supremum of the propositions expressed by its two conjuncts in the partial order of logical entailment.

Postulate 2c makes sure that we are dealing with systems that have at least one "answer" – whether right or wrong – to every "question" posed to the system.

Postulate 2d only allows for interpreted ordered systems which do not end up believing a contradiction when they receive a trivial or empty information (i.e., $\top$) as an input.

Finally, we are in the position to define what it means for a *nonmonotonic* conditional to be satisfied by an interpreted ordered system. Consider an arbitrary conditional $\varphi \Rightarrow \psi$ where φ and ψ are members of our language $\mathcal{L}$ from above, and where $\Rightarrow$ is a new nonmonotonic conditional sign. Then we say that a system satisfies $\varphi \Rightarrow \psi$ if, and only if, whenever the state that is associated with φ is fed into the system as an input, i.e., whenever the input represents a total belief in φ, the system will eventually end up believing ψ in its "answer states", i.e., the state that is associated with ψ is contained in all the states that are stable with respect to this input. If we collect all such conditionals $\varphi \Rightarrow \psi$ satisfied by the system, then we get what we call the 'conditional theory' corresponding to the system. In formal terms:

Definition 3 Let $\mathcal{S}_{\Im} = \langle S, ns, \leqslant, \Im \rangle$ be an interpreted ordered system:

1. $\mathcal{S}_{\Im} \models \varphi \Rightarrow \psi$ if and only if for every $\Im(\varphi)$-stable state s_{stab}: $\Im(\psi) \leqslant s_{stab}$.
2. $\mathcal{TH}_{\Rightarrow}(\mathcal{S}_{\Im}) = \{\varphi \Rightarrow \psi \,|\, \mathcal{S}_{\Im} \models \varphi \Rightarrow \psi\}$
   (the conditional theory corresponding to $\mathcal{S}_{\Im}$).

$\mathcal{TH}_{\Rightarrow}(\mathcal{S}_{\Im})$ may be interpreted as the set of "soft laws" or "normality laws" represented by the interpreted system. Leitgeb [40] gives an interpretation of the cognitive states that correspond to such conditionals in terms of so-called *conditional beliefs*, where conditional beliefs are to be distinguished conceptually from beliefs *in* conditionals.

Here is an example: consider again the simple qualitative network which we presented as a discrete ordered dynamical system above. In order to turn it into an *interpreted* ordered system, we have to equip it with an interpretation mapping $\Im$ that is defined on a propositional language $\mathcal{L}$. Let, e.g., $\mathcal{L}$ be determined by the set $\{b, f, w, p\}$ of propositional variables (for 'Tweety is a bird', 'Tweety is able to fly', 'Tweety has wings', 'Tweety is a penguin'). We choose the following interpretation mapping: let $\Im(b) = \{n_1\}$, $\Im(f) = \{n_1, n_2\}$, $\Im(w) = \{n_1, n_3\}$, $\Im(p) = \{n_1, n_4\}$, and $\Im(\neg\varphi) = 1 - \Im(\varphi)$, where the latter is to be understood in the way that whenever a node is active in $\Im(\varphi)$ then the same node is inactive in $\Im(\neg\varphi)$ and vice versa.[13]

[13] So 1 here is actually the constant 1-function, i.e., the function that maps each node to the activation value 1.

One can show that there is one and only one interpretation that has these properties and which also satisfies the postulates in Definition 2. Note that we have assumed $\Im(\neg\varphi) = 1 - \Im(\varphi)$ just for convenience, as it becomes easier then to pin down an interpretation for our example. It is not *implied* at all by Definition 2 that the pattern of active nodes that is associated with a negation formula $\neg\varphi$ is actually identical to the complement of the pattern of active nodes that belongs to the formula φ; this is merely the way in which we have set up our example. One consequence of this choice of $\Im$ is that, e.g., the following material conditionals turn out to be members of $\mathcal{TH}_{\Im}$: $p \supset b$, $(p \wedge w) \supset b$, $\neg b \supset \neg p$, and so forth.

Reconsidering our example from above, the dynamics of the system which we studied back then now turns out to have the following symbolic counterparts:

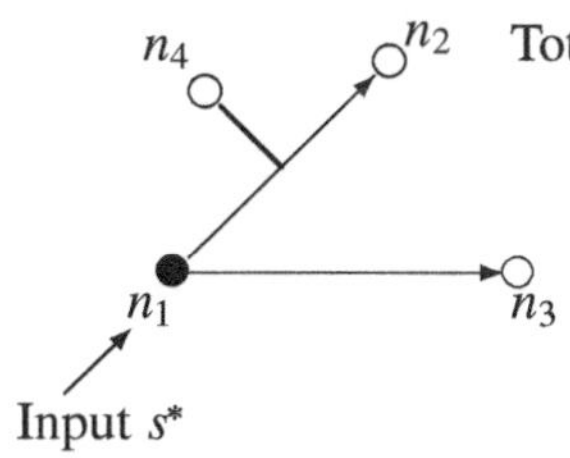

Total belief: b

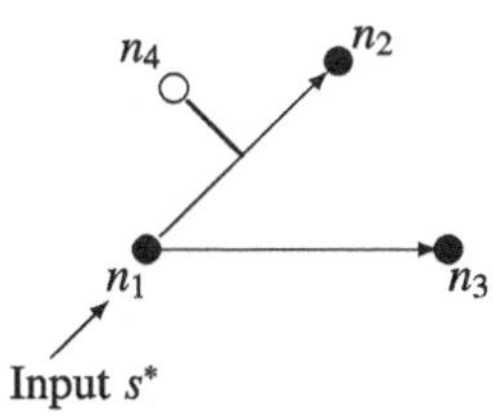

Believed: $b, f, w, (f \wedge w), (b \vee p), \ldots$
Not believed: $p, \neg b, \ldots$

Hence, $\mathcal{S}_{\Im} \models$
$b \Rightarrow f, b \Rightarrow w, b \Rightarrow (f \wedge w), (b \vee p) \Rightarrow f, \ldots$

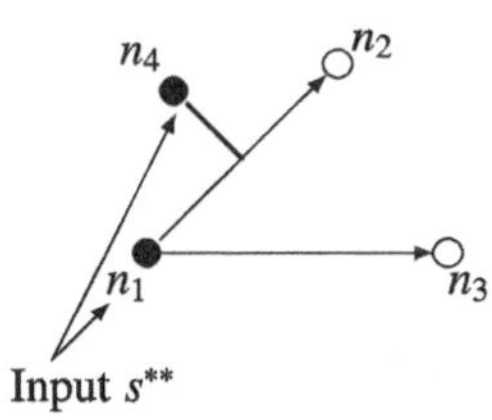

Total belief: $b \wedge p$

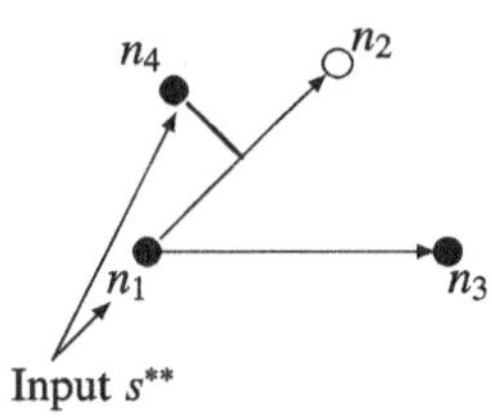

Believed: $b, p, w, (b \wedge p \wedge w), \neg f, \ldots$
Not believed: $f, \neg b, (f \vee \neg b), \ldots$

Hence, $\mathcal{S}_{\Im} \models$
$(b \wedge p) \Rightarrow \neg f, p \Rightarrow b, p \Rightarrow \neg f, \ldots$

Obviously, there will be lots of "soft" if-then "laws" about birds and penguins which this interpreted ordered system will get wrong. After all, it would be very surprising indeed if a little network with just four nodes were able to represent all of the systematic relationships between birds and penguins and flying and having wings faithfully. But the example should suffice to give a clear picture of how the definitions above are to be applied.

So we find that in this case $\mathcal{TH}_{\Rightarrow}(\mathcal{S}_{\mathfrak{I}})$ contains, e.g., $b \Rightarrow f$, $b \Rightarrow w$, $b \Rightarrow (f \wedge w)$, $(b \vee p) \Rightarrow f$, $(b \wedge p) \Rightarrow \neg f$, $p \Rightarrow b$, $p \Rightarrow \neg f$ without containing, e.g., $b \Rightarrow p$, $(b \vee p) \Rightarrow p$, $(b \wedge p) \Rightarrow f$. In particular, we see that $b \Rightarrow f \in \mathcal{TH}_{\Rightarrow}(\mathcal{S}_{\mathfrak{I}})$ while $(b \wedge p) \Rightarrow f \notin \mathcal{TH}_{\Rightarrow}(\mathcal{S}_{\mathfrak{I}})$.

What can be said in general terms about the conditional theories $\mathcal{TH}_{\Rightarrow}$ corresponding to interpreted dynamical systems? Here is the answer from the logical point of view:

Theorem 4 (Soundness of C)

Let $\mathcal{S}_{\mathfrak{I}} = \langle S, ns, \leqslant, \mathfrak{I} \rangle$ be an interpreted ordered system:

Then $\mathcal{TH}_{\Rightarrow}(\mathcal{S}_{\mathfrak{I}})$ is sound with respect to the rules of the system C of nonmonotonic conditional logic (see [32] *for details on this system), i.e.:*

1. *For all $\varphi \in \mathcal{L}$: $\varphi \Rightarrow \varphi \in \mathcal{TH}_{\Rightarrow}(\mathcal{S}_{\mathfrak{I}})$ (Reflexivity)*
2. *$\mathcal{TH}_{\Rightarrow}(\mathcal{S}_{\mathfrak{I}})$ is closed under the following rules: for φ, ψ, $\rho \in \mathcal{L}$,*
   $\frac{\mathcal{TH}_{\mathfrak{I}} \models \varphi \leftrightarrow \psi, \varphi \Rightarrow \rho}{\psi \Rightarrow \rho}$ *(Left Equivalence)*
   $\frac{\varphi \Rightarrow \psi, \mathcal{TH}_{\mathfrak{I}} \models \psi \supset \rho}{\varphi \supset \rho}$ *(Right Weakening)*
   $\frac{\varphi \Rightarrow \psi, \varphi \wedge \psi \Rightarrow \rho}{\varphi \Rightarrow \rho}$ *(Cautious Cut)*
3. *If $\mathcal{S}_{\mathfrak{I}}$ satisfies the uniqueness condition (remember Definition* 2*), then $\mathcal{TH}_{\Rightarrow}(\mathcal{S}_{\mathfrak{I}})$ is also closed under*
   $\frac{\varphi \Rightarrow \psi, \varphi \Rightarrow \rho}{\varphi \wedge \psi \Rightarrow \rho}$ *(Cautious Monotonicity)*
4. *$\mathcal{TH}_{\Rightarrow}(\mathcal{S}_{\mathfrak{I}})$ is consistent, i.e., $\top \Rightarrow \bot \notin \mathcal{TH}_{\Rightarrow}(\mathcal{S}_{\mathfrak{I}})$.*

So given the uniqueness assumption – an interpreted orderered system has a unique answer to each interpreted input – the class of conditionals it satisfies is closed under a well-known and important system of nonmonotonic conditional logic, namely the system C of *cumulative reasoning*, which is given by the rules listed above. Note that monotonicity, or strengthening of the antecedent, is *not* a valid rule for interpreted systems: as our example from above has shown, there may be formulas φ, ψ, ρ in $\mathcal{L}$, such that the conditional $\varphi \Rightarrow \psi$ is satisfied by a system but $\varphi \wedge \rho \Rightarrow \psi$ is not.

One can also show a corresponding completeness theorem for the system C with respect to this interpreted ordered systems semantics for $\Rightarrow$:

Theorem 5 (Completeness of C)

Let $\mathcal{TH}_{\Rightarrow}$ be a consistent theory of conditionals closed under the rules of C while extending a given classical theory $\mathcal{TH}$ as expressed by the Left Equivalence and the Right Weakening rules:

It follows that there is an interpreted ordered system $\mathcal{S}_{\Im} = \langle S, ns, \leqslant, \Im \rangle$, such that $\mathcal{TH}_{\Rightarrow}(\mathcal{S}_{\Im}) = \mathcal{TH}_{\Rightarrow}$, $\mathcal{TH}_{\Im} \supseteq \mathcal{TH}$, and $\mathcal{S}_{\Im}$ satisfies the uniqueness condition.

This means that whatever conditional theory you might be interested in, as long as it is closed under the rules of the system C, it is possible to find an interpreted ordered system which satisfies precisely the conditionals contained in that theory. These results can be found in Leitgeb [37].

It is also possible to extend these results into various directions. In particular, some interpreted ordered systems can be shown to have the property that each of their states s may be decomposed into a set of substates s_i which can be ordered in a way such that the dynamics for each substate s_i is determined by the dynamics for the substates $s_1, s_2, \ldots, s_{i-1}$ at the previous point of time. Such systems are called 'hierarchical' in Leitgeb [38]. We will not go into any details, but one can prove further soundness and completeness theorems for such *hierarchical* interpreted systems and the system CL = C + Loop of nonmonotonic conditional logic, where Loop is the following rule:

$$\frac{\varphi_0 \Rightarrow \varphi_1, \varphi_1 \Rightarrow \varphi_2, \ldots, \varphi_{j-1} \Rightarrow \varphi_j, \varphi_j \Rightarrow \varphi_0}{\varphi_0 \Rightarrow \varphi_j} \text{(Loop)}$$

Note that Loop is a weakened version of transitivity, whereas standard transitivity is *not* valid, just as the rule of cautious monotonicity above is a weakened version of monotonicity without standard monotonicity being valid anymore. (Consult [32] for more information on CL.)

In Leitgeb [36, 37], further soundness and completeness theorems can be found for more restricted classes of interpreted dynamical systems and even stronger logical systems for nonmonotonic conditionals. E.g., the important system P of so-called *preferential reasoning*, where P results from adding the rule

$$\frac{\varphi \Rightarrow \rho, \psi \Rightarrow \rho}{(\varphi \vee \psi) \Rightarrow \rho} \text{(Or)}$$

to the system CL, is sound and complete with respect to another particular class of interpreted dynamical systems. P coincides with Adams' [1] logical system for indicative conditionals as well as with the "flat" fragment of Lewis' [43] logic for subjunctive conditionals. ('Flat' means: iterations of subjunctive conditionals and other compositional constructions on their basis are excluded.) Moreover, various semantical systems for nonmonotonic reasoning have been found to "converge" on system P as their logical calculus.

As one can show, if artificial neural networks with weights are extended by an information ordering as well as an interpretation mapping along the lines explained above, then they turn out to be special cases of interpreted ordered systems. Furthermore, if the underlying artificial neural network consists of layers of nodes, such that the layers are arranged hierarchically, and all connections between nodes reach from one layer to the next one, then the interpreted ordered system is indeed a hierarchical one.

In more formal detail: $\langle U, W, A, O, NET, ex\rangle$ is an artificial neural network if and only if

1. U is a finite and nonempty set of nodes.
2. $W : U \times U \to \mathbb{R}$ assigns a weight to each edge between nodes.
3. A maps each node $u \in U$ to an activation mapping $A_u : \mathbb{R}^3 \to \mathbb{R}$ such that the activation state $a_u(t+1)$ of u at time $t+1$ depends on the previous activation state $a_u(t)$ of u, the current net input $net_u(t+1)$ of u, and the external input $ex(u)$ fed into u, i.e. $a_u(t+1) = A_u(a_u(t), net_u(t+1), ex(u))$.
4. O maps each node $u \in U$ to an output mapping $O_u : \mathbb{R} \to \mathbb{R}$ such that the output state $o_u(t+1)$ of u at time $t+1$ is solely dependent on the activation state $a_u(t+1)$ of u, i.e. $o_u(t+1) = O_u(a_u(t+1))$.
5. NET maps every node $u \in U$ to a net input (or propagation) mapping $NET_u : (\mathbb{R} \times \mathbb{R})^U \to \mathbb{R}$ such that the net input $net_u(t+1)$ of u at time $t+1$ depends on the weights of the edges leading from nodes u' to u, and on the previous output states of the nodes u', i.e. $net_u(t+1) = NET_u(\lambda u'.\langle W(u',u), o_{u'}(t)\rangle)$.[14]
6. $ex : U \to \mathbb{R}$ is the external input function.

We can view such networks as ordered dynamical systems when we define:

1. $S = \{s \mid s : U \to \mathbb{R}\}$.
2. $ns : S \to S$ with $ns(s)(u) := A_u(s(u), NET_u(\lambda u'.\langle W(u',u), O_{u'}(s(u'))\rangle), 0)$
   (so, in the definition of the internal next-state function, $ex(u)$ is set to 0).
3. $\leqslant \subseteq S \times S$ with $s \leqslant s'$ if and only if for all $u \in U$: $s(u) \leqslant s'(u)$.
   (Thus, $sup(s, s')$ is simply $max(s, s')$.)

$\langle S, ns, \leqslant\rangle$ is an ordered discrete dynamical system, such that $F_{s^*}(s) = sup(s^*, ns(s)) = max(s^*, ns(s))$ which entails that $F_{s^*}(s)(u) = max(s^*(u), ns(s)(u)) = max(s^*(u), A_u(s(u), NET_u(\lambda u'.\langle W(u',u), O_{u'}(s(u'))\rangle), 0))$, which corresponds to the assumption that the external input to a network interacts with the current activation state of the network by taking the maximum of both. Given this assumption, the dynamics of artificial neural networks and the dynamics of the corresponding ordered dynamical systems coincide. If the network is layered, then the corresponding ordered system is hierarchical. Stable states are regarded as the relevant "answer" states just as it is the case in the standard treatment of neural networks. If such networks are equipped with a corresponding interpretation

[14] $\lambda u'.\langle W(u',u), o_{u'}(t)\rangle$ is the function that maps u' to the pair $\langle W(u',u), o_{u'}(t)\rangle$.

mapping $\Im$ as defined above, they satisfy conditional theories which are closed under the rules of well-established systems of logic for nonmonotonic conditionals.

Furthermore, on the level of representation or interpretation we have:

- In interpreted ordered systems, *propositional formulas* are represented as total states s of the system; in particular, in interpreted neural networks, propositional formulas are represented as patterns of activity distributed over the nodes of the network.
- In interpreted ordered systems, *nonmonotonic conditionals* are represented through the overall dynamics of the system; in particular, in interpreted neural networks, nonmonotonic conditionals are represented by means of the network topology and the manner in which weights are distributed over the connections of the network. It is not single edges that correspond to conditionals, but the conditional theory that belongs to an interpreted network is a set of soft constraints that is represented by the network as a whole.

Thus, in contrast with the old McCulloch-Pitts idea, the representation of formulas in interpreted dynamical systems is distributed, as suggested by connectionists. At the same time, the set of conditionals satisfied by an interpreted dynamical system is closed under the rules of systems of nonmonotonic conditional logic that were introduced, and which have been studied intensively, by researchers in the tradition of the symbolic computation paradigm of cognitive science. Subsymbolic inference may be fundamentally nonmonotonic, as claimed by Smolensky (reconsider Sect. 7.2), but that does not mean that it could not be formalized in logical terms – it only means that the formalization has to be given in terms of systems of nonmonotonic reasoning or conditional logic.

The dynamical systems paradigm and the symbolic computation paradigm may thus be taken to yield complementary perspectives on the one and the same cognitive system. The precise meaning of this 'complementarity' is given by soundness and completeness theorems. Although these results only apply to highly idealized imitations of actual structures in the brain, the possibility of having such correspondences between symbolic and dynamic descriptions at all should be of great interest to philosophers of mind. Moreover, since nonmonotonic conditionals have been shown to have interpretations in terms of preference or similarity orderings of possible worlds or in terms of conditional probability measures (recall Sect. 7.3), the nonmonotonic conditionals that are satisfied by interpreted dynamical systems may be taken to represent aspects of some of these semantic structures. In this way, neural networks – artificial ones, but maybe even biological ones – may be understood as representing orderings of possible worlds or conditional probability measures, accordingly. This might pave the way for new interpretations and explanations of cognition done by neural networks, which should be relevant to cognitive scientists. Finally, as follows from the results above, conditionals in natural language, normality conditionals used by computer scientists, and the conditionals by which one may describe the dynamics of neural networks all seem to converge on more or less the same logic. This constitutes tentative evidence for two conclusions: first, the correspondence with normality conditionals in computer science indicates why conditionals in natural language might have the logical properties that they have – because we, much as the computer systems in artificial intelligence, need to be able to cope with exceptions. Secondly, the neural network semantics above suggests how we, natural language speakers, are capable of determining whether or not a conditional is acceptable to us – by feeding the information that is conveyed by its antecedent into a neural network that is run offline and the stable states of which are checked for whether they contain the information conveyed by the consequent of the conditional. Both of these tentative conclusions should be of obvious relevance to philosophers of language who are interested in conditionals.

7.5 Some Open Questions

Here is an (incomplete) to-do-list in this area of research:

Extending soundness/completeness results: How can all of the logical systems discussed by Kraus et al. [32], Lehmann and Magidor [34], and beyond be characterized in terms of connectionistically plausible and elegant constraints on interpreted dynamical systems? So far, there only seem to be partial answers to this question, sometimes relying on very restricted classes of dynamic systems. Which logical systems do we get if we drop the uniqueness assumption (see Definition 2)? How can full-fledged systems of conditional logic for subjunctive conditionals, for which nestings of conditionals and the application of propositional connectives to conditionals are well-defined, be represented by means of interpreted dynamical systems? The results achieved up to this point seem more suitable for indicative conditionals for which the meaningfulness of nesting and the application of propositional connectives are less plausible.

Characterizing learning in neural networks by logical rules: As we have seen, state transitions in a fixed (possibly, trained) neural network can be described by means of conditionals. However, it is as yet unknown how learning processes in networks – by which the weights in a network change under the influence of a learning algorithm and training data – can be represented by logical rules. Learning schemes such as Hebbian learning or backpropagation (by which the weights of connections between co-active nodes are increased) might translate into particular systems of inductive logic in which inferences can be drawn from factual training data and conditionals to learned conditionals. In order to facilitate this study, computer implementations of interpreted networks and their learning algorithms will be crucial.

Applying the theory to open problems in uncertain reasoning: The results achieved by the previous tasks are expected to feed back on open problems in uncertain reasoning. E.g.: The standard theory of belief revision was created as a theory for the "one-shot" revision of beliefs by a single piece of evidence.[15] It is well-known that belief revision and (preferential) nonmonotonic reasoning are more or less intertranslatable. Attempts of extending the theory of belief revision to iterated occurrences of evidence led to a multitude of suggestions lacking clear philosophical interpretation. By means of the results achieved in this area, it might be possible to understand evidence-induced changes of networks as iterated belief revisions. We hypothesise that different schemes of iterated revision correspond to, and can be understood as, different learning algorithms for neural networks.

Applying the theory in philosophy of science: In philosophy of science, it was realized early on that new empirical evidence can have the effect that previous hypotheses must be withdrawn, as a scientists might learn that what she had regarded likely is actually not. As Flach [20] argues, the same logics that govern valid commonsense inferences can be interpreted as logics for scientific induction, i.e., for data constituting incomplete und uncertain evidence for empirical hypotheses. Schurz [52] demonstrates that scientific laws are subject to normality or ceteris paribus restrictions that obey the logic of nonmonotonic reasoning. At the same time, the study of neural networks is expected to transform our philosophical understanding of science: Churchland [14] presents networks as models of scientific theories and regards prototype representations in networks as a system's explanatory understanding of its inputs. Bechtel [6] explains scientific model building in terms of the satisfaction of soft constraints represented in networks. Bird [8] observes: "The time is ripe for a reassessment of Kuhn's earlier work in the light of connectionist and neural-net research". Is it possible to throw some new light on these insights from the philosophy of science on the basis of new findings on logical accounts of neural networks and learning?

References

1. Adams, E. (1975). *The logic of conditionals: An application of probability to deductive logic* (Synthese library, Vol. 86). Dordrecht: Reidel.
2. Adams, E. (1998). *A primer of probability logic* (CSLI lecture notes). Stanford: Center for the study of language and information.
3. Bader, S., & Hitzler, P. (2005). Dimensions of neural-symbolic integration – a structured survey. In S. N. Artemov, H. Barringer, A. S. d'Avila Garcez, L. C. Lamb, & J. Woods (Eds.), *We will show them! Essays in honour of Dov Gabbay* (Federation for computational logic, pp. 167–194). London: College Publications, Int.
4. Balkenius, C., & Gärdenfors, P. (1991). Nonmonotonic inferences in neural networks. In J. A. Allen, R. Fikes, & E. Sandewall (Eds.), *Principles of knowledge representation and reasoning* (pp. 32–9). San Mateo: Morgan Kaufmann.
5. Bechtel, W. (1994). Natural deduction in connectionist systems. *Synthese, 101*(3), 433–463.
6. Bechtel, W. (1996). What should a connectionist philosophy of science look like? In R. M. McCauley (Ed.), *The Churchlands and their critics* (pp. 121–44). Massachusetts: Basil Blackwell.

[15] Gärdenfors [22] is the classic reference, and Hansson [25] is a nice textbook on belief revision.

7. Bennett, J. (2003). *A philosophical guide to conditionals*. Oxford: Clarendon Press.
8. Bird, A. (2002). What is in a paradigm? *Richmond Journal of Philosophy, 1*(ii), 11–20.
9. Blutner, R. (2004). Nonmonotonic inferences and neural networks. *Synthese, 142*, 143–74.
10. Boutsinas, B., & Vrahatis, M. N. (2001). Artificial nonmonotonic neural networks. *Artificial Intelligence, 132*, 1–38.
11. Brewka, G., Dix, J., & Konolige, K. (1997). *Nonmonotonic reasoning. An overview* (CSLI lecture notes, Vol. 73). Stanford: CSLI Publications.
12. Chalmers, D. J. (1990). Syntactic transformations on distributed representations. *Connection Science, 2*(1 & 2), 53–62.
13. Chen, C. -H., & Honavar, V. (1999). A neural-network architecture for syntax analysis. *IEEE Transactions on Neural Networks, 10*(1), 94–114.
14. Churchland, P. M. (1989). *A neurocomputational perspective: The nature of mind and the structure of science*. London: MIT Press.
15. d'Avila Garcez, A. S., Broda, K., & Gabbay D. M. (2001). Symbolic knowledge extraction from trained neural networks: A sound approach. *Artificial Intelligence, 125*, 153–205.
16. d'Avila Garcez, A. S., Broda, K. B., & Gabbay D. M. (2002). *Neural-symbolic learning systems*. London: Springer.
17. d'Avila Garcez, A. S., Lamb, L. C., & Gabbay, D. M. (2009). *Neural-symbolic cognitive reasoning*. Berlin: Springer.
18. d'Avila Garcez, A. S., Besold, T. R., de Raedt, L., Földiak, P., Hitzler, P., Icard, T., Kühnberger, K. -U., Lamb, L. C., Miikkulainen, R., & Silver, D. L. (2015). *Neural-symbolic learning and reasoning: Contributions and challenges*. Paper presented at the 2015 AAAI spring symposium series, Stanford.
19. Dietz, E. -A., Hölldobler, S., Palacios, L. (2015). *A connectionist network for skeptical abduction*. Paper presented at NeSy 2015, Buenos Aires.
20. Flach, P. A. (2000). Logical characterizations of inductive learning. In D. M. Gabbay & R. Kruse (Eds.), *Handbook of defeasible reasoning and uncertainty management systems* (Vol. 4, pp. 155–96). Dordrecht: Kluwer.
21. Fodor, J., & Pylyshyn, Z. (1988). Connectionism and cognitive architecture: A critical analysis. *Cognition, 28*, 3–71.
22. Gärdenfors, P. (1988). *Knowledge in flux*. Cambridge, MA: The MIT Press.
23. Gärdenfors, P. (1994). How logic emerges from the dynamics of information. In J. Van Eijck & A. Visser (Eds.), *Logic and information flow* (pp. 49–77). Cambridge: The MIT Press.
24. Ginsberg, M. L. (Ed.). (1987). *Readings in nonmonotonic reasoning* (pp. 1–23). Los Altos: Morgan Kaufmann.
25. Hansson, S. O. (1999). *A textbook of belief dynamics*. Dordrecht: Kluwer.
26. Hitzler, P., Hölldobler, S., & Seda, A. K. (2004). Logic programs and connectionist networks. *Journal of Applied Logic, 2*(3), 245–272.
27. Hölldobler, S. (1993). *Automated inferencing and connectionist models* (Post-doctoral thesis).
28. Hölldobler, S., & Kalinke, Y. (1994). Towards a massively parallel computational model for logic programming. In *Proceedings ECAI94 Workshop on Combining Symbolic and Connectionist Processing* (pp. 68–77). ECCAI.
29. Hölldobler, S. (2009). Cognitive science, computational logic and connectionism. In M. Adriani, et al. (Eds.), *Proceedings of the International Conference on Advanced Computer Science and Information Systems (ICACSIS)* (pp. 1–6).
30. Horgan, T., & Tienson, J. (1996). *Connectionism and the philosophy of psychology*. Cambridge: The MIT Press.
31. Kleene, S. C. (1956). Representation of events in nerve nets and finite automata. In C. E. Shannon & J. McCarthy (Eds.), *Automata studies* (pp. 3–42). Princeton: Princeton University Press.
32. Kraus, S., Lehmann, D., & Magidor, M. (1990). Nonmonotonic reasoning, preferential models and cumulative logics. *Artificial Intelligence, 44*, 167–207.
33. Legendre, G., Miyata, Y., & Smolensky, P. (1994). *Principles for an integrated connectionist/symbolic theory of higher cognition*. Hillsdale: L. Erlbaum.

34. Lehmann, D., & Magidor, M. (1992). What does a conditional knowledge base entail? *Artificial Intelligence, 55*, 1–60.
35. Leitgeb, H. (2001). Nonmonotonic reasoning by inhibition nets. *Artificial Intelligence, 128*(1–2), 161–201.
36. Leitgeb, H. (2003). Nonmonotonic reasoning by inhibition nets II. *International Journal of Uncertainty, Fuzziness and Knowledge-Based Systems, 11*(suppl., issue 2), 105–35.
37. Leitgeb, H. (2004). *Inference on the low level. An investigation into deduction, nonmonotonic reasoning, and the philosophy of cognition* (Applied logic series). Dordrecht: Kluwer/Springer.
38. Leitgeb, H. (2005). Interpreted dynamical systems and qualitative laws: From inhibition networks to evolutionary systems. *Synthese, 146*, 189–202.
39. Leitgeb, H. (2005). Reseaux de neurones capables de raisonner. *Dossier Pour la Science* (Special issue of the French edition of the *Scientific American*) October/December, 97–101.
40. Leitgeb, H. (2007). Beliefs in conditionals vs. conditional beliefs. *Topoi, 26*(1), 115–32.
41. Leitgeb, H. (2007) Neural network models of conditionals: An introduction. In X. Arrazola, J. M. Larrazabal, et al. (Eds.), *LogKCA-07, Proceedings of the First ILCLI International Workshop on Logic and Philosophy of Knowledge, Communication and Action* (pp. 191–223). Bilbao: University of the Basque Country Press.
42. Levesque, H. (1990). All I know: A study in autoepistemic logic. *Artificial Intelligence, 42*, 263–309.
43. Lewis, D. (1973). *Counterfactuals*. Oxford: Blackwell.
44. Makinson, D. (1994). General patterns in nonmonotonic reasoning. In D. M. Gabbay, C. J. Hogger, & J. A. Robinson (Eds.), *Handbook of logic in artificial intelligence and logic programming* (Vol. 3, pp. 35–110.). Oxford: Clarendon Press.
45. McCulloch, W. S., & Pitts, W. H. (1943). A logical calculus of the ideas immanent in nervous activity. *Bulletin of Mathematical Biophysics, 5*, 115–33; Reprinted in: W. S. McCulloch, *Embodiments of mind*. Cambridge, MA: The MIT Press (1965).
46. Ortner, R., & H. Leitgeb (2011). Mechanizing induction. In D. Gabbay, S. Hartmann, & J. Woods (Eds.), *Handbook of the history of logic. Volume 10: Inductive logic* (pp. 719–772). Oxford: Elsevier.
47. Peters, S., & Westerstahl, D. (2006). *Quantifiers in language and logic*. Oxford: Oxford University Press.
48. Pinkas, G. (1991). Symmetric neural networks and logic satisfiability. *Neural Computation, 3*, 282–291.
49. Pinkas, G. (1995). Reasoning, nonmonotonicity and learning in connectionist networks that capture propositional knowledge. *Artificial Intelligence, 77*, 203–247.
50. Rojas, R. (1996). *Neural networks – a systematic introduction*. Berlin: Springer.
51. Rumelhart, D. E., McClelland, J. L., & the PDP Research Group (1986). *Parallel distributed processing* (Vols. 1 and 2). Cambridge: The MIT Press.
52. Schurz, G. (2002). Ceteris paribus laws: Classification and deconstruction. *Erkenntnis, 57*(3), 351–72.
53. Schurz, G., & Leitgeb, H. (Eds.). (2005). Special volume on "Non-monotonic and uncertain reasoning in the focus of paradigms of cognition. *Synthese, 146*, 1–2.
54. Smolensky, P. (1988). On the proper treatment of connectionism. *Behavioral and Brain Sciences, 11*, 1–23.
55. Smolensky, P. (1990). Tensor-product variable binding and the representation of symbolic structures in connectionist networks. *Artificial Intelligence, 46*, 159–216.
56. Smolensky, P., & Legendre, G. (2006). *The harmonic mind: From neural computation to optimality-theoretic grammar*. Cambridge, MA: The MIT Press.
57. Stenning, K., & Lambalgen, M. van (2008). Human reasoning and cognitive science. Cambridge: The MIT Press.
58. van Benthem, J. (1984). Foundations of conditional logic. *Journal of Philosophical Logic, 13*(3), 303–49.
59. Van Gelder, T. J. (1998). The dynamical hypothesis in cognitive science. *Behavioral and Brain Sciences, 21*, 615–65.

60. Van Gelder, T. J. (1999). Distributed versus local representation. In R. Wilson & F. Keil (Eds.), *The MIT encyclopedia of cognitive sciences* (pp. 236–38). Cambridge: The MIT Press.
61. Wheeler, G. (2017). Machine epistemology and big data. In L. McIntyre & A. Rosenberg (Eds.), *The Routledge companion to philosophy of social science* (pp. 321–329). New York: Routledge.

Chapter 8
Proof Theory

Jeremy Avigad

Abstract Proof theory began in the 1920s as a part of Hilbert's program, which aimed to secure the foundations of mathematics by modeling infinitary mathematics with formal axiomatic systems and proving those systems consistent using restricted, finitary means. The program thus viewed mathematics as a system of reasoning with precise linguistic norms, governed by rules that can be described and studied in concrete terms. Today such a viewpoint has applications in mathematics, computer science, and the philosophy of mathematics.

8.1 Introduction

At the turn of the nineteenth century, mathematics exhibited a style of argumentation that was more explicitly computational than is common today. Over the course of the century, the introduction of abstract algebraic methods helped unify developments in analysis, number theory, geometry, and the theory of equations, and work by mathematicians like Richard Dedekind, Georg Cantor, and David Hilbert towards the end of the century introduced set-theoretic language and infinitary methods that served to downplay or suppress computational content. This shift in emphasis away from calculation gave rise to concerns as to whether such methods were meaningful and appropriate in mathematics. The discovery of paradoxes stemming from overly naive use of set-theoretic language and methods led to even more pressing concerns as to whether the modern methods were even consistent. This led to heated debates in the early twentieth century and what is sometimes called the "crisis of foundations."

In lectures presented in 1922, Hilbert launched his *Beweistheorie*, or Proof Theory, which aimed to justify the use of modern methods and settle the problem of foundations once and for all. This, Hilbert argued, could be achieved as follows:

J. Avigad (✉)
Department of Philosophy and Department of Mathematical Sciences, Carnegie Mellon University, Pittsburgh, PA 15213, USA
e-mail: avigad@cmu.edu

S. O. Hansson, V. F. Hendricks (eds.), *Introduction to Formal Philosophy*, Springer Undergraduate Texts in Philosophy, https://doi.org/10.1007/978-3-319-77434-3_8

- First, represent portions of the abstract, infinitary mathematical reasoning in question using formal axiomatic systems, which prescribe a fixed formal language and precise rules of inference.
- Then view proofs in these systems as finite, combinatorial objects, and prove the consistency of such systems—i.e. the fact that there is no way to derive a contradiction—using unobjectionable, concrete arguments.

In doing so, said Hilbert,

> ... we move to a higher level of contemplation, from which the axioms, formulae, and proofs of the mathematical theory are themselves the objects of a contentional investigation. But for this purpose the usual contentual ideas of the mathematical theory must be replaced by formulae and rules, and imitated by formalisms. In other words, we need to have a strict formalization of the entire mathematical theory.... In this way the contentual thoughts (which of course we can never wholly do without or eliminate) are removed elsewhere—to a higher plane, as it were; and at the same time it becomes possible to draw a sharp and systematic distinction in mathematics between the formulae and formal proofs on the one hand, and the contentual ideas on the other. [17]

Gödel's second incompleteness theorem shows that any "unobjectionable" portion of mathematics is insufficient to establish its own consistency, let alone the consistency of any theory properly extending it. Although this dealt a blow to Hilbert's program as it was originally formulated, the more general project of studying mathematical reasoning in syntactic terms, especially with respect to questions of algorithmic or otherwise concrete content, has been fruitful. Moreover, the general strategy of separating syntactic and semantic concerns and of maintaining a syntactic viewpoint where possible has become a powerful tool in formal epistemology. (See [34, 42] for more on Hilbert's program.)

Today, Proof Theory can be viewed as the general study of formal deductive systems. Given that formal systems can be used to model a wide range of types of inference—modal, temporal, probabilistic, inductive, defeasible, deontic, and so on—work in the field is varied and diverse. Here I will focus specifically on the proof theory of *mathematical* reasoning, but even with this restriction, the field is dauntingly broad: the 1998 *Handbook of Proof Theory* [9] runs more than 800 pages, with a name index that is almost as long as this article. As a result, I can only attempt to convey a feel for the subject's goals and methods of analysis, and help ease the reader into the broader literature. References are generally to surveys and textbooks, and results are given without attribution.

In the next section, I describe natural deduction and a sequent calculus for first-order logic, and state the cut-elimination theorem and some of its consequences. This is one of the field's most fundamental results, and provides a concrete example of proof-theoretic method. After that, I survey various aspects of proof-theoretic analysis, and, finally, in the last section, I discuss some applications.

8.2 Natural Deduction and Sequent Calculi

I will assume the reader is familiar with the language of first-order logic. Contemporary logic textbooks often present formal calculi for first-order logic with a long list of axioms and a few simple rules, but these are generally not very convenient for modeling deductive arguments or studying their properties. A system which fares better on both counts in given by Gerhard Gentzen's system of *natural deduction*, a variant of which we will now consider.

Natural deduction is based on two fundamental observations. The first is that it is natural to describe the meaning, or appropriate use, of a logical connective by giving the conditions under which one can *introduce* it, that is, derive a statement in which that connective occurs, and the methods by which one can *eliminate* it, that is, draw conclusions from statements in which it occurs. For example, one can establish a conjunction $\varphi \wedge \psi$ by establishing both φ and ψ, and, conversely, if one assumes or has previously established $\varphi \wedge \psi$, one can conclude either φ or ψ, at will.

The second observation is that it is natural to model logical arguments as taking place under the context of a list of hypotheses, either implicit or explicitly stated. If Γ is a finite set of hypotheses and φ is a first-order formula, the *sequent* $\Gamma \Rightarrow \varphi$ is intended to denote that φ follows from Γ. For the most part, these hypotheses stay fixed over the course of an argument, but under certain circumstances they can be removed, or *canceled*. For example, one typically proves an implication $\varphi \rightarrow \psi$ by temporarily assuming that φ holds and arguing that ψ follows. The introduction rule for implication thus reflects the fact that deriving ψ from a set of hypotheses Γ together with φ is the same as deriving $\varphi \rightarrow \psi$ from Γ.

Writing Γ, φ as an abbreviation for $\Gamma \cup \{\varphi\}$, the rules for natural deduction are shown in Fig. 8.1. The quantifier rules are subject to the usual restrictions. For example, in the introduction rule for the universal quantifier, the variable x cannot be free in any hypothesis. For intuitionistic logic, one also needs the rule *ex falso sequitur quodlibet*, which allows one to conclude $\Gamma \Rightarrow \varphi$ from $\Gamma \Rightarrow \bot$, where $\bot$ represents falsity. One can then define negation, $\neg\varphi$, as $\varphi \rightarrow \bot$. For classical logic, one adds *reductio ad absurdum*, or proof by contradiction, which allows one to conclude $\Gamma \Rightarrow \varphi$ from $\Gamma, \neg\varphi \Rightarrow \bot$.

For many purposes, however, *sequent calculi* provide a more convenient representation of logical derivations. Here, sequents are of the form $\Gamma \Rightarrow \Delta$, where Γ and Δ are finite sets of formulas, with the intended meaning that the conjunction of the hypotheses in Γ implies the *disjunction* of the assertions in Δ. The rules are as shown in Fig. 8.2. The last rule is called the *cut rule*: it is the only rule containing a formula in the hypothesis that may be entirely unrelated to the formulas in the conclusion. Proofs that do not use the cut rule are said to be *cut free*. One obtains a proof system for intuitionistic logic by restricting Δ to contain at most one formula, and adding an axiomatic version of *ex falso sequitur quodlibet*: $\Gamma, \bot \Rightarrow \varphi$. The cut-elimination theorem is as follows:

$$\frac{}{\Gamma, \varphi \Rightarrow \varphi}$$

$$\frac{\Gamma \Rightarrow \varphi \qquad \Gamma \Rightarrow \psi}{\Gamma \Rightarrow \varphi \wedge \psi} \qquad \frac{\Gamma \Rightarrow \varphi_0 \wedge \varphi_1}{\Gamma \Rightarrow \varphi_i}$$

$$\frac{\Gamma \Rightarrow \varphi_i}{\Gamma \Rightarrow \varphi_0 \vee \varphi_1} \qquad \frac{\Gamma \Rightarrow \varphi \vee \psi \qquad \Gamma, \varphi \Rightarrow \theta \qquad \Gamma, \psi \Rightarrow \theta}{\Gamma \Rightarrow \theta}$$

$$\frac{\Gamma, \varphi \Rightarrow \psi}{\Gamma \Rightarrow \varphi \rightarrow \psi} \qquad \frac{\Gamma \Rightarrow \varphi \rightarrow \psi \qquad \Gamma \Rightarrow \varphi}{\Gamma \Rightarrow \psi}$$

$$\frac{\Gamma \Rightarrow \varphi}{\Gamma \Rightarrow \forall y \; \varphi[y/x]} \qquad \frac{\Gamma \Rightarrow \forall x \; \varphi}{\Gamma \Rightarrow \varphi[t/x]}$$

$$\frac{\Gamma \Rightarrow \varphi[t/x]}{\Gamma \Rightarrow \exists x \; \varphi} \qquad \frac{\Gamma \Rightarrow \exists y \; \varphi[y/x] \qquad \Gamma, \varphi \Rightarrow \psi}{\Gamma \Rightarrow \psi}$$

Fig. 8.1 Natural deduction. Derivability of a sequent $\Gamma \Rightarrow \varphi$ means that φ is a consequence of the set of hypotheses Γ, and Γ, φ denotes $\Gamma \cup \{\varphi\}$

$$\frac{}{\Gamma, \varphi \Rightarrow \Delta, \varphi}$$

$$\frac{\Gamma, \varphi_i \Rightarrow \Delta}{\Gamma, \varphi_0 \wedge \varphi_1 \Rightarrow \Delta} \qquad \frac{\Gamma \Rightarrow \Delta, \varphi \qquad \Gamma \Rightarrow \Delta, \psi}{\Gamma \Rightarrow \Delta, \varphi \wedge \psi}$$

$$\frac{\Gamma, \varphi \Rightarrow \Delta \qquad \Gamma, \theta \Rightarrow \Delta}{\Gamma, \varphi \vee \theta \Rightarrow \Delta} \qquad \frac{\Gamma \Rightarrow \Delta, \varphi_i}{\Gamma \Rightarrow \Delta, \varphi_0 \vee \varphi_1}$$

$$\frac{\Gamma, \Rightarrow \Delta, \varphi \qquad \Gamma, \theta \Rightarrow \Delta}{\Gamma, \varphi \rightarrow \theta \Rightarrow \Delta} \qquad \frac{\Gamma, \varphi \Rightarrow \Delta, \psi}{\Gamma \Rightarrow \Delta, \varphi \rightarrow \psi}$$

$$\frac{\Gamma, \varphi[t/x] \Rightarrow \Delta}{\Gamma, \forall x \; \varphi \Rightarrow \Delta} \qquad \frac{\Gamma \Rightarrow \Delta, \psi[y/x]}{\Gamma \Rightarrow \Delta, \forall x \; \psi}$$

$$\frac{\Gamma, \varphi[y/x] \Rightarrow \Delta}{\Gamma, \exists x \; \varphi \Rightarrow \Delta} \qquad \frac{\Gamma \Rightarrow \Delta, \psi[t/x]}{\Gamma \Rightarrow \Delta, \exists x \; \psi}$$

$$\frac{\Gamma \Rightarrow \Delta, \varphi \qquad \Gamma, \varphi \Rightarrow \Delta}{\Gamma \Rightarrow \Delta}$$

Fig. 8.2 The sequent calculus

Theorem 2.1 *If $\Gamma \Rightarrow \Delta$ is derivable in the sequent calculus with cut, then it is derivable without cut.*

Gentzen's proof gives an explicit algorithm for removing cuts from a proof. The algorithm, unfortunately, can yield an iterated exponential increase in the size of proofs, and one can show that there are cases in which such an increase cannot be avoided. The advantage of having a cut-free proof is that the formulas in each sequent are built up directly from the formulas in the sequents above it, making it easy to extract useful information. For example, the following are two consequences of the cut-elimination theorem, easily proved by induction on cut-free proofs.

The first is known as *Herbrand's theorem*. Recall that a formula of first-order logic is said to be *existential* if it consists of a block of existential quantifiers followed by a quantifier-free formula. Similarly, a formula is said to be *universal* if it consists of a block of universal quantifiers followed by a quantifier-free formula. Herbrand's theorem says that if it is possible to prove an existential statement from some universal hypotheses, then in fact there is an explicit sequence of terms in the language that witness the truth of the conclusion.

Theorem 2.2 *Suppose $\exists \vec{x}\ \varphi(\vec{x})$ is derivable in classical first-order logic from a set of hypotheses Γ, where φ is quantifier-free and the sentences in Γ are universal sentences. Then there are sequences of terms $\vec{t}_1, \vec{t}_2, \ldots, \vec{t}_k$ such that the disjunction $\varphi(\vec{t}_1) \lor \varphi(\vec{t}_2) \lor \ldots \lor \varphi(\vec{t}_k)$ has a quantifier-free proof from instances of the sentences in Γ.*

For intuitionistic logic, one has a stronger property, known as the *explicit definability property*.

Theorem 2.3 *Suppose $\exists \vec{x}\ \varphi(\vec{x})$ is derivable in intuitionistic first-order logic from a set of hypotheses Γ in which neither $\lor$ nor $\exists$ occurs in a strictly positive part. Then there are terms $\vec{t}$ such that $\varphi(\vec{t})$ is also derivable from Γ.*

Theorem 2.2 provides a sense in which explicit information can be extracted from certain classical proofs, and Theorem 2.3 provides a sense in which intuitionistic logic is constructive. We have thus already encountered some of the central themes of proof-theoretic analysis:

- Important fragments of mathematical reasoning can be captured by formal systems.
- One can study the properties of these formal systems, for example, describing transformations of formulas and proofs, translations between formulas and proofs in different systems, and canonical normal forms for formulas and proofs.
- The methods provide information about the logic that is independent of the choice of formal system that is used to represent it.

For more on the cut-elimination theorems, see [11, 23, 31, 36, 38].

8.3 Methods and Goals

8.3.1 Classical Foundations

Recall that Hilbert's program, broadly construed, involves representing mathematical reasoning in formal systems and then studying those formal systems as mathematical objects themselves. The first step, then, requires finding the right formal systems. It is common today to view mathematical reasoning as consisting of a properly mathematical part that is used in conjunction with more general forms of logical reasoning, though there are still debates as to where to draw the line between the two. In any case, the following list portrays some natural systems of reasoning in increasing logical/mathematical strength:

1. pure first-order logic
2. primitive recursive arithmetic (denoted *PRA*)
3. first-order arithmetic (*PA*)
4. second-order arithmetic (PA^2)
5. higher-order arithmetic (PA^ω)
6. Zermelo-Fraenkel set theory (*ZF*)

Primitive recursive arithmetic was designed by Hilbert and Bernays to be a patently finitary system of reasoning. The system allows one to define functions on the natural numbers using a simple schema of primitive recursion, and prove facts about them using a principle of induction:

$$\varphi(0) \wedge \forall x\ (\varphi(x) \to \varphi(x+1)) \to \forall x\ \varphi(x).$$

In words, if φ holds of 0 and, whenever it holds of some number, x, it holds of $x+1$, then φ holds of every number. Here φ is assumed to be a quantifier-free formula. In fact, one can replace this axiom with a suitable induction *rule*, whereby primitive recursive arithmetic can be formulated without quantifiers at all. Surprisingly, via coding of finitary objects as natural numbers, this system is expressive and strong enough to develop most portions of mathematics that involve only finite objects and structures [3]. Peano arithmetic can be viewed as the extension of *PRA* with induction for all first-order formulas.

There is no effective axiomatization of second- or higher-order logic that is complete for the standard semantics, where second-order quantifiers are assumed to range over all subsets of the universe of individuals. As a result, one has to distinguish axiomatic second- and higher-order logic from the corresponding semantic characterization. Axiomatically, one typically augments first-order logic with comprehension rules that assert that every formula defines a set (or predicate):

$$\exists X\ \forall y\ (X(y) \leftrightarrow \varphi)$$

Here φ is a formula in which X does not occur, although φ is allowed to have other free variables in addition to y. One can augment these with suitable choice principles as well. Second-order arithmetic can be viewed as the extension of Peano arithmetic with second-order logic and second-order principles of induction, but one can, alternatively, interpret second-order arithmetic in second-order logic together with an axiom asserting the existence of an infinite domain. Similar considerations hold for higher-order logic as well.

Axioms for set theory can be found in any introductory set theory textbook, such as [22]. Of course, these axioms can be extended with stronger hypotheses, such as large cardinal axioms. For information on primitive recursive arithmetic, see [14, 38]; for first-order arithmetic, see [11, 15, 18]; for second-order arithmetic, see [35]; for higher-order arithmetic, see [36].

8.3.2 Constructive Foundations

Given the history of Hilbert's program, it should not be surprising that proof theorists have also had a strong interest in formal representations of constructive and intuitionistic reasoning. From an intuitionistic standpoint, the use of the excluded middle, $\varphi \vee \neg\varphi$, is not acceptable, since, generally speaking, one may not know (or have an algorithm to determine) which disjunct holds. For example, in classical first-order arithmetic, one is allowed to assert $\varphi \vee \neg\varphi$ for a formula φ that expresses the twin primes conjecture, even though we do not know which is the case. If one restricts the underlying logic to intuitionistic logic, however, one obtains *Heyting arithmetic*, which is constructively valid.

Stronger systems tend to be based on what has come to be known as the Curry-Howard-Tait *propositions as types* correspondence. The idea is that, from a constructive perspective, any proposition can be viewed as specifying a type of data, namely, the type of construction that warrants the claim that the proposition is true. A proof of the proposition is thus a construction of the corresponding type. For example, a proof of $\varphi \wedge \psi$ is a proof of φ paired with a proof of ψ, and so $\varphi \wedge \psi$ corresponds to the type of data consisting of pairs of type φ and ψ. Similarly, a proof of $\varphi \to \psi$ should be a procedure transforming a proof of φ into a proof of ψ, so $\varphi \to \psi$ corresponds to a type of functions. This gives rises to systems of *constructive type theory*, of which the most important examples are *Martin-Löf type theory* and an impredicative variant designed by Coquand and Huet, the *calculus of constructions*. Thus, our representative sample of constructive proof systems, in increasing strength, runs as follows:

1. intuitionistic first-order logic
2. primitive recursive arithmetic (*PRA*)
3. Heyting arithmetic (*HA*)
4. Martin-Löf type theory (*ML*)
5. the calculus of inductive constructions (*CIC*)

Good references for intuitionistic systems in general are [7, 39]. For more information on type theory, see [30]; for the calculus of inductive constructions in particular, see [8].

8.3.3 Reverse Mathematics

In the 1970s, Harvey Friedman observed that by restricting the induction and comprehension principles in full axiomatic second-order arithmetic, one obtains theories that are strong enough, on the one hand, to represent significant parts of ordinary mathematics, but weak enough, on the other hand, to be amenable to proof-theoretic analysis. He then suggested calibrating various mathematical theorems in terms of their axiomatic strength. Whereas in ordinary (meta)mathematics, one proves theorems from axioms, Friedman noticed that it is often the case that a mathematical theorem can be used in the other direction, namely, to prove an underlying set-existence principle, over a weak base theory. That is, it is often the case that a theorem of mathematics is formally *equivalent* to a set comprehension principle that is used to prove it.

In that years that followed, Friedman, Stephen Simpson, and many others worked to calibrate the axiomatic assumptions used in a wide range of subjects. They isolated five key theories along the way:

1. RCA_0: a weak base theory, conservative over primitive recursive arithmetic, with a *recursive comprehension axiom*, that is, a principle of comprehension for recursive (computable) sets.
2. WKL_0: adds *weak König's lemma,* a compactness principle, to RCA_0.
3. ACA_0: adds the *arithmetic comprehension axiom*, that is, comprehension for arithmetically definable sets.
4. ATR_0: adds a principle of *arithmetical transfinite recursion,* which allows one to iterate arithmetic comprehension along countable well-orderings.
5. $\Pi^1_1{-}CA_0$: adds the Π^1_1 *comprehension axiom*, that is, comprehension for Π^1_1 sets.

Simpson [35] provides the best introduction to these theories and the reverse mathematics program.

8.3.4 Comparative Analysis and Reduction

We have now seen a sampling of the many formal systems that have been designed to formalize various aspects of mathematics. Proof theorists have also invested a good deal of energy in understanding the relationships between the systems. Often, results take the form of *conservation theorems* which fit the following pattern, where T_1 and T_2 are theories and Γ is a class of sentences:

> Suppose T_1 proves a sentence φ, where φ is in Γ. Then T_2 proves it as well (or perhaps a certain translation, φ').

Such a result, when proved in a suitably restricted base theory, provides a foundational reduction of the theory T_1 to T_2, justifying the principles of T_1 relative to T_2. For example, such theorems can be used to reduce:

- an infinitary theory to a finitary one
- a nonconstructive theory to a constructive one
- an impredicative theory to a predicative one
- a nonstandard theory (in the sense of nonstandard analysis) to a standard one

For example:

1. Versions of primitive recursive arithmetic based on classical, intuitionistic, or quantifier-free logic all prove the same Π_2 theorems (in an appropriate sense) [38].
2. The Gödel-Gentzen double-negation interpretation and variations, like the Friedman-Dragalin A-translation, interpret a number of classical systems in intuitionistic ones, such as *PA* in *HA* [1, 10, 12, 38, 39].
3. There are various translations between theories in the language of (first-, second-, or higher-order) arithmetic and subsystems of set theory [27, 35].
4. Both $I\Sigma_1$, the subsystem of Peano arithmetic in which induction is restricted to Σ_1 formulas, and WKL_0, the subsystem of second-order arithmetic based on Weak König's Lemma, are conservative over primitve recursive arithmetic for the class of Π_2 sentences [2, 11, 15, 20, 33, 35, 38].
5. Cut elimination or an easy model-theoretic argument shows that a restricted second-order version, ACA_0, of Peano arithmetic is a conservative extension of Peano arithmetic itself. Similarly, Gödel-Bernays-von Neumann set theory *GBN*, which has both sets and classes, is a conservative extension of Zermelo-Fraenkel set theory. See, for example, [28, 35]. In general, proofs in ACA_0 may suffer an iterated exponential increase in length when translated to *PA*, and similarly for *GBN* and *ZF*, or $I\Sigma_1$ and *PRA*.
6. Theories of nonstandard arithmetic and analysis can be calibrated in terms of the strength of standard theories [19].
7. The axiom of choice and the continuum hypothesis are conservative extensions of set theory for Σ^2_1 sentences in the analytic hierarchy [22].

Such results draw on a variety of methods. Some can be obtained by direct translation of one theory into another. Many are proved using cut-elimination or normalization [11, 33]. The double-negation translation is a remarkably effective tool when it comes to reducing classical theories to constructive ones, and can often be supplemented by realizability, functional interpretation, or other arguments [1, 20, 37]. Model-theoretic methods can often be used, though they do not provide specific algorithms to carry out the translation [15, 18]. Even forcing methods, originally developed as a set-theoretic technique, can be fruitfully be applied in proof-theoretic settings [4, 22].

8.3.5 Characterizing Logical Strength

The results described in the previous section serve to characterize the strength of one axiomatic theory in terms of another. Showing that a theory T_2 is conservative over T_1 shows that, in particular, T_2 is consistent, if T_1 is. This provides a comparison of the *consistency strength* of the two theories.

But there are other ways of characterizing the strength of a theory. For example, the notion of an *ordinal* generalizes the notion of a counting number. Starting with the natural numbers, we can add an infinite "number," ω, and keep going:

$$0, 1, 2, 3, \ldots, \omega, \omega + 1, \omega + 2, \omega + 3, \ldots$$

We can then proceed to add even more exotic numbers, like $\omega \cdot 2$, ω^2, and ω^ω. The ordering on these particular expressions is computable, in the sense that one can write a computer program to compare any two them. What makes them ordinals is that they satisfy a principle of *transfinite induction,* which generalizes the principle of induction on the natural numbers. Ordinal analysis gauges the strength of a theory in terms of such computable ordinals: the stronger a theory is, the more powerful the principles of transfinite induction it can prove. See, for example, [26, 27, 36].

Alternatively, one can focus on a theory's *computational strength.* Suppose a theory T proves a statement of the form $\forall x \; \exists y \; R(x, y)$, where x and y range over the natural numbers, and R is a computationally decidable predicate. This tells us that a computer program that, on input x, systematically searches for a y satisfying $R(x, y)$ always succeeds in finding one. Now suppose f is a function that, on input x, returns a value that is easily computed from the least y satisfying $R(x, y)$. For example, $R(x, y)$ may assert that y codes a halting computation of a particular Turing machine on input x, and f may return the result of such a computation. Then f is a computable function, and we can say that the theory, T, proves that f is totally defined on the natural numbers. A simple diagonalization shows that no effectively axiomatized theory can prove the totality of every computable function in this way, so this suggests using the set of computable functions that the theory can prove to be total as a measure of its strength.

A number of theories have been analyzed in these terms. For example, by the results in the last section, the provably total computable functions of *PRA*, $I\Sigma_1$, RCA_0, and WKL_0 are all the primitive recursive functions. In contrast, one can characterize the provably total computable functions of *PA* and *HA* in terms of higher-type primitive recursion [5, 37], or using principles of primitive recursion along an ordinal known as ε_0 [27, 36]. Weaker theories of arithmetic can be used to characterize complexity classes like the polynomial time computable functions [11].

8.4 Applications

In this final section, I will describe some of the ways that proof theory interacts with other disciplines. As emphasized in the introduction, I am only considering applications of the traditional, metamathematical branch of proof theory. Formal deductive methods, more broadly, have applications across philosophy and the sciences, and the use of proof-theoretic methods in the study of these formal deductive systems is far too diverse to survey here.

8.4.1 Proof Mining

One way in which traditional proof-theoretic methods have been applied is in the process of extracting useful information from ordinary mathematical proofs. The reductive results of the twentieth century showed, in principle, that many classical proofs can be interpreted in constructive terms. In practice, these ideas have been adapted and extended to the analysis of ordinary mathematical proofs. Georg Kreisel described the process of extracting such information as "unwinding proofs," and Ulrich Kohlenbach has more recently adopted the name "proof mining" [20].

Substantial work is needed to turn this vague idea into something practicable. Ordinary mathematical proofs are not presented in formal systems, so there are choices to be made in the formal modeling. In addition, the general metamathematical tools have to be tailored and adjusted to yield the information that is sought in particular domains. Thus the work requires a deep understanding of both the proof-theoretic methods and the domain of mathematics in question. The field has already had a number of successes in fields like functional analysis and ergodic theory; see, for example, [20].

8.4.2 Combinatorial Independences

Yet another domain where a syntactic, foundational perspective is important is in the search for natural combinatorial independences, that is, natural finitary combinatorial principles that are independent of conventional mathematical methods. The Paris-Harrington statement [24] is an early example of such a principle. Since then, Harvey Friedman, in particular, has long sought to find exotic combinatorial behavior in familiar mathematical settings. Such work gives us glimpes into what goes on just beyond ordinary patterns of mathematical reasoning, and yields interesting mathematics as well. See the extensive introduction to [13] for an overview of results in this area.

8.4.3 Constructive Mathematics and Type Theory

As noted above, proof theory is often linked with constructive mathematics, for historical reasons. After all, Hilbert's program was initially an attempt to justify mathematics with respect to methods that are finitary, which is to say, syntactic, algorithmic, and impeccably constructive. Contemporary work in constructive mathematics and type theory draws on the following facts:

- Logical constructions can often be interpreted as programming principles.
- Conversely, programming principles can be interpreted as logical constructions.
- One can thereby design (constructive) proof systems that combine aspects of both programming and proving.

The references under Sect. 8.3.2 above provide logical perspectives on constructive type theory. For a computational perspective, see [25].

8.4.4 Automated Reasoning and Formal Verification

Another domain where proof-theoretic methods are of central importance is in the field of automated reasoning and formal verification. In computer science, researchers use formal methods to help verify that hardware and software are bug-free and conform to their specifications. Moreover, recent developments have shown that computational formal methods can be used to help verify the correctness of complex mathematical proofs as well. Both efforts have led to interactive approaches, whereby a user works with a computational proof assistant to construct a formal proof of the relevant claims. The have also led to more automated approaches, where software is supposed to carry out the task with little user input. In both cases, proof-theoretic methods are invaluable, for designing the relevant logical calculi, for isolating features of proofs that enable one to cut down the search space and traverse it effectively, and for replacing proof search with calculation wherever possible.

For more information on automated reasoning, see [16, 29]. For more information on formally verified mathematics, see [41], or the December 2008 issue of the *Notices of the American Mathematical Society*, which was devoted to formal proof.

8.4.5 Proof Complexity

Finally, the field of proof complexity combines methods and insights from proof theory and computational complexity. For example, the complexity class NP can be viewed as the class of problems for which an affirmative answer has a short (polynomial-size) proof in a suitable calculus. Thus the conjecture that NP is not equal to co-NP (which is weaker than saying P is not equal to NP) is equivalent to saying that in general there is no propositional calculus that has efficient proofs of every tautology. Stephen Cook has suggested that one way of building up to the problem is to show that *particular* proof systems are not efficient, by establishing explicit lower bounds. Such information is also of interest in automated reasoning, where one wishes to have a detailed understanding of the types of problems that can be expected to have short proofs in various calculi. The works [21, 28, 32, 40] provide excellent introductory overviews.

References

1. Avigad, J. (2000). Interpreting classical theories in constructive ones. *Journal of Symbolic Logic, 65*, 1785–1812.
2. Avigad, J. (2002). Saturated models of universal theories. *Annals of Pure and Applied Logic, 118*, 219–234.
3. Avigad, J. (2003). Number theory and elementary arithmetic. *Philosophia Mathematica, 11*, 257–284.
4. Avigad, J. (2004). Forcing in proof theory. *Bulletin of Symbolic Logic, 10*, 305–333.
5. Avigad, J., & Feferman, S. Gödel's functional ("Dialectica") interpretation. In [9] (pp. 337–405).
6. Barwise, J. (Ed.), (1977). *The handbook of mathematical logic*. Amsterdam: North-Holland. [Contains a number of introductory articles on proof theory and related topics.]
7. Beeson, M. J. (1985). *Foundations of constructive mathematics*. Berlin: Springer.
8. Bertot, Y., & Castéran, P. (2004). *Interactive theorem proving and program development: Coq'Art: The calculus of inductive constructions*. Berlin: Springer.
9. Buss, S. R. (Ed.). (1998). *The handbook of proof theory*. Amsterdam: North-Holland. [Provides a definitive overview of the subject.]
10. Buss, S. R. An introduction to proof theory. In Buss [9] (pp. 1–78).
11. Buss, S. R. First-order proof theory of arithmetic. In Buss [9] (pp. 79–147)
12. Feferman, S. Theories of finite type related to mathematical practice. In Barwise [6] (pp. 913–971).
13. Friedman, H. (to appear). *Boolean relation theory and incompleteness*. Cambridge University Press.
14. Goodstein, R. L. (1957). *Recursive number theory: A development of recursive arithmetic in a logic-free equation calculus*. Amsterdam: North-Holland.
15. Hájek, P., & Pudlák, P. (1993). *Metamathematics of first-order arithmetic*. Berlin: Springer.
16. Harrison, J. (2009). *Handbook of practical logic and automated reasoning*. Cambridge: Cambridge University Press.
17. Hilbert, D. (1922). Neubegründung der Mathematik. Erste Mitteilung. *Abhandlungen aus dem mathematischen Seminar der Hamburgischen Universität, 1*, 157–177. Translated by Ewald, W. (1996). As the new grounding of mathematics. First report. In Ewald, W. (Ed.), *From Kant to Hilbert: A source book in the foundations of mathematics* (Vol. 2, pp. 1115–1134) Oxford: Clarendon.
18. Kaye, R. (1991). *Models of Peano arithmetic*. Oxford: Clarendon.
19. Keisler, H. J. (2006). Nonstandard arithmetic and reverse mathematics. *Bulletin of Symbolic Logic, 12*, 100–125.
20. Kohlenbach, U. (2008). *Applied proof theory: Proof interpretations and their use in mathematics*. Berlin: Springer. [An introduction to proof mining.]

21. Krajíček, J. (1995). *Bounded arithmetic, propositional logic, and complexity theory*. Cambridge: Cambridge University Press.
22. Kunen, K. (1980). *Set theory: An introduction to independence proofs*. Amsterdam: North-Holland.
23. Negri, S., & von Plato, J. (2008). *Structural proof theory*. Cambridge: Cambridge University Press.
24. Paris, J., & Harrington, L. A mathematical incompleteness in Peano arithmetic. In [6] (pp. 1133–1142)
25. Pierce, B. (2004). *Advanced topics in types and programming languages*. Cambridge, MA: MIT Press.
26. Pohlers, W. Subsystems of set theory and second order number theory. In Buss [9] (pp. 209–335).
27. Pohlers, W. (2009). *Proof theory: The first step into impredicativity*. Berlin: Springer. [An introduction to ordinal analysis.]
28. Pudlák, P. The lengths of proofs. In [9] (pp. 547–637).
29. Robinson, J. A., & Voronkov, A. (Eds.). (2001). *Handbook of automated reasoning* (Vols. 1 and 2). Amsterdam/New York: Elsevier; Cambridge: MIT Press.
30. Sambin, G. (Ed.). (1998). *Twenty-five years of constructive type theory*. Oxford: Clarenden.
31. Schwichtenberg, H. Proof theory: Some aspects of cut-elimination. In Barwise [6] (pp. 867–895).
32. Segerlind, N. (2007). The complexity of propositional proofs. *Bulletin of Symbolic Logic, 13*, 417–481.
33. Sieg, W. (1985). Fragments of arithmetic. *Annals of Pure and Applied Logic, 28*, 33–72.
34. Sieg, W. (1999). Hilbert's programs: 1917–1922. *Bulletin of Symbolic Logic, 5*, 1–44.
35. Simpson, S. G. (1999). *Subsystems of second-order arithmetic*. Berlin: Springer
36. Takeuti, G. (1987). *Proof theory* (2nd ed.). Amsterdam: North-Holland.
37. Troelstra, A. S. Realizability. In [9] (pp. 407–473).
38. Troelstra, A. S., & Schwichtenberg, H. (2000). *Basic proof theory* (2nd ed.). Cambridge: Cambridge University Press. [An introductory text.]
39. Troelstra, A. S., & van Dalen, D. (1988). *Constructivism in mathematics: An introduction* (vols. 1 and 2). Amsterdam: North-Holland. [An overview of constructive mathematics.]
40. Urquhart, A. (1995). The complexity of propositional proofs. *Bulletin of Symbolic Logic, 1*, 425–467.
41. Wiedijk, F. (2006). *The seventeen provers of the world*. Berlin: Springer.
42. Zach, R. (2006). *Hilbert's program then and now*. In D. Jacquette (Ed.), *Philosophy of logic* (pp. 411–447). Amsterdam: Elsevier. [A nice historical overview.]

Chapter 9
Logics of (Formal and Informal) Provability

Rafal Urbaniak and Pawel Pawlowski

9.1 Introduction

Provability logics are, roughly speaking, modal logics meant to capture the formal principles of various provability operators (which apply to sentences) or predicates (which apply to sentence names). The first candidate for a provability logic was the modal logic **S4**, which contains as axioms all the substitutions of classical tautologies (in the language with $\Box$; throughout this survey when talking about instances or substitutions we'll mean instances and substitutions in the full language of the system under consideration), all substitutions of the schemata:

$$
\begin{array}{ll}
\text{(K)} & \Box(\varphi \rightarrow \psi) \rightarrow (\Box\varphi \rightarrow \Box\psi) \\
\text{(M)} & \Box\varphi \rightarrow \varphi \\
\text{(4)} & \Box\varphi \rightarrow \Box\Box\varphi
\end{array}
$$

and is closed under two rules of inference: *modus ponens* (from $\vdash \varphi$ and $\vdash \varphi \rightarrow \psi$ infer $\vdash \psi$), and *necessitation* (Nec): if $\vdash \varphi$, then $\vdash \Box\varphi$.

The principles of **S4** seem sensible when $\Box\varphi$ is read as 'it is provable that φ': if an implication and its antecedent are provable, then so is its consequent, whatever is provable should be true, and if something is provable, we can prove that it is (by simply displaying the proof). The system was used in 1933 by Gödel to interpret

R. Urbaniak (✉)
Centre for Logic and Philosophy of Science, Ghent University, Ghent, Belgium

Institute of Philosophy, Sociology and Journalism, University of Gdansk, Gdansk, Poland
e-mail: rafal.urbaniak@ugent.be

P. Pawlowski
Centre for Logic and Philosophy of Science, Ghent University, Ghent, Belgium
e-mail: pawel.pawlowski@ugent.be

S. O. Hansson, V. F. Hendricks (eds.), *Introduction to Formal Philosophy*, Springer Undergraduate Texts in Philosophy, https://doi.org/10.1007/978-3-319-77434-3_9

intuitionistic propositional calculus (which is closely related to reasoning about provability). Alas, **S4** turned out to be inadequate as a tool for modeling the behavior of *formal provability predicate* within axiomatic arithmetic, mostly due to the fact that (M), also (in the context of provability logics) called *local reflection*, while intuitively plausible, cannot be provable in a consistent sufficiently strong axiomatic arithmetic for the formal provability predicate of that arithmetic. Let us elaborate.

Let's fix our attention on the standard first-order axiomatic arithmetic called *Peano Arithmetic* (**PA**). With this system in the background, instead of talking about an arithmetical formula φ, we can use a coding to represent it by some natural number, denoted by $\ulcorner\varphi\urcorner$. Once we've done this, there is (a standard way to construct) an arithmetical formula $\mathtt{Prov}_{\mathbf{PA}}(x)$ true exactly about the codes of those formulas, which are provable in **PA**. This is the formal provability predicate of **PA**.

One crucial property of this predicate is stated by *Löb's Theorem*, according to which for any arithmetical φ, if $\mathbf{PA} \vdash \mathtt{Prov}_{\mathbf{PA}}(\ulcorner\varphi\urcorner) \rightarrow \varphi$, then already $\mathbf{PA} \vdash \varphi$. This means that reflection can hold only for those sentences which are already theorems of **PA**, and not universally for all sentences of arithmetic, and so **S4** cannot be the logic of formal provability predicate.

It turns out that another modal logic is the provability logic of formal arithmetical provability—it's the *Gödel-Löb logic* **GL**. Its axioms are all the substitutions of classical tautologies, all the substitutions of (K), all the substitutions of:

$$\text{(Löb)}\ \Box(\Box\varphi \rightarrow \varphi) \rightarrow \Box\varphi$$

and the rules are *modus ponens* and necessitation. Various modal logics similar to **GL** have been developed for various notions of provability related to the standard formal provability.

In the language of **GL** we can express claims such as 'p is provable', but we cannot express things such as 't is a proof of p' (that is, we cannot express *explicit provability statements*). The latter task can be achieved in the so-called *Logic of Proofs* (**LP**), whose language is much richer: it contains terms for proofs, ways of constructing complex terms for proofs, and a predicate '_ is a proof of _'. **LP** is an adequate logic of explicit provability. Various extensions of **LP** has been developed.

Somewhat independently of the research on the logic of the formal provability predicate, attempts have been made to develop a formal logic of informal mathematical provability, for which (M) holds. The challenge is to develop a sensible system which can be mixed with other parts of mathematics without running into inconsistency due to (Löb) or related reasons.

This survey discusses the developments described above in a bit more detail.

9.2 The Beginnings

9.2.1 *Modal Logic S4*

Formulas of *the language of a propositional modal logic* $\mathcal{L}_M$ are built from propositional variables $p_1, p_2, \ldots$, two propositional constants $\bot$ (contradiction) and $\top$ (logical truth), classical connectives $\neg, \wedge, \vee, \rightarrow, \equiv$, brackets, and unary modal connectives $\Box$ and $\Diamond$, in the standard manner. Sometimes, without loss of generality, we'll treat $\mathcal{L}_M$ as containing only a single classical connective and a single modal operator—this will shorten some definitions, and is enough to make all the other connectives definable. Given a formal language (not necessarily $\mathcal{L}_M$, the context will make the range of meta-variables clear on each occasion), we'll use lower case Greek letters $\varphi, \psi, \chi, \ldots$ as meta-variables for formulas of that language (sometimes, we'll also use σ as a metavariable for an arithmetical *sentence*).

A *normal modal logic* contains as axioms all the substitutions of formulas of $\mathcal{L}_M$ for propositional variables in classical tautologies, all substitutions (in $\mathcal{L}_M$) of the schema:

$$\Box(\varphi \rightarrow \psi) \rightarrow (\Box\varphi \rightarrow \Box\psi) \tag{K}$$

and is closed under two rules of inference: *modus ponens* (from $\vdash \varphi$ and $\vdash \varphi \rightarrow \psi$ infer $\vdash \psi$) and *necessitation* (Nec): if $\vdash \varphi$, then $\vdash \Box\varphi$. The weakest normal modal logic is called **K**, all other normal logics are its extensions.

The standard semantics of $\mathcal{L}_M$ involves *relational models* (also called *Kripke models*). A *frame* $\mathcal{F}$ is a tuple $\langle W, R\rangle$, where W is a non-empty set of possible worlds (or simply nodes, if you're not too much into bloated terminology) and R is a binary relation on W ('is a possible world from the perspective of'), often called an *accessibility relation*. A model **M** over $\mathcal{F} = \langle W, R\rangle$ is a triple $\langle W, R, \Vdash\rangle$, where $\Vdash$ is a *forcing* (or *satisfaction*) relation between W and the formulas of $\mathcal{L}_M$ (think about it as 'being true in'), satisfying the following conditions for any $w \in W$ and any $\varphi, \psi \in \mathcal{L}_M$:

$$
\begin{array}{ll}
w \nVdash \bot & w \Vdash \top \\
w \Vdash (\varphi \rightarrow \psi) & \text{iff } w \nVdash \varphi \text{ or } w \Vdash \psi \\
w \Vdash \Box\varphi & \text{iff for all } w' \in W, \text{ if } wRw', \text{ then } w' \Vdash \varphi
\end{array}
$$

It turns out that the class of formulas forced in every node in every frame is exactly the class of theorems of **K**. Sound and complete semantics for various other normal modal logics is obtained by putting further conditions on R.

One modal logic that will be of particular interest for us is **S4**, which (in one of the formulations) is obtained from **K** by adding as axioms all the instances of the following schemata:

$$\Box\varphi \rightarrow \varphi \tag{M}$$

$$\Box\varphi \rightarrow \Box\Box\varphi \tag{4}$$

(M) is sometimes called (T), but in what follows we'll often use **T** as a variable for an axiomatic theory, so to avoid confusion, we'll stick to (M). **S4** is sound and complete with respect to frames in which the accessibility relation is reflexive ($\forall w \in W\ wRw$) and transitive ($\forall w_1, w_2, w_3\ \in W(w_1 R w_2 \wedge w_2 R w_3 \rightarrow w_1 R w_3)$).

Modal connectives of various modal systems admit various interpretations. $\Box$ can be interpreted as logical necessity, metaphysical necessity, physical necessity, moral obligation, knowledge, etc.[1] Different modal systems are taken to capture principles essential for these various notions. In what follows, we'll be concerned with the reading on which $\Box\varphi$ means 'it is provable that φ' (this reading will need further specifications, as it will turn out). Now the question is: which modal logic captures adequately the formal principles that hold for this reading?

Prima facie, **S4** seems like a decent candidate. (K) holds, because the consequent of a provable implication whose antecedent is provable is also provable. (M) holds, because whatever is provable is true. Equation (4) holds, because if φ is provable, then by producing a proof of φ, by the same token, you are proving that it is provable (necessitation is reliable for pretty much the same reason). But are these considerations satisfactory? Not completely. First of all, we still don't know if there aren't any principles that hold for provability but are not provable in **S4**, because the argument so far was about the soundness of **S4** with respect to our intuitions about provability, not about completeness. Secondly, the argument is somewhat handwavy—it would be good to have a more precise explication of the notion of provability involved. Thirdly, even with such an explication in hand, we have to double-check if all principles of **S4** hold with respect to this explication. Things will turn out to be more complicated than one might initially expect.

9.2.2 Intuitionism and S4

S4 was first proposed as a logic of provability in the context of Brouwer's intuitionistic logic, which, very roughly speaking, results from replacing the notion of truth with that of constructive provability. The intuitionstic logic was formalized by Heyting [31] as Intuitionistic Propositional Calculus (**IPC**) (see also Troelstra and van Dalen [76]). On the intuitionistic approach, a mathematical claim is true just in case it has a proof, and false just in case there is a proof that it leads to contradiction. This idea inspired Heyting and Kolmogorov [32, 33, 45] to introduce the so-called *Brouver-Heyting-Kolmogorov* (BHK) semantics, which identifies truth with provability, falsehood with refutability, and further specifies:

[1]Notice however that different interpretations might make different principles plausible. For instance, (M) is not too convincing in the deontic reading, for unfortunately, not all that should be the case indeed is the case.

A proof of $\varphi \wedge \psi$ consists of a proof of φ and a proof ofψ.
A proof of $\varphi \vee \psi$ is provided by giving either a proof of φ or a proof of ψ.
A proof of$\varphi \rightarrow \psi$ is a construction of proofs of ψ from proofs of φ.
$\bot$ has no proof and $\neg\varphi$ means $\varphi \rightarrow \bot$.

Gödel [24] attempted to formalize the BHK semantics. He put forward **S4** as the logic of classical provability. Then, he suggested a translation t from the non-modal language of intuitionistic logic into $\mathcal{L}_M$ by taking a non-modal formula and putting a box in front of each of its subformulas (in fact, this translation is already mentioned in [61]). Gödel proved that if **IPC** $\vdash \varphi$, then **S4** $\vdash t(\varphi)$. The implication in the opposite direction has been later on proved by McKinsey and Tarski [52]. Thus, **IPC**, in a sense, can be taken to be about the classical provability, if, indeed, **S4** is the logic of classical provability (there are other modal logics into which IPC can be translated). Alas, an explicit provability semantics of $\Box$ in **S4** was missing, and so, the picture wasn't quite complete. One natural candidate for the interpretation of $\Box$ was a formal provability predicate in a standard axiomatized mathematical theory, to which we will now turn.

9.2.3 Arithmetical Provability Predicate

Considerations of formal provability predicate (or predicates) are usually developed in the context of an axiomatic arithmetic. This is the case for various reasons: via Gödel coding, instead of expressions, we can talk about numbers, standard arithmetical theories are usually strong enough to include a sufficiently reach theory of syntax (*modulo* coding), and arithmetic in general is a field where many results are already known and can be borrowed and applied to syntax.

For the sake of simplicity, we'll focus on one fairly standard axiomatic arithmetic: Peano Arithmetic (**PA**), although many results apply to other arithmetical theories, including some weaker ones (see for example Hájek and Pudlak [27] for details). The language of **PA**, $\mathcal{L}_{\mathbf{PA}}$, is a first-order language with identity and a few specific symbols: 0, S, $\times$ and $+$ (in the standard model of arithmetic $\mathbb{N}$ interpreted as referring to the number zero, the successor function, multiplication, and addition, respectively). For any number m, the *standard numeral* for m has the form $\underbrace{S \ldots S}_{m} 0$ and is abbreviated by $\overline{m}$. The specific axioms of **PA** consist of:

$$\forall x \ (0 \neq Sx) \tag{PA 1}$$

$$\forall x, y \ (Sx = Sy \rightarrow x = y) \tag{PA 2}$$

$$\forall x \ (x + 0 = x) \tag{PA 3}$$

$$\forall x, y \ (x + Sy = S(x + y)) \tag{PA 4}$$

$$\forall x\ (x \times 0 = 0) \tag{PA 5}$$

$$\forall x, y\ (x \times Sy = (x \times y) + x) \tag{PA 6}$$

and all the instances of the induction schema:

$$\varphi(0) \wedge \forall x\ (\varphi(x) \rightarrow \varphi(S(x))) \rightarrow \forall x\ \varphi(x) \tag{PA Ind}$$

Formulas of $\mathcal{L}_{\mathbf{PA}}$ can be classified according to their logical complexity. If t is a term not containing x, $\forall x \leq t\ \varphi(x)$ and $\exists x \leq t\ \varphi(x)$ abbreviate $\forall x\ (x \leq t \rightarrow \varphi(x))$ and $\exists x\ (x \leq t \wedge \varphi(x))$ respectively. Such occurrences of quantifiers are called *bounded*, and formulas whose all quantifiers are bounded are called Δ_0-formulas. The hierarchy proceeds in two "layers", that of Π_n and that of Σ_n formulas. $\Pi_0 = \Sigma_0 = \Delta_0$. Σ_{n+1}-formulas are of the form $\exists x_1, \ldots, x_k\ \varphi(x_1, \ldots, x_k)$, where $\varphi(x_1, \ldots, x_k)$ is Π_n. Π_{n+1}-formulas are of the form $\forall x_1, \ldots, x_k\ \varphi(x_1, \ldots, x_k)$, where $\varphi(x_1, \ldots, x_k)$ is Σ_n. Every formula of $\mathcal{L}_{\mathbf{PA}}$ is logically equivalent to a Σ_n formula and to a Π_m formula, for some n and m (and there always exist the least such n and m).

The class of Σ_1 formulas is of particular interest, because it turns out that a function is recursively enumerable (see Smith [68] for a nice introduction to the topic) just in case it is Σ_1-definable. This result, for instance, makes sure that an axiomatic system which is strong enough to handle Σ_1-sentences (in a sense to be specified) is strong enough to properly handle computable functions, including those related to syntactic manipulations, and so is strong enough to prove things about syntax of a formal language within it.

We say that an arithmetical theory $\mathbf{T}$ is Σ_1*-sound* just in case for any Σ_1-formula φ, if $\mathbf{T} \vdash \varphi$, then $\mathbb{N} \models \varphi$ (that is, φ is true in the standard model of arithmetic). The dual notion is that of Σ_1*-completeness*. $\mathbf{T}$ is Σ_1-complete just in case for any sentence $\varphi \in \Sigma_1$, if $\mathbb{N} \models \varphi$, then $\mathbf{T} \vdash \varphi$.

Fact 1.1 **PA** *is* Σ_1*-complete.*

There are various ways of *coding syntax*, effectively mapping syntactic objects, such as expressions, formulas, sentences and sequences thereof to natural numbers, so that each syntactic object τ of $\mathcal{L}_{\mathbf{PA}}$ is represented by its Gödel code $\ulcorner \tau \urcorner$. The details are unimportant here, so let's just focus on one of them and work with it (again, see Smith [68] for an accessible introduction).

Consider now any theory $\mathbf{T}$ in $\mathcal{L}_{\mathbf{PA}}$ extending **PA**. It is said to be *elementary presented* just in case there is an arithmetical Δ_0-formula $\mathtt{Ax}_{\mathbf{T}}(x)$ true of a natural number just in case it is a code of an axiom of $\mathbf{T}$. Such a formula can be further used in a fairly standard way to construct a Δ_0 arithmetical formula $\mathtt{Prf}_{\mathbf{T}}(y, x)$ which is the standard binary proof predicate of $\mathbf{T}$ such that it is true of natural numbers m and n just in case m is the code of a sequence of formulas which is a proof of the formula whose code is n (the details of the construction are inessential here). Moreover:

(Binumeration) If in the standard model $\mathtt{Prf}_{\mathbf{T}}(m, n)$, then $\mathbf{PA} \vdash \mathtt{Prf}_{\mathbf{T}}(\overline{m}, \overline{n})$
If in the standard model $\neg\mathtt{Prf}_{\mathbf{T}}(m, n)$, then $\mathbf{PA} \vdash \neg\mathtt{Prf}_{\mathbf{T}}(\overline{m}, \overline{n})$

$\mathtt{Prf}_{\mathbf{T}}(y, x)$ can be further used to define the so-called *standard provability predicate* (since we won't be talking about non-standard provability predicates, we'll simply talk about provability predicates, assuming they're standard) and the *consistency statement*:

$$\mathtt{Prov}_{\mathbf{T}}(x) := \exists y\ \mathtt{Prf}_{\mathbf{T}}(y, x)$$

$$\mathtt{Con}(\mathbf{T}) := \neg\mathtt{Prov}_{\mathbf{T}}(\ulcorner\bot\urcorner)$$

$\mathtt{Prov}_{\mathbf{T}}(y)$ is obtained from a Δ_0 formula by preceding it with an existential quantifier, and so, it is a Σ_1-formula. Therefore, by Σ_1-completeness, the first half of (Binumeration) holds for it (and the second one fails, for somewhat more complicated reasons):

$$\text{If in the standard model } \mathtt{Prov}_{\mathbf{T}}(n) \text{ is true, then } \mathbf{PA} \vdash \mathtt{Prov}_{\mathbf{T}}(\overline{n})$$

Note however, that even though the second half of (Binumeration) fails, $\mathtt{Prov}_{\mathbf{T}}(x)$ succeeds at *defining* provability, in the sense that $\mathtt{Prov}_{\mathbf{T}}(\overline{\ulcorner\varphi\urcorner})$ is true in the standard model of arithmetic just in case in fact $\mathbf{T} \vdash \varphi$ (by the way, from now on we'll skip using the bar above numbers coding of formulas, assuming it is normally there, that is, that in the formulas we'll mention, numerals of codes of formulas are standard).

Still assuming $\mathbf{T}$ is elementary presented, $\mathtt{Prov}_{\mathbf{T}}(x)$ satisfies the following so-called *Hilbert-Bernays conditions* [34, 48] for any arithmetical formulas φ, ψ:

$$\mathbf{T} \vdash \varphi \text{ iff } \mathbf{PA} \vdash \mathtt{Prov}_{\mathbf{T}}(\ulcorner\varphi\urcorner) \tag{HB1}$$

$$\mathbf{PA} \vdash \mathtt{Prov}_{\mathbf{T}}(\ulcorner\varphi \to \psi\urcorner) \to (\mathtt{Prov}_{\mathbf{T}}(\ulcorner\varphi\urcorner) \to \mathtt{Prov}_{\mathbf{T}}(\ulcorner\psi\urcorner) \tag{HB2}$$

$$\mathbf{PA} \vdash \mathtt{Prov}_{\mathbf{T}}(\ulcorner\varphi\urcorner) \to \mathtt{Prov}_{\mathbf{T}}(\ulcorner\mathtt{Prov}_{\mathbf{T}}(\ulcorner\varphi\urcorner)\urcorner) \tag{HB3}$$

In particular, the provability predicate of $\mathbf{T}$ can be taken to be that of $\mathbf{PA}$ itself. Also, keep in mind, that most of the results apply to certain theories weaker than $\mathbf{PA}$ and to elementary presented theories extending $\mathbf{PA}$, either of which we usually chose to ignore for the sake of simplicity.

Another important piece of the puzzle will be Gödel's *incompleteness theorems*, which we include here in a somewhat modernized statement:

Theorem 1.2 *If an elementary presented theory* $\mathbf{T}$ *extends* $\mathbf{PA}$ *and is consistent, then there is a sentence* $G \in \mathcal{L}_{\mathbf{PA}}$ *such that* $\mathbf{T} \nvdash G$ *and* $\mathbf{T} \nvdash \neg G$. *Moreover,* $\mathbf{T} \nvdash \mathtt{Con}(T)$.

Incompleteness follows from a more general result (which have been stated by Carnap [17]; see Gaifman [23] for a deeper historical discussion):

Lemma 1.3 (Diagonal Lemma) *For any formula* $\varphi(x) \in \mathcal{L}_{\mathbf{PA}}$ *there is a sentence* $\lambda \in \mathcal{L}_{\mathbf{PA}}$ *such that*

$$\mathbf{PA} \vdash \lambda \equiv \varphi(\ulcorner\lambda\urcorner)$$

The Diagonal Lemma, when we take $\varphi(x)$ to be $\neg\mathtt{Prov}_{\mathbf{PA}}(x)$, entails the existence of a sentence that can be used in the incompleteness proof, which provably satisfies the condition:

$$G \equiv \neg\mathtt{Prov}_{\mathbf{PA}}(\ulcorner G \urcorner)$$

Such a G is independent of **PA**. The result generalizes: if a theory satisfies certain requirements and is consistent, its Gödel sentence is independent of it.

Quite a few years later Henkin [30] asked a related question: what happens, however, with sentences such as:

$$H \equiv \mathtt{Prov}_{\mathbf{T}}(\ulcorner H \urcorner)? \qquad \text{(Henkin)}$$

The question was soon answered by Löb [48]:

Theorem 1.4 (Löb) *If the Diagonal Lemma applies to* **T**, *and the provability predicate of a theory* **T** *satisfies (his formulation of) the Hilbert Bernays conditions (HB1-3),* $\mathbf{T} \vdash \mathtt{Prov}_{\mathbf{T}}(\ulcorner \varphi \urcorner) \rightarrow \varphi$ *if and only if* $\mathbf{T} \vdash \varphi$.

For a given sentence φ, the formula '$\mathtt{Prov}_{\mathbf{T}}(\ulcorner \varphi \urcorner) \rightarrow \varphi$' is called *reflection for* φ *(over* ***T****)*, and Löb's theorem says that reflection is provable in **T** for all and only theorems of **T**. The theorem can be obtained fairly easily from the Diagonal Lemma applied to $\mathtt{Prov}_{\mathbf{T}}(x) \rightarrow \varphi$. For if the Diagonal Lemma applies to **T** (it is enough that **T** extends **PA**), the Lemma entails the existence of a ψ such that:

$$\mathbf{T} \vdash \psi \equiv (\mathtt{Prov}_{\mathbf{T}}(\ulcorner \psi \urcorner) \rightarrow \varphi) \qquad \text{(L)}$$

The rest of the reasoning is propositional.

1	$\mathbf{T} \vdash (\mathtt{Prov}_{\mathbf{T}}(\ulcorner \psi \urcorner) \rightarrow \varphi) \rightarrow \psi$	(L), Classical logic
2	$\mathbf{T} \vdash \psi \rightarrow (\mathtt{Prov}_{\mathbf{T}}(\ulcorner \psi \urcorner) \rightarrow \varphi)$	(L), Classical logic
3	$\mathbf{T} \vdash \mathtt{Prov}_{\mathbf{T}}(\ulcorner \psi \rightarrow (\mathtt{Prov}_{\mathbf{T}}(\ulcorner \psi \urcorner) \rightarrow \varphi) \urcorner)$	(HB1), 2
4	$\mathbf{T} \vdash \mathtt{Prov}_{\mathbf{T}}(\ulcorner \psi \urcorner) \rightarrow \mathtt{Prov}_{\mathbf{T}}(\ulcorner (\mathtt{Prov}_{\mathbf{T}}(\ulcorner \psi \urcorner) \rightarrow \varphi) \urcorner)$	(HB2), 3
5	$\mathbf{T} \vdash \mathtt{Prov}_{\mathbf{T}}(\ulcorner \psi \urcorner) \rightarrow (\mathtt{Prov}_{\mathbf{T}}(\ulcorner \mathtt{Prov}_{\mathbf{T}}(\ulcorner \psi \urcorner \urcorner)) \rightarrow \mathtt{Prov}_{\mathbf{T}}(\ulcorner \varphi \urcorner))$	(HB2), 4
6	$\mathbf{T} \vdash \mathtt{Prov}_{\mathbf{T}}(\ulcorner \psi \urcorner) \rightarrow \mathtt{Prov}_{\mathbf{T}}(\ulcorner \mathtt{Prov}_{\mathbf{T}}(\ulcorner \psi \urcorner \urcorner))$	(HB3)
7	$\mathbf{T} \vdash \mathtt{Prov}_{\mathbf{T}}(\ulcorner \psi \urcorner) \rightarrow \mathtt{Prov}_{\mathbf{T}}(\ulcorner \varphi \urcorner)$	5, 6
8	$\mathbf{T} \vdash \mathtt{Prov}_{\mathbf{T}}(\ulcorner \varphi \urcorner) \rightarrow \varphi$	Assumption
9	$\mathbf{T} \vdash \mathtt{Prov}_{\mathbf{T}}(\ulcorner \psi \urcorner) \rightarrow \varphi$	7, 8
10	$\mathbf{T} \vdash \psi$	Classical logic, 1, 8
11	$\mathbf{T} \vdash \mathtt{Prov}_{\mathbf{T}}(\ulcorner \psi \urcorner)$	(HB1), 10
12	$\mathbf{T} \vdash \varphi$	9, 11

9.2.4 The Inadequacy of S4 with Respect to Formal Provability

Coming back to the question of whether $\Box$ of **S4** can be sensibly interpreted as the formal provability predicate: what happens when we take $\Box\varphi$ to mean $\mathtt{Prov}_{\mathbf{T}}(\ulcorner\varphi\urcorner)$? As it turns out, things fall apart quite quickly. For the sake of simplicity we'll take the case where $\mathbf{T} = \mathbf{PA}$, but the point generalizes to consistent recursively axiomatizable extensions of **PA**.

Since $\mathbf{S4} \vdash \Box\varphi \rightarrow \varphi$ for any φ, the interpretation would require that for all $\varphi \in \mathcal{L}_{\mathbf{PA}}$, $\mathbf{PA} \vdash \mathtt{Prov}_{\mathbf{PA}}(\ulcorner\varphi\urcorner) \rightarrow \varphi$. But this, jointly with Löb's theorem, would entail that for any $\varphi \in \mathcal{L}_{\mathbf{PA}}$, $\mathbf{PA} \vdash \varphi$. So, if **PA** is consistent, **S4** is not the logic of the formal provability predicate of **PA**.

There is a somewhat different way to notice the inadequacy of **S4** in this context, already brought up by Gödel [24]. The formula expressing $\mathtt{Con}(PA)$ is $\neg\mathtt{Prov}_{\mathbf{PA}}(\ulcorner\bot\urcorner)$, which is logically equivalent to $\mathtt{Prov}_{\mathbf{PA}}(\ulcorner\bot\urcorner) \rightarrow \bot$. At the modal level, this is just an axiom of **S4**, since $\Box\bot \rightarrow \bot$ falls under schema (M). Thus, if **S4** was adequate, we would have $\mathbf{PA} \vdash \mathtt{Con}(PA)$, which would contradict Gödel's second incompleteness theorem. Moreover, necessitation would yield $\Box(\Box\bot \rightarrow \bot)$, and so in **S4** we would be able to derive the claim that the consistency claim is derivable, which again, contradicts Gödel's second incompleteness theorem.

At this stage, at least two questions remain open. What is the right modal logic of formal provability? What's the right provability semantics for S4?

9.3 Gödel-Löb Modal Logic (GL)

9.3.1 Axiomatizing GL

The inadequacy of **S4** was mainly due to Löb's theorem. How to proceed to obtain a modal logic better fit to the formal provability interpretation?

The first move results from noticing that it was (M) that was responsible for the trouble. Consequently, (M) has to be dropped. Another step is to notice that a formalized version of Löb's theorem can be proved in any elementary presented **T** extending **PA**, so that we have:

$$\mathbf{T} \vdash \mathtt{Prov}_{\mathbf{T}}(\ulcorner\mathtt{Prov}_{\mathbf{T}}(\ulcorner\varphi\urcorner) \rightarrow \varphi\urcorner) \rightarrow \mathtt{Prov}_{\mathbf{T}}(\ulcorner\varphi\urcorner)$$

So, our modal logic of provability should validate the corresponding modal principle:

$$\Box(\Box\varphi \rightarrow \varphi) \rightarrow \Box\varphi \tag{Löb}$$

GL (from Gödel-Löb) is a modal system resulting from these moves. Its axioms are all the substitutions of classical tautologies (in the language of $\mathcal{L}_M$), all the substitutions of (K), all the substitutions of (Löb), and the rules are *modus ponens* and necessitation.

Note that while (Nec) is a rule of **GL**, we cannot have $\varphi \rightarrow \Box\varphi$ as an axiom schema. While (Nec) is well-motivated (it says, in the intended interpretation, that any theorem is provably provable), the implication would say that anything *true* is provable, and that is far from obvious. In the arithmetical setting, we already have the formalized version of the second incompleteness theorem:

$$\mathbf{PA} \vdash \mathtt{Con}(\mathbf{PA}) \rightarrow \neg\mathtt{Prov}_{\mathbf{PA}}(\ulcorner\mathtt{Con}(\mathbf{PA})\urcorner)$$

so if we also had:

$$\mathbf{PA} \vdash \mathtt{Con}(\mathbf{PA}) \rightarrow \mathtt{Prov}_{\mathbf{PA}}(\ulcorner\mathtt{Con}(\mathbf{PA})\urcorner)$$

it would follow that $\mathbf{PA} \vdash \neg\mathtt{Con}(\mathbf{PA})$.

Now, is **GL** at least sound with respect to the formal provability interpretation? Well, the necessitation rule is the modal version of (HB1) and (K) is the modal version of (HB2). We can also prove in GL the modal version of (HB3), that is, $\mathbf{GL} \vdash (4)$, and so it can also be dropped when moving from **S4** to **GL**.

To get a better grasp of proofs in GL, let's see what the proof of this fact looks like. Before we give the proof we need two introductory steps. For one thing, since we have (Nec) and (K), we can easily move from $\mathbf{GL} \vdash (\varphi \rightarrow \psi)$ or from $\mathbf{GL} \vdash \Box(\varphi \rightarrow \psi)$ to $\mathbf{GL} \vdash \Box\varphi \rightarrow \Box\psi$. In what follows we'll make such moves without hesitation, sometimes calling them (Distr)—since they basically consist in distributing $\Box$ over material implication. If you're not convinced, the official argument starts with $\vdash \varphi \rightarrow \psi$. Use (Nec) to obtain $\vdash \Box(\varphi \rightarrow \psi)$. Then (K) tells you $\vdash \Box(\varphi \rightarrow \psi) \rightarrow (\Box\varphi \rightarrow \Box\psi)$, and so by detachment $\vdash \Box\varphi \rightarrow \Box\psi$. It's just too repetitive to go through these moves every time.

For another, we'll need this fairly straightforward fact:

Fact 3.1 $\mathbf{GL} \vdash \Box(\varphi \wedge \psi) \equiv (\Box\varphi \wedge \Box\psi)$

Proof Reason within **GL**. From left to right:

1. $\varphi \wedge \psi \rightarrow \varphi$	Tautology
2. $\varphi \wedge \psi \rightarrow \psi$	Tautology
3. $\Box(\varphi \wedge \psi) \rightarrow \Box\varphi$	(Distr), 1
4. $\Box(\varphi \wedge \psi) \rightarrow \Box\psi$	(Distr), 2
5. $\Box(\varphi \wedge \psi) \rightarrow (\Box\varphi \wedge \Box\psi)$	Classical logic, 3, 4

From right to left:

1. $\varphi \rightarrow (\psi \rightarrow (\varphi \wedge \psi))$	Tautology
2. $\Box\varphi \rightarrow (\Box\psi \rightarrow \Box(\varphi \wedge \psi))$	(Distr), 1
3. $(\Box\varphi \wedge \Box\psi) \rightarrow \Box(\varphi \wedge \psi)$	Classical logic, 2

□

Observe also that Fact 3.1 entails (left to right, together with conjunction elimination):

$$\mathbf{GL} \vdash \Box(\Box\varphi \wedge \varphi) \rightarrow \Box\Box\varphi \tag{3.2}$$

Finally we have:

Fact 3.3 **GL** $\vdash$ (4), *that is* **GL** $\vdash \Box\varphi \rightarrow \Box\Box\varphi$.

Proof Again, let's reason within GL.

1. $\varphi \rightarrow ((\psi \wedge \chi) \rightarrow (\chi \wedge \varphi))$	Tautology
2. $\varphi \rightarrow ((\Box\Box\varphi \wedge \Box\varphi) \rightarrow (\Box\varphi \wedge \varphi))$	Substitution, 1
3. $\varphi \rightarrow (\Box(\Box\varphi \wedge \varphi) \rightarrow (\Box\varphi \wedge \varphi))$	Fact 3.1, 2
4. $\Box\varphi \rightarrow \Box(\Box(\Box\varphi \wedge \varphi) \rightarrow (\Box\varphi \wedge \varphi))$	(Distr), 3
5. $\Box(\Box(\Box\varphi \wedge \varphi) \rightarrow (\Box\varphi \wedge \varphi)) \rightarrow \Box(\Box\varphi \wedge \varphi)$	(Löb)
6. $\Box\varphi \rightarrow \Box(\Box\varphi \wedge \varphi)$	Classical logic, 4, 5
7. $\Box\varphi \rightarrow \Box\Box\varphi$	(3.2), 6

□

We've shown that (4) is derivable in **GL**. But since (M) was the source of the problems, we also need to make sure it is not a derivable theorem schema for **GL**. Simply dropping it from the axiom schemata is not enough.

Fact 3.4 *If* **GL** $\nvdash \bot$*, it is not the case that for any* φ*,* **GL** $\vdash \Box\varphi \rightarrow \varphi$.

Proof Suppose the opposite holds. Then we have **GL** $\vdash \Box\bot \rightarrow \bot$, and so we can reason within GL:

1. $\Box\bot \rightarrow \bot$	Assumption
2. $\Box(\Box\bot \rightarrow \bot)$	(Nec), 1
3. $\Box(\Box\bot \rightarrow \bot) \rightarrow \Box\bot$	(Löb)
4. $\Box\bot$	Detachment, 2, 3
5. $\bot$	Detachment, 1, 4

□

The assumption of consistency of **GL** above is explicit not because we have any serious doubts about it. It's rather that when it is explicitly stated, the unprovability of reflection can be easily proven in a few steps, as we've just seen. Proof of a similar claim without this assumption is slightly more convoluted.

Fact 3.5 **GL** $\nvdash \bot$ *and* **GL** $\nvdash \Box p \to p$.

Proof The general structure of the argument is this. We show that all theorems of **GL** have a certain property, which $\bot$ and $\Box p \to p$ don't have, and so $\bot$ and $\Box p \to p$ are not theorems of **GL**. The property is: *being a classical propositional tautology under the following translation*. So now we need to define a translation t from $\mathcal{L}_M$ into the classical propositional language, which translates all theorems of **GL** into classical tautologies, but at the same time translates $\bot$ and $\Box p \to p$ into formulas whose negations are classically satisfiable. Let's start with the translation:

$$\begin{aligned} t(\bot) &= \bot \\ t(p) &= p \text{ (for all propositional variables)} \\ t(\varphi \to \psi) &= (\varphi)^\star \to (\psi)^\star \\ t(\Box\varphi) &= \top \end{aligned}$$

Clearly:

1. If φ is a substitution of a classical tautology, $t(\varphi)$ is a tautology. This is because the translation effectively is a substitution, and it gives a formula in the classical propositional language, in which all substitutions of tautologies are classical tautologies.
2. t(K) is $\top \to (\top \to \top)$, which is a classical tautology.
3. t(Löb) is $\top \to \top$, which also is a tautology.

We handled the axioms of **GL**, making sure their translations are classical tautologies. Now we need to take care of the inference rules.

4. Consider *modus ponens* (arguments for any classical propositional rule are pretty much the same). One can still apply *modus ponens* to $t(\varphi)$, $t(\varphi \to \psi) = (t(\varphi) \to t(\psi))$. So if **GL** $\vdash \varphi$, **GL** $\vdash \varphi \to \psi$, we know that **GL** $\vdash \psi$, and that the following are tautologies: $t(\varphi)$, $t(\varphi) \to t(\psi)$, and $t(\psi)$.
5. What about necessitation? Say **GL** $\vdash \varphi$ so that also **GL** $\vdash \Box\varphi$. Quite trivially $t(\Box\varphi) = \top$, which is a tautology.

Together, points 1–5 show that all theorems of **GL** translate into classical tautologies. Finally, we have to show that the translations of the formulas that we're interested in aren't tautologies.

6. $t(\Box p \to p) = \top \to p$, which is not a tautology.
7. $t(\bot) = \bot$, which also isn't a tautology.

Points 6–7 mean that these formulas are not theorems of **GL**, which completes the proof. □

9.3.2 Another Way Towards GL: K4LR

Another modal logic that might come to mind when one thinks of $\Box$ as provability is **K4LR**. Just as **GL**, it allows necessitation and classical consequence (for the modal language), and just as **GL** it has (K) as an axiom schema. But it keeps (4), drops (Löb), and admits the following *Löb's rule* (LR) instead:

$$\text{If } \vdash (\Box\varphi \rightarrow \varphi),\ \text{infer } \vdash \Box\varphi \tag{LR}$$

It turns out that **GL** and **K4LR** have the same theorems.

Fact 3.6 *If* **K4LR** $\vdash \varphi$, *then* **GL** $\vdash \varphi$.

Proof We need to check that **GL** proves the axioms of **K4LR** and that it is closed under its rules. As for the axioms, (K) is shared, and Fact 3.3 shows that **GL** $\vdash$(4). As for the rules (Nec) is shared, and we only need to show that **GL** is closed under (LR). This can be shown by the following reasoning within **GL**:

1. $\Box\varphi \rightarrow \varphi$	Assumption (as a GL-theorem)
2. $\Box(\Box\varphi \rightarrow \varphi)$	(Nec), 1
3. $\Box(\Box\varphi \rightarrow \varphi) \rightarrow \Box\varphi$	(Löb)
4. $\Box\varphi$	MP, 2, 3
5. φ	MP, 1, 4

□

Implication in the opposite direction also holds.

Fact 3.7 *If* **GL** $\vdash \varphi$, *then* **K4LR** $\vdash \varphi$.

Proof (K) and (Nec) and classical logic in the background are shared. The only thing that needs to be shown is **K4LR** $\vdash$(Löb), that is that (LR) in the context of **K4LR** is strong enough to give us the formula corresponding to the rule.

To see that this is not immediately obvious, note that, in principle, rules are weaker than corresponding implications, because they apply to theorems only. For instance, (Nec) is sensible because it says that any theorem is necessary, but the formula $p \rightarrow \Box p$ is not an axiom of any sensible standard modal logic, for it says, roughly, that *any truth is necessary*. Let's prove (Löb) within **K4LR**.

1. $\Box(\Box\varphi \rightarrow \varphi) \rightarrow (\Box\Box\varphi \rightarrow \Box\varphi)$	(K)
2. $\Box(\Box\varphi \rightarrow \varphi) \rightarrow \Box\Box(\Box\varphi \rightarrow \varphi)$	(4)
3. $\Box[\Box(\Box\varphi \rightarrow \varphi) \rightarrow \Box\varphi] \rightarrow [\Box\Box(\Box\varphi \rightarrow \varphi) \rightarrow \Box\Box\varphi]$	(K)
4. $\Box[\Box(\Box\varphi \rightarrow \varphi) \rightarrow \Box\varphi] \rightarrow [\Box(\Box\varphi \rightarrow \varphi) \rightarrow \Box\Box\varphi]$	PL, 1, 3
5. $\Box[\Box(\Box\varphi \rightarrow \varphi) \rightarrow \Box\varphi] \rightarrow [\Box(\Box\varphi \rightarrow \varphi) \rightarrow \Box\varphi]$	PL, 2, 4
6. $\Box(\Box\varphi \rightarrow \varphi) \rightarrow \Box\varphi$	(LR), 5

Line 1 applies (K) to the antecedent of (Löb). In line 2 we use (4) to modalize the antecedent of (Löb) even deeper. Line 3 applies axiom (K) to distribute necessity over the antecedent and the consequent of (Löb). By classical logic, lines 2 and 3 allow us to replace $\Box\Box(\Box\varphi \to \varphi)$ with $\Box(\Box\varphi \to \varphi)$ in the antecedent of the consequent of the formula in line 3. In line 5, thanks to line 1 we could remove one box in the last consequent of the formula in line 4. Now we notice that line 5 is just the premise for an application of (LR) and we apply this rule. □

9.3.3 GL vs. Logical Necessity

What difference does it make to read $\Box$ as 'it is provable in the system' rather than as 'it is logically necessary'? Well, (K), (4) and necessitation intuitively speaking hold for both readings. But there are some important differences.

One thing, already mentioned, is that **GL** $\nvdash \Box p \to p$, while all suitable candidates for a modal logic of logical necessity (the main one being S5) prove $\Box p \to p$. And rightly so, for it seems intuitively true that whatever is logically necessary holds.

What about (Löb)? It obviously is a theorem of **GL** (yes, we're sloppy about the distinction between formula and schemata, but this shouldn't cause any misunderstanding). But if we read $\Box$ as logical necessity, it is somewhat difficult to sort out our modal intuitions about (Löb). Notice that our intuitions about necessity validate reflection for all formulas. Among them is:

$$\Box\bot \to \bot$$

This seems right: if a contradiction is necessary, it is true (hopefully, it isn't). Also, the implication seems to be a logical truth itself, and as such should also be necessary. So we have:

$$\Box(\Box\bot \to \bot)$$

as an intuitive truth when $\Box$ is read as logical necessity. At the same time, we (with a few notable exceptions, such as Graham Priest) don't think there are true contradictions – so, a fortiori, we don't think there are necessary contradictions. This gives us:

$$\neg\Box\bot$$

Put those two things together, and we easily have:

$$\neg[\Box(\Box\bot \to \bot) \to \Box\bot]$$

So, when we read $\Box$ as logical necessity, we have an intuitively convincing formula which is the negation of an instance of (Löb)! And in general, it seems false that just because all necessary sentences are true, all sentences are true.

Another way to see why (Löb) is problematic when $\Box$ is read as necessity is this. Substitute $\neg p$ and apply a few trivial classical moves and the fact that $\Diamond\varphi$ is equivalent to $\neg\Box\neg\varphi$:

$$\Box(\Box\neg p \to \neg p) \to \Box\neg p$$

$$\Box\neg(p \wedge \Box\neg p) \to \Box\neg p$$

$$\neg\Box\neg p \to \neg\Box\neg(p \wedge \Box\neg p)$$

$$\Diamond p \to \Diamond(p \wedge \Box\neg p)$$

But the last formula says that if p is possible then it is possible that p and yet p is necessarily false. This surely isn't an intuitively convincing principle of logical necessity.

9.3.4 GL and Deontic Modalities

Multiple sources, when introducing **GL**, mention [67]—a paper titled "The Logical Basis of Ethics" as the source in which the Löb's theorem stated as a modal formula first occurred. However, to our knowledge, none of these sources actually explains what it was doing there. For instance, in Verbrugge [78] all that we can read about it is:

> Ironically, the first time that the formalized version of Lö's theorem was stated as the modal principle [...] was in a paper by Smiley in 1963 about the logical basis of ethics, which did not consider arithmetic at all.

We'd like to be more specific, and so a short digression about how (Löb) entered the stage follows.

Smiley is developing the ideas from Anderson [1], where an attempt is made to define a deontic modality O (*it is obligatory that*) in terms of an alethic one (*it is necessary that*). Anderson's basic idea is to define:

$$O\varphi =_{df} \Box(\neg\varphi \to S)$$

where S is an unspecified constant expressing the claim that some sanction is applied. Given this analysis, a modal logic of obligation is obtained via a translation from a modal logic of necessity (as far as Anderson's system is concerned, the resulting system is **S4** with O instead of $\Box$).

Smiley [67] contains an interesting discussion of a philosophical concern raised by Nowell-Smith and Lemmon [60] as to whether S is supposed to contain a deontic aspect, or is it supposed to be purely factive, and different difficulties arising in these two cases, but let's put these issues aside. Smiley's point, however, is that a contrapositive reformulation of Anderson's account makes the connection between the left-hand side and the right-hand side more intuitive. Instead of meeting a sanction if ϕ is not performed, Smiley talks about being a consequence of a moral code, so that:

$$O\varphi \equiv \Box(T \rightarrow \varphi)$$

where T is a moral code. Now, Smiley argues, we can ask what inferential principles hold for O no matter which particular T is chosen.

An important difference in Smiley's approach, however, is that the above equivalence is not definitional, and so the respective formulas aren't replaceable in all contexts. This is like treating the equivalence as an assumption (to which (Nec) cannot be applied) rather than as an axiom schema. This reading, he claims, validates (K), application of (Nec) to tautologous formulas, but not axioms (4) or (M). Accordingly, he conjectures that the right modal logic will comprise exactly these.

Further on, Smiley discusses another reading, on which $O\varphi$ holds if φ follows from the totality of obligatory propositions. But then, there is no single sentence expressing this totality, and so instead of using T as a sentential symbol, Smiley considers using it as an operator, so that $T\varphi$ means that φ belongs to this totality. While one might object that it is not clear what philosophical progress can be made by analysing being an obligatory sentence in terms of following from the totality of obligatory sentences (especially as, presumably, all obligatory sentences that follow from the totality of obligatory sentences are already in it), we can still ask what formal properties O thus defined would have. Smiley's reply is that given that such operators aren't treated in any modal logic, we should turn to arithmetic and the arithmetical provability predicate, which, arguably, might have similar formal properties as 'being a consequence of a moral code' (so the claim in [78] that the paper didn't consider arithmetic at all is a bit hasty). A this point, Smiley observes that on this arithmetical reading, all tautologous formulas are theorems, (K) and (4) hold, and so do *modus ponens* and (Nec). Then, Smiley mentions Löb's theorem saying:

> ... and Kripke has pointed out to me that this proof can itself be arithmetised to provide a proof of the formula $O(OA \rightarrow A) \rightarrow OA$.
> [67, 244]

So, indeed, Smiley does mention the formula in the context of ethics.

Now, just a few words about what *doesn't* happen in the paper. Smiley doesn't explain how (Löb) would be understood if the modality is interpret as a deontic modality, doesn't discuss any philosophical motivations for (Löb) in this interpretation (independent of the behavior of provability in arithmetic), and doesn't propose (Löb) as an additional axiom of a modal logic of obligation.

Come to think of it, (Löb) in the deontic reading, doesn't seem too plausible. On one hand, it should be the case that *whatever* should be the case happens (i.e.

obligations should be obeyed). On the other hand, it seems unintuitive that just because of that, simply anything *whatsoever* should be the case.

9.3.5 GL and Formal Provability

We know **S4** turned out inadequate with respect to formal provability predicate. **GL** does a much better job. To elaborate, we first need to explain the relation between $\mathcal{L}_M$ and $\mathcal{L}_{\mathbf{PA}}$ that will underlie what follows.

A mapping from propositional variables of $\mathcal{L}_M$ to the set of sentences of $\mathcal{L}_{\mathbf{PA}}$ is called an *arithmetical realization*. In a sense, an arithmetical realization tells us which variables are to be interpreted as which sentences of arithmetic. Given an elementary presented theory **T**, any arithmetical realization r can be extended to a **T**-interpretation $r_{\mathbf{T}}(\varphi)$ of a modal formula, by the following conditions:

$$r_{\mathbf{T}}(\bot) = \bot \quad r_{\mathbf{T}}(\top) = \top$$

$$r_{\mathbf{T}}(p) = r(p) \text{ for any variable } p$$

$$r_{\mathbf{T}}(\varphi \to \psi) = r_{\mathbf{T}}(\varphi) \to r_{\mathbf{T}}(\psi)$$

$$r_{\mathbf{T}}(\Box\varphi) = \mathtt{Prov}_{\mathbf{T}}(\ulcorner r_{\mathbf{T}}(\varphi)\urcorner)$$

If you worry that $\mathcal{L}_{\mathbf{PA}}$ doesn't really contain $\bot$ and $\top$, feel free to replace them with any $\mathcal{L}_{\mathbf{PA}}$-formulas that are, respectively, refutable and provable by pure logic. Let's call the set of all possible **T**-interpretations of $\varphi \in \mathcal{L}_M$ (under all possible realizations) $\varphi_{\mathbf{T}}$.

Given the correlation between the axioms and rules of **GL** and the Hilbert-Bernay's conditions and Löb's theorem, adequacy of **GL** at least in one direction is clear:

Fact 3.8 **GL** *is sound with respect to the arithmetical interpretation, that is:*

$$\textit{If } \mathbf{GL} \vdash \varphi, \textit{ then } \mathbf{PA} \vdash \varphi_{\mathbf{T}}.$$

(where by $\mathbf{PA} \vdash \varphi_{\mathbf{T}}$ *we mean that* **PA** *proves all the members of* $\varphi_{\mathbf{T}}$*).*

In fact, implication in the opposite direction also holds, provided that **T** is Σ_1-sound, so that the claim can be strengthened to equivalence [71]:

Theorem 3.9 (Solovay Completeness) *If* **T** *is* Σ_1*-sound, then for any* $\varphi \in \mathcal{L}_M$*:*

$$\mathbf{GL} \vdash \varphi \quad \textit{if and only if} \quad \mathbf{T} \vdash \varphi_{\mathbf{T}}.$$

This shows that given a sensible arithmetical theory, those principles of its formal provability predicate that are provable in arithmetic are adequately axiomatized by **GL**. The proof lies beyond the scope of this survey, but the general strategy can be quickly described. Assume **GL** $\nvdash \varphi$. Then, by the results to be described in Sect. 9.3.6 (feel free to read this passage again after reading that section) there is a finite transitive and reversely well-founded model such that for some w in it, $w \nVdash \psi$. Since the set of worlds in the model W is finite, we can safely identify W with an initial segment of natural numbers $= \{1, 2, \ldots, n\}$ with $w = 1$ and $1Ri$ just in case $1 < i \leq n$. The tricky part now, the part for which Solovay is deservedly famous, is using this arithmetical counterpart of W to construct an interpretation such that the arithmetical theory fails to prove the arithmetical interpretation of φ.

9.3.6 Relational Semantics for GL

We have drawn a connection between **GL** and the formal provability predicate. What about relational semantics for **GL**, though? As it turns out [63], there is a natural class of relational models with respect to which **GL** is sound and complete.

Theorem 3.10 **GL** *is sound and complete with respect to the class of finite frames in which R is transitive and irreflexive.*

There is a somewhat different class of frames with respect to which **GL** is sound and complete. We say that the accessibility relation R is *reversely well-founded* in W just in case every non-empty subset X of W has an R-maximal element (that is, a $w \in X$ such that $\neg\exists w' \in W\ wRw'$).

Theorem 3.11 **GL** *is sound and complete with respect to transitive and reversely well-founded frames.*

Notice that there is a connection between these two. Any reversely well-founded R is irreflexive, and a transitive R on a finite set is reversely well-founded just in case it is irreflexive. The result can be strengthened:

Theorem 3.12 **GL** *is sound and complete with respect to finite transitive and reversely well-founded frames.*

Since the proof employs a construction that given a formula to be checked gives an upper limit on the finite size of models to be checked, the proof by the same token proves the decidability of **GL**.

The full proof of weak completeness (that is, the one that applies to theoremhood, read on for details) is beyond the scope of this survey. To give you a taste, however, we'll run the following interesting part of the argument to the effect that if (Löb) holds in a frame, its accessibility relation is reversely well-founded. We'll argue by contraposition, by showing that if a frame isn't reversely well-founded, there is a possible world in it and a forcing relation over it, such that (Löb) fails there.

So assume R is not reversely well-founded. This means there is a set $X \subseteq W$ such that the elements of X constitute an infinite chain $w_1 R w_2 R w_3 \ldots$. Take $\Vdash$ such that $w \Vdash p$ for all $w \in W \setminus X$ and $w' \Vdash \neg p$ for all $w' \in X$. Pick an arbitrary $w \in X$. Now we want to show that the antecedent of (Löb), $\Box(\Box p \rightarrow p)$, holds in

w. This requires showing that $\Box p \to p$ holds in any world accessible from w. So assume wRv. We'll want to show $v \Vdash \Box p \to p$.

Either $v \in X$ or $v \notin X$. If the former, then v can access at least one world in the infinite chain. So for some $u \in X$, vRu. Since p is false in all elements of X we have $u \nVdash p$ and so $v \Vdash \Diamond\neg p$, that is $v \Vdash \neg\Box p$. But this classically entails $v \Vdash \Box p \to p$. If the latter, $v \Vdash p$, and classically $v \Vdash \Box p \to p$.

Either way, if wRv, $v \Vdash \Box p \to p$. Since our choice of v was arbitrary, and the only assumption was that wRv, this means that $w \Vdash \Box(\Box p \to p)$. This is the antecedent of (an instance) of (Löb). On the other hand, w is in a chain in X, and so it can access a world where p fails, and so $w \Vdash \neg\Box p$, which is the negation of (our instance of) (Löb).

Notice that the property of being conversely well-founded isn't first-order definable.[2] That is, there is no first-order formula containing the binary predicate letter R which holds in a model just in case R is conversely well-founded. For suppose there is such a formula ψ. Introduce infinitely many new constants $c_1, c_2, \ldots$. Consider the infinite set of formulas composed of ψ and $\{c_i R c_j \mid i < j\}$ (which jointly state the existence of an infinite chain). Each finite subset of that set is satisfiable in a model (take any finite conversely well-founded model, where there are more objects than constants under consideration). But then, by *compactness theorem for first-order classical languages* (which in one of its formulations says that if any finite subset of a set of first-order formulas is satisfiable, then so is the whole set), the whole set has a model. Among other things, this model makes ψ true (in which no new constants occur), and yet, R in it cannot be conversely well-founded, because it has to contain an infinite R-chain of objects corresponding to the new constants.

Compactness (to be elaborated on in Sect. 9.3.7), as used in the above argument, holds for first-order logic, and our argument was about first-order definability, so its use in this context is legitimate. This issue shouldn't be confused with the question of compactness of **GL**, because, as it will turn out, **GL** itself is not compact.

One remark: soundness and completeness in the above theorems is taken in the *weak* sense: ψ is valid in all finite, transitive and reversely well-founded frames just in case it is a theorem of **GL**. This sense is to be distinguished from strong soundness and completeness. To introduce this notion we have to define **GL**-*derivability* first. We say that ψ is **GL**-derivable from (a finite or infinite) premise set Γ ($\Gamma \vdash_{\mathbf{GL}} \psi$) just in case there is a proof of ψ from the axioms of **GL** and formulas belonging to Γ by the rules of **GL**, provided (Nec) is applied only to theorems of **GL** (alternatively, iff there is a proof of ψ from *theorems of* **GL** and elements of Γ by means of *modus ponens* only; the latter formulation has the advantage that the deduction theorem to the effect that $\vdash \varphi \to \psi$ just in case $\varphi \vdash \psi$ applies to derivability thus defined). Now, the relevant strong completeness claim is that $\Gamma \vdash_{\mathbf{GL}} \psi$ just in case in every possible world in every finite, transitive and reversely well-founded model, if all elements of Γ are true in it, then so is ψ. Alas, the claim is false—strong

[2]See [14] for an extensive introduction to issues related to definability of properties of frames.

completeness for **GL** fails, pretty much for the same reasons for which compactness fails for **GL**. We'll explain this in more detail soon.

Given the arithmetical soundness and completeness, relational semantics provides us with a handy tool for showing that a certain claim about provability is not provable in **PA**: to show this it is enough that its modal counterpart is not provable in **GL**, and given the relational semantics, to show this it is enough to construct a transitive reversely well-founded model making the claim false. A nice example [15] is that of

$$\Box(\Box p \vee \Box\neg p) \rightarrow (\Box p \vee \Box\neg p)$$

Is it provable in **GL**? The answer is negative. Consider a model composed of three possible worlds a, b, c such that $aRb, aRc, c \Vdash \neg p, b \Vdash p$. Since b and c are blind worlds, $b, c \Vdash \Box p, \Box\neg p, \Box p \vee \Box\neg p$, and so $a \Vdash \Box(\Box p \vee \Box\neg p)$. Yet, a sees a world where p and a world where $\neg p$, and so $a \nVdash \Box p \vee \Box\neg p$.

This means there is an arithmetical sentence σ which can be assigned to p by a realization, such that the negation of the resulting **PA**-interpretation of the formula in question can be consistently added to it.

In other words, it is consistent with **PA** that it is provable that either σ is provable or refutable, but nevertheless σ is neither provable nor refutable. And this clearly is a nice little fact that will score you some extra points in a late night conversation with a stranger in a pub.

9.3.7 Compactness Failure for GL

Compactness was already mentioned in Sect. 9.3.6. We say that a logic **L** is *compact* just in case for any set Γ of formulas of the appropriate formal language, if every finite subset of Γ has a model suitable for **L**, the whole Γ has such a model. In the case of **GL**, compactness would mean that for every set Γ of formulas of $\mathcal{L}_M$, if every finite subset of Γ has a transitive and conversely well-founded model, then so does Γ. Interestingly, compactness fails for **GL**.

To see why, first reflect on the meaning of $\Diamond$ in this context. $\Diamond\varphi$ is defined as $\neg\Box\neg\varphi$, and so while '$\Box\varphi$' is read as 'φ is provable', the intuitive reading of '$\Diamond\varphi$' is 'φ is not refutable' or 'the negation of φ is not provable'.

Now we can proceed with the reasoning. Take an infinite assembly of propositional variables $p_0, p_1, \ldots$, and the following infinite set of formulas:

$$C = \{\Diamond p_0\} \cup \{\Box(p_i \rightarrow \Diamond p_{i+1}) \mid i \in \mathbb{N}\}$$

Every finite subset of C has a transitive reversely well-founded model. There is no space for the general argument here, but to see why this is plausible consider the finite subset of C:

$$C_2 = \{\Diamond p_0, \Box(p_0 \to \Diamond p_1), \Box(p_1 \to \Diamond p_2)\}$$

Take (the transitive closure of) the frame composed of w, w_0, w_1, w_2 only, such that $wRw_0Rw_1Rw_2$. Whether propositional variables are true at w is irrelevant, for the other worlds take the forcing relation such that $w_i \Vdash p_i$ and $w_i \nVdash p_j$ if $i \neq j$. We'll argue that $w \Vdash C_2$.

- Since wRw_0 and $w_0 \Vdash p_0$, $w \Vdash \Diamond p_0$.
- $w_2 \nVdash p_0$, so $w_2 \Vdash p_0 \to \Diamond p_1$. For a similar reason, $w_1 \Vdash p_0 \to \Diamond p_1$. Moreover, since $w_1 \Vdash p_1$, $w_0 \Vdash \Diamond p_1$, and so $w_0 \Vdash p_0 \to \Diamond p_1$. This shows that $p_0 \to \Diamond p_1$ holds in all worlds accessible from w. So $w \Vdash \Box(p_0 \to \Diamond p_1)$.
- A perfectly analogous argument goes for $w \Vdash \Box(p_1 \to \Diamond p_2)$.

However, C doesn't have a transitive and conversely well-founded model. For suppose there is a model with a w such that $w \Vdash C$. Define:

$$X = \{v \mid wRv \wedge \exists i\ v \Vdash p_i\}$$

That is, collect all the possible worlds accessible from w where at least one p_i holds. We have $\Diamond p_0 \in C$, so $w \Vdash \Diamond p_0$, and X is non-empty, say wRw_0, $w_0 \Vdash p_0$. Since $w \Vdash \Box(p_0 \to \Diamond p_1)$, $w_0 \Vdash p_0 \to \Diamond p_1$. So $w_0 \Vdash \Diamond p_1$, and there is a $w_1 \in X$ such that $w_1 \neq w_0$ (R is irreflexive) such that w_0Rw_1 and $w_1 \Vdash p_1$. But $w \Vdash \Box(p_1 \to \Diamond p_2)$ and (by transitivity) wRw_1, and so $w_1 \Vdash p_1 \to \Diamond p_2$. Therefore $w_1 \Vdash \Diamond p_2$, and so w_1 has to see yet another member of X, etc. In short: X has to contain an infinite chain, which contradicts the assumption that R is conversely well-founded.

The example of C can be also used to explain why strong completeness fails for **GL**. We already know C has no transitive, conversely well-founded model. Another way to say this is that C semantically entails $\bot$ (with respect to this class of frames): $C \models \bot$. Yet, $\bot$ is not derivable from C, $C \nvdash \bot$—for any proof from C could only use a finite number of premises from C, and we already know that no finite subset of C entails (and so, by soundness, proves) $\bot$.

The fact that compactness fails is partially responsible for why the semantic completeness for **GL** is a bit more tricky than one for a more usual modal logic. Normally, in the proof, one constructs a canonical model by taking infinite sets of consistent formulas as possible worlds; for **GL**, however there are syntactically consistent sets of formulas which nevertheless aren't semantically coherent.

9.3.8 Letterless Sentences and the Normal Form Theorem for GL

A *letterless sentence* of $\mathcal{L}_M$ is a formula built from the classical and modal connectives, devoid of propositional variables, and containing only $\bot$ among its atomic formulas. Instead of preceding φ with n boxes, we'll write $\Box^n\varphi$.

One of the reasons why letterless sentences are interesting is because some of them formally certain fairly natural statements. For instance, $\neg\mathtt{Prov}_{\mathbf{T}}(\ulcorner\bot\urcorner)$ is (=formalizes) the consistency statement (hopefully true, but under standard conditions unprovable), $\mathtt{Prov}_{\mathbf{T}}(\ulcorner\neg\mathtt{Prov}_{\mathbf{T}}(\ulcorner\bot\urcorner)\urcorner)$ is the provability of consistency (false), $\neg\mathtt{Prov}_{\mathbf{T}}(\ulcorner\bot\urcorner) \rightarrow \neg\mathtt{Prov}_{\mathbf{T}}(\ulcorner\neg\mathtt{Prov}_{\mathbf{T}}(\ulcorner\bot\urcorner)\urcorner)$ expresses the second incompleteness theorem, etc.

One of the interesting uses of **GL** (to which we will move soon) relate to the existence of a certain decision procedure, whose description employs the notion of a normal form. By a *normal form of a letterless sentence* φ we mean a truth-functional combination of sentences of the form $\Box^i\bot$.

Theorem 3.13 (Normal form theorem (Boolos)) *For any letterless formula $\varphi \in \mathcal{L}_M$, there is a normal form letterless ψ such that* $\mathbf{GL} \vdash \varphi \equiv \psi$.

The normal form theorem, while at first it might seem abstract, will come handy quite soon, see Sect. 9.3.11.

9.3.9 ω-Consistency

An arithmetical theory $\mathbf{T}$ is *ω-inconsistent* just in case there is a formula $\psi(x)$ in the language of $\mathbf{T}$ such that for each $n \in \mathbb{N}$ we have $\mathbf{T} \vdash \psi(\overline{n})$, and yet, we also have $\mathbf{T} \vdash \exists x\ \neg\psi(x)$. $\mathbf{T}$ is *ω-consistent* iff it is not ω-inconsistent.

ω-inconsistency doesn't entail inconsistency *simpliciter*. After all, $\mathbf{T}$ can have a non-standard model, where all the standard numbers have the property expressed by ψ, and yet, some non-standard number (not named by a numeral) is a witness to $\exists x\ \neg\psi(x)$ (see [29, 43, 46] for more details on non-standard models of arithmetic). By the same token, consistency doesn't entail ω-consistency either.

Now, in the standard interpretation of **GL**, $\Box$ represents provability. What is represented by $\Diamond$? Well, by definition $\Diamond\varphi$ just in case $\neg\Box\neg\varphi$, and so $\Diamond\varphi$ holds just in case the negation of φ is not provable – that is, just in case φ is consistent (with the underlying axiomatic system of arithmetic). In this sense, **GL** can be thought of as a logic of consistency.

The question arises—what is the logic of ω-consistency? Can its propositional principles be axiomatized? The answer is easier than expected.

Let $\omega\mathtt{Con}_{\mathbf{T}}(\ulcorner\varphi\urcorner)$ be the arithmetical formula expressing the ω-consistency of φ with an elementary presented theory $\mathbf{T}$. If we think of it as $\Diamond\varphi$, then $\Box\varphi$ corresponds to $\neg\omega\mathtt{Con}_{\mathbf{T}}(\ulcorner\neg\varphi\urcorner)$. Accordingly, let's modify the definition of realisation so that:

$$r_{\mathbf{T}}(\Box\varphi) = \neg\omega\mathtt{Con}_{\mathbf{T}}\ulcorner\neg r_{\mathbf{T}}(\varphi)\urcorner.$$

Theorem 3.14 *Given the above conventions, the set of always provable formulas is axiomatized by* **GL**, *and the set of always true formulas is axiomatized by* **S** *(an extension of* **GL** *described in Sect.* 9.4.1*).*

9.3.10 Provability in Analysis

Roughly speaking, analysis is second-order arithmetic, that is, arithmetic with second-order logic available, where the infinity of instances of the induction schema is replaced with a single induction axiom (see however [66] for more details and variety of higher-order systems):

$$\forall P\ [(P(0) \wedge \forall x\ (Px \rightarrow P(Sx)) \rightarrow \forall x\ Px]$$

What is the logic of provability of analysis? Again, no surprises: it is **GL**.

Now imagine we want to strengthen the system with the so-called *ω-rule* will allows to infer $\forall x\ \psi(x)$ from $\psi(\overline{n})$ for all $n \in \mathbb{N}$. Note: ψ is ω-inconsistent with **T** just in case $\neg\psi$ is derivable from **T** by *one* application of the ω-rule.

It turns out that this move doesn't change the underlying logic of provability. Still, **GL** is the modal logic of provability in analysis with the ω-rule.

9.3.11 Applications of GL

Solovay completeness allows us to use **GL** to make inferences about provability predicates of elementary presented theories. Let's call a sentence of $\mathcal{L}_{\mathbf{PA}}$ a *constant sentence of* **PA** if it belongs to the least set of formulas containing $\bot$ (if you don't like $\bot$ being explicitly in $\mathcal{L}_{\mathbf{PA}}$ take it to be $0 = 1$), closed under classical connectives, such that if ψ is a constant sentence, then so is $\mathtt{Prov}_{\mathbf{PA}}(\ulcorner\psi\urcorner)$ (the notion generalizes to other theories). The notion was introduced by Harvey Friedman, who asked whether there is an effective decision procedure for evaluating the truth-value of constant sentences. The answer is positive and relies on Theorem 3.13. The procedure is this:

- Take the letterless $\varphi \in \mathcal{L}_{\mathbf{PA}}$ and find the letterless $\psi \in \mathcal{L}_M$ such that $r_{\mathbf{T}}(\psi) = \varphi$ (notice, for letterless sentences, the choice of r is irrelevant).
- Put ψ in the normal form.
- $\Box^i\bot$ has the same truth value as $\bot$, so delete all $\Box^i$ in front of $\bot$.
- We are left with a sentence in the non-modal language of propositional logic, devoid of propositional variables. Evaluate it. It is true just in case so is φ.

One nice general result about **GL** that has interesting consequences is De Jongh-Sambin fixed point theorem. To introduce it, some preliminaries are needed. A formula φ of $\mathcal{L}_M$ is said to be *modalized in the propositional variable p* just in case every occurrence of p in φ is within the scope of $\Box$ (this also applies to vacuous cases, so formulas devoid of p are also modalized in p). A formula is called a p*–formula* if it contains no variable other than p.

A formula φ is called a *fixed point* of a formula ψ with respect to variable p just in case φ contains only those sentence letters that occur in ψ, doesn't contain any occurrence of p, and:

$$\mathbf{GL} \vdash \Box(p \equiv \psi) \equiv \Box(p \equiv \varphi)$$

Theorem 3.15 (De Jongh-Sambin fixed point theorem) *If ψ is modalized in p, there is a fixed point φ for ψ relative to p.*

This form of the theorem is useful for eliminating apparent self-reference from arithmetical sentences: finding their provably equivalent counterparts which do not contain self-reference. For instance, if ψ is $\neg\Box p$, the fixed point is $\neg\Box\bot$. So, by fixed point theorem:

$$\mathbf{GL} \vdash \Box(p \equiv \neg\Box p) \equiv \Box(p \equiv \neg\Box\bot)$$

By arithmetical soundness of **GL**, for any arithmetical sentence χ, we have $\mathbf{PA} \vdash \chi \equiv \neg\mathtt{Prov}_{\mathbf{PA}}(\ulcorner\chi\urcorner)$ just in case $\mathbf{PA} \vdash \chi \equiv \neg\mathtt{Prov}_{\mathbf{PA}}(\ulcorner\bot\urcorner)$. So χ, equivalent to its own unprovability turns out to be also provably equivalent to the consistency statement. A few more examples. The fixed point of $\Box p$ is $\top$. So if we take χ provably equivalent to its own provability, the fixed point theorem tells us that we can equally well describe χ without self-reference, in the sense that:

$$\mathbf{PA} \vdash \chi \equiv \mathtt{Prov}_{\mathbf{PA}}(\ulcorner\chi\urcorner) \text{ iff } \mathbf{PA} \vdash \chi \equiv \top$$

and similarly:

$$\mathbf{PA} \vdash \chi \equiv \mathtt{Prov}_{\mathbf{PA}}(\ulcorner\neg\chi\urcorner) \text{ iff } \mathbf{PA} \vdash \chi \equiv \mathtt{Prov}_{\mathbf{PA}}(\bot)$$

$$\mathbf{PA} \vdash \chi \equiv \neg\mathtt{Prov}_{\mathbf{PA}}(\ulcorner\neg\chi\urcorner) \text{ iff } \mathbf{PA} \vdash \chi \equiv \bot$$

The utility of the fixed point theorem might perhaps become more clear if we look at a somewhat different formulation. Let $\psi(p)$ be a formula containing p among propositional variables occurring in it.

Theorem 3.16 (De Jongh-Sambin, second formulation) *For any $\psi(p) \in \mathcal{L}_M$ modalized in p, there is a formula $\varphi \in \mathcal{L}_M$ containing only variables from ψ, not containing p, such that:*

$$\mathbf{GL} \vdash \varphi \equiv \psi(\varphi)$$

Any fixed points of $\psi(p)$ are provably equivalent in **GL**.

In a sense, fixed point theorem is the modal counterpart of the Diagonal Lemma. This formulation makes it clear why the theorem is called a fixed-point theorem. Generally, a fixed point of a function f is an argument such that $f(x) = x$, and φ is the fixed point of ψ because $\psi(\varphi) \equiv \varphi$.

Moreover, the proof is effective, in the sense that it provides a recipe for constructing appropriate fixed points. Some examples of formulas and their fixed point are:

Formula	Fixed Point
$\Box p$	$\top$
$\Box\neg p$	$\Box\bot$
$\neg\Box p$	$\neg\Box\bot$
$\neg\Box\neg p$	$\bot$
$q \wedge \Box p$	$q \wedge \Box q$

Consider the third formula. $\neg\Box p$ says that p isn't provable. Its fixed point is $\neg\Box\bot$, and so, by the fixed point theorem, we have:

$$\mathbf{GL} \vdash \neg\Box\bot \equiv \neg\Box(\neg\Box\bot)$$

But the arithmetical realization of $\neg\Box\bot$ for **T** is $\texttt{Con}(\mathbf{T})$. So the above formula (from left to right):

$$\neg\Box\bot \rightarrow \neg\Box(\neg\Box\bot) \tag{G2}$$

represents the formalized version of Gödel's second incompleteness theorem: if the theory is consistent, it doesn't prove its own consistency.

In fact, Gödel's second incompleteness can be fairly easy reached in **GL** without the full power of the fixed point theorem:

1. $\Box(\Box\bot \rightarrow \bot) \rightarrow \Box\bot$	(L)
2. $\neg\Box\bot \rightarrow \neg\Box(\Box\bot \rightarrow \bot)$	contraposition, 1
3. $\neg\Box\bot \rightarrow \neg\Box(\neg\Box\bot)$	def. of $\neg$, 2

Second incompleteness is about not being able to prove the consistency claim. This, however, can be strengthened to the undecidability of consistency, because in **GL** it is also possible to prove that if the inconsistency is not provable, then neither is the inconsistency claim:

1. $\Box\bot \rightarrow \bot$	(M)
2. $\Box(\Box\bot \rightarrow \bot)$	(Nec), 1
3. $\Box\Box\bot \rightarrow \Box\bot$	(Distr), 2
4. $\neg\Box\bot \rightarrow \neg\Box\Box\bot$	Contraposition, 3

The modally formalized version of Gödel's first incompleteness theorem is:

$$\neg\Box\bot \rightarrow (\Box(p \equiv \neg\Box p) \rightarrow \neg\Box p) \tag{G1}$$

It also can be proved within **GL** (we'll use "CL" to mark moves made by classical propositional logic)

1.	$\Box p \rightarrow p$	(M)
2.	$\Box\neg p \rightarrow \neg p$	(M)
3.	$\Box p \wedge \Box\neg p \rightarrow \bot$	CL, 1, 2
4.	$(\Box p \equiv \Box\neg p) \wedge \Box p \rightarrow \Box p \wedge \Box\neg p$	CL
5.	$(\Box p \equiv \Box\neg p) \wedge \Box p \rightarrow \bot$	CL, 3, 4
6.	$(\Box p \equiv \Box\neg p) \wedge \Box p \rightarrow \Box\bot$	CL, 5
7.	$\Box(p \equiv \neg p) \rightarrow (\Box p \equiv \Box\neg p)$	(Distr)
8.	$\Box(p \equiv \neg p) \wedge \Box p \rightarrow \Box\bot$	CL, 6,7
9.	$\neg(\Box(p \equiv \neg p) \rightarrow \neg\Box p) \rightarrow \Box\bot$	CL, 8
10.	$\neg\Box\bot \rightarrow (\Box(p \equiv \neg p) \rightarrow \neg\Box p)$	CL, 9

Another argument which is quite easy to run with **GL** at hand, is for the claim that no sentence consistent with **PA** can imply all reflection principles. For suppose that for any φ

1. $\mathbf{GL} \vdash S \rightarrow (\Box\varphi \rightarrow \varphi)$	Assumption
2. $\mathbf{GL} \vdash S \rightarrow (\Box\neg S \rightarrow \neg S)$	Instance of 1
3. $\mathbf{GL} \vdash \Box\neg S \rightarrow \neg S$	CL, 2
4. $\mathbf{GL} \vdash \neg S$	(Löb), 3

The result means that in **PA** (and in any sensible **T** extending **PA**) we cannot finitely axiomatize reflection principles involving the provability predicate of any sensible $\mathbf{T}'$ extending **PA** (**PA** included).

Coming back to the fixed point theorem, observe that it doesn't hold for all formulas of $\mathcal{L}_M$. For instance, a fixed point of p (or $\neg p$) itself would be a letterless sentence S_p such that $\mathbf{GL} \vdash S_p \equiv p$ (or $\mathbf{GL} \vdash S_{\neg p} \equiv \neg p$), and it can be easily proven by induction on formula length that there is no such a letterless sentence.

9.4 Close Kins of GL

9.4.1 Modal Logic S

What happens, however, when instead of asking about those principles that are provable in the background arithmetic, we ask about those principles which are true in the standard model?

Modal system **S** is defined as the closure of **GL** together with (M) under *modus ponens* and substitution. Notice: (Nec) is inadmissible, apart from the job it does in generating the theorems of **GL**, included in **S**. Otherwise we could argue:

$$\begin{array}{ll} 1.\ \Box\bot \to \bot & \text{(M)} \\ 2.\ \Box(\Box\bot \to \bot) & \text{(Nec), 1} \\ 3.\ \Box\bot & \text{(Löb), 2} \\ 4.\ \bot & \text{CL, 1, 3} \end{array}$$

The failure of (Nec) means that **S** is not a normal modal logic. Interestingly, despite the failure of (Nec) in **S**, **S** validates the inference from $\mathbf{S} \vdash \varphi$ to $\mathbf{S} \vdash \Diamond\varphi$, which is not validated by **GL**. For reflection for $\neg\varphi$ gives $\mathbf{S} \vdash \Box\neg\varphi \to \neg\varphi$, contraposition gives $\mathbf{S} \vdash \varphi \to \neg\Box\neg\varphi$. The result then follows by the definition of $\Diamond$ and modus ponens.

Now, let $S(\varphi)$ be:

$$(\Box\varphi_1 \to \varphi_1) \wedge \cdots \wedge (\Box\varphi_k \to \varphi_k)$$

where $\Box\varphi_1, \ldots, \Box\varphi_k$ are all subformulas of φ of the form $\Box\chi$. The following holds:

Theorem 4.1 (II Solovay Completeness) *If* **T** *is sound (that is, for any* $\varphi \in \mathcal{L}_{\mathbf{PA}}$, *if* $\mathbf{T} \vdash \varphi$, *then* $\mathbb{N} \models \varphi$*), the following conditions are equivalent for any* $\psi \in \mathcal{L}_M$*:*

$$\mathbf{S} \vdash \psi$$

$$\mathbf{GL} \vdash S(\psi) \to \psi$$

$$\mathbb{N} \models \psi_{\mathbf{T}}$$

In a sense, Theorem 4.1 tells us that the only claims always true but not always provable are those which are required to make reflection hold.

Since **S** isn't a normal modal logic, it doesn't have straightforward relational models. A semantics for **S** in terms of the so-called *tail models* have been given by Visser [80].

9.4.2 Strong Provability and Grzegorczyk's Grz

Consider exending a realization r to the *Grzegorczyk* **T***-interpretation* $r^G_{\mathbf{T}}$, which differs from **T**-interpretation in what it does with the modal operator:

$$r^G_{\mathbf{T}}(\Box\varphi) = r^G_{\mathbf{T}}(\varphi) \wedge \mathtt{Prov}_{\mathbf{T}}(\ulcorner r^G_{\mathbf{T}}(\varphi)\urcorner)$$

Thus, while the standard interpretation reads the box as *provable*, Grzegorczyk interpretation reads it as *true and provable* [26]. The arithmetically complete logic of strong provability [15] is **Grz**, which is obtained from **S4** by adding:

$$\Box(\Box(\psi \rightarrow \Box\psi) \rightarrow \psi) \rightarrow \psi \tag{Grz}$$

There is an interesting connection between most of the axioms of **Grz** and the properties of strong provability provable in a somewhat weaker system **K4** (that is, **GL** without (Löb)) [70]. Define $[s]\varphi$ as $\Box\varphi \wedge \varphi$. Then, if **K4** $\vdash \varphi$, then also **K4** $\vdash [s]\varphi$. **K4** proves (all instances of) the following:

$$[s]\varphi \wedge [s](\varphi \rightarrow \psi) \rightarrow [s]\psi$$
$$[s]\varphi \rightarrow [s][s]\varphi, \quad [s][s]\varphi \rightarrow [s]\varphi$$
$$[s]\varphi \rightarrow \varphi$$

Note however, that if we replace $\Box$ in (Grz) with $[s]$, the result is not a **K4** theorem schema (it is invalidated by a single-world model, where ψ is false in the only possible world).

By the way, strong provability can be used in obtaining **GL** in yet another manner. For suppose you want to enrich **K4** with something that does the job of the diagonal lemma at the arithmetical level. To do this is, we need to say that if something is the consequence of a diagonalization, it should be a theorem. One way to capture this intuition is to add the *Diagonalization Rule*:

$$\text{If } \vdash [s](p \equiv \varphi(p)) \rightarrow \psi, \text{ then } \vdash \psi. \tag{DR}$$

De Jongh has proven that **GL** is closed under (DR), and Smoryński has shown that **K4**+(DR) coincides with **GL**.

Grz is another example of a logic of provability into which a translation of intuitionistic **IPC** can be given. Extensions of **Grz** to the context of the logic of proofs have been further studied in [58, 59].

9.4.3 *Provability of Σ_1-Sentences (GLV and GLSV)*

Σ_1 sentences have the specific property that for any such sentence σ

$$\mathbf{PA} \vdash \sigma \to \mathtt{Prov}_{\mathbf{PA}}(\ulcorner\sigma\urcorner).$$

Accordingly, the modal logics of the provability of Σ_1 sentences has been characterized by Visser by first taking **GLV** to be an extension of **GL** with all formulas of the form $p \to \Box p$ (it's crucial that p is a propositional variable, because Σ_1 aren't closed under arbitrary Boolean combinations). The rules of **GLV** are the same as those of **GL**—*modus ponens* and (Nec). Then, **GLSV** has as axioms all the theorems of **GLV** and all instances of reflection, and its only rule of inference is *modus ponens*.

A realization r is a Σ_1-*realization* if for any propositional variable p, $r(p)$ is a Σ_1-sentence. Call a relational model a GLV-model just in case it is finite, irreflexive, transitive and such that for all $w, v \in W$ and all propositional variables p:

$$wRv, w \Vdash p \Rightarrow v \Vdash p$$

(the last condition means that accessibility preserves the satisfaction of propositional variables).

Theorem 4.2 [82, 84]) **GLV** $\vdash \psi$ *just in case* ψ *is valid in all GLV-models just in case for all* Σ_1*-realization* r, **PA** $\vdash r^{PA}(\psi)$. **GLSV** $\vdash \psi$ *iff for all* Σ_1 *realizations* r, $r^{PA}(\psi)$ *is true in the standard model.*

9.5 The Logic of Proofs LP

9.5.1 *Motivations*

One of the reasons why thinking about provability is tricky, especially in the context of first-order theories, is that a universal quantifier is involved. Given that first-order arithmetical theories have non-standard models which contain non-standard numbers, this leads to certain troubles. In general, if we take a model M of an arithmetical theory, it might be the case that $\exists x\ \varphi(x)$ holds in that model, with no $\varphi(\overline{n})$ holding for $n \in \mathbb{N}$, because the existential formula has a non-standard witness.

In particular, this applies to $\mathtt{Prov}_{\mathbf{T}}(\ulcorner\varphi\urcorner)$, which in fact means $\exists x\ \mathtt{Prf}_{\mathbf{T}}(x, \ulcorner\varphi\urcorner)$. The problem is, this formula doesn't entail that for some $n \in \mathbb{N}$, $\mathtt{Prf}_{\mathbf{T}}(\overline{n}, \ulcorner\varphi\urcorner)$. This feature results in certain disparities in the behavior of $\mathtt{Prov}_{\mathbf{T}}(y)$ and $\mathtt{Prf}_{\mathbf{T}}(x, y)$.

The case where this is particularly visible is that of reflection. *Explicit reflection* has the form $\mathtt{Prf}_{\mathbf{T}}(\overline{n}, \ulcorner\varphi\urcorner) \to \varphi$. All instances of explicit reflection are provable in the underlying arithmetical theory **T** (satisfying our standard conditions). For either $\mathbb{N} \models \mathtt{Prf}_{\mathbf{T}}(\overline{n}, \ulcorner\varphi\urcorner)$ or $\mathbb{N} \not\models \mathtt{Prf}_{\mathbf{T}}(\overline{n}, \ulcorner\varphi\urcorner)$. In the former case, then indeed there is a proof of φ in **T**, and since $\mathbf{T} \vdash \varphi$, by classical logic, $\mathbf{T} \vdash \mathtt{Prf}_{\mathbf{T}}(\overline{n}, \ulcorner\varphi\urcorner) \to \varphi$. If, on the other hand, it's not the case that $\mathtt{Prf}_{\mathbf{T}}(\overline{n}, \ulcorner\varphi\urcorner)$, then (since it's a Δ_0 formula, and we assumed **T** to be sufficiently strong) $\mathbf{T} \vdash \neg\mathtt{Prf}_{\mathbf{T}}(\overline{n}, \ulcorner\varphi\urcorner)$, and again, by classical logic, $\mathbf{T} \vdash \mathtt{Prf}_{\mathbf{T}}(\overline{n}, \ulcorner\varphi\urcorner) \to \varphi$. Either way, an arithmetical theory satisfying the standard strength requirements proves explicit reflection, for any $n \in \mathbb{N}$ and any $\varphi \in \mathcal{L}_{\mathbf{PA}}$.

In contrast, due to Löb's theorem, given a consistent and sufficiently strong **T**, *local reflection* for $\mathtt{Prov}_{\mathbf{T}}(x)$:

$$\mathtt{Prov}_{\mathbf{T}}(\ulcorner\varphi\urcorner) \to \varphi$$

is provable only for those formulas, which are theorems of **T**. Indeed, at the level of **GL**, due to (Löb), one cannot consistently add reflection $\Box\varphi \to \varphi$, for otherwise, the reasoning already described in Sect. 9.4.1 can be used to derive contradiction.

Clearly, $\mathcal{L}_M$ lacks the resources to represent explicit reflection, because $\Box$ represents provability, and $\Box\varphi \to \varphi$ can be used to represent local reflection only. A more expressive language to achieve that goal has been devised to underlie the logic of proofs, **LP** [5, 7, 8].

9.5.2 The Language and Axioms of LP

The pure language of the logic of proofs extends the non-modal propositional language with new symbols:

- *proof variables* $(x, y, z, \dots)$ and *proof constants* $(a, b, c, \dots)$,
- three proof operation symbols: *binary application* $(\cdot)$, *binary union* $(+)$ and *unary proof checker* $(!)$,
- *is a proof of* symbol $(:)$.

A *proof polynomial* is either a proof variable, or a proof constant, or is built from proof polynomials by means of $\cdot$, $+$, or $!$. Binary application intuitively corresponds to *modus ponens*, in the sense that if t is a proof of $\varphi \to \psi$, and s is a proof of φ, then $t \cdot s$ is a proof of ψ. The union of two proofs t and s, $t + s$ proves anything that either t or s does. The proof checker operation checks whether a given proof t of φ is correct, and if it is, it yields a proof that t proves φ.

The *is a proof of* symbol is used to construct atomic formulas from proof polynomials and formulas. If t is a proof polynomial and φ is a formula, $t{:}\varphi$ is a formula saying that t *is a proof of* φ, or simply t *proves* φ.[3]

[3]A short historical remark: Gödel suggested the use of explicit proofs to interpret **S4** in a public lecture in Vienna, without describing the logic. The content of the lecture has been published in 1995, and the logic of proofs was formulated independently before that publication.

One rule of **LP** is *modus ponens*. The axioms of **LP** (on top of classical propositional logic) are:

$$t{:}(\varphi \to \psi) \to (s{:}\varphi \to (t \cdot s){:}\psi) \qquad \text{(Application)}$$

$$t{:}\varphi \to \varphi \qquad \text{(Reflection)}$$

$$t{:}\varphi \to !t{:}(t{:}\varphi) \qquad \text{(Proof checker)}$$

$$s{:}\varphi \to (s+t){:}\varphi,\ t{:}\varphi \to (s+t){:}\varphi \qquad \text{(Sum)}$$

Another rule of **LP**, allows to introduce $c{:}\varphi$ as a theorem, whenever φ is an axiom, where c is a new constant in a given proof. In this context, the **LP**-counterpart of (Nec) is a derivable rule.

Fact 5.1 *If* **LP** $\vdash \varphi$, *then for some proof polynomial* p, **LP** $\vdash p{:}\varphi$.

Notice that **LP** is not a normal modal logic: we can't simply treat $t{:}$ as we would treat $\Box$ in a normal modal logic. For instance, (K) for $t{:}$ fails, as this is not generally the case:

$$t{:}(p \to q) \to (t{:}p \to t{:}q)$$

9.5.3 Properties of LP

LP is decidable [53]. It is also sound and complete with respect to provability interpretation in **PA** (where proof polynomials are mapped to appropriate proof codes). Quite some time ago we seemed to have left **S4** behind, as arithmetically inadequate. One of the nice features of **LP** is that it brings **S4** back to the table. A *forgetful projection* of an **LP**-formula φ is a modal formula resulting from replacing all the occurrences of $t{:}(\varphi)$ with $\Box(\varphi)$.

Theorem 5.2 [6]) *The forgetful projection of* **LP** *is* **S4**.

Possible worlds semantics for **LP** (requiring the extension of the standard framework with the so-called *evidence function*) has been developed by Mkrtychev [53] and Fitting [21].

LP itself doesn't allow to express (and a fortiori) prove mixed statements about explicit proofs and provability, which nevertheless seem of indepedent interest. For instance, in the provability semantics the following *explicit-implicit principle* is valid:

$$\neg(t{:}\varphi) \to \Box\neg(t{:}\varphi).$$

9.5.4 *Mixed Logic of Explicit Proofs and Provability (B)*

Such claims can be proven in a mixed logic of explicit proofs and provability, **B** [5]. The axioms are that of **GL** enriched with:

$$t{:}\varphi \to \varphi$$

$$t{:}\varphi \to \Box(t{:}\varphi)$$

$$\neg(t{:}\varphi) \to \Box\neg(t{:}\varphi)$$

and the *Rule of reflection*, which allows to infer $\mathbf{B} \vdash \varphi$ from $\mathbf{B} \vdash \Box\varphi$. Artemov proved also the following:

Theorem 5.3 **B** *is sound and complete with respect to the semantics of proofs and provability in* **PA**.

For a survey of further studies and properties of **LP** and its extensions, see [10] and other developments in [58, 85, 86].

9.6 Formal Logics of Informal Provability

9.6.1 *Motivations*

Informal provability is closely related to what mathematicians do when they prove theorems, rather than to formal provability in an axiomatic system [72]. A sentence is informally provable if it is provable by any commonly accepted mathematical means. According to the proponents of *the standard view* there is no important difference between formal and informal proofs. Any informal proof, they say, is just a sloppy and incomplete version of a fully formalized proof in an appropriate formal theory. Thus, informal provability reduces to formal provability within some axiomatic system, usually to some version of set theory.

Yet, some people disagree with the above picture [37, 47, 51, 57]:

- It is not clear which axiomatic system we should choose to represent informal provability. It seems that the informal notion of provability is unified whereas in different formal systems different theorems are provable.
- It is not clear how to convert an informal proof into a formal one.
- It is not clear whether we should associate each informal proof with exactly one formal proof or with some abstraction class of formal representations of informal proofs, and if yes, how such a class is to be identified.
- It is not clear whether the conversion to formal proofs preserves identity laws for informal proofs. It may be the case, that two substantially different informal proofs are associated with exactly the same formal representation.

- Formal proofs are stated in a fully formalized language. Informal proofs on the other hand are stated in the natural language expanded with additional mathematical vocabulary.
- The role of axioms is different in proofs of these two kinds. In formal proofs axioms are simply one of the syntactically admissible ways of extending a given proof. In an informal proof, axioms partially or fully explicate the meaning of the notions involved.
- The justification of subsequent steps is of a different nature. In formal proofs it's purely syntactical. In informal proofs, mathematicians often refer to semantical notions such as truth-preservation or mathematical intuition.
- The reflection schema, which says that if something is (informally) provable then it is true, is intuitively compelling for informal provability. Yet, as already discussed, for any sensible notion of formal provability, we cannot have it.

Since Gödel , however, there is an agreement as to which principles are intuitively correct for informal provability: those are the principles of **S4**. So, if we were to produce an axiomatization of those principles, which intuitively hold for informal provability, the validity of all the instances of the reflection schema is crucial. For formal provability, by Löb's theorem, we know that the reflection schema is only provable for theorems – and there is no independent philosophical motivation for this restriction to be imposed on informal provability.

One way out might be to strengthen the underlying axiomatic system by brute force by adding all the instances of the reflection schema. One thing to observe, however is that even a small amount of reflection schema turns out to be arithmetically strong:

Fact 6.1 *Let* **T** *be a theory consisting of* **PA** *and all the instances of the reflection schema for* $\mathtt{Prov}_{\mathbf{PA}}(x)$ *restricted to* Π_1 *formulas. Then* $\mathbf{T} \vdash \mathtt{Con}(\mathbf{PA})$.

Proof Let φ be $\forall x\ x \neq x$. It is a Π_1 pure logical contradiction. Let's abbreviate it as $\bot$. By the assumption, $\mathbf{T} \vdash \mathtt{Prov}_{\mathbf{PA}}(\ulcorner\bot\urcorner) \rightarrow \bot$. By classical logic, $\mathbf{T} \vdash \neg\bot$. So we have $\mathbf{T} \vdash \neg\mathtt{Prov}_{\mathbf{PA}}(\ulcorner\bot\urcorner)$. □

But say we're not worried about intuitively obvious claims about provability turning out to be arithmetically strong. Another observation is that as far as formal provability is concerned, we can only consistently add reflection for the *old* formal provability predicate, thus obtaining a *new* formal theory with *new* formal provability predicate, for which reflection still fails to be provable.

Indeed, what we can consistently do still falls short of accepting reflection for the theory that we are working within. Suppose we extend $\mathcal{L}_{\mathbf{PA}}$ with an additional primitive symbol, a new provability predicate P, for which we add to our background arithmetical theory **T** (extending **PA**) all instances of the Hilbert-Bernays conditions and all the instances of the reflection schema, thus obtaining a new theory **TP**. We run into the following problem:

Fact 6.2 *No* **TP** *satisfying the conditions below is consistent.*

$$\mathbf{TP} \vdash P(\ulcorner\varphi\urcorner) \to \varphi$$

$$\textit{If } \mathbf{TP} \vdash \varphi, \textit{ then } \mathbf{TP} \vdash P(\ulcorner\varphi\urcorner)$$

$$\textit{(HB1-3) for P hold.}$$

This and related results motivated various attempts to develop a formal logic of informal provability (which formally captures inferential principles intuitively valid for informal provability, most notably reflection) while avoiding such pitfalls. The main idea is that instead of constructing a formal provability predicate within arithmetic, one develops a logic of informal provability by introducing a new symbol for provability and considering various axioms and rules that might apply to it. We'll now take a look at the main candidates, which come in two flavors. The first group treats informal provability as an operator not as a predicate, thus blocking those inferential moves with are available for predicates, but not for operators, and thus avoiding contradiction at the cost of limited expressivity. The second group of theories treats informal provability as a predicate, but limit the scope the Hilbert-Bernays conditions for the new provability predicate. At the end of this section we'll also take a look at two stray dogs which don't really fit into any of these groups.

9.6.2 Epistemic Arithmetic (EA)

Historically, the first theory of informal provability is Shapiro's *Epistemic Arithmetic* (**EA**) presented in [64] and developed by Goodman [25] and Flagg and Friedman [22]. The idea here is to extend the standard arithmetical language $\mathcal{L}_{\mathbf{PA}}$ to $\mathcal{L}_K$ by adding a unary operator K that applies to formulas. The underlying arithmetical theory is **PA**, and the behavior of K is characterized by the following rules:

KI If $\Gamma \vdash \varphi$ and every element of Γ is epistemic, then $\Gamma \vdash K(\varphi)$
KE $K(\varphi) \vdash \varphi$

where a formula φ is ontic iff it does not contain any occurrences of the operator K and is epistemic iff it has the form $K(\psi)$ for some formula ψ. So EA has all axioms of **PA** and the above two rules for K. Note, the above rules imply **S4** principles for K.

Unfortunately, the internal logic of **EA** (that is, what in **EA** is provably provable) is quite a weak theory – in a sense, it is an elementary extension of intuitionistic Heyting Arithmetic (**HA**). Define a translation V from $\mathcal{L}_{HA}$, the language of **HA**, into $\mathcal{L}_K$. We use $\bar{\varphi}$ to indicate that φ belongs to $\mathcal{L}_{HA}$ as follows:

1. For atomic formulas: $V(\bar{\varphi}) = K(\bar{\varphi})$,
2. $V(\overline{\varphi \wedge \psi}) = K(V(\overline{\varphi})) \wedge K(V(\overline{\psi}))$,
3. $V(\overline{\varphi \vee \psi}) = K(V(\overline{\varphi})) \vee K(V(\overline{\psi}))$,

4. $V(\overline{\varphi \to \psi}) = K(K(V(\overline{\varphi})) \to K(V(\overline{\psi})))$,
5. $V(\overline{\varphi \equiv \psi}) = K(K(V(\overline{\varphi})) \equiv K(V(\overline{\psi})))$,
6. $V(\overline{\neg\varphi}) = K(\neg K(V(\overline{\varphi})))$,
7. $V(\overline{\forall x\ \varphi(x)}) = K(\forall x\ V(\overline{\varphi(x)}))$,
8. $V(\overline{\exists x\ \varphi(x)}) = \exists x\ KV(\overline{\varphi(x)})$.

Just for the sake of simplicity we will write φ instead of $\bar{\varphi}$ whenever it does not lead to confusion. The above translation is sound and complete in the following sense:

Theorem 6.3 *For every* $\varphi \in \mathcal{L}_{HA}$, *if* **HA** $\vdash \varphi$, *then* **EA** $\vdash V(\varphi)$.

Theorem 6.4 [25]) *For every* $\varphi \in \mathcal{L}_{HA}$, *if* **EA** $\vdash V(\varphi)$, *then* **HA** $\vdash \varphi$.

EA, however, does have some interesting properties—we'll mention only two of them. The *numerical existence property* is that for any formula φ, if **EA** $\vdash \exists x\ K\varphi(x)$ then for some natural number n, **EA** $\vdash K\varphi(n)$. The *disjunction property* is that if **EA** $\vdash K(\varphi \vee \psi)$ then either **EA** $\vdash K(\varphi)$ or **EA** $\vdash K(\psi)$.

9.6.3 Modal Epistemic Arithmetic (MEA)

In Shapiro's **EA**, K is a primitive operator which cannot be further analyzed. Horsten [35] suggests that the provability operator is not primitive but complex. He distinguishes between two components of informal provability: the modal and the epistemic.

The modal component is associated with possibility. The epistemic component is explained in terms of a mathematical proof. Instead of just one operator K we have two unary operators applying to formulas: $\Diamond$ and P, where $\Diamond$ is interpreted as possibility and P intuitively stands for "some mathematician has a proof that…". In $\mathcal{L}_{\mathbf{PA}}$ extended with these two operators, $\mathcal{L}_{MEA}$, and following these intuitions we present the so-called Modal Epistemic Arithmetic (**MEA**) [35]. The axioms of **MEA** are as follows:

1. all the axioms of **PA** with induction for the extended language,
2. $\Diamond\varphi \to \varphi$ where φ is ontic i.e. $\varphi \in \mathcal{L}_{\mathbf{PA}}$,
3. $P(\varphi) \to \varphi$,
4. $P(\varphi) \to P(P(\varphi))$,
5. $(\Diamond P(\varphi) \wedge \Diamond P(\varphi \to \psi)) \to \Diamond P(\psi)$,
6. all axioms of the modal system **S5** for $\Diamond$,

and a rule of inference: if φ is a theorem, then so is $\Diamond P(\varphi)$.

Axioms 1 and 2 are some variants of the reflection principle which is provable for P for ontic sentences, and for $\Diamond$ for all sentences. It does not follow that reflection is provable for $\Diamond P$. Axioms 3 and 4 are standard axioms for provability ((HB3) and (HB1)). Note that (HB3) works for provability operator and (HB1) for $\Diamond P$. By a

$\Diamond P$-formula we will mean any formula φ where all subformulas of φ of the form $P\chi$ are immediately preceded with $\Diamond$.

Observation 6.5 *Let $\varphi \in \mathcal{L}_{MEA}$ be a $\Diamond P$-formula. Then the following claims hold:*

$$\mathbf{MEA} \vdash \Diamond P\varphi \rightarrow \varphi$$

$$\mathbf{MEA} \vdash \Diamond P\varphi \rightarrow \Diamond P \Diamond P\varphi$$

The above observation shows that we have a certain version of reflection schema and certain version of (HB3), at least for a restricted class of formulas.

9.6.4 Problems with Provability as an Operator

The main aim of treating provability as an operator is to circumvent the impossibility that arises for the formal provability predicate—that of having all HB conditions and all the instances of the reflection schema at the same time.

Theorem 6.6 (Montague's theorem) *Peano Arithmetic, if consistent, cannot contain (or be consistently extended to contain) a (possibly complex) predicate for which all Hilbert-Bernays conditions and all instances of the reflection schema hold.*

Proof Suppose that there is such a predicate, call it P. We will use natural deduction system. Argue inside the theory:

$1.\lambda \equiv P(\ulcorner \neg\lambda \urcorner)$	Diagonal lemma
1.1λ	Hypothesis
$1.2P(\ulcorner \neg\lambda \urcorner)$	Equivalence elimination: 1,1.1
$1.3\neg\lambda$	Modus ponens and reflection schema: 1.2
$2.\neg\lambda$	Reductio ad absurdum: 1.1 $\rightarrow$ 1.3
$3.P(\ulcorner \neg\lambda \urcorner)$	(HB1)
$4.\neg P(\ulcorner \neg\lambda \urcorner)$	CL, 1, 2
5.contradiction	CL, 3, 4

□

In order to prove Montague's theorem one applies the diagonal lemma to a certain formula involving provability predicate. But if provability is treated as an operator, we cannot use the diagonal lemma to generate this paradoxical formula.

MEA is capable of proving variants of reflection schema. It is an interesting result, for the name of the game here is to gather as many instances of reflection schema as possible without inconsistency. Unfortunately, the theory has some other philosophical problems:

1. The choice which rules are postulated for P and which are postulated for $\Diamond$ seems somewhat arbitrary. It is possible to consider different combination of those rules. For instance, to add axiom (K) directly for P.
2. The reflection schema is available only for $\Diamond P$. It is not clear why other types of reflection shouldn't be introduced. For instance, reflection restricted to Σ_1 formulas doesn't look completely insane.
3. Usually provability is treated as a predicate and not as an operator. There seems to be no motivation for using an operator, independent of blocking the Montague's theorem.
4. Both **EA** and **MEA** seem to be a bit too weak– there are translations to **HA** which preserve theorems.

9.6.5 PEA and Its Basis

Another strategy is to treat informal provability as a predicate and weaken some of the Hilbert-Bernays conditions for this predicate. Again, expand $\mathcal{L}_{\mathbf{PA}}$ with an additional predicate P for informal provability, thus obtaining a new language $\mathcal{L}_P$. The idea here is straightforward: we divide the set of problematic principles (HB conditions and the reflection schema) for the additional predicate P between two theories: PPEA and its basis. Then we add to PEA all the instances of the axiom saying that if something is derivable in the basis, it is informally provable.

We will start with a theory called the basis of **PEA** (**BPEA**) [36], which is defined by:

Basis Axiom 1 **PA** in extended language with induction extended to $\mathcal{L}_P$
Basis Axiom 2 $P(\ulcorner\varphi\urcorner) \to (P(\ulcorner\varphi \to \psi\urcorner) \to P(\ulcorner\psi\urcorner))$ for all $\varphi, \psi \in \mathcal{L}_P$
Basis Axiom 3 $P(\ulcorner\varphi\urcorner) \to P(\ulcorner P(\ulcorner\varphi\urcorner)\urcorner)$ for all $\varphi \in \mathcal{L}_P$

So, we have (K) and (4) for P. By $\mathtt{Prov}_B$ we mean the standard provability predicate of **BPEA**. **PEA** is given by the following axioms:

Axiom 1 **PA** in the extended language with induction extended to $\mathcal{L}_P$
Axiom 2 $P(\ulcorner\varphi\urcorner) \to \varphi$ for all $\varphi \in \mathcal{L}_P$
Axiom 3 $\mathtt{Prov}_B(\ulcorner\varphi\urcorner) \to P(\ulcorner\varphi\urcorner)$ for all $\varphi \in \mathcal{L}_P$

We have the reflection schema for P. Notice that we do not have (Nec) for P, but we have the implication $\mathtt{Prov}_B(\ulcorner\varphi\urcorner) \to P(\ulcorner\varphi\urcorner)$, which together with the reflection schema gives us $\mathtt{Prov}_B(\ulcorner\varphi\urcorner) \to \varphi$ which is a certain version of (Nec).

These theories are still under investigation. One of the nice things about **PEA**, apart from the reflection schema holding in it, is the fact that **PEA** has nice models.

Fact 6.7 *PEA has a model based on the standard model of arithmetic.*

However, it seems that the philosophical motivations underlying the system are somewhat lacking. While informal provability seems unified, this system clearly has two separate layers. The restrictions on the claims for which reflection can be used is still there—it's just that they're somewhat less visible, because they arise at the point in which a restriction is put on what can be provably provable (**Axiom 3**). Yes, **Axiom 2** guarantees that reflection is provable for any ϕ, but given that the internal logic of P is built starting from the forma provability predicate of **BPEA**, it holds universally at the price of being idle on many occasions.

9.6.6 Non-deterministic Many-Valued Approach

Another, rather non-standard approach [62] is to change the underlying logic and to build theories of informal provability where the notion is treated as a partial notion. The partition seems intuitive: some mathematical sentences are informally provable, others are informally refutable and some, it seems, are informally undecidable.

In order to model reasoning with a partial notion formally a three-valued logic comes handy. So think about partitions as values: 1 stands for *informally provable*, 0 for *informally refutable*, and n for *neither*.

One initial problem is that if we take a close look at disjunctions or conjunctions of mathematical sentences, it seems that their logical value depends not only on the values of their disjuncts or conjuncts. For instance, take two disjunction of two undecidable sentences. One is built from the Continuum Hypothesis and its negation. The other one is composed from the claim that the standard set theory (ZFC) is consistent, and the General Continuum hypothesis. The first one is informally provable, because it is a substitution of propositional tautology, whereas the other disjunction at least isn't known to be informally provable, and there is no contradiction in thinking that it is not informally provable.

In order to formally represent the above intuition consider a standard propositional language with one additional modal operator B ($\mathcal{L}_B$), intuitively read as *it is provable that*. Use the three values and the indeterminacy discussed above to define matrices for connectives and the operator B by:

φ	ψ	$\neg\varphi$	$B\varphi$	$\varphi \vee \psi$	$\varphi \wedge \psi$	$\varphi \rightarrow \psi$	$\varphi \equiv \psi$
1	1	0	1	1	1	1	1
n	1	n	$0/n$	1	n	1	n
0	1	1	0	1	0	1	0
1	n			1	n	n	n
n	n			$n/1$	$0/n$	$n/1$	$0/n/1$
0	n			n	0	1	n
1	0			1	0	0	0
n	0			n	0	n	n
0	0			0	0	1	1

where for two values x, y, when we write x/y we mean that either the formula has value x or y. The matrix is rather straightforward. The only interesting case is for B when φ has value n. Then either we cannot informally prove its undecidability then it remains n or we can do that, at the same time disproving $B(\varphi)$, hence value 0.

By an *assignment* we mean a function $v : Prop \mapsto Val$ from the set of propositional variables to the set of values. An *evaluation* is an extension of the assignment to complex formulas respecting conditions given above. The general phenomenon is that an assignment doesn't unambiguously determine unique evaluation: it only underlies a class of evaluations that extend it. For instance, if $v(p) = v(q) = n$, there will be one evaluation with $e_v^1(p \vee q) = n$ and another with $e_v^2(p \vee q = 1)$.

If we were to define a logic in terms of preservation of value 1, it would turn out to be too weak (for instance, conjunction and disjunction aren't even guaranteed to commute). We need one more requirement:

Definition 6.8 (Closure condition) For any $\mathcal{L}_B$-formulas $\varphi_1, \varphi_2, \cdots, \varphi_n, \psi$ such that

$$\varphi_1, \varphi_2, \cdots, \varphi_n \models \psi,$$

where $\models$ is the classical consequence relation for $\mathcal{L}_B$, for any e, if $e(B\varphi_i) = 1$ for any $0 < i \leq n$, then $e(B\psi) = 1$.

We will use $\acute{y}_C\, \varphi$ iff for all evaluations satisfying the closure condition φ has value 1. Similarly, we define $\Gamma \twoheadleftarrow_C \varphi$ as the preservation of value 1 in all evaluations satisfying the closure condition. This logic is called **CABAT**.

One of the most interesting features of this logic is the fact that under a certain translation of the standard provability predicate as B, the translation of Löb's theorem doesn't hold. This is a good sign. There was no initial intuition that Löb's theorem is correct principle for informal provability. It's a rather unwanted technical result.

If we look at Montague's theorem, after applying the diagonal lemma, the rest of the proof is done on the propositional level. We can translate all the premises of the theorem together with the formula resulting from application of diagonal lemma into CABAT language. It turned out that the theorem does not hold. This shows that using quite natural philosophical intuitions we can build a formal system with certain intuitive principles for provability which can be consistently extended with all the instances of the reflection schema.

The above proposal has its drawbacks. It is nothing but obscure how to build an arithmetical theory using this logic. It seems that the most common strategies for building partial models using three-valued logics will not work here. The second issue is that this proposal is still underdeveloped. The current semantics for CABAT is convoluted – it is not clear which evaluations are removed by the closure condition. Simpler and more intuitive semantics needs to be developed. Moreover, provability here is treated as an operator not as a predicate, and it is not clear what the consequences of moving to the predicate level with this logic would be.

9.6.7 Conditional Epistemic Obligation and Believability Logic

A somewhat different formal approach to our (at least prima facie) commitment to reflection arose in the context of formal axiomatic theories of truth built over arithmetic (see [18, 28, 38] for more details on the truth-theoretic aspects of the developments).

Truth-theoretic considerations aside, from the axioms and rules of **PA** local reflection for **PA** doesn't follow, and so we don't seem logically committed to local reflection for **PA** just because we accept the axioms and rules of the system. Yet, there seems something irrational about accepting **PA** without thinking that for any $\varphi \in \mathcal{L}_{\mathbf{PA}}$, indeed, if $\mathbf{PA} \vdash \varphi$, then φ (if we had a truth predicate available, we could use a single claim: that *whatever* **PA** *proves, is true*; but we're trying to avoid getting into a discussion of theories of truth). Assuming this is correct, the challenge is to explain why someone who accepts **PA** is rationally commited to reflection which nevertheless doesn't logically follow from the axioms of **PA** by the rules of **PA**. (Another example of a commitment of this sort is that to the Gödel sentence of **PA**, which seems true, even though it doesn't follow from **PA**.)

For such occasions, Ketland [44] introduced the notion of a *conditional epistemic obligation*. Ketland hasn't really explicated the notion, but only pointed out that once we accept a theory, we become conditionally epistemically obligated to accept some other claims in its language which nevertheless don't follow from the theory itself, and listed some examples such as that of reflection or the Gödel sentence.

A philosophically interesting explication of the notion of conditional epistemic obligation is not trivial (see [18, 19] for a discussion). But even putting this daunting task aside, the question arises whether we can achieve a more humble goal: that of describing the inferential behavior of the predicate expressing this sort of commitment by means of a formal system. Such an attempt can be found in the works of Cieślinski. For the sake of simplicity, we'll discuss the system as built over a particular arithmetical theory, **PA**.

Extend $\mathcal{L}_{\mathbf{PA}}$ with a new unary predicate symbol B, thus obtaining a new language $\mathcal{L}_B$. The goal is to describe the *theory of believability* built over **PA**. Let the result of taking the axioms of **PA** with induction extended to $\mathcal{L}_B$ be called **PAB**. Theory **Bel(PA)** extends **PAB** with the following axioms:

$$\forall \psi \in \mathcal{L}_B\ [\mathtt{Prov}_{PAB}(\ulcorner \psi \urcorner) \rightarrow B(\ulcorner \psi \urcorner)] \tag{A$_1$}$$

$$\forall \varphi, \psi \in \mathcal{L}_B\ [B(\ulcorner \varphi \urcorner) \wedge B(\ulcorner \varphi \rightarrow \psi \urcorner) \rightarrow B(\ulcorner \psi \urcorner)] \tag{A$_2$}$$

On top of the axioms, **Bel(PA)** has two additional rules of inference. (Nec) for B, which allows to infer

$$\mathbf{Bel(PA)} \vdash B(\ulcorner \psi \urcorner)$$

from

$$\mathbf{Bel(PA)} \vdash \psi,$$

and the *generalization rule* (Gen), which allows to infer

$$\mathbf{Bel(PA)} \vdash B(\ulcorner\forall x\ \psi(x)\urcorner)$$

from

$$\mathbf{Bel(PA)} \vdash \forall x\ B(\ulcorner\psi(x)\urcorner).$$

While the motivations for (Nec) and the axioms are quite straightforward (and not completely different from the considerations pertaining to **PEA**), one thing that makes the theory stand apart is (Gen). Normally, in **PA**, just because for every standard numeral $\overline{n}$ $\mathbf{PA} \vdash \psi(\overline{n})$, it doesn't follow that $\mathbf{PA} \vdash \forall x\ (\psi(x))$. This is because **PA** as a first-order theory admits non-standard models, and so can have a model in which there are non-standard numbers not denoted by any standard numeral. It is exactly (Gen) that allows our commitment (tracked by B, that is, the internal logic of **Bel(PA)**) to go beyond what **PA** already can prove, including reflection and the consistency of **PA**.

9.7 Further Topics

The scope of this survey (which is, admittedly, already quite long) is limited. In this section we list and briefly describe multiple further issues related to provability, which we couldn't properly cover. The list is, of course, far from complete.

9.7.1 Mixed Logic of Consistency and ω-Consistency

One interesting fact about the interaction between the notions of consistency and ω consistency is that the negation of the consistency of **PA**, while not being inconsistent with **PA**, is ω-inconsistent with **PA**. Now the question is: can we develop a propositional logic to reason about such claims?

For this purpose we need a *bimodal* logic with two modalities [39, 40]. Let $\Box\psi$ stand for the provability of ψ, and $\blacksquare$ stand for the ω-inconsistency of $\neg\psi$ with the background theory.

The axioms of **GLB** (**B** from *bimodal*) are all tautologies, all instances of (K) for $\Box$, all instances of (K) for $\blacksquare$, all instances of (Löb) for $\Box$ and all instances of (Löb) for $\blacksquare$, and all instances of the following schemata:

$$\Box\varphi \to \blacksquare\varphi$$

$$\neg\Box\varphi \to \blacksquare\neg\Box\varphi$$

The rules are *modus ponens* and (Nec) for $\Box$ (clearly, (Nec) for $\blacksquare$ follows by the first of the above formulas). Given a realization r for the language of **GLB**, it is extented to a **T**-interpretation by the standard conditions for classical connectives together with:

$$r_{\mathbf{T}}(\Box\varphi) = \mathtt{Prov}_{\mathbf{T}}(\ulcorner r_{\mathbf{T}}(\varphi)\urcorner)$$

$$r_{\mathbf{T}}(\blacksquare\varphi) = \neg\omega\mathtt{Con}_{T}(\ulcorner r_{\mathbf{T}}(\neg\varphi)\urcorner)$$

GLB is arithmetically sound and complete.

9.7.2 Provability Logic with Quantifiers

One way to extend $\mathcal{L}_M$ with quantifiers is to add *propositional quantifiers* binding propositional variables, which would allow to express claims such as 'some formula is not provable', or 'for every formula there is one which is provable just in case the former one isn't' (by $\exists p\ \neg\Box p$ and $\forall p\ \exists q\ (\Box q \equiv \neg\Box p)$. As proven by Shavrukov [65], the set of arithmetically valid sentences of this language is undecidable.

Another move to consider is moving to a first-order modal language and extending the intended semantics appropriately. Alas, the set of first-order formulas true in every realization is not effectively axiomatizable [2], and neither is the set of formulas provable in **PA** under any realization (the complexity of this set Π^0_2) [77]. Montagna et al. [56] showed moreover that quantified **GL** is not complete with respect to any class of Kripke frames, and that it doesn't have the fixed point property. Similarly, the first-order version of the logic of proofs has been proven to not be recursively enumerable [86].

9.7.3 Interpretability Logics

The notion of interpretability was introduced into meta-logic by Tarski et al. [75]. A theory **T** is interpretable in theory **U** just in case the language of **T** can be translated into that of **U** so that the translations of theorems of **T** become theorems of **U**. One example of interpretability is the relation between **PA** and the standard set theory **ZFC**. There is a translation from the language of arithmetic into the language of set theory, such that the translations of all theorems of **PA** are provable in **ZFC**.

The formal symbol representing interpretability was introduced into a language of a logic of provability by Švejdar [73]. The intended reading of $\varphi \rhd \psi$ if that for a sufficiently arithmetically rich theory **T** (such as **[PA]**), **T**$+\psi$ is interpretable in **T**$+\varphi$.

Interpretability logics were further studied by Visser [81, 83]. There is a sensible logic **IL** which axiomatizes interpretability principles valid in all sensible theories. It is **GL** expanded with:

$$\Box(\varphi \to \psi) \to \varphi \rhd \psi$$

$$(\varphi \rhd \psi \wedge \psi \rhd \chi) \to \varphi \rhd \chi$$

$$(\varphi \rhd \chi \wedge \psi \rhd \chi) \to (\varphi \vee \psi) \rhd \chi$$

$$\varphi \rhd \psi \to (\Diamond\varphi \to \Diamond\psi)$$

$$(\Diamond\varphi) \rhd \varphi$$

However, the class of all principles that hold in all realizations is sensitive to the choice of the underlying theory (see [83] for a comprehensive survey).

9.7.4 Generalization and Classifications

The notion of provability can be considered on a more general level, that of *provability logic of a given theory* **T** *relative to a metatheory* **U**—the notion was introduced by Artemov [4] and Visser [79]. Such a logic, denoted by $\mathbf{PL_T(U)}$, is the set of all propositional principles of provability for **T** that can be proven in **U**. From this perspective, **GL** is the provability logic $\mathbf{PL_{PA}(PA)}$, and **S** is $\mathbf{PL_{PA}}(Tr(\mathbb{N}))$, where $Tr(\mathbb{N})$ is the set of all $\mathcal{L}_{\mathbf{PA}}$-formulas true in the standard model of arithmetic. Much work has been done on the classification of logics $\mathbf{PL_T(U)}$, where **T** and **U** are known and independently studied extensions of **PA** [3, 4, 12, 41, 80].

9.7.5 Algebraic Approaches

Magari [49, 50] developed an algebraic approach to provability logic. The Magari algebra of **T**, is the set of **T**-sentences factorized modulo equivalence within **T**. Further applications of the algebraic toolkit to provability logics are Montagna [54, 55].

9.7.6 Connections with Other Domains

Provability logics have some use in computability theory. For instance, Beklemishev [12] uses them to investigate which computable functions can be proved to be total by means of restricted induction schemata. Another domain where provability logics find applications is proof theory [13]; see also Japaridze and Jongh [42] for a survey.

9.8 References and Further Readings

A historical account of the beginnings of the development of the logics of provability can be found in Boolos and Sambin [16]. For more introductory surveys of the logics of provability, read [74, 78]. For a short, but dense survey see Artemov [9]. For a survey focusing on self-reference read [70]. For material focusing on introducing the Logic of Proofs consult [10]. For more advanced surveys of the logic of provability, consult Japaridze and de Jongh [42] and Artemov and Beklemishev [11]. As for full blown book-long treatments, Boolos [15] is invaluable, and so is Smoryński [69].

References

1. Anderson, A. R. (1956). *The formal analysis of normative systems* (Technical report), DTIC Document.
2. Artemov, S. (1985). Nonarithmeticity of truth predicate logics of provability. *Doklady Akademii Nauk SSSR, 284*(2), 270–271.
3. Artemov, S. (1986). On modal logics axiomatizing provability. *Mathematics of the USSR-Izvestiya, 27*(3), 401–429.
4. Artemov, S. (1987). Arithmetically complete modal theories. *American Mathematical Society Translations, 2*(135), 39–54.
5. Artemov, S. (1994). Logic of proofs. *Annals of Pure and Applied Logic, 67*(1–3), 29–59.
6. Artemov, S. (1995). *Operational modal logic* (Technical report), Cornell University, MSI, pp. 95–29.
7. Artemov, S. (1998). *Logic of proofs: A unified semantics for modality and λ-terms* (Technical report), Cornell University, CFIS, pp. 98–06.
8. Artemov, S. (2001). Explicit provability and constructive semantics. *Bulletin of Symbolic logic, 7*, 1–36.
9. Artemov, S. (2006). Modal logic in mathematics. In P. Blackburn, J. van Benthem, & F. Wolter (Eds.), *Handbook of modal logic* (pp. 927–970). Burlington: Elsevier.
10. Artemov, S. (2007). On two models of provability. In D. M. Gabbay, S. S. Goncharov, & M. Zakharyaschev (Eds.), *Mathematical problems from applied logic II* (pp. 1–52). New York: Springer.
11. Artemov, S. N., & Beklemishev, L. D. (2005). Provability logic. In D. M. Gabbay & F. Guenthner (Eds.), *Handbook of philosophical logic* (2nd ed., pp. 189–360). Dordrecht: Springer.
12. Beklemishev, L. D. (1990). On the classification of propositional provability logics. *Mathematics of the USSR-Izvestiya, 35*(2), 247–275.
13. Beklemishev, L. D. (2003). Proof-theoretic analysis by iterated reflection. *Archive for Mathematical Logic, 42*(6), 515–552.
14. Blackburn, P., de Rijke, M., & Venema, Y. (2001). *Modal logic*. Cambridge: Cambridge University Press.
15. Boolos, G. (1993). *The logic of provability*. Cambridge: Cambridge University Press.
16. Boolos, G., & Sambin, G. (1991). Provability: The emergence of a mathematical modality. *Studia Logica, 50*(1), 1–23.
17. Carnap, R. (1934). *Logische Syntax der Sprache*. Berlin/Heidelberg: Springer.
18. Cieśliński, C. (2017). *The epistemic lightness of truth. Deflationism and its logic*. Cambridge: Cambridge University Press.

19. Cieśliński, C. (2016). Minimalism and the generalisation problem: On Horwich's second solution. *Synthese*, 1–25. https://doi.org/10.1007/s11229-016-1227-5.
20. Feferman, S., Dawson, J. W., Jr., Kleene, S. C., Moore, G. H., Solovay, R. M., & van Heijenoort, J. (Eds.). (1986). *Kurt Gödel: Collected works. Volume 1: Publications 1929–1936*. Oxford/New York: Oxford University Press.
21. Fitting, M. (2003). *A semantics for the logic of proofs* (Technical report, TR – 2003012). CUNY Ph.D. Program in Computer Science.
22. Flagg, R. C., & Friedman, H. (1986). Epistemic and intuitionistic formal systems. *Annals of Pure and Applied Logic, 32*(1), 53–60.
23. Gaifman, H. (2006). Naming and diagonalization, from cantor to Gödel to Kleene. *Logic Journal of the IGPL, 14*(5), 709–728.
24. Gödel, K. (1933). Eine Interpretation des intuitionistischen Aussagenkalkuls. [Reprinted in [20]].
25. Goodman, N. D. (1984). Epistemic arithmetic is a conservative extension of intuitionistic arithmetic. *Journal of Symbolic Logic, 49*(1), 192–203.
26. Grzegorczyk, A. (1967). Some relational systems and the associated topological spaces. *Fundamenta Mathematicae, 60*(3), 223–231.
27. Hájek, P. & Pudlak, P. (1993). *Metamathematics of first-order arithmetic* (Vol. 2, pp. 295–297). Berlin: Springer.
28. Halbach, V. (2011). *Axiomatic theories of truth*. Cambridge: Cambridge University Press.
29. Hedman, S. (2004). *A first course in logic: An introduction to model theory, proof theory, computability, and complexity*. Oxford/New York: Oxford University Press.
30. Henkin, L. (1952). A problem concerning provability. *Journal of Symbolic Logic, 17*(2), 160.
31. Heyting, A. (1930). *Die formalen Regeln der intuitionistischen Mathematik*. Verlag der Akademie der Wissenschaften.
32. Heyting, A. (1931). Die intuitionistische Grundlegung der Mathematik. *Erkenntnis, 2*(1), 106–115.
33. Heyting, A. (1934). *Mathematische Grundlagenforschung Intuitionismus, Beweistheorie*. Springer.
34. Hilbert, D., & Bernays, P. (1939). *Grundlagen der Mathematik II*. Springer.
35. Horsten, L. (1994). Modal-epistemic variants of Shapiro's system of epistemic arithmetic. *Notre Dame Journal of Formal Logic, 35*(2), 284–291.
36. Horsten, L. (1997). Provability in principle and controversial constructivistic principles. *Journal of Philosophical Logic, 26*(6), 635–660.
37. Horsten, L. (1998). In defence of epistemic arithmetic. *Synthese, 116*, 1–25.
38. Horsten, L. (2011). *The Tarskian turn. Deflationism and axiomatic truth*. Cambridge: MIT Press.
39. Ignatiev, K. N. (1993). On strong provability predicates and the associated modal logics. *The Journal of Symbolic Logic, 58*(01), 249–290.
40. Japaridze, G. K. (1985). The polymodal logic of provability. In *Intensional Logics and Logical Structure of Theories: Material from the Fourth Soviet–Finnish Symposium on Logic*, Telavi (pp. 16–48).
41. Japaridze, G. K. (1986). *The modal logical means of investigation of provability*. Ph.D. thesis, Moscow State University.
42. Japaridze, G., & de Jongh, D. (1998). The logic of provability. In S. R. Buss (Ed.), *Handbook of proof theory* (Vol. 137, pp. 475–550). Burlington: Elsevier.
43. Kaye, R. (1991). *Models of Peano Arithmetic*. Oxford: Oxford University Press.
44. Ketland, J. (2005). Deflationism and the Gödel phenomena: Reply to tennant. *Mind, 114*(453), 75–88.
45. Kolmogorov, A. (1932). Zur Deutung der Intuitionistischen Logik. *Mathematische Zeitschrift, 35*(1), 58–65.
46. Kossak, R. (2006). *The structure of models of Peano Arithmetic*. Oxford: Clarendon Press.
47. Leitgeb, H. (2009). On formal and informal provability. In O. Bueno & Ø. Linnebo (Eds.), *New waves in philosophy of mathematics* (pp. 263–299). New York: Palgrave Macmillan.

48. Löb, M. H. (1955). Solution of a problem of Leon Henkin. *The Journal of Symbolic Logic, 20*(02), 115–118.
49. Magari, R. (1975). The diagonalizable algebras (the algebraization of the theories which express Theor.:II). *Bollettino della Unione Matematica Italiana, 4*(12), 117–125.
50. Magari, R. (1975). Representation and duality theory for diagonalizable algebras. *Studia Logica, 34*(4), 305–313.
51. Marfori, M. A. (2010). Informal proofs and mathematical rigour. *Studia Logica, 96*, 261–272.
52. McKinsey, J. C., & Tarski, A. (1948). Some theorems about the sentential calculi of Lewis and Heyting. *The Journal of Symbolic Logic, 13*(01), 1–15.
53. Mkrtychev, A. (1997). Models for the logic of proofs. In S. Adian & A. Nerode (Eds.), *Logical foundations of computer science '97* (pp. 266–275). Berlin/Heidelberg: Springer.
54. Montagna, F. (1978). On the algebraization of a Feferman's predicate. *Studia Logica, 37*(3), 221–236.
55. Montagna, F. (1979). On the diagonalizable algebra of Peano Arithmetic. *Bollettino della Unione Matematica Italiana, 16*(5), 795–812.
56. Montagna, F., et al. (1984). The predicate modal logic of provability. *Notre Dame Journal of Formal Logic, 25*(2), 179–189.
57. Myhill, J. (1960). Some remarks on the notion of proof. *Journal of Philosophy, 57*(14), 461–471.
58. Nogina, E. (1994). *Logic of proofs with the strong provability operator* (Technical report). Institute for Logic, Language and Computation, University of Amsterdam, ILLC Prepublication Series ML-94-10.
59. Nogina, E. (1996). Grzegorczyk logic with arithmetical proof operators. *Fundamentalnaya i Prikladnaya Matematika, 2*(2), 483–499.
60. Nowell-Smith, P., & Lemmon, E. (1960). Escapism: The logical basis of ethics. *Mind, 69*(275), 289–300.
61. Orlov, I. E. (1928). The calculus of compatibility of propositions. *Mathematics of the USSR, Sbornik, 35*, 263–286.
62. Pawlowski, P., & Urbaniak, R. (2018). Many-valued logic of informal provability: A non-deterministic strategy. *The Review of Symbolic Logic, 11*(2), 207–223.
63. Segerberg, K. K. (1971). *An essay in classical modal logic* (The Philosophical Society in Uppsala). Uppsala: Uppsala University.
64. Shapiro, S. (1985). Epistemic and intuitionistic arithemtic. In *Intensional mathematics*. New York: North Holland.
65. Shavrukov, V. Y. (1997). Undecidability in diagonalizable algebras. *The Journal of Symbolic Logic, 62*(01), 79–116.
66. Simpson, S. G. (2009). *Subsystems of second order arithmetic* (Vol. 1). Cambridge: Cambridge University Press.
67. Smiley, T. J. (1963). The logical basis of ethics. *Acta Philosophica Fennica, 16*, 237–246.
68. Smith, P. (2007). *An introduction to Gödel's theorems*. Cambridge: Cambridge University Press.
69. Smoryński, C. (1985). *Self-reference and modal logic* (Universitext). New York: Springer.
70. Smoryński, C. (2004). Modal logic and self-reference. In D. Gabbay & F. Guenthner (Eds.), *Handbook of philosophical logic* (Vol. 11, pp. 1–53). Dordrecht: Springer.
71. Solovay, R. M. (1976). Provability interpretations of modal logic. *Israel Journal of Mathematics, 25*(3–4), 287–304.
72. Sundholm, G. (1998). Proofs as acts and proofs as objects: Some questions for Dag Prawitz. *Theoria, 64*(2–3), 187–216.
73. Švejdar, V. (1983). Modal analysis of generalized Rosser sentences. *The Journal of Symbolic Logic, 48*(04), 986–999.
74. Švejdar, V. (2000). On provability logic. *Nordic Journal of Philosophical Logic, 4*(2), 95–116.
75. Tarski, A., Mostowski, A., & Robinson, R. M. (1953). *Undecidable theories*. Amsterdam: Elsevier.

76. Troelstra, A. & van Dalen, D. (1988). *Constructivism in mathematics* (Vols. 1 and 2). Amsterdam: Elsevier.
77. Vardanyan, V. A. (1986). Arithmetic complexity of predicate logics of provability and their fragments. *Soviet Mathematics Doklady, 33*(3), 569–572.
78. Verbrugge, R. L. (2017). Provability logic. In *The Stanford encyclopedia of philosophy* (Fall 2017 Edition), Edward N. Zalta (ed.), https://plato.stanford.edu/archives/fall2017/entries/logic-provability/.
79. Visser, A. (1981). *Aspects of diagonalization and probability*. Ph.D. thesis, University of Utrecht.
80. Visser, A. (1984). The provability logics of recursively enumerable theories extending Peano Arithmetic at arbitrary theories extending Peano Arithmetic. *Journal of Philosophical Logic, 13*(1), 97–113.
81. Visser, A. (1990). Interpretability logic. In P. Petkov (Ed.), *Mathematical logic* (pp. 175–209). Boston: Springer.
82. Visser, A. (1997). A course on bimodal provability logic. *Annals of Pure and Applied Logic, 73*, 109–142 (1995); *The Journal of Symbolic Logic, 62*(02), 686–687.
83. Visser, A. (1998). An overview of interpretability logic. In M. Kracht, M. de Rijke, H. Wansing, & M. Zakharyaschev (Eds.), *Advances in modal logic* (Vol. 1, pp. 307–359). Stanford: CSLI Publications.
84. Visser, A. (2008). Propositional combinations of σ_1 sentences in Heyting's arithmetic. *Logic Group Preprint Series, 117*, 1–43.
85. Yavorskaya, T. (2001). Logic of proofs and provability. *Annals of pure and applied logic, 113*(1), 345–372.
86. Yavorsky, R. E. (2001). Provability logics with quantifiers on proofs. *Annals of Pure and Applied Logic, 113*(1), 373–387.

Part III
Metaphysics and Philosophy of Language

Chapter 10
Theory of Concepts

Erich Rast

Abstract The word 'concept' is sometimes used as a synonym for 'property', but many authors use it in a more specific sense, for example as standing for unsaturated entities whose extensions are sets and classes, for Fregean senses, or for abstract objects. Although there is no universal agreement on a definition of concepts, a viable theory of concepts has to address a number of formal issues: How to deal with counterfactual and possibly contradictory concepts, how to restrict comprehension schemes in higher-order logic to avoid semantic paradoxes like the Paradox of Predication, how to nominalize concepts, and how to express similarity and typicality of concepts. The article gives a brief survey of the most important problems in concept theory and their possible solutions.

10.1 Key Notions and Problems

Concept theories draw on a rich tradition, ranging from Plato and Aristotle over Leibniz to Frege. Two key aspects of a theory of concepts need to be distinguished. (i) The cognitive aspect regards the role of concepts in cognition and how these enable an epistemic agent to classify and categorize reality. A concept system is sometimes considered the cornerstone and starting point of a 'logic of thinking.' (ii) From a metaphysical point of view, concept theory must provide an explanation of the ontological status of universals, how these combine, whether there are different modes of predication, and what it means in general for an object to fall under a concept. Both aspects will be addressed in what follows. The survey starts with a brief overview of selected problems and positions.

The Demarcation Problem. There is no general agreement in the literature on what a concept is. Sometimes 'concept' is more or less used as a synonym for 'property', but many authors use it in a more specific sense, for example as standing

E. Rast (✉)
IFILNOVA Institute of Philosophy, New University of Lisbon
e-mail: erast@fcsh.unl.pt

S. O. Hansson, V. F. Hendricks (eds.), *Introduction to Formal Philosophy*, Springer Undergraduate Texts in Philosophy, https://doi.org/10.1007/978-3-319-77434-3_10

for unsaturated entities whose extensions are sets and classes (Frege), for Fregean senses (Church), or for abstract objects (Zalta). One goal shared by many authors, despite terminological differences, is to carve out the differences between closely related notions such as concepts, properties, abstract objects, Leibnizian concepts, or Fregean senses and make these notions more precise.

Nominalism, Realism, Cognitivism. A particular object is said to fall under a singular or individual concept and likewise a group of objects sharing some common trait is said to fall under a general concept. Being sorts of universals, different stances towards general concepts may be taken: According to strict nominalism there are only particulars; quantification over predicate expressions is not allowed at all or very limitedly. In this view general concepts do not exist in reality although they might play a role as thinking devices. In contrast to this, according to realism predicates denote universals either directly or whenever the predicate has been nominalized. There *are* universals in the sense that one may fully quantify over them although they might not be considered to exist in the narrow sense. Cognitivism is a mixed position. In this view, there are universals but only insofar as they are represented (or representable) by mental states.

Intensionality, Hyperintensionality, Contradictory Concepts. Having a heart and having a liver are often given as an example of two different concepts with the same extension. Modal logics have been used to account for this difference. Normal possible worlds semantics does not, however, provide the means to distinguish two different mathematical concepts with the same extension from each other. For example, two different ways of describing an equiangular triangle will determine the same set of objects in all possible worlds. To tackle this problem a stronger form of intensionality known as hyperintensionality is needed. Moreover, a person might erroneously believe that 37 is a prime number while not believing that $21 + 16$ is prime, might erroneously believe that $\sqrt{2}$ is a rational number, or might muse about round squares. To represent irrational attitudes and impossible objects a logic must in one way or another allow contradictory statements. Since in classical logic any formula can be derived from a contradiction (ex falso quod libet) a paraconsistent logic is needed; such a logic allows one to derive some, but not arbitrary consequences from a contradiction.

Similarity. A concept may be more or less similar to other concepts. For example, the concept of being a chair is similar to the concept of being a stool and both of them are more similar to each other than any of them is to the concept of being the back of a horse. From a cognitive perspective it is desirable to have a concept theory that allows for a measure of similarity between concepts and the objects falling under them.

Typicality. Typically chairs have four legs, but some have less. Typically birds can fly, but penguins cannot fly. How can this typicality be accounted for?

10.2 Preliminaries of Logical Concept Theory

In order to formulate a broadly-conceived logical theory of concepts it is necessary to quantify over concepts or corresponding abstract objects. Unless a very strict nominalism based on first-order logic is defended this naturally involves the use of second-order logic. For this reason results from mathematical logic need to be taken into account when developing a logical theory of concepts, some of which are addressed in what follows.

Henkin Models and Standard Models. There are two kinds of models for higher-order logic. In a standard model, first-order variables range over a domain D, second-order variables over $\mathcal{P}(D)$ for predicates and $\mathcal{P}(D_1 \times \cdots \times D_n)$ for n-ary relations, third-order predicate variables over $\mathcal{P}(\mathcal{P}(D))$, and so on. In a Henkin model (general model), only a fixed subset of the powerset is chosen respectively. So for instance the quantifier in $\forall F[F(a)]$ ranges over a fixed subset of $\mathcal{P}(D)$. Higher-order logic with Henkin models is essentially a variant of many-sorted first-order predicate logic [10]. It is complete, compact and the Löwenheim-Skolem theorems hold in it, but does not allow one to define certain mathematical structures categorically, i.e. in a way that is unique apart from differences captured by the notion of an isomorphism between models. In contrast to this, higher-order logic with standard models is not complete, not compact, and the Löwenheim-Skolem theorems do not hold in it. Lack of a full-fledged proof theory is compensated by the ability to categorically define important concepts such as countable vs. uncountable domains, quantifiers like 'most', and well-foundedness conditions. The distinction between higher-order logic and second-order logic with standard models is less important, since the former can be reduced to the latter without significant loss of expressivity [11]. For this reason many authors focus on second-order logic.

Box 10.1 Comprehension schemes and stratification

Stratification: Formula ϕ is homogeneously stratified iff there is a function $f(.)$ that maps terms and formulas of the language to natural numbers such that for any atomic formula $P(x_1, \ldots, x_n)$ in ϕ, $f(P) = \max[f(x_i)] + 1$ and $f(x_i) = f(x_j)\ for\ 1 \leq i, j \leq n$.

$$\exists F \forall \vec{x}[F(\vec{x}) \leftrightarrow \phi(\vec{x})] \quad \text{(Scheme A)}$$

$$\exists F \forall \vec{x}[F(\vec{x}) \leftrightarrow (G(\vec{x}) \wedge \phi(\vec{x}))] \quad \text{(Scheme B)}$$

Conditions: (I) $\vec{x} := x_1, \ldots, x_n$ are free in ϕ, i.e. bound in the whole scheme; (II) F is not free in ϕ, i.e. not bound in the whole scheme; (III) ϕ is homogeneously stratified.

① Unrestricted Comprehension: Scheme A + I
② Predicative Comprehension: Scheme A + I, II, III
③ Separation Axiom: Scheme B + I, II

Logical Paradoxes and Comprehension. Given some condition expressible in a formal language, what concepts are there? One way to answer this question is by specifying a comprehension scheme. Unrestricted comprehension asserts that there is a concept corresponding to any condition ϕ that can be formulated in the language (see Box 10.1, Principle ①). It allows one to introduce Russell's paradox of predication, the analogue to the well-known set-theoretic paradox. Take the predicate $P(x)$ that is not predicable of itself and is defined as $\neg x(x)$. Choosing $\phi := \neg x(x)$ and existential instantiation allows one to derive $\forall x[P(x) \leftrightarrow \neg x(x)]$ and by universal instantiation the contradiction $P(P) \leftrightarrow \neg P(P)$. Different provisions to avoid such inconsistencies lead to higher-order logics with varying expressive power that reflect different stances towards nominalism, cognitivism, and realism.

Predicativity vs. Impredicativity. A definition is impredicative iff it quantifies over a collection of objects to which the defined object belongs; otherwise it is predicative. Some mathematicians like Poincaré, Weyl, and Russell himself held the view that paradoxes arise because a logic with unrestricted comprehension allows for impredicative definitions. As a solution, the logic is made predicative. One way to achieve this is by assigning an order to all variables and prescribe that in any atomic formula $P(x_1, \ldots, x_n)$ in a condition ϕ formulated in the language the order of all x must be lower than the order of P (see Box 10.1, Principle ② and *stratification*). This makes $\neg x(x)$ ungrammatical. Church's influential Simple Type Theory (STT) is another way to define a predicative higher-order logic. Every term has a type with corresponding domain. Starting with finitely many base types, infinitely many compound types can be built. If α and β are types, then $(\alpha\beta)$ is the type of a function that takes an object of type β and yields an object of type α, where β and α may themselves be compound types. Predicates and relations are represented by several functions. This is called Currying or Schönfinkelization. For example, a unary predicate P is of type $(\sigma\iota)$, indicating a function that takes a term of type ι and yielding a truth-value of type σ, a second-order predicate is of type $(\sigma(\sigma\iota))$, and so on. (Another notation which was popularized by Montague uses e for objects, t for truth-values, and the order is reversed.)

Impredicativity does not automatically lead to paradoxes. On the contrary, many useful mathematical concepts such as the induction principle used for defining natural numbers are impredicative. For this reason some conceptual realists opt for impredicative second-order logics that give rise to larger mathematical universes. In these logics comprehension is restricted less radically than in predicative ones (see e.g. Box 10.1, Principle ③) or full comprehension is combined with a limited substitution principle in order to gain more expressivity while avoiding the paradoxes. The downside is that it is harder to ensure consistency in such systems than in purely predicative logics.

Philosophical Relevance. First-order logic and predicative higher-order logic with Henkin models reflect a strict nominalist stance as has been defended by Lésniewski, for example. Predicative higher-order logic with standard models may also be considered nominalist in spirit, because predicative comprehension reduces the existence of general concepts to conditions explicitly given in the language.

In contrast to this, impredicative higher-order logics with standard models clearly reflect a realist stance. More fine-grained distinctions can be found in Cocchiarella [5, 6].

10.3 Concepts as Abstract Objects

Possibilism. While the conceptual realist wants to talk about concepts it would be implausible to claim that concepts exist in the same sense as ordinary objects. Therefore, many conceptual realists distinguish, pace Quine, between quantification as a means of counting and quantification as a means of asserting existence. A logic in which non-trivial properties can be ascribed to nonexistent objects is possibilist or Meinongian, where the latter term is often used for metaphysical theories that allow one to talk about contradictory objects. In a classical setting, possibilism can be obtained by introducing two sorts of quantifiers. Actualist quantifiers are mere means of counting and run over the total domain, whereas possibilist quantifiers additionally assert existence and run only over a subset of the total domain. Alternatively, a unary existence predicate $E(x)$ may be introduced to which possibilist quantifiers are relativized, for instance $\forall^* x A := \forall x[E(x) \rightarrow A]$ and $\exists^* x A := \exists x[E(x) \wedge A]$.

Nominalization. One positive answer to the problem of universals is to assert that we cannot only quantify over concepts but are also able to talk about concepts like *being nice* as objects. Sometimes λ-abstraction is thought to fulfill this purpose. Semantically, a term of the form $\lambda x.P(x)$ is interpreted as the function that with respect to an assignment g takes an a within the domain of x and whose result is the same as $P(x)$ evaluated with respect to the modified assignment g' that is the same as g except that $g'(x) = a$. One might then consider $\lambda x.P(x)$ to stand for *being nice* if P stands for the predicate *nice*. However, λ-terms can be used instead of relations (as in STT) and the converse transformation is also possible in a logic with both functions and relations, and so λ-abstraction might not be considered a tool for nominalization understood in the narrow sense. Abstract object theory [18] and alternative ontologies such as trope theories [4, 12, 16] provide more elaborate nominalization mechanisms. Differing considerably in details and terminology, generally in these approaches nuclear and extranuclear properties are distinguished from each other [14], where the former are being constitutive of an object and the latter are not, and two different modes of predication are available: An object, which does not have to be concrete or existent, encodes a property if the property takes part of a description or listing of the object's essential features whereas it exemplifies a property if it has the property accidentally. For example, in bundle trope theories an object encodes a property if its constituting bundle of properties (viz., property moments also sometimes called qualitons) contains the property and exemplifies a property if it stands in a designated relation to the property. A concept is, in this view, a nonexistent non-concrete bundle of primitive properties or property moments. Analogously, in abstract object theory aF stands for the fact that the abstract (nonexistent) object a encodes property F. Care must be taken to restrict the range of properties that can be encoded. For example, forming an abstract object *existent red sphere* must either be disallowed or the existence-entailing predicate 'existent' must be interpreted in a derived, non-literal way in this construction.

10.4 Concepts and Intensionality

Modal Concepts and Intensionality. Modal operators may be added to higher-order logic in the same way as they are added to first-order logic, which in the second-order setting allows one to precisely express philosophical positions about the modal properties of concepts. For example, Anti-Essentialism may be expressed by adding the following axiom:

$$\forall F[\exists x \Box F(x) \rightarrow \forall x \Box F(x)] \tag{10.1}$$

which may be paraphrased as "if an object has an essential property, then any object has this essential property."

Hyperintensionality. Inspired by Frege's informal distinction between the sense and the denotation of an expression, there is a tradition of hyperintensional logics in which the following Axiom of Extensionality does not hold:

$$\forall F \forall G(\forall \vec{x}[F(\vec{x}) \leftrightarrow G(\vec{x})] \rightarrow \forall H[H(F) \rightarrow H(G)]) \tag{10.2}$$

This axiom states that if exactly the same objects fall under two concepts, then the concepts are identical in Leibniz' sense of having the same properties. Despite considerable differences in detail, hyperintensional logics generally invalidate this axiom by interpreting expressions over a domain of fine-grained intensions, which are in turn mapped to their extensions by an extension function [13]. Consequently, two notions of identity are available in such a logic: coarse-grained extensional identity and fine-grained intensional identity interpreted over intensions (Fig. 10.1). By interpreting functions and operators standing for notions like de dicto belief over intensions it is possible to distinguish having a heart from having a liver and deal with ordinary cases of referential opacity like Frege's Morning–Evening Star example. Additionally, strong intensions allow one to represent attitudes that are not closed under logical consequence, i.e. someone's believing that 37 is prime while not believing that $21 + 16$ is prime.

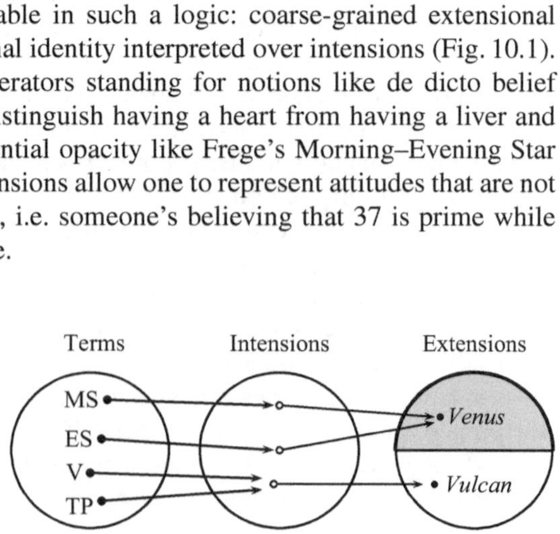

Fig. 10.1 Intensional versus extensional identity in a possibilist intensional logic. *ES* evening star, *MS* morning star, *V* Vulcan, *TP* the planet between Mercury and Sun

Contradictory Concepts. Representing irrational attitudes or contradictory concepts like being a round square requires substantial changes to the underlying logic. In a modal logical setting sometimes impossible worlds are introduced. At an impossible world 'anything goes'; arbitrary formulas, including contradictions, may be true at such a world by mere syntactic assignment. Another approach based on seminal work by Asenjo, da Costa, Anderson and Belnap is to use a 3-valued logic such as LP or RM3. These logics are paraconsistent and allow a contradictory formula to have a designated truth value that is interpreted as 'both true and false.' Paraconsistent logics have also been proposed as a way of dealing with the paradoxes, allowing the logic to mirror the philosophical position that there *are* real paradoxes and our talk about them is meaningful (Dialetheism).

The logical aspects of concept theory mentioned so far are well-known, but are not commonly combined into one all-encompassing metaphysical theory. Most authors focus on some of these aspects, such as how they can be used to answer the problem of universals, or logical reconstructions of historical positions such as Leibniz' Concept Calculus or Platonic Forms. References to further work are given in Sect. 10.6.

10.5 Geometrical Approaches

In this section some promising alternatives to the logical approach shall be mentioned, which are not metaphysical in the narrow sense. These broadly-conceived geometrical concept approaches fare particularly well with issues related to the cognitive aspects of concepts such as vagueness, typicality, and similarity and can either be combined with, or are thought to complement, logical theories.

Typicality. In a qualitative approach a preorder relation (preference relation) between all objects falling under a concept can be used to order objects falling under a given concept according to their typicality. The center represents a prototype and the nearer an object is to the center the more typical it is (Fig. 10.3a). In a logical setting this kind of typicality can be expressed in Preference Logics and related descendants of Lewis' Conditional Logic. Quantitative accounts induce a similar ordering by assigning a degree of typicality as a real number between 0 and 1 to each object as dependent on the concept it falls under, and despite some differences the two approaches can for many practical purposes be translated into each other. There are interconnections of these basic forms of typicality to non-monotonic logics for default reasoning, belief revision, ϵ-entailment, and plausibility and possibility measures.

Conceptual Spaces. Gärdenfors [9] proposes to model concepts not just on a symbolic, but also on a geometrical level. A conceptual space is an n-dimensional metric space with n quality dimensions, each of which represents a basic quality like *height*, *width*, *hue*, *saturation*, or *loudness*. A distance function allows for measuring the distance between any two points in such a space. In the simplest case of familiar n-dimensional Euclidean space this distance measure between two

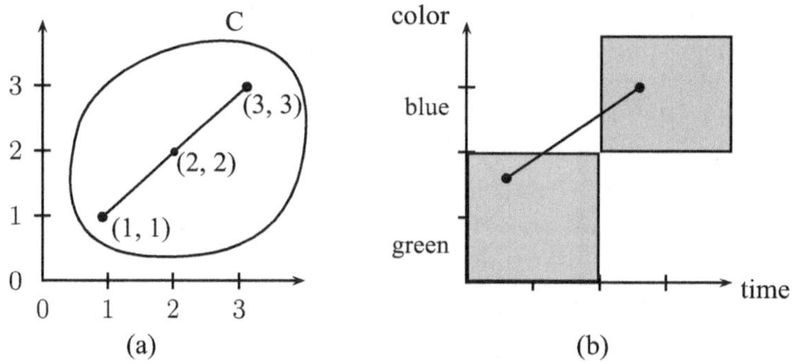

Fig. 10.2 (**a**) Convexity: for example for $t = \frac{1}{2}$ the point $\frac{1}{2} \cdot \langle 1, 1\rangle + (1 - \frac{1}{2}) \cdot \langle 3, 3\rangle = \langle \frac{1}{2}, \frac{1}{2}\rangle + \langle \frac{3}{2}, \frac{3}{2}\rangle = \langle 2, 2\rangle$ is also in C. (**b**) Goodman's concept *grue* is not convex

points $x = \langle x_1, \ldots, x_n\rangle$ and $y = \langle y_1, \ldots, y_n\rangle$ is defined as

$$d_E(x, y) = \sqrt{\sum_{i=1}^{n} w_i (x_i - y_i)^2} \tag{10.3}$$

where w_i represents the weight of the respective quality dimension. More general topological definitions of spaces allow for an adequate treatment of purely qualitative dimensions. Generally speaking, in a conceptual space objects are represented as vectors $x = \langle x_1, \ldots, x_n\rangle$ and concepts by regions in the space. Similarity between two objects in a conceptual space is defined as a function of their distance.

Gärdenfors has conjectured that natural concepts should be represented by convex regions. A region C of a space S is convex iff for any two points $x, y \in C$ any point $tx + (1 - t)y$, where $0 \leq t \leq 1$, on the line segment $\overline{xy}$ between x and y is also in C (Fig. 10.2a). One advantage of this assumption is that every convex region has a center, which may be interpreted as a prototypical object falling under the concept. Taking these centers $p_1, \ldots, p_k$ as starting points, concepts C_i can be defined around them by partitioning the space such that for each point $x \in C_i$, $d(p_i, x) \leq d(p_j, x)$ if $i \neq j$. The result is called a Voronoi diagram (Fig. 10.3b). The closer a point is to the center p_i of its concept C_i in such a partitioning, the higher is the degree of typicality of the object it represents. The convexity condition has also been taken as a first step toward distinguishing between natural and non-natural concepts. For example, with 'standard' quality dimensions Goodman's artificial concept *grue*, which is true of green objects before some point in time and of blue ones afterwards, is represented by a non-convex region (Fig. 10.2b). However, this solution depends on criteria for finding natural quality dimensions, as a natural concept may be turned into a non-natural one by changing the underlying dimensions and vice versa.

Formal Concept Analysis. In formal concept analysis a set M of attributes is associated with a set of objects G by a binary relation $I(x, y)$ read as "object x

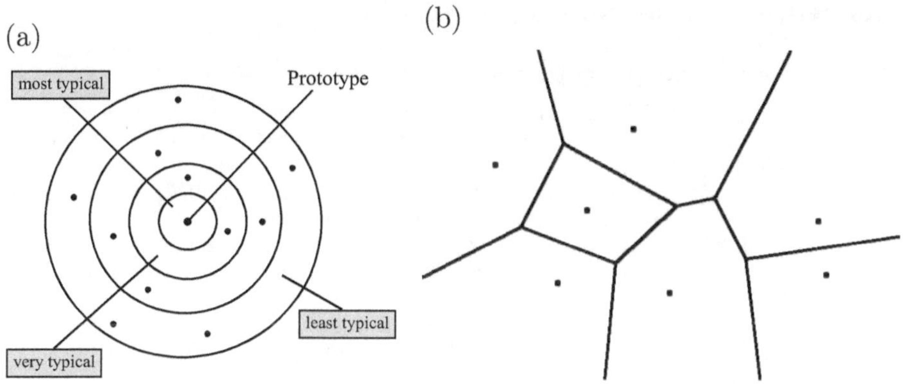

Fig. 10.3 (**a**) Typicality expressed as a preference ordering. (**b**) A Voronoi diagram with seven centers

has attribute G", where the triple $\langle M, G, I\rangle$ is called a context. Such a context may be thought of as a table with objects as rows and attributes as columns and a mark at the row-column intersection if the object at that row has the respective attribute. A formal concept is then a pair $\langle A, B\rangle$ of subsets $A \subseteq G$ and $B \subseteq M$ such that all objects in A share all the attributes in B. The formal concepts of a context can be ordered by a relation $(A, B) \leq (C, D)$ which is true iff $A \subseteq C$, false otherwise. Ordering all concepts in a context yields a *lattice* structure in which the least specific concept is at the bottom and the most specific one is at the top. Various methods and algorithms based on this representation have been used for data mining, machine learning, discovering new relationships between concepts, concept visualization, explaining human concept acquisition, and models of concept change.

10.6 Further Reading

Andrews [1] contains an introduction to type theory; reprints of original articles can be found in Benzmüller et al. [2]. Shapiro [17] is a comprehensive treatment of second-order logic. Burgess [3] discusses predicative and impredicative foundations of arithmetics with a focus on Frege. Metaphysical implications of different comprehension schemes are discussed at length in Cocchiarella [5, 6]. Priest [15] is a modern defense of possibilism and dialetheism; it may serve as a reference for further literature. Zalta [18] is the main work on abstract object theory and contains reconstructions of Platonic Forms and Leibniz' Concept Theory; many refinements can be found in Zalta's more recent works. Gärdenfors [9] is the seminal work on Conceptual Spaces. Ganter and Wille [7] and Ganter et al. [8] lay out formal concept analysis in a rigid manner.

References and Recommended Readings

Asterisks (∗) indicate recommended readings.

1. *Andrews, P. B. (2002). *An introduction to mathematical logic and type theory: To truth through proof.* New York: Academic Press. [Contains an introduction to higher-order logic suitable for beginners].
2. Benzmüller, C., Brown, C. E., Siekmann, J., & Stratman, R. (Eds.). (2008). *Reasoning in simple type theory* (Studies in logic, Vol. 17). London: College Publications.
3. Burgess, J. P. (2005). *Fixing Frege*. Princeton: Princeton University Press.
4. Castañeda, H. -N. (1989). *Thinking, language, experience*. Minneapolis: University of Minnesota Press.
5. Cocchiarella, N. B. (1994). Philosophical perspectives on formal theories of predication. In F. Guenthner & D. Gabbay (Eds.), *Handbook of philosophical logic* (Vol. 4, pp. 253–326). Dordrecht: Kluwer.
6. *Cocchiarella, N. B. (2007). *Formal ontology and conceptual realism* (Synthese library, Vol. 339). Dordrecht: Springer. [Intermediate-level text that lays out fine-grained philosophical distinctions between systems of higher-order logic].
7. Ganter, B., & Wille, R. (1998). *Formal concept analysis: Mathematical foundations*. Berlin/Heiderlberg/New York: Springer.
8. Ganter, B., Stumme, G., & Wille, R. (Eds.). (2005). *Formal concept analysis*. Berlin/New York: Springer.
9. *Gärdenfors, P. (2004). *Conceptual spaces: The geometry of thought*. Cambridge, MA: MIT Press. [Reference work on the theory of conceptual spaces].
10. Henkin, L. (1950). Completeness in the theory of types. *The Journal of Symbolic Logic, 15*(2), 81–91.
11. Hintikka, J. (1955). Reductions in the theory of types. In K. Jaakko & J. Hintikka (Eds.), *Two papers on symbolic logic* (Acta philosophica fennica, Vol. 8, pp. 57–115). Helsinki: Soc. Philosophica.
12. Mormann, T. (1995). Trope sheaves: A topological ontology of tropes. *Logic and Logical Philosophy, 3*, 129–150.
13. Muskens, R. (2007). Intensional models for the theory of types. *Journal of Symbolic Logic, 72*(1), 98–118.
14. Parsons, T. (1980). *Existence*. New Haven: Yale University Press.
15. *Priest, G. (2005). *Towards non-being: The logic and metaphysics of intentionality*. Oxford: Clarendon. [Seminal recent work on non-existence].
16. Rapaport, W. J. (1978). Meinongian theories and a Russellian paradox. *Noûs, 12*(2), 153–180.
17. Shapiro, S. (1991). *Foundations without foundationalism: A case for second-order logic*. Oxford: Clarendon Press.
18. *Zalta, E. (1983). *Abstract objects*. Dordrecht/Boston/Lancaster: D. Reidel. [Older, yet important work on abstract object theory].

Chapter 11
Categories

Jean-Pierre Marquis

Abstract Mathematical categories provide an abstract and general framework for logic and mathematics. As such, they could be used by philosophers in all the basic fields of the discipline: semantics, epistemology and ontology. In this paper, we present the basic definitions and notions and suggest some of the ways categories are starting to infiltrate formal philosophy.

11.1 Introduction

Mathematical categories were introduced in 1945 by the mathematicians Samuel Eilenberg and Saunders Mac Lane in order to define two concepts that were seen at that time to be more important and mathematically relevant, the concepts of functors and natural transformations. (See [10, 15, 16, 22, 23].) The concepts introduced were so clearly general that Eilenberg and Mac Lane could not refrain from picking from the philosophers' vocabulary.

> Now the discovery of ideas as general as these is chiefly the willingness to make a brash or speculative abstraction, in this case supported by the pleasure of purloining words from the philosophers: "Category" from Aristotle and Kant, "Functor" from Carnap, and "natural transformation" from then current informal parlance. [19, pp. 29–30.]

Eilenberg and Mac Lane themselves never made anything of the connection with the philosophical meaning of the term, although with hindsight, it is clear that they made a prescient choice of terminology. The original definition given by Eilenberg and Mac Lane does suggest that a mathematical category is a system of a uniform mathematical "kind" or type, e.g. the category of groups. Thus, many mathematical kinds form categories, allowing mathematicians to relate and compare those kinds

J.-P. Marquis (✉)
Département de philosophie, Université de Montréal Montréal, QC, Canada
e-mail: Jean-Pierre.Marquis@umontreal.ca

S. O. Hansson, V. F. Hendricks (eds.), *Introduction to Formal Philosophy*, Springer Undergraduate Texts in Philosophy, https://doi.org/10.1007/978-3-319-77434-3_11

via functors and natural transformations. Hence, categories could be thought of as some sort of organizational or classificatory tool for mathematical structures, although it certainly does not yield a proper classification in terms of a partition of mathematical structures. The *theory* of categories developed during the 1950s and 1960s and by the end of the 1960s and early 1970s, its connections to logic and the foundations of mathematics became clear. From the latter perspective, it can now be seen as a global framework encompassing both syntactical and semantical – in fact very often blurring the distinction between these two aspects – components of logical analysis, together with the links between these components.

It is my belief that mathematical categories are relevant to formal philosophy and metaphysics in at least two complementary ways. First, as formal tools, category theory and categorical logic can be seen to be generalizations of first-order logic and set theory. As such, they become fundamental frameworks to model, analyze and give a precise expression to philosophical concepts and theories. Thus, in as much as logical and set-theoretical tools can be used for philosophical purposes, categorical methods provide a more general and conceptual framework for the same purposes. By so doing, it often opens the door to unexpected connections and links between heretofore unrelated domains. Second, mathematical categories occupy a central and fundamental place in contemporary mathematics and the organization and foundations of contemporary mathematics can now be understood from this perspective. Thus, if one of the goals of metaphysics is to provide an understanding of various kinds of beings, in particular mathematical beings, then mathematical categories are more than relevant to this enterprise.

A caveat is in order. It will be impossible to do justice to the richness of the theory in such a short paper. In particular, some fundamental notions of the theory will simply be ignored. For instance, we will not say a word about adjoints, monads, toposes, categorical logic and higher-dimensional categories. This is nothing less than a shame and we certainly press the reader to look up the references in the text to learn about these notions too.

11.2 Categories: Definition

Informally, a mathematical category can be thought of as a network satisfying simple properties. The network is made of nodes $X, Y, Z, \ldots$ who exchange information. A sender can send multiple information and in various ways to a receiver. We can of course think of the sender X as representing a property and the information into a receiver Y as a way of transforming the given representation in X into a representation in Y. Thus, the information transmitted can tell us, for example, how one property varies with respect to another. One such way is represented by an arrow from the sender X to the receiver Y, namely by $X \xrightarrow{f} Y$. Notice that there can be many different arrows from one receiver to a sender as there are many different informations that X can send to Y. Also an information is automatically and always attached to a sender and a receiver, it is never "free floating" so to speak.

It is always originating from a definite sender and always received by a definite receiver and these data constitute an intrinsic part of the information transmitted. In a nutshell: information is *always* traceable. Naturally, there can be arrows going the opposite way and one and the same node can be the sender and the receiver of an information (as when I recall something to myself). It is also assumed that whenever a sender X sends an information to a receiver Y and the later sends an information to another receiver Z, then Z can retrieve the information that Y received from X but in this case, only via the message it received from Y. (I say in this case, since Z can also receive information directly from X.) In more technical terms, information composes. Furthermore, it is assumed that nodes are always "active", that is they always send information to themselves, which we will call the identity information and will denote by 1_X for a specific node X. The network is characterized by the following simple properties. Transfer of information is assumed to be associative: the information exchanged from W to Y via X and then sent to Z is the same as the information sent from W to X and then from X to Z via Y. Informally, this says that, in the end, the intermediary steps are irrelevant in the exchange of information: if you were to collect the information at one stage and verify it at that stage, you would get the same thing in the end if you were to verify it at a different step. A second simple property is that the identity information does not affect the information coming in nor the information coming out. A mathematical category can be thought of as such a simple network. As such it is hard to see why it could be of any interest. Before we go into this, we will give a formal definition of a category to fix the notation once and for all.[1]

Definition 1 A *category* **C** is a system made up of nodes or objects $X, Y, Z \ldots$ and arrows (or morphisms) $f, g, h, \ldots$ such that

- For each arrow f there are given nodes $dom(f)$, $cod(f)$, respectively the *domain* and the *codomain*, and we write

$$X \xrightarrow{f} Y \quad \text{or} \quad f : X \longrightarrow Y$$

whenever $X = dom(f)$ and $Y = cod(f)$;
- Given arrows $f : X \longrightarrow Y$ and $g : Y \longrightarrow Z$, there is a given arrow

$$g \circ f : X \longrightarrow Z$$

called the *composite* of f and g.

[1] We should emphasize that this is but one definition and that there are others (equivalent, of course). It very much depends on the background one wants to assume to start with and the goals one has in mind when using categories. If we were to assume that all mathematical entities have to be sets, we would give a slightly different definition. On the other hand, if we were to assume a purely formal set up, a fully specified formal framework, we would give a different definition still. For alternative definitions, see for instance [2, 19, 24].

- For each node X, there is an arrow:

$$1_X : X \longrightarrow X$$

called the *identity arrow* of X.

These data must satisfy the following properties:

- For all $f : W \longrightarrow X, g : X \longrightarrow Y, h : Y \longrightarrow Z$

$$h\circ(g\circ f) = (h\circ g)\circ f$$

that is, composition is *associative*.
- For all $f : X \longrightarrow Y$

$$f\circ 1_X = f = 1_Y\circ f$$

that is, the identity arrow is a unit.

The arrows of a category **C** automatically yield a criterion of identity for nodes.

Definition 2 An arrow $f : X \longrightarrow Y$ is said to be an *isomorphism* if it has an inverse, that is there is an arrow $g : Y \longrightarrow X$ such that $f\circ g = 1_Y$ and $g\circ f = 1_X$.

It can readily be shown that whenever such an inverse exists, it is unique and for that reason it is written f^{-1}.

Two nodes X and Y are *isomorphic*, written $X \simeq Y$, whenever there is an isomorphism between them. It is important to understand that isomorphic nodes are, from the point of view of the category in which they sit, indistinguishable. In our informal discussion, one could say that two isomorphic nodes contain the same information, possibly encoded differently. Thus, it might not be obvious that they are isomorphic and they can be isomorphic in more than one way.

Categories automatically come with a duality, via the notion of the opposite category, denoted by $\mathbf{C}^{op}$, of a category **C**. The nodes of $\mathbf{C}^{op}$ are the same as those of **C**, but the arrows go in the opposite direction. Thus, there is an arrow $f^o : Y \longrightarrow X$ in $\mathbf{C}^{op}$ if there is an arrow $f : X \longrightarrow Y$ in C. Composition is defined by

$$f^o \circ g^o = (g\circ f)^o.$$

Categories abound and are varied in nature. Philosophers have to keep in mind that the nodes and arrows can be interpreted in many different ways and in many different contexts. In other words, the nodes do not have to be entities of a specific kind, they do not have to be constituted in a uniform manner from one category to the next. The best way to illustrate this is probably by giving examples, to which we now turn.

11.3 Categories: Examples

Naturally, examples coming from mathematics arise immediately and are important. There are, however, two drawbacks to these examples in the present context. First, one has to rely on already known mathematical structures and functions, otherwise the examples are just as opaque as the definition of categories itself and one fails to see the shift from the already known concept, e.g. of groups, to the conceptually new level, namely the *category* of groups. Understandably, few philosophers are familiar with a large spectrum of mathematical structures, e.g. modules, groups, vector spaces, rings, fields, Banach spaces, Hilbert spaces, topological spaces, topological groups, Lie algebras, associative algebras, simplicial complexes, chain complexes, etc., in which case, one fails to see clearly the *variety* of cases naturally falling under the concept. Second, and almost in contradiction with what I have just said, the mathematical examples might lead the reader to believe that there is what might first appear to be a "deeper" unity to categories, namely that the nodes are in fact sets and the arrows are in fact functions between sets, that a category is merely a metamathematical device allowing mathematicians to put some order in their otherwise chaotic house of structures. This would be an undesirable impression and an unwarranted conclusion at this stage. Things are more subtle and delicate than what these cases suggest. So, let us start with examples coming from logic and abstract algebra and then give a few standard examples from mathematics.

(1) We will begin this enumeration with simple, abstract but useful examples. The category **0** has no node and no arrow. It is the empty category. The category **1** has one node and only the identity arrow. It can be pictured as follows:

The category **2** has two nodes and one arrow between them besides the identity arrows. Here is a picture of **2**:

Finally, the category **3** has three nodes and three non-trivial arrows, one of them being the composite of the other two. It can be represented thus:

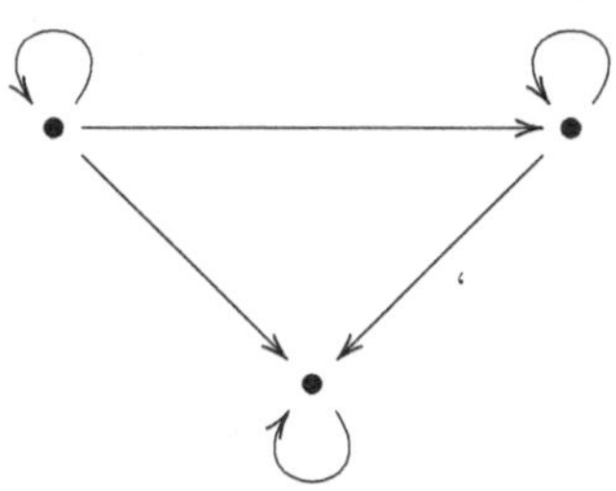

(2) Let $(P, \leq)$ be a pre-order, that is a reflexive and transitive relation. As a category, its objects are the points $p, q, r, \ldots$ of P and there is an arrow $f : p \longrightarrow q$ if and only if $p \leq q$. Thus there is at most one arrow between two nodes. It is easily verified that in this way one gets a category. In particular, any partial order is a category. This suggests immediately that category theory ought to be useful in mereology.

(3) Consider now propositions and logical relations between them, i.e. a deductive system. Thus the nodes are propositions $p, q, r, \ldots$ and there is an arrow $f : p \longrightarrow q$ if and only if $p \vdash q$ where "$\vdash$" denotes the usual consequence relation. Again, one immediately verifies that this is a category.

(4) Let T be a first-order theory (in a standard first-order language L, but in a categorical context, it is natural to work with many-sorted languages). It is possible to construct a category from T, denoted by C_T and called its *syntactic category* or also its *category of concepts*. Here is a sketch of the construction. (See, for instance, [20] or [14] for details.)

We first have to consider what is called a *formal set* $[\vec{x}; \varphi(\vec{x})]$, where $\vec{x}$ denotes a n-tuple of distinct variables containing all free variables of φ and φ is a formula of the underlying formal system L. Notice that a formal set is not a set as such. It is a purely syntactic construction. Two such formal sets, $[\vec{x}; \varphi(\vec{x})]$ and $[\vec{y}; \varphi(\vec{y})]$ are equivalent if one is the alphabetic variant of the other, that is if $\vec{x}$ and $\vec{y}$ have the same length and sorts and $\varphi(\vec{y})$ is obtained from $\varphi(\vec{x})$ by substituting $\vec{y}$ for $\vec{x}$ (and changing bound variables if necessary). This is clearly an equivalence relation and it is therefore possible to consider equivalence classes of such formal sets. A node of the syntactic category C_T is such an equivalence class of formal sets $[\vec{x}; \varphi(\vec{x})]$, where φ is a formula of the formal system L. The nodes of C_T are the equivalence classes of these formal sets, for *all* formulas of L. Notice: *all* formulas of the language are taken, not only those which appear in T. Thus, in a sense, the space of nodes is the collection of all possible properties and sentences expressible in that language, thus all possible theories in the given formal system. No logical relationship is considered at this stage, we have only identified the nodes. The next step will introduce the structure corresponding to the structure of that particular theory T and it is this step that will capture the particular features of T.

An arrow should be given by a formula of the theory T that defines a function, that is a formula $\theta(\vec{x}, \vec{y})$ of T that is provably functional. The only trick in the construction is to construct a morphism between two (equivalence classes of) formal sets $[\vec{x}; \varphi(\vec{x})]$ and $[\vec{y}; \psi(\vec{y})]$ in such a way that, when interpreted, it yields the *graph* of the function, in the standard set-theoretical sense of that expression, between the actual sets $\{(x_1, \ldots, x_n) \mid \varphi(\vec{x})\}$ and $\{(y_1, \ldots, y_m) \mid \psi(\vec{y})\}$. In fact, all definable functions in T will be represented by an arrow in C_T.

Needless to say, the verification that this is indeed a category requires more work than in the preceding cases.

What is especially clear in this case is that this is a purely syntactical construction. What makes it interesting is that the constructed category, an algebraic object usually denoted by C_T, embodies syntactical features of the theory T. It can be seen to be a generalization of the Lindenbaum-Tarski construction for propositional theories. We want to emphasize this aspect since categories are traditionally put in the semantical basket. In fact, categories have found a wide range of applications in proof theory. This example and the next show that categories are in both camps and provide a natural bridge between syntax and semantics.

(5) Instead of starting with a first-order theory, we could have started with a functional programming language L, e.g. Haskell. The nodes of the associated category are the data types of L and the arrows are the computable functions of L, e.g. the "programs". Again, this is a purely syntactic construction.

(6) Here are now some of the usual examples found in textbooks. The category **Set** has as its nodes sets and arrows, functions between sets. It is of course a very important category in mathematics. The category **Top** of topological spaces and continuous functions between them can easily be seen to be a category. So is the category **Grp** of groups and group homomorphisms. Readers acquainted with various mathematical structures will easily convince themselves that given a type of mathematical entities and structure-preserving functions between them, it is easy to verify that they form a category (or not!). See [19] for more standard examples.

(7) Any monoid M (or group for that matter) is a category. It has one object, which we might simply denote by a "•", and an arrow is simply an element of M. Composition of arrows correspond to composition of elements.

11.4 Basic Categorical Constructions

Category theory provides an extremely powerful language to express, explore and analyse concepts and theories. Some of these are directly relevant to exploration of formal metaphysics and formal ontology. We will here restrict ourselves to the most simple. Another interesting feature of category theory is that it does not presuppose that the objects it talks about, the nodes, are made up of points, elements or urelements. It has been argued that set theory could not be used in mereology, for instance, since one is then forced to assume that all objects are made up of elements, an assumption that does not fit with the ends of mereology. (See, for instance, [29].) It turns out that category theory does not suffer from this limitation at all. Quite the contrary. It can be used, for instance to express some of the basic concepts of mereology.

Let us start with what can certainly be considered to be one of the basic constructions of category theory: the notion of product of two objects X and Y of a category **C**. A *product* of X and Y in **C** is an object P *together with* two arrows $p_X : P \longrightarrow X$ and $p_Y : P \longrightarrow Y$, called the *projections* satisfying the following universal property: for all object Q with arrows $f : Q \longrightarrow X$ and $g : Q \longrightarrow Y$ there is a unique arrow $h : Q \longrightarrow P$ such that $p_X \circ h = f$ and $p_Y \circ h = g$. The definition holds readily in a simple commutative diagram, meaning that whenever you can go from one node to another node via two routes, then they are equal:

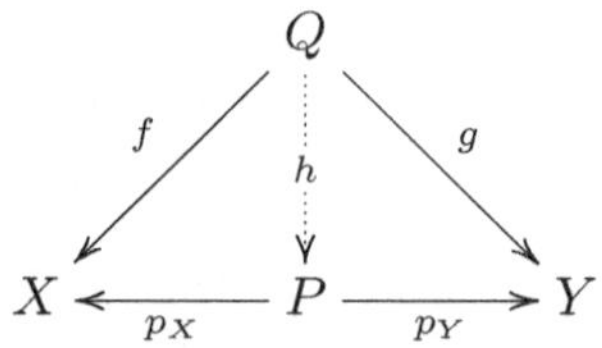

It can easily be shown that products are defined up to a unique isomorphism: if (P, p_X, p_Y) and (Q, q_X, q_Y) are products of X and Y, then there is a unique isomorphism $P \simeq Q$. For that reason, we talk about "the" product of X and Y and it is denoted by $X \times Y$ with the usual projections.

Using our informal analogy with information networks, a product of X and Y is a node that combines the information of X and the information of Y in a minimal manner, that is, any other node that can transmit at the same time information to X and to Y can actually be transmitted in a unique fashion through a product of X and Y. Of course, the interesting aspect of products is that once we have them, it is possible to consider how two kinds of information, e.g. coming from X and from Y, can be send together, that is from $X \times Y$. In the category **Set**, Cartesian products with the obvious projection functions are products in this sense. But here lies one of the strength of category theory: one can consider products in different categories and determine whether they exist and what they are in this context and here, the nodes do not have to be sets and the arrows do not have to be functions. Once again, the easiest example comes from logic: in a deductive system, a product of propositions p and q is the conjunction $p \wedge q$ (or any other proposition logically equivalent to it). Notice that it is not necessarily the case that all objects have a product in a category **C**. Whenever a category **C** does, we say that **C** has (binary) products. Notice also that if we take the empty product, we obtain an object, denoted by 1 such that for any object X of **C**, there is a unique arrow $X \longrightarrow 1$. The latter object is called the *terminal object* of **C**.[2]

The dual concept, usually called the coproduct, is also important. It is simply obtained by reversing the arrows in the foregoing definition. Thus, a *coproduct* of X and Y is an object S of **C** together with two arrows $i_X : X \longrightarrow S$, $i_Y : Y \longrightarrow S$, called the *inclusions*, such that for any object T with arrows $f : X \longrightarrow T$ and $g : Y \longrightarrow T$, there is a unique arrow $h : S \longrightarrow T$ with $f = h \circ i_X$ and $g = h \circ i_Y$. In diagrammatic form, we get:

[2]Notice that the latter concept can be defined directly in terms of arrows with no reference to the concept of product. Thus a category can have a terminal object without having all products.

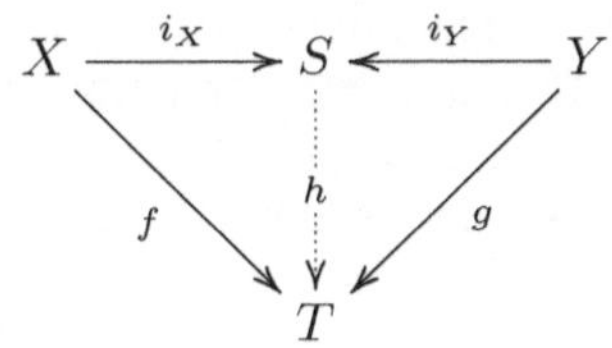

As with products, coproducts are defined up to a unique isomorphism and "the" coproduct of X and Y is usually denoted by $X \amalg Y$ with the usual inclusions. In **Set**, disjoint unions form coproducts. In a deductive system, the disjunction $p \vee q$ form a coproduct of p and q. Whenever coproducts exist for all pairs of objects of a category **C**, we say that **C** has (binary) coproducts. If we consider the empty coproduct, we get an object, denoted by 0, such that for any object X of **C**, there is an arrow $0 \longrightarrow X$. It is called the *initial object* of **C**.[3]

These constructions correspond to well-known concepts in various contexts. Algebraic topologists developed slightly more general concepts that find a natural expression in category theory and are directly relevant to mereology or mereotopology, namely the concepts of pullback and pushout.

Given the following diagram $X \xrightarrow{f} Z \xleftarrow{g} Y$ in a category **C**, a *pullback* of X and Y over Z is given by an object $X \times_Z Y$ together with two morphisms $X \times_Z Y \xrightarrow{p_X} X$ and $X \times_Z Y \xrightarrow{p_Y} Y$ such that $f \circ p_X = g \circ p_Y$ and satisfying the universal property: for all Q together with $Q \xrightarrow{l} X$ and $Q \xrightarrow{k} Y$ such that $g \circ k = f \circ l$, there is a unique arrow $Q \xrightarrow{h} X \times_Z Y$ such that $p_Y \circ h = k$ and $p_X \circ h = l$. In diagrammatic form:

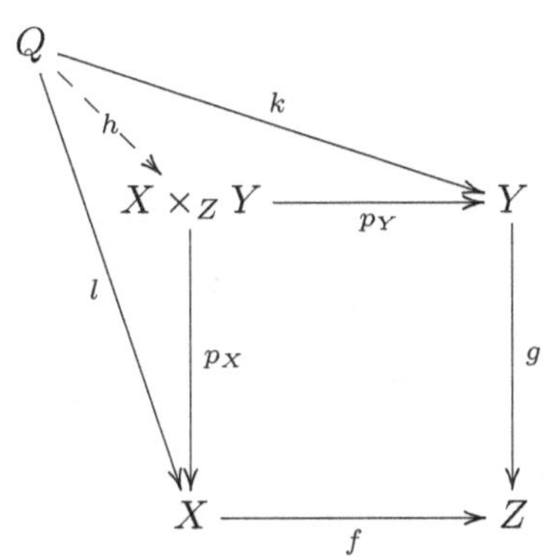

[3]The preceding remark concerning the possibility of defining the concept of terminal object directly applies *mutatis mutandis* to the concept of initial object.

Pullbacks can be understood as generalization of products. Indeed, it can easily be verified that the pullback of X and Y over the terminal object 1 is the same as the product of X and Y.

But there is an intuitive reading of pullbacks. It is the product of X and Y over the area of Z on which the arrows f and g agree. For instance, in the category **Set**, the pullback of $X \xrightarrow{f} Z \xleftarrow{g} Y$ can be described as the set of pairs $\{(x, y) | f(x) = g(y)\}$. Thus, one can think of a pullback as a local or parametrized product over an object.

The dual construction is just as important. Given the following diagram $X \xleftarrow{f} Z \xrightarrow{g} Y$ in a category **C**, a *pushout* of X and Y along Z is given by an object $X \amalg_Z Y$ together with two morphisms $X \xrightarrow[i_X]{} X \amalg_Z Y$ and $Y \xrightarrow[i_Y]{} X \amalg_Z Y$ such that $i_X \circ f = i_Y \circ g$ and satisfying the universal property: for all T together with $X \xrightarrow{p} T$ and $Y \xrightarrow{q} T$ such that $p \circ f = q \circ g$, there is a unique arrow $X \amalg_Z Y \xrightarrow{r} T$ such that $r \circ i_X = p$ and $r \circ i_Y = q$. In diagrammatic form:

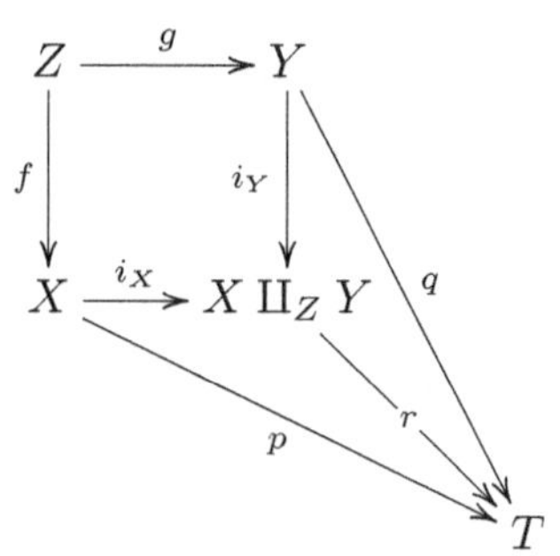

Notice that the notion of coproduct is a special case of a pushout: simply take the pushout over the initial object $X \xleftarrow{f} 0 \xrightarrow{g} Y$.

In the case of topological spaces, pushouts have a direct and intuitive interpretation. The pushout amounts to "gluing" together X and Y along the image of Z in both of them. Thus, for instance, if Z is the Euclidean unit disk $\mathbb{D}$ and X and Y are both unit spheres S^2, then the pushout $S^2 \amalg_{\mathbb{D}} S^2$ is a system of two spheres glued together along the images of $\mathbb{D}$ on both spheres. Figuratively, the two spheres are "smashed" together on the images of the disk.

It should be clear that these constructions can be iterated and combined in various ways. The foregoing constructions are all special cases of general constructions in categories, called *limits* and *colimits* in the literature. In the finite cases, one can show that finite limits and finite colimits can be constructed out of pullbacks and terminal object, on the one hand, and pushouts and initial object, on the other hand. But the situation is different when one considers arbitrary limits and colimits, concepts that are relevant to formal metaphysics.

A *diagram* D in a category $\mathbf{C}$ is a pattern of arrows (together with their domains and codomains). For instance, in the case of pullbacks, we had the diagram $X \xrightarrow{f} Z \xleftarrow{g} Y$. In the arbitrary case, we will denote an arrow of such a diagram by $D_i \xrightarrow{f_{ij}} D_j$.

A *cone* to a diagram D in $\mathbf{C}$ is an object C together with a family of arrows $c_i : C \longrightarrow D_i$, one for each D_i such that for each $f_{ij} : D_i \longrightarrow D_j$ of D, the following triangle commutes

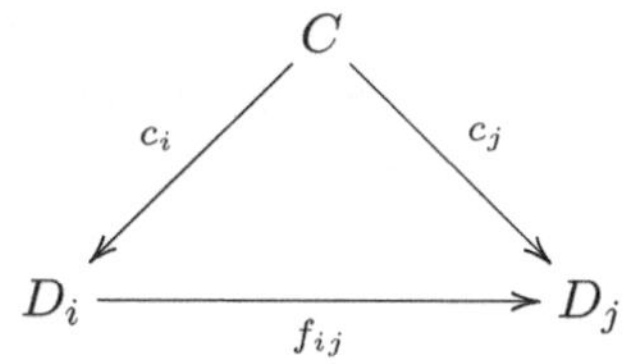

A *limit* for a diagram D is a cone, denoted by $\varprojlim D_j$ with arrows $\varprojlim D_j \xrightarrow{\pi_i} D_i$, to the diagram D satisfying the following universal property: for each cone C to D, there is a unique arrow $h : C \longrightarrow \varprojlim D_j$ such that for each arrow $f_{ij} : D_i \longrightarrow D_j$ of D, the following diagrams commute:

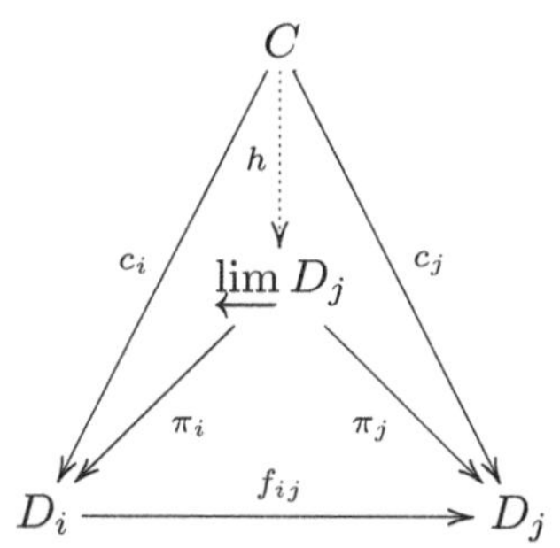

It is easy to see that products and pullbacks are special cases of (finite) limits. In the case of products, simply take the diagram consisting of two objects X and Y with no arrows between them. In the case of pullbacks, take the diagram $X \xrightarrow{f} Z \xleftarrow{g} Y$. In both cases, one verifies that the definition of limit yields the respective constructions. Informally, a limit can be thought of as a restricted product of the objects of the diagram, the restriction being given by the arrows of the diagram.

Naturally, the concept of limit has a dual, namely the concept of colimit.

A *cocone* from a diagram D in $\mathbf{C}$ is an object C together with a family of arrows $d_i : D_i \longrightarrow C$, one for each object D_i of the diagram D, such that for each $f_{ij} : D_i \longrightarrow D_j$, the following triangle commutes

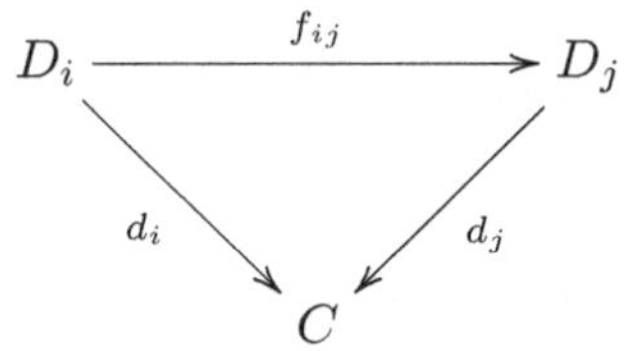

A *colimit* for a diagram D is a cocone, denoted by $\varinjlim D_j$ or simply colim D, with arrows $\iota_i : D_i \longrightarrow \varinjlim D_j$, one for each i, satisfying the following universal property: for each cocone C from D, there is a unique arrow $u : \varinjlim D_j \longrightarrow C$ such that for each arrow $f_{ij} : D_i \longrightarrow D_j$, the following diagrams commute:

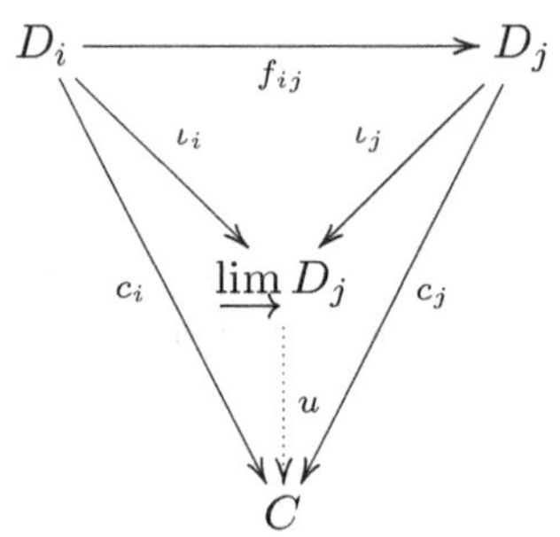

Informally, a colimit can be thought of as being a fusion – in the mereological sense of that expression – of the given pieces with the incidence relations provided by the arrows in the diagram.

Limits and colimits are powerful constructions. From a theoretical point of view, one wants to investigate how they combine, whether they commute, under what conditions, etc. For applications, one has to verify that the category in which one is working *has* limits and colimits.

11.5 Parts and Propert Parts in Category Theory

The notion of being a part of an object X receives an analysis that differs from the usual one offered by set theory and mereology. Let us first define the property of being monomorphic for an arrow: an arrow $i : A \longrightarrow X$ is said to be *monomorphic* or a *monomorphism* if it is left-cancellable, that is given any arrows $f, g : B \longrightarrow A$ such that $i \circ f = i \circ g$, then $f = g$. A monomorphism is usually denoted by $i : A \rightarrowtail X$.

In our informal analogy, we could say that the information coming out of A to X is never confused, i.e. distinct information remains distinct.

Informally, a monomorphism $i : A \longrightarrow X$ picks up a part of X. It is as if the information delivered by i would allow us to see clearly and distinctly a genuine part of X, even though i itself is not that part. Alternatively, one could think of i as yielding a copy of a part of X, namely the domain of i, A. In fact, these "copies" of parts of X are systematically related to one another as follows.

Two monomorphisms $i : A \rightarrowtail X$ and $j : B \rightarrowtail X$ are *equivalent* if there is an isomorphism $k : A \longrightarrow B$ such that $j \circ k = i$. This obviously defines an equivalence class of arrows and a *subobject* of X is defined to be such an equivalence class. Thus, technically speaking, category theorists identify a part of X as being an equivalence class of monomorphisms with codomain X.

Notice that there is a natural partial order on these subobjects of X: given $i : A \rightarrowtail X$ and $j : B \rightarrowtail X$, the subobject $[i]$ is included in the subobject $[j]$, $[i] \subset [j]$, if and only if there is an arrow $k : A \longrightarrow B$ such that $j \circ k = i$. (It can be shown that whenever such a k exists, it is a monomorphism and it is unique.)

We could say that a *part* of X is a monomorphism $i : A \rightarrowtail X$ isomorphic to a subobject of X.

How does this capture the notion of being a part? First, in the category **Set**, a monomorphism is the same thing as an injection and the image of an injection can be identified with a subset. (But this need not be the case in other categories.) In **Set**, a part of a set X is any isomorphic copy of a subset of X.[4] In **Top**, a part is automatically a subspace (the induced topology is taken care of by the arrows). In **Grp**, a part is a subgroup (again, a monomorphism in that category is automatically a homomorphism of group). Given a category C with products, a part $R \rightarrowtail X \times X$ is a relation. It is important, however, to understand that the notion of being a monomorphism is a generalization of the notion of being injective for a function. It therefore yields an analysis of parthood that is more general than the usual set-theoretical notion. We will come back to this point.

A *proper part* can be defined as follows: $i : A \rightarrowtail X$ is a *proper part* of X if $i : A \rightarrowtail X$ is a part of X but i is not an isomorphism.

An arrow $q : X \twoheadrightarrow Q$ is an *epimorphism* if it is right cancellable, that is given arrows $f, g : Q \rightarrow Y$ such that $f \circ q = g \circ q$, then $f = g$.

In **Set**, an epimorphism is a surjective function and epimorphisms are often the same as surjective homomorphisms. But not always. For instance, in the category of commutative rings with unit, epimorphisms are not necessarily surjective. In the case of Hausdorff spaces, epimorphisms are precisely continuous functions with a dense image. (See [4, pp. 28–29] for details.) Again, the notion of epimorphism is a natural generalization of the notion of surjective function, but it is context sensitive in a way the latter is not.

4This is shocking only if someone sticks firmly to the axiom of extensionality. From a categorical point of view, the axiom of extensionality is not the adequate criterion of identity for abstract sets.

We would like to show that the notion of a boundary between objects can be given a purely algebraic definition in a categorical framework, but in order to do so, we need to define two intermediate concepts, which are important in themselves.

First, given two objects X and Y of a category $\mathbf{C}$, one might want to consider the space of all arrows Y^X as an object of $\mathbf{C}$ itself. Of course, the latter does not always exist. Whenever it does, it can be defined via the structure of arrows of the category and provided $\mathbf{C}$ has products. Thus, given two objects X and Y of a category $\mathbf{C}$, an *exponentiation* of Y by X is an object Y^X together with an arrow $ev : Y^X \times X \longrightarrow Y$, called *evaluation*, such that for each Z and for each $f : Z \times X \longrightarrow Y$, there is a unique $\ulcorner f \urcorner : Z \longrightarrow Y^X$ such that $ev \circ (\ulcorner f \urcorner \times 1_X) = f$.[5] The diagram corresponding to the last property is:

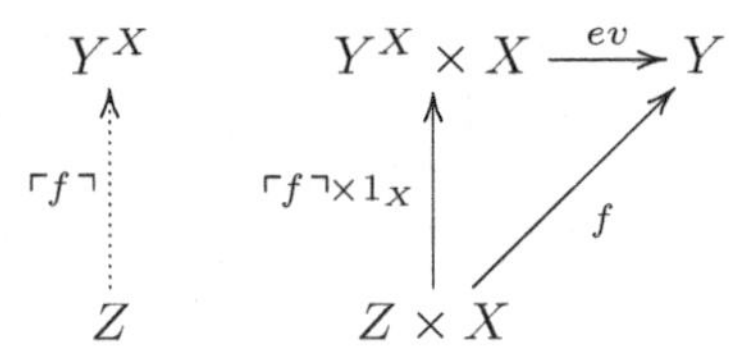

In **Set**, the exponentiation of Y by X is the set Y^X of all functions from X to Y. In a deductive system, Y^X is the implication $X \Rightarrow Y$.

When the category is a deductive system, the characterizing property of exponentiation benefits from a presentation in terms of deductive rules:

$$\frac{Z \vdash X \Rightarrow Y}{\overline{Z \wedge X \vdash Y}}$$

There are two rules: from top to bottom and from bottom to top.

The "dual" construction is less well known but just as important in our context. Given a category $\mathbf{C}$ with coproducts, the *subtraction* of X from Y, is an object written $Y \backslash X$, together with an arrow $i : Y \longrightarrow (Y \backslash X) \amalg X$ such that for each Z and each $f : Y \longrightarrow Z \amalg X$, there is a unique $\llcorner f \lrcorner : Y \backslash X \longrightarrow Z$ such that $(\llcorner f \lrcorner \amalg 1_x) \circ i = f$, i.e.:

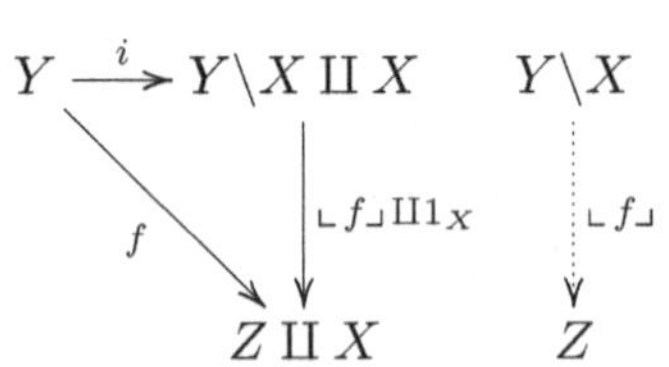

[5]Since this object is also defined up to a unique isomorphism, we immediately introduce the standard notation.

Again, when **C** is a deductive system, we can express the characterizing property in terms of deductive rules:

$$\frac{Y\backslash X \vdash Z}{\overline{Y \vdash Z \vee X}}$$

Whenever a category **C** has subtractions, products and a terminal object 1, we can define a *complement* X' of an object X by setting $X' = (1\backslash X)$. The *boundary* ∂X of an object X is then defined by[6]:

$$\partial X = X \wedge X'$$

It is important to understand that this definition of the boundary of an object is a genuine generalization of the definition usually found in topology.[7] It is the same definition when the category **C** is **Top**, but it applies to other cases just as well. We therefore have a general definition of boundary that applies to the usual cases but that could be used just as well in mereology, mereotopology and systems theory in general.

11.6 Functors and Natural Transformations

In a category, the information is carried by the arrows and the algebra of arrows. This situation should apply to categories themselves and, surprisingly perhaps, it does. An arrow between two categories **C** and **D** is called a *functor*.

Definition 3 A *(covariant) functor* $F : \mathbf{C} \longrightarrow \mathbf{D}$ assigns:

(1) to each node X of **C**, a node $F(X)$ of **D**
(2) to each arrow $f : X \longrightarrow Y$ of **C**, an arrow $F(f) : F(X) \longrightarrow F(Y)$ of **D**

in such a way that

- $F(g \circ f) = F(g) \circ F(f)$
- $F(1_X) = F_{(1_X)}$

In other words, a functor preserves identities and composition, i.e. the structure of a category.

[6]As far as I know, Bill Lawvere was the first to propose this general definition. For more on its properties, see [18].

[7]Notice that we do not have to take the closure of X since in the case of topological spaces, the operations define a coHeyting algebra, i.e. the algebra of closed sets.

A functor that reverses the order of composition, i.e. a functor $F : \mathbf{C} \longrightarrow \mathbf{D}$ such that $F(g \circ f) = F(f) \circ F(g)$ is said to be *contravariant*.

It can be asserted that contemporary mathematics is functorial: all basic concepts and constructions now introduced are functorial, i.e. they are functors. A list of examples could cover all fields of mathematics at all levels, from the most elementary to the most arcane. We will restrict ourselves to a few simple examples. The reader is urged to consult references for other relevant examples. (See, for instance, [2, 4, 19, 24].)

Examples:

(1) For each category **C**, there is an identity functor, $1_{\mathbf{C}}$, which sends each object to itself and each arrow to itself.
(2) The power-set operation $\wp : \mathbf{Set} \longrightarrow \mathbf{Set}$ is functorial. Its operation on objects is obvious. Given a function $f : X \longrightarrow Y$, for each subset $A \subset X$, $\wp(f)(A) = f[A]$, the image of A in Y under f. This is clearly a covariant functor.
(3) The power-set operation induces a second, different functor $\wp : \mathbf{Set}^o \longrightarrow \mathbf{Set}$, acting in the same way on objects but this time, for each subset $B \subset Y$, $\wp(f)(B) = f^{-1}[B]$, the inverse image of B in X. This is now a contravariant functor.
(4) A functor $F : \mathbf{2} \longrightarrow \mathbf{C}$ simply picks up an arrow of **C**.
(5) A functor $F : \mathbf{3} \longrightarrow \mathbf{C}$ picks up a commutative triangle of **C** or, in other terms, composable arrows of **C**.
(6) If M is a monoid, seen as a category, it can be verified that a functor $F : M \longrightarrow \mathbf{Set}$ picks up a set $F(\bullet) = X$ together with an (right) action, also called an M-*set* in the literature. More formally, an M-*set* can be described as a pair $(X, (f_m)_{m \in M})$ where, for each $m \in M$, $f_m : X \longrightarrow X$ such that for all m, n in M:

$$f_e = 1_X;$$

$$f_{mon} = f_m \circ f_n$$

where e is the unit of M and mon is the product in M. These readily describe systems that are in various states, the latter being given by the monoid M. It is worthwhile to notice the pattern here: the monoid M provides a schema of possible states whereas a functor $F : M \longrightarrow \mathbf{Set}$ or an M-**Set** is an actual system with an actual state space. Automata theory can be developped in this setting. See, for instance [1, 30].
(7) If P is a pre-order or a partial-order, seen as a category, then a functor $F : P \longrightarrow \mathbf{Set}$ sends each arrow $p \leq q$ of P to a function $F(p) \longrightarrow F(q)$ between sets. It is perhaps worth looking at the case when we have the linearly ordered set $(\mathbb{N}, \leq)$. In the case of the linearly ordered set, a functor $F : \mathbb{N} \longrightarrow \mathbf{Set}$ a sequence of sets

$$X_0 \longrightarrow X_1 \longrightarrow X_2 \longrightarrow \cdots$$

and functions $X_n \longrightarrow X_{n+1}$. Whereas in the previous case, we had a set in various states, here we have a sequence of sets seen as a whole, or evolving through time.

It is easy to see that functors compose and that therefore categories form a category.[8]

Definition 4 Given two parallel functors $F, G : \mathbf{C} \longrightarrow \mathbf{D}$, a *natural transformation* η from F to G, $\eta : F \longrightarrow G$, is given by the following data:

(1) For each object X of $\mathbf{C}$, an arrow $\eta_X : F(X) \longrightarrow G(X)$ of $\mathbf{D}$ such that:
(2) For each arrow $f : X \longrightarrow Y$ of $\mathbf{C}$, the following square commutes:

$$\begin{array}{ccc} F(X) & \xrightarrow{\eta_X} & G(X) \\ {\scriptstyle F(f)}\downarrow & & \downarrow{\scriptstyle G(f)} \\ F(Y) & \xrightarrow[\eta_Y]{} & G(Y) \end{array}$$

From the definition alone, natural transformations might seem to be odd beasts. As we have already said, they actually constitute the reason why Eilenberg and Mac Lane introduced functors and categories in the first place. In other words, natural transformations were noticed first as a mathematically significant phenomenon that deserved to be clarified. They turn out to pervade mathematics. Here are a few selected examples.

Examples:

(1) The power-set operation induces a third functor, contravariant again,

$$\wp : \mathbf{Set}^{op} \longrightarrow \mathbf{BA}$$

where **BA** is the category of Boolean algebras and Boolean homomorphisms. There is a parallel functor

$$\mathrm{Hom}(-, \mathbf{2}) : \mathbf{Set}^{op} \longrightarrow \mathbf{BA}$$

defined as follows: to each set X, $\mathrm{Hom}(X, \mathbf{2})$ is the set of all functions from X into the Boolean algebra $\mathbf{2}$, itself a Boolean algebra, and to each function $f : X \longrightarrow Y$, the functor $\mathrm{Hom}(f, \mathbf{2}) : \mathrm{Hom}(Y, \mathbf{2}) \longrightarrow \mathrm{Hom}(X, \mathbf{2})$ is defined by $\mathrm{Hom}(f, \mathbf{2})(g) = g \circ f$. There is a well-known isomorphism $\mathrm{Hom}(X, \mathbf{2}) \simeq \wp(X)$. In fact, this isomorphism is a natural transformation. This means that, for any function $f : X \longrightarrow Y$, the following square of Boolean algebras and Boolean homomorphisms commutes:

[8] There are obvious foundational issues arising at this point, but we will simply brush them under the carpet and ignore them altogether.

$$\begin{array}{ccc} X & & \mathrm{Hom}(X, \mathbf{2}) \overset{\cong}{\longrightarrow} \wp(X) \\ \Big\downarrow f & & \Big\uparrow \qquad\qquad \Big\uparrow \\ Y & & \mathrm{Hom}(Y, \mathbf{2}) \underset{\cong}{\longrightarrow} \wp(Y) \end{array}$$

(2) Here is a simple and perhaps shallow example. Consider the identity functor $1_{\mathbf{Set}} : \mathbf{Set} \longrightarrow \mathbf{Set}$ and the covariant power-set functor $\wp : \mathbf{Set} \longrightarrow \mathbf{Set}$. Consider now the function $s_X : X \longrightarrow \wp(X)$ defined by $s_X(x) = \{x\}$. It can easily be verified that it is a natural transformtion $s_{(-)} : 1_{\mathbf{Set}} \longrightarrow \wp$.

11.7 Functor Categories, Presheaves and Sheaves

Although the constructions presented in the preceding section illustrate the power and flexibility of category theory, it can be argued that one of the most striking tool of the trade is that of functor categories and the latter are certainly relevant to metaphysics, since they can be used to model possible worlds, various semantics, constructive mathematics and numerous other situations.

Given two categories $\mathbf{C}$ and $\mathbf{D}$, the functor category $\mathbf{D}^{\mathbf{C}^{op}}$ has as its objects functors $F : \mathbf{C}^{op} \longrightarrow \mathbf{D}$ and arrows natural transformations $\eta : F \longrightarrow G$.[9] Very often in practice, $\mathbf{D}$ is taken to be the category $\mathbf{Set}$, but it need not be. Notice also that the operation can be iterated, i.e. $\mathbf{D}$ can already be a functor category. It is in fact not unlikely to consider in practice, even in logic, categories of the form $(\mathbf{D}^{\mathbf{C}^{op}})^{\mathbf{E}^{op}}$. Here are some key examples of functor categories.

Examples:

(1) The functor category $\mathbf{C}^{\mathbf{1}}$ is obviously the same as the category $\mathbf{C}$.
(2) The functor category $\mathbf{C}^{\mathbf{2}}$ is the category of arrows of $\mathbf{C}$, that is an object of $\mathbf{C}^{\mathbf{2}}$ is an arrow of $\mathbf{C}$. An arrow of $\mathbf{C}^{\mathbf{2}}$ consists of a pair of arrows of $\mathbf{C}$ making the appropriate square commute.
(3) The objects of the functor category $\mathbf{Set}^{\mathbb{N}}$ have been given in the example 7 above. An arrow between two sequences of sets is a natural transformation $\eta : F \longrightarrow G$ such that all the squares in the following diagram commute:

[9] Notice that the functors go from the *opposite* category $\mathbf{C}^{op}$. This is related to theoretical aspects of category theory that we cannot explain in such a short paper.

$$\begin{array}{ccccc} X_0 & \longrightarrow & X_1 & \longrightarrow & \dots \\ \Big\downarrow{\scriptstyle \eta_0} & & \Big\downarrow{\scriptstyle \eta_1} & & \Big\downarrow{\scriptstyle \eta_2} \\ Y_0 & \longrightarrow & Y_1 & \longrightarrow & \dots \end{array}$$

(4) If we take the category $\mathbf{Set}^{\mathbb{N}^{op}}$ instead, we simply reverse the arrows of the objects as follows:

$$\begin{array}{ccccc} X_0 & \longleftarrow & X_1 & \longleftarrow & \dots \\ \Big\downarrow{\scriptstyle \eta_0} & & \Big\downarrow{\scriptstyle \eta_1} & & \Big\downarrow{\scriptstyle \eta_2} \\ Y_0 & \longleftarrow & Y_1 & \longleftarrow & \dots \end{array}$$

The foregoing examples are all cases of what are called *presheaves* in the literature. They have been used, for instance, to construct models of non-Boolean negations. (See [17].) Among functor categories, categories of sheaves occupy a central position, both because they have strong links to topology *and* logic. They have been used to obtain important results in various areas of logic, for instance in modal logic, see [3, 12, 21], in intuitionistic logic and intuitionistic mathematics in general, see [25], in constructive mathematics, see [11], and, recently, in mathematical physics, see [5–9]. Thomas Mormann has used sheaves to model an ontology of tropes. The next step here would be to investigate the logic of tropes since a category of sheaves has an internal logic. (See [26] and also [27, 28] for other applications.). A standard reference on sheaves in mathematics and logic is [20].

11.8 Conclusion

As we have already mentioned, we have barely scratched the surface in the foregoing sections. A proper treatment would have included a discussion of categorical logic, Grothendieck toposes and modal logic, process algebras and higher-dimensional categories. This in itself strongly suggests that it ought to be included in the toolbox of anyone interested in formal metaphysics.

And there is more. Category theory is now being applied in many different ways in theoretical computer science, from linear logic to ontology engineering. (For an introduction to the latter, see for instance [13].) As we have also mentioned, it is finding its way in the foundations of physics. It is my conviction that it is bound to occupy a central position in formal philosophy.

References

1. Adámek, J., & Trnková, V. (1990). *Automata and algebras in categories* (Mathematics and its applications (East European series), Vol. 37). Dordrecht: Kluwer.
2. (***)Awodey, S. (2007). *Category theory* (Oxford logic guides, Vol. 49). Oxford: Clarendon Press. [A nice introduction to category theory.].
3. Awodey, S., & Kishida, K. (2008). Topology and modality: The topological interpretation of first-order modal logic. *The Review of Symbolic Logic, 1*(Special Issue 02), 146–166.
4. Borceux, F. (1994). *Handbook of categorical algebra. 1* (Encyclopedia of mathematics and its applications, Vol. 50). Cambridge: Cambridge University Press.
5. Crane, L. (2009). Relational spacetime, model categories and quantum gravity. *International Journal of Modern Physics A, 24*(15), 2753–2775.
6. Döring, A., & Isham, C. J. (2008). A topos foundation for theories of physics. I. Formal languages for physics. *Journal of Mathematical Physics, 49*(5), 25
7. Döring, A., & Isham, C. J. (2008). A topos foundation for theories of physics. II. Daseinisation and the liberation of quantum theory. *Journal of Mathematical Physics, 49*(5), 26.
8. Döring, A., & Isham, C. J. (2008). A topos foundation for theories of physics. III. The representation of physical quantities with arrows $\breve{\delta}^{o}(A)\ \underline{\Sigma} \longrightarrow \underline{\mathbb{R}^{\succcurlyeq}}$. *Journal of Mathematical Physics, 49*(5), 777–825
9. Döring, A., & Isham, C. J. (2008). A topos foundation for theories of physics: IV. Categories of systems. *Journal of Mathematical Physics, 49*(5), 826–884.
10. Eilenberg, S., & Mac Lane, S. (1945). A general theory of natural equivalences. *Transactions of the American Mathematical Society, 58*, 231–294.
11. Geuvers, H., Niqui, M., Spitters, B., & Wiedijk, F. (2007). Constructive analysis, types and exact real numbers. *Mathematical Structures in Computer Science, 17*(1), 3–36.
12. Ghilardi, S., & Zawadowski, M. (2002). *Sheaves, games, and model completions: A categorical approach to nonclassical propositional logics* (Trends in logic – studia logica library, Vol. 14). Dordrecht: Kluwer.
13. Johnson, M., & Rosebrugh, R. (2010). Ontology engineering, universal algebra, and category theory. In R. Poli, M. Healy, & A. Kameas (Eds.), *Theory and applications of ontology: Computer applications* (pp. 565–576). Dordrecht: Springer.
14. Johnstone, P. T. (2002). *Sketches of an elephant: A topos theory compendium. Volume 2* (Oxford logic guides, Vol. 44). Oxford: The Clarendon Press Oxford University Press.
15. Krömer, R. (2007). *Tool and object. A history and philosophy of category theory* (Volume 32 of science networks. Historical studies). Basel: Birkhäuser Verlag.
16. Landry, E., & Marquis, J.-P. (2005). Categories in context: Historical, foundational and philosophical. *Philosophia Mathematica (3), 13*(1), 1–43.
17. La Palme Reyes, M., Macnamara, J., Reyes, G. E., Zolfaghari, H. (1999). Models for non-Boolean negations in natural languages based on aspect analysis. In D. M. Gabbay & H. Wansing (Eds.), *What is negation?* (Vol. 13, pp. 241–260). Dordrecht: Kluwer.
18. (***)Lawvere, F. W., & Rosebrugh, R. (2003). *Sets for mathematics*. Cambridge: Cambridge University Press. [A gentle introduction to category theory. A nice place to start for philosophers.].
19. (***)Mac Lane, S. (1998). *Categories for the working mathematician* (Graduate texts in mathematics, 2nd ed., Vol. 5). New York: Springer. [The standard reference in the field. However, it does presuppose familiarity with various mathematical theories.]
20. (***)Mac Lane, S., & Moerdijk, I. (1994). *Sheaves in geometry and logic. A first introduction to topos theory* (Universitext). New York: Springer. [A reference on sheaves and toposes. Not an introductory book.].
21. Makkai, M., & Reyes, G. E. (1995). Completeness results for intuitionistic and modal logic in a categorical setting. *Annals of Pure and Applied Logic, 72*(1), 25–101.
22. Marquis, J. -P. (2006). What is category theory? In G. Sica (Ed.), *What is category theory?* (pp. 221–255). Monza: Polimetrica.

23. Marquis, J. -P. (2009). *From a geometrical point of view: A study of the history and philosophy of category theory* (Logic, epistemology, and the unity of science, 1st ed., Vol. 14). New York: Springer.
24. McLarty, C. (1995). *Elementary categories, elementary toposes* (Oxford logic guides, Vol. 21). Oxford: The Clarendon Press.
25. Moerdijk, I. (1998). Sets, topoi and intuitionism. *Philosophia Mathematica (3), 6*(2), 169–177. Perspectives on intuitionism.
26. Mormann, T. (1995). Trope sheaves. A topological ontology of tropes. *Logic and Logical Philosophy, 3*, 129–150.
27. Mormann, T. (1997). Topological aspects of combinatorial possibility. *Logic and Logical Philosophy, 5*, 75–92.
28. Mormann, T. (1999). Neither mereology nor Whiteheadian account of space yet convicted. *Analysis (Oxford), 59*(3), 174–182.
29. Smith, B. (1998). The basic tools of formal ontology. In N. Guarino (Ed.), *Formal ontology in information systems* (pp. 19–28). Amsterdam: IOS Press.
30. Worththington, J. (2012). A bialgebraic approach to automata and formal language theory. *Annals of Pure and Applied Logic, 163*(7), 745–762.

Chapter 12
Can Natural Language Be Captured in a Formal System?

Martin Stokhof

Abstract The question whether natural language can be captured in a formal system has been argued at length, and both a positive and a negative answer has been defended. The paper investigates the main lines of argument for both, and argues that the stalemate that appears to have been reached is an indication that the question itself rests on a wrong conception of the relation between natural languages and formal languages, and hence of the methodological status of formal modelling of natural language.

12.1 Introduction

The question that this paper addresses has been answered with both an unqualified "Yes" and an unqualified "No", and both answers are not only straightforward but apparently, they can also both be justified. That suggests that perhaps something is wrong with the question, and that hence the best possible answer we can give is "It depends". In what follows we will proceed to explore these three options.

But before we start we must do some preliminary work. First of all, we should further specify the question by determining what we will take "natural language" to refer to. Is that the phonetic system of a natural language, its morphology, or its syntax? Although in these areas, too, the question of the possibilities and limitations of formalisation are interesting and important, we will concentrate on the "meaning" aspect of natural language, i.e., on its semantics and its pragmatics. The main reason for this restriction is that it is in this area that the question appears to have its main philosophical interest.

It should be noted at the outset that by distinguishing the question of the possible formalisation of meaning from that of, say, syntax, we allow for the possibility

M. Stokhof (✉)
ILLC/Department of Philosophy, Universiteit van Amsterdam, Amsterdam, The Netherlands

Department of Philosophy, Tsinghua University, Beijing, China
e-mail: M.J.B.Stokhof@uva.nl

S. O. Hansson, V. F. Hendricks (eds.), *Introduction to Formal Philosophy*, Springer Undergraduate Texts in Philosophy, https://doi.org/10.1007/978-3-319-77434-3_12

that the first question can be answered in the negative while at the same time the latter is answered positively. (As we shall see, the reverse possibility seems much harder to maintain.) Such a situation would be in accordance with the theoretical view that, indeed, there is a principled distinction between the syntax of natural languages on the one hand, and their semantics and pragmatics on the other, and that this distinction is reflected in the nature of linguistic competence. The former would correspond to Chomsky's conception of "the faculty of language in the narrow sense" (FLN), that is connected not with the communicative function of actual natural languages, but with an evolutionary and physiologically distinct computational system that is not unique for language. Semantics, and pragmatics in so far as it deals with meaning, would belong to a different system, "the faculty of language in the broad sense" (FLB), a conceptual system that draws on different cognitive capacities that are physiologically realised in different ways than those characteristic for the computational system.[1]

Most people working in semantics would tend not to agree with such a radical difference between systems, and one reason for this disagreement is that they conceive of semantics as a homogeneous domain, despite the customary distinction between lexical semantics and structural semantics. According to this view, the former deals with the meaning content of lexical expressions, whereas the latter is concerned with the semantic rules that govern the ways in which meaningful expressions are combined into larger, meaningful wholes. For an account of the meaning content of a complex expression we need both, hence semantics, at least in theory, should be homogeneous and unified, although in practice there may be a division of labour, of course. This assumed homogeneity does not sit well with the Chomskyan division between FLN and FLB, in particular because, according to the received wisdom of compositionality, the semantic rules that are the prime subject of structural semantics, in their turn should be aligned with the rules of syntax. But if Chomsky would be right, that would mean that there is a fundamental split between structural semantics and lexical semantics after all, the former being part of FLN and the latter belonging to FLB. This is certainly a possible point of view, one which is potentially agreeable to Chomsky, but nevertheless awkward for most semanticists, since it robs them of a homogeneous empirical domain. Whether the question of formalisation can shed further light on this issue is something we will return to later on.

So, we focus in what follows on the question of the formalisability of meaning, and hence the question addressed will be "Can natural language meaning be captured in a formal system?". Do note that the relevance of that question extends beyond semantics and pragmatics as parts of linguistic theory: if natural language meaning is non-formalisable, then this will arguably have implications for the potential of formalisation in epistemology, metaphysics, and other philosophical disciplines. Inasmuch as the concepts and the argumentative strategies employed in

[1]Cf., Hauser et al. [11]. The position that semantics is not part of grammar is one that Chomsky has defended throughout, cf., e.g., [4], Chapter 2.

those disciplines depend on, and are conducted in, natural language, the effects of natural language meaning being formalisable, or not as the case may be, will trickle down and affect what formalisation can reasonably be expected to achieve in these disciplines.

12.2 Yes

The wholeheartedly positive answer "Yes" started to be heard in philosophy, and a little later on in linguistics, only at the end of the sixties of the previous century. There had been people who had suggested that the tools of formal logic could be applied to the analysis of natural language meaning in a systematic and empirically explanatory way earlier on, but their voices were hardly heard at the time,[2] or their suggestions were brushed aside as irrelevant.[3] In analytic philosophy the two dominant schools of thought, logical positivism and ordinary language philosophy, both shied away from the idea of formalisation of natural language meaning, though for different reasons. For the positivists, natural language as such lacked sufficient systematicity and their semantic analyses were carried out by constructing interpreted formal languages and studying their logic. That same lack of systematicity for ordinary language philosophers also presented an obstacle to formalisation, but in their view, this was not so much a reason to switch to formal languages as to adhere to a more descriptive and informal methodology. As we shall see, both perspectives are still around.

It was only with the work of Davidson and Montague, and later on Lewis, Partee, Bartsch and a host of others, that the very idea of a formal treatment of natural language meaning came into its own. In view of Chomsky's rejection of Bar-Hillel's suggestion that logic and linguistics join forces (cf., footnote 3), it may come as a surprise that it is Chomsky who was often referred to as a major source of inspiration: his work in generative syntax inspired confidence that on that score doubts concerning a lack of systematicity could be considered as settled.[4] Syntax being amenable to a rigorously formal treatment meant that one condition

[2]The most prominent example perhaps being Hans Reichenbach, whose *Elements of symbolic logic*, which dates from 1947, contained a substantial part devoted to systematic treatment of, among other things, tenses and other temporal expressions in natural language, which became known and very influential only much later.

[3]Which is what happened to Yehoshua Bar-Hillel, who, inspired by Carnap's work in intensional logic, in the early fifties suggested that the formal methods of logic could be applied to the results of Chomsky's early work in generative syntax so as to provide a formal semantics for natural language (cf., [1]). The proposal met with a brusque and negative response from Chomsky (cf., [3]), and it took another decade for other people to take up on this idea.

[4]Cf., e.g., Davidson: 'Recent work by Chomsky and others is doing much to bring the complexities of natural languages within the scope of serious theory.' [7], and Montague: 'On this point [viz., that natural languages can be treated formally, MS] I differ from a number of philosophers, but agree, I believe, with Chomsky and his associates.' [21].

for a formal semantics was met. For, in some form or other, the compositionality principle, that says that the meaning of a complex expression is determined by the meanings of its component parts and the way these are combined, was leading, as it had been in philosophical logic since the days of Frege. It requires that the 'means of combination' of the expressions of a language, i.e., its syntactic rules, be determined in order to serve as the carrier of a compositional specification of their meaning. No guts no glory, no syntax no (formal) semantics.

Of course, the availability of a formal theory of syntactic structure fulfils a necessary condition, by itself it does not show that meaning can be formalised as well. This is where the enterprise of formal semantics starts. Looking back, we can discern a number of different trends. Some authors seem to aim at what can almost be called 'transcendental' arguments of the formalisability of natural language meaning, whereas others take a much more empirical stance, and proceed stepwise. The former approach starts from an assumption that can be epitomised in the following quote from Moerdijk and Landman: 'Things must be very strange if they cannot be captured in mathematical terms.'[5] It is that ultimately every phenomenon that is systematic and that can be understood by means of a rational inquiry, has sufficient structure and can be modelled formally. The subsequent task is then to find, or to develop, the appropriate formal tools and to apply them to real instances. That, to be sure, is indeed an empirical undertaking, and one that may run into problems or even fail. But formalisability as such seems less an empirical issue than a precondition for a phenomenon to be meaningful and subject to rational inquiry in the first place.

It comes as no surprise that this way of looking at things can be found primarily among the philosophers and logicians who were engaged in the development of formal semantics. An example to illustrate this. Richard Montague started his 'Universal grammar' with the following statement ([21], p. 373):

> There is in my opinion no important theoretical difference between natural languages and the artificial languages of logicians; indeed, I consider it possible to comprehend the syntax and semantics of both kinds of languages within a single natural and mathematically precise theory.

This claim functions more as a starting point than as a conclusion. In his 'Universal grammar' paper Montague proceeds to specify the form and content of such a 'single natural and mathematically precise theory', which consists of algebraic frameworks for the analysis of expressions, meanings and the meaning relation that associates (analysed) expressions with meaning in an explicitly formal way. What is outlined in 'Universal grammar' is a general framework, one that needs to applied to concrete phenomena, as Montague himself has done, for example, in his seminal paper 'The proper treatment of quantification in ordinary English' [22]. What is interesting about Montague's starting point is that it is conceptual, rather than empirical. The possibility of formalising natural language meaning is a starting

[5]Cf., Moerdijk and Landman [20]. Cf., also Cresswell's defence of his use of set theory as his metalanguage in his *Logics and languages* [5].

point, one that needs to be tested, but not as a specific claim but as something that defines an entire theoretical approach.

For Montague, this has consequences also for the empirical character of the subsequent application of the general framework. Thomason explains this in his introduction to *Formal philosophy*, the collection of Montague's papers on semantics, as follows (Thomason 1974, p. 2):

> According to Montague the syntax, semantics, and pragmatics of natural languages are branches of mathematics, not of psychology. The syntax of English, for example, is just as much a part of mathematics as number theory or geometry [. . .] This mathematical conception of semiotics does not imply that data are irrelevant to, for instance, the syntax of English. Just as mathematicians refer to intuitions about points and lines in establishing a geometrical theory, we may refer to intuitions about sentences, noun phrases, subordinate clauses, and the like in establishing a grammar of English. But this conception does presuppose agreement among theoreticians on the intuitions, and it renders statistical evidence about, say, the reactions of a sample of native speakers to "Mary is been by my mother" just as irrelevant to syntax as evidence about their reactions to "$7 + 5 = 22$" would be to number theory.

According to this characterisation, which seems to capture an influential early way of thinking about the nature of formal theories of natural language meaning, such theories are empirical in as much as they describe some form of idealised semantic competence, that plays out in the intuitions of skilled theoreticians.[6] The crucial question then becomes whether thus isolating meaning from use by relying on a distinction between competence and performance, is an independently motivated move and the formalisability of meaning an empirical hypothesis, or rather a precondition for conceiving of natural language meaning as formalisable to begin with.

Other pioneers as well deliver arguments for a formal treatment of natural language meaning that are primarily conceptual. Cf. the following passage from an early paper of Davidson (1965):

> I propose what seems to me clearly to be a necessary feature of a learnable language: it must be possible to give a constructive account of the meaning of the sentences in the language. Such an account I call a theory of meaning for the language, and I suggest that a theory of meaning that conflicts with this condition, whether put forward by philosopher, linguist, or psychologist, cannot be a theory of a natural language; and if it ignores this condition, it fails to deal with something central to the concept of a language.

The 'constructive account' that Davidson refers to in this passage, is, of course, a Tarski-style theory of truth: a formal theory that specifies in a rigorously formal manner the truth conditions of the well-formed sentences of a natural language.

---

[6]Thus, in that respect aligning formal semantics with the generative tradition. Cf., Stokhof [30] for a diagnosis of how that came about.

One final example, is provided by Lewis' early paper 'General semantics', which starts with the following claim (Lewis 1970, p. 18):

> On the hypothesis that all natural or artificial languages of interest to us can be given a transformational grammar of a certain not-very-special sort, it becomes possible to give very simple answers to the following questions:
>
> (1) What sort of thing is a meaning?
>
> (2) What is the form of the semantic rules whereby meanings of compounds are built up from the meanings of their constituent parts?
>
> It is not my plan to make any strong empirical claim about language. To the contrary: I want to propose a convenient format for semantics general enough to work for a great variety of logically possible languages. This paper therefore belongs not to empirical linguistic theory but to the philosophy thereof.

What we see expressed here resembles the views of Montague and Davidson in relevant respects: that both natural and formal languages can be analysed in a similar syntactic framework, and that the formalisability of syntax makes a compositional semantics possible. What is explicit in this passage is the status accorded to the theory that is based on these observations: the claims of general semantics are not empirical but philosophical.

Despite their abundance in the early works of formal semanticists, it would be inappropriate to mention only conceptual arguments such as these. Even the foundational and theoretical work of Montague, Davidson, Lewis, Cresswell and others contains applications of their ideas to concrete phenomena of natural language meaning. In fact, as was already mentioned above, it was not Montague's theoretical outline of his approach in 'Universal grammar' that served as a model for formal semantics in the early days, but the applied version in his 'The proper treatment of quantification in ordinary English' [22], that he used to analyse certain phenomena concerning quantification such as de dicto/de re ambiguities, and the like. This is characteristic of formal semantics as a branch of linguistics, of course. It deals with empirical phenomena, analysing and describing them by means of formal systems. That not just preaches formalisability of natural language meaning, it also practices it. Success and failure are measured by empirical adequacy and formal rigour. And whenever the enterprise succeeds, one might say, that constitutes ever so many arguments that meaning is indeed formalisable.

In that sense then, there is abundant evidence for a positive answer to the question under discussion: formal semantics has made great strides over the last four decades in capturing central aspects of natural language meaning. Quantification, anaphora, tense and aspect, modality and conditionals, mass nouns, plurals, questions, presupposition, focus and information structure, vagueness, and a host of other aspects of natural language meaning have been studied with formal means, by developing formal systems that describe the phenomena in question and provide systematic ways to make predictions about acceptability judgements of competent language users. Although initially most work was mono-lingual, increasingly cross-linguistic studies and typological investigations are being conducted in the framework of formal semantics as well. So, the scope of the formalisations has steadily increased, both in terms of the phenomena captured as well as in terms of the languages covered, and this has contributed in important ways to our understanding of the phenomena in question.

A particularly successful example (though certainly not the only one) is provided by work on generalised quantifiers (by Barwise & Cooper, Van Benthem, Higginbotham & May, Keenan & Faltz, and many others). Building on Montague's original analysis (in [22]) and on logical work by Mostowski and others, semanticists have come up with a small number of general, formal properties of generalised quantifiers that characterise the quantifiers we actually find in natural language: among the totality of logically perfectly possible quantifiers these properties determine the restricted set we actually find in natural language. Although there is further discussion about empirical details, this is indeed an impressive result, one that seems to vindicate the positive answer to our question, perhaps even in a more convincing manner than the more conceptual considerations with which formal semantics started out. Other empirical results, too, provide insights into underlying constraints on how natural languages express meaning and reveal complicated relations between apparently unrelated phenomena, and taken together they seem to build a convincing case for the formal nature of fundamental aspects of natural language meaning.

Then all is well and the "Yes"-answer goes unchallenged? Not quite. There are a number of concerns that may provide reason to think things over. The first is the lack of a unified formal framework, the second concerns the distinction between structural and lexical semantic features.

Let us start with the first concern. In the early days of formal semantics research tended to be carried out in a uniform framework. Of course, there were various candidates for such a framework around, with Davidsonians preferring first order, extensional systems, Montagovians making use of higher-order intensional type theory, and Cresswellians favouring set-theory. But within each 'school' researchers tended to carefully fit their analyses in the overall framework of their choice, taking care to make sure that their results were actually consistent with those of others by showing that they could be unified in the overall framework. Fairly quickly this gave way to a much more liberal use of formal systems, with little or no attention being paid to their compatibility. Browsing through the literature one may find applications of domain theory, property theories, belief revision systems, event calculus, different many-valued logics, various non-monotonic logics, dynamic logic, various forms of game theory, second-order type theory, Martin-Löf's type theory, untyped lambda-calculus, Boolean algebras, lattices of various kinds, set theory with or without ur-elements: basically everything in the book has been thrown at natural language phenomena at some point. And then there are the custom built formal systems, such as various versions of discourse representation theory, of situation theory, and so on.

Pluriformity as such presumably is no objection *per se*, but the existing variety of methods and frameworks does shed a different light on the question of the formalisability of natural language meaning, and the positive answer that is indirectly provided by empirically successful descriptions and analyses. Where in the early days, formal languages appeared to be used as *models* for natural languages, as is testified by the conceptual considerations which we illustrated above, the more piecemeal approach that has become characteristic of formal semantics suggests a more pragmatic stance, in which formal languages are regarded not as models but as *tools*. This is a crucially different way of looking at what we do when we apply formal methods in semantics, and hence a different type of positive answer. In the first case, the "Yes" is continued with "because natural languages are formal languages", whereas in the second case the reason given reads "because natural languages actually can be described in a formal manner". The practices may not be that different, but the underlying ideologies are really quite distinct.

Does that mean that the second, more pragmatic stance, which regards formal systems as tools rather than as models, is unobjectionable? Again, the answer seems qualified. In as far as it instantiates the general attitude that, indeed, "things must be very strange if they cannot be described in mathematical terms", it represents no stronger a claim than that, like any other natural phenomenon, natural languages are systematic and by that very fact should be amenable to systematic, formal investigation. By itself that seems a reasonable point of view, although as we shall see later on, it can be challenged. But does it also excuse the actual pluriformity of methods employed? In as much as these methods themselves carry different assumptions about the nature of what they are used to describe, the answer here must be negative, at least for the time being. Different formal systems may ascribe to the natural languages that they are used to analyse wildly different ontologies, they may be actually logically incompatible among themselves as they are based on incompatible logical principles, they may make different assumptions concerning the cognitive capacities of natural languages users, they may draw the line between semantics and pragmatics at different points, and so on. How to deal with such divergence is a concern that cannot be left unanswered, and it would seem that unless a satisfactory answer is given, the indirect positive answer to our question that empirically successful applications provide still needs additional justification.

Let us now to turn to the second consideration that may provide some reasons to suspend judgement on the positive answer, viz., the distinction between lexical and structural aspects of natural language meaning. Again, there is a distinct development to be discerned here. In the early days of formal semantics, the focus of research was on the meaning of expressions and constructions that play a structural role in the formation of the meanings of complex expressions. Quantifiers, tense and temporal expressions, modal expressions, and the like, appear to function very much like logical expressions, and it makes sense to focus on their analysis by taking suitable formal languages with appropriate counterparts, as models for their semantic behaviour. These are expressions that are systematic, invariant over different occasions of use, largely invariant over the linguistic context in which they occur, and their semantic content can be captured, it seems, in general principles of a more or less logical nature. On the other end of the spectrum we find the meaning of the majority of lexical items, which form the input of the semantic rules, but which do not play a structural role themselves. Their meanings often vary with linguistic and non-linguistic context, and it is often very difficult to distinguish between those aspects of their meaning that are properly linguistic and those that are intimately connected with various kinds of world knowledge. To be sure, there are generalizable features, semantic properties that are characteristic for a class of lexical items, but these are difficult to establish and their specification always underdetermines the full meaning of a specific lexical item. Yet, in the further development of formal semantics there was an increasing interest to 're-instate content', i.e., to try and capture as much of the meaning of lexical expressions in formal models.[7]

As was mentioned above, for some authors this has been a reason to give up on the idea that natural language meaning is a homogeneous phenomenon, and to assign the specification of lexical and structural aspects to completely different kinds of theories.[8] Most formal semanticists do not draw such a drastic conclusion and prefer to view their field of study as essentially homogeneous. Yet the distinction is real, and it seems to signal a definite limit to the formalisability of natural language meaning that has to be acknowledged.

12.3 No

The considerations mentioned at the end of the last section suggest at least a qualification of the positive answer outlined earlier on. We will return to them later on in this section, but first consider some of the reasons that people have given for a more principled negative answer to our leading question.

For clear examples of such "No"-answers, we can go back in time again, in this case, almost all the way (as far as the Western philosophical tradition is concerned). Complaints about the unsystematic and misleading character of natural language, with its vagueness, lack of precision, ambiguities and referential failures, are of all times. Of course, not all such complaints are made in the context of the question whether natural language meaning can be formalised. One would expect that point of view to become prominent only when the development of suitable formal languages had made formalisation of natural languages an option to begin with. To a large extent that is true, specific arguments concerning formalisation were developed in close conjunction with the development of modern logic in the nineteenth and early twentieth century, in the work of Peirce, Bolzano, Frege, Russell, and the early Wittgenstein. But also in a non-formal setting philosophers of diverse orientation (empiricists, rationalists, hermeneuticists and romanticists alike) voiced concerns about the adequacy of natural language as a medium for rigorous philosophical thought. In the seventeenth century, a whole movement originated around the idea of creating artificial languages, and to the extent that such artificial languages (as developed for example by Wilkins, Dalgarno, Leibniz, and others) were of a formal nature, the various arguments in favour of their creation and

---

[7]Cf., Kamp and Stokhof [16] for extensive discussion of this development.

[8]Cf., above on Chomsky's distinction between the computational and the conceptual system. Cf., Higginbotham 1997 for extensive discussion of the implications of such a move for formal semantics.

deployment and their suitability as alternatives to natural languages can be looked at as arguments against the formalisability of the latter.[9]

But it was at the end of the nineteenth and in the early twentieth century that the issue became more pronounced. After all, one reason for the interest in the development and application of formal languages in the analysis of reasoning was the assumed deficiency of natural language. Explicating why he found himself forced to develop his 'conceptual notation' by the difficulties he encountered trying to analyse reasoning rigorously using natural language, Frege said it as follows in his *Begriffsschrift* ([8], p. 5–6[10]):

> I found the inadequacy of language to be an obstacle; no matter how unwieldy the expressions I was ready to accept, I was less and less able, as the relations became more complex, to attain the precision that my purpose required. This deficiency led me to the idea of the present ideography.

Vagueness and ambiguity, the lack of an explicit and formal structure, lack of precision and context-dependence are some of the deficiencies that Frege, Russell, the early Wittgenstein and, at a later stage, some of the logical positivists, identified. Such deficiencies, they claimed, could be overcome only by improving on natural language, or by a wholesale replacement of it, for those purposes, such as logic and philosophical analysis, for which explicitness, rigour and precision were crucial.

But to what extent are these considerations really arguments against the formalisability of natural language meaning, rather than a mere rejection of the idea as such? As was already mentioned, the observations as such were hardly new. Many philosophers had already deplored for instance the fact that in typical Indo-European languages existence, predication and identity tend to be expressed grammatically by one and the same verb. What was new is that with the development of explicit formal languages there was for the first time a real alternative: philosophers and logicians did not have to settle for what they regarded as a deficient tool, they could employ better ones, and even, develop such tools themselves as need arose. But perhaps more importantly, the conditions under which formalisation is possible also became much clearer. In particular the necessity of a formal specification of the syntax for a compositional semantics was identified as a condition that natural languages apparently did not satisfy.[11] And as we have seen above, it was exactly when opinion as to the formalisability of natural language syntax started to change, that the possibility of a formal semantics became a serious option.

The question whether a formal semantics of natural language can be, or should be, a compositional one, is much debated, and this is not the place to review the

---

[9]Cf., Maat [19].

[10]Page references are to the English translation in Van Heijenoort.

[11]The *locus classicus* is Tarski's 1944 paper on the semantic conception of truth, where he writes: "The problem of the definition of truth obtains a precise meaning and can be solved in a rigorous way only for those languages whose structure has been exactly specified. For other languages – thus, for all natural, 'spoken' languages – the meaning of the problem is more or less vague, and its solution can have only an approximate character."

various positions and arguments.[12] What is important is that even if full compositionality is replaced by something like 'systematicity' the demands on the formal specification of the syntax are not really diminished.[13] In fact, compositionality may be somewhat orthogonal to the question that is under discussion here,[14] but it would go too far to explore that in any detail.

The formal nature of syntax being more or less universally agreed upon, the real challenge for the formalisability of natural language meaning comes from a different quarter. It is a line of thought that in a sense generalises some of the old objections, regarding vagueness and context-dependence, and regards such features both as much more pervasive, and as a virtue rather than as a vice.

In analytic philosophy, this perspective on natural language had been endorsed in particular by the ordinary language philosophers, such as Austin, Ryle, Warnock and others. In their view, it is precisely because of its pervasive context-dependence that natural languages are able to serve the purposes that they do. Exact definitions, strictly delineated concepts, and a precise formal structure are not just constraints that natural languages do not meet, they would in fact diminish their usefulness.

Recently, similar observations and arguments have been subject of intense debate between minimalists and (radical) contextualists. Both parties agree that context-dependence is a defining characteristic of natural language. In fact, this realisation goes back to Frege, who in 'Der Gedanke' [9], argues that in order to preserve determinate meaning, we need to re-analyse natural language sentences of which the truth value depends on context, as implicitly containing a specification of the relevant contextual parameters. This form of 'eternalism' postulates a substantial difference between what intuitively is the meaning of such sentences and what this approach construes it to be. Also, there are good arguments to think that this form of 'de-contextualisation' will not always work, as it cannot account for the role of essential indexicals in action explanation.[15] In formal semantics indexicality is usually dealt with by associating context-dependence expressions with two distinct but related semantic objects: one that constitutes the context-independent content and another that determines such a content in each context.[16] But such an approach only works for a limited set of indexical expressions, such as pronouns, temporal expressions, locatives, etc.. Radical contextualists argue that in fact all expressions are context-dependent, that no descriptive context is immune for contextual variation. In fact, they claim, there is no such thing as semantic content, i.e., natural language meaning cannot be specified independent of the use that is made of language in concrete contexts. As Charles Travis formulates it ([36], p.41):

---

[12]Cf., Pagin and Westerståhl [23, 24] for a comprehensive overview.

[13]Cf., Pullum and Scholz [27].

[14]In view of the fact that for example model-theoretic approaches to syntax (cf., [26]), though definite alternatives to generative ones, are committed to the formal specifiability of syntax just as well.

[15]Cf. Perry [25].

[16]The most well-known instance of such an approach is that of Kaplan, cf., [17].

> The core thesis of [radical contextualism] is that any way for things to be which an English (or etc.) open sentence speaks of admits of understandings as to when something would be that way. Any of many different things may thus be said of a given item in saying it to be that way. The same variety of different things may thus be said of it in using that open sentence of it.

But how would one argue for such a position? The arguments that radical contextualists adduce to support their view usually consist of ingenious examples that indicate that even such apparently stable descriptive predicates as the adjective ‘red’ can be used in radically different ways depending on the context. Such observations and constructions are certainly appealing for a sympathetic reader who might be willing to generalise from them to the systematic position that radical contextualism defends. But it is also true that such observations as such do not constitute a principled argument: suggestive as they may be, they do not force one to accept that there are no context-independent aspects of meaning whatsoever that could be captured in a formal model.

12.4 It Depends

Confronted with two such well-argued, seemingly firmly justified, yet diametrically opposite answers to a simple question, we should perhaps stop and step back and ask ourselves whether the question we asked, which initially looked simple, straightforward and clear, might not be so on closer inspection. After all, taking sides would be decidedly odd for it would mean to reject sound argumentation and reasonable observation and to declare it as somehow due to a deep misunderstanding of the issue at stake. More plausible, it seems, is that each party answered a slightly different question: those who favour the “Yes”-answer are concerned with structural aspects of natural language meaning, whereas those who defend a “No”-answer focus primarily on lexical content.

So, does that mean that the best answer we can give to the unqualified question is “It depends”? Although it may look like it, the “It depends”-answer really is not a way of dodging the issue, but neither does it represent a definite, contentful stance on what natural language meaning is and on how semantics therefore should (or should not) be done. Rather, it represents a more meta-level perspective on what it is that we do when we do things formally. It assumes that the question that this paper is concerned with, should not be construed as a factual one: there is no fact of the matter as to whether natural language meaning is something (an object, a complex entity with a certain structure) that is formal in nature, the structure of which can hence be explicated in some formalised description. Formal theories are not descriptions of formal objects, they are specific ways of interacting with a complex phenomenon, some aspects of which lend themselves to formal representation, whereas others do not. Arguments that purport to show that one grand unified formal theory must be possible because the nature of what gets formalised allows for it, and arguments that are supposed to prove that such is not the case because there are no formal properties of natural language meaning to begin with, both miss this simple, but profound point.

One may feel that the "It depends"-answer motivated along the lines just sketched is something that actually holds across the board for any type of scientific inquiry. Many phenomena that we encounter in reality are too complex to be fully captured in a formal model or theory, and abstracting away from real but to a certain extent irrelevant aspects is standard procedure and in many cases, saves the day. Why wouldn't the formal study of natural language meaning be yet another instance of this general feature of scientific inquiry? That is an objection that deserves a much longer answer that we can provide here.[17] Let us just mention one important reason to think that natural language, in particular natural language meaning, might be a different case. It is that there is a distinct dependency in the case of natural language meaning between what is captured in a formal theory and the ways and means by which we formulate such a theory and understand those constructed by others: the object understood and the medium of understanding are by and large the same. This goes beyond the straightforward observation that, ultimately, any formal theory can be understood only because of our natural linguistic abilities. And of course, this should not be taken to mean that such formal theories cannot extend our understanding and those abilities, because when successful they do, in fact, it is one of the criteria in terms of which success is measured. What makes the case of the formalisation of natural language meaning different is that what we seek to understand and what that understanding ultimately relies on are one and the same thing. And that can be taken as an indication that the relation between formal theories in this domain has a different status.

So, what are formal descriptions, formal theories of natural language? They are, to borrow an apt phrase from Wittgenstein,[18] 'übersichtliche Darstellungen ('perspicuous presentations'): insightful, lucid, surveyable presentations of particular aspects of natural language meaning; presentations that are explicit, that can be formally manipulated, and that lend themselves to implementation; presentations that by their very being formal allow us to see and do things that we could not see or do as easily using the expressions they formalise. But also, and this is the crux of the matter, presentations that themselves can be understood only in terms of what they present: not exclusively, because that would mean that they don't add to our knowledge and insight and they obviously do; but nevertheless essentially, since our understanding of such presentations, the very fact even that we have an interest in constructing them, can be understood only in the context of natural language itself. This is a kind of self-reference that facilitates all kinds of 'looping effects' between what is described by the formal theory and what makes our understanding and our use of that formal theory possible.

[17]Cf., Stokhof and Van Lambalgen [32, 33] for further discussion.

[18]Cf., Wittgenstein [37], section 122.

This is a point of view on the nature of formalisation of human language and kindred phenomena, on its usefulness and its limitations, that authors in quite different traditions have expressed as well. Wittgenstein having already been mentioned, perhaps it is apt to end with a quote from another prominent twentieth-century philosopher, Martin Heidegger, who in his essay 'The way to language' stated ([13], p. 422):

> There is no such thing as a natural language, a language that would be the language of a human nature at hand in itself and without its own destiny. Every language is historical, also in cases where human beings know nothing of the discipline of history in the modern European sense. Nor is language as information *the sole* language in itself. Rather, it is historical in the sense of, and written within the limits set by, the current age.

For Heidegger then, just as for Wittgenstein, the distinction between formal language and natural language is not a real opposition, but a reflection of a particular way of dealing with the world. What matters is a clear awareness of the perspective we take, and the pragmatic concerns that motivate it. In that sense, truly "it depends".

12.5 Conclusion

So where does this leave us with regard to the question we started out with? As is customary with such profound questions, final answers are hard to come by. That does not mean that we should not address them, of course. Such considerations as we have reviewed in the above do tell us something, albeit not something definite. First of all, it is clear that the conceptual motivation for a positive answer is in general insufficient. Not only is it constitutive of the enterprise rather than based on independent evidence or considerations, also it seems to steer a formal approach far too much in the direction of "modelling", something that runs into deep conceptual problems. If anything, more evidence based arguments, consisting of actual and successful attempts at formalisation in the end carry more weight. Second, this more pragmatic view comes with a focus on formal systems as tools, rather than models. That seems a much more realistic perspective, but it does come with its own set of questions, the most important of which is: what are the adequacy criteria for our choice of tools from the enormous toolbox that logic, computer science and mathematics have to offer? If we take a theory of natural language meaning to be a theory of semantic competence, then the multiplicity of the formal systems that are employed constitutes a serious challenge, one that still needs to be met.

References[19]

1. Bar-Hillel, Y. (1954). Logical syntax and semantics. *Language, 30*, 230–237.
2. * Cappelen, H., & Lepore, E. (2004). *Insensitive semantics. A defense of semantic minimalism and speech act pluralism*. Oxford: Blackwell. [Outspoken defence of semantic minimalism].
3. Chomsky, N. (1955). Logical syntax and semantics: their linguistic relevance. *Language, 31*, 36–45.
4. Chomsky, N. (1980). *Rules and representations*. Oxford: Blackwell.
5. Cresswell, M. (1973). *Logics and languages*. London: Methuen.
6. Davidson, D. (1965). Theories of meaning and learnable languages. In Y. Bar-Hillel (Ed.), *Proceedings of the 1964 international congress for logic, methodology, and philosophy of science*. Amsterdam: North-Holland.
7. * Davidson, D. (1967). Truth and meaning. *Synthese, 17*, 304–323. [One of the origins of formal semantics].
8. * Frege, G. (1879). *Begriffsschrift. Eine der arithmetischen nachgebildete Formelsprache des reinen Denkens*. Halle a.S: Louis Nebert. English translation in Van Heijenoort 1970. [Introduces the distinction between grammatical form and logical form].
9. Frege, G. (1918). Der Gedanke: eine logische Untersuchung. *Beiträge zur Philosophie des deutschen Idealismus, 2*, 58–77. English translation by Peter Geach in Frege 1977.
10. Frege, G. (1977). *Logical investigations*. Oxford: Blackwell.
11. Hauser, M. D., Chomsky, N., & Tecumseh Fitch, W. (2002). The faculty of language: What is it, who has it, and how did it evolve? *Science, 298*, 1569–1579.
12. van Heijenoort, J. (1970). *Frege and Gödel. Two fundamental texts in mathematical logic*. Cambridge, MA: Harvard University Press.
13. Heidegger, M. (1978). The way to language. In *Basic writings* (pp. 393–426). London: Routledge.
14. Higginbotham, J. (1993). Grammatical form and logical form. *Philosophical Perspectives, 7*, 173–196.
15. Higginbotham, J. (1997). Reflections on semantics in generative grammar. *Lingua, 100*, 101–109.
16. Kamp, H., & Stokhof, M. (2008). Information in natural language. In J. van Benthem & P. Adriaans (Eds.), *Handbook of philosophy of information* (pp. 49–112). Amsterdam: Elsevier.
17. Kaplan, D. (1979). On the logic of demonstratives. In P. French et al. (Eds.), *Contemporary perspectives in the philosophy of language* (pp. 401–413). Minneapolis: University of Minnesota Press.
18. Lewis, D. K. (1970). General semantics. *Synthese, 22*, 18–67.
19. Maat, J. (2004). *Philosophical languages in the seventeenth century: Dalgarno, Wilkins, Leibniz*. Dordrecht: Kluwer.
20. Moerdijk, I., & Landman, F. (1981). *Morphological features and conditions on rules in Montague grammar*. Amsterdam: Universiteit van Amsterdam.
21. * Montague, R. (1970). Universal grammar. *Theoria, 36*, 373–398. [One of the origins of formal semantics].
22. Montague, R. (1973). The proper treatment of quantification in ordinary English. In J. Hintikka, J. Moravcsik, & P. Suppes (Eds.), *Approaches to natural language* (pp. 221–242). Dordrecht: Reidel.
23. Pagin, P., & Westerståhl, D. (2010a). Compositionality I: Definitions and variants. *Philosophy Compass, 5*, 250–264.
24. Pagin, P., & Westerståhl, D. (2010b). Compositionality II: Arguments and problems. *Philosophy Compass, 5*, 265–282.

[19] Asterisks (*) indicate recommended readings.

25. Perry, J. (1979). The problem of the essential indexical. *Noûs, 13*, 3–21.
26. Pullum, G., & Scholz, B. (2001). On the distinction between model-theoretic and generative-enumerative syntactic frameworks. In P. de Groote et al. (Eds.), *Logical aspects of computational linguistics* (pp. 17–43). Berlin: Springer.
27. Pullum, G., & Scholz, B. (2007). Systematicity and natural language syntax. *Croatian Journal of Philosophy, VII*, 375–402.
28. * Recanati, F. (2005). Literalism and contextualism: Some varieties. In G. Preyer & G. Peter (Eds.), *Contextualism in philosophy: Knowledge, meaning and truth* (pp. 171–198). Oxford: Oxford University Press. [Overview of the contextualism – minimalism debate].
29. Reichenbach, H. (1947). *Elements of symbolic logic*. New York: Dover/McMillan.
30. Stokhof, M. (2007). Hand or hammer? On formal and natural languages in semantics. *The Journal of Indian Philosophy, 35*, 597–626.
31. Stokhof, M. (2011). Intuitions and competence in formal semantics. In B. Partee, M. Glanzberg, & J. Skilters (Eds.), *The baltic international yearbook of cognition, logic and communication. Volume 6: Formal semantics and pragmatics. Discourse, context and models* (pp. 1–23). Riga: University of Latvia Press.
32. Stokhof, M., & van Lambalgen, M. (2011a). Abstractions and idealisations: The construction of modern linguistics. *Theoretical Linguistics, 37*(1–2), 1–26.
33. Stokhof, M., & van Lambalgen, M. (2011b). Comments-to-comments. *Theoretical Linguistics, 37*(1–2), 79–94.
34. Tarski, A. J. (1944). The semantic conception of truth and the foundations of semantics. *Philosophy and Phenomenological Research, 4*, 341–375.
35. Thomason, R. H. (1974). Introduction. In R. H. Thomason (Ed.), *Formal philosophy. Selected papers of Richard Montague* (pp. 1–71). New Haven/London: Yale University Press.
36. * Travis, C. (2006). Insensitive semantics. *Mind and Language, 21*, 39–49. [Outspoken defence of contextualism].
37. * Wittgenstein, L. (1953). *Philosophical investigations*. Oxford: Blackwell. [Classical source of a practice-oriented approach to natural language meaning].

Chapter 13
Reference and Denotation

Robert van Rooij

Abstract According to Frege, the meaning of an expression is the description that helps language users to determine what its reference is. Natural as the view might seem, it gives rise to the *conceptual problem* that it presupposes that we already know the meaning of the terms used in the description (Wittgenstein, Quine), and it is *empirically incorrect* because 'having a correct description in mind' is neither a sufficient nor a necessary condition for successful reference (Kripke, Kaplan). Perhaps reference for at least some times is non-descriptive, and depends on context. Anaphora have a referential use as well, picking up the speaker's referent of an earlier used indefinite description. The challenge of this view is to provide a satisfactory analysis of so-called donkey-sentences.

13.1 The Descriptive Theory of Meaning and Its Problems

The perhaps most 'natural' conception of 'meaning', at least in its point of departure, identifies 'meaning' with *naming*. The meaning of an expression is that what the expression *refers to*, or *is about*. What meaning does is to establish a *correspondence* between expressions in a *language* and things in the (model of the) *world*. For simple expressions, this view of meaning is natural and simple. The meaning of a proper name like 'John' or definite description like 'the number of major planets', for instance, is the object or number denoted by it, while the meaning of a simple declarative sentence like 'John came' could then be the *fact* that John came. Beyond this point of departure, things are perhaps less natural. What, for example, should be the things out in the world that common nouns and a negated sentence like 'John didn't come' are about? This referential, or *Millian*, theory of meaning gives rise to a serious empirical difficult as well: the *substitution problem.*

R. van Rooij (✉)
Institute for Logic, Language and Computation (ILLC), University of Amsterdam, Amsterdam , Netherlands
e-mail: R.A.M.vanRooij@uva.nl

S. O. Hansson, V. F. Hendricks (eds.), *Introduction to Formal Philosophy*, Springer Undergraduate Texts in Philosophy, https://doi.org/10.1007/978-3-319-77434-3_13

Assuming, by the principle of compositionality, that the meaning of a complex sentence depends only on the meanings of its part and the way these parts are put together, it follows that if two expressions have the same meaning, one can substitute the one expression for the other in a complex sentence without change of meaning. But because there are 8 major planets in our solar system, on the theory of meaning at hand the expressions '8' and 'the number of major planets' refer to the same thing, and thus have the same meaning. Still, we cannot substitute the expressions 'the number of major planets' for the number 8 in the sentence 'It is necessary that 8 is bigger than 7' without changing its truth value. Frege [5] concluded that the *meaning* of an expression should not be identified with the *reference*, or denotation, of that expression. Instead, the meaning of 'the number of major planets' is the description itself, and the meaning of a noun or name is given by a set of properties associated with the expression that give necessary and sufficient conditions for objects or stuff to be in its denotation. Obviously, the above substitution puzzle does not arise on such a view.

However appealing and natural this cluster theory of reference might be, it gives rise to at least two problems, one conceptual and one empirical in nature. The **first conceptual problem** concerns the predicates, or properties, used in the description that is supposed to identify the referent. What is the meaning of those predicates? The standard theory of meaning doesn't seem to do more than explaining the meaning of one part of the language in terms of other parts – the predicates in terms of which the descriptions are given. One proposal to solve this problem would be that we indeed have a set of 'basic predicates', e.g. the predicates that refer to *natural kinds*. But how does this reference come about, if not via description?

A possible way out of the above regress problem is to propose to study meaning in terms of outward and observable correlates of language behavior. Perhaps motivated by such concerns, [24], for instance, proposed that we should study the meaning of expressions in terms of their use, and the logical positivists proposed their verificationalist' analysis of meaning mainly because they considered the way to verify a sentence as a particularly good way to get clear how a certain sentence is used. The verificationalist' reductive analysis of meaning failed, for one thing because it is difficult – if not impossible – to interpret the terms of a language individually. To determine the meanings of expressions we have to look simultaneously at a whole group (or perhaps all) of expressions of the language *as a whole* [1, 9]. One way to cash out such a *holistic* theory of meaning idea would be to claim that the terms refer to whatever things, properties, and relations that do the best job of making the set of sentences true that speakers in fact consider to be true. Unfortunately, generalizing Quine's (1960) well-known argument for the indeterminacy of reference, Putnam [18] showed that this picture as such is not constrained enough to fix the meaning of the expressions of a language in the intuitively correct way. Even if one knows the truth value of a sentence in every possible circumstance, this doesn't necessarily mean that one knows the intuitively correct meanings of its constituents. For instance, it is possible to formulate highly counterintuitive meanings for expressions like *cat* and *mat*, so that in the actual world they refer to trees and cherries, respectively, without affecting the meaning of *The cat is on the mat*. To determine the meaning of the terms of our language, knowing the truth value or meaning of a collection of sentences is not enough, because the terms of the language can be assigned weird and 'unintended' interpretations.

The **second problem** for the description theory of reference is **empirical** in nature. Donnellan [2, 3] and Kripke [12] have convincingly argued that this theory leads to counterintuitive results for proper names. They have shown that speakers can refer, and even can *intend* to refer, to particular individuals without being able to describe or identify those individuals. Ordinary people can, for instance, use the name *Feynman* to denote the physicist Feynman even though they have no uniquely identifying set of descriptions in mind. Kripke argued that uniquely fitting some set of descriptions that the speaker associates with a proper name is not a sufficient condition for its successful use either. Kripke [12] and Putnam [17] have similarly argued that the set of properties that speakers or agents associate with *natural kind terms* should also not be equated with the meaning of the noun. This is made very clear by the 'Twin Earth' stories given by Putnam [17] and others. In Putnam's story, the stuff that the inhabitants of the counterfactual situation call *water* is superficially the same as the stuff *we* call *water*, but its chemical structure is not H_2O, but XYZ. If, then, both the earthling and his twin assert 'Water is the best drink for quenching thirst', intuitively they have said something different. But how can this be if they associate exactly the same description with the word and if speaker's description determines reference?

13.2 The Causal Theory of Reference, and Context-Dependence

According to Kripke, Putnam and others, the meaning of at least proper names and natural kind terms is simply what they refer to. But this gives rise to the question of *why* these expressions have the references they in fact have. At this point, Kripke proposed his causal theory of reference. Kripke [12] argues that proper name 'N' can refer to a, only if, and because, a is the entity that is the *source* of the reference-preserving link from the initial baptism of the expression to the speaker's use of the name. This causal 'theory' of reference, or of meaning, is in accordance with a naturalistic philosophy and seems also the natural candidate to limit the possible interpretations of the expressions of 'our' language to solve Putnam's paradox (cf. [14]).

The causal account of meaning is not without problems. For instance, it is unclear how a causal theory could ever determine the meaning of functional words, or of prepositions like '*in*'. Moreover, it is not clear how to cash out the causal account in a completely naturalistic way and there are problems of how to account for our intuitions that we can have false beliefs. One way solve both of these problems involves making use of so-called 'normality conditions'. But in order for the resulting analysis to be wholly naturalistic, we need a naturalistic analysis of such conditions. A natural candidate that suggests itself to provide such an analysis is Millikan's [15] bio-semantics, but it is controversial whether this theory can do the full job. The main problem of the causal theory, however, is the original substitution problem: if the meanings of 'Hesperus' and 'Phosphorus' are just their referents, 'Hesperus is Phosporus' is predicted to express the necessary true proposition. But, then, how can we account for the fact that agents can seriously doubt that such statements are true?

It is an obvious observation that what is expressed by a sentence is *context-dependent*: in different contexts the same sentence can express different things. What is expressed by 'I am living in Amsterdam' depends on who is the speaker in that context. Kaplan's (1989) theory of context dependence allows us to distinguish different reasons why a sentence is 'necessarily' true. First, what a sentence expresses in context c can be true everywhere. A sentence like 'Hesperus is Phosphorus' is necessary in this way, given that the meaning of a proper name is just its referent. But it might also be the case that a sentence is true in every context in which it is expressed. For instance, an English sentence like 'I am here now' is necessarily true for this reason. But now consider John's uttering of 'I am John'. Though this sentence is necessarily true, the sentence can, intuitively, still be informative. This is because the hearer might be ignorant of the identity of the speaker, or at least doesn't know that he is called 'John'. This intuition can be accounted for in the theory by claiming that the speaker doesn't know in which context he is. One might now propose to use this analysis to account for the other problems as well: people can doubt whether the identity statement 'Hesperus is Phosphorus' is really true, because the referent of a proper name is context dependent, just like the referent of 'I'. And indeed, not only the reference of expressions like 'I' and 'you' depend on contingent features of the context, but this is also true – at least according to the causal theory of reference – for proper names and natural kind terms. Notice, though, that there is a difference between the sense in which the reference of these expressions depends on context. The expression 'I' is context dependent, because *in English*, 'I' always refers to the speaker in direct speech, and the same expression *of English* might be uttered by different speakers. The reference of 'Phosphorus' and 'water', on the other hand, are context dependent only because in different worlds they have a different meaning, or causal origin. But, of course, in that sense the meaning of 'I' is context dependent as well, and depends on the language we speak. Assuming that we speak a particular language, it follows that we sometimes don't know the meanings of the expressions (names and nouns) we use. Though this conclusion is natural to some, to others it feels like a contradiction in terms.

13.3 Indefinites and Anaphora

According to Quine [20], our learning and use of pronouns marks our ability to refer. If we may believe [6], scholastic philosophers held that pronouns can refer back to indefinites because indefinites are referential expressions. The indefinite refers to that object that the speaker intends to refer to by the use of the indefinite. Moreover, if a speaker uses a referential expression in his utterance, the proposition expressed by this utterance is object-dependent. Geach [6] has criticised this account. If John intends to refer to d by his use of the indefinite *an S*, and wants to say of d that he is P, even though d is not, John is not saying something false when he claims *An S is P*, according to Geach, if there actually is an S that is P. In order not to make such a prediction, according to Geach, it is better to represent an assertion like *An S is P* semantically simply by an existential formula, $\exists x[Sx \wedge Px]$. The specific/unspecific distinction belongs to pragmatics, which should be kept separate from semantics. To handle pronouns, we should follow Quine's insight and treat them as bound variables. A sequence of the form *Some S is P. It is Q* should, according to him, be translated as $\exists x[Sx \wedge Px \wedge Qx]$.

But there are well-known problems with this latter assumption. First, it leads to the unnatural consequence that we can interpret a sentence with an indefinite or other anaphoric initiator only at the end of the whole discourse: incrementality is given up. Second, if we want to interpret the pronouns in a donkey sentence like *If a farmer owns a donkey, he beats it* as bound variables, it seems we have to represent the indefinites in the antecedent as universal quantifiers to get the truth conditions right. But then it seems we have to give up compositionality. We cannot treat indefinites in all contexts in the same way. Finally, sometimes we cannot even get the truth conditions right by assuming that all pronouns should be treated as bound variables. This was shown by Evans [4] by sentences like *Tom owned some sheep and Harry vaccinated <u>them</u>*. According to a Geachian analysis of this sentence, we learn that Harry vaccinated *some* sheep that Tom owned if we accept what is expressed by the sentence; what we seem to learn, though, is that Harry vaccinated *all* of the sheep that Tom owned. Evans proposed that in a sequence of the form *Some S are P. They are Q*, the pronoun *they* goes proxy for the description *(all) the S such that P*.[1] Such pronouns he called *E-type pronouns*.

The above argument does not show that all pronouns are E-type pronouns. The pronouns occurring in sentences like 'Every man loves *his* cat', for instance, seems to function like the bound variables of quantification theory. Indeed, since Evans [4], proponents of the E-type approach normally make a distinction between *bound* and *unbound* pronouns, claiming that such a distinction can be made on purely syntactic grounds; and propose that only unbound pronouns should be treated as E-type pronouns.

However, if we use the term *unbound pronoun* in the above sense, it seems that not even all unbound pronouns go proxy for the definite or universal noun phrase recoverable from the antecedent clause and should be treated as E-type pronouns. Consider for instance *Yesterday, John met <u>some girls</u>. <u>They</u> invited him to their place*. We don't want to say that *they* needs to stand for all the girls John met yesterday. If we want to say that the pronoun is going proxy for a description

[1] Evans [4] claimed that the pronoun *rigidly refers to* (all) the S such that P. See Neale [16] for a motivation of the interpretation I have chosen.

recoverable from its antecedent, the relevant description should not be definite or universal, but *indefinite*: *some girls that John met yesterday*. To treat the pronoun as an abbreviation of an indefinite description also seems to be needed to get the right reading of a sentence like *Socrates owned a dog, and it bit him.* It seems that this sentence can be true if there was a dog that Socrates owned and it bit him, although at the same time there was also another dog that he owned that did not bite him.

A correct analysis for such discourses was given in Kamp's [10] 'Discourse Represention Theory', Heim's [9] 'File Change Semantics and Groenendijk and Stokhof's [8], 'Dynamic Predicate Logic'. These theories treat anaphoric pronouns simply as bound variables and indefinites as existential quantifiers. However, they interpret existential quantifiers dynamically in such a way that they introduce new objects, or discourse referents, that are available for reference. In this way, they solved Geach's incrementality problem. Moreover, they assure that with negation and conditionals, a universal quantification over assignment functions or sequences of individuals is involved, thereby accounting for donkey sentences and solving Geach's compositionality problem.

Note that, due to their use of *existential* closure, the anaphoric pronoun *he* in a sequence like *A man is walking in the park. He is whistling* is basically treated as an abbreviation of the *indefinite description* 'a man who is walking in the park'. But claiming that the pronoun is an abbreviation of an *indefinite* description would be very implausible. Pronouns are *definite* expressions. To quote Quine [19, p. 113], "'He', 'she', and 'it' are definite singular terms on a par with 'that lion' and 'the lion' " But if a singular pronoun cannot be treated as a definite description that (in extensional contexts) refers to (all) of the object(s) that verify the antecedent sentence, how then can a pronoun be treated as a definite expression? Some *empirical phenomena* also show/suggest that unbound anaphoric pronouns should in general have a more specific interpretation than the standard dynamic theories can offer.

One of those specific phenomena is the case of *pronominal contradiction*, originally due to Strawson [22]. When John asserts *A man is running through the park*, Mary may react by saying *He is not running, but just walking*. It is clear that in such examples the pronoun cannot be used as an abbreviation for the indefinite description *a man is running in the park*. The natural assumption to make here, however, is to say that in this case the pronoun is used referentially, referring back to the *speaker's referent* of the antecedent indefinite, the man the speaker had in mind for his use of the indefinite.[2]

Pronominal contradiction examples are well known to be problematic for the recently developed dynamic semantic theories. It is normally assumed that these

[2]Arguably, a type of pronominal contradiction is also involved in Geach's [7] notorious Hob-Nob sentences: *Hob believes that a witch blighted Bob's mare, and Nob believes that she killed Cob's cow.* Hob-Nob sentences are problematic for standard analyses because (i) witches do not exist, which rules out a *de re* analysis of the antecedent-anaphor relationship, and (ii) the discourse can be true in case Nob doesn't believe that the witch he talks about blighted Bob's mare, which rules out a descriptive E-type analysis.

problematic examples are rather special, though, and that for the 'standard' cases this notion of speaker's reference is irrelevant. The following example, however, suggests that anaphoric pronouns are usually used in this referential way. If John says *A man called me up yesterday*, it would be odd for John to reply to Mary's question *Did he have a gravel voice?* by uttering *That depends, if he called up in the morning he did, if he called up in the afternoon, he did not* if in fact two men called John up yesterday. It is not easy to see how this phenomenon can be explained if it is assumed that pronouns should simply be treated as variables bounded by dynamic existential quantifiers. A natural explanation can be given if it is assumed that for the use of the pronoun the speaker must have a specific object 'in mind'.

Note that according to this account we don't need to make use of existential closure to account for the non-exhaustive interpretation of pronouns, although the *definiteness* of pronouns can still be explained. A pronoun that takes the indefinite in a sentence of the form *Some S is/are P* as antecedent need not be interpreted exhaustively, i.e., need not refer to *all* the (relevant) *S*'s that have property *P*, because it only refers to the (all) speaker's referent(s) of the antecedent indefinite.

Of course, when we account for the anaphor-indefinite relation in terms of the notion of 'speaker's reference', we can no longer give the usual explanation for the asymmetry in acceptability between *John owns a donkey. Mary beats it* versus *John is a donkey owner. *Mary beats it*. Proponents of dynamic semantics, starting with Heim [9], explained the asymmetry solely in terms of the use of an explicit indefinite in the first, and the lack thereof in the second. Though there are problems with this view, a proponent of the alternative picture still has to explain the asymmetry.

According to Kripke [13] and Stalnaker [21], the speaker's reference is relevant to semantics, but only through pronominalisation. That is, it is irrelevant for what is expressed by the sentence (or clause) in which the indefinite occurs, but is truth-conditionally relevant for what is expressed by a later sentence with a pronoun that takes an indefinite as its syntactic antecedent.

How can we account for the *referential* treatment of pronouns on the one hand, and for the *existential* reading of indefinites on the other? One has to assume that possibilities should contain more information than is assumed in standard dynamic semantics, and that the dynamics is (relatively) independent of truth conditions (e.g. van Rooy, [23]). In particular, it should be clear in a possibility what the speaker's referent is of (occurrences of) indefinite expressions. Because the above sketched treatment of pronouns is exactly in line with [13] proposal, it is only to be expected that such an analysis, just like Kripke's, has problems to deal with *donkey sentences* like *If a farmer owns a donkey, he beats it*. The problem is that in such a sentence the indefinites don't seem to be used specifically, while the pronouns can arguably also not be treated as an abbreviation for *definite* descriptions, because it seems that the sentence can also be true in case one farmer owns more than one donkey. One way to solve the problem is to assume that a logical operator like *negation* is treated as an *intensional* operator, in that it allows part of the context, i.e. the choice function, to shift. One has to realize, however, that [11] would call such an 'intensional' treatment of negation a *monster*.

References

1. Davidson, D. (1967). Truth and meaning. *Synthese, 17*, 304–323.
2. Donnellan, K. (1966). Reference and definite descriptions. *Philosophical Review, 75*, 281–304.
3. Donnellan, K. (1978). Speaker reference, descriptions, and anaphora. In P. Cole (Ed.), *Syntax and semantics, volume 9: Pragmatics* (pp. 47–68). New York: Academic.
4. Evans, G. (1979). Pronouns, quantifiers and relative clauses (1). *The Canadian Journal of Philosophy, 7*, 467–536.
5. ***Frege, G. (1892). Über Sinn und Bedeutung. *Zeitschrift für philosophie und philosophische Kritik* (Vol. 50, pp. 25–50). [The classical paper making a distinction between meaning and reference].
6. Geach, P. (1962). *Reference and generality*. Ithaca: Cornell University Press.
7. Geach, P. (1967). Intentional identity. *Journal of Philosophy, 64*, 627–632.
8. Groenendijk, J., & Stokhof, M. (1991). Dynamic predicate logic. *Linguistics and Philosophy, 14*, 39–100.
9. Heim, I. (1982). *The semantics of definite and indefinite noun phrases*. Ph.D. dissertation, University of Massachusetts, Amherst.
10. ***Kamp, H. (1981). A theory of truth and semantic representation. In J. Groenendijk, et al. (Eds.), *Formal methods in the study of language* (pp. 277–322). Amsterdam: Mathematisch Centrum. [The first paper in the 'dynamic' tradition that treats indefinites uniformly as existential quantifiers, also in donkey sentences].
11. ***Kaplan, D. (1989). Demonstratives. In I. Almog, et al. (Eds.), *Themes from Kaplan*. Oxford: Oxford University Press. [The classic paper where the different roles of context of utterance and world of evaluation is forcefully argued for.]
12. ***Kripke, S. (1972/1980). Naming and necessity. In D. Davidson & G. Harman (Eds.), *Semantics of natural language* (pp. 253–355; 763–769). Dordrecht: Reidel. [The crucial paper which contains all the classical arguments against the descriptive theory of reference for proper names, and where the alternative causal theory is sketched.]
13. Kripke, S. (1977). Speakers reference and semantic reference. In: P. French et al. (Eds.), *Studies in the Philosophy of Language, Midwest Studies in Philosophy* (pp. 255–276). University of Minneapolis: Minnesota Press.
14. Lewis, D. (1984). Putnam's paradox. *The Australian Journal of Philosophy, 62*, 221–236.
15. Millican, R. (1984). *Language, thought and other biological categories*. Cambridge, MA: MIT Press.
16. Neale, S. (1990). *Descriptions*. Cambridge, MA: MIT Press.
17. ***Putnam, H. (1975). The meaning of meaning. In K. Gunderson (Ed.), *Language, mind and knowledge*. Minneapolis: University of Minnesota Press. [The classical paper where the twin-earth argument against the descriptive theory of reference of common nouns is introduced.].
18. Putnam (1981). *Reason, Truth, and History*. Cambridge: Cambridge University Press.
19. ***Quine, W. V. O. (1960). *Word and object*. New York/London: Technology Press/Wiley. [Contains the classical formulation of the indeterminacy of reference argument, and defends a holistic theory of meaning.].
20. Quine, W. V. O. (1974). *The roots of reference*. La Salle: Open Court.
21. Stalnaker, R. (1998). On the representation of context. *Journal of Logic, Language and Information, 7*, 3–19.
22. Strawson, P. (1952). *Introduction to logical theory*. London: Methuen.
23. van Rooy, R. (2001). Exhaustivity in dynamic semantics; referential and descriptive pronouns. *Linguistics and Philosophy, 24*, 621–657.
24. Wittgenstein, L. (1953). *Philosophical investigations*. Oxford: Blackwell.

Chapter 14
Indexicals

Philippe Schlenker

Abstract Indexicals are context-dependent expressions such as *I*, *you*, *here* and *now*, whose semantic value depends on the context in which they are uttered. They raise two kinds of questions. First, they are often thought to be scopeless – e.g. with *I* rigidly referring to the speaker – and to give rise to non-trivial patterns of inference – e.g. *I exist* seems to be *a priori* true despite the fact that *I necessarily exist* isn't. Second, indexicals may play a crucial role in the expression of irreducibly De Se thoughts, and both the existence of such thoughts and the ways in which they can be reported in indirect discourse must be elucidated. The Kaplanian picture posits that indexicals take their value from a distinguished context parameter, whose very nature is responsible for some entailments, and which remains fixed – hence the apparent scopelessness of indexicals. It further posits that while indexicals may serve to express irreducibly De Se thoughts, these may not be reported as such in indirect discourse (no 'De Se readings'). Both tenets have been criticized in recent research: there are a variety of constructions across languages in which the context

The initial version of this chapter was completed in 2010, and revised in 2014. It thus fails to take into account the most recent developments in indexical semantics; see Deal 2017 for an up-to-date discussion. Thanks to Pranav Anand for allowing me to copy-and-paste some examples from his dissertation, and to Paul Egré for helpful comments on an initial version of this chapter. I am particularly grateful to Isidora Stojanovic, for very detailed written comments, and to Paul Postal for catching some typos. The bibliography was prepared by Lucie Ravaux. This work was supported by a Euryi grant of the European Science Foundation ('Presupposition: a Formal Pragmatic Approach'), and then by an Advanced Grant of the European Research Council ('New Frontiers of Formal Semantics') under the European Union's Seventh Framework Programme (FP/2007–2013)/ERC Grant Agreement N°324,115–FRONTSEM (PI: Schlenker). Research was conducted at Institut d'Etudes Cognitives (ENS), which is supported by grants ANR-10-IDEX-0001-02 PSL* and ANR-10-LABX-0087 IEC.

P. Schlenker (✉)
Département d'Etudes Cognitives, Ecole Normale Supérieure, Institut Jean-Nicod (ENS – EHESS – CNRS), PSL Research University, Paris, France

Department of Linguistics, New York University, New York, NY, USA

S. O. Hansson, V. F. Hendricks (eds.), *Introduction to Formal Philosophy*, Springer Undergraduate Texts in Philosophy, https://doi.org/10.1007/978-3-319-77434-3_14

parameter appears to be shifted; and several types of indirect discourse (some of them involving context shift) do give rise to De Se readings.

Indexicals are context-dependent expressions such as *I*, *you*, *here* and *now*, whose semantic value depends on the context in which they are uttered (e.g. *I* denotes John if uttered by John, and Mary if uttered by Mary).[1] Indexicals in the strict sense (e.g. *I*, *here*, *now*) can be interpreted on the sole basis of the spatio-temporal properties of the speech act – in particular who is talking to whom, where and at what time. Demonstratives (e.g. uses of *he*, *she* or *that* without antecedent in the discourse) require in addition that one have access to the *referential intentions* of the speaker and/or to a notion of *salience*. In this chapter, we will focus on indexicals in the strict sense.

14.1 Foundational Questions

Indexicals raise several foundational questions for natural language semantics. For the sake of concreteness, we will start from a modal analysis in which the meaning of a sentence is assimilated to a function from world-time pairs to truth values. Thus a sentence *S* is evaluated relative to a an interpretation function [[.]] which takes as parameters a time t, a world w, and also an assignment function s (for individual variables), with $[[S]]^{s,t,w} = 1$ (for 'true') or 0 (for 'false'). We will see how this architecture must be modified to handle indexicals. But we start by stating five foundational questions that are raised by indexical expressions (see for instance [4, 40], Maier [19] and Schlenker [30] for other surveys, and [11] for a far-reaching synthesis of recent theoretical and empirical insights into shifted indexicals across languages).

14.1.1 Semantics and Logic: Context Dependency and Scopelessness.

Intuitively, the semantic value of an indexical is determined relative to the *context of a speech act*. But different speech acts – and hence different contexts – may co-occur in the same world and at the same time, hence world-time parameters as usually construed are insufficiently fine-grained to provide the value of indexicals. This immediately leads to a question about the general format of our semantic analysis:

[1]Here and throughout, italicization is used in the text for emphasis, or for quotation or quasi-quotation (but italicization is not used within formulas).

Q1. Which parameters should be added to the interpretation function to handle indexicals?

There is another side to this problem. Whichever answer is given to *Q1*, indexicals often seem to be special because they are 'scopeless', in the sense that they fail to interact scopally with other operators. To make the point concrete, let us compare the behavior of the word *I* to the apparently synonymous expression *the speaker*:

(1) a. The speaker is always boring.
a'. I am always boring.
b. The speaker is necessarily boring.
b'. I am necessarily boring.

Uttered by myself (= PS) at a conference, (1)a and (1)b have, among others, readings on which *the speaker* is semantically dependent on the operators *always* and *necessarily*, and thus need not refer to me, PS. Things are entirely different with (1)a'-b': *I* denotes ('rigidly') the speaker of the actual speech act, rather than whoever might be the speaker at other times or in other worlds. So our second question is:

Q2. Why do indexicals seem to be scopeless?

Indexicals give rise to *valid inferences* which are non-trivial to explain in logic. *I exist* or *I am here now* would seem to be *a priori* true, in the sense that *whenever these sentences are uttered they cannot fail to be true.* But these validities differ from 'normal' ones. In particular, it does not follow from their *a priori* status that the corresponding sentences prefixed with *necessarily* are true: *Necessarily, I exist* and *Necessarily, I am here now* are usually quite false (similar facts hold when *necessarily* is replaced with *always*). Tautologies, by contrast, are *a priori* true and remain so when they are prefixed with *necessarily* (e.g. *Necessarily, p or not p*). This leads to our third question:

Q3. How can indexicals give rise to a priori *true sentences which, when preceded by 'necessarily', can become false?*

14.1.2 Attitudes and Attitude Reports

What is the 'cognitive significance' of a sentence? Or to put it differently, what is the contribution of a sentence to the belief state of an agent who holds it to be true? A simple-minded view would take this cognitive significance to be given by the *information it provides about the world and time at which it is uttered.* This would make for an elegant connection with a semantic theory that countenances world and times. Let us call the *intension* or *content* of a sentence *S* the function given by:

(2) Content(S) = λt, w [[S]] $^{s, t, w}$

Here $\lambda t, w$ abbreviates $\lambda t\ \lambda w$, and thus Content(S) is a function which (given an assignment function s) associates to times t and worlds w the value that *S* has at t in w – namely [[S]] $^{s, t, w}$ (we abbreviate $\lambda t\ \lambda w$ as $\lambda t, w$ when we want to think of the arguments, which technically are taken 'one at a time', as pairs; here we think of this function as taking a pair of arguments <t, w > and returning the value that *S* has at t in w).

Here too, however, our initial picture is too simple. To take a well-known example, if David sees himself through a mirror, the cognitive significance of the sentence (A) *My pants are on fire* will be very different from that of (B) *His pants are on fire* – despite the fact that both sentences are about him, David, and are presumably true in the same world-time pairs (note that David is likely to take immediate action in (A), but not necessarily in (B)). A simpler case to analyze is provided by Perry's amnesiac example [17, 23]. Rudolf Lingens, an amnesiac, might have access to all available knowledge about the world (for instance because he is in a *very* well-furnished library at Stanford). He might know lots of things about Lingens, but he would still not be in a position to assert (A') *I am Rudolf Lingens* – though he would definitely be able to claim (B′) *Rudolf Lingens is Rudolf Lingens*. Hence our fourth question:

Q4: What determines the cognitive significance of sentences with indexicals?

There is another side to this question. If we ask how thoughts are *reported* in language, it often seems that the distinction between the two direct discourse sentences (A) and (B) gets lost in the report. For instance, the report in (3) is made equally true if David asserted (A) or if he asserted (B):

(3) David says that his pants are on fire.

It seems that the fine-grained semantic difference between (A) and (B) is not preserved in the report (note that it will not do to report (A) by saying: *David says that my pants are on fire*, which makes a claim about the speaker rather than about David). This observation is particularly important from the standpoint of a Fregean analysis of meaning. For Frege (1892), the same notion of Sense (or *Sinn*) accounts for (i) the cognitive significance of a sentence *and also* for (ii) its truth-conditional contribution in attitude reports (in possible worlds treatments, a Sense is reinterpreted as an intension, or function from world-time pairs to truth values). But in these examples the two roles ((i) and (ii)) seem to be fulfilled by different objects: the cognitive significance of a sentence with indexicals is directly tied to their context-dependency, whereas the truth-conditional contribution of a clause in an attitude report seems *not* to report the precise contribution of the indexicals that appeared in the original statement. This leads us to our fifth question:

Q5: Can the cognitive significance of thoughts expressed with indexicals be fully captured in attitude reports? If not, why is this not the case?

14.2 The Kaplanian Picture

14.2.1 Basic Analysis

Kaplan [15, 16] offered a unified answer to these questions, one that has proven very influential in the last 40 years. Technically, the basic idea is that expressions of a language are evaluated with respect to a *context* parameter in addition to whatever other parameters are needed for semantic evaluation. Contexts may be taken as primitive, in which case one must define various functions that output the agent [= speaker], hearer [= addressee], location, time and world of a context c, henceforth written as c_a, c_h, c_l, c_t and c_w.[2] Alternatively, contexts may be identified with tuples of the form <speaker, (addressee), time of utterance, world of utterance, etc>. The speaker, addressee, time and world of the context are sometimes called its 'coordinates'.[3]

To make things concrete, we assume – following Kaplan – that the form of the interpretation function is $[[.]]^{c, s, t, w}$: given a context of utterance c, an assignment function s, a time of evaluation t, and a world of evaluation w, an expression filling the slot of . receives a certain value. Concretely, we can provide the reference rules in (4) and the rules of composition in (5). The former indicate that when evaluated under a context c the words *I*, *you*, and *here* respectively denote the agent, hearer and location of c; the latter specify that *now* and *actually* have the effect of shifting the time and world of evaluation to the time and world of the context.[4]

(4) a. $[[I]]^{c, s, t, w} = c_a$
b. $[[you]]^{c, s, t, w} = c_h$
c. $[[here]]^{c, s, t, w} = c_l$

[2]Here and throughout, we will make the simplifying assumption that all contexts are contexts of *utterance*. As I. Stojanovic reminds us, Kaplan [15] was more careful and for this reason used the term *agent* rather than *speaker* of a context.

[3]The two approaches – primitive contexts, or contexts *qua* tuples – are semantically equivalent if there is an appropriate mapping between primitive contexts and the relevant tuples. For concreteness, assume that (i) each context c determines an agent c_a, a time c_t and a world c_w; and that (ii) for every triple of the form <x, t, w> comprising an individual x, a time t, and world w, there is at most one speech act that corresponds to it. Then we can equate the set {c: c is a context} with the set {<x, t, w>: x is an individual who is the agent of a speech act at time t in world w}. Note, however, that if the object language is endowed with context-denoting expressions, there might be important *syntactic* differences between context-denoting variables (e.g. c_1, c_2, etc) and syntactically represented triples (e.g. $<x_1, t_1, w_1>$, $<x_2, t_2, w_2>$, etc). See for instance Schlenker [27, 28] and Stechow [35, 36] for different representational choices in the syntax (e.g. with context variables in Schlenker [28], and triples in Schlenker [27] and Stechow [35, 36]).

[4]For readability, we give *now* and *actually* a syncategorematic treatment. Note that there are arguments in the literature that show that *actually* is not a *bona fide* indexical [9]. In fact, we do not know of a single case of a clear world indexical; we disregard this (potentially important) fact in this chapter.

(5) For any formula F,
a. $[[\text{now } F]]^{c, s, t, w} = [[F]]^{c, s, c_t, w}$
b. $[[\text{actually } F]]^{c, s, t, w} = [[F]]^{c, s, t, c_w}$

With these tools in hand, we can give the definition of truth in (6). It says roughly that a sentence *S* uttered in a context c is true if *S* is true according to our interpretation function, setting the context parameter to c *and the time and world parameters to the time and world of* c respectively.

(6) Truth
If a root sentence *F* is uttered in a context c, and if the assignment function s adequately represents the intentions of the speech act participants for the demonstratively used pronouns that appear in *F* (treated as free variables), then:

F is true in c just in case $[[F]]^{c, s, c_t, c_w} = 1$ (where c_t and c_w are the time and world of c respectively).

With this background in mind, we can proceed to answer the five questions we raised at the outset.

Q1. Which parameter should be added to the interpretation function to handle indexicals?

Clearly, this has to be a *context parameter*. In Kaplan's analysis, contexts are ontologically distinct from other parameters, and strictly more finely individuated than times or worlds (because distinct contexts can exist at the same time and in the same world).[5]

Q2. Why do indexicals seem to be scopeless?

There are two answers to this question in Kaplan's analysis. On a *technical level*, Kaplan's idea was that we happen to find in natural language operators that manipulate the various parameters, *except* the context parameter. In (7), we provide by way of example semantic rules for the operators *always* and *necessarily*, which shift the time and world parameters respectively.

[5]Other authors adopt frameworks in which at least one other parameter is of the same ontological type as contexts. This happens in particular if worlds are replaced with situations or events; in such a case, contexts can be taken to be situations or events of a particular sort, which blurs the ontological distinction between the two parameters – but does not make it unnecessary: for reasons discussed below, double indexing is crucial to obtain the right behavior for indexicals. As we emphasize below, when such a move is made one must give independent criteria for what counts as the context parameter, lest the discussion about 'context shift' should become rather confused.

(7) Let F be a clause.
a. [[always F]] $^{c, s, t, w}$ = 1 iff for every time t' accessible from t in w, [[F]] $^{c, s, t', w}$ = 1
b. [[necessarily F]] $^{c, s, t, w}$ = 1 iff for every world w'accessible from w at t, [[F]] $^{c, s, t, w'}$ = 1

The crucial observation is that in each case the context parameter remains unchanged. Thus if the sentences in (1)a-a' have the Logical Forms (i.e. the abstract syntactic representations) in (8)a-b respectively, we will obtain different truth conditions for them.

(8) a. Always [the speaker is boring]
b. Always [I am boring]

In both cases, we start by writing that [[always F]] $^{c, s, c_t, c_w}$ = 1 iff for every time t', [[F]] $^{c, s, t', c_w}$ = 1. In the case of (8)a, the latter condition becomes: ... [[the speaker is boring]] $^{c, s, t', c_w}$ = 1; in (8)b, it becomes: ... [[I am boring]] $^{c, s, t', c_w}$ = 1. Noun phrases may depend on the time of evaluation, which is why [[the speaker]] $^{c, s, t', c_w}$ denotes in this case the person who is speaking **at t'** in c_w. By contrast, in accordance with (4)a, [[I]] $^{c, s, t', w}$ always denotes the agent of c (= c_a), which is why the two claims end up making assertions about different people.

This explains why *always* fails to affect the interpretation of the indexical *I*. But couldn't one define other operators that manipulate the context parameter? Kaplan grants that his semantic framework makes it possible to define such operators, but he claims that they are never found in natural language (see also Lewis [18]). For this reason, he calls such operators 'monsters', and his empirical claim has come to be known as the 'Prohibition Against Monsters':

(9) Prohibition Against Monsters: No natural language operator manipulates the context parameter.

This empirical claim has been disputed in recent semantic research; we come back to this point in Sect. 14.3.2. But as we have presented things, the stipulation in (9) is needed to explain why indexicals fails to interact scopally with operators.

While the Prohibition Against Monsters is often taken as primitive (e.g. in standard linguistic accounts of indexicals), for Kaplan it was a derived property. His main philosophical claim was that indexicals display their unusual scopal behavior because they are expressions of 'direct reference'; it was direct reference, not the Prohibition Against Monsters, which motivated his account. In Kaplan's words, "directly referential expressions are said to refer directly without the mediation of a Fregean *Sinn*", which means that "the relation between the linguistic expression and the referent is not mediated by the corresponding propositional component, the content or what-is-said" ([15], p. 568). Kaplan did not mean by this that *nothing* mediates the relation between the linguistic expression and the individual. In fact, indexicals come with rules of use that establish a dependency between contexts and denotations. But these rules are, for him, quite different from *semantic contents*, which are just functions from world-time pairs (rather than contexts) to individuals or truth values. On the assumption that various operators (e.g. *necessarily* and *always*) only have access to the *content* of an expression, we derive the fact that indexicals cannot interact scopally with them. Importantly, however, Kaplan's formal framework can be adopted without accepting his views on direct reference; in such a case, the Prohibition Against Monsters needs to be taken as primitive if one wishes to derive the same predictions as Kaplan – unless one just abandons the Prohibition, as is now often done on empirical grounds – a point we will revisit in Sect. 14.3.2.

Let us turn to the question of *a priori* true vs. necessarily true sentences:

Q3. How can indexicals give rise to a priori *true sentences which, when preceded by 'necessarily', can become false?*

Given the definition of truth in (6), it seems natural to posit that a sentence is *a priori* true just in case it is true in every conceivable context:

(10) A sentence *F* is *a priori* true iff for each context c, *F* is true in the context c, i.e. (given (6)) iff $[[F]]^{c, s, c_t, c_w} = 1$.

Let us apply this definition to *I exist*. We assume, as is standard, that *exist* evaluated at a time t and a world w is true of precisely those individuals that exist at t in w. So to determine whether *I exist* is *a priori* true, we ask whether:

(11) for each context c, $[[\text{I exist}]]^{c, s, c_t, c_w} = 1$, i.e. c_a exists at c_t in c_w

Kaplan claims that the condition in (11) is satisfied *because of what contexts are*. Specifically, contexts obey (among others) the two conditions in (12):

(12) For any context c:
a. the agent c_a of c exists at the time c_t of c in the world c_w of c.
b. the agent c_a of c is at the location c_l of c at the time c_t of c and in the world c_w of c.

Thanks to (12)a, the condition in (11) is always met – which guarantees that *I exist* is indeed *a priori* true.

To obtain this result, we considered the value of our sentence in different contexts c – while setting the time and world parameters to the corresponding coordinates of c. But when we consider the sentence *Necessarily, I exist*, we only vary the world parameter. Thus (13) follows from the rule we posited for *necessarily* in (7)b.

(13) Uttered in a context c, *Necessarily I exist* is true iff $[[\text{necessarily I exist}]]^{c, s, c_t, c_w} = 1$, iff for every world w' accessible from c_w at c_t, $[[\text{I exist}]]^{c, s, c_t, w'} = 1$, iff for every world w' accessible from c_w at c_t, c_a exists at c_t in w'.

The latter condition has no reason to be satisfied, because for most relevant values of w', w' is not the world of the context c. Thus we have explained how a sentence can be *a priori* true even though it becomes false when prefixed with *necessarily*.

Let us turn to our questions about attitudes and attitude reports.

Q4: What determines the cognitive significance of sentences with indexicals?

We will start with a perspective which is in part foreign to Kaplan's analysis, but is standard in the semantic literature (e.g. [13, 30, 40]). Under what conditions does one believe that a sentence *S* is true? Just in case one believes that one is in a context in which *S* is true. In standard epistemic logic, an individual x is taken to believe that a sentence *S* is true just in case each world compatible with what x believes is one in which *S* is true. It is easy to adapt this analysis to the present case by replacing worlds with contexts:

(14) An individual x believes that a sentence *S* is true just in case each <u>context</u> compatible with what x believes is one in which *S* is true.

Given our definition of truth in (6), this condition can be rewritten as (15):

(15) An individual x believes a sentence *S* is true just in case for each context c compatible with what x believes, $[[S]]^{c,\, s,\, c_t,\, c_w} = 1$.

With this definition in hand, it can be seen that a sentence *S* is *a priori* true just in case it can be believed no matter what one's beliefs are – which seems intuitively reasonable. Thus there is both a conceptual and a technical connection between the analysis of belief and the analysis of *a priori* knowledge.

The condition in (15) immediately explains why (16)a has a very different cognitive significance from (16)b for the amnesiac Rudolf Lingens.

(16) a. I am Rudolf Lingens.
b. Rudolf Lingens is Rudolf Lingens.

According to our analysis, Lingens believes (16)a just in case each context c compatible with what he believes is one for which c_a = Rudolf Lingens – which is precisely not the case here, since he does not know which individual he is. By contrast, (16)b is trivial for him just as it is for everybody else, since for every such context c, Rudolf Lingens = Rudolf Lingens.

Kaplan develops a slightly different analysis. As we saw, it is crucial that expressions be evaluated with respect to a context parameter *in addition* to the 'usual' parameters – notably, the time and world parameters. Now Kaplan's idea is that an expression is *first* evaluated with respect to a context, which yields the *semantic content* of that expression. The content is *then* fed a world and time of evaluation to yield the denotation of the expression (for a referential expression, its denotation is an individual; for a sentence, it is a truth value). In this *façon de parler*, the meaning of an expression, called by Kaplan a 'character', is a function from contexts to contents; and a 'content' is just a function from world-time pairs to denotations (individuals or truth values).

(17) Character and Content

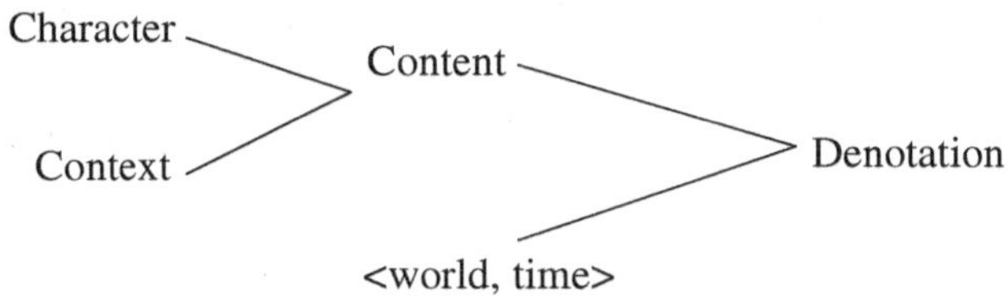

In this picture, what provides the cognitive significance of an expression is its *character*: it is because 'Lingens is at Stanford' and 'I am at Stanford' have different characters that Lingens can believe the former (because he has complete knowledge of the world he is in) without thereby believing the latter (because he does not know in which context he is located). By contrast, what provides the closest Kaplanian equivalent of Frege's notion of sense is the *content* of the sentence. The Prohibition Against Monsters entails that modal operators may only be sensitive to the content of an expression, not to its full character (more precisely: for any operator *Op* that is not monstrous, if *F* and *F′* have the same content but possibly different characters in a context c, *Op F* and *Op F′* must have the same value when evaluated in c). To take an example, on the assumption that the proper name *Lingens* is rigid and thus denotes the same individual in all possible worlds, the character of the sentence S = *I am Lingens* can be characterized as follows (using the notation $\lambda c\ \lambda t, w\ F$ in the meta-language to define a function from contexts to a function from world and times to truth values; as before, $\lambda t, w$ can be taken to abbreviate $\lambda t\ \lambda w$).

(18) $\text{Character(S)} = \lambda c\ \lambda t, w\ [c_a = \text{Lingens}]$

On the assumption that c* is a context whose agent is Lingens, the content of S in c* is:

(19) $\text{Content}_{c*}\text{(S)} = \text{Character(S)(c*)} = [\lambda c\ \lambda t, w\ c_a = \text{Lingens}]\text{(c*)} = \lambda t, w\ [\text{Lingens} = \text{Lingens}]$

Kaplan's analysis is compatible with the analysis we developed in (15), but it is not equivalent with it. It is *compatible* with it because it is possible to state (15) within a Kaplanian framework. To this end, an auxiliary notion is helpful, that of the *diagonal* $\delta(\chi)$ of a character χ, defined as follows:

(20) $\delta(\chi) = \lambda c\ \chi(c)(c_t)(c_w)$

If χ is the character of a clause *F*, the diagonal of χ can be identified with *the set of contexts c such that F uttered in c is true* according to the definition in (6). In effect, $\delta(\chi)$ is a proposition-like object – with the only difference that it corresponds to a set of contexts rather than to a set of worlds or world-time pairs. So we can refine Kaplan's analysis by granting that the cognitive significance of a sentence is provided by its character, but that the only thing that matters is whether the agent believes the *diagonal* of this character. Still, our initial theory is *not equivalent* to Kaplan's, because the latter does not provide a reductive analysis of what it means for someone to 'believe' a character; it leaves open the possibility that an agent x might believe a sentence *F* and disbelieve a sentence *F'* as long as they have different characters, *even if their diagonals* are identical. Our initial analysis precluded this possibility.

In order to determine whether an individual believes a sentence *S*, we must have access to the character (or at least to the diagonal of the character of *S*), rather than just to its content. But as we noted at the outset, attitude reports often seem to 'lose' the precise indexical nature of the attitudes they report, hence the question:

Q5: Can the cognitive significance of thoughts expressed with indexicals be fully captured in attitude reports? If not, why is this not the case?

As we had noted, there is an important difference between thinking *My pants are on fire* or *His pants are on fire*, even in case both possessive pronouns refer to the same individual. Still, in indirect discourse both situations can be reported by saying: *John thinks that his pants are on fire* (where *his* refers to John):

(21) John says: 'My pants are on fire' → John says that his pants are on fire

John says: 'His pants are on fire' → John says that his pants are on fire
(where 'his' refers to John)

Kaplan accounts for this observation by positing a semantics in which *John thinks that his pants are on fire* is true just in case there is *some* character which John asserts, and whose content in the context of John's thought act is that John's pants are on fire:

(22) *John says that his pants are on fire* (where *his* denotes John) is true at c*, t*, w* iff there is a character χ such that:
(i) the content of χ given the context of John's speech act (call it *c*) is that John's pants are on fire: $\chi(c) = \lambda t,w$ John's pants are on fire at t,w [= the content of the embedded clause], and
(ii) John asserts χ at t*, w*.

This analysis is of course compatible with Kaplan's two main claims: (i) the cognitive significance of sentences is given by their character; but (ii) attitude operators, like all other natural language operators, are only sensitive to the content of their argument. It immediately follows from (22) that two clauses that have the same content at the context utterance can be substituted *salva veritate* under *John says that* __.

There are two ways in which Kaplan's analysis could be extended: first, it could presumably be applied to other attitude verbs, such as *believe*, rather than just to verbs of saying; second, one may wish to give a reductive analysis of what it means to 'assert' or to 'believe' a character, using the diagonal operator defined above. Applied to belief reports, this extension leads to the following analysis:

(23) *John believes that his pants are on fire* (where *his* denotes John) is true c*, t*, w* iff there is a character χ such that:
(i) the content of χ given the context of John's thought act (call it *c*) is that John's pants are on fire: $\chi(c) = \lambda t, w$ John's pants are on fire at t in w, and
(ii) for each context c' compatible with what John believes at t* in w*, $[\delta(\chi)](c') =$ true, i.e. $\chi(c')(c'_t)(c'_w) = 1$.

Technical note. This analysis is not without problems. As Stechow and Zimmermann [37] show (following Crimmins [10]), this semantics makes the unfortunate prediction that *John believes that his pants are on fire* should be true as soon as John's pants really are on fire. Consider (24), calling its Character χ^* (where *actually* has the semantics defined in (5)b):

(24) It is either not so that John's pants are actually on fire now, or else John's pants are on fire.

The problem is that any rational individual can realize that (24) uttered in a context c and evaluated at the time c_t and in the world c_w of c is true. This is because $\chi^*(c)(c_t)(c_w)$ is true just in case: John's pants are not on fire at c_t in c_w, or John's pants are on fire at c_t in c_w – which is a tautology. Thanks to the *actually* and *now* operators, however, the *content* of χ^* in c is $\chi^*(c) = \lambda t, w$ [John's pants are not on fire at c_t in c_w or John's pants are on fire at t in w]. With the assumption that John's pants are in fact on fire at c_t in c_w, the first disjunct must be false, and thus we get: $\chi^*(c) = \lambda t, w$ [John's pants are on fire at t in w]. But this means that there *is* a character whose content is that John's pants are on fire, which is believed by John – χ^* is such a character. So the sentence *John thinks that his pants are on fire* should be true. But to reach this conclusion, we did not make reference to any non-trivial beliefs on John's parts! The analysis has gone wrong (but see Sect. 14.3 for an analysis of attitude reports that does *not* rely on quantification over characters).

14.2.2 Qualifications

While the technical picture we offered above is simple and appealing, not all of its components are essential – or empirically correct, for that matter. There is at least one important insight that should be preserved by any theory[6]:

[6]See Stojanovic [38] for a discussion of the minimal requirements on theories that aim to handle Kaplan's indexical examples.

(25) Double indexing
The semantic procedure must make it possible to evaluate expressions under at least two kinds of parameters: The context parameter, and whatever time and world parameters are otherwise necessary to deal with modal and temporal operators. Keeping the distinction is essential to capture the fact that time and world operators need not shift the context of evaluation of indexicals.

What about the other components of the Kaplanian picture? Their status is considerably less clear.

14.2.2.1 Direct Reference

As we saw at the outset, Direct Reference has the advantage of explaining why indexicals do not usually seem to interact scopally with other operators. But the Prohibition Against Monsters can derive (or rather stipulate) this fact within frameworks that accept Double Indexing but not Direct Reference. Furthermore, we will see in Sect. 14.3.2 that there are cases in which indexicals do in fact interact scopally with other operators, which casts doubt on a directly referential analysis.

14.2.2.2 Modal Logic

A relatively inessential property of the Kaplanian picture is that it involves an intensional system with one world parameter, one time parameter, and an assignment function that provides values to individual variables – with the crucial addition of a context parameter. As it happens, there is considerable evidence in semantics for the view that *independently of issues of indexicality* one needs to have simultaneously access to several world and time parameters ([9]; note that event/situation parameters could replace time or world parameters, but we would still need to have several of them). One way to implement the resulting system is to take the object language to include time and world variables, and to relativize the interpretation function to an assignment function that provides values not just to individual variables, but also to time and world (or situation/event) variables. When this step is taken, and combined with Kaplan's addition of a context parameter, the interpretation function takes the form $[[.]]^{c, s}$ rather than $[[.]]^{c, s, t, w}$ – with the important difference that in the first case the assignment function s provides values to individual as well as time and world variables, whereas in the second case it is only responsible for individual variables.

This technical refinement also opens a further technical possibility: we could postulate that the object language contains a distinguished context variable – call it $c*$ – whose value is also provided by the assignment function s. In effect, the interpretation function would then simply have the form $[[.]]^{s}$, and the word *I* would be represented as I_{c*} to guarantee that its value depends on the context s(c*).[7] In order to obtain an adequate definition of truth, we would need to stipulate that s(c*) denotes the context of the actual speech act. But stipulations of this sort are needed in any event for demonstratively used pronouns – when we analyze the sentence *He$_1$ [pointing] is smart but he$_2$ [pointing] is not*, we need to stipulate that the pronouns *he$_1$* and *he$_2$* refer to the 'right' individuals. This is the reason our definition of truth in (6) made explicit reference to "the intentions of the speech act participants"; in the case at hand, we would require that $s(x_1)$ and $s(x_2)$ be the individuals intended by the speaker when he uttered *he$_1$* and *he$_2$*.

Note that since assignment functions are just functions from variables (distinguished by integers) to objects, we can also write $[[.]]^{s}$ as in (26), where we have a long sequence with the value of $c*$, followed by the values of the individual variables $x_1, x_2, \ldots$, time variables $t_1, t_2, \ldots$, and world variables $w_1, w_2, \ldots$.

(26) $[[.]]^{s(c*), s(x_1), s(x_2), \ldots, (t_1), s(t_2), \ldots, s(w_1), s(w_2), \ldots}$

Thus an assignment function essentially makes it possible to relativize the interpretation function to an arbitrary number of individual, time, and world parameters – in addition to a context parameter.

14.2.2.3 Contexts

In Kaplan's analysis, contexts are primitive. This view contrasts with 'index theory', according to which an arbitrary number of *independently varying* parameters might become necessary when we analyze the semantics of more complex expressions (this view originated in Scott 1970; see Kaplan [15] and [14] for discussion). According to index theory, then, the interpretation function could take a form like $[[.]]^{x, x', x'', \ldots, t, t', t'', \ldots, w, w', w'', \ldots}$, which is immediately analogous to what we had in (26), except that no context parameter is present. We could add parameters for the agent, time and world of utterance, e.g. as $x*$, $t*$, $w*$, thus yielding:

(27) $[[.]]^{x*, t*, w*, x, x', x'', \ldots, t, t', t'', \ldots, w, w', w'', \ldots}$

Kaplan's objection against this implementation is that it misses some validities. The argument is as follows:

(i) A sentence is *valid* just in case it is true under all values of the parameters.
(ii) If x*, t* and w* are treated as *separate* parameters, in order to determine whether *I exist* is true we will have to evaluate it under values of these parameters that do *not* guarantee that x* exists at t* in w*; hence the sentence will not come out as valid.

[7] Alternatively, we could state a rule such as: $[[I]]^{s} = s(c*)_a$ – which is the counterpart in this system of the Kaplanian rule $[[I]]^{c, s, t, w} = c_a$.

(iii) Treating contexts as primitive avoids this problem – as long as we stipulate that: (a) for any context c*, *the agent of c* exists at the time of c* in the world of c** (in accordance with (12)a); (b) to determine whether a sentence is valid, we only evaluate it at parameters that are coordinates of the context parameter (in accordance with (6); this was precisely what we did for *I exist* in (11)).

A minimally different implementation of Kaplan's ideas would *reduce* contexts to triples of the form <x*, t*, w*>, with x* the agent of the speech act, t* its time, and w* its world. The interpretation function would then take the form $[[.]]_{<x^*, t^*, w^*>, x, x', x'', \ldots, t, t', t'', \ldots, w, w', w'', \ldots}$, which would avoid the problem faced by 'index theory' if (a) only triples <x*, t*, w* > that correspond to possible contexts are considered, and (b) we only evaluate the sentence at parameters that are coordinates of <x*, t*, w* >.[8]

But this raises a *further* possibility, which is to *stick to 'index theory', while revising our notion of validity.* Let us say that a sentence is *Kaplan-valid* for the interpretation function represented in (27) just in case it is valid for all values of the parameters for which (a) *<x*, t*, w*> is a possible context*, and (b) all other parameters are coordinates of <x*, t*, w*>. It is immediate that this would yield something equivalent to the preceding theory. In Kaplan's original analysis, we partly placed in the ontology – in what contexts *are* – the stipulations necessary to ensure that the correct inferences come out as valid. In the present reformulation, we directly define a notion of validity that captures the desired inferences.

Even within Kaplan's original framework, a non-standard notion of validity might be needed anyway. We already noted that when testing for validity, we must restrict attention to time and worlds parameters that are coordinates of the context (or else *I exist* and *I am here now* would not come out as valid). But there is a further problem that concerns contexts themselves. The argument is in two steps. First, we note with Predelli [26] that Kaplan's original analysis incorrectly predicts that (28) should be a contradiction.

(28) I am not here right now. (... Please leave a message after the tone.)

Since this sentence is perfectly coherent (e.g. as produced by an answering machine), there must be 'improper contexts', ones whose author is *not* located at the time of the context in the world of the context. We must thus enlarge Kaplan's original set to include improper contexts. Second, we note that once this step is taken we are left with the task of deriving Kaplan's original inferences: if there are improper contexts, how can *I am here now* come out as being 'normally' valid? The natural way to regain these inferences is to take (Kaplan-) valid sentences to be those that are true *with respect to the set of proper contexts.* But once this move is made, we can of course ask whether we couldn't just as well have started with index theory to define Kaplan-validity.

[8] See fn. 3. for further technical remarks.

More generally, Kaplan sought to derive certain *a priori* inferences by devising a system in which they came out as logical truths. But what counts as a logical inference is by no means a clear or settled question. Distinguishing between those inferences that are true by virtue of the meaning of the words from those that are true by virtue of world knowledge is, in this case as in others, a very difficult question, as Predelli's example makes clear.

14.2.2.4 Character and Content

As we showed in our discussion in Sect. 14.2.1, it is not the full character of a clause that is needed to assess its cognitive significance, but just its *diagonal*. But it is also unclear whether the notion of content as defined serves a useful purpose. As argued by various authors (see for instance Perry [24, 25], and also Stojanovic [38, 39]), there are a variety of notions of 'content' that could be argued to play a linguistic role, and Kaplan's notion is just one of them (we will see in the next section that Kaplanian contents are often inadequate to fulfill one of their main roles, which was to account for attitude reports). Furthermore, as shown by Ninan [21], a Kaplanian content can be defined on the basis of a semantics that is not based on Kaplan's parameters (for instance, within a semantics with time and world variables one can abstract over these variables to obtain the appropriate notion of content); and conversely, a semantics based on Kaplan's parameters need not give rise to Kaplan's notion of content (some of these parameters may be given the same status as the context parameter in Kaplan's analysis, so that they are not abstracted over in the computation of content).

14.3 De Se Reports and Shifted Indexicals

We will now show that there are quite a few cases across languages in which attitude operators manipulate the context of evaluation of indexicals. For all theories, this suggests that the Prohibition Against Monsters must be relaxed; in addition, these data pose a serious problem for the claim that indexicals are 'directly referential'.

14.3.1 De Se Reports

We start by showing that it is possible, contrary to the predictions of Kaplan's theory of indirect discourse, to preserve in indirect discourse the cognitive significance of indexicals. This is just a prelude, however, because the construction we consider does not use indexicals in the report; but in Sect. 14.3.2 we will show that the same semantic effect can in some languages be obtained by using in the report indexicals whose context of evaluation is 'shifted'.

The first observation is that what syntacticians call 'PRO', the unpronounced subject of an infinitive, is always understood to report a first person (or in some cases second person) thought when it is immediately embedded under an attitude verb [8, 20]. This is illustrated by the following scenario, in which *PRO* is inappropriate to report a third-person thought – by contrast with *he*, which is acceptable whether the thought to be reported was first- or third-personal.

(29) John is so drunk that he has forgotten that he is a candidate in the election. He watches someone on TV and finds that this person is a terrific candidate, and thinks: 'This guy should be elected'. Unbeknownst to John, the candidate he is watching on TV is John himself.
a. True: John hopes that he will be elected
b. False: John hopes PRO to be elected [28]
(by contrast, b. this is ok in a scenario in which the thought was: 'I should be elected')

Following the terminology of Lewis [8, 17], semanticists say that (29)b is a 'De Se' report because it is true only in case the agent has a first person thought. Interestingly, an artificial pronoun very much like *PRO*, called *he**, was posited by the philosopher Castañeda for purely conceptual reasons [5–7]. In effect, *PRO* embedded under an attitude verb is an English realization of Catañeda's *he**.[9]

Since Kaplan's analysis of indirect discourse was designed to *predict* that such distinctions cannot be drawn in indirect discourse, it is ill-suited to account for these contrasts. Inspired by Lewis [17], Chierchia [8] suggested that the semantics of attitude reports is more fine-grained than usually thought in possible worlds semantics. In essence, his idea was that the value of a clause embedded under an attitude verb may be as fine-grained as a set of triples of the form <individual, time, world>. It is immediate that such triples are homologous to contexts. Technically, however, no syntactic or morphological connection to indexicality was posited in Chierchia's treatment. Rather, it was assumed that a λ-operator could appear at the 'top' of the embedded clause to bind an individual variable. For simplicity, we represent this operator above an empty complementizer *C*, though this is just for notational convenience:

(30) John hopes λi C PRO_i to be elected

A crucial assumption is that, in attitude reports, *PRO* must always be bound by the closest λ-operator. To obtain an interpretable structure, we must still say what the role of the complementizer is. We will assume that it simply returns a proposition when applied to a clause (the same measure can be applied to the word *that*).

[9] So-called 'logophoric' person markers can also be seen as natural language realizations of Castañeda's *he**. See for Schlenker [30] for discussion, and [22] for a contrary view.

(31) a. $[[C\ F]]^{c, s, t, w} = [[that\ F]]^{c, s, t, w} = \lambda t'\ \lambda w'\ [[F]]^{c, s, t', w'}$
b. From (a), it follows that
$[[\lambda i\ C\ PRO_i\ be\text{-}elected]]^{c, s, t, w} = \lambda x'\ [[C\ PRO_i\ be\text{-}elected]]^{c, s[i \rightarrow x']10, t, w}$
$= \lambda x'\ \lambda t'\ \lambda w'\ [[PRO_i\ be\text{-}elected]]^{c, s[i \rightarrow x'], t', w'} = \lambda x'\ \lambda t'\ \lambda w'$ x' is elected at t' in w'.

We can think of the function defined in (31)b as associating truth values to sets of triples of the form <individual, time, world>. Since the latter are context-like objects, we can extend to the object-language operators *believe*, *hope*, etc., a homologue of the rule we used in Sect. 14.2.1 to explicate under what conditions an individual x believes that a sentence *S* is true. In (14)–(15), we had suggested that this is the case precisely if each context compatible with x's belief makes *S* true. Similarly, we will say that an individual x stands in the 'believe' relation to the denotation of an embedded clause just in case each context compatible with what x believes satisfies the embedded clause. Given the kind of denotation we have in (31)b, the rule must state that the *coordinates* of all such contexts make the embedded clause true.

(32) a. $[[believes^{De\ Se}]]^{c, s, t, w}(F)(x) = true$
iff for each context c' compatible with what x believes at t in w,
$F(c'_a)(c'_t)(c'_w) = true$
b. $[[hope^{De\ Se}]]^{c, s, t, w}(F)(x) = true$
iff for each context c' compatible with what x hopes at t in w,
$F(c'_a)(c'_t)(c'_w) = true$

The same semantics can be extended to the verb *hope,* as shown in (32)b.

An important consequence of this analysis is that *John hopes to be elected* is true just in case each context compatible with John's hope is one in which he could utter truly: 'I am elected'. Equivalently, *John hopes to be elected* is true just in case he stands in the 'hope' relation to the *diagonal* Δ of the character of *I am elected*. This result is just what is needed to account for the falsity of (29)b, since in our scenario John does *not* have a first person hope. The equivalence between *John hopes to be elected* and *John stands in the 'hope' relation to the diagonal of 'I am elected'* is stated in (33),[11] where we have assumed for convenience that δ was part of the object language.

---

[10] $s[i \rightarrow x']$ is that assignment function which is identical to s, with the possible exception that it assigns x' to *i*.

[11] For simplicity, we consider a variant of (29) in which John's first person hope is of the form 'I am elected' rather than 'I should be elected'.

(33) a. $[[\text{John hopes}^{\text{De Se}}\ \lambda i\ \text{C PRO}_i \text{ to be elected}]]^{c,\, s,\, t,\, w} = \text{true}$ iff for each context c' compatible with what John hopes at t in w, c'_a is elected at c'_t in c'_w.

b. Suppose that δ is part of the object language, with $[[\delta\ [\text{I be-elected}]]]^{c,\, s,\, t,\, w} = \lambda c'\ [[\text{I be-elected}]]^{c',\, s,\, c't,\, c'w}$ – which we call Δ. Then John stands in the 'hope' relation to Δ iff for each context c' compatible with what John hopes at t in w, $\Delta(c') = 1$, iff for each context c' compatible with what John hopes at t in w, c'_a is elected at c'_t in c'_w.

Of course in English δ does not seem to be part of the object language: *John hopes that I am elected* clearly does not allow the word *I* to be shifted (for if so it would intuively denote John). But things are different in other languages, as we will now see.

14.3.2 Shifted Indexicals in Indirect Discourse

We now suggest that there are constructions in which the diagonal δ does in fact appear in the object language. This will show that Kaplan's analysis was not just wrong about De Se readings, but also about monsters: sometimes attitude operators are Kaplanian monsters (a conclusion anticipated in Israel and Perry [14]; see Deal [11] for a distinct, and far more systematic, view of the cross-linguistic typology).

How can we establish the existence of monsters? We will discuss examples that have the form of (34), where <I> and <here> are indexicals:

(34) John says that ... <I> ... <here> ...

The argument has three steps.

(i) First, we argue that the presence of the diagonal operator in the embedded clause is *compatible* with the semantics of the sentence – in particular <*I*> should intuitively denote John, and <*here*> should intuitively denote John's location.[12]

(ii) Second, we exclude the possibility that the embedded clause is quoted. This is an essential step because on any theory it is unsurprising that *John says: 'I am a hero'* should attribute to John a claim about John himself (because in this case *say* establishes a relation between John and a string of words rather than with a proposition). In English, the presence of the word *that* rules out a quotative reading, but other languages could have quotative complementizers. Still, one can block quotative readings by observing that grammatical dependencies cannot normally 'cross' quotation marks. To illustrate, let us note that without explicit quotation marks *John says I like Mary* is ambiguous between *John says that I like Mary* and *John says 'I like Mary'*. But the second reading disappears in the more complex sentence *This is the person who [John says I like _]:* it

[12] It follows from the semantic analysis that both expressions are predicted to be read De Se.

cannot be interpreted as *this is the person who John says ['I like _']*, with a dependency between *who* and the object position of the most deeply embedded clause, marked as _. In technical syntax, *who* is said to be 'extracted' from this object position; and we see here that extraction cannot cross quotation marks. In this case, *I* behaves like a *bona fide* Kaplanian indexical: when quotation is excluded, *I* unambiguously refers to the actual speaker. As we will see, the facts are different in other languages.

(iii) Third, we want to exclude the possibility that the purported indexicals are in fact anaphoric elements. This is no trivial matter: anaphoric expressions can often have, among others, a deictic reading, whereby they pick their denotation from the context. What distinguishes such anaphoric elements from *bona fide* indexicals is that the latter can never have unambiguously anaphoric readings. For instance, the word *later* in *I will go for a walk later* may *appear* to be an indexical, because it can be understood to mean *later than now*. But other examples suggest that it is anaphoric – e.g. in *I met John yesterday morning; later he went for a walk, later* is understood as *later than the salient time* at which I met him at which I met John.

Following precisely this logic, Anand and Nevins [1] and Anand [2] conclude that there are clear cases of shifted indexicals in Zazaki. They show in particular that Zazaki indexicals can optionally shift in some constructions that rule out quotation – for instance (35), a Zazaki version of the English examples we just discussed.

(35) Extraction in Zazaki

i. **čɛnɛkɛ [kɛ Hɛseni va mɨ *t* paci kɛrda] rindɛka**
girl that Hesen said I *t* kiss did pretty.be-PRES
'The girl that Hesen said {Hesen, I} kissed is pretty.' (Anand and Nevins, 2004)

ii. **Piyaa-o [kɛ Rojda va kɛ mɨ *t* paci kɛrd] Ali biyo**
Person that Rojda said that I *t* kiss did Ali was
'Ali was the person that Rojda said {Rojda, I} kissed.' (Anand and Nevins, 2004)

Following the spirit of their proposal, we can handle these data within Kaplan's logic by postulating that the diagonal operator δ used in (33)b can optionally be found in the embedded clause, as shown in (36).

(36) John say δ I be a hero.

When this operator is present, it establishes a relation between John and the *diagonal* of the character of *I am a hero*, and thus attributes to him a claim that every context c compatible with his claim is one in which c_a is a hero at c_t in c_w. This result is derived using the techniques we saw at work in (33)b.

Anand and Nevins's analysis makes interesting fine-grained predictions. In particular, they predict that in Zazaki indirect discourse, *if one indexical is shifted under an attitude reports, then all the other indexicals are shifted as well* ('Shift Together'). The reason for this is that if one indexical gets shifted, then the δ operator must be present, and must thus shift the context parameter. Because there is a single context parameter, once it is shifted, the value of the original context is lost, and thus all indexicals in the same clause must be shifted as well. They show in detail that this and related predictions are borne out in Zazaki (see Deal [11] for a cross-linguistic analysis that makes systematic use of 'Shift Together').

Several other cases of shifting under attitude reports have been discussed in the literature. For instance, it was suggested in Schlenker [28] that sentences very much like (36) can be found in Amharic indirect discourse; and it was also claimed that in English *two days ago* is a shiftable indexical, while *the day before yesterday* is an unshiftable one (these data have been debated, however; see Anand [2] for a contrary view). One salient question in the literature is whether Anand and Nevins's treatment with a single context parameter is sufficient. Several examples have been discussed in which 'Shift Together' *fails* to hold (but see Deal [11] for a contrary view); in fact, data of precisely this type led Schlenker [28] to adopt a more expressive system in which there are *context variables* in the object language, which makes it possible to analyze many more readings than are predicted by Anand and Nevins. Such a system must still be able to account for the fact that in English *I* cannot be shifted; this was done by having a distinguished variable c^* which always denotes the actual speech act (as was done above in Sect. 14.2.2). As things stand, it would seem that 'Shift Together' holds true in some languages but not in others. Clearly, however, more research is needed to obtain a deeper understanding of this debate (see Schlenker [30] for further remarks, and Anand [2] and Deal [11] for an in-depth discussion).

What is clear, however, is that these data on indexical shift suggest that Kaplan's Prohibition Against Monsters needs to be revisited, and that theories of direct reference have serious challenges to address.

14.3.3 Shifted Indexicals in Free Indirect Discourse

Free Indirect Discourse is a type of reported speech, found primarily in literature, in which different indexicals are evaluated with respect to different contexts, *even in the absence of any (overt) attitude operator* (we use the sign # to mark semantic infelicity).

(37) a. Tomorrow was Monday, Monday, the beginning of another school week! (Lawrence, *Women in Love*; cited in Banfield [3])
b. #He thought: 'Tomorrow was Monday, Monday, the beginning of another school week!'
c. #He thought that tomorrow was Monday, Monday, the beginning of another school week!

The thought expressed in (37) is attributed to the character whose attitude is described rather than to the narrator; it can optionally be followed by a post-posed parenthetical, such as ..., *he thought* or ..., *he said*. Descriptively, Free Indirect Discourse behaves as a mix of direct and of indirect discourse: tenses and pronouns take the form that they would have in a standard attitude report (e.g. *She wondered where he was that morning*), while everything else – including *here, now, today, yesterday* and the demonstratives (e.g. *this*) – behaves as in direct discourse. In other words, a passage in Free Indirect Discourse may be obtained by changing the person and tense markers of a quotation to those of an indirect discourse embedded under an attitude verb in the desired person and tense.

Importantly, the indexicals that 'shift' in Free Indirect Discourse in English do not do so in standard indirect discourse. This fact alone shows that shifting in Free Indirect Discourse is not entirely reducible to the issues discussed in Sect. 14.3.2. There are two main types of extant analyses: some try to treat Free Indirect Discourse as a non-standard form of direct discourse (e.g. Schlenker [29]); while others treat it as a form of indirect discourse with a non-standard attitude operator (e.g. Sharvit [31, 32]). As things stand, the debate is wide open (see Eckardt [12] for a recent analysis).

14.4 Conclusion

We can now go back to the five questions we asked at the outset.

Q1 (Parameters). On most theories, indexicals are handled by relativizing semantic interpretation to a context-like parameter in addition to the parameters that are otherwise necessary to handle temporal and modal constructions. There are many options in the implementation, however (contexts can be taken as primitive, as in Kaplan's work; or they can be seen as tuples of coordinates; and there are even versions of 'index theory' that can emulate the results of context-based analyses).

Q2 (Scopelessness). The impression that indexicals are scopeless is in some cases incorrect: there are natural language constructions in which indexicals can be 'shifted' in attitude reports. Why does this rarely or never happen in English? For some theorists [1], this is simply because in English attitude verbs fail to embed the diagonal operator. For other theorists [28], this is because most English indexicals are specified as depending on a distinguished context variable which never gets bound. In either case, scopelessness is *not* invariably a property of expressions whose value is intuitively determined by a context of speech. We could *redefine* the terms 'indexical' and 'context' to ensure that (i) a context is, *by definition*, a parameter which is not manipulated by any operator; and (ii) an indexical (i.e. an expression whose value is determined by the context) can *by definition* never be monstrous (see Zimmermann [40] and Stalnaker [33, 34]).[13]

[13]Note that a consequence of this definitional move is that there is no context parameter, and hence no indexicals, in Zazaki as studied by Anand and Nevins. The reason is that according to them *all*

But Kaplan's analysis should not be equated with this definitional move; it had some empirical 'bite' – part of which seems to have been refuted.

Q3. (A priority and necessity). The fact that a sentence *S* can be *a priori* true while *Necessarily S* is false becomes unsurprising once the two notions are adequately explicated. The key is to ensure that *S* comes out as *a priori* true just in case for any context c, *S* is true in c, i.e. true when evaluated with respect to c and the corresponding coordinates of c. By contrast, *Necessarily S* is true at c just in case it is true when evaluated with respect to c and different values of the world parameter.

Q4 (Cognitive significance). The cognitive significance of a sentence *S* with indexicals is determined by the information it contains about the context in which it was uttered – it must be one of the contexts c such that *S* is true in c. Within post-Kaplanian frameworks, the cognitive significance of a sentence is given by the *diagonal* of its character, but here too there are many options for the implementation.

Q5 (Attitude reports). Contrary to what was predicted by Kaplan's theory of indirect discourse, the precise cognitive significance of sentences with indexicals can in some cases be faithfully reported in indirect discourse, thanks to expressions that are unambiguously De Se. *PRO*, the unpronounced subject of English infinitives, is a case in point. Shifted indexicals in constructions that allow them are another.

References and Recommended Readings[14]

1. Anand, P., & Nevins, A. (2004). Shifty operators in changing contexts. In: *Proceedings of semantics and linguistic theory (=SALT) 14.*
2. * Anand, P. (2006). De *De Se*. Ph.D. dissertation, MIT. [very detailed empirical study of indexicals and De Se reports in several languages, including Zazaki]
3. Banfield, A. (1982). *Unspeakable Sentences (Narration and Representation in the Language of Fiction)*. London: Routledge & Kegan Paul.
4. Braun, D. (2001). Indexicals. In *Stanford encyclopedia of philosophy,*http://plato.stanford.edu/entries/indexicals/
5. Castañeda, H.-N. (1966). He: A study in the logic of self-consciousness. *Ratio, 7*, 130–157.
6. Castañeda, H.-N. (1967). Indicators and quasi-indicators. *American Philosophical Quaterly, 4*(2), 85–100.
7. Castañeda, H.-N. (1968). On the logic of attributions of self-knowledge to others. *The Journal of Philosophy, 65*(15), 439–459.
8. Chierchia, G. (1987). Anaphora and attitudes *de se'*. In B. van Bartsch & E. van Boas (Eds.), *Language in context*. Foris: Dordrecht.
9. Cresswell, M. (1990). *Entities and indices*. Dordrecht: Kluwer.

parameters can in principle be shifted in that language (in particular, what we would otherwise call the context parameter is shifted by the diagonal operator).

[14]Asterisks (*) indicate recommended readings.

10. Crimmins, M. (1998). Hesperus and phosphorus: Sense, pretense, and reference. *The Philosophical Review, 107*, 1–47.
11. * Deal, A. R. (2017). Shifty asymmetries: Universals and variation in shifty indexicality. Manuscript. Berkeley: University of Californa.
12. Eckardt, R. (2014). *Semantics of free indirect discourse: How texts allow us to mind-read and eavesdrop*. Leiden: Brill.
13. Haas-Spohn, U. (1994). *Versteckte Indexicalität und subjective Bedeutung*. Ph. D. dissertation, Universität Tübingen.
14. Israel, D., & Perry, J. (1996). Where monsters dwell. In J. Seligman & D. Westerstahl (Eds.), *Logic, language and computation* (Vol. 1). Stanford: CSLI Publications.
15. * Kaplan, D. (1977/1989). Demonstratives. In P. Almog & Wettstein (Eds.) *Themes from Kaplan*. Oxford: Oxford University Press, 1989 [the main classic in the literature on indexicals]
16. * Kaplan, D. (1978). On the logic of demonstratives. *Journal of Philosophical Logic, 8*, 81–98. [a shorter presentation of some of the results in Kaplan 1977/1989]
17. * Lewis, D. (1979). Attitudes de dicto and de se. *Philosophical Review, 88*(4), 513–543. [main classic in the literature on attitudes De Se].
18. Lewis, D. (1980). Index, context, and content. In S. Kanger & S. Ohman (Eds.), *Philosophy and grammar* (pp. 79–100). Dordrecht: Reidel. Reprinted in Lewis 1998.
19. * Maier, E. (2006). *Belief in context: Towards a unified semantics of De Re and Se attitude reports*. Ph.D. dissertation, University of Nijmegen. [Includes a very clear survey of results on shifted indexicals and De Se readings]
20. Morgan, J. (1970). On the criterion of identity for noun phrase deletion. In *Proceedings of Chicago Linguistic Society (= CLS)* 6.
21. Ninan, D. (2010). *Semantics and the objects of assertion*. Manuscript, University of St. Andrews.
22. Pearson, H. (2012). *The sense of self: Topics in the semantics of De Se expressions*. Ph.D. dissertation, Harvard University.
23. Perry, J. (1993). The problem of the essential indexical. In *The problem of the essential indexical and other essays*. New York: Oxford University Press.
24. Perry, J. (1997). Reflexivity, indexicality and names. In W. Kunne et al. (Eds.), *Direct reference, indexicality and proposition attitudes*. Stanford: CSLI-Cambridge University Press.
25. Perry, J. (2001). *Reference and reflexivity*. Stanford: CSLI Publications.
26. Predelli, S. (1998). Utterance, interpretation and the logic of indexicals. *Mind and Language, 13*, 400–414.
27. Schlenker, P. (1999). *Propositional atittudes and indexicality: A cross-categorial approach*. Doctoral dissertation, MIT.
28. Schlenker, P. (2003). A Plea for monsters. *Linguistics and Philosophy, 26*, 29–120.
29. Schlenker, P. (2004). Context of thought and context of utterance (A note on free Indirect discourse and the historical present). *Mind and Language, 19*(3), 279–304.
30. * Schlenker, P. (2011). Indexicality and De Se reports. In von Heusinger, Maienborn, & Portner (Eds.), *Semantics* (Vol. 2, Article 61, pp. 1561–1604). Mouton de Gruyter. [longer survey than the present one, with more details on the linguistic side of things.]
31. Sharvit, Y. (2004). Free indirect discourse and *de re* pronouns. In R. Young (Ed.), *Proceedings of semantics and linguistic theory (= SALT) 14* (pp. 305–322). Ithaca: CLC Publications, Cornell University.
32. * Sharvit, Y. (2008). The puzzle of free indirect discourse. *Linguistics and Philosophy, 31*. 353–395. [a good source on the theoretical and empirical issues raised by Free Indirect Discourse; somewhat technical]
33. Stalnaker, R. (1981). Indexical belief. *Synthese, 49*, 129–151.
34. Stalnaker, R. (1999). *Context and content*. Oxford: Oxford University Press.
35. von Stechow, A. (2002). Binding by verbs: Tense, person and mood under attitudes. In H. Lohnstein & S. Trissler (Eds.), *The syntax and semantics of the left periphery* (Vol. 44). Berlin: de Gruyter.

36. von Stechow, A. (2003). Feature deletion under semantic binding: Tense, person, and mood under verbal quantifiers. In M. Kadowaki & S. Kawahara (Eds.), *NELS 33: Proceedings of the thirty-third annual meeting of the North East linguistic society, GLSA* (pp. 379–404). Amherst: University of Massachusetts.
37. von Stechow, A., & Zimmermann, T. E. (2005). A problem for a compositional treatment of de re attitudes. In G. Carlson & F. J. Pelletier (Eds.), *Reference and quantification: The partee effect* (pp. 207–228). Stanford: Center for the Study of Language and Information.
38. Stojanovic, I. (2008). *What is said: An inquiry into reference, meaning and content.* DM Verlag, Saarbrücken.
39. Stojanovic, I. (2009). Semantic content. *Manuscrito, 32*, 123–152.
40. Zimmermann, T. E. (1991). Kontextabhängigkeit. In A. von Stechow & D. Wunderlich (Eds.) *Semantik: ein internationales Handbuch der zeitgenössischen Forschung.* [Very detailed presentation of the formal and empirical issues as of the beginning of the 1990's.]

Chapter 15
Necessity and Possibility

Melvin Fitting

Abstract We give a basic introduction to modal logic. This includes possible world semantics, axiom systems, and quantification. Ideas and formal machinery are discussed, but all proofs (and meta-proofs) are omitted. Recommendations are given for those who want more.

15.1 Introduction

Modal operators qualify truth in some way: necessary truth, knowable truth, provable truth, eventual truth, and so on. All these have many formal properties in common while, of course, differing on others. One can abstract these properties and study them for their own sake just as elementary algebra abstracts algebraic equations from natural language problems about weights, measures, distances, and ages. The idea in all cases is that abstraction should provide us with a simple setting in which the formal manipulation of symbols according to precise rules will lead us to results that can be applied back to the complex 'real' world in which the problems arose.

If modal operators are many, what then formally constitutes a modal operator? We do not want to get into the infinite regress of philosophical debate here. A good working definition is, a modal operator is one we can investigate using the formal tools that have been developed for this purpose. Of course this is a time-dependent characterization—tools are human artifacts after all. Here we just consider the core of the subject, *normal modal logics*. These are the best understood using the simplest tools. They do not exhaust the subject.

Modal operators come in dual pairs. Dual to necessity is possibility: X is possibly true if it is not necessary that not-X is true, and X is necessarily true if it is not

M. Fitting (✉)
Professor emeritus, City University of New York, Graduate Center, Departments of Mathematics, Philosophy, and Computer Science, New York City, NY, USA
e-mail: melvin.fitting@gmail.com

S. O. Hansson, V. F. Hendricks (eds.), *Introduction to Formal Philosophy*, Springer Undergraduate Texts in Philosophy, https://doi.org/10.1007/978-3-319-77434-3_15

possible that not-X is true. Similarly knowability and consistency (with knowledge) are duals, and so on. Following custom we will use $\Box$ for any necessity-like modal operator and $\Diamond$ for its dual. It will not hurt if you read $\Box X$ as *necessarily* X and $\Diamond X$ as *possibly* X as in the title of this chapter, though you should not think these are the only readings.

(Propositional) formulas are built up from propositional letters using propositional connectives, just as in classical propositional logic, together with a rule of formation saying: if X is a formula so are $\Box X$ and $\Diamond X$. We will be informal about what the propositional connectives are, but it generally is some subset of $\wedge$ (and), $\vee$ (or), $\neg$ (not), $\supset$ (material implication), $\equiv$ (equivalence).

What sort of tools are available for formal modal investigation? Historically, axiom systems came first in modern times, with natural deduction systems, tableau systems, and such things following. Algebraic generalizations of truth tables came along in the 1940s. But ever since *possible world semantics* (relational semantics, Kripke semantics) was developed in the 1960s it has been the common starting point, and it is where we begin.

15.2 Possible World Semantics

What are possible worlds? Don't ask. This is generally a misleading question. One does not need to know what truth is in order to use truth tables—one just needs to know how it behaves with respect to the logical connectives. Likewise one does not need to know what constitutes domains of classical first-order models. It's whatever you like. You get to specify, according to intended application. The logical truths of first-order logic are those that hold no matter what the domain. Well, possible worlds are like that. You get to specify what possible worlds are, according to intended application, and the logical truths of modal logic are those that hold no matter what your specification might have been. For instance, if I am interested in what can be said about a coin flip there are plausibly two possible worlds, one in which the outcome is heads, one in which it is tails. Nothing else matters for this purpose. If I am interested in what is necessary given (our current understanding of) physical laws, possible worlds might be all ways the real universe could be, consistent with those laws. Or it could be all ways some particular experiment might come out. The choice is yours. The question is, what are the laws that hold across all such choices.

Besides possible worlds there is one more essential piece of machinery: an *accessibility relation*. For a particular intended application it may easily be that not all possible worlds are equally possible under all circumstances. For instance suppose the modal operator we have in mind is *from now on*. Then in evaluating the truth of a formula today we must take tomorrow into account, but we can ignore yesterday—tomorrow is accessible from today but yesterday is not. If the modal operator is *has always been* the situation is reversed—yesterday is relevant to today but tomorrow is not.

Definition 2.1 A *frame* is a pair $\langle \mathcal{G}, \mathcal{R} \rangle$ in which $\mathcal{G}$ is a non-empty set and $\mathcal{R}$ is a binary relation on $\mathcal{G}$.

When working with a frame $\langle \mathcal{G}, \mathcal{R} \rangle$ the members of $\mathcal{G}$ are commonly called *possible worlds* or *states*, and for $x, y \in \mathcal{G}$, if $x\mathcal{R}y$ one says that y *is accessible from* x or even x *sees* y. For the time being we will put no constraints on $\mathcal{R}$, though we will consider some later on.

When working with truth tables each line represents an assignment of truth values to propositional letters. We can choose what line to work with—that is arbitrary—but having made such a choice there are fixed rules for evaluating the truth or falsity of more complex formulas. Modal models are like this too, except that truth values for propositional letters can be different at different possible worlds.

Definition 2.2 A *(possible world) model* is a triple, $\mathcal{M} = \langle \mathcal{G}, \mathcal{R}, \mathcal{V} \rangle$ where $\langle \mathcal{G}, \mathcal{R} \rangle$ is a frame and $\mathcal{V}$ assigns a truth value to each propositional letter at each possible world (a *valuation*). That is, if P is a propositional letter and $w \in \mathcal{G}$ then $\mathcal{V}(P, w) \in \{\mathit{true}, \mathit{false}\}$. We say the model $\langle \mathcal{G}, \mathcal{R}, \mathcal{V} \rangle$ is *based on* the frame $\langle \mathcal{G}, \mathcal{R} \rangle$.

Given a model, truth values for complex formulas are calculated, world by world, according to certain set rules. At each possible world, propositional connectives behave in their usual truth-table way. But also, $\Box X$ is taken to be true at possible world w if X is true at all possible worlds accessible from w. Similarly $\Diamond X$ is taken to be true at w if X is true at some possible world accessible from w. Thus necessary truth is truth at all possible worlds that are relevant, while possible truth is truth under at least one relevant alternative. Here are the evaluation rules stated precisely. Assume $\mathcal{M} = \langle \mathcal{G}, \mathcal{R}, \mathcal{V} \rangle$ is a model—we write $\mathcal{M}, w \Vdash X$ to indicate that formula X is true at possible world w of model $\mathcal{M}$, and $\mathcal{M}, w \nVdash X$ to indicate that it is not. We give one representative propositional connective case—the others are similar.

$$\begin{aligned}
\mathcal{M}, w \Vdash P &\iff \mathcal{V}(P, w) = \mathit{true}, \text{ for } P \text{ a propositional letter} \\
\mathcal{M}, w \Vdash X \supset Y &\iff \mathcal{M}, w \nVdash X \text{ or } \mathcal{M}, w \Vdash Y \\
\mathcal{M}, w \Vdash \Box X &\iff \mathcal{M}, z \Vdash X \text{ for every } z \in \mathcal{G} \text{ with } w\mathcal{R}z \\
\mathcal{M}, w \Vdash \Diamond X &\iff \mathcal{M}, z \Vdash X \text{ for some } z \in \mathcal{G} \text{ with } w\mathcal{R}z
\end{aligned}$$

Call a formula K-*valid* if it evaluates to *true* at every possible world of every model. The 'K' refers to Kripke, since this is the logic that is given by all possible world models (Kripke models), without any special conditions or restrictions. Typical examples of validities of K are: $\Box(A \wedge B) \equiv (\Box A \wedge \Box B)$, $\Diamond(A \vee B) \equiv (\Diamond A \vee \Diamond B)$, and $(\Box A \wedge \Diamond B) \supset \Diamond(A \wedge B)$. Typical examples of non-validities are: $\Box(A \vee B) \supset (\Box A \vee \Box B)$ and $(\Diamond A \wedge \Diamond B) \supset \Diamond(A \wedge B)$. Think about what these are saying when the modal operators are interpreted in various ways (necessity, knowability, and so on) and you will see that these validities and non-validities are as they ought to be.

15.3 Adding Conditions

Let P be a propositional letter and consider the formula $\Box P \supset P$. One would naturally assume that whatever is necessary is certainly true, so this formula should be valid. But recall that $\Box$ represents many different modalities. Suppose we read $\Box$ as 'is true starting tomorrow.' For this we would not want $\Box P \supset P$, and in fact it is not K-valid. Consider the model $\mathcal{M}_t = \langle \mathcal{G}, \mathcal{R}, \mathcal{V} \rangle$ in which $\mathcal{G}$ consists of two possible worlds; let us call them today and $\mathsf{tomorrow}$, and in which we take $\mathsf{today}\,\mathcal{R}\,\mathsf{tomorrow}$, so that $\mathcal{R}$ represents the passage of time (in a narrow way, of course). Let $\mathcal{V}(P, \mathsf{today}) = \mathit{false}$ and $\mathcal{V}(P, \mathsf{tomorrow}) = \mathit{true}$. We have $\mathcal{M}_t, \mathsf{today} \Vdash \Box P$ because the only possible world accessible from today is $\mathsf{tomorrow}$, and we have $\mathcal{M}_t, \mathsf{tomorrow} \Vdash P$ because $\mathcal{V}(P, \mathsf{tomorrow}) = \mathit{true}$. On the other hand, $\mathcal{M}_t, \mathsf{today} \nVdash P$ because $\mathcal{V}(P, \mathsf{today}) = \mathit{false}$. It follows that $\mathcal{M}_t, \mathsf{today} \nVdash \Box P \supset P$, and so indeed $\Box P \supset P$ is not K-valid. If we want $\Box P \supset P$ to hold throughout a model, additional restrictions must be imposed.

Suppose $\Box P$ is read, 'P is known'. One cannot know false things, so we would expect P to be so if $\Box P$ is. But if $\Box P$ is read, 'P is believed,' we would not have the same expectation. One way of thinking about knowledge and belief, involving possible worlds, was explored in detail by Hintikka. We are ignorant in varying ways about the actual state of the world—we may not know if it is snowing at the South Pole, or if it is not, for instance. We may say that the actual world is just one among several possible worlds; in some it is snowing at the South Pole and in others it is not, and so we do not know whether it is or not. But in all of these possible worlds either it is snowing or it is not snowing, and so we know that disjunctive fact. What we know is what is true in all the possible worlds accessible to us. Roughly speaking, the range of relevant possible worlds is a representation of our ignorance. Then the difference between knowledge and belief is that for knowledge the actual world must be among those that are accessible, while for belief it need not be. Beliefs need not be tied to actual facts, merely to possible facts.

A binary relation $\mathcal{R}$ is called *reflexive* if $x\mathcal{R}x$ holds for every x for which the relation is meaningful. Call a frame reflexive if its accessibility relation is reflexive, and likewise for models based on reflexive frames. It would be a good exercise for the reader to show that $\Box P \supset P$ is true at every possible world of any reflexive model. Conversely, if a frame is *not* reflexive, some model based on it will falsify $\Box P \supset P$ at some possible world. The argument goes as follows. Suppose $\langle \mathcal{G}, \mathcal{R} \rangle$ is not reflexive; say for a particular $w \in \mathcal{G}$ we do not have $w\mathcal{R}w$. Let $\mathcal{V}$ be the valuation given by $\mathcal{V}(P, w) = \mathit{false}$ and $\mathcal{V}(P, x) = \mathit{true}$ for all $x \in \mathcal{G}$ where $x \neq w$. In this model, $\Box P$ is true at w because P is only false at w, which is not accessible from w, while p is true at all other possible worlds, which thus includes all possible worlds accessible from w. But by construction, P is false at w and hence so is $\Box P \supset P$.

Call a model a T-*model* if it, or more properly its frame, is reflexive, and let us say a formula is T-*valid* if it evaluates to *true* at every possible world of every T-model.

Then $\Box X \supset X$ is T-valid for all formulas X and not just for propositional letters. T-validity is appropriate if $\Box$ represents necessity, knowability, provability, being true at every time, and many other modalities. It is not appropriate for believability, obligatory, being true at some time, and so on.

The most commonly investigated propositional modal logics can be captured by putting various conditions on the accessibility relation of frames, just as we did above. These logics have names that are historical—do not look for a pattern. But here are those that are most commonly considered.

$\mathcal{R}$ is *serial* if, for every x there is some y such that $x\mathcal{R}y$. A formula is D-valid if it is true at all possible worlds of every serial model. A typical D validity is $\Box X \supset \Diamond X$. It is easy to see that every T-frame is also a D frame, and so this formula is also a T-validity.

$\mathcal{R}$ is *transitive* if $x\mathcal{R}y$ and $y\mathcal{R}z$ implies $x\mathcal{R}z$. A formula is K4-valid if it is true at all possible worlds of every transitive model. Typical K4 validities are $\Box X \supset \Box\Box X$ and $\Diamond\Diamond X \supset \Diamond X$.

A formula is S4-valid if it is true at all possible worlds of every model that is both reflexive and transitive. Typical S4 validities are $\Box X \equiv \Box\Box X$ and $\Diamond\Diamond X \equiv \Diamond X$.

$\mathcal{R}$ is *symmetric* if $x\mathcal{R}y$ implies $y\mathcal{R}x$. A formula is KB-valid if it is true at all possible worlds of every symmetric model. Typical KB validities are $X \supset \Box\Diamond X$ and $\Diamond\Box X \supset X$.

A formula is S5-valid if it is true at all possible worlds of every model that is reflexive, symmetric, and transitive. Notice that with S4 validity, two consecutive $\Box$ occurrences are equivalent to one. S5 has the property that every string of mixed $\Box$ and $\Diamond$ operators collapses to its last member. For example, $\Diamond\Box\Diamond\Box X \equiv \Box X$ is an S5 validity. This is a very strong property and may or may not be desirable—it depends on the application you have in mind. Investigators in game theory commonly assume the knowledge possessed by agents meets the S5 conditions, for example.

These are hardly all the conditions that have been imposed on models. Further, it is by no means the case that all modal logics that are of interest can be characterized by imposing conditions on frames. Nonetheless, this works for the modal logics that are most commonly used, and it provides a good entry point to a broader subject.

15.4 Axiomatics

Axiomatic formulations of modal logics were investigated long before possible world semantics came along, but today it is common to reverse the historical order. Among the first important results concerning modal semantics was Kripke's proof that several familiar axiomatically formulated modal logics corresponded to logics that had simple semantic characterizations. Here is an outline, for the record. We formulate our axiom systems using axiom *schemes*, without a rule of substitution.

Basic Axiom Schemes

- all classical tautologies (or enough of them)
- $\Box(X \supset Y) \supset (\Box X \supset \Box Y)$
- $\Diamond X \equiv \neg\Box\neg X$

Rules of Inference

- $\frac{X, X \supset Y}{Y}$ *(modus ponens)*
- $\frac{X}{\Box X}$ *(necessitation)*

As usual, an axiomatic proof is a finite sequence of formulas each of which is an instance of an axiom scheme or follows from earlier lines by one of the rules of inference. A proof proves its last line. Call the axiom system above K. It can be shown that the formulas provable in K are exactly the formulas that are K-valid, as defined semantically. We omit the proof here, but it is not difficult.

The modal logics discussed in Sect. 15.3 can be axiomatized by adding schemes to the system for K. We present this as a table. Many more logics can be handled in a similar way—these are merely meant to be representative.

Validity	Axiom Schemes
D	$\Box X \supset \Diamond X$
T	$\Box X \supset X$
K4	$\Box X \supset \Box\Box X$
S4	$\Box X \supset X, \Box X \supset \Box\Box X$
KD	$X \supset \Box\Diamond X$
S5	$\Box X \supset X, \Box X \supset \Box\Box X, X \supset \Box\Diamond X$

In addition to axiom systems many modal logics (including most common ones) have natural deduction systems, tableau (tree) proof systems, and Gentzen sequent calculi. However, there are many modal logics that have axiom systems but not (known) proof systems of these other kinds. Details can be tricky here.

15.5 Quantification

In classical logic, quantification is relatively straightforward. The language is enhanced by adding individual variables and relation symbols (and perhaps also constant and function symbols). Quantifiers, $\forall$ and $\exists$ are also added, along with relatively uncomplicated rules of formation. Classical models are introduced consisting of a non-empty domain and an interpretation of relation symbols by relations on that domain. Then machinery is introduced that has the effect of making a universally quantified formula true if every value from the domain makes the formula being quantified true, and an existentially quantified formula true if some value from the domain does so. There is basically only one way of doing all this.

Modally things are not as simple. If one understands quantification from an *actualist* point of view, quantifiers range over what actually exists (whatever that means), while from a *possibilist* point of view quantifiers range over what might exist. (Corresponding temporal versions are *presentist* and *eternalist*.) These differing conceptions can be represented using possible world models in a rather direct way. Of course non-empty domains are involved, and relation symbols are interpreted by relations on domains, but the interpretation might vary from possible world to possible world. For an actualist quantificational semantics we associate with each possible world of a model a non-empty domain, where different worlds can have different domains. These might be disjoint, overlap, or have some other more complex relationship. When evaluating a quantified formula at a possible world, the quantifier is understood to range over the things in the domain of that world only. For a possibilist quantificational semantics, we can think of these separate domains as being combined into one single set, with quantifiers ranging over it no matter at what possible world the truth of a quantified statement is being evaluated. Possibilist semantics validates $(\forall x)\Box A(x) \equiv \Box(\forall x)A(x)$ while actualist semantics does not; in fact it validates neither $(\forall x)\Box A(x) \supset \Box(\forall x)A(x)$ nor $\Box(\forall x)\Box A(x) \supset (\forall x)\Box A(x)$.

The different semantic versions are commonly referred to as *constant domain* (possibilist) and *varying domain* (actualist). It is not the case that one is right and the other wrong, but rather each represents a distinct notion of what quantification in a modal setting is about. In a sense, modal machinery does not dictate, but rather it tells you the consequences of a choice which is made for reasons of philosophical position, taste, or just convenience.

One can even have a formal language with both actualist and possibilist quantifiers, and investigate interactions between the two of them and modal operators. Alternately one could introduce an *existence predicate*, say $E(x)$. Then one could work with an underlying possibilist semantics, think of the things that E is true of at a possible world as the actual existents there, and understand actualist quantification as possibilist quantification relativized to E. The machinery is versatile.

A treatment of equality can be added to the formal machinery. This is closely related to what one imagines the 'things' in quantifier domains to be. Suppose we understand domains to consist of objects in some concrete sense, chairs, tables, beer mugs. Objects do not, so to speak, split apart, so we would want the validity of $(x = y) \supset \Box(x = y)$. Likewise neither do they combine so we would also want the validity of $\neg(x = y) \supset \Box\neg(x = y)$. It is rather straightforward to achieve this. On the other hand, we might think of our 'things' more intensionally. Are "the tallest tower in Paris" and "the tallest structure built by Eiffel" equal or not? They are in the actual world, but one could certainly create alternate possible worlds in which they are different, or even in which one but not the other is non-existent. These are *non-rigid* designators and their behavior is more complex then that of the objects mentioned earlier. It is possible to introduce quantification over such things too, but it requires more care and nuance.

The quantificational semantics described so far descends directly from the work of Saul Kripke. There is an alternative version due to David Lewis. According to Lewis things cannot exist in the domains of more than one possible world, but they can have *counterparts* in other worlds. Indeed, something existing in one world can have one, many, or no counterparts in an alternate world. Formally, models consist of possible worlds, an alternativeness relation, and a counterpart relation relating quantifier domains. Roughly speaking, something has a property necessarily at a world if at all alternative worlds, all its counterparts have the property. This is an extremely flexible semantics, with relationships to the Kripke-style version. Once again, the machinery provides an array of tools, but it is up to the user to decide what tools to make use of.

15.6 Concluding Comments

The formal machinery of possible worlds, and the accompanying proof procedures, are remarkably plastic. Different conditions on necessity and possibility can be accommodated. Different concepts of existence and quantification can be modeled. Different approaches to identity can be investigated. One should not think of the machinery as settling philosophical problems, but rather as clarifying them. A philosophical hypothesis that can be formalized is coherent. A formalization makes explicit the consequences of adopting that hypothesis. By itself no formalization can say a philosophical position is correct, merely that it is understandable. But to be understandable is almost as good as being true.

Recommended Readings

1. Blackburn, P., de Rijke, M., & Venema, Y. (2001). *Modal logic* (Tracts in theoretical computer science). Cambridge: Cambridge University Press. [A definitive treatment for those who want (much) more].
2. Blackburn, P., Van Benthem, J., & Wolter, F. (Eds.). (2007). *Handbook of modal logic*. Amsterdam: Elsevier.
3. Fitting, M. (2004). First-order intensional logic. *Annals of Pure and Applied Logic, 127*, 171–193.
4. Fitting, M. (2006, revised 2015). Intensional logic. *The Stanford Encyclopedia of Philosophy* (Summer 2015 Edition), Edward N. Zalta (ed.), https://plato.stanford.edu/archives/sum2015/entries/logic-intensional/ [Easily available, detailed, and free].
5. Fitting, M., & Mendelsohn, R. (1998). *First-order modal logic*. Dordrecht/Boston: Kluwer. [Aimed at philosophers, provides proofs, discusses tableau systems].
6. Garson, J. (2000, revised 2016). Modal logic. *The Stanford Encyclopedia of Philosophy* (Summer 2016 Edition), Edward N. Zalta (ed.), https://plato.stanford.edu/archives/sum2015/entries/logic-modal/
7. Garson, J. (2006). *Modal logic for philosophers*. Cambridge: Cambridge University Press.
8. Girle, R. (2000). *Modal logics and philosophy*. Teddington: Acumen.

9. Hughes, G. E., & Cresswell, M. J. (1996). *A new introduction to modal logic*. London: Routledge. [A classic].
10. Kripke, S. (1980). *Naming and necessity*. Cambridge: Harvard University Press. [Does not use formal machinery, but it is always in the background.].
11. Lewis, D. (1986). *On the plurality of worlds*. Oxford: Blackwell.

Chapter 16
Bivalence and Future Contingency

Gabriel Sandu, Carlo Proietti, and François Rivenc

Abstract This work presents an overview of four different approaches to the problem of *future contingency* and *determinism* in temporal logics. All of them are bivalent, viz. they share the assumption that propositions concerning future contingent facts have a determinate truth-value (true or false). We introduce *Ockhamism*, *Peirceanism*, *Actualism* and $T \times W$ semantics, the four most relevant bivalent alternatives in this area, and compare them from the point of view of their expressiveness and their underlying metaphysics of time.

16.1 Introduction

A major problem for *schoolmen* was to reconcile *divine foreknowledge* with *future contingency*, the latter being a prerequisite for human *free choice*. In modern times, when theological concerns have become less pressing, the so-called *future contingents problem* has shifted back to the more mundane Aristotelian question of how to accommodate the latter with the principle of *bivalence*, i.e. the thesis that all propositions, including those concerning future contingent facts, are either true or false. Both problems amount to the same if one assumes that only *true* propositions may be known (*nihil scitum nisi verum*) and that God has a *full science* about the future. But if one doesn't care much about God's omniscience, then this puzzle

G. Sandu
Department of Philosophy, History, Culture and Art Studies, University of Helsinki , Finland
e-mail: gabriel.sandu@helsinki.fi

C. Proietti (✉)
Department of Philosophy, Lund University, Lund, Sweden
e-mail: Carlo.proietti@fil.lu.se

F. Rivenc
University of Paris 1 Panthéon - La Sorbonne, Paris, France
e-mail: francois.rivenc@orange.fr

S. O. Hansson, V. F. Hendricks (eds.), *Introduction to Formal Philosophy*, Springer Undergraduate Texts in Philosophy, https://doi.org/10.1007/978-3-319-77434-3_16

becomes less urgent and one may just solve the dilemma by discarding one of its horns, i.e. the principle of bivalence. This is what most of the contemporary approaches to future contingency do (see [2, 10, 18]).[1]

Nonetheless, there are many reasons for preserving bivalence. Logical simplicity is perhaps the most instrumental of them. Others are related to language expressivity and the fact that non-bivalent approaches seem mostly unable to distinguish *simple* truths about the future from *settled* truths about it.

There are several bivalence-preserving solutions to the future contingents problem. Many of them were already known to the scholastics (see [13]). We will present those which have preserved their relevance up to nowadays: *Ockhamism*, *Peirceanism* (both formulated by Prior), *Actualism* and $W \times T$ semantics. All of them (except maybe Peirceanism) have been inspired by the medieval tradition. The advantage of contemporary tensed-logical approaches lies in their rigor and their comparability, mostly due to the fact that they all have the same semantic format.

There are no shared desiderata for a best choice among these solutions. Metaphysical considerations, tacit or explicit, about the "real" structure of time play a major role in the discussion and may easily turn into an "ideological" debate. Nonetheless, it is instructive to compare how the different approaches account not only for the openness of the future, but also for some additional intuitions about time and truth. One of these is *retrogradation of truth*. When one evaluates *ex post* a sentence like "there will be a sea-battle tomorrow", she is driven to assign a determinate truth-value to it and say, for example, that this sentence *was* true (in case a *sea battle* is actually taking place).[2] Related to retrogradation is a more general concern about expressivity: the formalism should account for the intuitive meaning of different tensed constructions in natural language. This means that the language and its semantics must be able to express the different truth-conditions of propositions like the following.

(1) There will be a sea-battle tomorrow.
(2) Laws of physics will hold tomorrow.
(3) There is a sea-battle, so it was true yesterday (but not settled) that there would be a sea-battle.
(4) The coin will come up heads. It is possible though, that it will come up tails, and then later it will come up tails again (though at that moment it could come up heads), and then, inevitably, still later it will come up tails yet again.[3]
(5) There is a sea-battle, but there could have been none.

In the next section we will present in detail Ockham's analysis (reconstructed by Prior) of the *future contingents problem*. In Sects. 16.3, 16.4, 16.5 and 16.6 we will

[1]The forerunner of all these solutions has been considered by many scholars (but not all of them) to be Aristotle in chapter IX of *On Interpretation*.

[2]This is MacFarlane's *determinacy intuition* (see [10], p. 322) as opposed to the *indeterminacy intuition* (future contingent sentences are neither true nor false at the moment of utterance).

[3]This example is taken from Belnap and Green [2].

introduce the four mentioned bivalent logical systems for solving it and discuss how they fare with respect to retrogradation and sentences like (1)–(5). A. Prior deserves the merit for having formulated two of them, in Chapter VII of his *Past, Present and Future*. He also deserves huge credit for introducing, in the same place, their common branching-time semantics, even though, as we will explain later, he did not grant them a major philosophical relevance.

16.2 Ockham's Argument

We freely adapt Ockham's version of the argument leading from *divine foreknowledge* to the *necessity of the future* as exposed in his *Tractatus de Praedestinatione* (1320 ca.). Ockham carefully reconstructs the argument in order to isolate two fundamental premises of it and to eventually reject one of them. The first premise is

(P1) Necessarily, if God knew in the past that p, then p.

which is on a par with the standard epistemic principle that knowledge implies truth, formulated by the medievals as *nihil scitum nisi verum*. The second premise is

(P2) If it has been the case that p, then *necessarily* it has been the case that p,

that we can represent in a temporal language[4] as

$$Pp \to \Box Pp$$

and which goes under the name of the *principle of necessitation of the past* (PNP): *quod fuit, non potest non fuisse*. The kind of necessity involved here is not *logical* but *historical* necessity, or necessity *per accidens* as the medievals called it: what has been the case is *(historically) necessary*, for it is not any longer possible for it not to have been the case.

If we apply (P2) to divine foreknowledge we get as a first conclusion:

(C1) If God knew in the past that p, then *necessarily* God knew in the past that p.

A third premise is derived from the modal schema, $\Box(p \to q) \to (\Box p \to \Box q)$, known as schema **K**, which states that "if a conditional and its antecedent are *necessary*, then the consequent is also *necessary*". A special instance of it is

[4]Our language consists of atomic formulas $p, q, \ldots$ (to be read as "pure" present-tense sentences such as "there is a sea battle") and recursively built on Boolean operators $\neg$ ("not"), $\wedge$ ("and"), $\vee$ ("or"), $\to$ ("if - then"), the temporal operators F ("it will be the case that") and P ("it has been the case that") and an additional operator $\Box$ to be read as "it is necessary that". We will also make use of dual operators like $G := \neg F \neg$ ("it will always be the case that"), $H := \neg P \neg$ ("it has always been the case that") and $\Diamond := \neg \Box \neg$ ("it is possible that").

(P3) If (necessarily, if God knew that p, then p), then (if necessarily God knew that p then necessarily p).

By *Modus Ponens* from (P1) and (P3) we obtain

If necessarily God knew that p then necessarily p.

and finally, by (C1) and *transitivity*:

If God knew that p, then necessarily p.

If p is a future-tensed statement, such as "I will be sitting tomorrow" (or Ockham's favorite example "Peter will be chosen"), then future-tensed statements are necessary and determinism follows – by assuming divine foreknowledge or bivalence, which here amount to the same.

Ockham points out that this argument lies essentially on **(PNP)**:

> This argument is based on the proposition that a singular proposition true about the past is necessary. Therefore if "this is white" is true now, "this will be white was true" is necessary. Consequently, it is necessary that it happens, and it cannot come about otherwise.[5]

Ockham's solution touches precisely on this point: he does not reject the principle but suggests a restriction of it. On the other hand, he maintains that God knows already, or from the beginning of time, which future events are going to happen. Again, since knowledge implies truth, saying that God knows that p will be the case amounts to saying that it is true now that p will be the case. Thus, propositions about the future already have a truth-value, even if we ignore which one, and the principle of *bivalence* is preserved. Indeed, throughout his *Tractatus* Ockham maintains that bivalence is the rationale of *divine foreknowledge*.

Ockham observes that one can block determinism and preserve the contingency of the future by limiting the universality of (PNP). This principle should only hold for the past and present tensed propositions which are not "equivalent" with any future tensed ones.[6] Formally speaking, we should not be allowed to derive, from propositions like

$$Pp \rightarrow \Box Pp$$

instances like

$$PFp \rightarrow \Box PFp$$

by unrestricted substitution. Blocking such a free substitution and invalidating formulas like the last one is precisely what qualifies a logical solution as Ockhamist.

---

[5] See Ockham [12, p. 99].

[6] Equivalence is to be understood in the same sense in which "it was the case yesterday that I will quit smoking in two days" is equivalent with "I will quit smoking tomorrow".

16.3 Prior's Ockhamism

Chapter VII of Prior [16] offers a first axiomatization of an *Ockhamist* temporal logic. One of the axioms of this system is the formula $p \to \Box p$, of which (PNP) is an instance. But this schema does not allow substitution of formulas containing the F operator, i.e. we may derive from it instances like $Pp \to \Box Pp$ but not $PFp \to \Box PFp$.[7]

In chapter VII we also find the first formulation of a *sound* semantics for this system: the nowadays universally adopted *tree-like models* for branching time. These models represent time as

> ... a line without beginning or end which may break up into branches as it moves from left to right (i.e. from past to future), though not the other way; so that from any point there is only one route to the left (into the past) but possibly a number of alternative routes to the right.[8]

From Prior's point of view this semantics is just a heuristic or pedagogical device and was not intended to constitute an alternative representation of the Ockhamist logic.[9] On the contrary, the proof-theoretic approach was meant to replace and absorb the fictional representation and *reification* of time which is carried by a model-theoretic representation. Nonetheless, as we said, these structures have nowadays become such a universal tool that, with the risk of being injust to Prior, we will base our analysis on them. We therefore define the *Ockhamist logic* **O** as the set of all formulas which are valid in the class of the *Ockhamist models* that we are going to present.

Central to Prior's definition is the notion of a *tree-like structure* $\mathcal{T}$, like the one depicted in Fig. 16.1, which is a pair $\langle T, <\rangle$, where T is a set of moments $m, m' \ldots$ and $<$ is a strict ordering relation (i.e. irreflexive, transitive and asymmetrical) over T, where the $<$-predecessors of any point m are *totally ordered* by $<$ and where the intuitive meaning of $m < m'$ is "*m precedes m'*". A *history* h is a *maximal chain* in T for the relation $<$. The set of histories $h_1, h_2, \ldots$ in T will be denoted by $H(T)$. Given a moment m, H_m will designate the set of all histories containing it. Note already that if $m < m'$ then $H_{m'} \subseteq H_m$.

A history h represents a specific and well determined course of events, relative to which every proposition is true or false at m, including those about the future. We can formally represent that by an evaluation function V, which assigns a subset of

---

[7]To be precise, Prior uses here a more expressive temporal language with metric operators F_n ("it will be the case in n intervals of time") and P_m ("it was the case m intervals ago"), where n and m are two quantifiable variables to be interpreted with (rational or real) non-negative numbers measuring intervals of time. For the sake of simplicity we will avoid using metric operators, since F, P and $\Box$ are sufficient for the points we need to make.

[8]See Prior [16, p. 126].

[9]This is probably one reason why Prior does not even face the question of *completeness*.

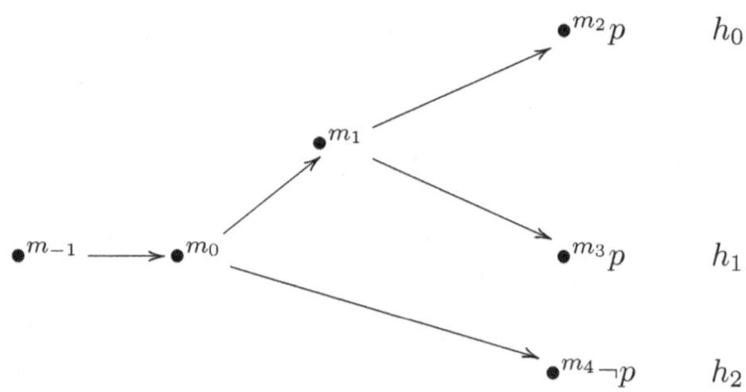

Fig. 16.1 A model for branching time

$T \times H(T)$ to every propositional variable p (see Fig. 16.1). A further requirement is that, given a moment m, V does not varies with the different histories in H_m, i.e. we have

(Uniqueness) $\langle m, h\rangle \in V(p)$ if and only if for all $h' \in H_m$, $\langle m, h'\rangle \in V(p)$

We can then define an *Ockhamist model* $\mathcal{M} = \langle T, <, V\rangle$ for our tensed language by extending V in the following way:

$\mathcal{M}, \langle m, h\rangle \models p$ iff $\langle m, h\rangle \in V(p)$
$\mathcal{M}, \langle m, h\rangle \models \neg\phi$ iff $\mathcal{M}, \langle m, h\rangle \not\models \phi$
$\mathcal{M}, \langle m, h\rangle \models \phi \wedge \psi$ iff $\mathcal{M}, \langle m, h\rangle \models \phi$and$\mathcal{M}, \langle m, h\rangle \models \psi$
$\mathcal{M}, \langle m, h\rangle \models P\phi$ iff $\exists m' < m$ such that $\mathcal{M}, \langle m', h\rangle \models \phi$
$\mathcal{M}, \langle m, h\rangle \models F\phi$ iff $\exists m' > m$ such that $\mathcal{M}, \langle m', h\rangle \models \phi$
$\mathcal{M}, \langle m, h\rangle \models \Box\phi$ iff $\forall h'(h' \in H_m \Rightarrow \mathcal{M}, \langle m, h'\rangle \models \phi)$

Evaluating a future tensed proposition Fp w.r.t. a moment and a history amounts to checking if p is satisfied "later on" in the same history. The general idea behind this is that when we talk about the future we actually pick a *prima facie* course of events h as being the most plausible candidate among all possible futures. Historical necessity is instead equated with "truth in all histories" and, given the *uniqueness* condition, it is easy to check that present and past-tensed propositions (e.g. p, Pp, PPp etc.) are, if true, necessary.

O can easily distinguish among *contingent* and *settled* truths about the future. Indeed, contingent sentences like Fp ("there will be a sea-battle") may very well be true but not *necessary*: in our model $\langle m_0, h_0\rangle \models Fp$ but $\langle m_0, h_0\rangle \models \neg\Box Fp$. Ockhamist semantics also respects the intuition that some sentences about the future, like (2) (see introduction), can be true in a *stronger* sense, i.e. also necessary, when they hold in all possible branches. This is a fortiori the case of logical tautologies $\top$: both $\top$ and $\Box\top$ are valid in Ockhamist models, for tautologies are true at every pair $\langle m, h\rangle$.

It is easy to verify that (PNP) does not hold in general in this semantics and in particular, as claimed by Ockham, it fails for sentences containing a reference to the future. Indeed, as the reader may check, in the model of Fig. 16.1 at the moment m_0 we have $\langle m_0, h_0\rangle \models PFp$ but $\langle m_0, h_0\rangle \models \neg\Box PFp$. Nevertheless, in accordance

Fig. 16.2 Heads and tails

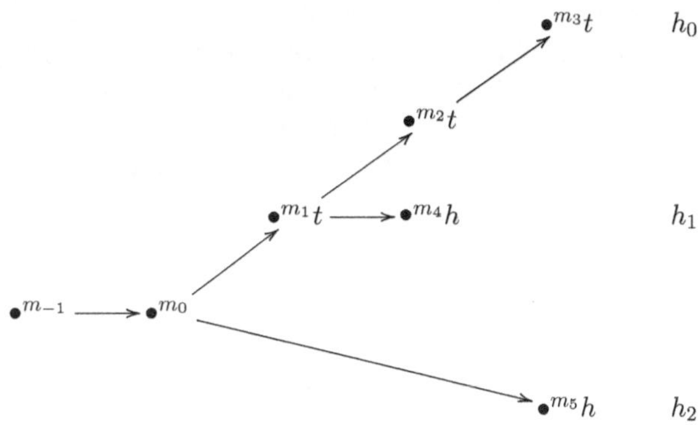

with Ockham, (PNP) is valid for propositions which are "not equivalent to future-tensed ones", in our case those not containing any operator F.[10]

We may notice that *retrogradation of truth* is secured by the fact that $p \to PFp$ is valid in the Ockhamist semantics. More generally, we can easily account for sentences like (3) in Sect. 16.1, which can be translated as $p \wedge PFp \wedge P\neg\Box Fp$ and which are true at $\langle m_2, h_0 \rangle$ in our model.

Complex propositions like (4) make plural references to different possible futures at different points in the tree. Here too, Ockhamism is powerful enough to express its truth conditions. For example, (4) can be translated by the formula $Fh \wedge \Diamond F(t \wedge \Diamond Fh \wedge F(t \wedge \Box Ft))$.[11] This formula is satisfied at $\langle m_0, h_2 \rangle$ by the model in Fig. 16.2.

The intuition behind (5) of Sect. 16.1 is that we should also be able to refer to *this* precise moment in courses of events which are, properly speaking, no more possible: this is the sense of a counterfactual with a false antecedent. In the Ockhamist semantics this can be expressed in many cases by moving back and forth along the branches. The truth conditions of (5) can be "mimicked" by $p \wedge P\Diamond F\neg p$,[12] which is indeed satisfied in the model of Fig. 16.1 at $\langle m_2, h_0 \rangle$. Nonetheless, not all counterfactuals seem to be expressible, as (5), by simple combinations of F, P and $\Box$. We will come back to this point in Sect. 16.6.

To resume, Prior's Ockhamism is a very expressive framework that enables the distinction between contingent and settled propositions about the future. But there is a major philosophical objection against it, which concerns the notion of a *prima facie* course of events. Since, at m, all histories in H_m are equally possible, it is not clear how one should be able to single out any one of them. However, according to many, when talking about future events, we need to refer to *the* actual future. But in

[10]Finer-grained distinctions are induced in Prior's actual system by the use of metric operators.

[11]Where h means "the coin lands head" and t stands for "the money lands tail".

[12]This translation is not completely faithful. A metric language can better express (5) with $p \wedge P_n \Diamond F_n p$.

this semantics (and also in Prior's view) there is no such a thing. The main problem with it seems to be its neutrality between two opposite views: one according to which there is no designated course of events, and the other which, on the contrary, allows one to refer to the actual course of events. Peirceanism and Actualism are meant to bear these opposite stands in a more radical way.

16.4 The Peircean System

Restricting (PNP) is not the only way to block arguments for determinism. As one may evince from Ockham's argument, it is also crucial that God is able to know in the past what will happen later. This is only possible if we assume that $p \rightarrow PFp$ is valid. The latter is an uncontroversial principle of minimal temporal logics, but not of Prior's *Peircean logic*.[13] The Peircean system **P** was favored by Prior over **O** as the only one which fleshes out the intuition that the future is not "real" until it becomes present,[14] the only exception being represented by that parcel of the future which is already present in its causes.

Prior introduces the idea behind **P** as a variant of the traditional solution (rejecting bivalence to save indeterminism), where a different interpretation of the F operator plays, in some peculiar sense, the role usually ascribed to a third truth-value or a *truth-value gap*. A *Peircean model* is easily obtained from an Ockhamist one by modifying the clause for F as follows:

$$\mathcal{M}, \langle m, h\rangle \models F\phi \text{ iff } \forall h'((h' \in H_m) \Rightarrow \exists m'(m < m' \wedge \mathcal{M}, \langle m', h'\rangle \models \phi))$$

Again, we identify **P** with the set of all formulas valid in the class of Peircean models. Fp now means something like "given any course of events, it will be the case that p". The intuition is that, speaking about the future, it does not make sense to pick up any *prima facie* designated history, since all possible futures stand on a par from the present standpoint. F has now the same meaning as the expression $\Box F$ in **O**: indeed **P** can be seen as a fragment of it. Thereby, **P** is also bivalent and the law of excluded middle holds also for future contingent propositions, i.e. $Fp \vee \neg Fp$ is valid. But, contrary to the Ockhamist semantics, $Fp \vee F\neg p$ can very well fail as well as $\neg Fp \rightarrow F\neg p$ (but its converse holds).[15] The Peircean solution has some counterintuitive backups: future "necessary" propositions like (2) are still true, but future contingent ones like (1) are now simply *false* (consider the model

[13]See Prior [16] chap. VII p. 132.

[14]See Prior [15].

[15]It should also be noticed that the Peircean sense of "it will always be the case that" is no more expressed by the combination $\neg F \neg$, thus G has to be defined as a new primitive operator by the following clause

$$\mathcal{M}, \langle m, h\rangle \models G\phi \text{ iff } \forall h' \forall m'((h' \in H_m \wedge m < m') \Rightarrow \mathcal{M}, \langle m', h'\rangle \models \phi).$$

in Fig. 16.1 as a Peircean model). Nonetheless, one may still distinguish between necessarily false propositions, those ϕs for which both $\neg F\phi$ and $F\neg\phi$ are true, and contingently false ones, those ϕs for which $\neg F\phi$ is true but $F\neg\phi$ is not.

It is easy to check that $p \rightarrow PFp$ is no more valid. Thus *retrogradation of truth* is undermined. In general, propositions like (3), saying that something "was going to be the case" are regarded simply as (bad) *façons de parler* to express the fact that something is *now* the case. Similar problems arise for (4) and (5) and many other examples. In general, since **P** is a proper fragment of **O**, it seems that the Peircean is committed to a "deflationist" view about temporal truth, according to which many sentences we commonly express in natural language are simply misleading paraphrases.

16.5 Actualism and *TRL* Semantics

All along his *Tractatus* Ockham seems to presuppose that there is, among all possible future courses of events, a designed *actual future*, a sort of *thin red line* among all other branches,[16] that God already knows from all eternity. This designed history should be, contrary to Prior's claims, not only a *prima facie* one. Adherence to Ockham's word is not the only reason to stipulate such a special history. It seems that we often refer to this unique entity in order to make sense of peculiar sentences such as "Tomorrow I will quit smoking, even if all evidence speaks against that".[17]

The Actualist view has been encoded by means of the so-called TRL semantics (from "thin red line"). There are different possible ways of defining a TRL semantics in a branching structure (see Barcellan and Zanardo [1] and Braüner et al. [3]), but all of them must fulfill some natural requirements. First of all, looking at sentences like (4), it seems clear that a model should not only specify a designated branch corresponding to "the true history", but also many others: one for every counterfactual moment t. Following Barcellan and Zanardo [1], we define a TRL semantics on the basis of an Ockhamist model via a function $\mathcal{A}(t)$ from T to $H(T)$, which picks the actual future at a moment m. Then we define an actual future operator f_A with the following clause:

$$\mathcal{M}, \langle m, h\rangle \models f_A\phi \text{ iff } \exists m' \in \mathcal{A}(m)(m < m' \wedge \mathcal{M}, \langle m', \mathcal{A}(m)\rangle \models \phi)$$

This function is supposed to respect some natural constraints, the most immediate being

TRL1 $\quad m \in \mathcal{A}(m)$

and the second being the condition of "coherence"

TRL2 $\quad \forall m_1, m_2(m_1 < m_2 \rightarrow \mathcal{A}(m_1) = \mathcal{A}(m_2))$

[16] This famous expression was coined by Belnap and Green [2].

[17] For a more accurate discussion of this point see Hasle and Øhrstrom [6] and Øhrstrom et al. [14].

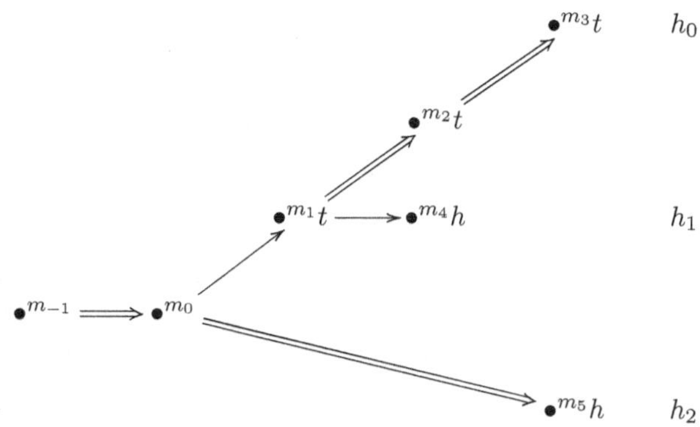

Fig. 16.3 Failure of $\phi \rightarrow Pf_A\phi$. Double arrows indicate each moment's actual future

According to Belnap and Green [2], such conditions generate some serious problems of inconsistency for the Actualist conception. If we put together **TRL1** and **TRL2** the order $<$ is forced to be linear. On the other hand, they claim, if we discard **TRL2** we obtain "unreasonable results", e.g. we invalidate many natural principles such as (a) $PP\phi \rightarrow P\phi$, (b) $f_A f_A\phi \rightarrow f_A\phi$ and (c) $\phi \rightarrow Pf_A\phi$.[18]

TRL2 is actually a strong coherence condition; Barcellan and Zanardo [1] showed that we can instead reasonably opt for the weaker

TRL2* $\quad \forall m_1, m_2(m_1 < m_2 \wedge m_2 \in \mathcal{A}(m_1) \rightarrow \mathcal{A}(m_1) = \mathcal{A}(m_2))$

and escape most of the "unreasonable results". They also add the further condition

TRL3 $\quad$ there exists an m^* such that for all $m < m^*$, $\mathcal{A}(m) = \mathcal{A}(m^*)$

where $\mathcal{A}(m^*)$ defines the unique "real" history of the model.[19] It is possible to check that this definition preserves many temporal laws such as (a) and (b). The formula (c) $\phi \rightarrow Pf_A\phi$ is not valid instead – as an example, consider the failure of $t \rightarrow Pf_At$ at m_1 in the model in Fig. 16.3 – but is nonetheless satisfied at any moment of $\mathcal{A}(m^*)$. An additional problem for this semantics is that it cannot properly block (PNP) for, as one may easily check,

$$f_A\phi \rightarrow \Box f_A\phi$$

is a valid formula, as well as its converse. From this point of view, TRL semantics are not completely Ockhamist. In order to make (PNP) fail and express (1)–(5) one should enrich the language with other future tense operators.[20]

To summarize, the major "logical" inconvenience of the Actualist operator f_A is that when we combine it with P and $\Box$ many "natural" principles seem to fail

---

[18]See Belnap and Green [2] p. 380.

[19]For a proof of uniqueness see Barcellan and Zanardo [1] p. 5.

[20]Barcellan and Zanardo use the peircean operators of Sect. 16.2 as primitives.

and we have to recur to other future operators to adjust them. But it is fair to notice that failures of "intuitive" principles are not specific of f_A and that they at least do not seem to lead to an "inconsistency" of the Actualist conception.[21] From a more metaphysical standpoint, the most common objection to Actualism, in this or other forms, is that it involves a commitment to facts "that do not supervene upon any physical, chemical or psychological states of affairs" [2].

16.6 T × W Semantics

Branching time semantics are not the only possible "technical" solution for preserving future contingency and bivalence. Another option is represented by $T \times W$ semantics, introduced in Thomason [19].[22] Whereas branching-time is based on the idea of *overlapping* histories, $T \times W$ starts from the intuition of there being a plurality of separated possible courses of events (or worlds) which may have "equivalent" past histories up to a point and *diverge* afterwards.[23] The models of Fig. 16.4 represent this difference.

For a formal definition, we need a set T of moments, an irreflexive linear order $<$ on it, a set W of possible worlds and a family $\{\sim_t \mid t \in T\}$ of equivalence relations among them, intuitively denoting sameness up to a certain point in time t. A frame is a tuple $\langle T \times W, <, \{\sim_t\}_{t \in T}\rangle$ where

- $T \times W$ is the set of $\langle t, w\rangle$ such that $t \in T$ and $w \in W$
- for all $t \in T$, $\sim_t$ is an equivalence relation
- for all $t' \in T$, if $w \sim_t w'$ and $t' < t$ then $w \sim_{t'} w'$

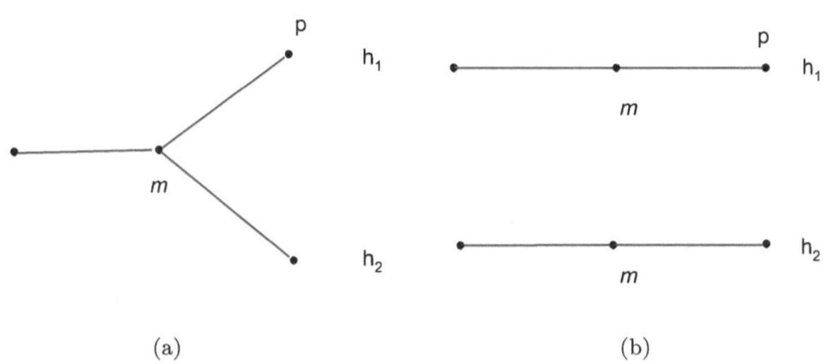

Fig. 16.4 Ockhamist models. (**a**) A tree-like model. (**b**) A $T \times W$ model

[21] For a more articulated defence of Actualism see Øhrstrom [13].

[22] Complete logical systems for this semantics have been formulated later by von Kutschera [20] and Di Maio and Zanardo [5].

[23] For the notions of *overlap* and *divergence* see the famous Lewis [9] pp. 198–209.

Given a valuation V, assigning to every p a subset of $W \times T$, a model is obtained by expanding V to a satisfaction relation in the usual way for Boolean and temporal operators (e.g. $\langle t, w\rangle \models F\phi$ iff for some $t' > t$, $\langle t', w\rangle \models \phi$) and defining the $\Box$-clause as:

$$\langle t, w\rangle \models \Box\phi \text{ iff for all } w' \text{ such that } w \sim_t w', \langle t, w'\rangle \models \phi$$

Necessity at $\langle t, w\rangle$ means truth at the "same" moment in all other equivalent histories. We get the Ockhamist notion of *historical* necessity by an adequate specification of $\sim_t$ as "sharing the same past up to t" i.e.

$$w \sim_t w' \text{ iff for all } t' \leq t, \langle t', w'\rangle \text{ and } \langle t', w\rangle \text{ satisfy the same propositional letters.}$$

It is relevant to notice that under some specific conditions a branching *Ockhamist* model can be *tranformed* into a $T \times W$-model in a truth preserving way.[24] This happens when we have a *synchronized tree*, i.e. a tree whose branches are all *isomorphic*.[25] Under this condition the $T \times W$ semantics is at least as expressive as the Ockhamist semantics of Sect. 16.3, i.e. we can account in the same way for the truth conditions of (1)–(5), and even more.[26] Indeed, in $T \times W$ necessity operators are defined via a more arbitrary equivalence relation among histories, which does not force the *uniqueness* condition (see Sect. 16.3). Therefore, one is free to define new necessity and possibility operators by relaxing or making more accurate the equivalence relations. Relaxing the equivalence relation allows to quantify over histories that diverge even before a given moment m. By this means, it is possible to handle propositions like "for all that I know it could have been raining last night", where the construction "for all that I know ..." is to be read as an *epistemic*

[24]Full equivalence between the logic of general Ockhamist structures and $T \times W$ does not hold. A famous counterexample is provided by the *Burgess formula* (see Burgess [4] and Reynolds [17]), which is valid for the first class but not for the second.

[25]More precisely, a *synchronized* tree is a tree-like structure where it is possible to define a partition I (the "instants") of the set T that satisfies the following conditions (see also Wölfl [21]):

(a) For every $i \in I$ and every $h \in H(T)$ there is exactly one $m_{i,h} \in T$ with $m_{i,h} \in i \cap h$
(b) For all $i, i' \in I$ and all $h, h' \in H(T)$, from $m_{i,h} < m_{i',h}$ it follows that $m_{i,h'} < m_{i',h'}$

Given a *synchronized* Ockhamist model $\mathcal{T} = \langle T, <, V\rangle$ we can define a $T \times W$ model $\mathcal{T}' = \langle T' \times W', <', \{\sim_t\}_{t\in T}, V'\rangle$ by taking:

- $T' = I$ and $W' = H(T)$
- $i <' i'$ iff $m_{i,h} < m_{i',h}$ for some $h \in H(T)$
- $h \sim_i h'$ iff $m_{i,h} = m_{i,h'}$
- $\langle i, h\rangle \in V'(p)$ iff $\langle m_{i,h}, h\rangle \in V(p)$

and it is straightforward to check that $\mathcal{T}, \langle m, h\rangle \models \phi$ if and only if $\mathcal{T}', \langle m, h\rangle \models \phi$.

[26]The situation is more complex if the tree is not synchronized. For an accurate study of the relationships between branching-time semantics and $T \times W$ see Wölfl [21].

possibility operator. Here, indeed, we are driven to consider as epistemic alternatives more histories than those which share the same past.[27] The case of *counterfactuals* presents analogous features.[28]

16.7 Conclusions

We have presented four logical systems which deal with a perennial philosophical problem: the problem of *future contingents.* Apart from tackling the problem in a rigorous way, the four logical approaches have the same model-theoretical format. This makes the solutions comparable and allows us to see what are the gains and losses in terms of expressivity, the relation between future contingents and the principle of bivalence, and the metaphysical commitments we make.

The system $T \times W$ has at least the same expressive power as the Ockhamist semantics, but it has received scarce attention or has even been fiercely opposed. Thomason himself dismissed it in the very same paper in which he introduced it [19]. Most of the reasons for this attitude are grounded in metaphysical considerations. Whereas branching time is regarded as an almost an adequate representation of McTaggart's A-series conception, $T \times W$ is instead associated with the B-series conception and seems to commit to a *reification* of time.[29] Moreover, quantification over *non actual and non overlapping* histories is seen by many as an additional commitment to modal *realism* (i.e. the philosophical thesis that non actual worlds are *real* or *exist* on a par with the actual one). To many, overlap seems more faithful to an intuitive notion of *causality*: at any moment m there is just one past that we cannot change or influence and many possible futures we can "act upon" and "decide which one to take". In $T \times W$, at any point, there is just one future; contingency and

[27] See also Iacona [7].

[28] The same "redefinitions" of necessity and possibility operators can of course be carried out, in principle, also in an Ockhamist model. However, this goes against one of the philosophical motivations behind the branching time semantics, according to which all tensed constructions ought to be expressed with reference to points in time that are connected to the present point of evaluation – by some (back and forth) path over the temporal tree. This requirement may be too restrictive when we need to consider, e.g., fictional alternatives or histories diverging in the far past.

[29] The notions of A-series and B-series were introduced by [11]. The A-series conception of time, also called the dynamic view, resumes the way we experience time by being "in a flux" and opens up to *presentism* – a view that McTaggart himself did not endorse – where only the (constantly changing) "now" properly exists. According to this conception past, present, and future tenses are primitive concepts for referring to events in time. Other temporal concepts such as instants in time and the earlier-later relation between them, are to be derived from the formers. On the other hand, according to the B-series conception – which accounts for a "bird-eye view" of time and according to which the entire series of instants exists – instants and their earlier-later relation are the primitive concepts and tenses are derived from them.

causal influence on the future seem to be definable only in terms of a counterfactual dependence,[30] i.e. in terms of what would happen if the actual course of events were different.

In response to the criticisms against $T \times W$ one may point out that it is not clear how the choice of a particular semantics should commit us to a certain ontology. Additionally, it does not seem that other bivalent approaches like Ockhamism and Actualism are safe from these problems: if bivalence holds and truth is relative to a particular course of events then we are just one step far from admitting that other courses of events are fictional ones, and that the metaphor of branching seems just an unsuccessful compromise. $T \times W$ keeps the order of truth and the order of causality on two separate plans. Peirceanism, with a radically different definition of truth for future tensed propositions, seems to be the only radical alternative. However, defenders of Peirceanism face at least two burdens: they should deal with a less expressive language and have to find a justification for the strange asymmetry which makes it that future contingents are just *false*. Non-bivalent approaches admitting truth-value gaps for future contingents, as the one defined in [18] and which Prior hoped for,[31] seem to be the only possible way to fully preserve symmetry between truth and falsity.

References and Recommended Readings

1. Barcellan, B., & Zanardo, A. (1999). Actual futures in Peircean branching-time logic. citeseer.ist.psu.edu/326930.html.
2. Belnap, N., & Green, M. (1994). Indeterminism and the thin red line. In J. Tomberlin (Ed.), *Philosophical perspectives 8: Logic and language* (pp. 217–244). Atascadero: Ridgeview Publishing Company.
3. Braüner, T., Øhrstrom, P., & Hasle, F. (2000) Determinism and the origins of temporal logic. In H. Barringer, M. Fisher, D. Gabbay, & G. Gough (Eds.), *Advances in temporal logics* (pp. 185–206). Dordrecht: Kluwer.
4. Burgess, J. (1978). The unreal future. *Theoria, 44*(3), 157–179.
5. Di Maio, M., & Zanardo, A. (1998). A Gabbay-rule free axiomatization of T×W validity. *Journal of Philosophical Logic, 27*(5), 435–487.
6. Hasle, P., & Øhrstrom, P. (1995). *Temporal logic from ancient ideas to artificial intelligence*. Dordrecht: Kluwer Academic.
7. Iacona, A. (2009). Commentary: Combinations of tense and modality by R. Thomason. *Humana Mente, 8*, 185–190.
8. Lewis, D. (1979). Counterfactual dependence and time's arrow. *Nous, 13*, 455–76.
9. Lewis, D. (1986). *On the plurality of worlds*. Oxford: Blackwell.
10. MacFarlane, J. (2003). Future contingents and relative truth. *The Philosophical Quarterly, 53*, 321–336.
11. McTaggart, J. M. E. (1908). The unreality of time. *Mind, 187*, 457–474.
12. Ockham, W. (1983). *Predestination, God's foreknowledge, and future contingents*. Indianapolis: Hackett.

[30] See [8].

[31] See [16] p. 137.

13. Øhrstrom, P. (2009). In defence of the thin red line: A case for ockhamism. *Humana mente, 8*, 17–32.
14. Øhrstrom, P., Hasle, P., & Braüner, T. (1998). Ockhamistic logics and true futures of counterfactual moments. In *Proceedings of the Fifth International Workshop on Temporal Representation and Reasoning (TIME-98)* (Vol. 18).
15. Prior, A. N. (1966). Postulates for tense-logic. *American Philsophical Quarterly, 3*, 153–161.
16. Prior, A. N. (1967). *Past, present and future*. Oxford: Oxford University Press.
17. Reynolds, M. (2002). Axioms for branching time. *Journal of Logic and Computation, 12*, 679–697.
18. Thomason, R. (1970). Indeterminist time and truth-value gaps. *Theoria, 36*(3), 264–281.
19. Thomason, R. H. (1984). Combinations of tense and modality. In F. G. D. Gabbay (Ed.), *Handbook of philosophical logic* (pp. 205–234). Dordrecht: Kluwer Academic.
20. von Kutschera, F. (1997). $T \times W$ completeness. *Journal of Philosophical Logic, 26*, 241–250.
21. Wölfl, S. (2002). Propositional Q-logic. *Journal of Philosophical Logic, 31*, 387–414.

Part IV
Epistemology

Chapter 17
Epistemic Logic and Epistemology

Wesley H. Holliday

Abstract This chapter provides a brief introduction to propositional epistemic logic and its applications to epistemology. No previous exposure to epistemic logic is assumed. Epistemic-logical topics discussed include the language and semantics of basic epistemic logic, multi-agent epistemic logic, combined epistemic-doxastic logic, and a glimpse of dynamic epistemic logic. Epistemological topics discussed include Moore-paradoxical phenomena, the surprise exam paradox, logical omniscience and epistemic closure, formalized theories of knowledge, debates about higher-order knowledge, and issues of knowability raised by Fitch's paradox. The references and recommended readings provide gateways for further exploration.

17.1 Introduction

Once conceived as a single formal system, epistemic logic has become a general formal approach to the study of the structure of knowledge, its limits and possibilities, and its static and dynamic properties. In the twenty-first century there has been a resurgence of interest in the relation between epistemic logic and epistemology [6, 19, 34, 37, 41]. Some of the new applications of epistemic logic in epistemology go beyond the traditional limits of the logic of knowledge, either by modeling the dynamic process of knowledge acquisition or by modifying the representation of epistemic states to reflect different theories of knowledge. In this chapter, we begin with basic epistemic logic as it descends from Hintikka [22] (Sects. 17.2 and 17.3), including multi-agent epistemic logic (Sect. 17.4) and doxastic logic (Sect. 17.5),

W. H. Holliday (✉)
Department of Philosophy, University of California, Berkeley, CA, USA
e-mail: wesholliday@berkeley.edu

S. O. Hansson, V. F. Hendricks (eds.), *Introduction to Formal Philosophy*, Springer Undergraduate Texts in Philosophy, https://doi.org/10.1007/978-3-319-77434-3_17

followed by brief surveys of three topics at the interface of epistemic logic and epistemology: epistemic closure (Sect. 17.6), higher-order knowledge (Sect. 17.7), and knowability (Sect. 17.8).

17.2 Basic Models

Consider a simple formal language for describing the knowledge of an agent. The sentences of the language, which include all sentences of propositional logic, are generated from atomic sentences $p, q, r, \ldots$ using boolean connectives $\neg$ and $\wedge$ (from which $\vee$, $\rightarrow$, and $\leftrightarrow$ are defined as usual) and a knowledge operator K.[1] We write that the agent knows that p as Kp, that she does *not* know that p *and* q as $\neg K(p \wedge q)$, that she knows *whether or not* q as $Kq \vee K\neg q$, that she knows that she does not know that *if* p, *then* q as $K\neg K(p \rightarrow q)$, and so on.

We interpret the language using a picture proposed by Hintikka [22], which has since become familiar in philosophy. Lewis [27] describes a version of the picture in terms of *ways the world might be*, compatible with one's knowledge:

> The content of someone's knowledge of the world is given by his class of *epistemically accessible* worlds. These are the worlds that might, for all he knows, be his world; world W is one of them iff he knows nothing, either explicitly of implicitly, to rule out the hypothesis that W is the world where he lives. (27)

The first part of the picture is that whatever is true in at least one of the agent's epistemically accessible worlds *might, for all the agent knows, be true in his world,* i.e., he does not know it to be false. The second part of the picture is that whatever is true in *all* of the agent's epistemically accessible worlds, the agent knows to be true, perhaps only implicitly (see [27, §1.4]).

Here we talk of "scenarios" rather than worlds, taking w, v, u, ... to be scenarios and W to be a *set* of scenarios.[2] For our official definition of epistemic accessibility, call a scenario v epistemically accessible from a scenario w iff everything the agent knows in w is true in v [41, §8.2].

Consider an example. A spymaster loses contact with one of his spies. In one of the spymaster's epistemically accessible scenarios, the spy has defected (d). In another such scenario, the spy remains loyal ($\neg d$). However, in all of the spymaster's epistemically accessible scenarios, the last message he received from the spy came a month ago (m). Hence the spymaster knows that the last message he received from the spy came a month ago, but he does not know whether or not the spy has defected, which we write as $Km \wedge \neg(Kd \vee K\neg d)$.

---

[1]To reduce clutter, I will not put quote marks around symbols and sentences of the formal language, trusting that no confusion will arise.

[2]In our formal models, "scenarios" will be unstructured points at which atomic sentences can be true or false. We are not committed to thinking of them as Lewisian possible worlds.

We assess the truth of such sentences in a *model* $\mathcal{M} = \langle W, R_K, V \rangle$, representing the epistemic state of an agent.[3] W is a nonempty set, the set of scenarios. R_K is a binary relation on W, such that for any w and v in W, we take wR_Kv to mean that scenario v is epistemically accessible from scenario w. Finally, V is a valuation function assigning to each atomic sentence p a subset of W, $V(p)$, which we take to be the set of scenarios in which p holds.

Given our definition of epistemic accessibility, and the fact that everything an agent *knows* is true, our intended models are ones in which R_K is *reflexive*: wR_Kw for all w in W. We call such models *epistemic models*.

Let φ and ψ be any sentences of the formal language. An atomic sentence p is *true* in a scenario w in a model $\mathcal{M} = \langle W, R_K, V \rangle$ iff w is in $V(p)$; $\neg\varphi$ is true in w iff φ is *not* true in w; $\varphi \wedge \psi$ is true in w iff φ and ψ are true in w; and finally, the modal clause matches both parts of the picture described above:

(MC) $K\varphi$ is true in w iff φ is true in every scenario v such that wR_Kv.

We say that a sentence is *satisfiable* iff it is true in some scenario in some model (otherwise *unsatisfiable*) and *valid* iff it is true in all scenarios in all models. We may also relativize these notions to a restricted class of models, such as the intended class of epistemic models in which R_K is reflexive. A sentence is *satisfiable in the class* iff it is true in some scenario in some model in the class and *valid over the class* iff it is true in all scenarios in all models in the class.

Figure 17.1 displays a simple epistemic model for the spymaster example, where we draw a circle for each scenario (with all atomic sentences true in the scenario indicated inside the circle), and we draw an arrow from a scenario w to a scenario v iff wR_Kv. Observe that $Km \wedge \neg(Kd \vee K\neg d) \wedge d$ is true in w_1: d is true in w_1 by description; yet neither Kd nor $K\neg d$ is true in w_1, because neither d nor $\neg d$ is true in all scenarios epistemically accessible from w_1, namely w_2 and w_1 itself; however, Km is true in w_1, since m is true in all scenarios epistemically accessible from w_1. We could construct a more complicated epistemic model to represent the spymaster's knowledge and ignorance of other matters, but this simple model suffices to show that $Km \wedge \neg(Kd \vee K\neg d) \wedge d$ is satisfiable.

Let us now consider a sentence that is unsatisfiable in epistemic models. In a twist on Moore's [29] paradox, Hintikka [22, §4.17] considers what happens if I tell you something of the form *you don't know it, but the spy has defected*, translated as $d \wedge \neg Kd$. This may be true (as in w_1), but as Hintikka observes, you can never know it. You can never know that the spy has defected but you don't know it.

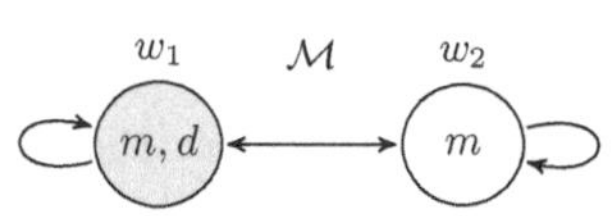

Fig. 17.1 A simple epistemic model

[3]Hintikka presented his original formal framework somewhat differently. Such details aside, we use the now standard relational structure semantics for normal modal logics.

Formally, $K(d \wedge \neg Kd)$ cannot be true in any scenario in an epistemic model; it is unsatisfiable, as we show in Sect. 17.3 below. It follows that $\neg K(d \wedge \neg Kd)$ is true in every scenario, so it is valid over epistemic models.

Since we take $wR_K v$ to mean that everything the agent knows in w is true in v, one might sense in (MC) some circularity or triviality. As a technical matter, there is no circularity, because R_K is a primitive in the model, not defined in terms of anything else. As a conceptual matter, we must be clear about the role of the epistemic model when paired with (MC): its role is to represent the content of one's knowledge, *what one knows*, not to analyze *what knowledge is* in terms of something else.[4] (As we discuss in Sects. 17.6 and 17.7, with richer epistemic structures we can also formalize such analyses of knowledge.) Finally, (MC) is not trivial because it is not neutral with respect to all theories of knowledge.[5]

---

[4]It is important to draw a distinction between epistemic accessibility and other notions of indistinguishability. Suppose that we replace R_K by a binary relation E on W, where our intuitive interpretation is that wEv holds "iff the subject's perceptual experience and memory" in scenario v "exactly match his perceptual experience and memory" in scenario w [28, 553]. Suppose we were to then define the truth of $K\varphi$ in w as in (MC), but with R_K replaced by E. In other words, the agent knows φ in w iff φ is true in all scenarios that are experientially indistinguishable from w for the agent. (Of course, we could just as well reinterpret R_K in this way, without the new E notation.) There are two conceptual differences between the picture with E and the one with R_K. First, given the version of (MC) with E, the epistemic model with E does not simply represent the content of one's knowledge; rather, it commits us to a particular view of the conditions under which an agent has knowledge, specified in terms of perceptual experience and memory. Second, given our interpretation of E, it is plausible that E has certain properties, such as *symmetry* (wEv iff vEw), which are questionable as properties of R_K (see Sect. 17.7). Since the properties of the relation determine the valid principles for the knowledge operator K (as explained in Sects. 17.3 and 17.7), we must be clear about which interpretation of the relation we adopt: epistemic accessibility, experiential indistinguishability, or something else. Here we adopt the accessibility interpretation.

Finally, note that while one may read $wR_K v$ as "for all the agent knows in w, scenario v might be the scenario he is in," one should *not* read $wR_K v$ as "in w, the agent considers scenario v possible," where the latter suggest a subjective psychological notion. The spymaster may not subjectively consider it possible that his spy, whom he has regarded for years as his most trusted agent, has defected. It obviously does not follow that he *knows* that his spy has not defected, as it would according to the subjective reading of R_K together with (MC).

[5]For any theory of knowledge that can be stated in terms of R_K and (MC), the rule RK of Sect. 17.3 must be sound. Therefore, theories for which RK is not sound, such as those discussed in Sect. 17.6, cannot be stated in this way. Given a formalization of such a theory, one can always define a relation R_K on scenarios such that $wR_K v$ holds iff everything the agent knows in w according to the formalization is true in v. It is immediate from this definition that if φ is not true in some v such that $wR_K v$, then the agent does not know φ in w. However, it is *not* immediate that if φ is true in all v such that $wR_K v$, then the agent knows φ in w. It is the right-to-left direction of (MC) that is not neutral with respect to all theories of knowledge.

17.3 Valid Principles

The reflexivity of R_K guarantees that the principle

$$\mathsf{T} \quad K\varphi \to \varphi$$

is valid.[6] For if $K\varphi$ is true in a scenario w, then by (MC), φ is true in all epistemically accessible scenarios, all v such that wR_Kv. Given wR_Kw by reflexivity, it follows that φ is true in w. (Conversely, if a relation R_K on a nonempty set W is not reflexive, then one can construct a model $\mathcal{M} = \langle W, R_K, V\rangle$ in which an instance of T is false. Thus, T *corresponds* to reflexivity.) It is also easy to verify that

$$\mathsf{M} \quad K(\varphi \wedge \psi) \to (K\varphi \wedge K\psi)$$

is valid over all models, simply by unpacking the truth definition. Using propositional logic (PL), we can now show why sentences of the Moorean form $p \wedge \neg Kp$ cannot be known:

(0) $K(p \wedge \neg Kp) \to (Kp \wedge K\neg Kp)$ instance of M;
(1) $K\neg Kp \to \neg Kp$ instance of T;
(2) $K(p \wedge \neg Kp) \to (Kp \wedge \neg Kp)$ from (0)–(1) by PL;
(3) $\neg K(p \wedge \neg Kp)$ from (2) by PL.

The historical importance of this demonstration, now standard fare in epistemology, is that Hintikka explained a case of unknowability in terms of logical *form*. It also prepared the way for later formal investigations of Moorean phenomena (see [10] and refs. therein) in the framework of *dynamic epistemic logic*, discussed in Sect. 17.8.

To obtain a deductive system (**KT**) from which all and only the sentences valid over our reflexive epistemic models can be derived as theorems, it suffices to extend propositional logic with T and the following rule of inference:

$$\mathsf{RK} \quad \frac{(\varphi_1 \wedge \cdots \wedge \varphi_n) \to \psi}{(K\varphi_1 \wedge \cdots \wedge K\varphi_n) \to K\psi} \quad (n \geq 0).$$

We interpret the rule to mean that if the sentence above the line is a theorem of the system, then the sentence below the line is also a theorem. Intuitively, RK says that the agent knows whatever follows logically from what she knows.

The soundness of RK shows that basic epistemic models involve a strong idealization. One can interpret these models as representing either the idealized (implicit, "virtual") knowledge of ordinary agents, or the ordinary knowledge of idealized agents (see [37] and refs. therein). There is now a large literature on alternative models for representing the knowledge of agents with bounded

[6] Throughout we use the nomenclature of modal logic for schemas and rules.

rationality, who do not always "put two and two together" and therefore lack the *logical omniscience* reflected by RK (see [18] and refs. therein). As we discuss in Sects. 17.6 and 17.7, however, the idealized nature of our mathematical models can be beneficial in some philosophical applications.[7]

17.4 Multiple Agents

The formal language with which we began in Sect. 17.2 is the language of *single-agent* epistemic logic. The language of *multi-agent* epistemic logic contains an operator K_i for each agent i in a given set of agents. (We can also use these operators for different time-slices of the same agent, as shown below.) To interpret this language, we add to our models a relation R_{K_i} for each i, defining the truth of $K_i\varphi$ in a scenario w according to (MC) but with R_{K_i} substituted for R_K.

Suppose that the spymaster of Sect. 17.2, working for the KGB, is reasoning about the knowledge of a CIA spymaster. Consider two cases. In the first, although the KGB spymaster does not know whether his KGB spy has defected, he does know that the *CIA spymaster*, who currently has the upper hand, knows whether the KGB spy has defected. Model $\mathcal{N}$ in Fig. 17.2 represents such a case, where the solid and dashed arrows are the epistemic accessibility relations for the KGB and CIA spymasters, respectively. The solid arrows for the KGB spymaster between w_1 and w_2 indicate that his knowledge does not distinguish between these scenarios, whereas the absence of dashed arrows for the CIA spymaster between w_1 and w_2 indicates that her knowledge does distinguish between these scenarios, as the KGB spymaster knows. In the second case, by contrast, the KGB spymaster is uncertain not only about whether his KGB spy has defected, but also about whether the CIA spymaster knows whether the KGB spy has defected. Model $\mathcal{N}'$ in Fig. 17.2 represents such a case. The KGB spymaster does not know whether he is in one of the upper scenarios, in which the CIA spymaster has no uncertainty, or one of the lower scenarios, in which the CIA spymaster is also uncertain about whether the KGB spy has defected. While $K_{\text{KGB}}(K_{\text{CIA}}d \vee K_{\text{CIA}}\neg d)$ is true in w_1 in $\mathcal{N}$, it is false in w_1 in $\mathcal{N}'$.

Let us now turn from the representation of what agents know about the world and each other's knowledge, using multi-agent epistemic models, to formalized reasoning about such knowledge, using multi-agent epistemic logic.

For a sample application in epistemology, consider the *surprise exam paradox* (see [33] and refs. therein). A tutor announces to her student that she will give him a surprise exam at one of their daily tutoring sessions in the next n days, where an exam on day k is a surprise iff the student does not know on the morning of day k that there will be an exam that day. The student objects, "You can't wait until the last day, day n, to give the exam, because if you do, then I'll know on the morning

[7] For additional ways of understanding idealization in epistemic logic, see [45].

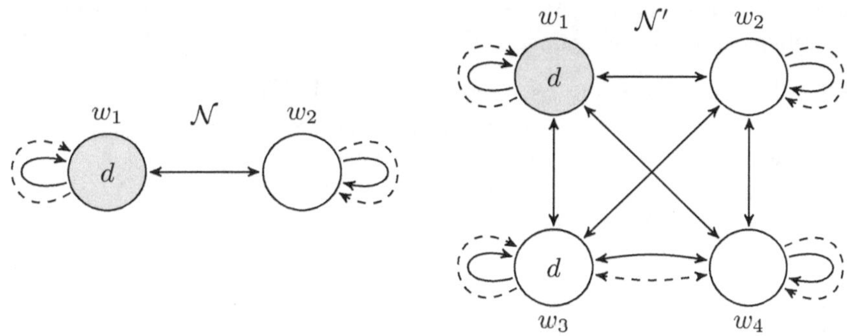

Fig. 17.2 Multi-agent epistemic models

of day n that the exam must be that day, so it won't be a surprise; since I can thereby eliminate day n, you also can't wait until day $n-1$ to give the exam, because if you do, then I'll know on the morning of day $n-1$ that the exam must be that day, so it won't be a surprise...." Repeating this reasoning, he concludes that the supposed surprise exam cannot be on day $n-2$, day $n-3$, etc., or indeed on any day at all. His reasoning appears convincing. But then, as the story goes, the tutor springs an exam on him sometime before day n, and he is surprised. So what went wrong?

Consider the $n = 2$ case. For $i \in \{1, 2\}$, let e_i mean that the exam is on day i, and let $K_i\varphi$ mean that the student knows on the morning of day i that φ, so our "multiple agents" are temporal stages of the student.[8] The tutor's announcement that there will be a surprise exam can be formalized as $(e_1 \wedge \neg K_1 e_1) \vee (e_2 \wedge \neg K_2 e_2)$. Now consider the following assumptions:

(A) $K_1((e_1 \wedge \neg K_1 e_1) \vee (e_2 \wedge \neg K_2 e_2))$;
(B) $K_1(e_2 \rightarrow K_2 \neg e_1)$;
(C) $K_1 K_2(e_1 \vee e_2)$.

Assumption (A) is that the student knows that the tutor's announcement of a surprise exam is true. Assumption (B) is that the student knows that he has a good memory: if the tutor waits until day 2 to give the exam, then the student will remember that it was not on day 1. Assumption (C) is that the student knows that he will also remember on the morning of day 2 that there was or will be an exam on one of the days (because, e.g., this is a school rule). The last assumption is that the student is a perfect logician in the sense of RK from Sect. 17.3. Let RK_i be the rule of inference

[8] A similar formalization applies to the *designated student paradox* [33, 317], a genuinely multi-agent version of the surprise exam paradox.

just like RK but for the operator K_i. Then we can derive a Moorean absurdity from assumptions (A), (B), and (C)[9]:

(4) $\big(K_2(e_1 \vee e_2) \wedge K_2\neg e_1\big) \rightarrow K_2e_2$ using PL and RK_2;
(5) $K_1\big(\big(K_2(e_1 \vee e_2) \wedge K_2\neg e_1\big) \rightarrow K_2e_2\big)$ from (4) by RK_1;
(6) $K_1(K_2\neg e_1 \rightarrow K_2e_2)$ from (C) and (5) using PL and RK_1;
(7) $K_1\neg(e_2 \wedge \neg K_2e_2)$ from (B) and (6) using PL and RK_1;
(8) $K_1(e_1 \wedge \neg K_1e_1)$ from (A) and (7) using PL and RK_1.

We saw in Sect. 17.3 that sentences of the form of (8) are unsatisfiable in epistemic models, so we must give up either (A), (B), (C), or RK_i.[10] In this way, epistemic logic sharpens our options. We leave it to the reader to contemplate these options. There is much more to be said about the paradox (and the $n > 2$ case), but we have seen enough to motivate the interest of multi-agent epistemic logic.

The multi-agent setting also leads to the study of new epistemic concepts, such as *common knowledge* [39], but for the sake of space we return to the single-agent setting in the following sections.

17.5 Knowledge and Belief

The type of model introduced in Sect. 17.2 can represent not only the content of one's knowledge, but also the content of one's *beliefs*—and how these fit together. Let us extend the language of Sect. 17.2 with sentences of the form $B\varphi$ for belief and add to the models of Sect. 17.2 a *doxastic* accessibility relation R_B. We take wR_Bv to mean that everything the agent *believes* in w is true in v, and the truth clause for $B\varphi$ is simply (MC) with $K\varphi$ replaced by $B\varphi$ and R_K replaced by R_B. (For richer models representing *conditional* belief, see [8, 36].)

How do epistemic and doxastic accessibility differ? At the least, we should not require that R_B be reflexive, since it may not be that everything the agent believes

[9] We skip steps for the sake of space. E.g., we obtain (4) by applying RK_2 to the tautology $((e_1 \vee e_2) \wedge \neg e_1) \rightarrow e_2$. We then obtain (5) directly from (4) using the special case of RK_1 where $n = 0$ in the premise $(\varphi_1 \wedge \cdots \wedge \varphi_n) \rightarrow \psi$, known as Necessitation: if ψ is a theorem, so is $K_1\psi$. It is important to remember that RK_i can only be applied to *theorems* of the logic, not to sentences that we have derived using undischarged assumptions like (A), (B), and (C). To be careful, we should keep track of the undischarged assumptions at each point in the derivation, but this is left to the reader as an exercise. Clearly we have not derived (8) as a theorem of the logic, since the assumptions (A), (B), and (C) are still undischarged. What we have derived as a theorem of the logic is the sentence abbreviated by $((A) \wedge (B) \wedge (C)) \rightarrow$ (8).

[10] We can derive (8) from (A), (B), and (C) in a doxastic logic (see Sect. 17.5) without the T axiom, substituting B_i for K_i. Thus, insofar as $B_1(e_1 \wedge \neg B_1e_1)$ is also problematic for an ideal agent, the surprise exam paradox poses a problem about belief as well as knowledge.

in a scenario w is true in w. Instead, it is often assumed that R_B is *serial*: for all w, there is some v such that wR_Bv, some scenario where everything the agent believes is true. Given seriality, it is easy to see that the principle

$$\mathsf{D} \quad B\varphi \to \neg B\neg\varphi$$

is valid, in which case we are considering an agent with consistent beliefs. (Indeed, D *corresponds* to seriality in the same way that T corresponds to reflexivity, as noted in Sect. 17.3.) With or without seriality, the analogue of RK for belief,

$$\mathsf{RB} \quad \frac{(\varphi_1 \wedge \cdots \wedge \varphi_n) \to \psi}{(B\varphi_1 \wedge \cdots \wedge B\varphi_n) \to B\psi} \quad (n \geq 0),$$

is also sound, an idealization that can be interpreted in ways analogous to those suggested for RK in Sect. 17.3, although RK raises additional questions (see Sect. 17.6).

How are epistemic and doxastic accessibility related? At the least, if whatever one knows one believes, then every scenario compatible with what one believes is compatible with what one knows: wR_Bv implies wR_Kv. Assuming this condition, $K\varphi \to B\varphi$ is valid; for if φ is true in all v such that wR_Kv, then by the condition, φ is true in all v such that wR_Bv. Other conditions relating R_B and R_K are often considered, reflecting assumptions about one's knowledge of one's beliefs and beliefs about one's knowledge (see [37]).

It is noteworthy in connection with Moore's [29] paradox that if we make no further assumptions about the relation R_B, then $B(p \wedge \neg Bp)$ is *satisfiable*, in contrast to $K(p \wedge \neg Kp)$ from Sect. 17.3. In Sect. 17.7, we will discuss an assumption about R_B that is sometimes made and is sufficient to render $B(p\wedge\neg Bp)$ unsatisfiable.[11]

17.6 Epistemic Closure

The idealization that an agent knows whatever follows logically from what she knows raises two problems. In addition to the logical omniscience problem with RK noted in Sect. 17.3, there is a distinct objection to RK that comes from versions of the *relevant alternatives* (RA) [11] and *truth-tracking* [30] theories of knowledge. According to Dretske's [11] theory, RK would fail even for "ideally astute logicians" who are "fully appraised of all the necessary consequences... of every proposition" (1010); even if RB were to hold for such an ideal logician, nonetheless RK would not hold for her in general. Nozick's [30] theory leads to the same result. The reason is that one may satisfy the conditions for knowledge (ruling out the relevant alternatives, tracking the truth, etc.) with respect to some propositions and yet not with respect to all logical consequences of the set of those propositions, *even if* one

[11] In fact, the sentence $\neg B(p \wedge \neg Bp)$ precisely corresponds to a condition on R_B, namely that for every w, there is a v such that wR_Bv and for every u, vR_Bu implies wR_Bu.

has explicitly deduced all of the consequences. Hence the problem of epistemic closure raised by Dretske and Nozick is distinct from the problem of logical omniscience.

Dretske and Nozick famously welcomed the fact that their theories block appeals to the *closure of knowledge under known implication*,

$$\mathsf{K} \quad (K\varphi \wedge K(\varphi \to \psi)) \to K\psi,$$

in arguments for radical skepticism about knowledge.[12] For example, according to K, it is a necessary condition of an agent's knowing some mundane proposition p (Kp), e.g., that what she sees in the tree is a Goldfinch, that she knows that all sorts of skeptical hypotheses do not obtain ($K\neg\mathrm{SH}$), e.g., that what she sees in the tree is not an animatronic robot, a hologram, etc., assuming she knows that these hypotheses are incompatible with p ($K(p \to \neg\mathrm{SH})$). Yet it seems difficult or impossible to rule out every remote possibility raised by the skeptic. From here the skeptic reasons in reverse: since one has not ruled out every skeptical possibility, $K\neg\mathrm{SH}$ is false, so given K and the truth of $K(p \to \neg\mathrm{SH})$, it follows by PL that Kp is false. Hence we do not know mundane propositions about birds in trees—or almost anything else, as the argument clearly generalizes.

Rejecting the skeptical conclusion, Dretske and Nozick hold instead that K can fail. However, K is only one closure principle among (infinitely) many. Although Dretske [11] denied K, he accepted other closure principles, such as closure under conjunction elimination, $K(\varphi \wedge \psi) \to (K\varphi \wedge K\psi)$, and disjunction introduction, $K\varphi \to K(\varphi \vee \psi)$. Nozick [30] was prepared to give up even closure under conjunction elimination, but not closure under disjunction introduction. More generally, one can consider any closure principle of the form $(K\alpha_1 \wedge \cdots \wedge K\alpha_n) \to (K\beta_1 \vee \cdots \vee K\beta_m)$, such as $(Kp \wedge Kq) \to K(p \wedge q)$, $(K(p \vee q) \wedge K(p \to q)) \to Kq$, $K(p \wedge q) \to K(p \vee q)$, $K(p \wedge q) \to (Kp \vee Kq)$, etc.

To go beyond case-by-case assessments of closure principles, we can use an epistemic-logical approach to formalize theories of knowledge like those of Dretske, Nozick, and others, and then to obtain general characterizations of the valid closure principles for the formalized theories. To the extent that the formalizations are faithful, we can bring our results back to epistemology. For example, Holliday [24] formalizes a family of RA and "subjunctivist" theories of knowledge using richer structures than the epistemic models in Sect. 17.2. The main Closure Theorem identifies exactly those closure principles of the form given above that are valid for the chosen RA and subjunctivist theories, with consequences for the closure debate in epistemology: on the one hand, the closure failures allowed by these theories spread far beyond those endorsed by Dretske and Nozick; on the other hand, some closure principles that look about as useful to skeptics as K turn out to be valid according to these theories. While this result is negative for the theories in question,

[12]Note that the K axiom is derivable from the RK rule with the tautology $(\varphi \wedge (\varphi \to \psi)) \to \psi$.

the formalization helps to identify the parameters of a theory of knowledge that affect its closure properties, clarifying the theory choices available to avoid the negative results.

As a methodological point, it is noteworthy that the results about epistemic closure in [24], which tell us how RK fails for certain RA and subjunctivist theories of knowledge, apply to an agent whose beliefs satisfy full *doxastic closure* in the sense of RB. Thanks to this idealization, we can isolate failures of epistemic closure due to special conditions on knowledge, posited by a given epistemological theory, from failures of closure due to an agent's simply not "putting two and two together." This is an example of the beneficial role that idealization can play in epistemic logic, a point to which we return in Sect. 17.7.

17.7 Higher-Order Knowledge

Just as the T axiom $K\varphi \rightarrow \varphi$ corresponds to the reflexivity of R_K, other epistemic principles correspond to other conditions on R_K. In this way, our models give us another perspective on these principles via properties of accessibility.

First, consider *symmetry*: wR_Kv iff vR_Kw. Williamson [41, §8.2] observes that this assumption plays a crucial role in some arguments for radical skepticism about knowledge. Suppose that in scenario w, the agent has various true beliefs about the external world. The skeptic describes a scenario v in which those beliefs are false, but the agent is systematically deceived into holding them anyway. How does one know that one is not in such a scenario? Uncontroversially, it is compatible with everything the agent knows in the skeptical scenario v that she is in the ordinary scenario w. Given this, the skeptic appeals to symmetry: it must then be compatible with everything the agent knows in w that she is in v, which is to say that everything she knows in w is true in v. But since everything the agent believes in w about the external world is *false* in v, the skeptic concludes that such beliefs do not constitute knowledge in w.

If we require with the skeptic that R_K be symmetric, then the principle

$$\mathsf{B} \quad \neg\varphi \rightarrow K\neg K\varphi$$

is valid according to (MC).[13] Although this is often assumed for convenience in applications of epistemic logic in computer science and game theory, the validity of B is clearly too strong as a matter of epistemology (see [41]).[14]

[13]Assume $\neg\varphi$ is true in w, so φ is not true in w. Consider some v with wR_Kv. By symmetry, vR_Kw. Then since φ is not true in w, $K\varphi$ is not true in v by (MC), so $\neg K\varphi$ is true in v. Since v was arbitrary, $\neg K\varphi$ is true in all v such that wR_Kv, so $K\neg K\varphi$ is true in w by (MC).

[14]Note that if we reject the requirement that R_K be symmetric in *every* epistemic model, we can still allow models in which R_K is symmetric (such as the model in Fig. 17.1), when this is appropriate to model an agent's knowledge. The same applies for other properties.

It is easy to check that symmetry follows if R_K is both reflexive and *Euclidean*: if wR_Kv and wR_Ku, then vR_Ku. The latter property guarantees that

$$5 \quad \neg K\varphi \to K\neg K\varphi$$

is valid according to (MC). Hence if we reject the symmetry requirement and the validity of the B axiom, which corresponds to symmetry, then we must also reject the Euclidean requirement and the validity of the 5 axiom, which corresponds to Euclideanness. Additional arguments against the 5 axiom come from considering the interaction of knowledge and belief (recall Sect. 17.5).[15]

While the rejection of B and 5 is universal among epistemologists, there is another principle of higher-order knowledge defended by some. Corresponding to the condition that R_K is *transitive* (if wR_Kv and vR_Ku, then wR_Ku) is the principle

$$4 \quad K\varphi \to KK\varphi.^{16}$$

Similarly, corresponding to the condition that R_B is transitive is the principle $B\varphi \to BB\varphi$. Assuming the latter, $B(p \wedge \neg Bp)$ is unsatisfiable, which is the fact at the heart of Hintikka [22, §4.6–4.7] analysis of Moore's paradox.[17]

Hintikka [22, §5.3] argued that 4 holds for a strong notion of knowledge, found in philosophy from Aristotle to Schopenhauer. The principle has since become known in epistemology as "KK" and in epistemic logic as "positive introspection." Yet Hintikka [22, §3.8–3.9, §5.3–5.4] rejected arguments for 4 based on claims about agents' introspective powers, or what he called "the myth of the self-illumination of certain mental activities" (67). Instead, his claim was that for a strong notion of knowledge, *knowing that one knows* "differs only in words" from *knowing*. His arguments for this claim [22, §2.1–2.2] deserve further attention, but we cannot go into them here (see [36, §1]).

As Hintikka assumed only reflexivity and transitivity for R_K, his investigation of epistemic logic settled on the modal logic of reflexive and transitive models, **S4**, obtained by extending propositional logic with RK, T, and 4. Some objected to this proposal on the grounds that given $K\varphi \to B\varphi$, 4 implies $K\varphi \to BK\varphi$,

---

[15] Assuming $K\varphi \to B\varphi$, D, and 5, the principle $BK\varphi \to K\varphi$ is derivable (see [16, §2.4]). Given the same assumptions, if an agent is a "stickler" [30, 246] who believes something only if she believes that she knows it ($B\varphi \to BK\varphi$), then one can even derive $B\varphi \leftrightarrow K\varphi$ (see [26] and [17]). Given $K\varphi \to B\varphi$, D, B, and $B\varphi \to BK\varphi$, one can still derive $B\varphi \to \varphi$ (see [17, 485]).

[16] To see that 4 is valid over the class of transitive models, assume that $K\varphi$ is true in w in such a model, so by (MC), φ is true in all v such that wR_Kv. Consider some u with wR_Ku. Toward proving that $K\varphi$ is true in u, consider some v with uR_Kv. By transitivity, wR_Ku and uR_Kv implies wR_Kv. Hence by our initial assumption, φ is true in v. Since v was arbitrary, φ is true in all v such that uR_Kv, so $K\varphi$ is true in u by (MC). Finally, since u was arbitrary, $K\varphi$ is true in all u such that wR_Ku, so $KK\varphi$ is true in w by (MC).

[17] Assuming D, 4, and M for B, we have: (i) $B(p \wedge \neg Bp)$, assumption for reductio; (ii) $Bp \wedge B\neg Bp$, from (i) by M for B and PL; (iii) $BBp \wedge B\neg Bp$, from (ii) by 4 for B and PL; (iv) $\neg B\neg Bp \wedge B\neg Bp$, from (iii) by D and PL; (v) $\neg B(p \wedge \neg Bp)$, from (i)-(iv) by PL.

which invites various counterexamples (see the articles in *Synthese*, Vol. 21, No. 2, 1970). Rejecting these objections, Lenzen [26, Ch. 4] argued from considerations of the combined logic of knowledge and belief (and "conviction") that the logic of knowledge is at least as strong as a system extending **S4** known as **S4.2** and at most as strong as one known as **S4.4**. Others implicated 4 in the surprise exam paradox, while still others argued for 4's innocence (see [41, Ch. 6] and [33, Ch. 7–8]).

In addition to approaching questions of higher-order knowledge via properties of R_K, we can approach these questions by formalizing substantive theories of knowledge. While the relevant alternatives and subjunctivist theories mentioned in Sect. 17.6 are generally hostile to 4, other theories are friendlier to 4. For example, consider what Stalnaker [36] calls the *defeasibility analysis*: "define knowledge as belief (or justified belief) that is stable under any potential revision by a piece of information that is in fact true" (187). Like others, Stalnaker [37] finds such stability too strong as a necessary condition for knowledge; yet he finds its sufficiency more plausible. (Varieties of belief stability have since been studied for their independent interest, e.g., in [4], without commitment to an analysis of knowledge.) Formalizing the idea of stability under belief revision in models encoding agents' *conditional* beliefs, Stalnaker [36, 37] shows that under some assumptions about agents' access to their own conditional beliefs, the formalized defeasibility analysis validates 4.[18]

The most influential recent contribution to the debate over 4 is Williamson [40, 41, Ch. 5] *margin of error* argument, which we will briefly sketch. Consider a perfectly rational agent who satisfies the logical omniscience idealization of RK and hence K, setting aside for now the additional worries about closure raised in Sect. 17.6. Williamson argues that even for such an agent, 4 does not hold in general. Suppose the agent is estimating the height of a faraway tree, which is in fact k inches. Let h_i stand for *the height of the tree is i inches*, so h_k is true. While the agent's rationality is perfect, his eyesight is not. As Williamson [41, 115] explains, "anyone who can tell by looking at the tree that it is not i inches tall, when in fact it is $i + 1$ inches tall, has much better eyesight and a much greater ability to judge heights" than this agent. Hence for any i, we have $h_{i+1} \rightarrow \neg K \neg h_i$. In contrapositive form, this is equivalent to:

(9) $\forall i (K \neg h_i \rightarrow \neg h_{i+1})$.[19]

Now suppose that the agent reflects on the limitations of his visual discrimination and comes to know every instance of (9), so that the following holds:

(10) $\forall i (K(K \neg h_i \rightarrow \neg h_{i+1}))$.

[18] Stalnaker shows that the epistemic logic of the defeasibility analysis as formalized is **S4.3**, which is intermediate in strength between Lenzen's lower and upper bounds of **S4.2** and **S4.4**.

[19] Note that the universal quantifiers in (9), (10), (15), and (21) are not part of our formal language. They are merely shorthand to indicate a schema of sentences.

Given these assumptions, it follows that for any j, if the agent knows that the height is not j inches, then he also knows that the height is not $j+1$ inches:

(11) $K\neg h_j$ assumption;
(12) $KK\neg h_j$ from (11) using 4 and PL;
(13) $K(K\neg h_j \to \neg h_{j+1})$ instance of (10);
(14) $K\neg h_{j+1}$ from (12) and (13) using K and PL.

Assuming the agent knows that the tree's height is not 0 inches, so $K\neg h_0$ holds, by repeating the steps of (11)–(14), we reach the conclusion $K\neg h_k$ by induction. (We assume, of course, that the agent has the appropriate beliefs implied by (11)–(14), as a result of following out the consequences of what he knows.) Finally, by T, $K\neg h_k$ implies $\neg h_k$, contradicting our initial assumption of h_k.

Williamson concludes that this derivation of a contradiction is a reductio ad absurdum of 4. Rejecting the transitivity of epistemic accessibility, he proposes formal models of knowledge with non-transitive accessibility to model limited discrimination [40]. (For discussion, see *Philosophy and Phenomenological Research*, Vol. 64, No. 1, 2002, and a number of recent papers by Bonnay and Egré, e.g., [9]. Williamson [43] goes further and argues that an agent can know a proposition p even though the probability on her evidence that she knows p is as close to 0 as we like.) Since Williamson's argument assumes that the agent satisfies the idealization given by RK in Sect. 17.2, if it is indeed a reductio of 4 in particular, then it shows that 4 fails for reasons other than bounded rationality. As Williamson suggests (see [21, Ch. 25]), this shows how idealization in epistemic logic can play a role analogous to that of idealization in science, allowing one to better discern the specific effects of a particular phenomenon such as limited discrimination.

17.8 Knowability

We now turn from questions about epistemic closure and higher-order knowledge to questions about the limits of what one may come to know. As we will see, these questions lead naturally to a *dynamic* approach to epistemic logic.

Fitch [14] derived an unexpected consequence from the thesis, advocated by some anti-realists, that *every truth is knowable*. Let us express this thesis as

(15) $\forall q(q \to \Diamond Kq)$,

where $\Diamond$ is a *possibility* operator. Fitch's proof uses the two modest assumptions about K used for (0)–(3) in Sect. 17.3, T and M, together with two modest assumptions about $\Diamond$. First, $\Diamond$ is the dual of a *necessity* operator $\Box$ such that $\neg\Diamond\varphi$ follows from $\Box\neg\varphi$. Second, $\Box$ obeys the rule of Necessitation: if φ is a theorem, then $\Box\varphi$ is a theorem. For an arbitrary p, consider the following:

(16) $(p \wedge \neg Kp) \to \Diamond K(p \wedge \neg Kp)$ instance of (15).

Since we demonstrated in Sect. 17.3 that $\neg K(p \wedge \neg Kp)$ is a theorem, we have:

(17) $\Box\neg K(p \wedge \neg Kp)$ from (0)–(3) by Necessitation;
(18) $\neg\Diamond K(p \wedge \neg Kp)$ from (17) by duality of $\Diamond$ and $\Box$;
(19) $\neg(p \wedge \neg Kp)$ from (16) and (18) by PL;
(20) $p \to Kp$ from (19) by (classical) PL;
(21) $\forall p(p \to Kp)$ from (16)–(20), since p was arbitrary.

From the original anti-realist assumption in (15) that every truth is *knowable*, it follows in (21) that every truth is *known*, an absurd conclusion.

There is now a large literature devoted to this "knowability paradox" (see, e.g., [41, Ch. 12], [12], [33, Ch. 4] and [32]). There are proposals for blocking the derivation of (21) at various places, e.g., in the step from (19) to (20), which is not valid in *intuitionistic* logic, or in the universal instantiation step in (16), since it allegedly involves an illegitimate substitution into an intensional context. Yet another question raised by Fitch's proof concerns how we should interpret the $\Diamond$ operator in (15).

van Benthem [5] proposes an interpretation of the $\Diamond$ in the framework of dynamic epistemic logic (see [7] and the chapter of this Handbook by Baltag and Smets for refs.). As we state more formally below, the idea is that $\Diamond K\varphi$ is true iff *there is a possible change in one's epistemic state* after which one knows φ. Contrast this with the *metaphysical* interpretation of $\Diamond$, according to which $\Diamond K\varphi$ is true iff there is a possible world where one knows φ.

In the simplest dynamic approach, we model a change in an agent's epistemic state as an *elimination of epistemic possibilities*. Recall the spymaster example from Sect. 17.2. We start with an epistemic model $\mathcal{M}$ and an actual scenario w_1, representing the spymaster's initial epistemic state. Although his spy has defected, initially the spymaster does not know this, so $d \wedge \neg Kd$ is true in w_1 in $\mathcal{M}$. Suppose the spymaster then learns the news of his spy's defection. To model this change in his epistemic state, we *eliminate* from $\mathcal{M}$ all scenarios in which d is *not* true, resulting in a *new* epistemic model $\mathcal{M}_{|d}$, displayed in Fig. 17.3, which represents the spymaster's new epistemic state. Note that Kd is true in w_1 in $\mathcal{M}_{|d}$, reflecting the spymaster's new knowledge of his spy's defection.

The acquisition of knowledge is not always as straightforward as just described. Suppose that instead of learning d, the spymaster is informed that *you don't know it, but the spy has defected*, the familiar $d \wedge \neg Kd$. The resulting model $\mathcal{M}_{|d\wedge\neg Kd}$, obtained by eliminating from $\mathcal{M}$ all scenarios in which $d \wedge \neg Kd$ is false (namely w_2) is the same as $\mathcal{M}_{|d}$ in this case. However, while $d \wedge \neg Kd$ is true

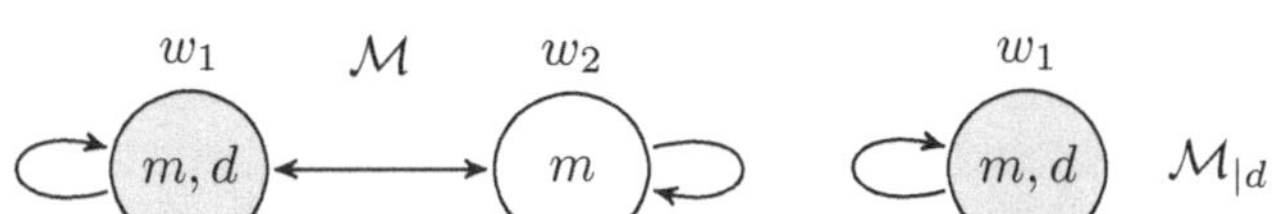

Fig. 17.3 Modeling knowledge acquisition by elimination of possibilities

in w_1 in $\mathcal{M}$, it becomes *false* in w_1 in $\mathcal{M}_{|d\wedge\neg Kd}$, since Kd becomes true in w_1 in $\mathcal{M}_{|d\wedge\neg Kd}$. As Hintikka [22] observes of a sentence like $d \wedge \neg Kd$, "If you know that I am well informed and if I address the words … to you," then you "may come to know that what I say *was* true, but saying it in so many words has the effect of making what is being said false" (68f).[20] Since $d \wedge \neg Kd$ is false in w_1 in $\mathcal{M}_{|d\wedge\neg Kd}$, so is $K(d \wedge \neg Kd)$.

Returning to the knowability paradox, van Benthem's proposal, stated informally above, is to interpret the $\Diamond$ in (15) such that $\Diamond K\varphi$ is true in a scenario w in a model $\mathcal{M}$ iff *there exists* some ψ true in w such that $K\varphi$ is true in w in the model $\mathcal{M}_{|\psi}$, obtained by eliminating from $\mathcal{M}$ all scenarios in which ψ is false. For example, in Fig. 17.3, $\Diamond Kd$ is true in w_1 in $\mathcal{M}$, since we may take d itself for the sentence ψ; but $\Diamond K(d \wedge \neg Kd)$ is false, since there is no ψ that will get the spymaster to know $d \wedge \neg Kd$. As expected, (15) is not valid for all sentences on this interpretation of $\Diamond$. Yet we now have a formal framework (see [3]) in which to investigate the sentences for which (15) *is* valid. In addition, this dynamic epistemic logical approach has inspired alternative frameworks for the analysis of knowability and other epistemic paradoxes (see [25]).

A much-discussed proposal by Tennant (see [32, Ch. 14]) is to restrict (15) to apply only to *Cartesian* sentences, those φ such that $K\varphi$ is consistent, in the sense that one cannot derive a contradiction from $K\varphi$. This restriction blocks the substitution of $p \wedge \neg Kp$, given (0)–(3) in Sect. 17.3. However, van Benthem [5] shows that (15) is not valid for all Cartesian sentences on the dynamic interpretation of $\Diamond$, which imposes stricter constraints on knowability. Another conjecture is that the sentences for which (15) is valid on the dynamic interpretation of $\Diamond$ are those that one can always learn without *self-refutation*, in the sense of Hintikka's remark above. Surprisingly, this conjecture is false, as there are sentences φ such that whenever φ is true, one can come to know φ by being informed of *some* true ψ, but one cannot always come to know φ by being informed of φ *itself* [5]. A syntactic characterization of the sentences for which (15) is valid on the dynamic interpretation of $\Diamond$ is currently unknown, an open problem for future research (see [10] for another sense of "everything is knowable"). We conclude by observing that while Fitch's proof may make trouble for anti-realism, reframing the issue in terms of the dynamics of knowledge acquisition opens a study of positive lessons about knowability (see [5, §8]; cf. [41, §12.1]).

[20] For discussion of such "unsuccessful" announcements in the context of the surprise exam paradox, see [15].

17.9 Conclusion

This survey has given only a glimpse of the intersection of epistemic logic and epistemology. Beyond its scope were applications of epistemic logic to epistemic paradoxes besides the surprise exam (see [35]), to debates about fallibilism and contextualism in epistemology (see refs. in [24]), to Gettier cases [42], and to social epistemology. Also beyond the scope of this survey were systems beyond basic epistemic logic, including quantified epistemic logic,[21] justification logic [2], modal operator epistemology [19], and logics of group knowledge [39]. For a sense of where leading figures in the intersection foresee progress, we refer the reader to Hendricks and Roy [21] and Hendricks and Pritchard [20]. Given the versatility of contemporary epistemic logic, the prospects for fruitful interactions with epistemology are stronger than ever before.[22]

References and Recommended Readings

1. Aloni, M. (2005). Individual concepts in modal predicate logic. *Journal of Philosophical Logic, 34*(1), 1–64. *** For a modern treatment of classic puzzles of quantified epistemic logic, go here next.
2. Artemov, S. (2008). The logic of justification. *The Review of Symbolic Logic, 1*(4), 477–513
3. Balbiani, P., Baltag, A., van Ditmarsch, H., Herzig, A., Hoshi, T., & de Lima, T. (2008). 'Knowable' as 'Known after an Announcement'. *The Review of Symbolic Logic, 1*, 305–334.
4. Baltag, A., & Smets, S. (2008). A qualitative theory of dynamic belief revision. In G. Bonanno, W. van der Hoek, & M. Wooldridge (Eds.), *Logic and the foundations of game and decision theory* (Texts in logic and games, Vol. 3, pp. 13–60). Amsterdam: Amsterdam University Press.
5. van Benthem, J. (2004). What one may come to know. *Analysis, 64*(2), 95–105. ***For further philosophical introduction to dynamic epistemic logic, go here next. A longer version appears as Chapter 9 of [32].
6. van Benthem, J. (2006). Epistemic logic and epistemology: The state of their affairs. *Philosophical Studies, 128*(1), 49–76.
7. van Benthem, J. (2011). *Logical dynamics of information and interaction*. New York: Cambridge University Press.
8. Board, O. (2004). Dynamic interactive epistemology. *Games and Economic Behavior, 49*, 49–80.
9. Bonnay, D., & Egré, P. (2009). Inexact knowledge with introspection. *Journal of Philosophical Logic, 38*, 179–227.
10. van Ditmarsch, H., van der Hoek, W., & Iliev, P. (2011). Everything is knowable – how to get to know *whether* a proposition is true. *Theoria, 78*(2), 93–114.

---

[21] See [22, Ch. 6], [16, §5], and [1] for discussion of quantified epistemic logic. Hintikka [23] has proposed a "second generation" epistemic logic, based on *independence-friendly* first-order logic, aimed at solving the difficulties and fulfilling the epistemological promises of quantified epistemic logic.

[22] For helpful discussion and comments, I wish to thank Johan van Benthem, Tomohiro Hoshi, Thomas Icard, Alex Kocurek, Eric Pacuit, John Perry, Igor Sedlár, Justin Vlasits, and the students in my Fall 2012 seminar on Epistemic Logic and Epistemology at UC Berkeley.

11. Dretske, F. (1970). Epistemic operators. *The Journal of Philosophy, 67*(24), 1007–1023. ***A classic source of current debates about epistemic closure.
12. Edgington, D. (1985). The paradox of knowability. *Mind, 94*(376), 557–568.
13. Ègre, P. (2011). Epistemic logic. In L. Horsten & R. Pettigrew (Eds.), *The continuum companion to philosophical logic* (Continuum companions, pp. 503–542). New York: Continuum. ***For an introduction to further topics in epistemic logic, go here next.
14. Fitch, F. B. (1963). A logical analysis of some value concepts. *The Journal of Symbolic Logic, 28*(2), 135–142.
15. Gerbrandy, J. (2007). The surprise examination in dynamic epistemic logic. *Synthese, 155*, 21–33.
16. Gochet, P., & Gribomont, P. (2006). Epistemic logic. In D. M. Gabbay & J. Woods (Eds.), *Handbook of the history of logic* (Vol. 7). Amsterdam: Elsevier.
17. Halpern, J. Y. (1996). Should knowledge entail belief? *Journal of Philosophical Logic, 25*, 483–494.
18. Halpern, J. Y., & Pucella, R. (2011). Dealing with logical omniscience: Expressiveness and pragmatics. *Artificial Intelligence, 175*, 220–235.
19. Hendricks, V. F. (2005). *Mainstream and formal epistemology*. New York: Cambridge University Press.
20. Hendricks, V. F., & Pritchard, D. (Eds.). (2008). *Epistemology: 5 questions*. Copenhagen: Automatic Press.
21. Hendricks, V. F., & Roy, O. (Eds.). (2010). *Epistemic logic: 5 questions*. Copenhagen: Automatic Press.
22. Hintikka, J. (1962). *Knowledge and belief: An introduction to the logic of the two notions*. Ithaca: Cornell University Press. ***The founding text of modern epistemic logic, still worth reading.
23. Hintikka, J. (2003). A second generation epistemic logic and its general significance. In V. F. Hendricks, K. F. Jørgensen, & S. A. Pedersen (Eds.), *Knowledge contributors* (Synthese library, Vol. 322). Dordrecht: Kluwer.
24. Holliday, W. H. (2015). Epistemic closure and epistemic logic I: Relevant alternatives and subjunctivism. *Journal of Philosophical Logic, 44*(1), 1–62.
25. Holliday, W. H. (2017). Knowledge, time, and paradox: Introducing sequential epistemic logic. In H. van Ditmarsch & G. Sandu (Eds.), *Jaakko Hintikka on knowledge and game theoretical semantics* (Outstanding contributions to logic, Vol. 12). New York: Springer.
26. Lenzen, W. (1978). Recent work in epistemic logic. *Acta Philosophica Fennica, 30*, 1–219.
27. Lewis, D. (1986). *On the plurality of worlds*. Oxford: Basil Blackwell.
28. Lewis, D. (1996). Elusive knowledge. *Australasian Journal of Philosophy, 74*(4), 549–567.
29. Moore, G. E. (1942). A reply to my critics. In P. A. Schilpp (Ed.), *The philosophy of G.E. Moore* (pp. 535–677). Evanston: Northwestern University Press.
30. Nozick, R. (1981). *Philosophical explanations*. Cambridge, MA: Harvard University Press.
31. Pacuit, E. (2013). Dynamic epistemic logic I and II. *Philosophy Compass, 8*(9), 798–833.
32. Salerno, J. (Ed.). (2009). *New essays on the knowability paradox*. New York: Oxford University Press.
33. Sorensen, R. (1988). *Blindspots*. Oxford: Clarendon Press. ***Filled with fascinating material relevant to epistemic logic.
34. Sorensen, R. (2002). Formal problems about knowledge. In P. K. Moser (Ed.), *The Oxford handbook of epistemology* (pp. 539–595). New York: Oxford University Press.
35. Sorensen, R, "Epistemic Paradoxes", *The Stanford Encyclopedia of Philosophy* (Fall 2017 Edition), Edward N. Zalta (ed.), https://plato.stanford.edu/archives/fall2017/entries/epistemic-paradoxes/.
36. Stalnaker, R. (1996). Knowledge, belief and counterfactual reasoning in games. *Economics and Philosophy, 12*, 133–162.
37. Stalnaker, R. (2006). On logics of knowledge and belief. *Philosophical Studies, 128*(1), 169–199. ***For more on the interaction of knowledge and belief, go here next.

38. Stine, G. C. (1976). Skepticism, relevant alternatives, and deductive closure. *Philosophical Studies, 29*(4), 249–261.
39. Vanderschraaf, Peter and Sillari, Giacomo, "Common Knowledge", *The Stanford Encyclopedia of Philosophy* (Spring 2014 Edition), Edward N. Zalta (ed.), https://plato.stanford.edu/archives/spr2014/entries/common-knowledge/.
40. Williamson, T. (1999). Rational failures of the KK principle. In C. Bicchieri, R. C. Jeffrey, & B. Skyrms (Eds.), *The logic of strategy* (pp. 101–118). New York: Oxford University Press. ***For payoffs of applying epistemic logic to epistemology, start here or with one of Williamson's other papers.
41. Williamson, T. (2000). *Knowledge and its limits*. New York: Oxford University Press.
42. Williamson, T. (2013). Gettier cases in epistemic logic. *Inquiry, 56*(1), 1–14.
43. Williamson, T. (2014). Very improbable knowing. *Erkenntnis, 79*(5), 971–999.
44. Williamson, T. (2015). A note on Gettier cases in epistemic logic. *Philosophical Studies, 172*(1), 129–140.
45. Yap, A. (2014). Idealization, epistemic logic, and epistemology. *Synthese, 191*(14), 3351–3366.

Chapter 18
Knowledge Representation for Philosophers

Richmond H. Thomason

Abstract This article provides an overview of the subfield of Artificial Intelligence known as "Knowledge Representation and Reasoning." This field uses the techniques of philosophical logic, but aims at providing a theoretical basis for the management of declarative information in automated reasoning systems. Three topics are singled out here for attention: planning and reasoning about actions, description logics, and nonmonotonic logics.

18.1 Philosophical Logic and Logical AI

Formal philosophy seeks to use formalized languages and their metatheory to illuminate philosophical problems. In its earlier stages (roughly, until around 1960), most work in this area relied on classical logics and philosophical analysis, and so is difficult to distinguish from the broader area of analytic philosophy. But in its later stages, many practitioners of formal philosophy became convinced that classical logics were inadequate for some philosophical purposes, and the later work typically involves the formalization of a language, the development of its logical properties, and informal and philosophical discussion of its significance for philosophy.

A philosophical project of this sort uses what Alonzo Church [13], pp. 47–58) called the *logistic method:* that is, it selects a target domain—an area of inquiry with characteristic forms of reasoning, and constructs a theory of the reasoning by providing a formalized logical system, including an axiomatization and model theory. Church had in mind mathematical domains and the sort of reasoning found in mathematical proofs, but philosophical logicians have used the method to study tense, modality, nondeclarative sentences, propositional attitudes such as knowledge and belief, contrary-to-fact conditionals, and many other linguistic constructions of philosophical interest.

R. H. Thomason (✉)
Philosophy Department, University of Michigan, Ann Arbor, MI, USA
e-mail: rthomaso@umich.edu

S. O. Hansson, V. F. Hendricks (eds.), *Introduction to Formal Philosophy*, Springer Undergraduate Texts in Philosophy, https://doi.org/10.1007/978-3-319-77434-3_18

Artificial Intelligence is an eclectic field, and harbors many methodologies. But since the publication of [42], and largely because of John McCarthy's subsequent influence, the logistic method has been used to understand the domains that AI seeks to create. The targets include reasoning about time and action (and, in particular, planning), reasoning about other agents, about space and material objects, and many other topics.

Logical AI is continuous with earlier work in philosophical logic and makes explicit use of it; it is, in fact, best to think of philosophical logic and logical AI as a single field. Logical AI is now much larger and more active than its philosophical parent discipline, and by now many of the most important trends are being pursued by professional computer scientists. Nevertheless, most of this work is as relevant to philosophy as the earlier work that was published in the philosophical journals, and philosophers who value the usefulness of logic should be aware of the computational literature.

18.2 The Emergence of Knowledge Representation in AI

The process that led to the emergence of Knowledge Representation as a subfield of Artificial Intelligence should be of interest to philosophers. In both philosophy and AI formal techniques are available, but their value and appropriateness can be questioned. In AI, however, the foundational debate was limited to a few years, and resulted in a clearcut outcome. There are subfields of AI that can avoid reasoning about propositional attitudes and using formalisms that incorporate intensional constructions. But in those that do consider this sort of reasoning, the value of explicit representations, and of logical theory as a source of these representations, is no longer at issue.

The recognition of Artificial Intelligence as a field goes back to a conference held at Dartmouth College in 1956. (See [47].) A few years later, in [42], John McCarthy explicitly proposed an approach to AI that would attempt to represent an agent's declarative knowledge explicitly, employing logical rules as an inference mechanism. The need to formalize common-sense knowledge, and the appropriateness of logic for this purpose, is a continuing theme in McCarthy's later work; see the papers collected in [45].

However, McCarthy's early proposals were not very influential and, through the 1970s, much of the work in AI—and certainly, work that involved implemented systems—either ignored declarative representations entirely or, in some cases, developed ad hoc representation systems that had little or nothing to do with the logical tradition. (Marvin Minsky's "frame-based" representations are an example; see [49].)

During the phase of AI (roughly, dating through the 1970s and well into the 1980s), when researchers were concentrating on small to medium-sized reasoning problems, the role of logic was debated. See, for instance, [27, 48]; there is a retrospective discussion of the issues in [37, Section 1.5]. A thorough history of AI, and of the ideas in play during this period, remains to be written. But it is clear that during the latter part of the 1980s and the early 1990s, this conflict was decided in favor of the logicians—not so much by explicit debate, but by widespread recognition among AI practitioners of the importance and usefulness of logical representations.

I believe that the following factors played an important part in these developments:

1. **Software engineering considerations.** Beyond a certain size, it is difficult or impossible to maintain software systems without a modular design, and without a clear, explicit understanding of the meaning of the representations. As reasoning systems became larger and more ambitious, these considerations provided a powerful motivation for using logical representations when possible. As programs become larger and more complex, you need not only a comprehensive, detailed account of what the program is supposed to do; even better, you want a proof that the algorithm is correct.
   You also need modularity. The software engineering reasons for modular representation of declarative knowledge are well documented in [60, Chapter 3]. Stefik is eclectic about representation systems, but that brings me to my next point.
2. **Universality of logic for declarative representations.** Gradually, it became realized that various alternatives to logical representations that had been proposed could be formalized as logics, and that treating them as such would deliver improved insights.[1]
3. **Decoupling theories from implementations.** Theorem proving is seldom the best way to approach the reasoning problems that arise in AI. But, as the AI community learned, logical modeling doesn't commit an AI researcher to a theorem-proving implementation. Theorem proving is not the only algorithm associated with logic—for instance, model construction is useful for many purposes—and the relationship between a logical theory and an implementation informed by it can be tenuous. At one extreme, logical modeling helps to understand the reasoning problem, and although the implementation is inspired by the logic, it is hard to say what the relationship is. At another extreme, it can be hard to distinguish the theory from the implementation.
4. **Computer science graduate education.** Computer science began to produce graduate students in large numbers in the 1980s. As these students entered the AI research pool, the comfort level of AI researchers with logic grew. Computer science departments provide training in theory, and

[1] In [26], Patrick Hayes argued that frame-based representations, which had widely been taken to be an alternative to logical representations, could be reproduced in a first-order logic with a mechanism for formalizing defaults. (Hayes used an epistemic operator for this purpose.) The later history vindicated this idea, as ideas about frames and semantic nets were transformed into *description logics*—representation services that can be embedded in first-order logic, or in well understood extensions of first-order logic. See Sect. 18.4.2, below, for more about description logics.

theoretical computer science is almost entirely a branch of logic. Many of the younger AI researchers at this time were accomplished logicians.

5. Small-scale successes. Some early special-purpose uses of logic were successful and influential. Examples are James Allen's interval logic for reasoning about time [2], and McCarthy's Situation Calculus for reasoning about actions [43].

The first major collection devoted to knowledge representation, [9], appeared in 1985. At this point the field had begun to move rapidly, and many of the papers in the collection were already outdated, although the volume covers ideas that were to become important themes in the future. At this point, the area gained popularity: a significant number of papers devoted to knowledge representation began to appear at the major AI meetings.

A series of international conferences devoted entirely to knowledge representation began in Toronto in 1989; the twelfth in this series took place in 2010. By 1989, the field was well-established, and from now on I'll refer to it as "KRR". As we'll see, the second 'R' is important.

18.3 The Scope and Subject Matter of KRR

18.3.1 The Importance of Reasoning

The title of the first KR proceedings, [3], is "Principles of Knowledge Representation and Reasoning." The clause 'and reasoning' was added intentionally, and marks a significant difference between KRR and the closely related field of philosophical logic.

A typical project in philosophical logic will formalize some topic, hopefully providing a model-theoretic semantics as well as good motivation for the formalization. Usually, there are forms of reasoning associated with the domain that is formalized, and a philosophical logician will recognize this by taking into account examples of reasoning that intuitively are good or bad and using this to justify the validities delivered by the theory. Presumably, the intuitions about validity are closely associated with our expertise in the associated reasoning. But the connection of the project to reasoning doesn't go further than this. A philosophical logician will hope that the structures that make formulas true and false will deliver new insights into topics of traditional philosophical interest, and the philosophical impact of the project will mostly depend on the quality of thes insights. Except for a sample of valid and invalid inferences, reasoning is absent from these picture.

But many robust and complex forms of reasoning are associated with the domains that philosophical logicians have explored. For instance, consider the problem of reading a narrative and—if the narrative is temporally coherent—figuring out how to order the events that are mentioned in the narration into a temporal sequence, and maintaining this timeline as more of the narrative is read. Temporal logicians working in the philosophical logic tradition hardly ever consider issues of this sort.

But such questions are crucially important in knowledge representation, because their answers may provide connections to useful, implementable reasoning services.

Also—and this is a new consideration—the design of the formal language may be influenced by the intended reasoning application. In [35], Hector Levesque and Ronald Brachman argue that there is a potential tradeoff between expressive adequacy of the language and the computational complexity of the reasoning. A formalized language that is ambitious expressively may be less useful when the reasoning application is taken into account, because the reasoning associated with it—for instance, theorem proving or satisfaction checking—is more complex.

In fact, some of the most successful projects in KRR involve the discovery of useful compromises between the competing factors in the Brachman-Levesque tradeoff. Edmund Clarke's use in [14] of a restricted temporal language for software validation is an example, as well as the temporal language of [2]. But good solutions to the tradeoff are difficult to find. (The application to description logics that Levesque and Brachman had in mind did not quite work out as they had hoped.)

18.3.2 Topics in KRR

The coverage of the 12 KRR proceedings that appeared by 2010[2] provide a useful guide to the major topics, as well as an indication of their importance for the field. Among these topics are: Planning, Description logics, Abduction, Multiagent Systems, Nonmonotonic Logic, Planning Agents, Spatial Logics, Belief Revision, Ontologies, and Preferences. Not so well represented at the KRR meetings, but important for philosophers, is the work on formalizing common sense reasoning.

The next sections will go into more detail about a few of these topics; references to the others can be found in the works cited in Sect. 18.5.

18.4 Details About Selected KRR Topics

18.4.1 Planning and Reasoning About Actions

By any measure, the most active area of research in KRR consists of logics for planning. Planning itself, or means-end reasoning, was recognized quite early as an important area of AI (see [59]), and in the earliest phases the paradigmatic examples of planning were taken from gamelike domains. Simon's paper contains the fundamental idea that *actions* are available to the planning agent, which when executed will change the state of the world (for instance, the state of a partially

---

[2]These are [1, 3, 4, 11, 15, 16, 19–21, 23, 40, 51].

filled-in crossword puzzle), and the idea that planning is a matter of searching a problem space for a series of operators that will achieve a given *goal*.

It is certainly possible to implement a planning system without a logical formalization of the reasoning. (And many AI researchers did this, thinking only of the problem of how, using heuristic search, to efficiently find a plan in the very large search spaces that arise even in simple problems.) But some AI researchers, following John McCarthy, took a logical approach. McCarthy's Situation Calculus, first presented in [43] and mentioned above in Sect. 18.2, was offered as a formalization of means-end reasoning. The ideas in this paper (originally published in 1963) are elaborated in [46], which is usually cited as the source of the Situation Calculus.

There is a continuous history of research on the logic of action, within the Situation Calculus, from the early 1970s to the present, through which insights have deepened, and the theory has been generalized to more challenging planning domains.

Here, I will concentrate only on the basic ideas of the Situation Calculus formalism and on the immediate logical problems that it generates. For other expositions of this topic, see [39, 58, 62].

The Situation Calculus is a many-sorted first-order theory, with designated sorts for *situations* and for *actions*. ('Situation' is McCarthy's term for 'state'; actions are taken to be individuals.) Many predicates of the Situation Calculus, then, will have a single argument place of situational type: these are called *fluent predicates*, and the values they take in models are called *fluents*. Actions are treated as primitives: they are individuals, and there is a designated sort of actions.

There is a special 3-place predicate ***Result*** expressing a relation between situations, actions, and situations; the idea is that $\boldsymbol{Result}(s_1, a, s_2)$ is true iff performing the action denoted by a in the situation denoted by s_1 leads to the situation denoted by s_2. We will assume that the outcome of performing an action is unique:

$$\forall s \forall a \exists s_1 \forall s_2 [\boldsymbol{Result}(s, a, s_2) \leftrightarrow s_2 = s_1].$$

A *causal theory* in the Situation Calculus provides *causal axioms* for each action. A causal axiom for an action a is supposed to say what will happen if the action is performed in appropriate circumstances. At the very least, then, a causal axiom for a will entail a conditional relating a *precondition* for the action to its *effects*. The purpose of a causal theory is to characterize precisely what the result of performing each action will be. Without this, it would be impossible to tell in general what state would result from performing a series of actions, so it would be impossible to produce a plan supported by a proof that the goal will be reached.

In the crossword puzzle domain, for instance, for each letter and cell of the puzzle there is an action of putting that letter in the cell. We define an empty cell to be one that contains nothing:

$$\forall x \forall s [\boldsymbol{Empty}(x, s) \leftrightarrow \forall z \neg \boldsymbol{In}(z, x, s)].$$

Consider the action of putting ‘t’ in a cell. This action’s precondition is that the cell must be empty; its effect is that ‘t’ is the cell. The following simple causal axiom captures these things nicely:

$$\text{(SCA) } \forall s \forall x \forall s' \forall y[\textbf{\textit{Result}}(s, \textbf{\textit{PutIn-t}}, s') \rightarrow [\textbf{\textit{Empty}}(x, s) \rightarrow \textbf{\textit{In}}(\textbf{\textit{t}}, x, s')]].$$

(Here, ***t*** is a constant denoting the letter ‘t’.)

This causal axiom allows too many models. The problem is that it doesn’t say anything about what happens in cells other than the single cell that is affected. Of course, nothing happens in these other cells. We want *causal inertia* to prevail—the other cells stay put, but (SCA) doesn’t entail this.

The fact that causal inertia will in general call for many more axioms than are required for causal change is, in its purest form, the *Frame Problem*. In a more general form, the Frame Problem is the question of how to axiomatize causal inertia.[3]

A *monotonic* approach to the Frame Problem states the inertial rules explicitly as axioms. An economical way to do this is to write axioms giving necessary and sufficient conditions under which a fluent holds in an arbitrary resultant situation. In the crossword domain, the axiom for the letter ‘t’ would look like this:

$$\begin{aligned}\text{(MCA) } \forall s \forall a \forall s' \forall x[\,[\textbf{\textit{Action}}(a) \wedge \textbf{\textit{Result}}(s, a, s') \wedge \textbf{\textit{In}}(\textbf{\textit{t}}, x, s')] \leftrightarrow \\ [\,[a = \textbf{\textit{PutIn-t}} \wedge \textbf{\textit{Empty}}(x, s)] \vee [a \neq \textbf{\textit{PutIn-t}} \wedge \textbf{\textit{In}}(\textbf{\textit{t}}, x, s)]\,]\,].\end{aligned}$$

This axiom guarantees that ‘t’ appears in a cell of a noninitial situation if and only if it was already there, or was just put there. It requires quantification over actions, but this is unproblematic, since actions can (and probably should) be treated as individuals.

It’s a bit disappointing that monotonic solutions to the Frame Problem are perfectly workable (in [55], for instance, Ray Reiter hows how monotonic solutions can be deployed in challenging and complex planning environments), because the nonmonotonic solutions are so much more interesting from a logical standpoint. These solutions require a logic that somehow supports exceptions to axioms. (See Sect. 18.4.3, below.) Given such a logic, causal inertia can be expressed as a simple, global default: “nothing changes.” Causal axioms then provide constraints that override the inertial default.

The discovery of anomalies in the nonmonotonic solutions [25] led to increasingly complex logical solutions which, since they appeal to causality, are of considerable philosophical interest. See, for instance, [38, 63].

This is by no means the end of the logical challenges. The *Ramification Problem* has to do with indirect effects of actions; the *Qualification Problem* has to do with the difficulty of stating universally correct preconditions for actions. The

[3]Philosophers should take note. In the philosophical literature, the Frame Problem has been widely misunderstood and wildly overgeneralized. See [58, Section 1.12].

literature on both these problems is extensive, and contains much material that is philosophically interesting; some entry points to the literature are [38, 41, 61].

Many other problems arise in the process of extending simple planning theories to more noisy and challenging domains. What about multiple agents? What about agents who are uncertain about the current state of the world? How can the discrete, action-based accounts of change used in planning be combined with continuous theories of natural change based on differential equations? Many authors have discussed these issues, but [55] is perhaps the best beginning point.

Before turning to other matters, let's consider how the logical approach to planning allows declarative information (knowledge) to be separated from the heuristics and procedures that may be involved in reasoning with the knowledge. The causal axioms, other domain axioms, and the causal theory constitute a declarative theory that can be formalized in a logic with well understood properties. It can be validated by checking it against intuitions and evidence about the domain. It is relatively easy to update. And it is independent of any particular implementation. It can be combined with any algorithm for finding a plan, and with any heuristics for narrowing the search, that an implementer chooses to use. Each of these things is an advantage, from a software engineering standpoint. Taken together, the case for this modular approach, separating out the declarative knowledge and using a logic to formalize it, is quite compelling.

18.4.2 Description Logics

Many AI applications, as well as planning, will need a separate representation of the knowledge used by the system. This creates a need for a plug-in KR service that is relatively easy to learn and to use, and that can reliably and efficiently deliver the conclusions that are needed by the system. Description logics fill this niche better than any other KR service.

There are many description logics, so I will confine myself here to the basic recipe for a description logic: separate general from factual information. Insist that general information takes the form of concept definitions, and include in the KR language useful constructs for forming such definitions. Restrict the form of the factual information; for instance, do not allow disjunctive formulas.

Definitions in a description logic might look like this:

$$\begin{array}{l}
\textbf{\textit{Mother}} : \text{HAS-AT-LEAST-1}(\textbf{\textit{child}}) \text{ AND } \textbf{\textit{Female}}. \\
\textbf{\textit{Employed}} : \text{HAS-AT-LEAST-1}(\textbf{\textit{employer}}). \\
\textbf{\textit{Working-Mother}} : \textbf{\textit{Mother}} \text{ AND } \textbf{\textit{Employed}}. \\
\textbf{\textit{Orphan}} : \text{HAS-0}(\textbf{\textit{parent}}). \\
\textbf{\textit{parent}} : \text{INVERSE}(\textbf{\textit{child}}).
\end{array}$$

Given these definitions, a description logic would be able to infer that Agnes is a working mother if it is told that Agnes is a mother and is self-employed. You would hope that it would be able to infer an inconsistency if it is told that Bert is Agnes' child and Bert is an orphan.

The reasoning algorithms for many description logics are well understood and deliver reliable results, often with excellent efficiency. Good documentation is available for many of these systems. Much work has been devoted to extending the expressiveness of description logics, and many of these extensions—for instance, attempts to include temporal reasoning—are philosophically interesting.

See [6] for a recent survey of this topic, with many references; [5] has many details, including descriptions of some of the leading systems.

18.4.3 Nonmonotonic Logics and Nonmonotonic Reasoning

Nonmonotonic logic might well have been developed earlier, by philosophical logicians, but in fact this topic emerged from logical AI. Monotonicity is a property of the consequence relation $\vdash$: if $\Gamma \vdash \phi$ then $\Gamma \cup \{\psi\} \vdash \phi$. This says that adding a new axiom to an axiomatic system produces more theorems. A consequence relation is nonmonotonic if it fails to have the monotonicity property.

Common-sense reasoning is full of examples of nonmonotonicity. For instance, let Γ be the set of observations that assumptions that are in play for for a practical agent—a person or a robot, and let Δ be the set of conclusions that the agent draws from these observations. Suppose the agent observes, in situation s_1, that a certain cup is on a certain table. Then the formula

$$\text{(A1) } \boldsymbol{On(cup87, table15}, s_1)$$

will be in Γ. The agent has no reason to think the cup will be disturbed in the span of time between s_1 and, say, s_2. Then our agent will suppose (A2), which will therefore belong to Δ, because it is concluded from observations and the agent's causal theory.

$$\text{(A2) } \boldsymbol{On(cup87, table15}, s_2)$$

Suppose the agent daydreams, receiving no new information between s_1 and s_2. Observing the table, the agent learns that, contrary to expectations, the cup is gone; the new observation (A3) is the negation of (A2).

$$\text{(A3) } \neg\boldsymbol{On(cup87, table15}, s_2)$$

In a monotonic logic, we get an inconsistent theory if we add (A3) to Γ. In a nonmonotonic logic, the conclusion is retracted when the addition is made and $\Gamma \cup \{(A3)\}$ is consistent. The generalization that produced the incorrect conclusion—

in this case, a nonmonotonic axiom of causal inertia—is retained. The conclusion that is withdrawn marks an exception to the axiom.

John McCarthy's Circumscription Theory, proposed in [44], involves a relatively simple modification of first-order logic. Since it can be fairly easily explained, we will use it to illustrate the workings of a nonmonotonic logic. The language of Circumscription Theory is first-order, but special *abnormality* predicates are introduced to mark exceptions.

An exception-tolerant Causal Inertia axiom for the crossword puzzle domain would be stated as follows.

$$\text{(NMCI) } \forall x \forall y \forall s \forall a[[\boldsymbol{Cell}(x) \wedge \boldsymbol{In}(y, x, s) \wedge \boldsymbol{Result}(s, a, s') \wedge \neg \boldsymbol{Ab}(x, y, s)] \rightarrow \boldsymbol{In}(y, x, s')$$

The axiom guarantees that what is in a cell stays put through a change unless there is an abnormality involving the cell, its occupant, and the change. (Another inertia axiom would be needed to ensure that empty cells stay empty unless there is an exception.)

We obtain a nonmonotonic logic by taking account in the definition of logical consequence only models of a theory Γ in which the extensions of the various abnormality predicates are minimized. With just one abnormality predicate, say the 3-place predicate in (NMCI), the definition is simple. A model M is *better* than a model M', $M \prec M'$, iff the extension Ab of $\boldsymbol{Ab}$ in M is a proper subset of the extension Ab$'$ of $\boldsymbol{Ab}$ in M'. A model M of Γ is *minimal* iff no model of Γ is better than M. Finally $\Gamma \vdash \phi$ iff every minimal model of Γ satisfies ϕ.

One advantage of Circumscription, from the standpoint of formalizing domains, is that it is possible to write axioms about abnormalities—about what to expect when things go wrong. These axioms themselves may involve abnormalities. See [36] for details. This article is also an excellent introduction to Circumscription Theory; and treatments can be found in any of the books on nonmonotonic logic cited in the bibliography to this paper, as well as in [12].

A very large body of work on nonmonotonic logic and its applications has accumulated over the last 30 years, most of it appearing in the AI journals, but some in philosophical venues.

The two other leading approaches to nonmonotonic logic are *Default Logic* and modal theories such as *Autoepistemic Logic*. Default Logic, originating in [54], takes a more proof-theoretic approach to the topic. A default theory consists of two components: the monotonic component is a set of ordinary first-order axioms, and the nonmonotonic component is a set of *default rules*.

In the special case I'll consider here,[4] a (normal) default rule

$$\phi \;/\; \psi$$

[4] I will ignore general default rules in this exposition, and only consider normal defaults.

looks like an ordinary rule of inference, but has a different interpretation: the conclusion ψ can be inferred from the premise ϕ as long as it is consistent to do so.

A default rule like

$$\text{(DR1)}\ \textbf{\textit{TurnSwitch}}\ /\ \textbf{\textit{LightOn}}$$

could be read "Infer that the light will go on if you turn the switch."

If the monotonic axioms are $\{\textbf{\textit{TurnSwitch}}\}$ and the only default rule is (DR1), then ***LightOn*** can be concluded. But if the monotonic axioms are $\{\textbf{\textit{TurnSwitch}}, \neg\textbf{\textit{LightOn}}\}$, then ***LightOn*** cannot be concluded.

Just as the rules of first-order logic allow a theory to be derived from first-order axioms, Reiter assumes that *extensions* can be derived from a default theory: an extension is a set of formulas that a perfect reasoner might infer from the monotonic axioms and default rules of the theory.

But defaults can *conflict*, and this makes things complicated. The standard example is the *Nixon Diamond*: the monotonic theory is $\{\textbf{\textit{Quaker}}(\textbf{\textit{Nixon}}),\textbf{\textit{Republican}}(\textbf{\textit{Nixon}})\}$ and there are two default rules:

$$\textbf{\textit{Quaker}}(\textbf{\textit{Nixon}})\ /\ \textbf{\textit{Pacifist}}(\textbf{\textit{Nixon}})\ \text{and}\ \textbf{\textit{Republican}}(\textbf{\textit{Nixon}})\ /\ \neg\textbf{\textit{Pacifist}}(\textbf{\textit{Nixon}}).$$

Here, it is not clear what to say: both default rules can be consistently applied separately to the monotonic axioms, but not both. Reiter associates *two* extensions with the Nixon Diamond: one concludes that Nixon is a pacifist, the other that Nixon is not a pacifist. Given the information in the default theory, and forgetting whatever else we know about Nixon, there is no way to choose between these two extensions.

Theorists differ about how to think about this, but the most interesting interpretation, from a logical standpoint, is that the relation between premises and their logical consequences is not unique in nonmonotonic logic: perfect reasoners can draw different sets of conclusions from the same default theory. This idea is particularly attractive in metaethical applications of nonmonotonic logic; see John F. Horty's work, cited in the bibliography.

Reiter's main technical achievement in [54] consists of two definitions of the extension relation, and a proof that the definitions are equivalent. For more about default logic, see [7, 18, 32, 57], and any of the general treatments of nonmonotonic logic listed in the bibliography.

The thought behind autoepistemic logic is that a default rule applies unless the reasoning agent knows something to the contrary; this suggests that defaults can be formalized using an epistemic modal operator. For more about this approach, see [33] and (again) any of the general references to nonmonotonic logic in the bibliography.

The field of *Argumentation Theory* is only tenuously connected to KR, but it uses ideas from nonmonotonic logic and is potentially important for philosophy. The idea is to treat arguments abstractly, constructing a theory of notions like the relations of *attack* and *defeat* between arguments, and attempts to develop a notion of *extension* for arguments analogous to Reiter's extensions for default logic. The literature on Argumentation Theory is by now rather extensive. It has been applied to many reasoning domains, including the law, but has not yet gotten the attention it deserves from philosophers.[5]

For general discussions of Argumentation Theory. see [8, 53]. [22] is an early, influential paper in the field. For applications to the law, see, [34, 56].

Although I will only mention the programming language PROLOG and its extensions briefly here, there is a strong, continuous tradition of work in this area in KRR. With its declarative programming style, PROLOG programming offers a distinctive and important compromise between declarative transparency and implementability. PROLOG's negation-as-failure provides a connection to nonmonotonic reasoning.

And PROLOG can be extended in interesting ways. Some of these extensions become large-scale projects that attract research groups, and offer KRR services for important areas of reasoning. I have already mentioned Ray Reiter's extension in [55] of PROLOG into a language for cognitive robotics. Stable models and answer sets are another area of this kind; see [24].

Horty's work combining nonmonotonic logic and deontic logic provides a good example of how ideas originating in KRR can be fruitfully applied in metaethics; see [28–31]. Certainly, these ideas can be applied in many other areas of philosophy as well.

18.5 How Can a Philosopher Access the Field?

Much of the literature in KRR is technical, but this should not be a problem for formal philosophers—especially since so much of it overlaps with philosophical logic. For those who want a systematic introduction, [10] is an excellent resource. For those interested in specific topics, as well as references to the literature, [64] is a very useful resource. For commonsense reasoning, see [17, 50]. If anyone wants to get a comprehensive, detailed sense of the research in this field, there is nothing like the KRR proceedings listed above in Footnote 2. There is a great deal of material there, but the quality is very high.

Bibliography

1. Aiello, L. C., Doyle, J., & Shapiro, S. (Eds.). (1996). *KR'96: Principles of Knowledge Representation and Reasoning*. San Francisco: Morgan Kaufmann.
2. Allen, J. (1983). Maintaining knowledge about temporal intervals. *Communications of the ACM, 26*(11), 832–843.

[5]John Pollock's work, however, is an exception. Pollock developed a theory of nonmonotonic reasoning that is closely related to Argumentation Theory. See, for instance, [52].

3. Allen, J. F., Fikes, R., & Sandewall, E. (Eds.). (1989). *KR'89: Principles of Knowledge Representation and Reasoning*. San Mateo: Morgan Kaufmann.
4. Allen, J. F., Kautz, H. A., Pelavin, R., & Tennenberg, J. (Eds.). (1991). *KR'91: Principles of Knowledge Representation and Reasoning*. San Mateo: Morgan Kaufmann.
5. Baader, F., Calvanese, D., McGuinness, D. L., Nardi, D., & Patel-Schneider, P. (Eds.). (2003). *The description logic handbook: Theory, implementation and applications*. Cambridge: Cambridge University Press.
6. Baader, F., Horrocks, I., & Sattler, U. (2008). Description logics. In F. van Harmelen, V. Lifschitz, & B. Porter (Eds.), *Handbook of knowledge representation* (pp. 135–179). Amsterdam: Elsevier.
7. Besnard, P. (1992). *Default logic*. Berlin: Springer.
8. Besnard, P., & Hunter, A. (2008). *Elements of argumentation*. Cambridge, MA: The MIT Press.
9. Brachman, R. J., & Levesque, H. J. (Eds.). (1985). *Readings in knowledge representation*. Los Altos: Morgan Kaufmann.
10. Brachman, R. J., & Levesque, H. (2004). *Knowledge representation and reasoning*. Amsterdam: Elsevier.
11. Brewka, G., & Lang, J. (Eds.). (2008). *KR2008: Proceedings of the Eleventh International Conference*. Menlo Park: AAAI Press.
12. Brewka, G., Niemelä, I., & Truscyński, M. (2008). Nonmonotonic reasoning. In F. van Harmelen, V. Lifschitz, & B. Porter (Eds.), *Handbook of knowledge representation* (pp. 239–284). Amsterdam: Elsevier.
13. Church, A. (1959). *Introduction to mathematical logic* (Vol. 1). Princeton: Princeton University Press.
14. Clarke, E. M., Grumberg, O., & Peled, D. A. (1999). *Model checking*. Cambridge, MA: The MIT Press.
15. Cohn, A. G., Schubert, L., & Shapiro, S. C. (Eds.). (1998). *KR'98: Principles of Knowledge Representation and Reasoning*. San Francisco: Morgan Kaufmann.
16. Cohn, A. G., Giunchiglia, F., & Selman, B. (Eds.). (2000). *KR2000: Principles of Knowledge Representation and Reasoning*. San Francisco: Morgan Kaufmann.
17. Davis, E. (1991). *Representations of common sense knowledge*. San Francisco: Morgan Kaufmann.
18. Delgrande, J. P., & Schaub, T. (2000). The role of default logic in knowledge representation. In J. Minker (Ed.), *Logic-based artificial intelligence* (pp. 107–126). Dordrecht: Kluwer Academic Publishers.
19. Doherty, P., Mylopoulos, J., & Welty, C. A. (Eds.). (2006). *KR2006: Proceedings, Tenth International Conference on Principles of Knowledge Representation and Reasoning*. Menlo Park: AAAI Press.
20. Doyle, J., Sandewall, E., & Torasso, P. (Eds.). (1994). *KR'94: Principles of Knowledge Representation and Reasoning*. San Francisco: Morgan Kaufmann.
21. Dubois, D., Welty, C., & Williams, M.-A. (Eds.). (2004). *KR2004: Principles of Knowledge Representation and Reasoning*. AAAI Press.
22. Dung, P. M. (1995). On the acceptability of arguments and its fundamental role in nonmonotonic reasoning, logic programming, and n-person games. *Artificial Intelligence, 77*(2), 321–257.
23. Fensel, D., Giunchiglia, F., McGuinness, D., & Williams, M.-A. (Eds.). (2002). *KR2002: Principles of Knowledge Representation and Reasoning*. San Francisco: Morgan Kaufmann.
24. Gelfond, M. (2008). Answer sets. In F. van Harmelen, V. Lifschitz, & B. Porter (Eds.), *Handbook of knowledge representation* (pp. 285–316). Amsterdam: Elsevier.
25. Hanks, S., & McDermott, D. (1986). Default reasoning, nonmonotonic logics and the frame problem. In T. Kehler & S. Rosenschein (Eds.), *Proceedings of the Fifth National Conference on Artificial Intelligence* (pp. 328–333), Los Altos. Morgan Kaufmann: American Association for Artificial Intelligence.
26. Hayes, P. (1981). The logic of frames. In B. Webber & N. J. Nilsson (Eds.), *Readings in artificial intelligence* (pp. 451–458). Los Altos: Morgan Kaufmann.

27. Hayes, P. (1987). A critique of pure reason. *Computational Intelligence, 3*, 179–185.
28. Horty, J. F. (1994). Moral dilemmas and nonmonotonic logic. *Journal of Philosophical Logic, 23*(1), 35–65.
29. Horty, J. F. (1995). Deontic logic and nonmonotonic reasoning. In D. Nute (Ed.), *Essays in defeasible deontic logic*. Dordrecht: Kluwer Academic Publishers.
30. Horty, J. (1997). Nonmonotonic foundations for deontic logic. In D. Nute (Ed.), *Defeasible deontic logic* (pp. 17–44). Dordrecht: Kluwer Academic Publishers.
31. Horty, J. F. (2001). *Agency and deontic logic*. Oxford: Oxford University Press.
32. Horty, J. F. (2007). Defaults with priorities. *Journal of Philosophical Logic, 36*(4), 367–413.
33. Konolige, K. (1994). Autoepistemic logic. In D. Gabbay, C. J. Hogger, & J. A. Robinson (Eds.), *Handbook of logic in artificial intelligence and logic programming, volume 3: Nonmonotonic reasoning and uncertain reasoning* (pp. 217–295). Oxford: Oxford University Press.
34. Kowalski, R. A., & Toni, F. (1996). Abstract argumentation. *Artificial Intelligence and Law, 4*, 275–296.
35. Levesque, H. J., & Brachman, R. J. (1995). A fundamental tradeoff in KR and reasoning. In R. J. Brachman & H. J. Levesque (Eds.), *Readings in knowledge representation*. Los Altos: Morgan Kaufmann.
36. Lifschitz, V. (1988). Circumscriptive theories: A logic-based framework for knowledge representation. *Journal of Philosophical Logic, 17*(3), 391–441.
37. Lifschitz, V., Morgenstern, L., & Plaisted, D. (2008). Knowledge representation and classical logic. In F. van Harmelen, V. Lifschitz, & B. Porter (Eds.), *Handbook of knowledge representation* (pp. 3–88). Amsterdam: Elsevier.
38. Lin, F. (1995). Embracing causality in specifying the indirect effects of actions. In C. Mellish (Ed.), *Proceedings of the Fourteenth International Joint Conference on Artificial Intelligence* (pp. 1985–1991). San Francisco: Morgan Kaufmann.
39. Lin, F. (2008). Situation calculus. In F. van Harmelen, V. Lifschitz, & B. Porter (Eds.), *Handbook of knowledge representation* (pp. 649–669). Amsterdam: Elsevier.
40. Lin, F., Sattler, U., & Truszczynski, M. (Eds.). (2010). *KR2010: Proceedings of the Twelfth International Conference*. Menlo Park: AAAI Press.
41. McCain, N., & Turner, H. (1995). A causal theory of ramifications and qualifications. In C. Mellish (Ed.), *Proceedings of the Fourteenth International Joint Conference on Artificial Intelligence* (pp. 1978–1984). San Francisco: Morgan Kaufmann.
42. McCarthy, J. (1959). Programs with common sense. In *Proceedings of the Teddington Conference on the Mechanization of Thought Processes* (pp. 75–91), London. Her Majesty's Stationary Office.
43. McCarthy, J. (1969). Situations, actions, and causal laws. In M. Minsky (Ed.), *Semantic information processing* (pp. 410–417). Cambridge: The MIT Press. Originally published in 1963 as a technical report.
44. McCarthy, J. (1980). Circumscription—a form of non-monotonic reasoning. *Artificial Intelligence, 13*(1–2), 27–39.
45. McCarthy, J. (1990). *Formalizing common sense: Papers by John McCarthy*. Norwood: Ablex Publishing Corporation. Edited by Vladimir Lifschitz.
46. McCarthy, J., & Hayes, P. J. (1969). Some philosophical problems from the standpoint of artificial intelligence. In B. Meltzer & D. Michie (Eds.), *Machine intelligence 4* (pp. 463–502). Edinburgh: Edinburgh University Press.
47. McCarthy, J., Minsky, M. L., Rochester, N., & Shannon, C. E. (2006). A proposal for the Dartmouth summer research project on artificial intelligence. *The AI Magazine, 27*(4), 12–14.
48. McDermott, D. (1987). A critique of pure reason. *Computational Intelligence, 3*, 151–160.
49. Minsky, M. (1981). A framework for representing knowledge. In J. Haugeland (Ed.), *Mind design* (pp. 95–128). Cambridge, MA: The MIT Press. Originally published in 1974 as an MIT technical report.
50. Mueller, E. T. (2006). *Commonsense reasoning*. Amsterdam: Elsevier.
51. Nebel, B., Rich, C., & Swartout, W. (Eds.). (1992). *KR'92: Principles of Knowledge Representation and Reasoning*. San Francisco: Morgan Kaufmann.

52. Pollock, J. L. (2001). Defeasible reasoning with variable degrees of justification. *Artificial Intelligence, 133*(1–2), 233–282.
53. Prakken, H., & Vreeswijk, G. (2001). Logics for defeasible argumentation. In D. M. Gabbay & F. Guenthner (Eds.), *Handbook of philosophical logic, volume IV* (2nd ed., pp. 219–318). Amsterdam: Kluwer Academic Publishers
54. Reiter, R. (1980). A logic for default reasoning. *Artificial Intelligence, 13*(1–2), 81–132.
55. Reiter, R. (2001). *Knowledge in action: Logical foundations for specifying and implementing dynamical systems*. Cambridge, MA: The MIT Press.
56. Rissland, E. L., Ashley, K. D., & Loui, R. P. (2003). AI and law: A fruitful synergy. *Artificial Intelligence, 150*(1–2), 1–15.
57. Schaub, T. (1998). The family of default logics. In D. M. Gabbay & P. Smets (Eds.), *Handbook of defeasible reasoning and uncertainty management systems, volume 2: Reasoning with actual and potential contradictions* (pp. 77–134). Dordrecht: Kluwer Academic Publishers.
58. Shanahan, M. (1997). *Solving the frame problem*. Cambridge, MA: The MIT Press.
59. Simon, H. A. (1966). *On reasoning about actions* (Technical Report Complex Information Processing Paper #87), Carnegie Institute of Technology, Pittsburgh.
60. Stefik, M. J. (1995). *An introduction to knowledge systems*. San Francisco: Morgan Kaufmann.
61. Thielscher, M. (2001). The qualification problem: A solution to the problem of anomalous models. *Artificial Intelligence, 131*(1–2), 1–37.
62. Thomason,R. H. (2003). Logic and artificial intelligence. Stanford Encyclopedia of Philosophy. http://plato.stanford.edu/archives/fall2003/entries/logic-ai/.
63. Turner, H. (2008). Nonmonotonic causal logic. In F. van Harmelen, V. Lifschitz, & B. Porter (Eds.), *Handbook of knowledge representation* (pp. 759–776). Amsterdam: Elsevier.
64. van Harmelen, F., Lifschitz, V., & Porter, B. (Eds.). (2008). *Handbook of knowledge representation*. Amsterdam: Elsevier.

Chapter 19
Representing Uncertainty

Sven Ove Hansson

Abstract Our uncertainty about matters of fact can often be adequately represented by probabilities, but there are also cases in which we, intuitively speaking, know too little even to assign meaningful probabilities. In many of these cases, other formal representations can be used to capture some of the prominent features of our uncertainty. This is a non-technical overview of some of these representations, including probability intervals, belief functions, fuzzy sets, credal sets, weighted credal sets, and second order probabilities.

19.1 Uncertainty in Decisions

Many decisions are difficult because we do not know the effects of our alternatives. Consider the following examples:

- You have been offered a free two-day hang-gliding course, but you partner is worried, and says: "You must first find out how dangerous it is."
- You are tempted to buy a lottery ticket. The top prize would solve all your financial problems, but the ticket is quite expensive. Should you buy it?
- You are offered to bet on a horse but you do not know its chances.
- The authorities have to decide if a new chemical can be used, but they do not know what effects it may have on human health and the environment.

In the first two cases it is likely that probabilities can be of some help (but the decision may still be a very difficult one). In the last two cases it is not obvious that probabilities can at all be used. Can formal methods nevertheless be used to throw light on decision problems like these?

The formal representation of uncertainty has mostly been discussed in contexts of decision-guidance, but the topic is interesting also apart from applications to decision-making. Formal representations can be used to distinguish between

S. O. Hansson (✉)
Division of Philosophy, Royal Institute of Technology (KTH), Stockholm, Sweden
e-mail: soh@kth.se

S. O. Hansson, V. F. Hendricks (eds.), *Introduction to Formal Philosophy*, Springer Undergraduate Texts in Philosophy, https://doi.org/10.1007/978-3-319-77434-3_19

different kinds of ignorance or lack of knowledge, and they can also contribute to our understanding of phenomena such as belief change and conditional beliefs.

19.2 Possibility Sets and Weight Functions

You can be uncertain about various matters, for instance about the state of the world, how to describe it, what options are available to you in a decision, your evaluation of various outcomes, and your own moral and philosophical principles. This chapter is devoted to uncertainty about the state of the world and about the correctness of our descriptions of it.[1] The simplest representation of such uncertainty is what we may call a *possibility set*, a set consisting of the alternatives one is uncertain between. Suppose that I know that either Ann, Bob, or Cai baked the cake, but I do not know which of them. Then the cake was baked by a member of the possibility set {Ann, Bob, Cai}.

In many cases, our uncertainty concerns a number, i.e. the value of a numerical variable. Then the possibility set will be a set of numbers, a *numerical possibility set*. In the most common applications, such sets have the form of intervals. Even if you do not know how much money you have on your account, you may know that you have between €500 and €1000. Then the number of euros on your account is an element of [500, 1000], the set of real numbers not lower than 500 and not higher than 1000. The common rules of arithmetic can be extended to intervals, as outlined in Box 19.1 and illustrated in the following examples [22, pp. 11–14]:

> She has between €1000 and €2000 and he has between €500 and €700. Thus, together they have between €1500 and €2700, and she has between €300 and €1500 more than he has.
>
> There will be between 10 and 20 participants in the competition, and each of them will use between 3 and 5 fishing-rods. Therefore, between 30 and 100 fishing-rods will be used.

Box 19.1 Some rules of interval arithmetic

$$[a, b] + [c, d] = [a + c, b + d]$$
$$[a, b] - [c, d] = [a - d, b - c]$$
$$[a, b] \times [c, d] = [\mathrm{min}(a \times c, a \times d, b \times c, b \times d), \mathrm{max}(a \times c, a \times d, b \times c, b \times d)]$$
$$[a, b] \times [c, d] = [a \times c, b \times d] \text{ if } a, b, c, \text{ and } d \text{ are all non-negative.}$$

---

[1]It is sometimes unclear whether an agent's uncertainty in a particular matter concerning herself is attributable to her lack of factual information or to the fact that she has not made some decision that could have resolved the uncertainty. This type of "ambiguous" uncertainty underlies several of the well-known decision-theoretical paradoxes [15].

Possibility sets provide a simple but also very limited form of uncertainty representation. We often have reasons to differentiate between the elements, and put more weight on some than on the others. If I know that Cai is very fond of baking, then I may give more weight to the possibility of her being the baker than to Ann or Bob. This can be expressed by assigning numbers to each of them. I may for instance introduce a function f such that $f(\text{Ann}) = 0.25$, $f(\text{Bob}) = 0.25$, and $f(\text{Cai}) = 0.50$. This is our second major form of uncertainty representation, a *weighted possibility set*. It consists of a possibility set and a *weight function* that assigns a weight to each of its elements. We will assume that the weights are non-negative and that they add up to 1.

The weight function is often construed as a probability function. It is important to recognize that the notion of probability has a precise mathematical definition. A probability function is a function that satisfies the laws of probability, the laws obeyed by random events in the real world, such as throws of dice and coins. These laws are elegantly summarized in the Kolmogorov axioms that are given in Box 19.2. A mathematical entity that does not satisfy these laws should not be called "probability".

There are two major interpretations of probabilities. First, as in Fig. 19.1, we can think of them as mental entities. In that case, if you assign the probability 1/6 to the dice yielding a six, you make a report on your own state of mind, a "subjective" probability. Alternatively, as in Fig. 19.2, we can think of probabilities as properties of the physical world, existing independently of our minds. In that case your report on the dice is a statement about (tendencies in) the world, an "objective" probability. These two interpretations lead us to quite different developments in the representation of uncertainties. Let us begin with the former.

19.3 The Subjective Interpretation

Many proponents of the subjective interpretation are Bayesians. They maintain that in order to be rational, a person's subjective degrees of belief have to comply with the probability axioms. According to this view, rational uncertainty can always be represented by a probability function. If you are uncertain about whether Bern is the capital of Switzerland, then there must be some definite probability lower than 1 that you can assign to the statement "Bern is the capital of Switzerland". For the Bayesian, uncertainty is just another name of probability.

The major argument for this position is also an argument for the maximization of expected (i.e. probability-weighted) utility. It can be shown that a person who does not abide by this decision rule, or whose probability assignments violate the probability axioms, can have a Dutch book made against her. By this is meant a bet that she would be sure to lose whatever the outcome of the game would be. (See Box 19.3.) Since this is irrational behaviour, it can then be concluded that in order to be rational we should all make our decisions in accordance with a subjective probability assignment that obeys the axioms [16, pp. 381–382].

Box 19.2 Axioms for degree-of-belief representations

The Kolmogorov axioms for probability

Let Ω be the set of all events under consideration. $A, B, \ldots$ are sets of events, i.e. subsets of Ω. Let p be a function from sets of events to real numbers. Then p is a *probability function* if and only if it satisfies the following three axioms:

1. $p(A) \geq 0$ for all $A \subseteq \Omega$. (non-negativity)
2. $p(\Omega) = 1$ (normalization)
3. If $A \cap B = \varnothing$, then $p(A \cup B) = p(A) + p(B)$. (finite additivity)

The above formulation can only be used if the number of events (elements of Ω) is finite. In the more general case, axiom 3 has to be reformulated, and the following should hold for an infinite series $A_1, A_2, \ldots$ of events such that no two of them have an element in common:

3'. $p(A_1 \cup A_2 \cup \ldots) = \sum_{n=1}^{\infty} A_n$ (additivity)

Axioms for other degree-of-belief representations

Alternative degree-of-belief representations usually satisfy the first two but not the third of the Kolmogorov axioms. Instead of (3), the belief function Bel of Dempster-Shafer theory satisfies the following:

4. If $A \cap B = \varnothing$, then $\mathsf{Bel}(A \cup B) \geq \mathsf{Bel}(A) + \mathsf{Bel}(B)$. (finite superadditivity)

In possibility theory, (3) is replaced by the following requirement on the possibility measure Poss:

5. $\mathsf{Poss}(A \cup B) = \max(\mathsf{Poss}(A), \mathsf{Poss}(B))$

Importantly, this argument refers to the requirements of rational decision-making, not those of rational belief. The demands of rational action (practical rationality) and those of rational belief (theoretical rationality) need not coincide. Holding certain beliefs may be rational in the sense of furthering the achievement of practical goals without being rational in the ratiocinative sense. The Dutch book argument does not tell us what we should believe, only what we should act as believing.

Apart from that, empirical studies have shown that human behaviour commonly violates the Kolmogorov axioms [5, 10]. Therefore, if we are looking for a representation of actual human uncertainty, then we may have good reasons to look for one that does not satisfy the axioms. And of course, if we are not convinced

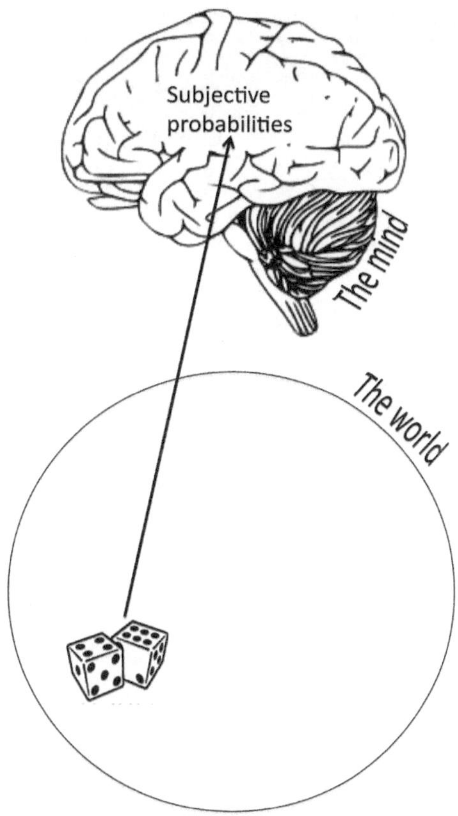

Fig. 19.1 Subjective probabilities

by the Dutch book argument, then we are free to choose rationality principles for (decision-guiding) beliefs based on other criteria.

Several numerical representations of subjective degrees of belief have been presented that are weaker, i.e. require less, than the Kolmogorov axioms. Among the most important of these are the so-called *belief functions* of Dempster-Shafer theory [2, 25]. The characteristic difference between probability functions and belief functions is that the latter can assign more weight to a set than the sum of what it assigns to its elements. For an example, consider the two mutually exclusive events "The Nigerian team will defeat the South African one tomorrow" (N) and "The South African team will defeat the Nigerian team tomorrow" (S). First suppose that your degrees of belief are represented by a probability function p that assigns the value 0.1 to each of the two events, i.e. $p(N) = p(S) = 0.1$. It then follows that $p(\{N, S\}) = 0.2$, i.e. the probability of either team winning is the sum of the probabilities of each of them winning. Next, suppose that instead, your degrees of belief are represented by a belief function Bel, also such that $\mathsf{Bel}(N) = \mathsf{Bel}(S) = 0.1$. From this we cannot conclude what the value of $\mathsf{Bel}(\{N, S\})$ is. It may be equal to 0.2, but it may also be higher, for instance 0.8. This is sensible under some

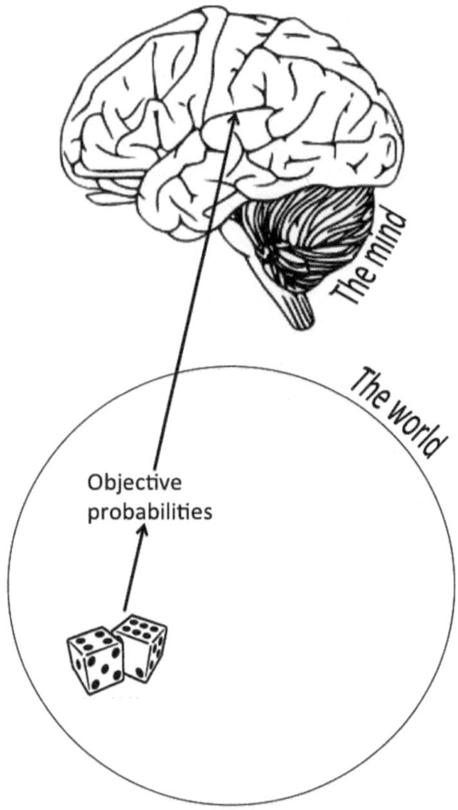

Fig. 19.2 Objective probabilities

interpretations of Bel, since you may have reasons to believe that either N or S will happen, without having much evidence in favour of either of the two alternatives.

All this has been based on the assumption that the subjective uncertainty that we intend to represent reflects our lack of knowledge about the physical world. However, there are also other sources of uncertainty, in particular vagueness and ambiguities in our language.

ADILAH: Look at him! How would you describe him? Is he bald or not?
MIAHUA: I am inclined to call him bald, but I do not know for sure.

Miahua is not uncertain about what the man's head looks like, but she is uncertain about how to use the word "bald". She may assign weights to the two alternatives, perhaps 0.6 to "bald" and 0.4 to "not bald". However, these numbers need not be interpretable as probabilities. Another option is to interpret them as indicating *fuzzy set* membership.

Box 19.3 An example of a Dutch book

Bob entertains the following beliefs about a dice:

The probability that it yields "1" is 1/6.
The probability that it yields "4" is 1/6.
The probability that it yields "6" is 1/6.
The probability that it yields a prime number is 4/6.

Martha offers Bob the following bets:

If you pay €2 you get €13 if it yields "1".
If you pay €2 you get €13 if it yields "4".
If you pay €2 you get €13 if it yields "6".
If you pay €8 you get €13 if it yields a prime number.

Since Bob believes all these bets to be favourable he accepts them all. As a result of this, he pays €14, and whatever the outcome of the throw he will receive €13 back. This is a sure loss, in other words a Dutch book. The reason for this failure is that his probability assignments do not satisfy the Kolmogorov axioms. (See Box 19.2.)

In common ("crisp") set theory, there is always a definite answer to whether or not an object is an element of a given set. A set can be represented by an indicator function (membership function, element function). Let μ_A be the indicator function of a set A. Then for all x, $\mu_A(x)$ is equal to 1 if x is an element of A, and equal to 0 if it is not. The function does not take any other value than 0 or 1. In contrast, the indicator function of a fuzzy set can take any value in the interval [0, 1]. If $\mu_A(x) = .5$, then x is "half member" of A, and if $\mu_A(x) = .6$, then it is somewhat more member than non-member [28]. Fuzzy sets can be used to represent vagueness, such as the vagueness that made Miahua uncertain whether the man was bald or not [26, p. 27]. Fuzzy set membership does not satisfy Kolmogorov's axiom system.

Possibility theory is a variant of fuzzy set theory in which a function Poss with the properties of an indicator function replaces the probability function as a measure of uncertainty [3, 29].

19.4 Credal Sets

Let us return to Fig. 19.2, and to the account of probabilities that treats them as properties of the physical world. This is an interpretation with a strong intuitive appeal. Suppose I tell you that the probability that a certain dice will yield a six is 1/6. You investigate the dice and find that it is loaded and yields a six in about 1 in 3 throws. When telling me this, you expect me to say "Oh, then I was wrong", rather than "But what I said was right, since I reported a probability, and probabilities are states of mind". In cases like these we take probabilities to represent the (counterfactual) frequency that would be recorded if a similar triggering event were repeated under similar circumstances a large number of times. The probability that an atom of Fermium-257 will decay within the next 100 days is close to 0.5. This is a property of the natural world, not merely a subjective belief.

But things do not end here. If there are objective probabilities, then we can be (subjectively) uncertain about which these probabilities are. Consider the following example:

> There are two urns in the room. One of them contains 5 red and 95 black balls. The other contains 95 red and 5 black ones. Someone puts one of the urns – you do not know which – in front of you and asks: "If you draw a ball from this urn, what is the probability that it is red?"

A quite natural answer would be "It is either 0.05 or 0.95". More precisely, you hesitate between two probability functions, p_1 and p_2, which differ in the probabilities they assign to the event (R) that the ball you draw is red. We have $p_1(R) = 0.05$ and $p_2(R) = 0.95$. Your uncertainty can then be expressed with the simplest uncertainty representation introduced above, namely a possibility set $\{p_1, p_2\}$. A possibility set that has probability functions as its elements is called a *credal set* ([20]; cf. [4]). This way of representing uncertainty has been favoured by many philosophers, and also developed in considerable technical detail by statisticians [1]. As one example of its use, climatologists studying the effects of climate change on the probability of extreme weather events employ a range of credible models that generate different probability functions. The outcome of their calculations can then be expressed as a credal set, containing a range of probability functions [9].

In many cases, uncertainty about a specific probability can be expressed as a *probability interval*. Summarizing the outcomes of climate modelling, we may say for instance: "Around 2050, the yearly probability of this whole valley being flooded will be between 1% and 4%." In general, if we have a credal set $\{p_1, \ldots, p_n\}$, then for each event A we can find the minimal probability that it can assign to an event A, i.e. the lowest value of $p_k(A)$ for any p_k. This is the lower probability generated by the credal set, often denoted $\underline{p}(A)$. The upper probability $\overline{p}(A)$ is defined analogously, and together they form the probability interval.

Probability intervals are an intuitively accessible and often quite useful way to express uncertainties [11]. However, it must be observed that information is lost when we simplify a credal set $\{p_1, \ldots, p_n\}$ to the pair $\langle \underline{p}(A), \overline{p}(A) \rangle$ [27]. Consider the following example:

> A dollar coin has been found among the property of a deceased cardsharp. We suspect that it may be unfair. We do not know in which direction it would then be biased, but we know that the most biased coins available to him yield either heads or tails with a frequency of 90%.

Here, the probability of heads in a single throw can be represented by a credal set containing all the probability functions assigning probabilities between 0.1 and

0.9 to heads in a single throw (H). Therefore, $\underline{p}(H) = 0.1$ and $\overline{p}(H) = 0.9$. At first sight it seems as if we can perform further calculations using just $\underline{p}(H)$ and $\overline{p}(H)$. For example, the probability of three heads in a row (HHH) is anywhere in the interval [0.001, 0.729], so that the lower limit coincides with $(\underline{p}(H))^3$ and the upper with $(\overline{p}(H))^3$. However, this pattern cannot be generalized. The probability of getting heads in the first but not in the second throw (HT) is anywhere in the interval [0.09, 0.25]. This interval cannot be calculated from $\underline{p}(H)$ and $\overline{p}(H)$. In fact it is not difficult to construct a credal set that has the same values of $\underline{p}(H)$ and $\overline{p}(H)$ but a different value of $\overline{p}(HT)$.[2] As this example shows, the use of probability intervals instead of the full credal set is as mathematically precarious as it is intuitively accessible, and great care must therefore be exercised in the calculative use of such intervals.

19.5 Weighted Credal Sets

The elements of a credal set may differ in their credibility. To represent these differences we can assign weights to them, thus obtaining a weighted credal set. Most commonly, these weights are probabilities, which leads to a model with probabilities on two levels, as illustrated in Fig. 19.3.

The use of two levels of probabilities (first- and second-order probabilities) has been put in doubt by philosophers since David Hume [17, pp. 182–183] who have worried that if we allow for two levels, then there is no way to avert an infinite regress of higher and higher orders of probability. However, since the two levels in this model represent different types of entities, the process that took us from the first to the second level cannot necessarily be repeated. If it cannot, then no such regress gets started.

Another common criticism is that the two levels of probabilities are unnecessary since they can always be reduced to one. The following variant of our cardsharp example can be used to assess that argument:

> Among the property of the deceased cardsharp we find a dime that may or may not be biased. All biased dimes that have been made have either a 0.1 or a 0.9 probability of heads. Our credal set is $\{p_1, p_2, p_3\}$, where $p_1(H) = 0.1$, $p_2(H) = 0.5$, and $p_3(H) = 0.9$. Furthermore, our subjective probability function $\mathfrak{p}$ over the credal set is such that $\mathfrak{p}(p_1) = 0.25$, $\mathfrak{p}(p_2) = 0.5$, and $\mathfrak{p}(p_3) = 0.25$.

We are now going to throw the dime once. Given all this information, what probability should we assign to it landing heads? The obvious answer is 0.5. We obtain this answer by reducing the two levels to one [23, p. 58]. Letting $\hat{\mathfrak{p}}$ denote the resulting reductive probability function we have:

---

[2]Let the credal set consist of all probability functions that assign to $p(H)$ a value in either of the two intervals [0.1, 0.2] and [0.8, 0.9]. Then we still have $\underline{p}(H) = 0.1$ and $\overline{p}(H) = 0.9$, but $\overline{p}(HT) = 0.16$.

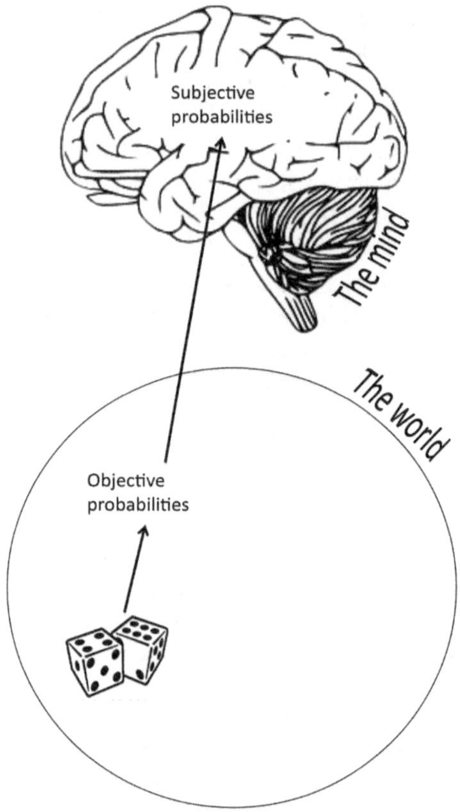

Fig. 19.3 Two-levelled probabilities

$$\hat{\mathfrak{p}}(H) = \mathfrak{p}(p_1) \times p_1(H) + \mathfrak{p}(p_2) \times p_2(H) + \mathfrak{p}(p_3) \times p_3(H) = 0.5$$

At this point, a critic of the two-levelled model might well ask: "So what is the point? Why all this trouble when you could instead, directly, have assigned the subjective probability 0.5 to H?" But there is a good answer to that. Suppose that we also want to know the probability of obtaining three heads in a row when the coin is tossed thrice. That probability is:

$$\hat{\mathfrak{p}}(HHH) = \mathfrak{p}(p_1)\times p_1(HHH)+\mathfrak{p}(p_2)\times p_2(HHH)+\mathfrak{p}(p_3)\times p_3(HHH) = 0.245$$

which is different from $(\hat{\mathfrak{p}}(H))^3 = 0.125$. Thus, if we had followed the advice of the critic we would have obtained the wrong answer to problems involving iterated tosses of the coin. In order to get the right answer in a one-levelled model we would have to give up the assumption that the different tosses in a series are independent events, which is the precondition for deriving the probability of HHH from that of H in the standard way [13]. Hence, the reduction of two-levelled to one-levelled probabilities comes at a high price, namely that we cannot treat events as independent which we intuitively perceive as such. In order to make sense of the complex world that we live in we have to treat some events as similar but independent. This applies to both everyday and scientific reasoning. It is for instance essential for the conceptualization of a repeatable scientific experiment. Although two-levelled probabilities can in principle always be reduced to a single level, the reduced account often has much less explanatory value.

The two-levelled model also has the advantage of providing a lucid account of how empirical information makes us change our views on objective (physical) probabilities. Suppose that we throw the above-mentioned coin five times, and get a series of five heads ($HHHHH$ or in short H^5). This makes us revise $\mathfrak{p}$ and, in particular, increase the value of $\mathfrak{p}(p_3)$, but how much? This we can find out with standard conditionalization on the second level, i.e.

$$\mathfrak{p}(p_3 \mid H^5) = \frac{\mathfrak{p}(H^5 \mid p_3) \times \mathfrak{p}(p_3)}{\mathfrak{p}(H^5 \mid p_1) \times \mathfrak{p}(p_1) + \mathfrak{p}(H^5 \mid p_2) \times \mathfrak{p}(p_2) + \mathfrak{p}(H^5 \mid p_3) \times \mathfrak{p}(p_3)}$$
$$\approx 0.90$$

and similarly $\mathfrak{p}(p_1 \mid H^5) \approx 2 \times 10^{-5}$ and $\mathfrak{p}(p_2 \mid H^5) \approx 0.10$. We now have a revised second-level probability function, $\mathfrak{p}(\ \mid H^5)$, which directly gives rise to a new reductive probability function $\hat{\mathfrak{p}}(\ \mid H^5)$. With its help we can answer the question "Given what we know after observing H^5, what is now our best estimate of the probability that a toss of this coin yields heads?" The answer is $\hat{\mathfrak{p}}(H \mid H^5) \approx 0.86$. This reductive conditionalization is of course very different from standard conditionalization. An ordinary probability function p yields $p(H \mid H^5) = 1$ since given that H^5 took place it is certain that H also took place [13].

The following example illustrates the practical importance of this type of probability revision:

> Two estimates have been made of the probability that a major explosion will take place in the first year's operation of a new explosives factory (E). According to one estimate, that probability is $p_1(E) = 0.0001$, and according to the other it is $p_2(E) = 0.01$. We believe the former estimate to be much more reliable than the latter, and we have $\mathfrak{p}(p_1) = 0.999$ and $\mathfrak{p}(p_2) = 0.001$. Hence the (reductive) probability of an explosion is $\hat{\mathfrak{p}}(E) \approx 0.00011$, which means that we can almost disregard p_2.
>
> But after a couple of months a major explosion takes place. We therefore update our second-level probability and obtain $\mathfrak{p}(p_1 \mid E) \approx 0.91$ and $\mathfrak{p}(p_2 \mid E) \approx 0.09$. These estimates, rather than the original ones, should be used for instance when we consider whether to open another factory of exactly the same type.

Revisions of this type are an essential mechanism for learning from experience. The change in an epistemic agent's estimate of the (objective) probability of an event E that ensues after learning that E has occurred can be used as a numerical measure of the agent's uncertainty concerning the probability of E [14].

We have focused in this section on the use of probability functions on the second level in a two-levelled system. However, proposals have also been made to assign weights other than probabilities to credal sets, such as fuzzy sets or other measures of "epistemic reliability" that do not satisfy the Kolmogorov axioms [7].

19.6 Decision-Making Under Uncertainty

Decision rules for uncertainty can be classified according to the information that they require about the options that are available to be chosen between.[3]

Possibility sets. Suppose that for each of the options that we can choose between we have a possibility set telling us what outcomes can follow, but we know nothing about the comparative plausibility of these outcomes. In this case we have to base the decision on the values of the outcomes. The most common decision rule for this purpose is the *maximin* rule. It tells us to identify for each option its security level, i.e. the value of the worst outcome that it can result in. We should then choose one of the options with the highest security level. According to a variant of this rule, the *leximin rule* (lexicographic maximin), if there is more than one alternative with the highest security level, then the one with the highest second-worst outcome should be chosen. If there is more than one alternative with the highest value of the second-worst outcome, then the third-worst outcomes are compared, etc. [24].

The maximin rule is maximally cautious. The other extreme is represented by the *maximax rule* according to which we should choose one of the alternatives with the highest hope level (level of the best outcome). If the values of the outcomes can be expressed in numbers, then a middle road can be chosen with the help of *Hurwicz's index*. This is a number α between 0 and 1 that expresses the degree of cautiousness (not the degree of pessimism, although that is how it is usually described). The recommendation is to choose an option that maximizes the value of

$$\alpha \times \mathfrak{min}(A) + (1 - \alpha) \times \mathfrak{max}(A)$$

where $\mathfrak{min}(A)$ is the security level and $\mathfrak{max}(A)$ the hope value [18].

Credal set. Given a credal set, we can apply the *maximin expected utility rule* (MMEU). Then for each option we have to find the probability function that gives rise to the lowest expected utility (probability-weighted utility). This is the probabilistic security level of the option in question. The rule requires that we choose an option with the highest possible probabilistic security level [6]. An alternative would be to calculate both the probabilistic security level and the analogous probabilistic hope level, and then apply Hurwicz's index to find a compromise between the two.

Weighted credal set. If we have a weighted credal set, then we can calculate the weighted average of the probabilities, assigning the appropriate weight to each probability. Each weighted average is then multiplied with the corresponding utility, in the usual manner of expected utility calculations. If the weights are second-order

[3]For more information on decision rules, see Chap. 34.

probabilities, then this amounts to using reductive probabilities to calculate expected utilities. The same method can also be used for weights that do not satisfy the probability axioms.

Daniel Ellsberg [4] proposed an adjustment of this rule to make it more cautious. Suppose that we have a measure (such as $\hat{\mathfrak{p}}$) that represents the best probability estimate. We can then weigh it against the probabilistic security level, using an index of the same type that was proposed by Hurwicz. The resulting value can be described as a cautionized variant of expected utility.

Hence, even if we have limited information about probabilities, or none at all, formal representations of uncertainty make it possible to consistently apply decision criteria such as degrees of cautiousness. But of course, when more information is available, it is mostly advisable to apply a decision procedure that makes use of it.

Acknowledgements I would like to thank Richard Bradley and Karin Edvardsson Björnberg for very useful comments on an earlier version of this text.

References and Proposed Readings

Asterisks (∗) indicate recommended readings.

1. Berger, J. O. (1994). An overview of robust Bayesian analysis. *Test, 3*, 5–59.
2. Dempster, A. P. (1968). A generalization of Bayesian inference. *Journal of the Royal Statistical Society. Series B (Methodological), 30*, 205–247.
3. *Dubois, D., & Prade, H. (2001). Possibility theory, probability theory and multiple-valued logics: a clarification. *Annals of Mathematics and Artificial Intelligence, 32*, 35–66. [Provides a philosophical background to possibility theory and some related approaches.].
4. Ellsberg, D. (1961). Risk, ambiguity, and the savage axioms. *Quarterly Journal of Economics, 75*, 643–669.
5. Emby, C. (2009). A comparison of elicitation methods for probabilistic multiple hypothesis revision. In V. Arnold (Ed.), *Advances in accounting behavioral research* (Vol. 12, pp. 85–108). Bingley: Emerald Group Publishing.
6. Gärdenfors, P. (1979). Forecasts, decisions and uncertain probabilities. *Erkenntnis, 14*, 159–181.
7. Gärdenfors, P., & Sahlin, N. -E. (1982). Unreliable probabilities, risk taking, and decision making. *Synthese, 53*, 361–386.
8. *Gärdenfors, P., & Sahlin, N. -E. (Eds.). (1988). *Decision, probability, and utility: Selected readings*. Cambridge: Cambridge University Press. [Excellent collection with many of the classic texts, in particular on second-order measures.].
9. Ghosh, S., & Mujumdar, P. P. (2009). Climate change impact assessment: Uncertainty modeling with imprecise probability. *Journal of Geophysical Research Atmospheres, 114*, D18113.
10. Grinblatt, M., & Keloharju, M. (2009). Sensation seeking, overconfidence, and trading activity. *Journal of Finance, 64*, 549–578.
11. *Hall, J., Twyman, C., & Kay, A. (2005). Influence diagrams for representing uncertainty in climate-related propositions. *Climatic Change, 69*, 343–365. [An introduction to probability intervals with useful examples of how they can be used in practice.]
12. *Halpern, J. (2003). *Reasoning about uncertainty*. Cambridge, MA: MIT Press. [An excellent book-length introduction to the topic of this chapter.]

13. Hansson, S. O. (2008). Do we need second-order probabilities? *Dialectica, 62*, 525–533
14. Hansson, S. O. (2009). Measuring uncertainty. *Studia Logica, 93*, 21–40.
15. Hansson, S. O. (2017). Uncertainty and control. *Diametros, 53*, 50–59.
16. Harsanyi, J. C. (1977). On the rationale of the Bayesian approach: Comments on professor Watkin's paper. In R. E. Butts & J. Hintikka (Eds.). *Foundational problems in the special sciences* (pp. 381–392). Dordrecht: Reidel.
17. Hume, D. ([1739] 1888). *A treatise of human nature*, Reprinted from the original edition in three volumes and edited, with an analytical index, by L.A. Selby-Bigge. Oxford: Clarendon Press.
18. Hurwicz, L. (1951). *The generalized Bayes-minimax principle: A criterion for decision-making under uncertainty* (Cowles Commission Discussion Paper, 355). Chicago. http://cowles.econ.yale.edu/P/ccdp/st/s-0355.pdf.
19. *Kyburg, H. E., Jr. (1992). Getting fancy with probability. *Synthese, 90*, 189–203. [Brief but informative discussion of probability and uncertainty.].
20. Levi, I. (1980). *The enterprise of knowledge*. Cambridge, MA: MIT Press.
21. *Miranda, E. (2008). A survey of the theory of coherent lower previsions. *International Journal of Approximate Reasoning, 48*, 628–658. [Overview on probability intervals and related approaches.].
22. Moore, R. E. (1979). *Methods and applications of interval analysis*. Philadelphia: SIAM.
23. Savage, L. J. (1972). *The foundations of statistics* (2nd Ed.). New York: Dover.
24. Sen, A. K. (1970). *Collective choice and social welfare*. San Francisco: Holden-Day.
25. Shafer, G. (1976). *A mathematical theory of evidence*. Princeton: Princeton University Press.
26. Unwin, S. D. (1986). A fuzzy set theoretic foundation for vagueness in uncertainty analysis. *Risk Analysis, 6*, 27–34.
27. *Walley, P. (2000). Towards a unified theory of imprecise probability. *International Journal of Approximate Reasoning, 24*, 125–148. [A thorough introduction to credal sets and related approaches.].
28. Zadeh, L. (1965). Fuzzy sets. *Information and Control, 8*, 338–353.
29. Zadeh, L. (1978). Fuzzy sets as the basis for a theory of possibility. *Fuzzy Sets and Systems, 1*, 3–28.

Chapter 20
Belief Change

Sven Ove Hansson

Abstract All formal models of belief change involve choices between different ways to accommodate new information. However, the models differ in their loci of choice, i.e. in what formal entities the choice mechanism is applied to. Four models of belief change with different loci of choice are investigated in terms of how they satisfy a set of important properties of belief contraction and revision. It is concluded that the locus of epistemic choice has a large impact on the properties of the resulting belief change operation.

20.1 Requirements on Belief Change

The rationality of beliefs can be discussed either in a static or a dynamic perspective. In a static perspective, we discuss what rationality requires of a person's state of beliefs, taken as a snapshot at a particular point in time. In a dynamic perspective, the topic is instead how one should rationally change one's state of belief in response to new information. The logical investigation of belief change came to light in the 1980s and is therefore a comparatively new area of formal philosophy.

In the static approach to belief rationality we can distinguish between on the one hand substantial requirements of rationality and on the other hand formal (or structural) requirements. For instance, suppose that after his university studies in biology, Donald still believe that dinosaurs and humans have once lived side by side. Then we would consider his belief system to be irrational, for substantial rather than formal (structural) reasons. If he believes that all snakes are poisonous, and he also believes that the black-tailed python is a non-poisonous snake, then this provides us with another, structural, type of reason to consider his current belief system to be irrational. The crucial criterion why this is a structural rather than a substantial failure of rationality is that we can discern the irrationality without even knowing

S. O. Hansson (✉)
Division of Philosophy, Royal Institute of Technology (KTH), Stockholm, Sweden
e-mail: soh@kth.se

S. O. Hansson, V. F. Hendricks (eds.), *Introduction to Formal Philosophy*, Springer Undergraduate Texts in Philosophy, https://doi.org/10.1007/978-3-319-77434-3_20

the meaning of the three terms "snake", "black-tailed python" and "poisonous". Structural rationality falls within the purview of logic and will therefore be at our focus here.

Can structural requirements be made also in the dynamic approach to belief rationality? More precisely, are there sensible rationality requirements on the process of belief change? The following examples are intended to show that there may indeed be such requirements, but also that these requirements may not be uncontroversial:

> *Example 1* For many years, Derek was confident that his wife was faithful to him. But one day a neighbour told him stories that convinced him that she was cheating on him. When he confronted her, she could explain everything, and he regained his previous strong belief in her faithfulness. But something strange happened. He never regained his belief that she loved him. He could not explain why. All misunderstandings had been cleared, and everything else was as before, but still he was unable to believe in her love any more.

Derek's pattern of belief change contradicts the following, seemingly quite compelling requirement on rational belief change:

> *The recovery principle:* If the agent first gives up and then fully regains a belief, then she will also regain all the beliefs that she had before this sequence of loss and regain took place.

But obvious as it may seem, the recovery principle is far from uncontroversial. The following example has been put forward to show that it does not hold in general:

> *Example 2* [10] First I believed that Cleopatra had a son, and therefore of course also that she had a child. But then a person whom I rely on told me that the book in which I learned this was a historical novel. Accepting this, I gave up my belief that Cleopatra had a child. But soon afterwards I heard a highly respected scholar mention in passing that Cleopatra was a mother. Then I again believed that Cleopatra had a child. However, I did not regain my previous belief that she had a son.

The pattern exhibited in this example contradicts the principle of recovery. After first losing and then taking back my belief that Cleopatra had a child, I still lack one of the beliefs that I had originally, namely that she had a son. There has been a considerable debate on whether examples like this disprove the recovery principle or they are so untypical that the principle is still a useful idealization [8, 15].

Our next example concerns the acquisition of new beliefs:

Example 3

LOGICIAN: Yesterday you told me that you had no idea whatsoever whether Mohammed has any children. Now you profess to be firmly convinced that he is a father. What has happened?

JESSICA: Yesterday evening I saw him in the supermarket with baby food in his shopping cart.

LOGICIAN: I don't see how that can prove him to be a father. He could for instance be shopping for a friend or a relative. Is this all the new information that made you change your mind?
JESSICA: Yes it is.
LOGICIAN: Is there any way in which you can logically derive that he is a father from the fact that he bought baby food, perhaps in combination with something else you knew before?
JESSICA: No, I don't think so.
LOGICIAN: You disappoint me. You received new information that does not contradict what you believed before. Then you can conclude whatever follows logically from the new information in combination with your previous beliefs. But that is all. If you go beyond that, how can you be trusted as a rational thinker?
JESSICA: I'm sorry if I disappoint you, but given what I saw, it seems so plausible that he is a father that I can't help believing it.

Our Logician advances the following guideline for rational belief change:

> *The principle of deductivism:* If the agent adopts a new belief that does not contradict her previous beliefs, then she comes to believe in everything that follows logically from the combination of her old beliefs and the new belief, but nothing beyond that.

As should be clear from the example, this is a contestable principle since it requires, essentially, that we refrain from making any non-deductive inferences.

The theory of belief change is concerned with how human beings change their beliefs. Since our brains (and minds) are finite, we should expect them only to have room for a finite number of beliefs. However, since our language is unlimited, we have to be careful about how we express the requirement of finitude. For instance, I believe in each of the sentences on the following infinite list:

Beethoven completed less than 10 symphonies.
Beethoven completed less than 11 symphonies.
Beethoven completed less than 12 symphonies.
...
Beethoven completed less than 1.000 symphonies.
...

Each of these sentences differs in meaning from all the others, so it follows from my assent to all of them that my set of beliefs contains infinitely many sentences that are unique in terms of meaning. But obviously, this does not make my set of beliefs infinite in any interesting way. All of these sentences follow from the first. Instead of requiring that the set of beliefs be finite we should require it to be *finite-based*, i.e. everything that it contains should follow logically from some finite set of beliefs.

Finite-basedness is of course a static requirement. The corresponding dynamic requirement is that finite-basedness should be preserved under belief change. This should apply both when we add a new belief and when we remove an old one. Thus the following two principles should both apply:

The principle of finite-based contraction: If an agent with a finite-based set of beliefs gives up one of these beliefs, then her new set of beliefs is also finite-based.

The principle of finite-based revision: If an agent with a finite-based set of beliefs adopts a single new belief, then her new set of beliefs is also finite-based.

We now have four principles of rationality for belief change that are all expressible in natural language: Recovery, Deductivism, Finite-based contraction, and Finite-based revision. In the remainder of this chapter, we will introduce some formal approaches to belief change and use these four postulates to compare their properties.

20.2 A Basic Framework for Belief Change

To begin with, we need a general framework in which we can express different approaches to belief change. Following the well-established tradition in the field, we will assume that the agent's (static) belief state at each point in time is represented by a set of sentences, called the *belief set* and denoted K. The belief set is logically closed (closed under logical consequence), by which is meant that it contains everything that it logically implies. This can be expressed with a consequence relation Cn, such that for any set X of sentences, $\mathrm{Cn}(X)$ is the set of its logical consequences. Our requirement that K is logically closed can then be expressed with the simple formula $K = \mathrm{Cn}(K)$.

Logical closure is an idealization, and obviously not a realistic property of a person's set of beliefs. No one has sufficient logical and mathematical competence to believe in everything that follows logically from what she believes. However, as an idealization it is quite useful, since it allows us to work with much simpler formal models than what we would otherwise have needed.

In the logic of belief change, all changes result from inputs. An input usually consists in a sentence and an instruction saying what to do with that sentence. Standardly there are three such instructions, namely "remove this sentence", "add this sentence", and "add this sentence and retain consistency".

The instruction "remove this sentence" is performed with an operation of *contraction*, usually denoted $\div$. Thus $K \div p$ is the outcome of removing p from K. Contraction is assumed to satisfy the following postulates:

$$K \div p \subseteq K \text{ (inclusion)}$$

$$K \div p = \mathrm{Cn}(K \div p) \text{ (closure)}$$

$$p \notin K \div p, \text{ unless } p \text{ is a logical truth (success)}$$

The instruction "add this sentence" is performed with the operation of *expansion*, denoted $+$. It is a simple set-theoretical operation, defined as follows:

$$K + p = \mathrm{Cn}(K \cup \{p\})$$

Expansion has the virtue of simplicity, but it also has the damaging property of bringing us to inconsistency whenever we assimilate some information that contradicts what we believed before. (If $\neg p \in K$ then $K + p$ is inconsistent.) Therefore we need the more sophisticated operation of *revision* that corresponds to the instruction "add this sentence and retain consistency". Revision is denoted $*$ and assumed to have the following properties:

$K * p = \mathrm{Cn}(K * p)$ (closure)

$p \in K * p$ (success)

$K * p$ is consistent if p is consistent (consistency)

We now have sufficient notation to express the four conditions from Sect. 20.1 in formal language:

$K \subseteq K \div p + p$ (Recovery)
If $\neg p \notin K$ then $K * p = K + p$ (Deductivism)
If K is finite-based, then so is $K \div p$ (Finite-based contraction)
If K is finite-based, then so is $K * p$ (Finite-based revision)

Let us now turn to the construction of belief change operations. Expansion is set-theoretically defined, but how should contraction and revision be constructed? Beginning with contraction, there are many ways to remove a sentence p from a belief set K that contains it. Obviously, p itself has to be thrown out, but that is not enough. We also have to make sure that we do not preserve elements of K that together imply p. For instance, if $q \in K$, then we also have $q \rightarrow p \in K$ (since $p \in K$ and K is logically closed). Since $q \in K$ and $q \rightarrow p \in K$ together imply p, at least one of them has to go in the construction of $K \div p$.

Similar considerations apply to revision. If $\neg p$ is in K, then it has to be removed in the construction of $K * p$ in order to achieve consistency. For every sentence q in K we also have $q \rightarrow \neg p$ in K, and either q or $q \rightarrow \neg p$ has to be removed in the construction of $K * p$.

Thus, both contraction and revision involve choices. We can frame these choices in various ways: as a choice which sentences to retain, a choice which sentences to remove, a choice among possible outcomes, etc. In the next four sections we will investigate four alternative frames for the choices involved in belief change. These frames turn out to induce different properties in the operations, as we will see by testing them against our four postulates from Sect. 20.1.

20.3 AGM: The Standard Approach

In 1985, Carlos Alchourrón (1931–1996), Peter Gärdenfors, and David Makinson published a paper that became the starting-point of modern research on the logic of belief change. The model they proposed is usually called "AGM" after their initials. When constructing contraction, $K \div p$, they started with the observation that among the many subsets of K not implying p, some are inclusion-maximal, i.e. they are as large as they can be without implying p. A set X is an inclusion-maximal p-excluding subset of K (in short: a p-remainder of K) if and only if it is a subset of K that does not imply p, but if we extend it with any additional element from K, then it will imply p. The set of p-remainders of K is denoted $K \perp p$.

Intuitively, we want to retain as much of K as we can without obtaining p. That could lead us to take one of the elements of $K \perp p$ as the contraction outcome. However, it may be impossible to single out one of these elements as better than all the others. If several p-remainders share the top position, then our post-contraction beliefs should be those that are held in all these top remainders. Formally, this is achieved by introducing a selection function γ that selects a subset $\gamma(K \perp p)$ of $K \perp p$ that consists, intuitively speaking, of the "best" elements of $K \perp p$. The outcome of contracting K by p is the intersection of all elements of $\gamma(K \perp p)$, i.e.

$$K \div p = \bigcap \gamma(K \perp p).$$

This construction is called *partial meet contraction.* One way to construct γ is to base it on a transitive relation covering all subsets of K that are x-remainders for some sentence x. If γ selects the elements of $K \perp p$ that are highest ranked according to such a relation, then the resulting contraction is a *transitively relational partial meet contraction.*

The AGM paper reported axiomatic characterizations of these operations. A sentential operation on a belief set K is a partial meet contraction if and only if it satisfies the following six axioms:

$K \div p = \text{Cn}(K \div p)$ (closure)
$K \div p \subseteq K$ (inclusion)
If $p \notin K$ then $K \div p = K$ (vacuity)
$p \notin K \div p$, unless p is a logical truth (success)
If $p \leftrightarrow q$ is a logical truth then $K \div p = K \div q$ (extensionality)
$K \subseteq (K \div p) + p$ (recovery)

Furthermore, such an operation is transitively relational if and only if, in addition, it satisfies the following two axioms:

$(K \div p) \cap (K \div q) \subseteq K \div (p\&q)$ (conjunctive overlap)
If $p \notin K \div (p\&q)$ then $K \div (p\&q) \subseteq K \div p$ (conjunctive inclusion)

The construction of revision in AGM is based on the simple observation that if p cannot be consistently added to K, then that is because $\neg p$ is in K. ($K + p$ is inconsistent if and only if K implies $\neg p$.) Therefore, all we have to do to make p consistently addable is to first remove $\neg p$. This line of reasoning (which can also be found in earlier work by Isaac Levi) gives rise to the following construction of revision in terms of contraction and expansion:

$$K * p = (K \div \neg p) + p \text{ (the Levi identity)}$$

It turns out that if revision is defined in this way, then the contraction operation on which the revision operation $*$ was based can be regained as follows:

$$K \div p = K \cap (K * \neg p) \text{ (the Harper identity)}$$

An operation is called a *partial meet revision* if and only if it is obtained via the Levi identity from a partial meet contraction, and it is a *transitively relational partial meet revision* if and only if it is obtained in that way from a transitively relational partial meet contraction. The AGM trio showed that partial meet revision is exactly characterized by the following six axioms:

$K * p = \text{Cn}(K * p)$ (closure)
$K * p \subseteq K + p$ (inclusion)
If $\neg p \notin K$ then $K + p \subseteq K * p$ (vacuity)
$p \in K * p$ (success)
If $p \leftrightarrow q$ is a logical truth then $K * p = K * q$ (extensionality)
If p is consistent then so is $K * p$ (consistency)

In order to characterize transitively relational partial meet revision, the following two axioms have to be added:

$K * (p \& q) \subseteq (K * p) + q$ (superexpansion)
If $\neg q \notin K * p$ then $(K * p) + q \subseteq K * (p \& q)$ (subexpansion)

Let us now return to the four principles that we introduced in Sect. 20.1 and reformulated as formal postulates in Sect. 20.2. As we saw above, Recovery is among the postulates for contraction, so it is satisfied. Deductivism is also satisfied; it follows directly from two of the revision postulates (inclusion and vacuity). However, both the remaining two postulates, Finite-based contraction and Finite-based revision, fail for the AGM operations.[1]

[1]See [12] for a proof that Finite-based contraction does not hold. The proof that Finite-based revision does not hold has not been published, and is therefore given here: Let S be an infinite set of logical atoms in the language, let p be another such atom, and let $K = \text{Cn}(\{\neg p\})$. Then $\{\neg p \vee s \mid s \in S\}$ is a subset of K that does not imply $\neg p$. It follows from compactness and the axiom of choice that there is some X such that $\{\neg p \vee s \mid s \in S\} \subseteq X \in K \bot \neg p$ [1]. Let γ be a selection function such that $\gamma(K \bot \neg p) = \{X\}$ and let $*$ be the revision based on γ. Then $\{\neg p \vee s \mid s \in S\} \cup \{p\} \subseteq K * p$, and since $K * p$ is logically closed we have $S \subseteq K * p$ and consequently $K * p$ is not finite-based.

20.4 Choosing Among Possible Worlds

A set of sentences is maximally consistent if and only if it is consistent but there is no sentence in the language that can be added to it without making it inconsistent. Maximally consistent sets are often called possible worlds since their structure is considered suitable for a total description of a state of the world.

A belief set K is compatible with a possible world if and only if nothing in the belief set contradicts it. Due to the special properties of possible worlds, this is equivalent with the requirement that the belief set is a subset of the possible world. It can also be shown that every belief set is equal to the intersection of all possible worlds that includes it. We can therefore replace belief sets by sets of possible worlds in our deliberations. The agent's belief state is then represented by a set of possible worlds (whose intersection is equal to the belief set).

A simple geometrical representation can be used to aid our intuitions [9]. In Fig. 20.1, think of each point in the square as a possible world. The circle in the middle contains exactly those possible worlds that are compatible with the current belief state. The area covered by the parabola represents those possible worlds in which p holds. This representation is quite intuitive, once you get accustomed to the fact that a smaller area represents a larger belief set, not the other way around.

Belief revision has a remarkably simple representation in this framework. In our example, consider the revision $K * p$. Its outcome should be a set of possible worlds in which p is true. Since we want to change as little as possible, the obvious solution is to let it consist of the intersection of the circle and the parabola, i.e. of those of the currently unrejected worlds in which p is true.

But this was a simple case, in which the new information was compatible with what was already believed. What should we do if the parabola (p) and the circle (K) have an empty intersection? Well, since there are no p-worlds in K we will then have to do with p-worlds that are as close, or similar, to K-worlds as possible. For that purpose we can think of K as surrounded by a system of spheres, with the worlds most similar to it in the sphere closest to K itself, those second-most similar

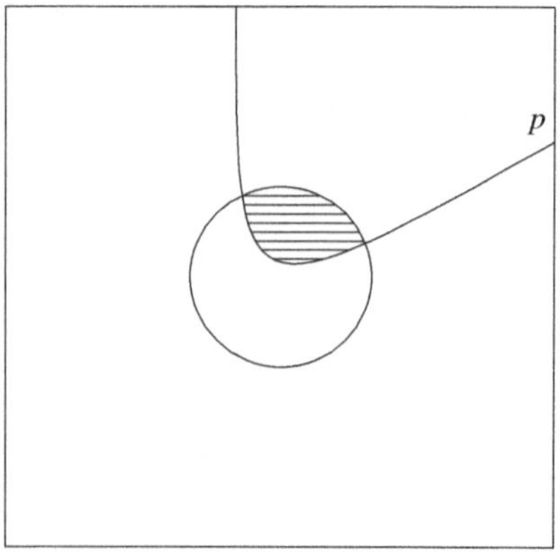

Fig. 20.1 Revision by a sentence p that is compatible with the present belief set

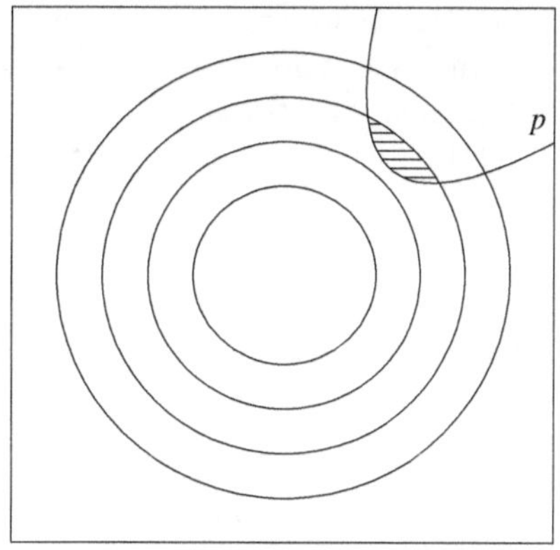

Fig. 20.2 Revision by a sentence p that is incompatible with the present belief set

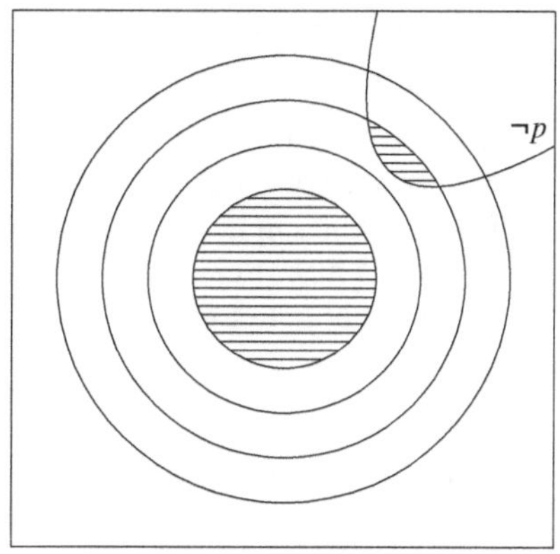

Fig. 20.3 Contraction by p

in the next sphere, etc., as in Fig. 20.2. The outcome of revision by p is then equal to the set of p-worlds in the innermost sphere that contains some p-worlds. (Such a system of spheres corresponds, of course, to an ordering of the possible worlds.)

Contraction is somewhat less intuitive than revision in possible world models. To contract by p means to allow for the possibility that $\neg p$, i.e. to allow for some possible worlds in which $\neg p$ holds. In a spheres model, these should be the $\neg p$-worlds that are closest to the belief set, i.e. those that are situated in the closest sphere that contains some $\neg p$-worlds. The contraction outcome will then be the union of these $\neg p$-worlds and the original belief set, as shown in Fig. 20.3.

Although the possible worlds construction is quite different from the partial meet construction, the two ways to construct contraction and revision turn out to yield exactly the same result. In other words, an operation on a belief set K is a transitively relational partial meet contraction if and only if it can be constructed in the way indicated in Fig. 20.3, and it is a transitively relational partial meet revision if and only if it can be constructed in the way indicated in Fig. 20.2. This surprising result is based on a one-to-one correspondence called "Grove's bijection" between remainders and possible worlds [9], [11, pp. 53–55].

Hence, it makes no difference for the belief change operations if we apply a choice mechanism to remainders or to possible worlds.[2] This might give the impression that it makes no difference what we apply the choice mechanism to. That, however, is an unjustified generalization, as we will see in the following two sections.

20.5 Belief Bases

In an infinite language, belief sets are very large entities. Since they contain all logical consequences of what the agent believes, they contain a lot of sentences that provide logical connections between epistemically unrelated beliefs. This can have rather strange consequences:

> I believe that the earth is (approximately) spherical (e). I also believe that I have my house key in my left trouser pocket (k). Consequently, I believe that the earth is spherical if and only if my house key is in my left pocket ($e \leftrightarrow k$). I put my hand in the pocket to pick up the key. It is not there! I have to give up my belief in k. I cannot then, on pain of inconsistency, retain both my belief in e and my belief in $e \leftrightarrow k$.

Both e and $e \leftrightarrow k$ are elements of the belief set. Therefore, when I find out that k is false, I have to choose between retaining e and retaining $e \leftrightarrow k$. The option of keeping $e \leftrightarrow k$ and giving up e is not excluded automatically, but has to be excluded by the selection mechanism. This appears inappropriate, since $e \leftrightarrow k$ is a merely derived belief that should arguably disappear automatically when k is given up.

Considerations like this led to the construction of belief change operations in which the actual choice takes place among "real" beliefs (like e and k), and the "merely derived" beliefs (like $e \leftrightarrow k$) have no role in the selection process. The crucial construction is a *belief base* consisting of the "real" beliefs, from which the rest of the belief set can be derived. The belief base is denoted B. It satisfies the criterion $\mathrm{Cn}(B) = K$, and contrary to K it does not have to be logically closed. For most purposes we assume that B is finite.

Partial meet contraction and revision can be performed on belief bases in the same way as for belief sets: We define $B \perp p$ as the set of inclusion-maximal subsets of B that do not imply p and γ as a selection function that selects a non-empty subset of each such remainder set. This gives rise to the partial meet contraction $B \div p = \bigcap \gamma(B \perp p)$. Partial meet revision is defined as $(B \div \neg p) \cup \{p\}$.

For any given belief set K, we obtain *base-generated* partial meet contraction and revision (denoted $\widehat{\div}$ and $\widehat{*}$) by assigning to it a belief base B and a selection function for that belief set:

[2] AGM is also equivalent to a construction based on selection among the sentences in K, namely epistemic entrenchment [6, 7, 16, 17]. It is also close to equivalent to another such construction, safe contraction [2, 18].

$$K \mathbin{\widehat{\div}} p = \mathrm{Cn}(B \div p)$$

$$K \mathbin{\widehat{*}} p = \mathrm{Cn}(B * p)$$

These operations have been rather carefully investigated and also axiomatically characterized. It is easy to show that base-generated partial meet contraction does not satisfy Recovery. Let p, q, and r be logically independent sentences, and let $K = \mathrm{Cn}(\{p, q, r\})$. We can assign to it a belief base $B = \{p, q, r\}$. Any selection function γ for B will yield $\gamma(B \perp (q \vee r)) = \{\{p\}\}$, thus $K \mathbin{\widehat{\div}} (q \vee r) = \mathrm{Cn}(\{p\})$. It follows that $K \mathbin{\widehat{\div}} (q \vee r) + (q \vee r) = \mathrm{Cn}(\{p, q \vee r\})$, from which we can see that Recovery does not hold.

It is equally easy to show that Deductivism is satisfied. If $\neg p \notin K$ then it holds for any belief base B for K and any partial meet contraction $\div$ on B that $B \div \neg p = B$, and we can conclude that $K \mathbin{\widehat{*}} p = \mathrm{Cn}(B \cup \{p\}) = K + p$.

When we assign a belief base to a finite-based belief set, then we can always choose a finite belief base, and it would indeed be difficult to justify doing otherwise. Provided that we use finite belief bases when that is possible, both Finite-based contraction and Finite-based revision hold for the base-generated partial meet operations. We can conclude that in terms of these four postulates it makes a big difference whether we apply selection functions to the remainders of a belief set or to the remainders of a belief base for it.

20.6 Descriptor Revision

Since we now know that it is important what we apply the choice mechanism to, we have reason to ponder what are the appropriate objects of choice. From the viewpoint of cognitive realism, possible worlds seem unsuitable since they are large structures beyond our grasp. Remainders of belief sets are problematic for much the same reason. Even if the original belief set K is finite-based, if p is a non-tautologous element of K, then the remainder set $K \perp p$ has an infinite number of elements, none of which is finite-based.[3] Remainders of a finite belief base are somewhat more plausible objects of choice, but there is an additional problem with the partial meet construction that applies to belief sets and belief bases alike: It is difficult to justify that we intersect the remainders chosen by the selection function. If the remainders are the top candidates for being the contraction outcome, then their intersection is not one of the top candidates.[4] Then why should it be chosen? [19].

[3]Provided that the language has an infinite number of non-equivalent sentences. For a proof, see [12].

[4]More precisely: if there is more than one top candidate and the top candidates are all p-remainders for some sentence p, then their intersection is not itself a top candidate.

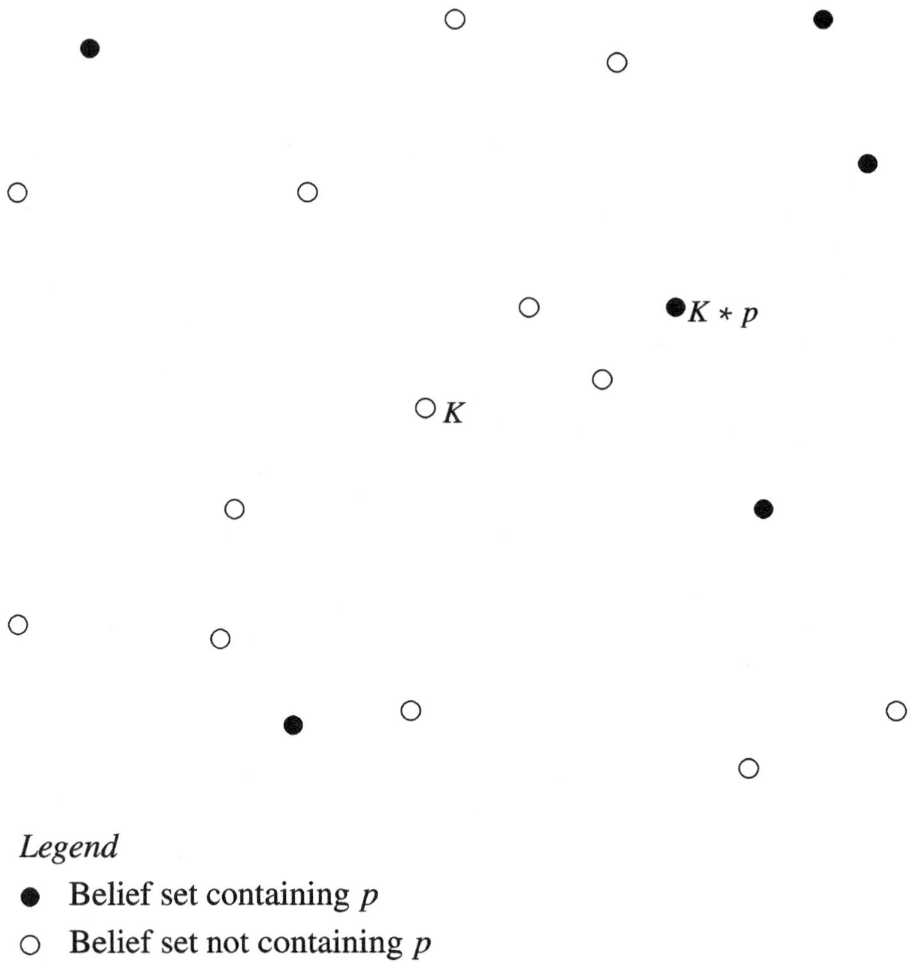

Fig. 20.4 Descriptor revision. $K * p$ is the belief set closest to K among those that contain p. Note that the circles denote belief sets, not possible worlds

Another alternative is to apply the selection mechanism directly to the set of possible outcomes of change. Presumably, not all logically closed sets are suitable contraction outcomes. We can therefore assume that there is an *outcome set* ($\mathbb{X}$) consisting of all the belief sets that can be reached by an operation of change. Intuitively, we can think of $\mathbb{X}$ as consisting of all those belief sets that are coherent, stable, and/or plausible enough to be suitable as outcomes of an operation of belief change. In a cognitively realistic model, all elements of $\mathbb{X}$ should be finite-based.

In addition to this we need a selection mechanism that always singles out exactly one element of the outcome set. One plausible such mechanism is a distance measure. We can think of the elements of the outcome set $\mathbb{X}$ as dispersed in some kind of metric space, as illustrated in Fig. 20.4. $K * p$ is then found as the belief set closest to K in which p holds. (We have to assume that distances are always dissimilar, or that there is some other mechanism to arbitrate ties.)

Contraction can be obtained in the same way: We can identify $K \div p$ as the belief set that is closest to K among those elements of $\mathbb{X}$ that are subsets of K and do not contain p. Furthermore, this model opens up for a more general approach to belief change called *descriptor revision* [13, 14]. To introduce it we need a belief operator $\mathfrak{B}$ such that $\mathfrak{B}p$ denotes that p is believed. We can use this notation to express a wide variety of success conditions for belief change. For instance, $\mathfrak{B}p \vee \mathfrak{B}q$ means that either p or q is believed and $\mathfrak{B}p \vee \neg\mathfrak{B}q$ means that either p is believed or q disbelieved. (The expressions formed with $\mathfrak{B}$ in this way are called belief descriptors.)

Descriptor revision has a single operation of change, denoted $\circ$. It can be applied to all types of descriptors. For instance, in a distance-based framework, the operation $K \circ \neg\mathfrak{B}p$ takes us from K to the belief set closest to K that does not contain p.[5] Similarly, the operation $K \circ (\mathfrak{B}p \vee \mathfrak{B}q)$ produces as outcome the closest belief set in which $\mathfrak{B}p \vee \mathfrak{B}q$ is satisfied, i.e. the closest belief set containing either p or q. Common revision (by sentences) is a special case of descriptor revision, since we can identify $K * p$ with $K \circ \mathfrak{B}p$. Both general descriptor revision ($\circ$) and its restriction to sentential revision ($*$) have been axiomatically characterized. One of the advantages of descriptor revision is that it can easily be extended to iterated change. Hence, in a distance-based framework we obtain $K \circ (\mathfrak{B}p \vee \mathfrak{B}\neg p) \circ (\mathfrak{B}q \vee \mathfrak{B}\neg q)$ by going first from K to the closest belief set containing either p or $\neg p$, and then from there to the closest belief set containing either q or $\neg q$. (This corresponds to making up one's mind first about p and then about q.)

It is a rather straight-forward exercise to show that neither Recovery nor Deductivism holds in this framework. However, provided that the elements of $\mathbb{X}$ are finite-based, both Finite-based contraction and Finite-based revision hold for descriptor revision.

20.7 Conclusion

The outcome of this investigation is summarized in Table 20.1. We have found that the properties of operations of change depend to a large degree on what formal structures we use as objects of choice. Two important philosophical questions are involved here: When we choose rationally what to believe, what are the objects that our choices should be applied to? And what structural properties should a rational agent's choice patterns comply with? Both these are questions that can

[5] In the limiting case when the outcome set contains no element that satisfies the descriptor, nothing is changed, i.e. the original belief set is the outcome of the operation.

Table 20.1 Summary of how the application of the choice mechanism to different objects impacts the satisfaction of four postulates of belief change

Objects of choice	Satisfaction of postulates			
	Recovery	Deductivism	Finite-based contraction	Finite-based revision
Remainders of belief sets	+	+	−	−
Possible worlds	+	+	−	−
Remainders of belief bases	−	+	+	+
Potential outcomes	−	−	+	+

be asked in an informal language, but we need a formal language to perform the logical analysis that shows how closely the two questions are connected with each other.

References and Proposed Readings

1. Alchourrón, C. E., & Makinson, D. (1981). Hierarchies of regulation and their logic. In R. Hilpinen (Ed.). *New studies in deontic logic* (pp. 125–148). Dordrecht: D. Reidel Publishing Company.
2. Alchourrón, C. E., & Makinson, D. (1985). On the logic of theory change: Safe contraction. *Studia Logica, 44*, 405–422
3. *Alchourrón, C. E., Gärdenfors, P., & Makinson, D. (1985). On the logic of theory change: Partial meet contraction and revision functions. *Journal of Symbolic Logic, 50*, 510–530. [The major starting-point of the whole research area.].
4. *Fermé, E., & Hansson, S. O. (2011). AGM 25 years. Twenty-five years of research in belief change. *Journal of Philosophical Logic, 40*, 295–331. [A comparatively brief and non-technical overview of the research area.].
5. *Fermé, E., & Hansson, S. O. (In press). *Belief change. Introduction and overview*. Springer 2018. [A comparatively non-technical overview of the area.].
6. *Gärdenfors, P. (1988). *Knowledge in flux. Modeling the dynamics of epistemic states*. Cambridge: The MIT Press. [A philosophical account of the AGM model by one of its originators.].
7. Gärdenfors, P., & Makinson, D. (1988). Revisions of knowledge systems using epistemic entrenchment. In M. Y. Vardi (Ed.). *Proceedings of the Second Conference on Theoretical Aspects of Reasoning About Knowledge* (pp. 83–95). Los Altos: Morgan Kaufmann.
8. Glaister, S. M. (2000). Recovery recovered. *Journal of Philosophical Logic, 29*, 171–206.
9. Grove, A. (1988). Two modellings for theory change. *Journal of Philosophical Logic, 17*, 157–170.
10. Hansson, S. O. (1991). Belief contraction without recovery. *Studia Logica, 50*, 251–260.
11. *Hansson, S. O. (1999). *A textbook of belief dynamics. Theory change and database updating*. Dordrecht: Kluwer. [Detailed introduction with a strong focus on AGM and belief base operations. Contains proofs of all the major theorems.].
12. Hansson, S. O. (2008). Specified meet contraction. *Erkenntnis, 69*, 31–54.
13. Hansson, S. O. (2014). Descriptor revision. *Studia Logica, 102*, 955–980.
14. *Hansson, S. O. (2017). *Descriptor revision. Belief change through direct choice*. Cham: Springer. [A full treatment of descriptor revision.].

15. Makinson, D. (1997). On the force of some apparent counterexamples to recovery. In E. G. Valdés, et al. (Eds.). *Normative systems in legal and moral theory, Festschrift for Carlos Alchourrón and Eugenio Bulygin* (pp. 475–481). Berlin: Duncker and Humblot.
16. *Rott, H. (2001). *Change, choice and inference: A study of belief revision and nonmonotonic reasoning*. Oxford: Oxford University Press. [An exposition of AGM theory with a strong emphasis on its connections with the study of rational choice.].
17. Rott, H. (2003). Basic entrenchment. *Studia Logica, 73*, 257–280.
18. Rott, H., & Hansson, S. O. (2014). Safe contraction revisited. In S. O. Hansson (Ed.). *David Makinson on classical methods for non-classical problems* (pp. 35–70). Dordrecht: Springer.
19. Sandqvist, T. (2000). On why the best should always meet. *Economics and Philosophy, 16*, 287–313.

Chapter 21
Probability Theory

Darrell P. Rowbottom

Abstract This chapter covers the epistemic or information-based interpretations of probability: logical, subjective, objective Bayesian, and group level. It explains how these differ from aleatory or world-based interpretations of probability, presents each in detail, and then discusses its strengths and weaknesses.

21.1 The Ubiquity of Probability Talk

Rarely does one get through a day without encountering a reference to probability or one of its relatives. Meteorological reports purport to tell us the chance of rain, and meteorologists take this to reflect a particular kind of probability. Bookies offer betting odds, which we take up according to how probable we take particular events to be. And when we're asked whether we'll attend an event, we often answer "Probably!" or "Probably not!"

Probabilities are regularly appealed to in philosophy as well. For example, it's often taken for granted in informal philosophy that the more probable of two mutually exclusive and jointly exhaustive events is the better one to bet on when offered even odds (to further the end of winning). But this isn't as obvious at it may initially seem; in fact, I hold it to be wrong under some interpretations of probability. And sometimes I encounter discussions in which it is boldly proclaimed that inductive inferences have something to do with probabilities, although what these probabilities are is never discussed. In fact, I have read several papers like this. Whole arguments are constructed on probability claims, but the understanding

D. P. Rowbottom (✉)
Lingnan University, Tuen Mun, Hong Kong
e-mail: darrellrowbottom@ln.edu.hk

S. O. Hansson, V. F. Hendricks (eds.), *Introduction to Formal Philosophy*, Springer Undergraduate Texts in Philosophy, https://doi.org/10.1007/978-3-319-77434-3_21

of probability under discussion is never explained. Perhaps an intuitive notion is supposed to be operating? My view is that this will not do for serious philosophy.

Let me try to convince you. Imagine a weather forecaster says that the chance of rain tomorrow is zero, but that it nevertheless rains. Should we be angry with the forecaster? Must her methods be fundamentally flawed? Against intuitive expectations, the answer may lie in the negative if she is operating with a *frequency in the limit* interpretation of probability. All that would be required, for her statement to be correct, is that in identical meteorological circumstances, were these to be repeated infinitely many times, rain would occur the next day with a frequency of zero. To see that this is compatible with rain occurring, consider the following set, with infinitely many members, in which *R* represents rain and *N* represents no rain:

$$\{N, R, N, N, R, N, N, N, R, N, N, N, N, R, \ldots\}$$

The set continues in the same pattern, with five *N*s to the next *R*, then six after the subsequent *N*, and so on. Mathematically, the ratio of *R*s to *N*s is zero in the limit.

I could spend more space trying to convince you that it's important to understand how probability talk may be interpreted, but this may be poorly spent given that you've already made the effort to consult this piece. So I'll move on to the meat.

21.2 Epistemic Versus Aleatory Interpretations of Probability

Probability is a mathematical notion, and the issue of its interpretation typically arises only when we seek to apply it. In fact, mathematicians have had interpretative difficulties too, especially when it comes to understanding which kinds of arguments for their successful predictions were more fundamental. You can read about this elsewhere; see Hacking [15] and Shafer and Vovk [44]. The best place for us to start is to recognize the most fundamental interpretative rift if we begin with a measure-theoretic view of probability (based, for example, on Kolmogorov's or Popper's axioms of probability).[1] This is between *aleatory* and *epistemic* views.

The basic distinction is not so hard to grasp. Roughly, *aleatory* probabilities are 'out there' in the world; indeed '*alea*' is the Latin word for 'die', although one should not conclude that gambling concerns only *aleatory* probabilities. *Epistemic*

[1] Shafer and Vovk [44] argue that we should not begin by understanding probability in a measure-theoretic way, but instead in a game-theoretic way. As such, their interpretative strategy is different from those considered here (although the Dutch Book argument, which we will cover in due course, is game-theoretic in nature).

probabilities concern us more intimately, and roughly relate to our epistemic states.[2] To give a flavour of how individuals thinking on either side of this divide might apply probabilities, consider the following dialogue:

Philosopher: Here is a normal two-sided coin. What is the probability that it will land on heads when I flip it?
Mr Epistemic: One half.
Ms Aleatory: I don't know.
Mr E: What do you mean, you don't know?
Ms A: It might be biased. We'd need to see the experiment repeated several times before forming a reasonable opinion.
Mr E: Yes, it could be biased! But you've no reason to think it's biased one way, rather than another, so why not just assign the two possibilities the same probability, namely one half?
Ms A: Well that's just guessing.
Mr E: But isn't that a reasonable betting strategy? What odds would you accept on a 'heads' bet? That way we can work out how probable you think it is . . .
Ms A: If I did bet, I'd use my knowledge about similar circumstances, i.e. similar coins being flipped by similar people, and the frequencies from those.

This should be enough to give a flavour of the disagreements that can occur in this dimension. I experience such disagreements frequently, when I pose the same question as the philosopher, in the dialogue, to students.

'*Objective*' is often used in place of '*aleatory*'. I avoid the former term because probabilities might be thought to be 'out there' *in the non-material world* on some epistemic views of probability. Most notably, probabilities may be construed as logical relations between propositions, which might be taken to exist, *qua* abstracta, even in possible worlds containing no beings capable of grasping them.

Now since this chapter appears in a section of the handbook on 'Epistemology', I will cover only the epistemic interpretations of probability in what follows. Although it is true that one of the aleatory interpretations has been used, in the past, in epistemic contexts—specifically, inductive probabilities have been construed, e.g. by Reichenbach, in terms of relative frequencies of truth of consequences given the truth of the premises (in the limit)—this is now widely considered to have been in error. I will not rehearse the arguments here, for lack of space. See Rowbottom ([31], pp. 39–41) for more discussion.

Each of the interpretations covered below has variants. So the interpretative possibility space is much more complex than the normal philosophical taxonomy makes apparent. The best I can do, in what follows, is to provide an overview of the positions on the standard taxonomy and to flag any areas where there are subtleties of interpretative difference that are easy to miss.

[2]I call the two kinds of probability 'world-based' and 'information-based' in Rowbottom [35]; I think this is better terminology, but it's non-standard.

21.3 The Logical Interpretation

The basic idea behind the logical interpretation of probability is that there are degrees of partial entailment, between propositions or groups thereof, in addition to entailment relations.[3] (The talk of 'degrees of partial entailment' is from Carnap [5]; Keynes [21], the architect of the logical interpretation, used 'logical relations'.) Consider a situation where p entails q. Then the probability of q given p, or the *conditional probability* of q on p, is equal to unity; $\mathrm{P}(q, p) = 1$. And similarly, if p and q are logically inconsistent then $\mathrm{P}(p, q) = 0$ and $\mathrm{P}(q, p) = 0$. Now if we allow for other kinds of logical links, of varying strengths, we can imagine that, in general, $\mathrm{P}(p, q) = r$ where r is any real number between zero and one. It is worth noting that Popper explained the logical interpretation—or more accurately his variant thereof—rather differently, by appealing to the notion of logical content. Specifically, he claimed that $\mathrm{P}(a, b)$, interpreted logically, measures: 'the degree to which the statement a contains information which is contained by b' ([26], p. 292). (A hint that this is problematic may be gleaned by thinking of the rule of 'or introduction' in natural deduction. If a is not entailed by b, then a will have infinitely many consequences that b does not. And *vice versa*.)

How about *unconditional* probabilities? Keynes ([21], pp. 6–7) declared that: 'No proposition is in itself either probable or improbable, just as no place can be intrinsically distant; and the probability of the same statement varies with the evidence presented, which is, as it were, its origin of reference.' So in general, on the logical view, talk of unconditional probabilities is understood as elliptical. When I talk of the probability of the next president of the USA being a Democrat, for instance, I assume some things are true; that the Democratic Party is not a figment of my imagination, that the USA is a real country, and so forth. But we may nevertheless define unconditional probabilities in terms of a special class of conditional probabilities, following the suggestion of Popper ([26], pp. 284–285). The idea is that $\mathrm{P}(p)$ may be understood to represent $\mathrm{P}(p, T)$, where T represents any tautology. This gives a probability that doesn't depend on anything other than the laws (or axioms) of logic being true.

So far I have mentioned only logical relations. But as this is an *epistemic* interpretation of probability, the reader may be wondering how *we* come into the picture. The short answer is that most advocates of the logical interpretation have held that our personal degrees of belief—this is a technical term, about which I will say more shortly—should map on to the equivalent logical relations in order to be rational. Again, an analogy with entailment helps. If p entails q and I'm certain that p, then I ought to be certain that q (subject to some other appropriate conditions

[3] There is also a 'classical interpretation', which predates the logical one. On the classical view, probabilities are defined (roughly) in terms of the ratio of favourable outcomes to possible outcomes. The problem with this view is that it appears to require that each outcome be *equipossible*. So it could not handle a biased coin; e.g. in calculating the probability of heads on one flip, we'd always arrive at one half. (And someone might very well take the coin landing on its edge to be possible, and thereby be forced to conclude that the probability of heads was a third.) For more on the classical interpretation, see Gillies [13] and Rowbottom [35].

obtaining, e.g. that I desire to know whether q and recognise that p entails q).[4] More generally, let p entail q to a specific degree: $P(q, p) = r$. My degree of belief in q given p—$D(q, p)$—will be rational only if it maps on to the appropriate logical relation, i.e. only if $D(q, p) = P(q, p) = r$. This was the view of Keynes ([21], p.4):

> What particular propositions we select as the premises of our argument naturally depends on subjective factors peculiar to ourselves; but the relations, in which other propositions stand to these, and which entitle us to probable beliefs, are objective and logical.

I must reiterate, because it is a common error to think otherwise, *that the logical view does not say that probabilities are rational degrees of belief*. As Keynes ([21], p. 11) clearly stated, 'probability':

> In its most fundamental sense... refers to the logical relation between two sets of propositions... Derivative from this sense, we have the sense in which... the term *probable* is applied to the degrees of rational belief.

I emphasise this issue because accepting a logical interpretation of probability does not entail accepting that a degree of belief is rational if and only if it maps on to the appropriate logical relation (in the way described above). Take Popper as a case in point. One of his most striking theses was that the logical probability of any synthetic universal statement is zero relative to any finite evidence, and plausibly any evidence that we might possess; see Popper ([25], Appendix *vii) and Rowbottom ([31], Section 2.3; [32]). But he did not conclude that it is irrational to believe in such laws.

The main problem with the logical interpretation is that it is hard to see how such logical probabilities may be measured (or how to define them *operationally*).[5] And if they cannot be measured, then it seems reasonable to doubt that they exist. Keynes [21] had a rather complicated position on this issue.[6] Roughly—see O'Donnell [24] and Rowbottom [30] for more—he thought that some relations can be grasped by intuition, but that others can only be calculated by employing an *a priori* synthetic principle, namely the principle of indifference.[7] In essence, his idea was that this principle is applicable when our intuition fails, as an extension of logical proof to non-demonstrative cases:

---

[4]How to connect reasons for belief and entailment is much more complicated than it may first appear. See, for example, Streumer [45].

[5]In the words of De Finetti ([7], p. 23): 'For any proposed interpretation of Probability, a proper operational definition must be worked out: that is, a device apt to measure it must be constructed.' One could argue with this, of course, but it seems odd to want to posit a kind of probability that isn't generally measurable! What purpose would it serve?

[6]Keynes also believed in non-numerical probabilities, which complicates matters further, but we can put this to one side for present purposes.

[7]This was earlier called 'the principle of non-sufficient reason', and goes back (although not under the same name) to Bernoulli [2].

> If the truth of some propositions, and the validity of some arguments, could not be recognised directly, we could make no progress... [T]he method of logical proof... enables us to know propositions to be true, which are altogether beyond the reach of our direct insight... ([21], p. 53, f.1).

So just arriving at the correct degree of belief in p given q, say as a result of brainwashing, is not sufficient to have a rational degree of belief in p given q. (It is only *necessary*, for any rational degree of belief, that $D(q, p) = P(q, p) = r$.) One must have some suitable procedure for arriving at $P(p, q)$. And this brings us to the principle of indifference:

> [T]hat equal probabilities must be assigned to each of several arguments, if there is an absence of positive ground for assigning *unequal* ones ([21], p. 42).

The problem with the application of the principle is that it leads to paradoxical results when there is more than one way to 'carve up' the space of possibilities. The most famous and widely discussed of these are the water-wine paradox (see van Fraassen [46], pp. 304–305 and Mikkelsson [23]) and Bertrand's ([3], p. 4) geometrical paradox (see Rowbottom [34]). But an even simpler example will suffice.

"If I flip a coin twice, what is the probability of getting two heads?" One respondent might take there to be three different possibilities—no heads, one head, and two heads—and conclude, assigning each an equal probability by the principle of indifference, that the answer is one third. Another might take there to be four different possibilities—no heads, one head and one tail (in order), one tail and one head (in order), and two heads—and instead conclude that the answer is a quarter. Which is right?

Several readers may think that the latter is clearly correct, and indeed Keynes ([21], p. 60) introduced an indivisibility criterion in order to argue for this; in short, the possibility of 'one head' is divisible into the possibilities of 'one head and one tail (in order)' and 'one tail and one head (in order)'. It might surprise those readers to learn that Carnap ([5], pp. 562–565) instead argued that one should think of possibilities in the first way in the coin flip example. His motivation was that learning by Bayesian conditionalisation, on the basis of the behaviour of the coin in repeats of the experiment, would otherwise not be possible. See Gillies ([13], p. 45–46) for more discussion of this.

In any event, the whole point of paradoxes such as Bertrand's is that they concern continuous cases and hold even if one introduces an indivisibility criterion. So despite occasional attempts to defend the principle of indifference or similar strategies—see Jaynes [18], Marinoff [22]—the more widespread view—see van Fraassen [46], Gillies ([13], p. 49), Shackel [43], and Rowbottom [34]—is that the paradox is insoluble.[8]

[8]This said, some putative solutions have recently appeared. See Aerts and Sassoli de Bianchi [1], and Gyenis and Rédei [14]. Both papers are rather technical in character.

21.4 The Subjective Interpretation

The subjective interpretation eschews talk of logical relations, to focus instead on personal degrees of belief.[9] (We'll come to the ontology of 'degrees of belief' a little later.) Sometimes it is said that in the subjective interpretation, probabilities *just are* degrees of belief. But this is a mistake, albeit one supported by a casual reading of the likes of De Finetti.[10] The reason is that not all degrees of belief need satisfy the probability calculus; and as a matter of fact, even clever people like logicians and mathematicians can have inconsistent beliefs. Thus the subjective interpretation relies on the idea that it is *necessary* for a person's degrees of belief to satisfy the probability calculus in order for those degrees of belief to be rational. So following Keynes ([21], p. 20), advocates of the subjective view hold that: 'Belief, whether rational or not, is capable of degree'. In addition, they hold that *rational* degrees of belief do not violate the axioms of probability; they range between 0 and 1, and so on.

In fact, Ramsey and De Finetti hailed it as a virtue of the subjective interpretation that the axioms of probability can be *derived* from a consideration of the rules that degrees of belief ought to obey, via a consideration of rational betting behaviour. This is usually known as the Dutch Book Argument, and sometimes as the Ramsey-De Finetti theorem. The basic idea is simple. Put someone in a betting scenario. If they bet in such a way as to be susceptible to losing whatever happens, then they must have accepted bets with betting quotients—these will be explained below—that fail to satisfy the axioms of probability. Now assume that those betting quotients reflect the person's degrees of belief and that being susceptible to losing whatever happens is irrational. The conclusion is that rational degrees of belief satisfy the axioms of probability.

One way to set up such a scenario is as follows. T (a tester) asks B (a bettor) to choose a number q, her betting quotient on E, on the understanding that T will then choose a stake S, that B will pay T the sum of qS, and that B will receive S if E occurs. If E does not occur, T will keep the sum of qS. Note that T may choose a positive of negative value for S. (Giving a sum of qS, in the event that S is positive, is equivalent to receiving a sum of $q|S|$ in the event that S is negative, and so forth. So if S is negative, then T will pay B the sum of $q|S|$, in return for $|S|$ if E occurs; else, B will keep $q|S|$.) A final stipulation is that B does not know which way she will be betting (i.e., whether T will choose a positive or negative value for S).

---

[9]In the words of Ramsey ([27], pp. 72–73): 'the kind of measurement of belief with which probability is concerned is . . . of belief *qua* basis of action . . . with beliefs like my belief that the earth is round . . . which would guide my action in any case to which it was relevant.' Ramsey's example seems somewhat odd, however, because it seems that all conceivable beliefs can be relevant to action in appropriate cases. For example, I could be asked what I believe about some obscure philosophical issue and desire to express the truth. An asseveration I made in response would be guided by that belief.

[10]The following statement, for example, is misleading: 'only subjective probabilities exist—i.e. the degree of belief in the occurrence of an event attributed by a given person at a given instant with a given set of information.' (De Finetti [8], pp. 3–4)

You may already have thought of several problems with this scenario. If the stake is too small, such as \$0.01, then B might give any old answer. Or B may fear paying T the sum of qS in the event that T selects S positive if E concerns the far future because she'll then be out of pocket for a long time—and select a value of q with the intent to entice B to select a negative value for S. These problems can solved by refining the scenario; the magnitude of the stake can be set so that it's significant, the betting procedure can be altered such that money is always given to the tester (or the bets can be on utilities rather than cash), and so forth. But there are other problems that appear insoluble. For example, the fact that B does not know which way T is going to bet does not mean that she should not have an opinion about how T will bet. And recognising this leads to the rather remarkable result that a bettor might rationally select a betting quotient of zero for an event that she is sure will occur; see Rowbottom [28]. According to the axioms of probability, though, the probability of such an event should be unity! Similar problems have been discussed in considerable depth elsewhere, e.g. by Seidenfeld et al. [42] and Hájek [16].[11]

Given the problems with the Dutch book argument, it is surprising that a much more promising alternative, proposed by De Finetti [7], is less well known. This is to use a scoring rule. The idea is to put the bettor—or the person whose degrees of belief you want to elicit—into a situation in which she believes that she will be penalised by a particular loss, dependent on her forecast (which is a number similar to a betting quotient in the gambling case above) and what happens. Now we can proceed only by assuming that she wishes to minimise her expected loss, and need not worry about how she anticipates that someone else—a tester (or bookie)—will behave. For more on this strategy, see Schervish et al. [37].

It is crucial to note that there are several different understandings of degrees of belief. De Finetti, for instance, thought of these in an *operational* and *behavioural* manner, such that they are *identical* to betting quotients or forecasts. Most plausibly, degrees of belief might be understood, in such a vein, as dispositions to bet or forecast in particular ways. By contrast, the more popular view at present is that degrees of belief are *credences* or degrees of confidence. Such credences may not be reflected in betting behaviour, which is an advantage. One worry, however, is how to measure them (as something above and beyond, say, forecast dispositions). Perhaps the most obvious route is to appeal to some kind of personal awareness, like strength of feeling. But this is a dubious move. As Ramsey ([27], p. 71) pointed out:

> This view would be very inconvenient, for it is not easy to ascribe numbers to the intensities of feelings; but apart from this it seems to me observably false, for the beliefs which we hold most strongly are often accompanied by practically no feeling at all; no one feels strongly about things he takes for granted.[12]

[11] A good place to read more about Dutch Books is Hájek [17].

[12] Note, in particular, the comment about ascribing numbers to intensities of feeling. The idea that people can have precise degrees of belief corresponding to any rational number between 0 and 1—or perhaps beyond, if we're discussing degrees of belief that don't satisfy the probability

So just as there are different accounts of belief, in the philosophy of mind, there are different accounts of degrees of belief in the philosophy of probability (or formal epistemology). In fact, here is an area in which there is the potential for a valuable exchange between these two areas of philosophy. This is suggested by my recent debate with Eric Schwitzgebel—see Schwitzgebel [38–40] and Rowbottom [29, 36]—on whether degrees of belief can explain cases of apparent 'in-between believing'. De Finetti's behavioural approach seems to have an analogue in the dispositional approach to belief (although De Finetti was concerned with only a narrow range of dispositions, at best). So are credences typically construed as mental representations, i.e. as beliefs are on a representational account in the philosophy of mind? (And might they not instead be construed as dispositional but not limited to betting and/or forecast scenarios?) For more on the nature of degrees of belief, see the discussion of Eriksson and Hájek [10]. They suggest that degrees of belief are primitive, and point to the success that decision theory has had without providing an analysis of the notion. But they do not engage with contemporary philosophy of mind.

21.5 Objective Bayesianism

The basic idea behind objective Bayesianism—which is championed by Jaynes [20] and Williamson [47, 48], for example—is that one can start in the same way as an advocate of the subjective view does, e.g. with a betting or forecast scenario, and then show that there are constraints on rational degree of belief which subjectivists do not consider. It is therefore unsurprising to find that Gillies ([12], Sect. 2) characterises 'objective Bayesianism' as follows:

> [The] approach could be called the 'topping-up' version of the logical interpretation of probability. The idea is to start with purely subjective degrees of belief. We then add one rationality constraint (coherence) to obtain the axioms of probability. However, this might be 'topped-up' by further rationality constraints derived from logical or inductive intuition. Thus the choice of different probabilities allowed by the subjective theory would be narrowed down, and eventually it might be possible to get back to a single rational degree of belief as in the original logical theory.

To be more specific, objective Bayesians think that degrees of belief should (a) be probabilities (in the sense of satisfying the calculus), (b) reflect the evidence of their possessors, especially in so far as this concerns observed frequencies and/or estimates of aleatory probabilities, and (c) otherwise be maximally non-committal. Williamson [48] calls these the (a) *probability*, (b) *calibration*, and (c) *equivocation*

calculus—is clearly an idealisation. It's more realistic to think that they lie in particular intervals. There is a related literature on imprecise probabilities, and in fact the idea of working with intervals was discussed at considerable length by Keynes [21]. For more on the notion, which is growing in popularity, see http://www.sipta.org and Bradley [4].

norms. With regard to (c), a principle called 'the maximum entropy principle' is used in place of the principle of indifference. Jaynes ([19], p. 623) described this as 'an extension of the principle of insufficient reason [i.e. indifference]', and declared 'that it can be asserted for the positive reason that it is uniquely determined as the one which is maximally noncommittal with regard to missing information'. It is superior to the principle of indifference at least in so far as it has greater generality.

How close the objective Bayesian view *really* is to the logical one is contested, and the disagreement rests to some extent on how one reads Keynes; see Rowbottom [30] and Williamson ([48], p. 23).[13] One potential advantage is the avoidance of the posit of logical relations 'out there'; instead rational degrees of belief may be *defined* in terms of degrees of belief that satisfy the aforementioned norms. But it is fair to say that some of the strongest criticisms of objective Bayesianism have a similar flavour to the strongest criticisms of the logical view. In particular, they focus on the successor to the principle of indifference, namely the maximum entropy principle. One suggestion is this is just as paradoxical as its forerunner; see Seidenfeld [41].

The phrase 'objective Bayesian' might also give the false impression that (learning by) Bayesian conditionalisation plays a central role in the interpretation of probability being proposed. But as Williamson [47, 48] makes clear, objective Bayesians may hold that individuals *should not* update their degrees of belief by Bayesian conditionalisation. This brings us on to the next section.

21.6 'Degree of Belief' Interpretations

It might be preferable to oppose the logical interpretation of probability to *degree of belief interpretations of probability*. (This is new terminology, but it seems fitting.) Compare the subjective and objective Bayesian interpretations, or what we might now call 'subjectivism' and 'objectivism'. On a *pure* subjective view, having degrees of belief that satisfy the axioms of probability is sufficient for rationality. But some introduce further rationality requirements, e.g. that degrees of belief should reflect observed frequencies where appropriate, as 'top ups'. Ultimately, it is best to understand this in terms of a spectrum; strong objectivists (like some 'objective Bayesians') hold that personal probabilities should always have special unique values, whereas pure subjectivists only require that they lie within a particular range. Those in the middle of the spectrum think that personal probabilities should have unique values in some contexts, but not in others, or think the range is narrower than pure subjectivists do.

Williamson [48], for instance, holds that sometimes there are different permissible ways to *equivocate*, or to be non-committal. And when this is the case, he

[13]For example, I argue that Keynes did hold that observed frequencies should constrain our degrees of belief, or at least that his interpretation could easily accommodate this idea. I also dispute the view that the principle of indifference is not as well motivated as the maximum entropy principle.

thinks there's a free choice about how to equivocate. As such, he holds that in some circumstances there is no particular personal probability distribution that one ought to adopt, although some distributions will nevertheless be wrong (in so far as they don't equivocate). Think back to the earlier example of the two flips of the coin. One can opt to use a sample space of {no heads, one head, two heads} or instead of {no heads, one head and one tail, one tail and one head, two heads}. If one equivocates on the first space, the probability of no heads will be one third. If one equivocates on the second space, the probability of no heads will be one quarter. So Williamson might suggest that the probability one has for 'no heads' ought to be one of these two values, other than anything else.[14]

21.7 Group Level Interpretations

One final kind of interpretative strategy is to consider *group*, rather than personal, degrees of belief. The idea behind this approach, which was first proposed by Gillies [11], is that a group can have a Dutch Book made against it if its members do not have the same degree of belief assignments, i.e. reach consensus, and those assignments do not satisfy the probability calculus. Imagine a married couple, with pooled financial resources. Romeo bets £100 that it will rain tomorrow, at even odds. Juliet simultaneously bets £150, at three to one on, that it will not rain. No matter what happens, they will lose £50. (Note that this depends on the individuals having degrees of belief too; so a group level interpretation is *supplemental* to a personal level one.)

This idea has not really caught on, but it can be developed in several ways as shown in Rowbottom ([33]; [35], Chapter 6). For example, group degrees of belief may be interpreted differently from personal degrees of belief; the latter might be understood as credences, while the former might be understood as agreed betting quotients (along with the appropriate dispositions). So a group may be understood to reach consensus about how to bet without sharing individual credences.

Furthermore, it is possible to consider a spectrum of group interpretations, ranging from purely intersubjective to 'interobjective'. On a pure intersubjective account, it does not matter how consensus is reached; it is enough that it is present. On an interobjective account, by contrast, particular procedures are also required in order to form consensus, e.g. critical discussion with input from all members of group who have relevant degrees of belief, and/or relevant expertise. There will therefore be some scenarios, at least, in which group probabilities have unique values.

---

[14]Williamson would presumably insist that the sample space in *this* case should be the latter. However, when the sample spaces are continuous, e.g. in the paradox of Bertrand [3], he thinks that it is allowable to equivocate on the basis of different sample spaces. In short, the idea is that the sample space to use is not clearly specified in the way the problem is set up.

Whatever else might be said about the merits of such approaches, it is sometimes the case that group decisions are better than individual ones; and talk of probabilities at the group level may therefore prove useful in the context of confirmation theory. (Often we're interested in the recommendation of a group, on the basis of the union of the background knowledge of the members. And it seems natural to talk about what the groups thinks, its degrees of confidence, and so forth.) But this research programme is still in its infancy.

Recommended Further Reading

Rowbottom [35] is the most accessible introduction to the interpretation of probability, and requires no mathematical background. It also covers the significance of the interpretation of probability in several contexts: philosophical, social scientific, and natural scientific.

Childers [6] is an intermediate-level introduction. It is especially noteworthy for its extended discussion of the maximum entropy principle, which lies at the heart of objective Bayesianism.

Gillies [13] is the classic textbook on the interpretation of probability. It is more advanced in character than the aforementioned books, and is a very rewarding read for those with solid mathematical backgrounds.

Eagle [9] is a useful collection of classic work on the philosophy of probability (rather than only 'contemporary readings', as its title unfortunately suggests). It is at an advanced, research, level.

Acknowledgements Work on this chapter was supported by a General Research Fund grant, 'Computational Social Epistemology and Scientific Method' (#341413), from Hong Kong's Research Grants Council. Thanks to Teddy Seidenfeld, Glenn Shafer, and the editors for comments on earlier versions.

References

1. Aerts, D., & Sassoli de Bianchi, M. (2014). Solving the hard problem of Bertrand's paradox. *Journal of Mathematical Physics, 55*, 083503.
2. Bernoulli, J. (1713 (2006)). *The art of conjecturing*. (E. D. Sylla, Trans.). Baltimore: Johns Hopkins University Press.
3. Bertrand, J. 1889. *Calcul des Probabilités* (3rd ed, c.1960). New York: Chelsea.
4. Bradley, S. (2014). Imprecise probabilities. In E. N. Zalta (Ed.), *Stanford encyclopedia of philosophy*. URL: https://plato.stanford.edu/entries/imprecise-probabilities/.
5. Carnap, R. (1962). *The logical foundations of probability* (2nd ed.). Chicago: University of Chicago Press.
6. Childers, T. (2013). *Philosophy and probability*. Oxford: Oxford University Press.
7. De Finetti, B. (1972). Subjective or objective probability: Is the dispute undecidable?', *Symposia Mathematica, 9*, 21–36.
8. De Finetti, B. (1974). *Theory of probability: A critical introductory treatment (1970)*. London: Wiley.
9. Eagle, A. (2009). *Philosophy of probability: Contemporary readings*. London: Routledge.

10. Eriksson, L., & Hájek, A. (2007). What are degrees of belief? *Studia Logica, 86*, 183–213.
11. Gillies, D. A. (1991). Intersubjective probability and confirmation theory. *British Journal for the Philosophy of Science, 42*, 513–533.
12. Gillies, D. A. (1998). Confirmation theory. In D. M. Gabbay & P. Smets (Eds.), *Handbook of defeasible reasoning and uncertainty management systems* (Vol. 1, pp. 135–167). Dordrecht: Kluwer.
13. Gillies, D. A. (2000). *Philosophical theories of probability*. London: Routledge.
14. Gyenis, Z., & Rédei, M. (2015). Defusing Bertrand's paradox. *British Journal for the Philosophy of Science, 66*, 349–373.
15. Hacking, I. (1975). *The emergence of probability*. Cambridge: Cambridge University Press.
16. Hájek, A. (2005). Scotching Dutch books. *Philosophical Perspectives, 19*, 139–151.
17. Hájek, A. (2008). Dutch book arguments. In P. Anand, P. Pattanaik, & C. Puppe (Eds.), *The Oxford handbook of rational and social choice*. Oxford: Oxford University Press.
18. Jaynes, E. T. (1973). The well posed problem. *Foundations of Physics, 3*(4), 477–492.
19. Jaynes, E. T. (1957). Information theory and statistical mechanics. *The Physical Review, 106*, 620–630.
20. Jaynes, E. T. (2003). *Probability theory: The logic of science*. Cambridge: Cambridge University Press.
21. Keynes, J. M. (1921). *A treatise on probability*. London: Macmillan.
22. Marinoff, L. (1994). A resolution of Bertrand's paradox. *Philosophy of Science, 61*, 1–24.
23. Mikkelson, J. M. (2004). Dissolving the wine/water paradox. *British Journal for the Philosophy of Science, 55*, 137–145.
24. O'Donnell, R. (1990). The epistemology of J. M. Keynes. *British Journal for the Philosophy of Science, 41*, 333–350.
25. Popper, K. R. (1959). *The logic of scientific discovery*. New York: Basic Books.
26. Popper, K. R. (1983). *Realism and the aim of science*. London: Routledge.
27. Ramsey, F. P. (1926). Truth and probability. In H. E. Kyburg, & H. E. Smokler (Eds.), *Studies in subjective probability* (pp. 61–92). New York: Wiley, 1964.
28. Rowbottom, D. P. (2007a). The insufficiency of the Dutch book argument. *Studia Logica, 87*, 65–71.
29. Rowbottom, D. P. (2007b). In-between believing and degrees of belief. *Teorema, 26*, 131–137.
30. Rowbottom, D. P. (2008). On the proximity of the logical and 'Objective Bayesian' interpretations of probability. *Erkenntnis, 69*, 335–349.
31. Rowbottom, D. P. (2010). *Popper's critical rationalism: A philosophical investigation*. London: Routledge.
32. Rowbottom, D. P. (2013a). Popper's measure of corroboration and P(*h*,*b*). *British Journal for the Philosophy of Science, 64*, 739–745.
33. Rowbottom, D. P. (2013b). Group level interpretations of probability: New directions. *Pacific Philosophical Quarterly, 94*, 188–203.
34. Rowbottom, D. P. (2013c). Bertrand's paradox revisited: Why Bertrand's "solutions" are all inapplicable. *Philosophia Mathematica, 21*, 110–114.
35. Rowbottom, D. P. (2015). *Probability*. Cambridge: Polity Press.
36. Rowbottom, D. P. (2016). How might degrees of belief shift? On actions conflicting with professed beliefs. *Philosophical Psychology, 29*, 732–742.
37. Schervish, M. J., Seidenfeld, T., & Kadane, J. B. (2009). Proper scoring rules, dominated forecasts, and coherence. *Decision Analysis, 6*, 202–221.
38. Schwitzgebel, E. (2001). In-between believing. *The Philosophical Quarterly, 51*, 76–82.
39. Schwitzgebel, E. (2002). A phenomenal, dispositional account of belief. *Nous, 36*, 249–275.
40. Schwitzgebel, E. (2010). Acting contrary to our professed beliefs, or the gulf between occurrent judgment and dispositional belief. *Pacific Philosophical Quarterly, 91*, 531–553.
41. Seidenfeld, T. (1986). Entropy and uncertainty. *Philosophy of Science, 53*, 467–491.
42. Seidenfeld, T., Schervish, M. J., and Kadane, J. B. (1990). When fair betting odds are not degrees of belief. PSA: Proceedings of the Biennial Meeting of the Philosophy of Science Association (pp. 517–524).

43. Shackel, N. (2007). Bertrand's paradox and the principle of indifference. *Philosophy of Science, 74*, 150–175.
44. Shafer, G., & Vovk, V. (2001). *Probability and finance: It's only a game*. New York: Wiley.
45. Streumer, B. (2007). Reasons and entailment. *Erkenntnis, 66*, 353–374.
46. Van Fraassen, B. C. (1989). *Laws and symmetry*. Oxford: Clarendon Press.
47. Williamson, J. O. D. (2005). *Bayesian nets and causality: Philosophical and computational foundations*. Oxford: Oxford University Press.
48. Williamson, J. O. D. (2010). *In defence of objective Bayesianism*. Oxford: Oxford University Press.
49. Williamson, J. O. D. (2011). Objective Bayesianism, Bayesian conditionalisation, and voluntarism. *Synthese, 178*, 67–85.

Chapter 22
Bayesian Epistemology

Erik J. Olsson

Abstract Bayesian epistemology provides a formal framework within which concepts in traditional epistemology, in particular concepts relating to the justification of our beliefs, can be given precise definitions in terms of probability. The Bayesian approach has contributed clarity and precision to a number of traditional issues. A salient example is the recent embedding of the so-called coherentist theory of epistemic justification in a Bayesian framework shedding light on the relation between coherence and truth as well as on the concept of coherence itself. Starting with the early work of Condorcet, the calculus of probability has proved to be a useful tool in the study of social aspects of knowledge as it is pursued in social epistemology.

22.1 Two Problems of Probabilistic Coherence

Let us start by examining the two concepts involved in the term "Bayesian epistemology".

First, we have the term "Bayesian" which in this context denotes a plethora of theories and approaches that make use of probability in the elucidation of phenomena having to do with our beliefs about the world. One aspect of the Bayesian approach, also called Bayesianism, is the representation of a state of belief as an assignment of probabilities to a set of propositions. Typically, Bayesians feel uncomfortable in assigning any empirical proposition probability 0 or 1. Rather, they recommend assigning probabilities strictly between 0 and 1. One reason for this is the so-called betting interpretation of probabilities according to which assigning a probability means that you are willing to accept certain bets. Assigning probability 1 to a proposition means then that you are willing to bet everything – your life, your

E. J. Olsson (✉)
Lund University, Lund, Sweden
e-mail: erik_j.olsson@fil.lu.se

S. O. Hansson, V. F. Hendricks (eds.), *Introduction to Formal Philosophy*, Springer Undergraduate Texts in Philosophy, https://doi.org/10.1007/978-3-319-77434-3_22

family etc. – on p being true. Since we are rarely willing to bet everything on a given (empirical) proposition being true, we should avoid assigning probability 1 to any such proposition, the Bayesian concludes.

This is the static aspect of Bayesianism. There is also a dynamic aspect enshrined in the recommendation that a rational inquirer should update her beliefs by conditionalizing on the new evidence in accordance with Bayes' rule. Let us first define the conditional probability of the hypothesis h given evidence e: $P(h \mid e) = P(e \mid h)P(h)/P(e)$. This equation still does not state anything about how the inquirer's probabilities should change given new evidence. Bayes' rule, sometimes also referred to as the principle of conditionalization, states that the inquirer should, upon receiving new evidence e, update her probability in h so that the latter corresponds to the conditional probability of h given e. In other words, $P^*(h) = P(h \mid e)$, where $P^*(h)$ is the new probability of h given evidence e. These two fundamental assumptions of Bayesianism have inspired a huge debate in philosophy of science and statistical theory, as well as in economics and decision theory. The reader is referred to Talbott [28] for an overview.

Let us proceed now to the term "epistemology" or "theory of knowledge". Epistemology is concerned with various aspects of knowledge. What is the nature of knowledge and how should the concept be defined? What sources give rise to knowledge? How far does our knowledge extend – are there limits in principle? Do we have knowledge at all – or do we have to accept some form of skepticism? If we know, do we know that we do? And so on. Traditionally, the answer to the first question – about the nature of knowledge – has been that knowledge amounts to justified true belief. If you have a belief and you entertain that belief with justification, then you know, provided of course that the proposition in question is true. Believing means in this context being sure or fully convinced of the truth of the proposition.

These standard characterizations of Bayesianism and epistemology reveal that it is not unproblematic to coherently combine the two into "Bayesian epistemology". Just to raise one question: How does the fact that knowledge requires full conviction square with the Bayesian recommendation not to assign 1 to any given empirical proposition? Does not skepticism about empirical knowledge ensue? Perhaps unsurprisingly, recent texts on "Bayesian epistemology" in fact do not address in any great detail problems in traditional epistemology but rather use the term as roughly synonymous with "Bayesianism" (e.g. Talbott [28]). Even though there are certain tensions to be overcome, a Bayesian approach can in fact be very effective in the elucidation of the justification part of the traditional concept of knowledge. This holds in particular of so-called coherentist accounts of justification.

The Truth Conduciveness of Coherence Pre-systematically, coherence is a good thing. If a set of beliefs is coherent, we tend to think that it is plausibly true, and that a more coherent set is more likely to be true than a less coherent one. Consider however the following example from Klein and Warfield ([18], 130–131):

> A detective has gathered a large body of evidence that provides a good basis for pinning a murder on Mr. Dunnit. In particular, the detective believes

> that Dunnit had a motive for the murder and that several credible witnesses claim to have seen Dunnit do it. However, because the detective also believes that a credible witness claims that she saw Dunnit two hundred miles away from the crime scene at the time the murder was committed, her beliefs set is incoherent (or at least somewhat incoherent). Upon further checking, the detective discovers some good evidence that Dunnit has an identical twin whom the witness providing the alibi mistook for Dunnit.

Let the original belief system of the detective contain the beliefs that (1) Dunnit had a motive; (2) several credible witnesses report that they saw Dunnit commit the murder; (3) a single credible witness reports that she saw Dunnit far away from the crime scene at the time of the murder. Let the extended belief system contain the same believes plus the additional beliefs that (4) Dunnit has an identical twin and (5) Dunnit did it. Then we would say that the extended system is more coherent than the original belief system. So we should expect the former to be more likely to be true. However, the extended system contains more propositions than the original system, and hence the probability of the conjunction of the propositions in the extended system must be lower than the probability of the conjunction of the propositions in the original system: disregarding some trivial special cases, the probability of a bigger conjunction is lower than the probability of a smaller conjunction. So, despite being more coherent, the extended system is actually less likely to be true. So, coherence is after all not correlated with plausible truth.

Defining Coherence There have been few convincing proposals for how to define coherence in traditional epistemology. The attempt to spell our coherence in purely logical terms, e.g. by A. C. Ewing [9], was soon seen to be too restrictive. Most other proposals suffer from serious incompleteness or imprecision. A case in point is the account due to Laurence BonJour [3], who regards coherence to be a concept with a multitude of different aspects, corresponding to the following *coherence criteria* (ibid.: 97–99):

1. A system of beliefs is coherent only if it is logically consistent.
2. A system of beliefs is coherent in proportion to its degree of probabilistic consistency.
3. The coherence of a system of beliefs is increased by the presence of inferential connections between its component beliefs and increased in proportion to the number and strength of such connections.
4. The coherence of a system of beliefs is diminished to the extent to which it is divided into subsystems of beliefs which are relatively unconnected to each other by inferential connections.
5. The coherence of a system of beliefs is decreased in proportion to the presence of unexplained anomalies in the believed content of the system.

Now it could well happen that one system S is more coherent than another system T in one respect, whereas T is more coherent than S in another. Perhaps S contains more inferential connections than T, which is less anomalous than S. If so, which system is more coherent in an overall sense? A difficulty pertaining to theories of coherence that construe coherence as a multifaceted concept is to specify how the different aspects are to be amalgamated into one overall coherence judgment. Bonjour's theory remains silent on this important point and, as we shall see, in several other regards as well.

22.2 A Bayesian Analysis of the Dunnit Example

Let us state the argument in more precise terms. A central claim in the coherence theory which has strong intuitive backing is the following:

(A) The more coherent a set is, the more probable it is.

Let us say that an extension K' of a set K is non-trivial if some of the beliefs that are K' but not in K neither follow logically from K, nor have a probability of 1. Klein and Warfield's argument against (A) rests on the following premises:

(B) Any non-trivial extension of a belief system is less probable then the original system.
(C) There exist non-trivial extensions of belief systems that are more coherent than the original system.

But, so the argument goes, (B) and (C) taken together contradict (A).

Let us look at the support for (B) and (C). While (B) is taken for granted, (C) is supported by the above Dunnit example. It is difficult to question (C). Intuitively the members of the extended set in the Dunnit example hang better together than the elements of the original set. Also, the original set contains an anomaly which is resolved through the introduction of the beliefs that Dunnit did it and has an identical twin who the witness providing the alibi mistook for Dunnit. Because no new anomaly is thereby introduced, it follows from Bonjour's fifth criterion that the extended set is more coherent.

But what about (B)? It derives support from its similarity with.

(B´) Any non-trivial extension of a set of propositions is less probable than the original set.

That claim follows directly from the laws of probability and is therefore entirely innocent. But notice that (B´) is about sets of propositions, whereas (B) is about belief systems. What is the difference? A belief system is not any old set of propositions but a set of propositions believed to be true by a subject. Hence, whereas the probability of a set of propositions is the probability that these propositions are all true, the probability of a belief system is the probability that these propositions are true, given that they are believed by the person in question. The former is an unconditional and the latter a conditional probability.

Let $S = \{p_1, \ldots, p_m\}$ and $S' = \{p_1, \ldots, p_m, p_{m+1}, \ldots, p_n\}$. Moreover, let B be a belief system corresponding to S and B' be a belief system corresponding to S'. Formally, (B´) can be expressed as follows:

(B´*) If S' is a non-trivial extension of S, then $P(p_1, \ldots, p_m, p_{m+1}, \ldots, p_n) < P(p_1, \ldots, p_m)$.

The claim (B) should rather be understood as follows:

(B*) If B' is a non-trivial extension of B, then $P(p_1, \ldots, p_m, p_{m+1}, \ldots, p_n \mid \text{belp}_1, \ldots, \text{belp}_m, \text{belp}_{m+1}, \ldots, \text{belp}_n) < P(p_1, \ldots, p_m \mid \text{belp}_1, \ldots, \text{belp}_n)$,

where belp_i states that the subject believes proposition p_i.

For the Dunnit argument it is (B*) that needs to hold, not (B´*). It can be shown however that (B*) is false. There can be non-trivial extensions of a belief system that are more probable than the original belief system. Suppose again that a robbery has been committed. A detective wishing to find out whether Dunnit did it (call that proposition r) consults independent witnesses that have a track-record of being sufficiently reliable so that the detective can routinely trust their reports. This reminds us of Bonjour's "cognitively spontaneous beliefs" which play a crucial role in his epistemology. We assume that the detective believes something just in case a witness has said so.

Suppose that the first witness reports that Dunnit was driving his car away from the crime scene at high speed (c) and the second that Dunnit is in the possession of a gun of the relevant type (g). The original belief system contains the propositions c and g. Now a new witness steps forward, claiming that Dunnit deposited a large sum of money in his bank the day after the robbery (m). The extended belief system contains the propositions c, g and m. The key notions of reliability and witness independence can be expressed in probability theory. For instance, that a given witness is a reliable belief producer can be expressed as follows:

$$P\,(\text{beli}|\text{i}) = \mathbf{p} \text{ and } P\,(\text{beli}|\text{not}-\text{i}) = 1 - \mathbf{q} \text{ for } \mathbf{p}, \mathbf{q} \approx 1 \text{ and i} = \text{c, g, m}.$$

Hence, the probability that you form the belief, if it is true, should be high, and the probability that you form the belief, if it is false should be low.

That the beliefs are independently held means that they there is no direct influence between the testimonies upon which they were based. This can be captured by saying that the detective's routinely acquired belief about some item of evidence is probabilistically independent of any other item of evidence or any other of his routinely acquired beliefs, conditional on the that item of evidence. We express this formally for two items of evidence using the notation of Dawid [7] for the propositional variables *c*, *g*, *r*, *belc* and *belg*. (The values of the propositional variable *c* are the propositions c and its negation not-c and similarly for the other propositional variables.)

$$belc \perp g, belg \mid c \text{ and } belg \perp c, belc \mid g$$

The first part of this statement is read *belc* is independent of *g* and *belg* given *c*, which is sometimes expressed by saying that *c* "screens off" *belc* from *g* and *belg*. This implies for instance that belc is independent of not-g and belg given not-c.

With a few additional assumptions it can now be proved that the extended belief system is more probable than the original system:

$$P(c, g|\text{belc}, \text{belg}) < P(\text{c}, \text{g}, \text{m}|\text{belc}, \text{belg}, \text{belm})$$

For more details and a proof, see Bovens and Olsson [6].

22.3 Bayesian Accounts of Coherence

Let us now return to the problem of how to define coherence. Bonjour's account serves to illustrate another general difficulty. The third criterion stipulates that the degree of coherence increases with the number of inferential connections between different parts of the system. As a system grows larger the probability is increased that there will be relatively many inferentially connected beliefs. For a smaller system, this is less likely. Hence, there will be a positive correlation between system size and the number of inferential connections. Taken literally, Bonjour's third criterion implies, therefore, that there will be a positive correlation between system size and degree of coherence. But this is not obviously correct.

Here is another general challenge for those wishing to give a clear-cut account of coherence. Suppose a number of eye witnesses are being questioned separately concerning a robbery that has recently taken place. The first two witnesses, Robert and Mary, give exactly the same detailed description of the robber as a red-headed man in his forties of normal height wearing a blue leather jacket and green shoes. The next two witnesses, Steve and Karen, also tell exactly the same story but only succeed in giving a very general description of the robber as a man wearing a blue leather jacket. So here we have two cases of exact agreement. In one case, the agreement concerns something very specific and detailed, while in the other case it concerns a more general proposition. This raises the question of which pair of reports is more coherent. Should we say that agreement on something specific gives rise to a higher degree of coherence, perhaps because such agreement seems more "striking"? Or should we rather maintain that the degree of coherence is the same, regardless of the specificity of the thing agreed upon?

The rich literature on Bayesian coherence measures provides various answers to these questions. Here are the two most discussed measures:

$$C_1(\text{p}_1, \ldots, \text{p}_\text{n}) = P(\text{p}_1 \wedge \cdots \wedge \text{p}_\text{n}) / P(\text{p}_1) \times \cdots \times \text{P}(\text{p}_\text{n})$$

$$C_2(\text{p}_1, \ldots, \text{p}_\text{n}) = P(\text{p}_1 \wedge \cdots \wedge \text{p}_\text{n}) / P(\text{p}_1 \vee \cdots \vee \text{p}_\text{n})$$

C_1 was put forward in Shogenji [27] while C_2 was tentatively proposed in Olsson [20] and, independently, in Glass [12]. As the reader can verify, C_1 is sensitive to size as well as to specificity, while this is not so for C_2. It has been suggested, therefore, that these two measures actually measure two different things. While C_2 captures the degree of agreement of the proposition in a set, C_1 is more plausible as a measure of how *striking* the agreement is. See Olsson [20] and also Bovens and Olsson [5] for a discussion of agreement vs. striking agreement. Since the appearance of these two measure, a large number of other alternative measures have been proposed, many of which are considered in Olsson and Schubert [24].

One influential thought in traditional epistemology is that coherence is somehow linked with "mutual support". The Bayesian way of thinking of support is in terms of a confirmation measure. Douven and Meijs [8] have proposed a general scheme for defining coherence measures given a measure S of degree of confirmation. For two propositions p and q, their suggestion takes the following form:

$$C_3\left(\mathrm{p},\mathrm{q}\right)=\frac{1}{2}\left(S\left(\mathrm{p},\mathrm{q}\right)+S\left(\mathrm{q},\mathrm{p}\right)\right)$$

Thus, the degree of coherence of a set of two propositions depends on how much they confirm each other on the average. In order to turn this scheme into a definite measure of coherence, we have to specify a particular measure of confirmation, of which there is no shortage in the Bayesian literature. Douven and Meijs's preferred choice is the difference measure advocated by Gillies [11] and others:

$$C_4\left(\mathrm{p},\mathrm{q}\right)=P\left(\mathrm{p}|\mathrm{q}\right)-P\left(\mathrm{p}\right)$$

Plugging in this measure in Douven and Meijs's recipe yields the following formula:

$$C_5\left(\mathrm{p},\mathrm{q}\right)=\frac{1}{2}\left(P\left(\mathrm{p}|\mathrm{q}\right)-P\left(\mathrm{p}\right)+P\left(\mathrm{q}|\mathrm{p}\right)-P\left(\mathrm{q}\right)\right)$$

But there are of course a whole range of other confirmation measures that could just as well have been employed, e.g., the ratio measure preferred by Schlesinger [25] and others:

$$C_6\left(\mathrm{p},\mathrm{q}\right)=P\left(\mathrm{p}|\mathrm{q}\right)/P\left(\mathrm{p}\right)$$

As is easily seen, the ratio measure of confirmation coincides with the Shogenji measure of coherence for the case of two propositions.

22.4 Impossibility Results for Coherence and Truth

The paper by Klein and Warfield and also Michael Huemer [17] spurred an intense debate on the relation between coherence and truth or high probability, a debate which is still on-going. The most thought-provoking results concern the possibility of finding a measure of coherence that is *truth conducive* in the following sense: if a set of beliefs A is more coherent than another set of beliefs B, then the probability of A is higher than the probability of B. Here it is assumed that the beliefs in question are somewhat reliable and independently held. Finding such a measure was first stated as an open problem in Olsson [20]. An impossibility result to that effect was first proved by Luc Bovens and Stephan Hartmann in their [4] book. A different impossibility theorem was proved in Olsson [21].

These impossibility results give rise to a mind-boggling paradox. How can it be that we trust and rely on coherence reasoning, in everyday life and in science, when in fact coherence is not truth conducive? Since the impossibility results were published a number of proposals have been made for how to avoid the anomaly they present us with. Olsson and Schubert [24] observed that, while coherence falls short of being truth conducive, it can still be "reliability conducive," i.e. more coherence, according to some measures, entail a higher probability that the sources are reliable, at least in a paradigmatic case. For a further development of this idea, see Schubert [26]. Staffan Angere [1, 2] has argued, based on the results of computer simulations, that the fact that coherence fails to be truth conducive in the sense just referred to does not prevent it from being connected with truth in a weaker, defeasible sense: almost all coherence measures that have an independent standing in the literature satisfy the condition that *most* cases of higher coherence are also cases of higher likelihood. Other researchers have proposed other ways of reconciling the impossibility results with our ordinary reliance on coherence. For an up-to-date overview of the debate, see Olsson [23].

22.5 Bayesian Social Epistemology

Following C. I. Lewis [19], most Bayesian coherence theorists take as their paradigm case a scenario involving a number of witnesses giving coherent testimonies. This is then taken to be analogous to the situation upon which traditional coherence theorists have been most interested: the coherence of one person's beliefs. It is perfectly possible to by-pass the second issue so as to focus only on witness scenarios, in which case the study falls under the area known as social epistemology. Bovens and Hartmann [4] elaborate on witness coherence and their book contains further references. A closely related topic is the Bayesian study of voting and the famous Condorcet Jury Theorem which states, roughly, that if voters are independent and somewhat reliable, the majority is more likely to have the right answer than anyone in the minority. Moreover, the chance that the majority is right approaches 1 as more voters are added. See for instance Goodin and List [14] for more on this.

The Jury Theorem belongs, more generally, to what Alvin I. Goldman [13] calls veritistic social epistemology which aims to evaluate social practices, jury voting being but one case, in terms of their veritistic outputs, where veritistic outputs includes states like knowledge, error and ignorance. Goldman focuses on the tendency of practices to produce *true belief* in the participants, true belief representing in his view a weak form of knowledge. Thus, states of true belief have *fundamental* veritistic value or disvalue, whereas practices have *instrumental* veritistic value insofar as they promote or impede the acquisition of fundamental veritistic value.

Let us now turn to the very concept of veritistic value. Goldman's main proposal is that degrees of belief (DB) have veritistic value relative to a question Q, so that any DB in the true answer to Q has the same amount of V-value as the strength of the DB. Goldman represents strength of belief as subjective probability. In Goldman's terminology, V-value of $DB_{\mathbf{x}}(\text{true}) = \mathbf{x}$. Suppose, for example, that Mary is interested in the question whether it will rain tomorrow. If the strength of Mary's belief that it will rain tomorrow is .8, and it will in fact rain tomorrow, then the V-value of Mary's state of belief vis-à-vis the rain issue is .8.

Suppose that a question begins to interest agent S at time t_1, and S applies a certain practice π in order to answer the question. The practice might consist, for instance, in a certain perceptual investigation or in asking a friend. If the result of applying π is to increase the V-value of the belief states from t_1 to t_2, then π deserves positive credit. If it lowers the V-value it deserves negative credit. If it does neither, it is neutral with respect to instrumental V-value. There is more complexity to come, however. In evaluating the V-value of a *practice*, we usually cannot focus merely on the one agent scenario. As Goldman notes, "[m]any social practices aim to disseminate information to multiple agents, and their success should be judged by their propensity to increase the V-value of many agents' belief states, not just the belief states of a single agent" ([13], 93). This is why we should be interested in the *aggregate* level of knowledge, or true belief, of an entire community (or a subset thereof).

Consider a small community of four agents: S_1–S_4. Suppose that the question of interest is whether p or not-p is true, and that p is in fact true. At time t_1, the several agents have DBs vis-à-vis p as shown in the corresponding column (see Table 22.1). Practice π is then applied, with the result that the agents acquire new DBs vis-à-vis P at t_2 as shown in the column under t_2.

At t_1 the group's mean DB in p is .55, so that .55 is their aggregate V-value at t_1. At t_2, the group's mean DB in p is .75, so that this is their new aggregate V-value. Thus the group displays an increase of .20 in its aggregate V-value. Hence the practice π displays positive V-value in this application.

A further issue is that there is a need to consider not just one application of a practice but many such applications. In evaluating a practice, we are interested in its performance across a wide range of applications. In order to determine the

Table 22.1 Individual degrees of belief for a community of four inquirers before and after applying a practice

	t_1	t_2
S_1	DB(p) = .40	DB(p) = . 70
S_2	DB(p) = .70	DB(p) = .90
S_3	DB(p) = .90	DB(p) = .60
S_4	DB(p) = .20	DB(p) = .80

V-value of the practice π in our example we would have to study how well it fares in other applications as well. This would presumably mean, among other things, varying the size of the population of inquirers as well as allowing it to operate on other initial degrees of belief. Once we have isolated the relevant set of applications against which the practice is to be measured, we can take its average performance as a measure of its V-value.

It follows from these considerations that, when assessing the V-value of a practice, we need to "average" twice. For each application A_i of the practice, we need to assess the average effect E_i it had on the degrees of belief of the members of the society. The V-value of the practice is then computed as the average over all the E_is.

As one can imagine, the task to compute the V-value of a social practice can become quite complicated in practice. For that reason, researchers have been interested in delegating it to computers. See Olsson [22] for a description of the simulation framework Laputa which allows V-values to be computed automatically for a variety of social practices.

22.6 The Value of Bayesian Epistemology

Pursuing Bayesian epistemology, as understood here and arguably in Bovens and Hartmann [4], means translating concepts and ideas from epistemology into the language of probability, especially concepts that relate to the way in which our beliefs are justified. This brings with it a number of advantages, many of which pertain to the use of formal methods generally. One has already been highlighted: by means of formalization vague or ambiguous concepts can be made precise and different senses distinguished. This was amply illustrated in our discussion of various ways of defining the concept of coherence – the central concept in the coherentist theory of justification – in probabilistic terms. Further, once a problem has been translated into probability theory, it can be handled in a more objective fashion than was previously possible. Our Bayesian treatment of the Dunnit example due to Klein and Warfield illustrates this advantage allowing it to be rigorously proved that one of their premises is false. The same example pinpoints another virtue of formalization: the possibility of making and upholding delicate distinctions that are difficult to express and sustain in ordinary language, i.e., the distinction between any old propositions and propositions that are believed to be true by some inquirer, and the implications of that difference for the probability of a set. See Hansson [16] for an illuminating discussion of the value of formalization.

Finally, formalization in a standard formal framework, probability being no exception, furthers the important scientific virtues of unity and integration. Thus, the marriage between coherence and probability has led to a tighter connection between epistemology and other areas of philosophy and science in which probability plays a major role. As we saw, authors have explored the rather obvious connection to confirmation theory, including Branden Fitelson [10]. Links to artificial intelligence – Bayesian networks and fuzzy logic respectively – are established in Bovens and Olsson [5] and Glass [15].

References and Proposed Readings[1]

1. Angere, S. (2007). The defeasible nature of coherentist justification. *Synthese, 157*(3), 321–335.
2. Angere, S. (2008). Coherence as a heuristic. *Mind, 117*(465), 1–26.
3. Bonjour, L. (1985). *The structure of empirical knowledge*. Cambridge, MA: Harvard University Press.
4. * Bovens, L., & Hartmann, S. (2003) *Bayesian epistemology*. Oxford: Clarendon Press. [A technically competent survey of applications of Bayesianism to problems in informal philosophy, including an impossibility result for coherence.]
5. Bovens, L., & Olsson, E. J. (2000). Coherentism, reliability and Bayesian networks. *Mind, 109*, 685–719.
6. Bovens, L., & Olsson, E. J. (2002). Believing more, risking less: On coherence, truth and non-trivial extensions. *Erkenntnis, 57*, 137–150.
7. David, A. P. (1979). Conditional independence in statistical theory. *Journal of the Royal Statistical Society, Ser B, 41*(1), 1–31.
8. Douven, I., & Meijs, W. (2007). Measuring coherence. *Synthese, 156*(3), 405–425.
9. Ewing, A. C. (1934). *Idealism: A critical survey*. London: Methuen.
10. Fitelson, B. (2003). A probabilistic measure of coherence. *Analysis, 63*, 194–199.
11. Gillies, D. (1986). In defense of the Popper-Miller argument. *Philosophy of Science, 53*, 110–113.
12. Glass, D. H. (2002). Coherence, explanation and Bayesian networks. In M. O'Neill, R. F. E. Sutcliffe, et al. (Eds.), *Artificial intelligence and cognitive science*, *Lecture Notes in Artificial Intelligence 2464* (pp. 177–182). Berlin: Springer.
13. Goldman, A. I. (1999). *Knowledge in a social world*. Oxford: Oxford University Press.
14. Goodin, R. E, & List, C. (2001). Epistemic democracy: Generalizing the Condorcet Jury theorem. *Journal of Political Philosophy, 9*(3), 277–306.
15. Glass, D. H. (2006). Coherence measures and their relation to fuzzy similarity and inconsistency in knowledge bases. *Artificial Intelligence Review, 26*(3), 227–249.
16. Hansson, S. O. (2000). Formalization in philosophy. *Bulletin of Symbolic Logic, 6*(2), 162–175.
17. Huemer, M. (1997). Probability and coherence justification. *Southern Journal of Philosophy, 35*, 463–472.
18. Klein, P., & Warfield, T. A. (1994). What price coherence? *Analysis, 54*, 129–132.
19. * Lewis, C. I. (1946). *An analysis of knowledge and valuation*. LaSalle: Open Court. [An early classic on probabilistic coherence theory.]
20. Olsson, E. J. (2002). What is the problem of coherence and truth? *The Journal of Philosophy, 99*, 246–272.
21. * Olsson, E. J. (2005). *Against coherence: Truth, probability, and justification*. Oxford University Press. [Applies Bayesian concepts to informal coherence theory, including an impossibility theorem.]
22. Olsson, E. J. (2011). A simulation approach to veritistic social epistemology. *Episteme, 8*(2), 127–143.

[1] Asterisks (*) indicate proposed readings.

23. * Olsson, E. J. (2017). Coherentist theories of epistemic justification. *The Stanford Encyclopedia of Philosophy* (Spring 2017 edn., Edward N. Zalta, (Ed.), URL= https://plato.stanford.edu/archives/spr2017/entries/justep-coherence/. [A regularly updated overview over the debate over the coherence theory of justification with particular emphasis on Bayesian approaches.]
24. Olsson, E. J., & Schubert, S. (2007). Reliability conducive measures of coherence. *Synthese, 157*(3), 297–308.
25. Schlesinger, G. (1995). Measuring degrees of confirmation. *Analysis, 55*, 208–212.
26. Schubert, S. (2011). Coherence and reliability: The case of overlapping testimonies. *Erkenntnis, 74*(2), 263–275.
27. Shogenji, T. (1999). Is coherence truth-conducive? *Analysis, 59*, 338–345.
28. Talbott, W. (2016). Bayesian epistemology. In *Stanford encyclopedia of philosophy* (Winter 2016 Edition, Edward N. Zalta (Ed.)). URL = http://plato.stanford.edu/entries/epistemology-bayesian/ [A regularly updated overview of Bayesianism and its problems.]

Chapter 23
Coherence

Sven Ove Hansson

Abstract We encounter the notion of coherence in many branches of philosophy. This overview introduces some basic distinctions that can be used to characterize concepts of coherence. After that, two formal frameworks for the analysis of coherence are introduced. The first of these is based on the logic of support relations. It is used to show that coherentism and foundationalism may be combinable rather than antithetical. The second framework assumes that coherence comes in degrees and that it can be measured in probability-based units. The properties of such measures is discussed, and so are the difficulties in constructing a measure of coherence that satisfies intuitively reasonable constraints.

23.1 Coherence Is Everywhere

We encounter the notion of coherence in many branches of philosophy.

In the theory of knowledge, coherentists claim that our beliefs all justify each other. Their adversaries, the foundationalists, maintain that a limited subset of the beliefs, the basic beliefs, provide the justification for all the others.

According to Bayesian epistemologists, a rational subject's beliefs must be probabilistically coherent, that is, comply with the laws of probability.

In the philosophy of science, internal tensions (incoherence) in a scientific theory or paradigm are seen as driving forces for its replacement by something better.

In metaphysics, coherentists about truth claim that the truth of a proposition consists in its coherence with other propositions. According to its main rival, correspondence theory, the truth of a proposition is constituted by its correspondence to objective features of the world.

S. O. Hansson (✉)
Division of Philosophy, Royal Institute of Technology (KTH), Stockholm, Sweden
e-mail: soh@kth.se

S. O. Hansson, V. F. Hendricks (eds.), *Introduction to Formal Philosophy*, Springer Undergraduate Texts in Philosophy, https://doi.org/10.1007/978-3-319-77434-3_23

Consistency in logic and mathematics is often described as a form of coherence.

Ethicists such as John Rawls have emphasized that our ethical principles and our judgments in practical ethical issues should form a coherent system (be in a "reflective equilibrium").

In decision theory and action theory, it is usually assumed that a rational plan has to be coherent [17].

In recent years, legal scholars have increasingly emphasized that the law and its interpretation must form a coherent system, preferably based on some common principles.

In spite of the ubiquity of coherence in philosophy, surprisingly few attempts have been made to clarify the general meaning of this term in precise, formal terms.[1] Most formal treatments of coherence have focused on only one application area (usually epistemology), and consequently they lack in generality. The formalization of coherence is still at an early stage, and no consensus has been reached on what criteria a good model should satisfy, or how it should be constructed. In the following section, some distinctions that are essential for the formalization of coherence will be introduced. After that we will have a look at two formalizations, one that is quite general and a more specialized one that is often referred to in epistemology.

23.2 Distinctions That We Need

Some things come in degrees but are nevertheless often discussed in all-or-nothing terms. Temperature is one of these. Although it (literally) comes in degrees we can say: "Yesterday it was hot but today it is not." Coherence is another:

"Her talk was more coherent than his." (*comparative coherence*)
"Her talk was coherent. His was not." (*absolute coherence*)

A model of coherence can treat it in either of these two ways. The absolute version is simpler and may be more clear for some purposes, but of course the comparative version has room for more nuances.

Another important distinction is that between, on the one hand, cohering with something else, and on the other hand, being coherent in itself [3]:

"Her views on capital punishment do not cohere with her more general moral views." (*relational coherence*)
"Her moral standpoints are remarkably incoherent." (*systemic coherence*)

[1]This has been pointed out repeatedly, for instance by Bender [3], Bonjour [5], Bartelborth [2], Olsson [20, pp. 12–13] and Moretti and Akiba [18].

Furthermore, systemic coherence can be treated in two different ways that it is important to distinguish between, although the distinction is somewhat intricate. We normally see the coherence of a system as a matter of how well its parts hang or hold together. This interpretation is also recorded in dictionaries; to be coherent means to "stick or cling firmly together" according to the Oxford English Dictionary. Thus consider the following set of three sentences:

(1) {"Life is sacred", "All murderers should be executed", "It is soon five o'clock"}

This has the appearance of being an incoherent set, since the two first sentences do not fit well together. But now consider the following set:

(2) {"Life is sacred and all murderers should be executed", "It is soon five o'clock"}

This set consists of only two sentences, and there is no conflict between the two. If we consider the coherence of a set as something to be determined solely by relations among its elements, then (2) must be deemed much more coherent than (1). In similar fashion, any incoherent set could be made more coherent by merging its most diverging elements into a single element. But presumably most of us would see such an operation as a way to hide the incoherence rather than reduce it. The reason for this is that when we see a set such as (2), we do not accept its elements as representing the actual parts of that which the set represents.

When two sets, such as (1) and (2), have the same contents, we tend to assume that they also have the same degree of coherence. The underlying assumption is that coherence is a property of the contents of the set, rather than a property of the collection of elements that is used to present the contents. In order to determine the coherence of the contents, we identify its "actual" constituent parts (which may be different from those that were presented to us). We then investigate to what extent these "actual" constituents hang together. When thinking in this way we apply a *presentation-insensitive* notion of systemic coherence.

But we should not exclude the possibility of evaluating the coherence of a particular presentation of a set. I once listened to an exposition of a new legislation that was correct but nevertheless confusing because of the disorganized order of presentation. I could then have said: "The material he presented in his talk was coherent, but the presentation was incoherent." Such a comparison would involve both a presentation-insensitive and a *presentation-sensitive* notion of systemic coherence.

Finally, we have to distinguish between the different *types of cohesive and repulsive forces* that operate in different systems whose coherence we want to analyze. In logic and mathematics, coherence depends on the forces of logical implication [7]. In other areas there is a wider variety of forces conferring or constraining coherence. In epistemology, different types of inferential relations (in a wide sense) can be at play, giving rise for instance to explanatory, evidential, justificatory, or probabilistic coherence [22, p. 144], [4, p. 96]. Ethical coherence can be construed for instance in terms of derivability from common underlying principles, presence of redundant support from several moral principles, or absence of conflicting statements.

23.3 The Support Relations Model

In this section we are going to introduce a simple model of systemic coherence [11]. Its assumption is that we have a system to be evaluated with respect to its coherence, and that this system is represented by a set. (If we are studying presentation-insensitive coherence, then this set has to be composed of the "actual" components of the system.) A support relation S represents the coherence-conferring forces in the system. If a and b are elements of the set, then aSb denotes that a supports b.[2] For illustration we can use diagrams in which the elements are represented by points and S by arrows. An arrow from a to b denotes that a supports b. See Fig. 23.1.

This is in several respects a highly simplified representation of the forces that make coherent systems stick together. First, support is treated as an all-or-nothing affair although we know that support comes in degrees. Secondly, only positive contributions to coherence are covered. In our example (1) above, what made the set incoherent was not just the lack of coherence-conferring relations among the elements but the presence of a conflictual relationship between the first two elements. Thirdly, the model cannot deal adequately with cases where two or more elements in combination provide a support that none of them confers alone, as the first two sentences do to the third in the following example:

(3) {"Amy's father is Chinese", "Amy's mother is British", "Amy is bilingual"}

These limitations can easily be removed. We can replace the binary relation by a function s on pairs of elements, such that $s(a, b)$ is a number representing the degree to which a supports b. Negative values can represent disconnecting forces. In this way we get rid of the first two limitations. To get rid of the third we just need to extend the function to cover expressions such as $s(\{a, b\}, c)$ in which the first argument is a set of sentences.

However, simple binary all-or-nothing support relations are suitable for illustrating certain essential properties of support relations, and they will therefore be used here. One of their major advantages is that they provide us with a convenient representation not only of the interconnected relations in a coherent set but also of the one-sided support relations from the base (basic beliefs) to the rest of the set

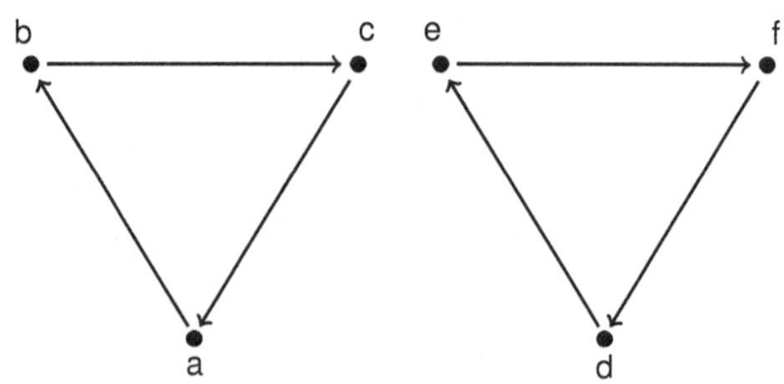

Fig. 23.1 Support relations among the six elements of the set $\{a, b, c, d, e, f\}$. Universal supportedness and Universal supportingness are both satisfied

[2] S is irreflexive, i.e. $\neg(xSx)$ holds for all x.

that are assumed to hold in a foundationalist framework. We can therefore use this simple model to clarify the relationship between coherentism and foundationalism.

Some more notation is needed. The set whose coherence or foundations we are going to investigate needs a name. We can call it E. We also need quantifiers. Unless otherwise specified, $\forall x$ means "for all $x \in E$" and $\exists x$ means "for some $x \in E$". We will also have use for the ancestral S^* of S. It denotes a chain of S-relations that connects two elements of E, hence:

aS^*b holds if and only if either aSb or there is a finite series of elements $x_1, \ldots, x_n$ such that aSx_1, x_kSx_{k+1} for all $1 \leq k < n$, and x_nSb.

Coherentism has been explicated by Ernest Sosa as meaning that "a body of knowledge is a free-floating raft every plank of which helps directly or indirectly to keep all the others in place, and no plank of which would retain its status with no help from the others" [30, p. 24]. It follows from this that nothing is unsupported and that everything supports something else. In the formal language, this means that the following two conditions should be satisfied:

$(\forall x)(\exists y)(ySx)$ (*Universal supportedness*)
$(\forall x)(\exists y)(xSy)$ (*Universal supportingness*)

These are two reasonable conditions, but they are not sufficient to define even a minimal notion of coherence. This can be seen from Fig. 23.1. In this diagram, both conditions are satisfied, but it would be strange to claim that the set $\{a, b, c, d, e, f\}$ is coherent. As several authors have pointed out, a coherent system should not have any isolated subsystem or part that is unconnected with the rest of the system [4, p. 97], [31]. This is ensured by the following simple condition:

$(\exists x)(\forall y)(xS^*y)$ (*Non-fragmentation*)

Figure 23.2 shows a case in which *Universal supportedness*, *Universal supportingness*, and *Non-fragmentation* are all satisfied. The combination of these three conditions ensures at least a minimal degree of coherence.[3]

Let us now turn to foundationalism. According to Ernest Sosa, it means that "every piece of knowledge stands at the apex of a pyramid that rests on stable and secure foundations whose stability and security does not derive from the upper stories or sections" [30, p. 24]. This condition refers to a proper, non-empty subset of E. Let us call it B. Then $E \setminus B$ denotes the superstructure, i.e. the set of

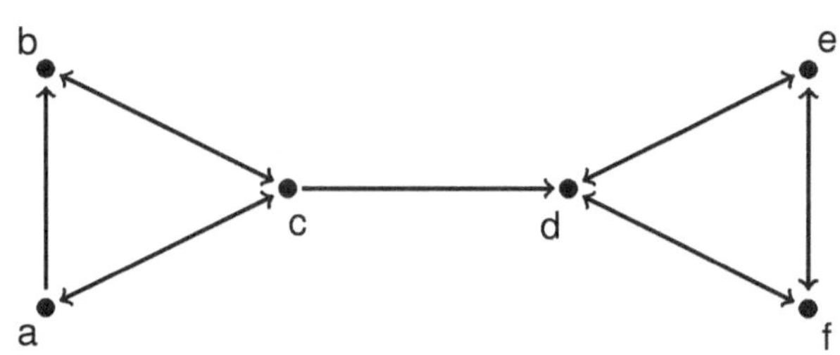

Fig. 23.2 In this case Universal supportedness, Universal supportingness and Non-fragmentation are all satisfied

[3]Alternative, stronger conditions are discussed in [11].

elements of E that are not also elements of B. The quotation from Sosa provides us with two conditions on B. First, it should not be supported by any element of the superstructure, i.e. $(\forall y \in E \setminus B)(\forall x \in B)\neg(ySx)$. Secondly, it should support all the elements of the superstructure. However, this support may be indirect. Therefore we do not need to require that $(\forall y \in E \setminus B)(\exists x \in B)(xSy)$; it is sufficient to require that $(\forall y \in E \setminus B)(\exists x \in B)(xS^*y)$. All this adds up to the following combined requirement:

There is a set B with $\varnothing \neq B \subset E$ such that
$(\forall y \in E \setminus B)(\forall x \in B)\neg(ySx)$ and
$(\forall y \in E \setminus B)(\exists x \in B)(xS^*y)$. (*Base*)

This is a reasonable definition of a base and therefore of foundationalism. It says, essentially, that the base supports the rest of the system but is not supported by it. See Fig. 23.3 in which this criterion is satisfied. This figure also illustrates that the base will not in all cases be uniquely defined. There are in fact no less than eleven alternative bases in this diagram, namely $\{a\}$, $\{a, d\}$, $\{a, d, e\}$, $\{a, d, e, f\}$, $\{a, b\}$, $\{a, b, d\}$, $\{a, b, d, e\}$, $\{a, b, d, e, f\}$, $\{a, b, c\}$, $\{a, b, c, d\}$, and $\{a, b, c, d, e\}$. This may appear disturbing, since we presumably want the base to be well-defined. But the problem can be solved fairly easily. One of the alternative bases, namely $\{a\}$, is uniquely inclusion-minimal, i.e., it is a subset of all the others. We can take it to be the genuine base, of which the other ten are mere extensions.

But now consider Fig. 23.4. Condition *Base* is satisfied here as well. But in this case there is no uniquely inclusion-minimal base, i.e., no base that is a subset of all the others. (To see this, just note that both $\{a, b\}$ and $\{a, b, c\}$ can serve as bases.) From this example we can conclude that contrary to common assumptions, the base of a foundationalist system need not be well-defined.

Fig. 23.3 A case in which Base is satisfied

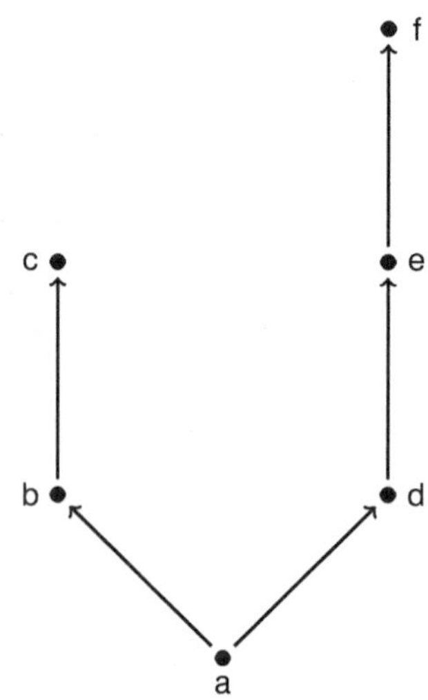

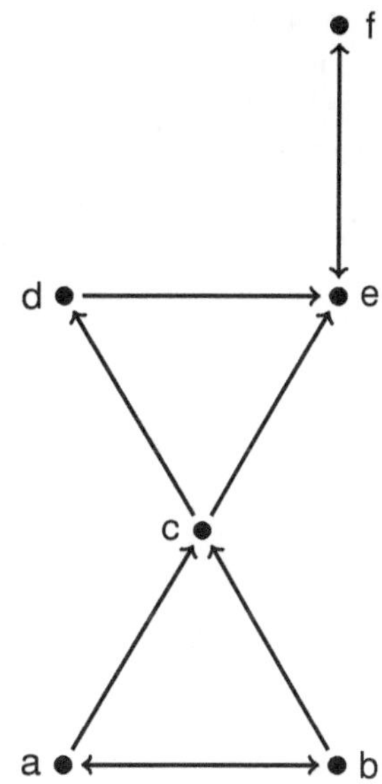

Fig. 23.4 Universal supportedness, Universal supportingness, Non-fragmentation, and Base are all satisfied. The example therefore has both coherentist and foundationalist properties

Figure 23.4 has another, even more important property. We have already seen that it satisfies *Base*. We can easily verify that it also satisfies our three conditions for coherentism, namely *Universal supportedness*, *Universal supportingness*, and *Non-fragmentation*. This example shows that the minimal conditions of coherentism and those of foundationalism are compatible with each other. This opens up the possibility of considering structures of support that are intermediate between, or combine, the classic notions of coherentism and foundationalism. The traditional dichotomy can then be replaced by descriptive models that recognize a wider variety of types of support structures. This more nuanced picture emerges, characteristically, from endeavours to express these notions in formal language so that their implications can be spelt out in full detail.

23.4 Probabilistic Measures of Coherence

Let us now turn to another formal representation of coherence that has attracted much attention lately, namely probabilistic coherence measures. Just as in the previous approach, it is assumed that we have a collection of objects, and that the coherence we are looking for is a property of this collection. In this more specified approach, it is assumed that the elements are propositions to which we can assign probabilities. However, the collection is not represented in the formal language by a set but by a sequence. The reason for this is that duplicates may be of interest. The collection $\langle a, a, b \rangle$ contains two reports that a, whereas $\langle a, b \rangle$ only has one such report. Consequently, these two collections may differ in their degrees of probabilistic coherence. Since the sets $\{a, a, b\}$ and $\{a, b\}$ are identical, they cannot be used to represent such distinctions.

A probabilistic coherence measure is a numerical function that takes us from such a sequence of propositions to a number that represents its degree of coherence. The measure is supposed to depend only on the probabilities of the elements of the sequence and of their Boolean combinations [20, p. 100].

Lewis [15, p. 338] proposed a definition of coherence in terms of probability. Beginning in 1999, veritably dozens of probabilistic coherence measures have been put forward. The following two are probably the ones most often referred to:

Shogenji's [28] coherence measure:

$$C_S(\langle A_1, \ldots, A_n \rangle) = \frac{P(A_1 \& \ldots \& A_n)}{P(A_1) \times \ldots \times P(A_n)}$$

Olsson's [19] coherence measure:

$$C_O(\langle A_1, \ldots, A_n \rangle) = \frac{P(A_1 \& \ldots \& A_n)}{P(A_1 \vee \ldots \vee A_n)}.$$

Shogenji's measure is the ratio between how probable it is that all the sentences are true and how probable this would have been if they had been probabilistically independent of each other. Olsson's measure has been called a measure of overlap. It is the ratio between the probability of all sentences being true and the probability of at least one of them being so. Other measures have been proposed for instance by Fitelson [9, 10], Meijs [16], Douven and Meijs [8], and Siebel and Wolff [29]. Bovens and Hartmann [6] have proposed an arguably somewhat more sophisticated approach in which the numerical measure is replaced by an incomplete ordering. Then sets can be incomparable in terms of their degrees of probabilistic coherence. However, consideration of the two measures already mentioned is sufficient to give a picture of the general nature of probabilistic coherence measures.

Shogenji's measure has been criticized for yielding counter-intuitive results for agreeing reports. Suppose that we initially have one report saying that a coin that was thrown yesterday yielded heads. This is represented by a sequence $\langle A \rangle$ with only one element, namely the proposition A with probability 0.5. Shogenji's measure yields the coherence $C_S(\langle A \rangle) = 1$. But then we receive another report saying the same thing. This increases the degree of coherence to $C_S(\langle A, A \rangle) = 2$. A third report will yield $C_S(\langle A, A, A \rangle) = 4$, etc. According to Fitelson [9] who pointed out this, the measure behaves counterintuitively since we should expect coherence to remain constant when more and more agreeing reports are received. Olsson's measure fares much better here; indeed we have $C_O(\langle A \rangle) = 1$, $C_O(\langle A, A \rangle) = 1$, etc. for any number for agreeing reports.

On the other hand, Olsson's measure has another disputable feature. Based on an insight first reported by Bovens and Hartmann [6], Koscholke and Schippers [14] showed that according to this measure, a set's degree of coherence cannot be increased by adding another proposition to the set. Examples are easily found in which this runs counter to intuition:

m Haris and Rosarita are going to marry.
n Haris and Rosarita have no romantic relationship.
r Rosarita is a refugee for whom a marriage is the only chance not to be sent back to a war zone.
h Haris is a pro-refugee activist.

It would seem plausible to claim in this case that the set $\{m, n, r, h\}$ is more coherent than its subset $\{m, n, r\}$, which is in its turn more coherent than $\{m, n\}$.

Shogenji's and Olsson's measures have a feature in common for which they have been much criticized: They both yield different results for logically equivalent sets. Akiba [1] said: "Obviously the coherence of two beliefs B_1 and B_2 should be no different from the coherence of one conjunctive belief $B_1\&B_2$." Therefore, he said, the two sets $\{B_1, B_2\}$ and $\{B_1\&B_2\}$ should have the same degree of coherence. This, however, is not the case. Let B_1 denote that the next throw of a fair dice will yield an odd number and B_2 that it will yield a prime. Then we have $C_S(\{B_1, B_2\}) = 4/3$ and $C_S(\{B_1\&B_2\}) = 1$. Similarly, $C_O(\{B_1, B_2\}) = 1/2$ and $C_O(\{B_1\&B_2\}) = 1$. Hence neither of these two measures satisfies Akiba's criterion.

In reply, Olsson has questioned Akiba's assumption that sets with the same contents should always have the same degree of coherence, irrespective of how these contents are distributed among sentences [20, p. 102]. Responding to this, Moretti and Akiba [18] showed that a wide variety of probabilistic coherence measures assign different degrees of coherence to logically equivalent sets of propositions. They call this the "problem of belief individuation". However, the seeming problem can be resolved with help of the distinction between presentation-sensitive and presentation-insensitive coherence that was introduced above. Different presentations of one and the same information can differ in their degrees of coherence, even though the coherence of the respective information is the same. Shogenji's and Olsson's coherence measures are presention-sensitive, and there is nothing wrong in them being so.[4] A presentation-insensitive measure would have to replace a given presentation by some standard presentation of the same information before measuring the probabilistic relations among the elements.

A considerable number of probabilistic coherence measures have been proposed. For instance, Koscholke [13] reviewed eighteen of them. Schippers [24, 25] has shown that it is impossible to construct a coherence measure that satisfies a small set of intuitively plausible properties. Based on this, he proposed that a pluralist approach to coherence measures may be appropriate [25, p. 972].

An issue of much interest for coherentist epistemology is whether a probabilistic coherence measure can be truth-conducive. By this is meant that a more coherent sequence is more likely to be true [12]. Several impossibility results have been put forward, indicating that no informative coherence measure can be truth-conducive [6, 12, 20, 21]. The debate is still on-going, and attempts to save

[4]Olsson introduced his measure in a framework where coherence is a property of reports, typically coming from different sources. (Bovens and Hartmann did the same.) Presentation-sensitivity is thus explicitly assumed.

truth-conduciveness have been made [23, 26, 27]. However, it is difficult to see how truth-conduciveness could be achieved with presentation-sensitive measures (like the ones currently under discussion). Expectedly, a truth-conducive measure should treat two sequences equally if they have the same relation to the truth, which they have if they convey the same information.

As mentioned at the outstart, the formal treatment of coherence is still at a surprisingly early stage, given the importance of this concept in several branches of informal philosophy. There is a need for new and innovative measures and models.

Acknowledgements I would like to thank Erik Olsson for very useful comments on an earlier version of this text.

References and proposed readings

Asterisks (∗) indicate recommended readings.

1. Akiba, K. (2000). Shogenji's probabilistic measure of coherence is incoherent. *Analysis, 60*, 356–359.
2. Bartelborth, T. (1999). Coherence and explanations. *Erkenntnis, 50*, 209–224.
3. Bender, J. W. (1989). Coherence, justification, and knowledge: The current debate. In J. W. Bender (Ed.), *The current state of the coherence theory* (pp. 1–14). Dordrecht: Kluwer.
4. Bonjour, L. (1985). *The structure of empirical knowledge*. Cambridge, MA: Harvard University Press.
5. ∗Bonjour, L. (1999). The dialectic of foundationalism and coherentism. In J. Greco & E. Sosa (Eds.), *The Blackwell guide to epistemology* (pp. 117–142). Malden: Blackwell. [A rigorous but informal introduction to coherence and foundationalism.]
6. ∗Bovens, L., & Hartmann, S. (2003). *Bayesian epistemology*. Oxford: Oxford University Press. [An excellent introduction that connects the issue of probabilistic coherence with other issues in epistemology and the philosophy of science.]
7. Daya, K. (1960). Types of coherence. *Philosophical Quarterly, 10*, 193–204.
8. Douven, I., & Meijs, W. (2007). Measuring coherence. *Synthese, 156*, 405–425.
9. Fitelson, B. (2003). A probabilistic theory of coherence. *Analysis, 63*, 194–199.
10. Fitelson, B. (2004). Two technical corrections to my coherence measure. Online paper, http://fitelson.org/coherence2.pdf.
11. ∗ Hansson, S. O. (2007). The false dichotomy between coherentism and foundationalism. *Journal of Philosophy, 104*, 290–300. [The support relations model.]
12. Klein, P., & Warfield, T. A. (1994). What price coherence? *Analysis, 54*, 129–132.
13. ∗Koscholke, J. (2016). Evaluating test cases for probabilistic measures of coherence. *Erkenntnis, 81*, 155–181. [Evaluates eighteen probabilistic coherence measures against a number of test cases.]
14. Koscholke, J., & Schippers, M. (2016). Against relative overlap measures of coherence. *Synthese, 193*, 2805–2814.
15. Lewis, C. I. (1946). *An analysis of knowledge and valuation*. La Salle: Open Court.
16. Meijs, W. (2006). Coherence as generalized logical equivalence. *Erkenntnis, 64*, 231–252.
17. Millgram, E. (2000). Coherence: The price of the ticket. *Journal of Philosophy, 97*(2), 82–93.
18. Moretti, L., & Akiba, K. (2007). Probabilistic measures of coherence and the problem of belief individuation. *Synthese, 154*, 73–95.
19. Olsson, E. J. (2002). What is the problem of coherence and truth? *Journal of Philosophy, 94*, 246–272.

20. ∗ Olsson, E. J. (2005). *Against coherence. Truth, probability, and justification.* Oxford: Clarendon Press. [An excellent overview with a particular emphasis on probabilistic measures and truth conduciveness.]
21. Olsson, E. J. (2005). The impossibility of coherence. *Erkenntnis, 63*, 387–412.
22. Rescher, N. (1979). *Cognitive systematization: A systems-theoretic approach to a coherentist theory of knowledge*. Oxford: Blackwell.
23. Roche, W. (2014). On the truth-conduciveness of coherence. *Erkenntnis, 79*, 647–665.
24. Schippers, M. (2014). Probabilistic measures of coherence: From adequacy constraints towards pluralism. *Synthese, 191*(16), 3821–3845.
25. ∗ Schippers, M. (2015). Towards a grammar of Bayesian coherentism. *Studia Logica, 103*, 955–984. [Proposes a set of intuitively reasonable constraints on a probabilistic coherence measure, and shows that for purely mathematical reasons, no measure can satisfy them all.]
26. Schippers, M. (2016). The problem of coherence and truth redux. *Erkenntnis, 81*, 817–851.
27. Schupbach, J. N. (2008). On the alleged impossibility of Bayesian coherentism. *Philosophical Studies, 141*, 323–331.
28. Shogenji, T. (1999). Is coherence truth-conducive? *Analysis, 59*, 338–345.
29. Siebel, M., & Wolff, W. (2008). Equivalent testimonies as a touchstone of coherence measures. *Synthese, 161*, 167–182.
30. Sosa, E. (1980). The raft and the pyramid: Coherence versus foundations in the theory of knowledge. *Midwest Studies in Philosophy, 5*, 3–25.
31. Spohn, W. (1999). Two coherence principles. *Erkenntnis, 50*, 155–175.

Part V
Philosophy of Science

Chapter 24
Computational Models in Science and Philosophy

Paul Thagard

Abstract Computer models provide formal techniques that are highly relevant to philosophical issues in epistemology, metaphysics, and ethics. Such models can help philosophers to address both descriptive issues about how people do think and normative issues about how people can think better. The use of computer models in ways similar to their scientific applications substantially extends philosophical methodology beyond the techniques of thought experiments and abstract reflection. For formal philosophy, computer models offer a much broader range of representational techniques than are found in traditional logic, probability, and set theory, taking into account the important roles of imagery, analogy, and emotion in human thinking. Computer models make possible investigation of the dynamics of inference, not just abstract formal relations.

Computer models are ubiquitous in the natural and social sciences, but are still rare in philosophy. This chapter will discuss the valuable contributions that such models make in the sciences and show how similar benefits can be gained in philosophy. Formal methods in philosophy have been limited to a relatively small set of tools such as predicate logic, set theory, and probability theory. But there are other branches of mathematics that are at least as relevant to central concerns in epistemology and metaphysics, including differential calculus, linear algebra, dynamic systems theory, and theory of computation. Computational models that draw on these kinds of mathematics can be highly valuable for understanding the structure and growth of knowledge and for grasping the nature of mind and reality.

P. Thagard (✉)
University of Waterloo, Waterloo, ON, Canada
e-mail: pthagard@uwaterloo.ca

S. O. Hansson, V. F. Hendricks (eds.), *Introduction to Formal Philosophy*, Springer Undergraduate Texts in Philosophy, https://doi.org/10.1007/978-3-319-77434-3_24

24.1 Computer Models in Scientific Applications

The development of digital computers and programs in the 1940s transformed many areas of science, starting with physics and later extending to biology, economics, cognitive psychology, and other fields. Physicists began to use computers to model the behavior of sub-atomic particles in nuclear fission and fusion ([13], ch. 8). To build bombs, physicists needed to understand how neutrons fission and scatter, but detailed experiments were not feasible and mathematical theory generated unsolvable equations. Hence John von Neumann and others employed the new tool of vacuum tube computers to recreate physical processes by modeling a sequence of random scatterings using what came to be called *Monte Carlo* methods. The differential equations in physical theory that assumed continuous quantities could be approximated by difference equations expressed in computer instructions. The new method replaced crude estimates of criticality by simulations that enable physicists to determine how detonations occur. Even the very primitive early computers could carry out calculations that would have taken humans hundreds of years. Now, some computers can perform quadrillions of operations per second, providing enormous capacity for simulating very complex systems.

Computer models are now widely used in fields of physics ranging from fluid dynamics to quantum mechanics [56]. Computational biology began in the 1960s and is now applied to many systems from cells to evolutionary development [23]. With the development of huge data bases in genomics and related fields, computers are used for bioinformatic purposes such as determining the function of model genes [18]. Computational chemistry is used to calculate the properties and changes of molecules and solids, with applications to the design of new drugs and materials [8]. Economists have long used computers to implement mathematical models of financial phenomena and are now turning to more realistic systems that model the interactions of somewhat intelligent agents (e.g. [4]). I will shortly give a more detailed account of the nature of computational models in science based on my own experience in developing models in cognitive psychology and neuroscience.

From the perspective of some traditional philosophical approaches, the use of computer models may seem puzzling. Consider the classical hypothetic-deductive method according to which theories consist of axiomatized hypotheses from which observations can be deduced. Why not just use mathematics to state the hypotheses and formal logic to deduce their consequences? There are many reasons why the logic-based version of hypothetico-deductivism is impractical.

First, scientific theories are rarely formalized so rigorously that deductions of the sort found in systems such as predicate logic can be made. Second, predicate logic is undecidable in the sense that there is no effective procedure for determining whether a formula is a consequence of a set of axioms. Third, more practically, theorists in physics and other fields have long known that calculating the consequences of their assumptions is mathematically very difficult. For example, it was already known in the eighteenth century that determining the motions of three bodies was difficult for Newtonian mechanics. Fourth, in the 1960s when computer models were newly used in meteorology, Edward Lorenz discovered that atmospheric systems are chaotic in that small differences in initial conditions can have very large effects in long-range predicted behavior. For these reasons, the logic-based view of hypothetico-deductive systems used to generate predictions and explanations in science does not capture well the actual practice of science. Computer models provide a powerful alternative to human deductions, generating valuable extensions to scientific methods [21].

I now give a more detailed description of how computer models work in science, drawing on my own experience building them for applications in psychology and neuroscience [47]. The methods I will describe are very common in the cognitive sciences, and are similar in many ways to how computer models operate in the natural and social sciences. I will note the relevant differences shortly. All computer models in science require ways of describing both conditions and changes. In physics, the conditions are usually represented by the values of variables, and the changes are represented by differential equations that describe how the values transform over time.

The first prominent computer model in psychology was the rule-based account of problem solving developed in the 1950s by Newell and Simon [26, 27], and this methodology expanded rapidly through the 1970s when cognitive science emerged as a recognized interdisciplinary field. I began building computer models in the 1980s in order to get a better understanding of analogical and other kinds of inference relevant to the discovery and acceptance of scientific theories [19, 20, 36]. The aim of computer modeling in psychology is to develop and test theories about how the mind works.

Since cognitive psychology supplanted behaviorism in the 1950s and 1960s, a psychological theory is an account of the structures and processes that enable minds to carry out such functions as perception, problem solving, learning, and language use. Candidate structures include propositions, concepts, images, Bayesian graphs, and neural networks [42]. Whereas many philosophers take propositions and concepts to be abstract entities, in cognitive science such structures are assumed naturalistically to be physical entities operating in brains and/or computers. Computer models of mind are different from computer models in physics and biology because of the fertile hypothesis developed in the 1950s that thinking is at least analogous to computation and perhaps more strongly is even a kind of computation. In contrast, computational models in physics and biology do not usually assume that entities such as atoms and non-neural cells are actually performing computations themselves.

Following the analogy between thinking and computing, mental structures can be modeled in computer programs via data structures, which are ways in which programming languages store and organize information for efficient use. Programming languages include a variety of data structures such as numbers, variables, strings, lists, and arrays. A high-level programming language such as LISP or Prolog contains extended ways of representing more complex information including propositions and concepts. Then a psychological theory about what kinds of representations the mind uses can be translated into a computer model with analogous kinds of data structures. A computer program is sometimes described just as a set of step-by-step instructions, but the instructions need to have data structures on which to operate, just as an inference needs propositions as well as rules of inference. Hence it is more accurate to describe computer programs and models as combinations of data structures and algorithms, which are effective methods expressed as finite steps of instructions.

It is surprisingly difficult to define more precisely what an algorithm is (see the Wikipedia article "Algorithm Characterizations"). For scientific purposes, algorithms are specified in order to capture changes taking place in the natural system being modeled. In the cognitive sciences, the algorithms specified serve to model the processes proposed in the psychological theory. For example, in rule-based psychological theories such as those of Newell and Simon [27] and Anderson [1], the algorithms specify how applying a rule to propositions can lead to inference to new propositions. This process is similar to use of modus ponens in formal logic, but much more complicated because many non-logical considerations such as past usefulness influence the algorithms that select what rules to fire. The data structures and algorithms of the computer program that implement the computational model correspond to the representations and processes that the psychological theory hypothesizes.

In computer modeling, it is important to distinguish between theories, models, and programs. Theories are general accounts of things, relations, and interactions that produce change. Computer models use data structures to characterize the things and relations, and use algorithms to capture the changes that result from the interactions. Computer programs are packages of code written in a specific language that implement the model and thereby provide a way of testing the theory. It is sometimes said that programs are theories, but programs contain myriad details particular to the programming language used. More accurately, programs implement models that approximate the claims made by theories. Cognitive scientists do not always move from theories to models to programs, because thinking about how to write a program in a familiar language can be a very useful way of developing ideas that can be used to specify models and programs. Computer modeling is a method for generating hypotheses as well as for testing them.

Psychological theories are not easy to test directly against experimental results, because their deductive implications are often unclear. When a theory, however, is specified in a model and implemented in a program, it becomes much easier to determine the implications of the theory. Unless the theory, model, and program are ridiculously simple, building a program and getting it to perform in psychologically realistic ways are highly non-trivial tasks. As the field of artificial intelligence has repeatedly found since its origin in the mid-1950s, computational implementation of functions such as perception and inference reveals unexpected difficulties. Some algorithms are computationally intractable in that the resources required increase exponentially with the size of the problem to be solved. For example, using truth tables to check for consistency in propositional logic is fine for very small numbers of sentences, but since n sentences require 2^n rows this method is not practical even on large computers for the millions of beliefs held by people and computer data bases. Hence implementing a theory in a running computer program provides preliminary evidence that the representations and processes postulated by the theory are physically realizable.

Given realizability, a computer program provides a way of testing a theory by examining whether the running program behaves in ways similar to how people behave in psychological experiments. There should be at least a qualitative fit between what the program does and what people do: the program performs roughly the same tasks in roughly the same ways. Ideally, there will also be a quantitative fit between program and human behavior, with statistics describing what the program does matching closely statistics generated in human experiments. Of course, even quantitative fit between program and experiments does not demonstrate that the original psychological theory is true, but it does provide some support. As in scientific theorizing in general, evaluation requires a full assessment of how well a theory compares to alternative theories in its ability to explain the full range of available evidence.

Computer modeling in the rule-based Newell and Simon tradition is still an important part of cognitive science, but it has been supplemented by approaches that more directly model the brain. In the 1980s neural networks models became prominent, also known by the terms *connectionist* (because they represent information by the connections between neurons) and *parallel distributed processing* [31]. These models are very different from rule-based and logic-based models in their data structures and algorithms. Instead of viewing problem solving and other cognitive tasks as a series of inferences applied to linguistic structures, neural network models adopt simpler data structures - artificial neurons and the links between them - and parallel algorithms that describe how activation (neural firing) spreads through populations of neurons. Current models in computational neuroscience are much more biologically realistic than connectionist models in employing much larger numbers of spiking neurons organized into populations that correspond to actual brain regions [7, 10–12, 28].

Although neural network models approach the mind very differently from the views of psychological operations found in folk psychology, formal logic, and rule-based systems, their use still fits with the general methodology I already described for computer modeling. Programs still consist of data structures and algorithms, although the structures are strange from the commonsense ones suggested by introspection and examination of written texts. Speech and writing are serial processes in which words, sentences, and inferences are generated one at a time using large structures such as concepts and propositions. From the perspective of computational neuroscience, concepts and propositions are patterns of activation in populations of thousands or millions of neurons (defenses and illustrations include [46, 47, 54]). For describing such patterns and exploring their operations, computer modeling is indispensable.

24.2 Philosophical Applications

Decades ago, Aaron Sloman [33] wrote that it was only a matter of time that any philosophers unfamiliar with computational modeling would be deemed incompetent! Currently, however, computer models are still rare in philosophy, although they have been used to study such topics as logic, causal reasoning, social evolution, ethical development, scientific reasoning and coherence. Specific examples will be provided below.

The key question is how computer models can be relevant to philosophical problems concerning the nature of knowledge, reality, and morality. On some views of philosophy, there would be no relevance. If the main goal of philosophy were to generate transcendent, a priori truths, then computer models would have little to contribute. Or if the main goal of philosophy were to analyze people's everyday concepts, then attention to language would obviate computer models. I think, however, that there are no significant a priori truths, and that philosophy should be like science in aiming to improve concepts rather than to analyze existing ones [45, 46, 48, 50]. Philosophy does not reduce to science, because its concerns have a degree of generality and normativity not found in any scientific field. But a naturalistic approach as pursued by Aristotle, Bacon, Locke, Hume, Mill, Peirce, Quine and many other philosophers, sees scientific results as highly relevant to philosophical issues, and hence opens the possibility that computational models might provide a useful philosophical methodology.

First consider epistemology. If one abandons as hopeless the traditional empiricist and rationalist goals to find an indubitable foundation for knowledge, then epistemology can reorient toward the much more interesting and accomplishable task of understanding the structure and development of knowledge. This task is very similar to the goal of cognitive psychology to understand how the mind/brain processes information about the world. Quine [30] also saw an alliance between epistemology and psychology, but was hampered by the theoretical and experimental limitations of the behaviorist psychology of his day. Current psychology has the intellectual resources to help address many key philosophical concerns about the nature of knowledge and inference. Here are some illustrations.

The main alternative to foundationalist epistemology is coherentism, according to which interlocking beliefs can be justified if they form a coherent set. Most philosophical discussions of coherence have only vaguely suggested how it can be objectively assessed. However, coherence can be made much more precisely calculable by considering it as a kind of constraint satisfaction problem of the sort naturally approached using neural network algorithms [37, 38, 40]. Moreover, coherence from this perspective can be formalized to an extent that enables proof that the problem of coherence is NP-hard, i.e. in a class of problems for which a guaranteed solution is unlikely to be found [55]. However, computer experiments show that connectionist and other algorithms can be used to model very large examples of scientific reasoning. Such modeling does not "prove" that coherentism is the best approach to epistemology, but it provides evidence that it can adequately characterize important aspects of belief evaluation.

The main alternative to coherentism in non-foundationalist epistemology is Bayesianism, which uses the tools of probability theory to analyze the structure and growth of knowledge. Merely assuming that probability theory provides answers to epistemological problems does not take one very far, but highly sophisticated computational tools for modeling Bayesian inference have been developed by philosophers, psychologists, and computer scientists (e. g. [14–16, 29, 34]). These computational tools have made possible the testing of Bayesian models as both accounts of actual human inference and as means of making accurate probabilistic inferences.

One advantage of formalizing philosophical ideas about inference in computational models is that it makes possible head-to-head comparison of their relative merits. For example, Thagard [41] compared coherence and Bayesian accounts of legal inference and argued that coherence is superior both descriptively and normatively. Epistemology, obviously, is concerned not just with the descriptive task concerning how people do think but also with the normative task of determining how people can think better. Normative concerns are not alien to science, as there are branches of applied science such as engineering and educational psychology that are as much concerned with improving the world as describing it. Computer models can contribute to normative deliberation by providing a means to explore the consequences of different ways of understanding the nature of knowledge. They are thus much more useful than thought experiments, in which philosophers' own intuitive reactions to stories they have made up are mysteriously used as evidence for the philosophers' preconceptions. As in science, computer models provide a link between theory and data, where the data can be actual cases of human knowledge development of the sort that occur in laboratory experiments and the history of science.

Computer models have other kinds of epistemological applications. For example, there is an old debate in the philosophy of science about whether there could be a "logic of discovery" [17]. This debate has been enriched by the development of various computer models of aspects of scientific discovery including generation of concepts, hypotheses, and descriptions of mechanisms (e.g. [3, 24, 39, 54]). Peirce's still-influential idea of abduction as a kind of inference involving both the generation and acceptance of explanatory hypotheses has been computationally explored using many techniques ranging from formal logic to neural networks (e.g. [22, 52]). Analogical inference can also be productively investigated using computational models [20, 35]. More traditional philosophical approaches involving formal logic can also be enhanced by computational modeling. In sum, computer modeling is as valuable a tool for epistemology as it is for cognitive psychology and other areas of science.

One might naturally suspect, however, that computer models are irrelevant to metaphysical questions about the fundamental nature of reality. As for epistemology, however, the potential arises within a naturalistic view of metaphysics that views it as continuous with science. For example, metaphysical questions about the nature of space and time might be informed by physical theories that are tested via computational models, although I do not know of any specific examples. But such models are clearly relevant to another central metaphysical question, the relation of mind and body.

Idealism, materialism, and dualism are the classic positions in the philosophy of mind. I think that evidence is rapidly mounting for a materialist resolution of the mind-body problem ([46, 48]; see also [2, 5, 6, 25]). Rather than pursuing inconclusive and prejudicial thought experiments, philosophers can examine evidence both for and against the hypothesis that mind events are brain events. This hypothesis is no different from many identity hypotheses that have come to be supported by large amounts of scientific evidence: water is H_2O, air is a mixture of gases, combustion is oxygenation, lightning is electricity, heat is motion of molecules, and so on. Support for mind-brain identity requires consideration of how well brain processes can explain the full range of psychological functions such as perception, inference, language, emotion, and conscience.

As my earlier discussion of computational neuroscience indicated, computer models are an important part of developing and testing neurocognitive theories. Philosophers can of course wait and watch for models most relevant to metaphysical concerns to be developed by scientists, but can accelerate progress by possessing the skills to build models themselves. For example, I had been investigating emotional thinking as a brain process [43], and was aware that conscious experience is a key part of emotion that according to some philosophers requires a non-materialist explanation. Hence I decided to develop a model of emotional consciousness, parts of which have been implemented computationally [49]. This model integrates two theories of emotion (cognitive appraisal and physiological perception) that have been taken as competitors by both philosophers and psychologists. Without computational tools that facilitate thinking of the brain as a parallel processor interconnecting both cognitive and bodily information, it would have been difficult to construct this model. By providing an evidence-based neural explanation of one important kind of consciousness, the model is highly relevant to the philosophical question of the relation between mind and body. Later work draws on new ideas from computational neuroscience to develop an improved theory of emotion [53].

I predict that further progress in computational neuroscience, along with rapidly growing evidence from brain scans and other experimental techniques, will provide further evidence for materialist metaphysics. Of course, those who favor dualism or idealism may see these developments as grounds for just ignoring scientific evidence and the computational models that connect them with theory. Ignorance is bliss.

Besides epistemology and metaphysics, the third major area of philosophy is ethics. Most computer modeling relevant to ethics has been performed by theorists interested in questions concerning the evolution of ethical strategies as modeled by game theory [9, 32]. I prefer a less abstract, more naturalistic approach to ethics that attempts to reach moral conclusions by developing coherent judgments about fundamental human needs [46, 48, 51]. From this perspective, moral intuitions are not a priori judgments achieving transcendent truths, but rather are the result of brain processes for emotional coherence. It follows that the model of emotional consciousness already described is highly relevant to understanding ethical judgments. The model provides a way of seeing how such judgments can be both cognitive and emotional, undercutting debates about emotivism that have exercised ethicists since the 1930s. Hence computer models can be highly relevant to ethical theory. Neuropsychological theories rooted in computational models can also be relevant to explaining puzzling ethical lapses such as conflicts of interest and self-deception [44].

In sum, computer models provide formal techniques that are highly relevant to philosophical issues in epistemology, metaphysics, and ethics. Such models can help philosophers to address both descriptive issues about how people do think and normative issues about how people can think better. The use of computer models substantially extends philosophical methodology beyond the timeworn techniques of thought experiments and abstract reflection.

For formal philosophy, computer models offer a much broader range of representational techniques than are found in traditional logic, probability, and set theory, allowing expansion to take into account the important roles of imagery, analogy, and emotion in human thinking. Just as significant, computer models make possible investigation of the dynamics of inference, not just abstract formal relations. Far from being oxymoronic, computational philosophy offers powerful new tools for investigating knowledge, reality, and morality.

References[1]

1. Anderson, J. R. (2007). *How can the mind occur in the physical universe?* Oxford: Oxford University Press.
2. *Bechtel, W. (2008). *Mental mechanisms: Philosophical perspectives on cognitive neuroscience*. New York: Routledge.
3. Bridewell, W., Langley, P., Todorovski, L., & Dzeroski, S. (2008). Inductive process modeling. *Machine Learning, 71*, 1–32.
4. Chen, S., Jain, L., & Tai, C. (2010). *Computational economics: A perspective from computational intelligence*. Hershe: Idea Group.
5. Churchland, P. M. (2007). *Neurophilosophy at work*. Cambridge: Cambridge University Press.
6. Churchland, P. S. (2002). *Brain-wise: Studies in neurophilosophy*. Cambridge, MA: MIT Press.
7. Churchland, P. S., & Sejnowski, T. (1992). *The computational brain*. Cambridge, MA: MIT Press.
8. Cramer, C. J. (2002). *Essentials of computational chemistry*. New York: Wiley.
9. Danielson, P. (1992). *Artificial morality: Virtuous robots for virtual games*. New York: Routledge.
10. Dayan, P., & Abbott, L. F. (2001). *Theoretical neuroscience: Computational and mathematical modeling of neural systems*. Cambridge, MA: MIT Press.
11. *Eliasmith, C. (2013). *How to build a brain: A neural architecture for biological cognition*. Oxford: Oxford University Press.

[1] Asterisks (*) indicate recommended readings.

12. Eliasmith, C., & Anderson, C. H. (2003). *Neural engineering: Computation, representation and dynamics in neurobiological systems*. Cambridge, MA: MIT Press.
13. Galison, P. (1997). *Image & logic: A material culture of microphysics*. Chicago: University of Chicago Press.
14. Glymour, C. (2001). *The mind's arrows: Bayes nets and graphical causal models in psychology*. Cambridge, MA: MIT Press.
15. Glymour, C. & Danks, D. (2008). Reasons as causes in Bayesian epistemology. *Journal of Philosophy, ADD*.
16. Griffiths, T. L., Kemp, C., & Tenenbaum, J. B. (2008). Bayesian models of cognition. In R. Sun (Ed.), *The Cambridge handbook of computational psychology* (pp. 59–100). Cambridge: Cambridge University Press.
17. Hanson, N. R. (1958). *Patterns of discovery*. Cambridge: Cambridge University Press.
18. Haubold, B., & Wiehe, T. (2006). *Introduction to computational biology: An evolutionary approach*. Basel: Birkhäuser.
19. Holland, J. H., Holyoak, K. J., Nisbett, R. E., & Thagard, P. R. (1986). *Induction: Processes of inference, learning, and discovery*. Cambridge, MA: MIT Press/Bradford Books.
20. Holyoak, K. J., & Thagard, P. (1995). *Mental leaps: Analogy in creative thought*. Cambridge, MA: MIT Press/Bradford Books.
21. Humphreys, P. (2004). *Extending ourselves: Computational science, empiricism, and scientific method*. Oxford: Oxford University Press.
22. Josephson, J. R., & Josephson, S. G. (Eds.). (1994). *Abductive inference: Computation, philosophy, technology*. Cambridge: Cambridge University Press.
23. Kitano, H. (2002). Computational systems biology. *Nature, 420*, 206–210.
24. Langley, P., Simon, H., Bradshaw, G., & Zytkow, J. (1987). *Scientific discovery*. Cambridge, MA: MIT Press/Bradford Books.
25. McCauley, R. N., & Bechtel, W. (2001). Explanatory pluralism and the heuristic identity theory. *Theory & Psychology, 11*, 736–760.
26. Newell, A., Shaw, J. C., & Simon, H. (1958). Elements of a theory of human problem solving. *Psychological Review, 65*, 151–166.
27. Newell, A., & Simon, H. A. (1972). *Human problem solving*. Englewood Cliffs: Prentice-Hall.
28. O'Reilly, R. C., & Munakata, Y. (2000). *Computational explorations in cognitive neuroscience*. Cambridge, MA: MIT Press.
29. Pearl, J. (2000). *Causality: Models, reasoning, and inference*. Cambridge: Cambridge University Press.
30. Quine, W. V. O. (1969). *Ontological relativity and other essays*. New York: Columbia University Press.
31. *Rumelhart, D. E., & McClelland, J. L. (Eds.). (1986). *Parallel distributed processing: Explorations in the microstructure of cognition*. Cambridge MA: MIT Press/Bradford Books.
32. Skyrms, B. (1996). *The dynamics of rational deliberation*. Cambridge, MA: Harvard University Press.
33. *Sloman, A. (1978). *The computer revolution in philosophy*. Atlantic Highlands: Humanities Press.
34. Spirtes, P., Glymour, C., & Scheines, R. (1993). *Causation, prediction, and search*. New York: Springer-Verlag.
35. Steinhart, E. (2001). *The logic of metaphor: Analogous parts of possible worlds*. Dordrecht: Kluwer.
36. Thagard, P. (1988). *Computational philosophy of science*. Cambridge, MA: MIT Press.
37. Thagard, P. (1989). Explanatory coherence. *Behavioral and Brain Sciences, 12*, 435–467.
38. Thagard, P. (1992). *Conceptual revolutions*. Princeton: Princeton University Press.
39. Thagard, P. (1998). Computation and the philosophy of science. In T. W. Bynum & J. H. Moor (Eds.), *The digital phoenix: How computers are changing philosophy* (pp. 48–61). Oxford: Blackwell.
40. Thagard, P. (2000). *Coherence in thought and action*. Cambridge, MA: MIT Press.

41. Thagard, P. (2004). Causal inference in legal decision making: Explanatory coherence vs. Bayesian networks. *Applied Artificial Intelligence, 18*, 231–249.
42. Thagard, P. (2005). *Mind: Introduction to cognitive science* (2nd ed.). Cambridge, MA: MIT Press.
43. Thagard, P. (2006). *Hot thought: Mechanisms and applications of emotional cognition*. Cambridge, MA: MIT Press.
44. Thagard, P. (2007). The moral psychology of conflicts of interest: Insights from affective neuroscience. *Journal of Applied Philosophy, 24*, 367–380.
45. Thagard, P. (2009). Why cognitive science needs philosophy and vice versa. *Topics in Cognitive Science, 1*, 237–254.
46. Thagard, P. (2010). *The brain and the meaning of life*. Princeton: Princeton University Press.
47. *Thagard, P. (2012). *The cognitive science of science: Explanation, discovery, and conceptual change*. Cambridge, MA: MIT Press.
48. Thagard, P. (forthcoming). *Natural philosophy: From social brains to knowledge, reality, morality, and beauty*. Oxford: Oxford University Press.
49. Thagard, P., & Aubie, B. (2008). Emotional consciousness: A neural model of how cognitive appraisal and somatic perception interact to produce qualitative experience. *Consciousness and Cognition, 17*, 811–834.
50. Thagard, P., & Beam, C. (2004). Epistemological metaphors and the nature of philosophy. *Metaphilosophy, 35*, 504–516.
51. Thagard, P., & Finn, T. (2011). Conscience: What is moral intuition? In C. Bagnoli (Ed.), *Morality and the emotions* (pp. 150–159). Oxford: Oxford University Press.
52. Thagard, P., & Litt, A. (2008). Models of scientific explanation. In R. Sun (Ed.), *The Cambridge handbook of computational psychology* (pp. 549–564). Cambridge: Cambridge University Press.
53. Thagard, P., & Schröder, T. (2014). Emotions as semantic pointers: Constructive neural mechanisms. In L. F. Barrett & J. A. Russell (Eds.), *The psychological construction of emotions* (pp. 144–167). New York: Guilford.
54. Thagard, P., & Stewart, T. C. (2011). The Aha! Experience: Creativity through emergent binding in neural networks. *Cognitive Science, 35*, 1–33.
55. Thagard, P., & Verbeurgt, K. (1998). Coherence as constraint satisfaction. *Cognitive Science, 22*, 1–24.
56. Thijssen, J. M. (2007). *Computational physics*. Cambridge: Cambridge University Press.

Chapter 25
Models of the Development of Scientific Theories

Gerhard Schurz

Abstract The three basic kinds of theory development are expansion, contraction and revision by empirical evidence (Sect. 25.1). Under empiricist assumptions, the history of scientific evidence can be represented by a sequence of true and cumulatively increasing evidence sets which in the limit determine the complete structure of the world (Sect. 25.2). Under these assumptions it turns out that purely universal hypotheses are falsifiable with certainty, but verifiable only in the limit, ∀-∃-hypotheses are falsifiable in the limit but not verifiable in the limit, and ∀-∃-∀-hypotheses are neither nor (Sect. 25.3). In the consequence, hypotheses with complex quantification structure can only be confirmed probabilistically (Sect. 25.4). While these results are based on "empiricist" assumptions, the revision of theories which contain theoretical concepts requires either a given partition of possible hypotheses out of which the most promising one is chosen (rational choice paradigm), or it requires steps of abductive belief revision (construction paradigm) (Sect. 25.5). Revision of scientific theories is based on a Lakatosian preference structure, following the idea that in case of a conflict between theory and data, only peripherical parts of the theory are revised, while the theory's core is saved from revision as long as possible (Sect. 25.6). Surprisingly, the revision of a false theory by true empirical evidence does not necessarily increase the theory's truthlikeness (Sect. 25.7). Moreover, increase in empirical adequacy does not necessarily indicate progress in theoretical truthlikeness; a well-known attempt to justify this inference is Putnam's no-miracles argument (Sect. 25.8).

G. Schurz (✉)
University of Düsseldorf, Düsseldorf, Germany
e-mail: schurz@phil-hhu.de

S. O. Hansson, V. F. Hendricks (eds.), *Introduction to Formal Philosophy*, Springer Undergraduate Texts in Philosophy, https://doi.org/10.1007/978-3-319-77434-3_25

25.1 Basic Notions of Theory Development

Three different kinds of theory development are expansions, contractions and revisions of theories (compare ch. 6/5 in *this volume*). These and other distinctions require *formal tools* for their reconstruction. We assume the logical framework of a 1st order language with the standard logical symbols ($\neg,\wedge,\vee,\rightarrow,\leftrightarrow,\forall,\exists,=$) and x_i ($i = 1, 2,\ldots$) for individual variables, a_i for (individual constants), P_i, R_i ... for n-ary predicates. $\mathcal{L}$ is our language and $S(\mathcal{L})$ the set of $\mathcal{L}$-sentences. Small (arabic) letters $s_i \in S(\mathcal{L})$ denote arbitrary sentences, h_i hypotheses, e_i evidence statements, T_i stands for theories which are sets of sentences, E_i for sets of evidence statements, and S_i for arbitrary sets of statements. $||$— denotes logical inference and Cn logical consequence, i.e. $\mathrm{Cn}(S){:=}\ \{s_i{\in}S(\mathcal{L}){:}\ S\ ||\!\!-\ s_i\}$ (":=" means "identity by definition"). In this framework (which is widespread in formal philosophy of science) a theory T_i is reconstructed as a consistent and logically closed set of sentences of a given set of characteristic axioms A_i of T_i, i.e. $T_i = \mathrm{Cn}(A_i)$. A_i is also called the (axiomatic) *base* of T_i.[1]

Within this formal background the (ordinary) *expansion* of a theory T by a new and T-compatible piece of information s is denoted as $T + s$ and defined as $T + s{:}= \mathrm{Cn}(T \cup\{s\})$. If $T = \mathrm{Cn}(A)$ an equivalent definition is $T + s = \mathrm{Cn}(A\cup\{s\})$. If s contradicts T ($T\ ||\!\!-\neg s$), then one must first contract T by $\neg s$ before one can expand T by s. The so-called *contraction* of T by s is denoted as $T\div s$ and intended to be some preferred T-subset which does no longer entail s. Different methods of defining contraction operations have been suggested (e.g. via intersections of maxichoice contractions, or via epistemic ordering or ranking functions over sentences or over corresponding possible worlds (cf. e.g. [11, 33], and ch. 6/5 *this volume*). Finally, the (ordinary) *revision* of T by a T-incompatible proposition s is denoted as $T{*}s$ and defined via the so-called Levi-identity as $T{*}s{:} = (T\div\neg s) + s$ (so-called after Isaac Levi in [10]). In other words, one revises T by a new piece of information s by first contracting T by $\neg s$ and then expanding this contraction by s. Note that for a T-compatible s, expansion coincides with revision ($T{*}s = T + s$).

The information by which scientific theories are typically expanded or revised are *evidences*, i.e. observations or measurement results; they are (typically) expressed by so-called *basic statements*. These are unnegated or negated atomic statements (also called "literals") of an assumed *empirical* (or non-theoretical) *sublanguage* $\mathcal{L}_e$ of $\mathcal{L}$ whose concepts are directly observable or measurable (i.e. $S(\mathcal{L}_e) \subseteq S(\mathcal{L})$).

[1]An alternative to the *sentential* representation of a theory T as a logically closed set of sentences is the *model-theoretic* (or "structuralist") representation of T by the set of *semantic models* or *possible worlds* which verify T (cf. e.g. [9, 17]).

Important for philosophy of science is the distinction between *AGM-contraction* after Alchourrón et al. [1] and *base-contraction* introduced by Hansson [13]. While in AGM-contraction the contraction-operation is applied to the entire set of T's consequences, whether basic or derived, in base-contraction this operation is only applied to T's axioms A, i.e. $T \div s$ is defined as $\mathrm{Cn}(A \div s)$. The difference is this. If an axiom $s \in A$ of a theory $T = \mathrm{Cn}(A)$ is removed, then in base-contraction T-consequences whose justificational support depends on axiom s have to be removed from T, too. In AGM-contraction this need not be so: one may retain consequences from T even after their premises have been removed. While base-revision is reasonable from a *foundation-oriented* viewpoint, AGM-revision makes sense if one adopts a *coherentistic* position (cf. also [33], ch. 3).

Beyond these ordinary notions of expansion and revision (be it AGM- or base-) we introduce in Sect. 25.5 the stronger notions of *abductive* expansion and revision.

25.2 Theory Development Under Empiricist Assumptions

The empiricist view of scientific theories has been defended by classical empiricists (e.g. John Locke) and early logical empiricists (e.g. [4]). This view makes two assumptions: (i) that evidences are certain, and (ii) that all non-logical concepts of scientific theories can be defined by empirical concepts, or in other words, that the scientific language contains no genuinely theoretical concepts, i.e. $\mathcal{L} = \mathcal{L}_e$. We let $E(\mathcal{L}_e)$ stand for the set of all basic (i.e. evidence) statements of $\mathcal{L}_e$. Each maximally consistent subset of $E(\mathcal{L}_e)$ (a so-called "diagram") is denoted by EW_k and represents a possible empirical world or "$\mathcal{L}_e$-world" over a given countably infinite domain of named individuals (i.e. every individual has a standard name in $\mathcal{L}_e$). We formalize the history of scientific evidences *over* a given $\mathcal{L}_e$-world EW in the form of an *evidence stream* $(e) = (e_0, e_1, \ldots)$, such that $EW = \{e_i: i \in \omega\}$ and the e_i are pairwise distinct. Given the empiricist assumptions, EW determines the *complete true* theory $T(EW)$, which is semantically given as the set of all true $\mathcal{L}_e$-statements in EW and syntactically as the uniquely determined maximally consistent and ω-complete extension of EW in $\mathcal{L}_e$.[2] Given this formal reconstruction, empiricist theory-development has the following properties:

1. The accumulation of evidences is *cumulative* because evidences are true and are never taken back. Formally, this means that the e_i are mutually compatible.
2. At every finite time point n, only a finite subset of EW is known. Therefore $T(EW)$ is never known with certainty. As Popper [28] has stressed, no purely *universal hypotheses* (e.g. $\forall xPx$) is verifiable by finitely many evidences, although

[2]A sentence set $S \subseteq S(\mathcal{L})$ is maximally consistent iff S is consistent and has no consistent proper extension in $S(\mathcal{L})$ and S is ω-complete iff whenever $\varphi[a_i] \in S$ for all individual constants a_i, then $\forall x\varphi[x] \in S$ (for $\varphi[a_i]$ an arbitrary $\mathcal{L}$-formula containing a_i).

it is falsifiable (namely by $\neg Pa$); and dually, no purely *existential hypothesis* (e.g. $\exists xPx$) is falsifiable by finitely many evidences, although it is verifiable (namely by Pa). Later it was discovered that quantificationally mixed hypotheses such as $\forall x \exists x Rxy$ are neither verifiable nor falsifiable by finitely many evidences. It follows that even under empiricist assumptions it need not be that theory development is cumulative at the level of hypotheses: false universal theories may be held for an arbitrarily long time before they get falsified; and theories with mixed quantifiers will never get verified or falsified.

25.3 Convergence in the Limit and Formal Learning Theory

The empiricist setting of Sect. 25.2 is the major framework of formal learning theory (cf. ch. 6/6, *this volume*). Evidence streams are called *data streams* in Kelly [15]. Formal learning theorists are aware that there exist theories with theoretical terms to which this setting doesn't apply. Still their setting may be regarded as a legitimate idealization in all contexts in which one may assume an unproblematic background knowledge (e.g. concerning measurement techniques) by which one can determine the truth value of all basic statements of the language in which theories are formulated. From now on we understand "$\mathcal{L}_e$" in this extended sense. Verification and falsification *with certainty* (in the sense of Carnap [4] and Popper [28]) are defined as follows:

(1) $h \in S(\mathcal{L}_e)$ is *verifiable* (or falsifiable, respectively) *with certainty* over an $\mathcal{L}_e$-world EW iff for every evidence stream (e) over EW there exists a time point n at which e_n entails h (or entails $\neg h$, resp.).

Purely existential hypotheses are verifiable and purely universal hypothesis are falsifiable with certainty; but quantificationally mixed hypotheses are neither nor. In view of this negative results formal learning theorists suggest to use the *weaker* epistemic standard of verifiability [viz. falsifiability] in the limit. Let $SQ(\mathcal{L}_e)$ be the set of all evidence sequences over $\mathcal{L}_e$-worlds; let (e_{1-n}) denote the sequence of the first n elements of (e); and let $E_n := \{e_i: 1 \le i \le n\}$ be the corresponding evidence set at time n. An assessment function for the hypotheses $h \in H$ in a given set of hypotheses H is a function $\alpha: H \times \{(e_{1-n}): (e) \in SQ(\mathcal{L}_e), n \in \omega\} \rightarrow$ {true, false} which conjectures at every time point n of any evidence stream (e) whether h is true or false. Then:

(2) A hypothesis $h \in S(\mathcal{L}_e)$ is *verifiable* (or falsifiable, resp.) *in the limit* iff there is an assessment function such that for every $\mathcal{L}_e$-world EW which verifies (or falsifies, resp.) h and evidence stream (e) over EW there exists a time point n after which α conjectures the correct truth value of h in EW forever (i.e. $\alpha(h,(e_{1-m})) =$ true for all $m \ge n$).

One of the major results of formal learning theory is the following:

(3) An ∃-∀-hypothesis (e.g. $\exists x \forall y Rxy$) is verifiable but not falsifiable in the limit. Dually a ∀-∃-hypotheses (e.g. $\forall x \exists y Rxy$) is not verifiable but falsifiable in the limit.

In particular, a ∀-hypothesis ($\forall x Fx$) is verifiable in the limit and falsifiable with certainty (and dually for ∃-hypotheses, with "verifiable" and "falsifiable" exchanged). Let us explain the basic idea underlying result (3). Define Dom(E_n) to be the *subdomain* of those individuals which appear in E_n. Then an assessment method α for $\exists x \forall y Rxy$ can be defined as follows: conjecture "true" as long as the so far observed evidence set E_n does not falsify $\exists x \forall y Rxy$ over Dom(E_n); otherwise conjecture "false". Then if h is true in EW there exists an individual a_k which appears at some time $t(a_k)$ and for which $\forall y Ra_k y$ is true, whence after time $t(a_k)$ α will conjecture "true" forever. However, if $\exists x \forall y Rxy$ is false in a given EW, then for every assessment method α one may construct a "demonic" evidence stream over EW such that α's conjectures don't stabilize but switch endlessly between "true" and "false" (for the detailed construction cf. [15], 51ff).

Although the method α for $\exists x \forall y Rxy$ is guaranteed to stabilize to the conjecture "true" after *some finite* time, one can never know *when* this time is reached, and hence, one can never know whether one has achieved the truth or not. Even if this intrinsic weakness of "verification in the limit" is accepted, the other bad news is that already hypotheses with three alternating quantifiers (∃-∀-∃ or ∀-∃-∀) are neither verifiable nor falsifiable in the limit. Kelly shows that these hypotheses are at least *gradually* verifiable (or falsifiable) in the limit, which is a still weaker property (whose definition is omitted here; cf. [ibid., 66ff]). However, for hypotheses with four alternating quantifiers even gradual verification (or falsification) fails.

Most contemporary philosophers of science would argue that even evidence statements may fail: perceptions may be erroneous and measurement devices may be malfunctioning. If we drop the infallibility assumption for evidences, then evidence statements in (e) may mutually contradict each other. It follows that evidence streams are no longer cumulative: $E_n \subseteq E_{n+1}$ does not always hold and we cannot define $E_n = \{e_i: 1 \leq i \leq n\}$. We rather have to construct the *evidence history* (E) as a sequence of evidence sets (E):= (E_0, E_1, E_2, ...) that is defined by revision (or expansion) operations: $E_{n+1} = E_n * e_{n+1}$. The axiom of success, $e_n \in E_n$, is no longer mandatory for revisions over contradicting evidence statements. To retain the positive results concerning verifiability and falsifiability it is necessary to set up the following *constraint of stable error-correction:* every false evidence in (e) is corrected once-and-forever after some finite time. This implies that for every true evidence e in EW (the given empirical world) there exists a time n such that for all $m \geq n$, $e \in E_m$.[3]

[3]The constraint of *error-correction in the limit* would not be sufficient.

25.4 Inductive Confirmation and Convergence with Probability

The major conclusion of the previous section is this: even under empiricist idealizations and in regard to the *weak* "in-the-limit" notions of verification and falsification, the range of those hypotheses which are verifiable or falsifiable is very *restricted*. As soon as one assumes *inductive confirmation* as a legitimate justification principle, things get better. A simple qualitative definition may run as follows:

(4) A hypothesis h is inductively *confirmed* by a finite evidence set E_n iff h is not falsified by E_n over the subdomain Dom(E_n), and this the confirmation is the stronger, the greater that part of Dom(E_n) which verifies h over Dom(E_n).[4]

However, since David Hume it is known that induction is not a generally reliable inference: it may fail in worlds (or event sequences) which are *non-uniform*, i.e. in which the future differs radically from the past. What one can only show is that (under mild conditions) induction is an *optimal* strategy (cf. [35]).

Often confirmation principles are formalized in the framework of *Bayesian* (subjective) *probabilities*, i.e. rational degrees of beliefs (cf. ch. 6/8 and 6/9, *this volume*). If $P{:}S(\mathcal{L}_e) \to [0,1]$ is a Bayesian probability function over the total set of statements, then the (posterior) probability of a hypothesis h_k given evidence e is given by the famous Bayes-formula as

(5) $P(h_k|\, e) = P(e|\, h_k) \cdot P(h)/\Sigma_{1\leq i\leq n} P(e|\, h_i) \cdot \mathrm{P}(h_i)$

where $\{h_1, \ldots, h_n\}$ is a (pragmatically given) partition of possible hypotheses that contains P_k, $P(h_i)$ is the prior probability of h_i and $P(e|h_i)$ the probability (or "likelihood") of e on the assumption that h_i. Formula (5) shows that the posterior probability of a hypothesis depends not only on its relation to the evidence (the likelihood), but also on its prior probability. Since most contemporary Bayesians agree that prior probabilities are *subjective*,[5] this dependency seems to constitute a counterargument to Bayesian confirmation theory. Bayesians counter that the dependency of the posterior probabilities of hypotheses on the priors becomes smaller and smaller the more evidences come in. Bayesians cite here *convergence* theorems ([8], 58), for example the following general convergence theorem:

(6) Gaifman and Snir [12]: Under the assumption that P is (not only finitely but) *countably additive* it holds for every possible $\mathcal{L}_e$-hypothesis h and evidence stream over a world EW with *probability* 1 that h's posterior $P(h|E_n)$ converges to the truth value of h in EW for $n \to \infty$.

Probabilistic convergence theorems of this sort are restricted in three ways: (1.) they hold only under the *empiricist* assumption that $\mathcal{L} = \mathcal{L}_e$ (Gaifman and Snir

[4]*Example:* If $E_2 = \{Raa, Rab, Rbc\}$, then $h = \exists x \forall y Rxy$ is neither falsified nor verified by E_2 over Dom(E_2) = $\{a,b,c\}$, though it is verified over $\{a,b\}$.

[5]The older view of Carnap [5] on "logically given" prior probabilities is hardly tenable; cf. Howson/Urbach ([14], 60); Earman ([8], 15).

express this by their requirement that $S(\mathcal{L}_e)$ "separates" the set of all possible worlds over $\mathcal{L}_e$); (2.) they hold only with probability 1 – whence convergence may fail in an uncountably infinite subset of the uncountably many $\mathcal{L}_e$-worlds, and (3.) they hold only under the condition of countable additivity, which involves inductive assumptions (this is demonstrated in ([15], 321ff) and [39]). Stronger convergence properties require stronger inductive assumptions (cf. [8], 108, and Schurz [41], ch. 4.7).

25.5 The Rational Choice and the Construction Paradigm of Theory Discovery: Learning Sentences and Abductive Theory Revision

In the previous sections we have studied the development of *scientific assessments* of hypotheses in the face of an evidence stream, but not the *discovery* of hypotheses. For Popperians theory discovery is a matter of psychology, not of logic – there are no rules for theory discovery. Formal learning theory, however, provides also rules for theory discovery. These rules assume that there exists a countably enumerable set H of possible (not necessarily disjoint) hypotheses in $\mathcal{L}_e$ which contains at least one true hypothesis, and an infinitely repetitive ordering $(h) = (h_0, h_1, \ldots,)$ of the hypotheses in H, i.e. every hypothesis occurs in (h) infinitely many often ([15], 224f). With such an enumeration at hand, an assessment method $\alpha(h_i,(e_{1-n}))$ can be transformed into a discovery method γ that assigns to each initial subsequence (e_{1-n}) $(n \in \omega)$ a hypothesis in H, recursively defined as follows:

(7) $\gamma((e_0)) = h_0$, and for each time n, if $\alpha(\gamma((e_{1-n}),(e_{1-n})) =$ "true", then $\gamma((e_{1-(n+1)})) = \gamma((e_{1-n}))$, and otherwise $\gamma((e_{1-(n+1)}))$: = the next hypothesis h^* in (h) behind $\gamma((e_{1-n}))$ for which $\alpha(h^*, (e_{1-(n+1)})) =$ "true".

The so-defined discovery method γ stabilizes to conjecturing the first true hypothesis in H for which α stabilizes to the assessment "true".

This discovery rule of formal learning theory is an example of the *rational choice paradigm* of theory discovery. Here one assumes that a list H which contains all possible and interesting hypotheses is given *in advance*, i.e. *before* any evidences have been received. Often this list H is assumed to form a *partition*, i.e. the hypotheses in H are pairwise incompatible and exhaustive (at least in relation to an assumed background knowledge K). Theory development consists in choosing the optimal (e.g. best confirmed) hypotheses in the face of the received evidence set E_n.

Although in some historical phases theory development proceeds according to the rational choice paradigm, this paradigm has its limitation in the fact that scientists rarely possess a list of all interesting hypotheses in advance from which they choose. Usually new interesting hypotheses are *constructed* in science from given evidence by inductive or abductive learning mechanisms and are then put to subsequent empirical tests ([34], §1; [37], §1.3). We call this view the *construction* paradigm of theory development. To fill the construction paradigm with content we need heuristics and/or algorithms which tell us how to construct and revise plausible theories in the face of an ongoing evidence stream.

Neither the AGM- nor the base-version of ordinary belief revision contains such learning mechanisms, although on different reasons. AGM-revision is extremely liberal: it allows $T{*}e$ to be any logically closed sentence set which lies between $\mathrm{Cn}(\{e\})$ and some maximally consistent extension of $\{e\}$ and which is "preferred"; but the AGM-axioms for preferences don't decide which one of these sets is preferred. Belief base revision, on the other hand, is purely *corrective* in the sense that here one cannot generate *new* quantified hypotheses from evidences (for details cf. [37], §1.2). In scientific theory development, however, the revision process is typically *creative* in the sense that it constructs new hypotheses.

In line with the two paradigms there have been suggested two ways of extending the theory of belief (base) revision in order to cover theory discovery. One way is based on the rational choice paradigm; it has been introduced by Levi (1980, 35f; 1991, 71ff, 146) and is further developed by Rott [33] and Olsson and Westlund (2006). Besides input-driven (or "routine") revisions, Levi introduces a second kind of revisions, so-called *deliberate* revisions, which result from an act of will and consist in choosing a hypothesis h from a given partition of possible hypotheses and adding it to the accepted beliefs.

The second way is based on the construction paradigm and consists in enriching the accepted beliefs by creative learning mechanisms. This idea has been implemented in two ways, by learning sentences and by abductive revision operations. The method of *learning sentences* has been introduced by Martin and Osherson (1998) (the name "learning sentences" comes from me). Assume a cumulative sequence of evidence sets $(E) = (E_0, E_1, \ldots)$. Let $(T){:} = (T_0, T_1, \ldots)$ be an *associated* sequence of theories (you may also say "belief system" instead of "theory") which by definition arises from (E) and an initial theory T_0 by *revision* operations: $T_n{:} = T_{n-1}{*}E_n$. Note that we revise with E_n (instead with only the new evidences in E_n) because (a) iterated revision is not always order-independent and (b) this definition works also if (E) is not cumulative. Assume the underlying empirical world EW makes the universal hypotheses $\forall xFx$ true. The problem is that ordinary belief base revision can never generate $\forall xFx$ in the face of the evidence stream (E) as long as T_0 is empty or contains only singular statements (proof in [37], §1.2). Martin and Osherson overcome this problem by adding a learning sentence of the form $Fa_i \rightarrow \forall xFx$ to the initial theory T_0 (for an arbitrary constant a_i). If $\forall xFx$ is true, $\forall xFx$ will enter the theory sequence at the first time n at which Fa_i occurs in the evidence stream, and will remain there forever, while if $\forall xFx$ is false, then $\neg\forall xFx$ will enter the theory sequence at the first time n at which a sentence of the form $\neg Fa_j$ has shown up in the evidence stream.

In the case of $\exists$-$\forall$-hypotheses learning sentence are more complicated than simple implications from evidences to hypotheses. More generally Martin and Osherson prove the following *theorem*:

(8) [Martin and Osherson 1998, (63), p. 153] For each problem of the form "which of the hypotheses in partition $\{h_1,\ldots,h_n\}$ is true in *EW*" that is solvable by a discovery algorithm γ (in the explained sense) there exists a set of learning sentences *L* such stringent belief base revision[6] applied to a cumulative evidence sequence (*E*) over *EW* and an initial belief set T_0 containing $\{\vee_{1\leq i\leq n}h_i\} \cup L$ will after some finite time *n* produce belief sets T_n, $T_{n+1},\ldots$ which contain the *true* element h^* of $\{h_1,\ldots,h_n\}$ forever (i.e., $h^* \in T_m$ for all $m\geq n$).

Martin and Osherson develop a fascinating combination of formal learning theory and belief revision. However, their account has two problems. First, it is restricted to the empiricist assumption $\mathcal{L}_e = \mathcal{L}$. Second, learning sentences are somehow unnatural: we do not literally believe "if this (and this ...) raven is black, then all ravens are black".

An alternative way of implementing learning mechanism is by closing the revision operation under non-deductive inferences. If these non-deductive inferences include abductions to conclusions with theoretical concepts, hypothesis creation transcends the empiricist assumption. Abductive belief revision has been elaborated, among others, by Pagnucco [27], Aliseda [2], Schurz [37] (see also Langley et al. [20]). Abductive belief expansion can be defined as follows (cf. [37], §3.2):

(9) Let $T(\mathcal{L})$ be the set of all possible theories in the given total language $\mathcal{L}$ (which may now contain theoretical terms; i.e. $S(\mathcal{L}_e) \subset S(\mathcal{L})$). The *abductive expansion* of a theory $T \in T(\mathcal{L})$ by a *T*-compatible evidence *e* is denoted by $T +_{\mathrm{a}} e$ and defined as follows: $(T +_{\mathrm{a}} e) = (T + e) + \mathrm{abd}(T,e)$, where "+" is the ordinary expansion function.

In this context, $abd(T,e)$: $T(\mathcal{L}) \times E(\mathcal{L}_e) \rightarrow S(\mathcal{L})$ is an abductive expansion function that assigns to each theory *T* and evidence *e* a consistent sentence $s{:} = abd(T,E)$ which is either a tautology or *explains E* within *K*.

This definition allows for the case in which no explanatory hypothesis for *E* is found; in that case $abd(T,e)$ is identified with a tautology (and the abductive expansion is called "improper").

Abductive expansion can be broken up into an ordinary expansion and an abductive inference step (this is in line with what ([33], 84) calls the "direct mode of foundationalist base revisions"). The same is not always possible for abductive revision. If an element *h* of belief set *T* explains a conjunction of evidences *E*, and *e* is a new piece of evidence contradicting *h*, then it is inefficient to remove *h* and generate an alternative hypothesis h^* from scratch. Scientists rather try to obtain the revised hypothesis h^* by a *direct* revision of the old hypothesis *h* given $E\cup\{e\}$ into some new hypothesis h^* which explains *e* and at the same time preserves the old explanations of the evidences in *E*. For example, assume *h* is a quantitative hypothesis saying that gas pressure is proportional to the gas temperature, and new

[6] A base contraction function $\div$ is stringent iff for each *T* and *e*, $T\div e$ is a preferred maximal *T*-subset not implying *e*.

data tell us that for low temperatures, the gas pressure is lower than predicted by *h*. Then scientists will not simply remove *h* from their belief set, but revise *h* by adding a new non-linear term to the linear relationship predicted by *h*. This leads to the following general definition of abductive belief revision "$*_a$":

(10) $T*_a\, e := T*e + \mathrm{rev}(h, e, T \div \neg e)$ (where "*" is the ordinary revision operator).

Here $h := h_{T,e}$ is the "explanatory loss" caused by the contraction by *e*, defined as the conjunction of all explanatory hypotheses which are in *T* but not in $T \div \neg e$. Moreover, rev: $T(\mathcal{L}) \times E(\mathcal{L}_e) \times \mathrm{Pow}(S(\mathcal{L})) \rightarrow S(\mathcal{L})$ is an abductive revision function which assigns to the "lost" hypothesis *h*, *h*-incompatible evidence *e* and theory-contraction $T \div \neg e$ a revised hypothesis $h^* := \mathrm{rev}(h, e, T \div \neg e)$ that is consistent with $T*e$ and is either a tautology or explains $E_{T,h} \cup \{e\}$, where $E_{T,h}$ is the set of all evidences which were explained by *h* in *T*.

In the process of hypothesis-revision, the revised hypothesis h^* is not only a function of the contracted theory $T \div \neg e$ and the new evidence *e* – which it would have to be according to Levi identity – but also a function of the old hypothesis *h* which has been removed from $T \div \neg e$. Schurz ([37], §3.2) concludes from this fact that Levi-identity fails for abductive belief revision.

25.6 Theoretical Concepts, Lakatosian Research Programs and Refined Falsificationism

Most scientific theories contain *theoretical concepts* such "electron" or "magnetic force" which do not occur in the evidence stream but go beyond the observable. For theoretical hypotheses – i.e. hypotheses containing theoretical terms – the empiricist assumption fails: their truth value is not determined by the evidence stream, even not in the limit; moreover, they are not obtainable from evidences by inductive generalizations from finite evidence sets. How are theoretical hypotheses confirmed at all?

Popper [28] has pointed out that usually scientific theories entail observational consequences and are, though not being verifiable, at least falsifiable via the rule of *Modus tollens* (if $T \Vdash E$ then $\neg E \Vdash \neg T$). Popper's falsificationist account of theory development rests on the idea that theories which are falsified by the actual evidences are laid aside. Popper's account of "instantaneous rejection" was criticized by Kuhn [16] and Lakatos [19]. Kuhn showed that in the history of science theories which contradict data are not rejected or laid aside; scientists rather introduce additional *auxiliary assumptions* which save the theory core or theory "paradigm" from being falsified. It is well known that Kuhn's criticism involved some more radical points, for example concerning the theory-dependence of evidence and the resulting irrationality of paradigm changes (so-called "scientific revolutions"). Many (and probably most) contemporary philosophers of science did not follow these radical aspects of Kuhn.

The less radical part of Kuhn's criticism was elaborated by Lakatos' account of *refined* falsificationism which significantly improved Popper's "naive account". Scientific hypotheses are never assessed in isolation. They form theories, which are *systems* of statements together with an *epistemic ranking* (preference ordering) over them. This ordering decides which elements are to be given up when conflicts between theory and data arise. Lakatos speaks of theories as consisting of a *theory core* that is surrounded by a *periphery* which contains less and less important parts, the more one moves from inside into outside layers of the theory. The outermost layer of a theory contains auxiliary assumptions, which assert the existence or non-existence of *disturbing factors*. They figure like a *protective belt* because by introducing new disturbing factors the theory core can always be protected from falsification. For example, in 1846, when J. Adams and U. Le Verrier discovered a considerable discrepancy between the predicted and the actually observed orbit of the planet Uranus, they postulated the existence of a yet undiscovered planet, Neptune, whose gravitational effect on Uranus was assumed to be responsible for its divergence from the predicted orbital path. Later on Neptune's existence was indeed independently confirmed by telescopic observations. But a similar scenario happened around 1856 when Le Verrier observed a divergence of the planet Mercury from its predicted orbit and postulated the existence of a yet smaller planet named *Vulcan*, which despite tenacious attempts could never be found by telescopes.

Similar accounts of theory development had already been given by Duhem [7] and Quine [32]. Duhem's thesis of the *holism of theory falsification* says that if a particular *version* T of a theory – consisting of T's core plus a particular periphery – contradicts a datum e, then all what one knows (by Modus Tollens) is that some part of T is false, but logic alone doesn't dictate which part of T should be given up. Lakatos [19], however, provided an answer to this question: the theory-parts which are given up should be as peripherical and unimportant as possible (this is a version of "prioritized base contraction" in the sense of Rott ([33], 40ff). Although it is logically speaking always possible to solve conflicts with data ("anomalies") by peripherical theory-revisions, Lakatos sets up an important rationality constraint to steps of this sort: a theory-revision should be *theoretically progressive*, by which Lakatos means that the new theory contains all the confirmed empirical content of the old theory plus some additional new empirical "excess" content. Moreover, Lakatos calls the new theory version *empirically progressive* if part of this excess content has been independently confirmed. If, on the other hand, theory revisions *reduce* the empirical content of a theory they are called *degenerative*.

In Lakatos' account of research programmes a theory is a *historical* entity: if it changes its periphery, it is merely another *version* of the same theory; only if it changes its core, it is a *new* theory. Formally speaking, a *theory history* in the sense of Lakatos is a sequence $(T) = (T_0, T_1, \ldots)$ which is associated with a sequence of evidence sets $(E) = (E_0, E_1, \ldots)$ in the sense defined in the pervious section. Each theory $T_i = \mathrm{Cn}(A_i)$ is now itself a ranked system of statements ($T_i(0)$, $T_i(1)$, $T_i(2), \ldots$), where $T_i(0) \subset A_i$ is the *core* of theory T_i and $T_i(n)$ ($n \geq 1$) are less and less important subsets of T-axioms. Building on the previous section we assume that each T_n is given as the abductive revision of the background theory T_0 by E_n (T_n: $= T_0 * E_\mathrm{n}$).

In line with Popper and Lakatos we do not assume that evidence statements are infallible, i.e. the evidence sequence (*E*) is not necessarily cumulative. Of course we make the Lakatosian assumption that the evidence sequence – though not being absolutely theory-neutral – is at least *theory-neutral* in regard to all those theories whose success is being compared. Only under that assumption can one have the *same* evidence sequence for all theories in (*T*), even for those with a different core. This is the decisive difference to Kuhn's relativistic account in which different theories with different cores would have *their own* evidence sequences, and rational comparisons of them were hardly possible. On the other hand, the decisive difference of Lakatos' account to neo-empiricist accounts is that the theories in (*T*) contain theoretical concepts, whence their truth is not determined by an empirical world *EW* over $\mathcal{L}_e$.

Theory-subsequences of (*T*) whose theories share the same theory core are called (in line with Kuhn) *normal* periods of science; while theory-successions (or pairs) in which the theory core changes are called scientific *revolutions*. We define the following Lakatosian criteria for the rational evaluation of theory-development (where "$\subset$" stands for proper and "$\subseteq$" for proper-or-improper set-inclusion):

(11) We define $E(T) = T \cap S(\mathcal{L}_e)$ as the set of all (confirmed or unconfirmed) empirical consequences of *T*, and $EC_n(T) := E(T) \cap E_n$ as the set of *T*'s confirmed empirical consequences at time *n*.
(12) A theory succession (T_n, T_{n+1}) is called

- *theoretically progressive* iff $E(T_n) \subset E(T_{n+1})$ and $EC(T_n) \subseteq EC(T_{n+1})$;
- *empirically progressive* iff $E(T_n) \subset E(T_{n+1})$ and $EC(T_n) \subset EC(T_{n+1})$;
- *stagnating* iff $E(T_n) = E(T_{n+1})$ and $EC(T_n) = EC(T_{n+1})$, and finally
- *degenerative* iff either $E(T_n) \supset E(T_{n+1})$ or $EC(T_n) \supset EC(T_{n+1})$.

In conclusion, Lakatos' model of theory development allows for the rational evaluation of theory development and the rational assessment of theory progress even if one allows for fallible evidences and for theories whose truth value is not determined by the complete empirical truth.

25.7 Verisimilitude and Truth-Approximation by Theory-Revision

Popper (1963, 233f) has argued that the main goal of science consist in progress in *verisimilitude* or *truthlikeness*. For although most scientific theories involve idealizations and hence are strictly speaking false, some of them are much *closer* to the truth than others. According to Popper's intuitive idea, a theory T_1 is closer to the truth than another theory T_2 iff T_1 has more true and less false consequences than T_2. It is well known that Popper's original definition of verisimilitude turned out to be inadequate (cf. ch. 7/2, *this volume*). In the following period two major families of accounts to verisimilitude have been developed which cured the mistake in Popper's original definition; they have been called the *disjunctive* and the *conjunctive* approach to truthlikeness (cf. [6, 18, 40]). Without being able to explain the precise definitions of verisimilitude in these accounts, we mentioned some major results concerning the connection between verisimilitude and belief revision:

(12) Niiniluoto [25]: Neither the expansion nor the revision of a false theory T by a true evidence leads always to an increase of T's verisimilitude.

Schurz [38] demonstrates that this holds even if the evidence is a single true basic statement (instead of a disjunction of a true and a false basic statement, as assumed in Niiniluoto's example). The following example makes this clear: assume the hypothesis $h{:} = b \rightarrow f_1 \wedge \ldots \wedge f_n$ is an implication leading from a true but unknown basic statement b to a conjunction of false basic statements, and the new input by which the theory $T{:} = \{h\}$ is expanded is b. Then $T + b = \mathrm{Cn}(\{b, f_1, \ldots, f_n\})$. Since the verisimilitude-loss due to the addition of n false basic statements f_i $(1 \leq i \leq n)$ may be much greater than the verisimilitude-gain due to the addition of the true b, the verisimilitude of $T + b$ may be much smaller than that of T. However, the volume Kuipers and Schurz [38] contains a lot of results which show that under restricted conditions positive connections between increase in verisimilitude and revision by true evidences can be obtained. In conclusion, truth-approximation can still be upheld as a major goal of theory development, although the paths towards this goal may have intermittent phases in which theories move away from the truth.

25.8 From Progress in Empirical Success to Progress in Theoretical Truth: Instrumentalism Versus Realism

Scientific realism is the view that the empirical success of a theory is a reliable indicator of the (approximate) truth of the theory, including the truth of its theoretical superstructure. In contrast, *scientific instrumentalism* holds that the theoretical superstructure of a theory has merely the instrumental purpose of entailing the evidences in a most simple and unifying way, but there is no reason to assume that this theoretical superstructure corresponds to an unobservable external reality. While for scientific realists the decisive progress in theory development consists in progress in truth approximation at the theoretical level, for instrumenalists or empiricists such as [42] scientific progress is confined to progress in *empirical success* (or empirical adequacy).

The standard justification of scientific realism is the *no-miracles* argument, which goes back to Putnam (1975, 73) and has been used in various ways as a defense of scientific realism (cf. Boyd 1984; [30]). This argument says that the empirical success of contemporary scientific theories would be a sheer miracle if we would not assume that their theoretical superstructure, or ontology, is approximately true in the sense of scientific realism. However, the no-miracles argument is beset by two strong counterarguments, an empirical and a theoretical counterargument.

The empirical counterargument is the *pessimistic meta-induction* argument of Laudan (1981). This argument points to the fact that in the history of scientific theories one can recognize radical changes at the level of theoretical superstructures, although there was continuous progress at the level of empirical success. On simple inductive grounds one should expect therefore that the theoretical superstructures of our presently accepted theories will also be overthrown in the future, and hence can in no way be expected to be approximately true.

The theoretical counterargument to the no-miracles argument is the *no-speculation* argument (cf. [34], §7.1). It points out that for every set of possible observations E one may construct *ex-post* and *ad-hoc* some speculative theory T which just entails ("explains") E, but has no independent empirical consequences. The empirical success of such speculative ad-hoc theories is in no way a reliable indicator of their approximate truth.

In Schurz [36] a justification of the inference from empirical success to partial theoretical truth has been suggested which does not presuppose the questionable NMA. It is based on relations of *correspondence* between historically consecutive theories T and T^* with nondecreasing empirical success, which hold if the following (simplified) conditions are satisfied:

(C1): The predecessor theory T speaks about a partition of circumstances $\{A_1, \ldots, A_n\}$ and contains a theoretical expression φ for which it entails a set of bilateral reduction sentences $\{B_i{:}1\leq i\leq n\}$ of the form

(B_i): $\forall x\forall t(A_i xt \rightarrow (\varphi(x) \leftrightarrow R_i xt))$ – in words: for all systems x and times t, under empirical circumstances A_i the presence of φ in system x is indicated by the empirical phenomenon R_i.

(C2): Every empirical prediction of the form $\exists t(A_i xt \wedge R_i xt) \rightarrow \forall t(A_j xt \rightarrow R_j xt)$ which follows from $\{B_i{:}1\leq i\leq n\}$ is entailed by the successor theory T^* in a *T^*-dependent* way, which means by definition that there exists a theoretical *mediator* description $\varphi^*_{i,j}x$ in T^* such that $\forall x\exists t(A_i xt \wedge R_i xt \rightarrow \varphi^*_{i,j}x)$ as well as $\forall x(\varphi^*_{i,j}x \rightarrow \forall t(A_j xt \rightarrow R_j xt))$ follows from T^*.

In addition it is required that the two theories T and T^* are *causally normal* in the sense that the circumstances A_i are described in terms of theory-independent empirical parameters. While condition (C2) is mild, condition (C1) on the predecessor theory T is a crucial constraint which requires that the theoretical concept φ of T figures as a common cause of several empirical regularities. The collection of the bilateral reduction sentences $\{B_i{:}1\leq i\leq n\}$ entailed by T is denoted as B(T,φ). (C1) excludes pre-scientific ad-hoc speculations from the application range of the correspondence theorem. Given these conditions it is possible to infer that also a part of the theoretical structure of T is preserved in T^*:

(13) *Correspondence theorem* [36]: Let T be a causally normal predecessor theory satisfying condition (C1) and T^* a causally normal successor theory satisfying condition (C2). *Then* T^* contains a theoretical expression $\varphi^*(x)$ such that $B(T,\varphi)\cup T^*$ is consistent and implies a *correspondence relation* of the form $\forall x\forall t(A_1 xt\vee \ldots \vee A_n xt \rightarrow (\varphi(x) \leftrightarrow \varphi^*(x))$. In words: whenever a system x

is exposed to one of the circumstances A_i, then x satisfies the T-theoretical description φ iff x satisfies the T^*-theoretical description φ^*.

Based on the correspondence theorem (13) one can argue that $\varphi(x)$ *refers indirectly* to the theoretical state of affairs described by $\varphi^*(x)$ – provided one assumes that $\varphi^*(x)$ refers and T^* is at least partially true. An example which is extensively discussed in Schurz [36] is the transition from the *phlogiston* theory to the modern *generalized oxidation* theory of combustion and saltification. For this theory-transition the correspondence theorem generates the following correspondence: *phlogistication* of a substance x corresponds to the acceptance of electrons by positively charged x-ions from their bonding partner, and *dephlogistication* of x corresponds to the *donation of electrons* of x's atoms to their electronegative bonding partner.

The correspondence theorem allows that T and T^* are mutually incompatible. However, that part of T which is needed to derive the correspondence, namely $B(T,\varphi)$ is always compatible with T^*. If one assumes that the observed phenomena are caused by an external reality whose structure can possibly be represented by an ideal but unknown "super-theory" T^+ which is causally normal and satisfies condition (C2), then (13) implies that also our presently accepted theories, as long as they are causally normal and satisfy condition (C1), must have got something right at their theoretical level. In conclusion, the correspondence theorem justifies a weak kind of scientific realism which is not based on the questionable no-miracles argument.

References[7]

1. Alchourrón, C. E., Gärdenfors, P., & Makinson, D. (1985). On the logic of theory change. *Journal of Symbolic Logic, 50*, 510–530.
2. Aliseda, A. (2006). *Abductive reasoning*. Dordrecht: Springer.
4. Boyd, R. (1984). The current status of scientific realism. In J. Leplin (Ed.), *Scientific realism* (pp. 41–82). Berkeley: University of California Press.
4. Carnap, R. (1928/2003). *The logical structure of the world and pseudoproblems in philosophy*. Chicago: Open Court.
5. Carnap, R. (1950). *Logical foundations of probability*. Chicago: University of Chicago.
6. Cevolani, G., & Festa, R. (2009). Scientific change, belief dynamics and truth approximation. *La Nuova Critica, 51*(52), 27–59.
7. Duhem, P. (1908/1991). *The aim and structure of physical theory*. Princeton: Princeton University Press.
8. Earman, J. (1992). *Bayes or bust?* Cambridge, MA: MIT Press.
9. French, S. (2008). The structure of theories. In S. Psillos & M. Curd (Eds.), *The Routledge Companion to Philosophy of Science* (pp. 269–280). London: Routledge.
10. Gärdenfors, P. (1981). An epistemic approach to conditionals. *American Philosophical Quarterly, 18*, 203–211.

[7]Asterisks (*) indicate recommended readings.

11. Gärdenfors, P. (1988). *Knowledge in flux*. Cambridge, MA: MIT Press.
12. Gaifman, H., und Snir, M. (1982). Probabilities over rich languages. *Journal of Symbolic Logic, 47*, 495–548.
13. * Hansson, S. O. (1999). *A textbook of belief dynamics*. Dordrecht: Kluwer. [Introduction in contemporary accounts of belief revision]
14. * Howson, C., und Urbach, P. (1996). *Scientific reasoning: The bayesian approach* (2nd ed.). Chicago: Open Court. [Introduction to probability theory with a focus on Bayesianism]
15. * Kelly, K. T. (1996). *The logic of reliable inquiry*. New York: Oxford University Press. [Textbook in formal learning theory]
16. Kuhn, T. S. (1962). *The structure of scientific revolutions*. Chicago: University of Chicago Press (3rd edition 1996).
17. Kuipers, T. A. F. (2000). *From instrumentalism to constructive realism*. Dordrecht: Kluwer.
18. Kuipers, T. A. F. & Schurz, G. (2011). Belief revision aiming at truth approximation. *Erkenntnis, 75*, 2011. (Guest-edited volume).
19. Lakatos, I. (1970). Falsification and the methodology of scientific research programmes. In I. Lakatos & A. Musgrave (Eds.), *Criticism and the growth of knowledge* (pp. 91–196). London: Cambridge University Press.
20. Langley, P., et al. (1987). *Scientific discovery. computational explorations of the creative process*. Cambridge, MA: MIT Press.
21. Laudan, L. (1981). *A confutation of convergent realism* (pp 107–138). Reprinted in D. Papineau (Ed.), (1996). *The philosophy of science*. Oxford: Oxford University Press.
22. Levi, I. (1980). *The enterprise of knowledge*. Cambridge, MA: MIT Press.
23. Levi, I. (1991). *The fixation of belief and its undoing*. Cambridge, MA: Cambridge University Press.
24. Martin, E., & Osherson, D. (1998). *Elements of scientific inquiry*. Cambrigdge, MA: MIT Press.
25. Niiniluoto, I. (1999). Belief revision and truthlikeness. In B. Hansson et al. (Eds.), *Internet Festschrift for Peter Gärdenfors*. http://www.lucs.lu.se/spinning
26. Olsson, E. J., & Westlund, D. (2006). On the role of research agenda in epistemic change. *Erkenntnis*, 65, 165–183.
27. Pagnucco, M. (1996). *The role of abductive reasoning within the process of belief revision* (Dissertation). Sydney: University of Syndey.
28. Popper, S. K. (1935/2002). *Logic of discovery*. London: Routledge, 2002.
29. Popper, K. (1963). *Conjectures and refutations*. London: Routledge.
30. * Psillos, S. (1999. *Scientific realism. How science tracks truth.* London/New York: Routledge. [Comprehensive overview of the contemporary debate in scientific realism]
31. Putnam, H. (1975). What is mathematical truth? In H. Putnam (Ed.), *Mathematics, matter and method* (pp. 60–78). Cambridge: Cambridge University Press.
32. Quine, W. v. O. (1951). Two dogmas of empiricism. *Philosophical Review, 60*, 20–43.
33. Rott, H. (2001). *Change, choice and inference: A study in belief revision and nonmonotonic reasoning*. Oxford: Clarendon Press.
34. Schurz, G. (2008a). Patterns of abduction. *Synthese, 164*(2008), 201–234.
35. Schurz, G. (2008b). The Meta-Inductivist's winning strategy in the prediction game: A new approach to Hume's problem. *Philosophy of Science, 75*, 278–305.
36. Schurz, G. (2009). When empirical success implies theoretical reference: A structural correspondence theorem. *British Journal for the Philosophy of Science, 60*(1), 101–133.
37. Schurz, G. (2011a). Abductive belief revision in science. In E. Olsson & S. Enqvist (Eds.), *Belief revision meets philosophy of science* (pp. 77–104). New York: Springer.
38. Schurz, G. (2011b). Verisimilitude and belief revision. With a focus on the relevant element account. *Erkenntnis, 75*(2011), 203–221.

39. Schurz, G., & Leitgeb, H. (2008). Finitistic and frequentistic approximations of probability measures with or without sigma-additivity. *Studia Logica, 89*(2), 258–283.
40. Schurz, G., & Weingartner, P. (2010). Zwart and Franssen's impossibility theorem holds for possible-world-accounts but not for consequence-accounts to verisimilitude. *Synthese, 172*, 415–436.
41. * Schurz, G. (2013). *Philosophy of science: a unified approach.* New York: Routledge. [Cromprehensive overview of contemporary philosophy of science, divided into introductory and advanced parts]
42. * Van Fraassen, B. (1980). *The scientific image.* Oxford: Clarendon Press. [Representative work of contemporary empiricism]

Chapter 26
Space and Time

John Byron Manchak

Abstract Here, formal tools are used to pose and answer several philosophical questions concerning space and time. The questions involve the properties of possible worlds allowed by the general theory of relativity. In particular, attention is given to various causal properties such as "determinism" and "time travel".

26.1 Introduction

It is no surprise that formal methods have proven to be quite useful in the philosophy of space and time. With them, great progress has been made on the question, heavily debated since Newton and Leibniz, of whether space and time are absolute or relational in character. And there is a related problem which has also been clarified considerably: whether or not various geometrical properties and relations are matters of convention [3, 6, 20].

These topics, interesting as they are, will not be considered here. Rather, the focus will concern the "global structure" of space and time. General relativity (our best large-scale physical theory) will be presupposed. But the investigation of global structure will allow us to step away from the complex details of this theory and instead examine space and time with an eye towards a number of fundamental features (e.g. topology, causal structure) [9].

An elegant mathematical formalism is central to the subject. So too are the associated space-time diagrams. Using these tools, questions of physical and philosophical interest can be posed and answered. A small subset of these questions are examined below. A number of others are discussed elsewhere [4].

I wish to thank David Malament and Steve Savitt for helpful comments.

J. B. Manchak (✉)
University of California, Irvine, CA, USA
e-mail: jmanchak@uci.edu

S. O. Hansson, V. F. Hendricks (eds.), *Introduction to Formal Philosophy*, Springer Undergraduate Texts in Philosophy, https://doi.org/10.1007/978-3-319-77434-3_26

26.2 Possible Worlds

General relativity determines a class of cosmological models. Each model represents a physically possible world which is compatible with the theory. We take such a model (also called a *spacetime*) to be an ordered pair (M, g). Here, M is a connected smooth, n-dimensional manifold ($n \geq 2$) and g is a smooth, Lorentzian metric on M. (Usually n is taken to be four, but possible worlds with other dimensions are also considered. For ease of presentation, a number of two-dimensional models will be examined here.)

The manifold M captures the shape (topology) of the universe and each point in M represents a possible event. From our experience, it seems that any event (a first kiss, for example) can be characterized by n numbers (one temporal and $n-1$ spatial coordinates). So naturally, the local structure of M is characterized by an n-dimensional Cartesian coordinate system. But globally, M need not have the same structure. Indeed, M can have a variety of possible shapes. A number of two-dimensional manifolds are familiar to us: the plane, the sphere, the cylinder, the torus. Note too that any manifold with any closed set of points removed also counts as a manifold. For example, the sphere with the "North pole" removed is a manifold.

A manifold does a fine job of representing the totality of possible events but more structure is needed to capture exactly how these possible events are related. The Lorentzian metric g provides this extra structure. We can think of g as a type of function which assigns a length to any vector at any point in M. But it is crucial that, at every point in M, the metric g not only assign some positive lengths but also some zero and negative lengths as well. In this way, g partitions all vectors at a point in M into three non-empty classes: the timelike (positive length), the lightlike (zero length), and the spacelike (negative length). The result is a light cone structure at each point in M (see Fig. 26.1). Physically, the light cone structure demarcates the upper bound to the velocities of massive particles (it is central to relativity theory that nothing can travel faster than light).

The light cone structure can certainly change smoothly from point to point. But it need not. In fact, a number of interesting physically possible worlds, and all of the examples considered below, have a light cone structure which remains constant (a metric with this property is said to be *flat*).

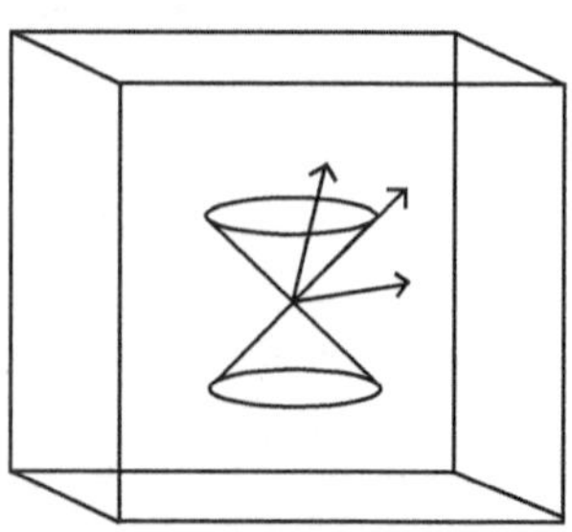

Fig. 26.1 The three-dimensional possible world (M, g). A representative light cone is depicted. Timelike vectors are inside the light cone; spacelike vectors, outside. Lightlike vectors are on the boundary

Now that we have given a characterization of physically possible worlds, we are in a position to ask a somewhat interesting question.

Question Given any shape, is there a physically possible world with that shape? (Answer: No.)

First we translate the question into the formalism: Given any n-dimensional manifold M, can a Lorentzian metric g be put on M? Next, we can get a grip on the question by noticing that a manifold admits a Lorentzian metric if and only if it admits a (non-vanishing) timelike vector field [9]. But an n-dimensional sphere does not admit a non-vanishing timelike vector field if n is even (this essentially follows from the famous "hairy ball theorem" of Brouwer). So, the answer to our question is negative. There is no physically possible world with an even number of dimensions (including our own) shaped like a sphere.

26.3 Orientability

Consider a physically possible world (M, g). The light cone which g assigns to any event in M has two lobes. And at any given event, we can certainly label one lobe as "future" and the other as "past". But can we do this for every event in M in a way that involves no discontinuities? If such a labeling is not possible, there could be no proper distinction between particles traveling "forward" and "backward" in time. If, on the other hand, such a labeling *is* possible, then we could, in a globally consistent way, give time a direction. A natural question arises here.

Question Can time be given a direction in all physically possible worlds? (Answer: No.)

We know from the previous section that any spacetime (M, g) must admit a timelike vector field on M. The question above amounts to whether any spacetime (M, g) must admit a *continuous* timelike vector field as well. A bit of thought produces a simple counterexample: a physically possible world shaped like the Möbius strip with a flat metric (see Fig. 26.2). In such a world, global notions of "past" and "future" are not meaningful.

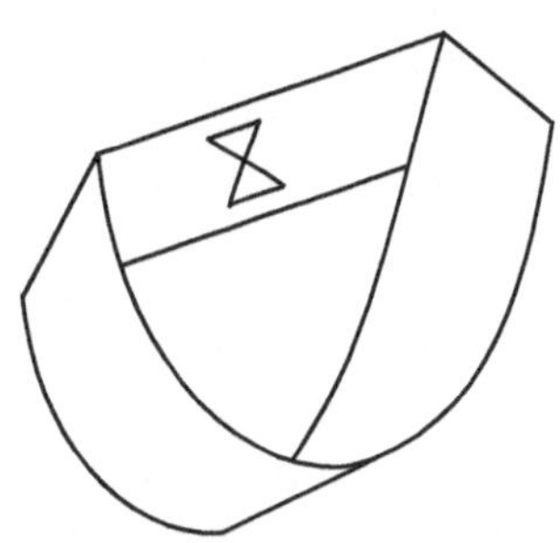

Fig. 26.2 The two-dimensional possible world (M, g) which does not admit a continuous nonvanishing timelike vector field. Here M is a Möbius strip

Let us say that a spacetime which does admit a continuous (non-vanishing) timelike vector field is *temporally orientable*. Because many other global conditions presuppose temporal orientability, it is customary to consider only spacetimes with this property. In what follows, we adhere to the custom.

26.4 Chronology

Suppose some physically possible world (M, g) is temporally orientable and that an orientation has been given. The next geometric object of study is the *future directed timelike curve* (sometimes called a *worldline*). It is simply a smooth curve on the manifold M such that all its tangent vectors are timelike and point to the future. A future directed timelike curve represents the possible life history of a massive particle; if there is a future directed timelike curve from some event p to some other event q, it must be, in principle, possible for a massive particle to travel from the one to the other.

(A *future directed lightlike curve* is defined analogously. A *future directed causal curve* is a smooth curve on the manifold such that all its tangent vectors are either timelike or lightlike and point to the future.)

We now are in a position to define a (two-place) relation $\ll$ on the events in M. We write $p \ll q$ if there exists a future directed timelike curve from p to q. (An analogous relation $<$ can be defined using future directed causal curves.) It is not difficult to prove that the relation $\ll$ is transitive: for any events p, q, and r, if $p \ll q$ and $q \ll r$, then $p \ll r$. At first, it also seems as though the relation cannot allow for distinct events p and q to be such that both $p \ll q$ and $q \ll p$ (which, by transitivity, would imply that $p \ll p$). In that case, a massive particle may travel from one event to another and then back again undergoing "time travel" of a certain kind. We ask the following question.

Question Is there a physically possible world which allows for time travel? (Answer: Yes.)

Let M be a two-dimensional cylindrical manifold and let the metric g be flat and such that timelike curves are permitted to loop around the cylinder (see Fig. 26.3). Clearly time travel is permitted since the relation $\ll$ holds between any two points in M. Due to their paradoxical time structures, physically possible worlds which allow for time travel have received a great deal of attention from philosophers [10, 21]. In what follows, we will say that a spacetime (M, g) satisfies the *chronology* condition if time travel is not permitted.

There is an interesting result concerning the shapes of physically possible worlds which satisfy the chronology condition. We say a manifold is *compact* if every sequence of its points has an accumulation point. (The sphere and torus are both compact while the plane is not.) One can show that if a spacetime (M, g) is such that M is compact, (M, g) must violate chronology [13]. However, the converse is false: Gödel spacetime is one counterexample.

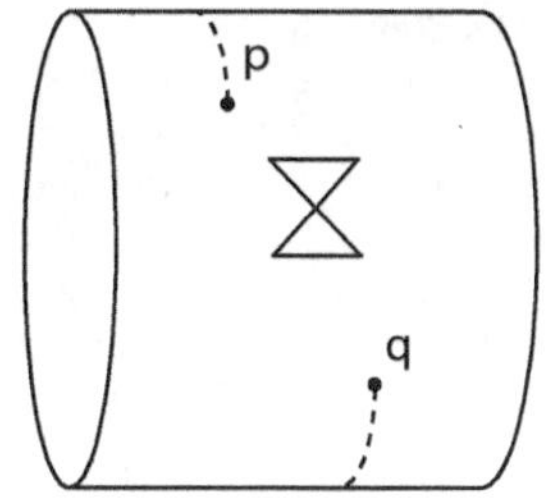

Fig. 26.3 The two-dimensional possible world (M, g). Timelike curves are permitted to loop around the cylinder M so that $p \ll q$ for all events p and q

26.5 Distinguishability

Given a physically possible world (M, g) and any event p in that world, we next can consider the collection of events in M which could have possibly influenced p. We call such a set the *past* (or *past domain of influence*) of p and formally it is defined as $I^-(p) \equiv \{q \in M : q \ll p\}$. In words, an event is a member of the past of p if there is a future directed timelike line from that event to p. Analogously, we can consider the collection of events in M which p may possibly influence. We call this set the *future* of p and define it as $I^+(p) \equiv \{q \in M : p \ll q\}$. (Analogous sets $J^-(p)$ and $J^+(p)$ can be defined using the $<$ relation.)

Are there physically possible worlds which allow distinct events to have identical pasts? Futures? There are. The example considered in the previous section (recall Fig. 26.3) is such that for any event p, $I^-(p) = I^+(p) = M$. So, clearly, we have for any distinct events p and q, $I^-(p) = I^-(q)$ and $I^+(p) = I^+(q)$. Following standard practice, let us say that any physically possible world which allows distinct events to have identical pasts is not *past distinguishing*. Analogously, let us say that any physically possible world which allows distinct events to have identical futures is not *future distinguishing* [13].

The example in the previous section was neither past nor future distinguishing but it also did not satisfy the chronology condition. Perhaps there is some connection.

Question If a physically possible world allows for time travel, must it allow different events to have influence over precisely the same set of future events? (Answer: Yes.)

To see the connection, assume that a spacetime (M, g) allows for time travel. So there must be distinct events p and q in M such that $p \ll q$ and $q \ll p$. From $p \ll q$, we know that $I^+(q) \subseteq I^+(p)$. From $q \ll p$, we know that $I^+(p) \subseteq I^+(q)$. Thus, $I^+(q) = I^+(p)$. So we conclude that every physically possible world which violates chronology also must violate future distinguishability. (An analogous result holds for past distinguishability.) Now, does the implication go in the other direction?

Question Must a physically possible world allow for time travel if it allows different events to have influence over precisely the same set of future events? (Answer: No.)

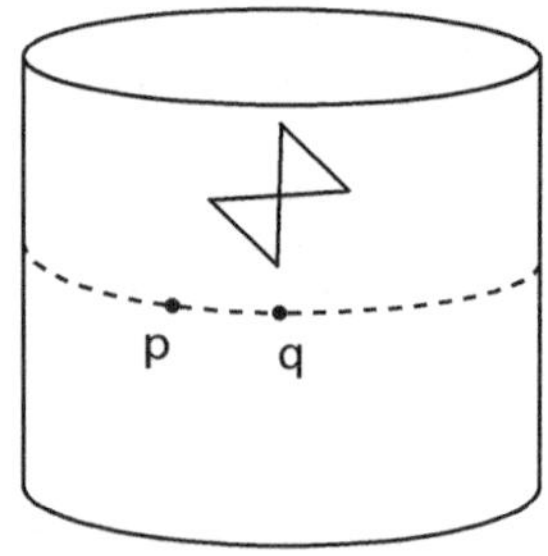

Fig. 26.4 The two-dimensional possible world (M, g). Chronology is not violated but the distinct events p and q have identical futures (the region above the dotted line) and also identical pasts (the region below the dotted line)

A counterexample is not too hard to find. Let M be a two-dimensional cylindrical manifold and let the metric g be flat and such that only lightlike curves are permitted to loop around the cylinder (see Fig. 26.4). This allows for the spacetime to satisfy chronology while allowing the points p and q to have the same futures (and also the same pasts).

There is a notable theorem concerning physically possible worlds which are both future and past distinguishing: Any two such worlds must have the same shape if they have the same causal structure. Formally, if (M, g) and (M', g') are past and future distinguishing and if there is a bijection $\varphi : M \to M'$ such that, for all p and q in M, $p \ll q$ if and only if $\varphi(p) \ll \varphi(q)$, then M and M' have the same topology [16].

26.6 Stability

Although the example in the previous section (recall Fig. 26.4) did not allow for time travel, it "almost" did. If the light cones were opened at each point, by even the slightest amount, chronology would be violated. So, there is a sense in which the example is (arbitrarily) "close" to worlds which allow for time travel. Physically possible worlds with this property are said to be not *stably causal*. Spelling out with precision the condition of stable causality requires a bit more formalism than we have available to us here. But fortunately there is an equivalent condition which is much easier to state.

We say a spacetime (M, g) admits a *global time function* if there is a smooth function $t : M \to \mathbb{R}$ such that, for any distinct events p and q, if $p \in J^-(q)$, then $t(p) < t(q)$. It is a fundamental result that this condition which guarantees the existence of "cosmic time" is both necessary and sufficient for stable causality [11]. We are now in a position to investigate how stable causality is connected to past (or future) distinguishability.

Question Is there a physically possible world which allows different events to have influence over precisely the same set of future events and yet is not close to any worlds which allow for time travel? (Answer: No.)

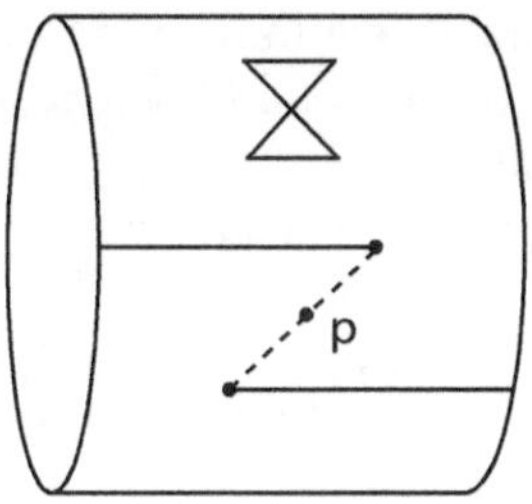

Fig. 26.5 The two-dimensional possible world (M, g). Because of the removed strips, the future and past distinguishability conditions hold. But stable causality is violated; if the light cones were to be opened by even the slightest amount at each point, it would be the case that $p \ll p$

Although it is not immediate, a violation of future (or past) distinguishability does indeed imply a violation of stable causality [13]. (The much weaker result, that a violation of chronology implies a violation of stable causality, should be clear.) Does the implication go in the other direction?

Question Is there a physically possible world in which different events always have influence over different sets of future events and yet is close to a world which allows for time travel? (Answer: Yes.)

To construct a spacetime which satisfies future (and past) distinguishability but violates stable causality, begin with the two-dimensional cylindrical manifold and let the metric be flat and such that timelike curves are permitted to loop around the cylinder (recall Fig. 26.3). Next, remove two strips which just prevent causal curves from connecting (see Fig. 26.5). The result is a spacetime, call it (M, g). One can verify that for any distinct points p and q in M, $I^-(p) \neq I^-(q)$ and $I^+(p) \neq I^+(q)$. But although there *is* a function $t : M \to \mathbb{R}$ such that t increases along every future directed causal curve, no such function exists which is also smooth. So, the spacetime fails to be stably causal.

26.7 Determinism

What does it mean to say that a physically possibly world is deterministic? Roughly the idea is that, in such a world, all events must depend upon the events at any one time. Let us make this precise.

Consider a spacetime (M, g) and let S be any subset of M. We define the *domain of dependence* of S, $D(S)$, to be the set points p in M such that every causal curve through p, without a past or future "end point", intersects S. The set $D(S)$ represents those events in M which depend entirely upon the events in S. Next, we say that a set S is *achronal* if, for any events p and q in S, it is not the case that $p \ll q$. A set of events which are thought to be happening at any one time must certainly be achronal.

Finally, we say that a spacetime has a *Cauchy surface* if there is an achronal set S in M such that $D(S) = M$. (In an n-dimensional spacetime, a Cauchy surface S necessarily has $n - 1$ dimensions.)

There are a number of theorems which can be interpreted as stating that what happens on a Cauchy surface fully determines what happens throughout the entire spacetime [1]. And although there are subtleties involved, for our purposes a physically possible world with a Cauchy surface (also called a *globally hyperbolic* spacetime) will be considered deterministic [2]. With determinism clearly defined, one might wonder about the following.

Question If a physically possible world is close to a world which allows for time travel, must it be indeterministic? (Answer: Yes.)

That determinism implies stable causality is non-trivial. But the result can even be strengthened: In any globally hyperbolic spacetime (M, g), a global time function $t : M \to \mathbb{R}$ can be found such that each surface of constant t is a Cauchy surface. Also, the shape of the Cauchy surfaces are all the same [7]. Does the implication go in the other direction?

Question If a physically possible world is indeterministic, must it be close to another world which allows for time travel? (Answer: No.)

A counterexample is easy to construct. Let the manifold M be the two-dimensional plane with one point removed. Let the metric g be flat. The resulting spacetime (M, g) admits a global time function $t : M \to \mathbb{R}$ but for any achronal set of events S the set $D(S)$ is not M (see Fig. 26.6). (If the point were not removed, the spacetime *would* be globally hyperbolic.) With the answer to this question, we can now note that chronology, future (or past) distinguishability, stable causality, and global hyperbolicity form a hierarchy of causal conditions (see Table 26.1).

There is a sense in which determinism is connected to the absence of "holes" in spacetime. We say a spacetime (M, g) is *internally causally compact* (i.e. it has no holes) if, for all events p and q, $J^-(p) \cap J^+(q)$ is compact. Note that a spacetime with a point removed is never internally causally compact since one can

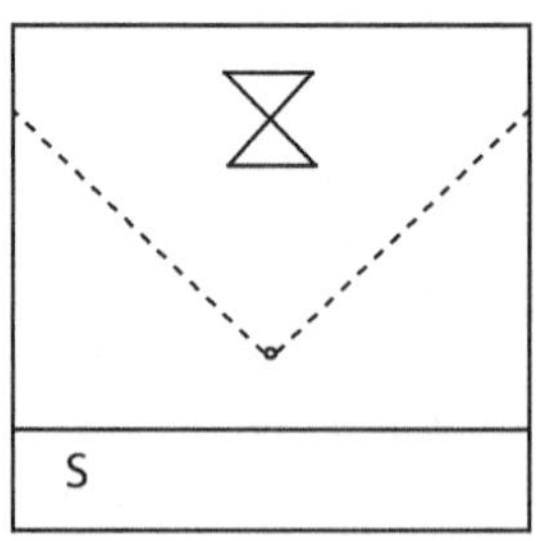

Fig. 26.6 The two-dimensional possible world (M, g). Stable causality is not violated but because of the removed point, any achronal surface S will be such that its domain of dependence (the region below the dotted line) is not all of M

Table 26.1 Causal Hierarchy

Chronology	$\Leftarrow$	Future (or Past)	$\Leftarrow$	Stable	$\Leftarrow$	Global
	$\not\Rightarrow$	Distinguishability	$\not\Rightarrow$	Causality	$\not\Rightarrow$	Hyperbolicity

find a sequence of points without accumulation point in $J^-(p) \cap J^+(q)$ if p and q are chosen so that $J^-(p) \cap J^+(q)$ "contains" the missing point. Now one can certainly show that global hyperbolicity implies internal causal compactness. What is fascinating is that the converse is also true if future (or past) distinguishability is assumed [7]. Thus, putting various results together, one can understand determinism to be equivalent to the conjunction of a weak causal condition and the requirement of no holes.

26.8 Reasonable Worlds

The discussion so far has concerned physically possible worlds. One is also interested in a subset of these worlds: the reasonable ones. However, what counts as a physically reasonable world is not always clear and often depends upon the context. Here, we briefly mention two questions concerning physically reasonable worlds.

The early results of global structure concerned "singularities" of a certain kind. The idea was to show, using fairly conservative assumptions, that all physically reasonable worlds must necessarily contain spacetime singularities. The project culminated in a number of general theorems [14]. And these theorems eventually led to serious worries concerning determinism. Indeed one natural question, still investigated today, is the following [19].

Question Is every physically reasonable world deterministic?

The question has different answers depending on how it is interpreted formally. And there is certainly much interpretive disagreement among physicists and philosophers [4]. If we can agree, for the time being, that not every physically reasonable spacetime is deterministic, then there is another question of interest.

Question Is there a physically reasonable world which allows for time travel?

Under some formal interpretations, the question has a negative answer [12]. Under others, the question is still open [5]. As before, the entire debate hinges on the details concerning the meaning of a "physically reasonable" world [18]. And of course, such details are best articulated and explored with the formalism.

References

Asterisks (∗) indicate recommended readings.

1. Choquet-Bruhat, Y., & Geroch, R. (1969). Global aspects of the Cauchy problem in general relativity. *Communications in Mathematical Physics, 14*, 329–335.
2. Earman, J. (1986). *A primer on determinism*. Dordrecht: D. Reidel.
3. ∗ Earman, J. (1989). *World enough and space-time*. Cambridge: MIT Press. [A complete treatment of absolute versus relational theories of space and time.]
4. ∗ Earman, J. (1995). *Bangs, crunches, whimpers, and shrieks*. Oxford: Oxford University Press. [A survey of topics in global spacetime structure.]
5. Earman, J., Wüthrich, C., & Manchak, J. (2016). "Time machines". In E. N. Zalta (Ed.). The Stanford encyclopedia of philosophy (Winter 2016 Edition). https://plato.stanford.edu/archives/win2016/entries/time-machine/.
6. Friedman, M. (1983). *Foundations of space-time theories*. Princeton: Princeton University Press.
7. Geroch, R. (1970). Domain of dependence. *Journal of Mathematical Physics, 11*, 437–449.
8. ∗ Geroch, R. (1978). *General relativity from A to B*. Chicago: University of Chicago Press. [An accessible introduction to general relativity.]
9. ∗ Geroch, R., & Horowitz, G. (1979). Global structure of spacetimes. In S. Hawking, & W. Israel (Eds.). *General relativity: An Einstein centenary survey* (pp. 212–293). Cambridge: Cambridge University Press. [An accessible presentation of topics in global spacetime structure.]
10. Gödel, K. (1949). An example of a new type of cosmological solutions of Einstein's field equations of gravitation. *Reviews of Modern Physics, 21*, 447–450.
11. Hawking, S. (1969). The existence of cosmic time functions. *Proceedings of the Royal Society A, 308*, 433–435.
12. Hawking, S. (1992). The chronology protection conjecture. *Physical Review D, 46*, 603–611.
13. Hawking, S., & Ellis, G. (1973). *The large scale structure of space-time*. Cambridge: Cambridge University Press.
14. Hawking, S., & Penrose, R. (1970). The singularities of gravitational collapse and cosmology. *Proceedings of the Royal Society A, 314*, 529–548.
15. ∗ Huggett, N. (1999). *Space from Zeno to Einstein*. Cambridge: MIT Press. [An accessible introduction, with primary source material, to various topics in space and time.]
16. Malament, D. (1977). The class of continuous timelike curves determines the topology of spacetime. *Journal of Mathematical Physics, 18*, 1399–1404.
17. ∗ Malament, D. (2006). Classical general relativity. In J. Butterfield, & J. Earman (Eds.), *Philosophy of physics* (pp. 229–274). Amsterdam: Elsevier. [A survey of topics in relativity theory.]
18. Manchak, J. (2013). Global spacetime structure. In R. Batterman (Ed.), *The Oxford handbook of philosophy of physics* (pp. 587–606). Oxford: Oxford University Press.
19. Penrose, R. (1979). Singularities and time-asymmetry. In S. Hawking & W. Israel (Eds.), *General relativity: An Einstein centenary survey* (pp. 581–638). Cambridge: Cambridge University Press.
20. Sklar, L. (1974). *Space, time, and spacetime*. Berkeley: University of California Press.
21. Stein, H. (1970). On the paradoxical time-structures of Gödel. *Philosophy of Science, 37*, 589–601.
22. ∗ Wald, R. (1984). *General relativity*. Chicago: University of Chicago Press. [A complete treatment of general relativity.]

Part VI
Value Theory and Moral Philosophy

Chapter 27
Formal Investigations of Value

Sven Ove Hansson

Abstract We can express values in three major ways: in terms of classification ("good", "bad", "best", etc.), comparison ("better", "at least as good", "equal in value"), and quantity (numbers are assigned). The interrelations among these three types of value expressions are surveyed, with a particular emphasis on relations of interdefinability. Furthermore, interrelations between value terms and terms expressing norms or choices are explored. Several of these connections have been surprisingly little studied, and further investigations may possibly lead to the discovery of additional connections among the different formal representations of value and value-related concepts.

27.1 Introduction

Example 1
CUSTOMER: Can you say something about the quality of these two wines, the Argentinian and the South-African one?
WAITER: Well the Argentinian wine is quite good but the South-African one is better.
CUSTOMER: So the South-African wine is the best of the two?
WAITER: No, that is not what I said. The Argentian wine is best of the two.
CUSTOMER: I am sorry but I cannot make sense of what you are saying.

Example 2
"I need to buy a new car. There are three options that I choose between, a Volkswagen, a Volvo, and a Peugot. I compared the first two and found the Volkswagen to be better than the Volvo. Then I compared the Volkswagen to the Peugot and concluded that the Peugot was better than the Volkswagen. But then I started to think about the Volvo again, and I couldn't avoid the

S. O. Hansson (✉)
Division of Philosophy, Royal Institute of Technology (KTH), Stockholm, Sweden
e-mail: soh@kth.se

S. O. Hansson, V. F. Hendricks (eds.), *Introduction to Formal Philosophy*, Springer Undergraduate Texts in Philosophy, https://doi.org/10.1007/978-3-319-77434-3_27

conclusion that the Volvo is better than the Peugot. So which car should I buy? I just can't make up my mind."

Example 3
UNHAPPY WIFE: Now that I have told you about all the problems in my marriage, do you recommend me to divorce?
MARRIAGE COUNSELLOR: No, my advice is to stay with your husband in spite of his faults.
UNHAPPY WIFE: So you think that it would be better for me to stay with him than to apply for a divorce.
MARRIAGE COUNSELLOR: No, it would be worse. But I nevertheless think that it is what you ought to do.

As these examples show, we have expectations that our value statements should cohere with each other. The second example also shows that we expect a rational person's choices to cohere with her values, and the third that we expect her norms and her values to form a coherent whole. These coherence issues are also important in moral philosophy. As one example of this, utilitarians and deontologists have different views on the exact nature of the required coherence between norms and values.

Formal representation has turned out to be indispensable if we wish to account in a precise way for coherence in issues such as these. However, it must be emphasized that the values held by a human being are inseparably connected with other components of her mind, such as her beliefs and her emotions. The very process of isolating her values from the rest of her mind involves a considerable idealization, and when these isolated values are expressed in a formal language we take the idealization one step further. Therefore, we should not expect to find a single, correct formalization. Instead, we should expect different formalizations to be suitable for capturing various features of what may be called the value-component of her state of mind.

This being said, there are a number of well-established representations, in particular preference relations, value functions, and choice functions, that have turned out to be useful for a wide variety of purposes. These devices are used not only in philosophy but even more in economics, psychology, and the decision sciences. Their most common use is to express what rationality demands of a person's values (and similarly of her norms and her choices). This chapter will provide an overview of these and some other representations, with a strong emphasis on how they relate to each other and in particular on whether they can be defined in terms of each other.

27.2 Values, Facts, and Norms

The separation of facts from values, and the principle that no "ought" can be derived from an "is", belong to the standard messages of elementary philosophy teaching. This exemplifies a general type of logical issues that can be raised for any two categories of statements. We can ask whether two such categories are logically separable, so that no element of one of them can be logically derived from elements of the other. Contrariwise, we may ask whether the two categories are interdefinable, so that for any element of one of them there is a logically equivalent element of the other. Obviously, it is also possible for such definability to go only in one direction.

Two categories for which rather subtle issues of this nature arise are those of norms and values. A normative expression such as "You ought to exercise two hours every day" prompts or commends some course of action. An evaluative expression such as "The best you can do is to exercise two hours every day" does not prompt or commend. Doing the best may for instance be a too demanding recommendation. An evaluative sentence may contextually imply advice or requirements, but that is not part of what it inherently means [11], [13, p. 143].

Terminological ambiguity often makes it difficult to uphold these distinctions. The terms "norm" and "normative" are sometimes used to cover both types of moral expressions. There is nothing wrong with such a terminological practice, as long as the distinction is made by some other linguistic means.[1]

Once this distinction has been made, it will be seen that the fact–value and is–ought delimitations do not coincide. Although "fact" and "is" denote the same category, "ought" refers to norms which is a separate category from that denoted by "value". Interesting issues of logical relationships arise among all three of these categories.

27.3 Varieties of Definability

George Edward Moore [20, pp. 172–173] pointed out that in spite of being different in meaning, a normative and an evaluative expression may be extensionally equivalent. In particular, a moral theory may imply a specific connection between values and moral requirements. However, it is important to distinguish between those relationships among concepts that hold according to a particular moral theory and those that hold conceptually. Moral standpoints may be supported by different kinds of arguments, but we should not expect substantial normative conclusions to follow from the structure of our concepts. In this chapter, the focus will be on conceptual connections that do not depend on the types of standpoints that tend to differ among moral theories. However, even on that general level it is important to distinguish between what we can call *definability* and *determinability*. The difference is that definability requires intensional equivalence whereas determinability only requires extensional equivalence. The word "bachelor" is definable in terms of "married" and "man" since there is an expression with these two words whose meaning coincides with that of "bachelor". The word "Stockholm" is determinable, but not definable in terms of "capital of Sweden" since these two expressions only have the same reference (for contingent reasons) but not the same meaning.

[1]The notion of supererogation, i.e. doing more good than what is morally required, is a particularly interesting case. It appears to be a composite concept that cannot be adequately explained without reference to both values and norms [15].

Bidirectional definability	$\subseteq$	Bidirectional determinability
$\rotatebox{90}{$\subseteq$}$		$\rotatebox{90}{$\subseteq$}$
Unidirectional definability	$\subseteq$	Unidirectional determinability

Fig. 27.1 The different types of definability and determinability referred to in the text

Since we are concerned with sets of expressions, such as value expressions, normative expressions etc., this distinction will have to be explicated for such sets. Let A and B be two sets of expressions. Then A is definable in B if and only if for every expression in A there is some expression in B that has the same meaning. Furthermore, A is determinable in B if and only if for every expression in A there is some expression in B that has the same reference. Obviously, definability implies determinability, but not the other way around.

For a simple example, let A be the English insect names and B the (scientific) Latin insect names. Then (if we disregard some minor ambiguities) A is definable in terms of B since for every English insect name there is a Latin insect name with the same meaning. However, the relationship does not go in the other direction since quite a few Latin insect names lack an English equivalent. This is a case of *unidirectional* definability. By *bidirectional* definability is meant that each of two sets of expressions is definable in terms of the other. A similar distinction can be drawn between unidirectional and bidirectional determinability.

Obviously, interdefinability is the ideal, and one might well ask whether definitions not complying with that standard should at all be considered. However, a connection between two categories of expressions (such as norms and values) can be philosophically interesting and/or practically useful although it is not derivable from purely conceptual knowledge, or works only in one direction. Therefore, all the four types of relations of definability and determinability specified in Fig. 27.1 are useful in philosophical investigations.

This chapter has its focus on value statements. After some basic specifications have been introduced in Sect. 27.4, three major types of value statements will be introduced in Sect. 27.5 in the form of a "value triangle". The three types are discussed in somewhat more detail in Sects. 27.6, 27.7, 27.8 and their logical interrelations are investigated in Sects. 27.9, 27.10, 27.11, and 27.12. After that, the logical interrelations between value expressions (of all three types) and statements about choice are studied in Sect. 27.13, and their interrelations with statements about norms are treated in Sect. 27.14. The general picture that emerges from these investigations will be briefly summarized in Sect. 27.15.

27.4 Three Basic Specifications

Most of our value statements in everyday life are ambiguous or at least unspecified in several ways. If I tell you that salmon is the best food fish, you may have to ask several questions in order to find out what I mean: Best for whom? Best in comparison to what? (Among the fish we can buy in our local store, or among those that are available anywhere in the world?) Best from what point of view? (Taste, nutritional value, etc.) Let us have a closer look at these three types of specifications.

The subject. Values can be related to persons in at least two ways. We may refer either to what is good or better *according to* a person or to what is good or better *for* that person. The distinction is not always made with sufficient clarity, but it is crucial in many contexts. One example is medical ethics where increasing emphasis on the patient's autonomy has led to a shift from arguments based on what is good for the patient to arguments based on what is good according to her. Both modes of speaking can also be applied to collective agents. In addition, value terms can be used in an impersonal way (that may at least sometimes be interpreted as "good for everyone").

Instead of saying "This is better according to him" we can say simply: "He prefers this." Logicians have often used the term "logic of betterness" when referring to values that are impersonal or assumed to hold *for* a person. The more common term "logic of preference" usually refers to values held *by* persons. However, no logical or otherwise structural differences seem to have been detected between the two types of connections between betterness and a person. A major reason for this is that the logical discourse on preferences does not usually refer to the preferences that actual people have but to the preferences of (idealized) rational agents. This is also the type of preferences that is usually discussed for instance in economics and decision theory.

This practice should be understood against the background that it would be difficult to identify any structural properties of the preferences (or other values) of agents who do not satisfy at least minimal requirements of rationality. We can assume that a rational agent does not both claim that Wagner's music is better than Verdi's and that the music of Verdi is better than that of Wagner. However, irrational agents can be expected to violate this and presumably any other structural requirement that we may wish to impose. This makes the values of (idealized) rational agents much more interesting than those of actual agents. Of course, we need not assume that agents are rational in all respects, only that they have reflected enough on their value statements to avoid certain structural features that further deliberation would show to be untenable.

The objects of evaluation. Most value statements have an (at least implicit) comparison class. It is one thing to say that Emma is a very good sprinter when you are discussing members of the local running club, but quite another thing to say so when discussing who should represent her country in the upcoming Olympic Games. Both in formal and informal accounts of values we need to keep track of the comparison class (also called alternative set). Quite a few pseudoparadoxes in value theory have their background in unmentioned shifts in the comparison class [12]. But on the other hand, carefully performed and described such shifts can be used to account for changes in values, and as we will see such shifts can also be used as a mechanism for interdefinability between different types of value statements.

Comparison classes can have interesting structural properties. A particularly important such property is *mutual exclusivity*. By this is meant that no two elements can be combined. The comparison class:

{dog owner, cat owner}

does not satisfy mutual exclusivity since it is possible to have both a cat and a dog. The more precisely described comparison class:

{dog owner but not cat owner, cat owner but not dog owner, both a dog owner and a cat owner, neither a dog owner nor a cat owner}

satisfies mutual exclusivity. It also satisfies *exhaustiveness*, i.e. it covers all possibilities. For most purposes, formal work is simplified by the use of exhaustive and mutually exclusive comparison classes.

We also need to determine what types of entities the comparison class consists of. Two approaches are common in the philosophical literature. One is to regard the elements as primitive, which means that they have no structural connections with each other. The other is to assume that they are sentences. This is often convenient since sentences representing states of affairs provide a highly versatile representation of both philosophical and mundane subject matter. In what follows, the letters x, y, z will be used to represent elements of the comparison class if they are taken as primitive. When these elements are assumed to be sentences they will instead be denoted by the letters p, q, r.

The evaluative viewpoint. Value statements can be made from different points of view, and they are therefore always ambiguous to the extent that the point of view has not been specified. The best car on sale is not necessarily the best car for me to buy. A good philosopher may be a bad mother, etc. There are at least three major ways in which such standards can be specified.

Many such specifications can be interpreted as positing a goal, such that the value terms refer to the achievement of that goal. We can for instance say that something is good from an ethical, economical, environmental, or aesthetic point of view. Something is "morally good" if it is good for satisfying our moral commitments and aspirations, "economically good" if it is good for achieving economic goals, etc. Such goals can be specified to different degrees and in different directions. The best car from the viewpoint of fuel economy may not be best from the viewpoint of total cost per kilometre.

Another way to disambiguate an evaluation is to mention one of the categories to which the evaluated object belongs. I have a friend who can be described as a good pianist but a bad driver. The two expressions refer to the same person, but evaluated according to our criteria for different categories that she belongs to, namely those of pianists respectively drivers. Such, category-specified value statements are quite common, and they are precise to the extent that we have determinate criteria for the categories in question [14].

As a limiting case, value statements may be intended to include all aspects, i.e., represent an evaluation that takes everything into account ("synoptic" values). It is contentious whether moral values and synoptic values coincide or whether the synoptic values are a broader or more over-arching category that includes non-moral values as well.

If vacillation between value criteria is allowed, then counter-examples can be constructed against any structural condition for value terms that we may think of. ("Rocky is the best saddle horse in the village, but not the best workhorse." Therefore, something may be both best and not best at the same time.) For formal analysis to be meaningful, we have to assume *criterial constancy*, i.e. the viewpoint of evaluation should be the same for all evaluations under consideration.

27.5 The Triangle of Value Concepts

In his book about the logic of probability, Rudolf Carnap distinguished between three major types of empirical descriptive terms. A *classificatory* concept such as "warm" divides objects into mutually exclusive classes. A *comparative* concept such as "warmer" compares two objects to each other. Finally, a *quantitative* concept such as "temperature" characterizes objects by assigning numerical values to them [3], [cf. 17]. The same three categories can be used to classify the value terms.

Among the classificatory value expressions we find those articulated with terms such as "good", "very bad", "almost worst", "fairly good", and "worst", all of which have a single referent that they identify as element of a class. In the formal language they are represented by monadic (one-place) predicates, such as G for "good" and B for "bad". The formula Gx means "x is good", and Bx means "x is bad".

Comparative value expressions such as "better", "worse", and "equal in value to" describe the relation between two referents. In the formal language they are expressed with dyadic (two-place) predicates. In what follows we will use $>$ for "better", $\sim$ for "equally good as", and $\geq$ for "at least as good as". Thus $x > y$ means that x is better than y, and $y \sim z$ that y and z are equally good. (A common alternative notation uses P instead of $>$, I instead of $\sim$, and R instead of $\geq$.)

Quantitative value expressions represent a referent's amount of value in numerical terms, i.e. in numbers saying "how good" something is. Quantitative value is expressed by a numerical function v that takes us from objects of evaluation to real numbers. Thus $v(x) = 3$ means that x has the value represented by the number 3.

In everyday life, moral statements are usually expressed with classificatory or comparative expressions. In moral theory, quantitative valuations are important, primarily since they are required in utilitarianism.

Table 27.1 The three major types of value expressions and their formal representations

Type of value expression	Formal representation	Examples
Classificatory	Monadic predicate	Good, bad, best
Comparative	Dyadic predicate	Better than, equally god as, at least as good as
Quantitative	Numerical function	Utility

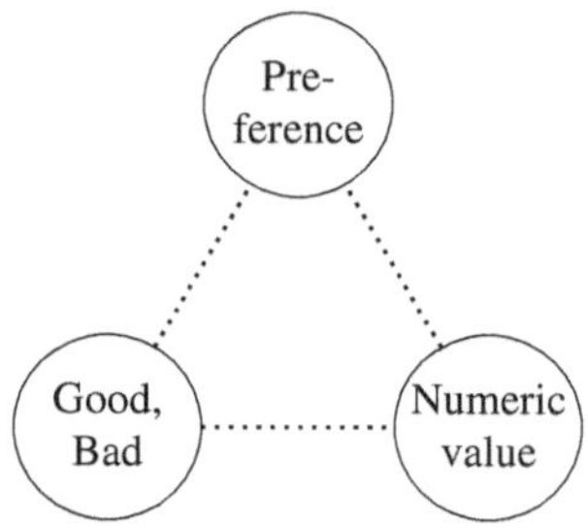

Fig. 27.2 The value triangle, representing the three major types of value statements

The three types of value terms are summarized in Table 27.1 and in the value triangle depicted in Fig. 27.2. The next three sections are devoted to the structural properties of each of these three types. For expository reasons we will begin with the comparative terms.

27.6 Comparative Value Concepts

Preference logic, the logic of the dyadic value predicates, is the most well-developed part of the logic of value concepts.[2] It has a long history. Aristotle discussed structural properties of preferences in Book III of his *Topics*. Representations in modern logical language were developed by Sören Halldén [9] and Georg Henrik von Wright [28].

The two fundamental comparative value concepts are "better" ($>$, strict preference) and "equal in value to" ($\sim$, value equality). The former of these represents both betterness and converse worseness, hence $x > y$ is taken to mean both "x is better than y" and "y is worse than x".

The relation $\geq$, "at least as good as" (weak preference) can be defined in terms of the two fundamental concepts:

$x \geq y$ if and only if either $x > y$ or $x \sim y$.

The three expressions "x is better than y", "y is better than x", and "x is equal in value to y" are usually taken to be mutually exclusive, i.e. no two of them can hold at the same time. It is also assumed that everything is equal in value to itself and

[2]For a more detailed exposition, see Chap. 29.

that equality in value always works in both directions. These assumptions add up to the following four constitutive properties of the comparative notions:

$x > y \rightarrow \neg(y > x)$ (*asymmetry of preference*)
$x \sim y \rightarrow y \sim x$ (*symmetry of indifference*)
$x \sim x$ (*reflexivity of indifference*)
$x > y \rightarrow \neg(x \sim y)$ (*incompatibility of preference and indifference*)

A much more controversial principle is completeness, according to which it holds for any two objects of the comparison class that either one of them is better than the other, or else they are equal in value. This property can be expressed in either of the following two equivalent ways:

$x > y \ \vee \ x \sim y \ \vee \ y > x$ (*completeness*)
$x \geq y \ \vee \ y \geq x$ (*completeness, alternative formulation*)

By far the most discussed postulate for comparative value is transitivity, according to which two steps of weak preference can be combined into one:

$x \geq y \geq z \ \rightarrow \ x \geq z$ (*transitivity*)

(To simplify the notation, we contract series of dyadic predicate expressions, thus writing $x \geq y \geq z$ for $x \geq y$ & $y \geq z$.)

Transitivity is often regarded as an essential rationality criterion.[3] The same applies to various weakened versions of it, such as:

$x > y > z \ \rightarrow \ x > z$ (*quasi-transitivity*)
$x_1 > x_2 > \ldots > x_n \ \rightarrow \neg(x_n > x_1)$ (*acyclicity*)

27.7 Classificatory Value Concepts

There is a wide variety of classifying value predicates: "good", "best", "bad", "very good" etc. Here, the focus will be on "good" and "bad" that are denoted by G respectively B. "Good" and "bad" are usually taken to be mutually exclusive, i.e. they cannot consistently both be applied to one and the same object of evaluation. If someone says that a particular novel is both good and bad, then this is perceived as paradoxical. We expect a resolution that typically assigns different evaluation criteria to the two statements, for instance: "The plot is good, but the language is bad". Due to our assumption of criterial constancy we can presume that goodness and badness are mutually exclusive:

$\neg(Gx \ \& \ Bx)$ (*mutual exclusiveness*)

If the objects of evaluation (elements of the comparison class) are represented by sentences, then additional logical principles can be introduced. In particular, we can express the intuition that a state of affairs and its negation are not (from the

[3]See Chaps. 29 and 31.

same point of view) both good or both bad. If you say "It is good to be married, and it is also good to be unmarried", then you typically mean that matrimony and bachelorhood are good in different respects or according to different criteria. Something similar can be said about the dismal pronouncement "It is bad to be married, and it is also bad to be unmarried." Such equivocations are excluded by the following principles:

$\neg(Gp \ \& \ G\neg p)$ (*non-duplicity of good*)
$\neg(Bp \ \& \ B\neg p)$ (*non-duplicity of bad*)

Two other potential postulates are $Gp \rightarrow B\neg p$ and (symmetrically) $Bp \rightarrow G\neg p$. (For both of them to hold it is sufficient and necessary that $B\neg p \leftrightarrow Gp$ holds.) However, it is easy to show that neither of them is a plausible postulate.

> My uncle is a great music lover. It would be good if I give him a recording of *Das Wohltemperierte Klavier* for his birthday. However, it would not be bad if I do not give him such a recording. This is because not doing so is compatible with giving him some other nice present that he will appreciate.

> Maria is an alcoholic who consumes different brands of whiskey every evening. It is bad that she drank Hazelburn whiskey yesterday. However, it would not have been much of a good thing if she had not done so, since then she would in all probability have taken some other whiskey instead.

Without further devices it seems difficult to obtain any plausible postulates for "good" and "bad" in addition to mutual exclusiveness and non-duplicity. There are at least two devices that we can use to obtain further postulates: shifts in the comparison class and the insertion of "good" and "bad" into a language that also contains a preference relation.

It is easy to find examples in which our usage of "good" and "bad" depends on the context. Jennifer and Robert are both members of the local chess club. Jennifer is one of its best players, but Robert seldom wins a game. When discussing members of the club it would be reasonable to say "Jennifer is a good player, but Robert is not". Suppose that they both join a large competition with several thousand participants, most of whom neither Jennifer nor Robert has much of a chance to defeat. In such a context it would be more natural to count neither Jennifer nor Robert as a good player.

To express this in the formal language we will use capital letters such as A and D to denote comparison classes. These letters can be attached as indices to the monadic value predicates G and B. Thus $G_A x$ means that x is good among the elements of A and $B_A x$ that x is bad among the elements of A. Johan van Benthem [26] has proposed the following postulates for such indexed monadic value predicates:

If $G_A x \ \& \ \neg G_A y$, then there is no comparison class D such that $G_D y \ \& \ \neg G_D x$ (*non-reversal of good*)
If $B_A x \ \& \ \neg B_A y$, then there is no comparison class D such that $B_D y \ \& \ \neg B_D x$ (*non-reversal of bad*)

G differentiates between x and *y* in *A* if and only if either $G_A x$ & $\neg G_A y$ or $G_A y$ & $\neg G_A x$. Furthermore, *G differentiates within A* if and only if there are $x, y \in A$ such that *G* differentiates between *x* and *y* in *A*. The corresponding definitions apply to the badness predicate *B*. With these definitions, the following postulates, also proposed by van Benthem, can be introduced:

If *G* differentiates between *x* and *y* in *D*, and $\{x, y\} \subseteq A \subseteq D$, then *G* differentiates between *x* and *y* in *A*. (*downward difference of good*)
If *B* differentiates between *x* and *y* in *D*, and $\{x, y\} \subseteq A \subseteq D$, then *B* differentiates between *x* and *y* in *A*. (*downward difference of bad*)
If $A \subseteq D$ and *G* differentiates within *A*, then it differentiates within *D*. (*upward difference of good*)
If $A \subseteq D$ and *B* differentiates within *A*, then it differentiates within *D*. (*upward difference of bad*)

The other device for obtaining postulates for "good" and "bad" is to include the dyadic and monadic value predicates in one and the same framework. This is intuitively plausible, since our classificatory and comparative concepts appear to be closely connected to each other. This was implicitly recognized already by Aristotle, when he said that "if one thing exceeds while the other falls short of the same standard of good, the one which exceeds is the more desirable" (*Topics*, III:3), which can be interpreted as a statement that:

$Gx \,\&\, \neg Gy \rightarrow x > y$ (*negation-sensitivity of good*)

Other, at least seemingly plausible, connections between the monadic and dyadic predicates include:

$\neg Bx \,\&\, By \rightarrow x > y$ (*negation-sensitivity of bad*)
$Gx \,\&\, By \rightarrow x > y$ (*bivalent sensitivity*)
$x > y \rightarrow Gx \vee By$ (*closeness*)
$Gx \,\&\, y \geq x \rightarrow Gy$ (*positivity of good*)
$Bx \,\&\, x \geq y \rightarrow By$ (*negativity of bad*)
$Gx \,\&\, Gz \,\&\, x \geq y \geq z \rightarrow Gy$ (*continuity of good*)
$Bx \,\&\, Bz \,\&\, x \geq y \geq z \rightarrow By$ (*continuity of bad*)
$Gx \,\&\, x \sim y \rightarrow Gy$ (*indifference-sensitivity of good*)
$Bx \,\&\, x \sim y \rightarrow By$ (*indifference-sensitivity of bad*)

27.8 Quantitative Value Concepts

A numerical function is any function that takes us from some objects to real numbers. In measurement theory, numerical functions are classified according to how much information they carry. Football teams have shirts with numbers on them.

A function that assigns to each football player the number on his or her shirt, for instance $v(\text{Ronaldinho}) = 10$, carries no other information than any other label that could be used for the same purpose. It is called a *nominal* function. Such functions have no use in the representation of values.

Other numerical functions represent an order or rank, so that something can be learnt from which of two objects is assigned the highest value. These are called *ordinal scales*. The ranking of tennis players is an example. The player ranked number 1 is presumably better than that ranked number 2, etc., but the differences on the scale have no significance. Thus the difference between the 1st and the 2nd player cannot be inferred, and it may be very different from that between the 200th and the 201st.

An *interval scale* has uniform differences. A common temperature scale (°C or °F) exemplifies this. The difference between 4 and 5 °C is the same as that between 40 and 41 °C. However, 10 °C is not ten times hotter than 1 °C. Ratios on an interval scale do not carry any meaningful information.[4]

Finally, on a *ratio scale* ratios are also meaningful. Length is measured on a ratio scale. Thus, 10 mm is ten times longer than 1 mm. These lengths stand in the same proportion to each other as 10 to 1 km (which is useful to know when reading a map with the scale 1:1,000,000). The scientific temperature scale is also a ratio scale, thus 300 °K (27 °C) is twice as hot as 150 °K (−123 °C).

The requirements on a numerical function that represents values depends on its intended use. For the purposes of a utilitarian moral theory a ratio scale will be necessary. This makes it possible to add values and to compare the values of aggregated wholes to each other. Other types of moral theories may be less demanding on the value function.

With these definitions in place we can now investigate interdefinabilities among the three categories of value statements. We will begin with the left side of the triangle of Fig. 27.2.

27.9 From Comparative to Classificatory Value

Several proposals have been put forward that define "good" and "bad" in terms of the dyadic predicates. The first such proposal was made by Albert P. Brogan [2], according to whom "good" means "better than its negation" and "bad" means "worse than its negation".

$Gp \leftrightarrow p > \neg p$ (negation-related good)
$Bp \leftrightarrow \neg p > p$ (negation-related bad)

[4]More precisely: The information that we can extract from knowing the exact values of ratios coincides with the information we can extract from just knowing for each ratio whether it is higher than, equal to, or less than 1.

This definition has a strong intuitive appeal, but of course it only works for relata that have a sentential structure so that they can be negated. Another disadvantage is that if G and B are defined in this way, then they do not always satisfy positivity, respectively negativity. For an example, let $\neg q \sim q \sim p > \neg p$. Then $Gp, q \geq p$ and $\neg Gq$, contrary to positivity.

Another major tradition is based on the identification of some neutral object or group of objects. Then "good" can be defined as "better than something neutral" and "bad" as "worse than something neutral". As a general recipe this works for non-sentential as well as sentential objects of comparison:

$Gx \leftrightarrow x > n$ (neutrality-related good)
$Bx \leftrightarrow n > x$ (neutrality-related bad)

Several proposals have been made on how to specify the neutral object(s). Most of these proposals require the objects to be represented by sentences. Some authors have recommended that the neutral propositions should be tautologies [6, p. 37] or contradictions [29, p. 164]. Writing $\top$ for an arbitrary tautology and $\perp$ for an arbitrary contradiction we then have:

$Gp \leftrightarrow p > \top$ (tautology-related good)
$Bp \leftrightarrow \top > p$ (tautology-related bad)
$Gp \leftrightarrow p > \perp$ (contradiction-related good)
$Bp \leftrightarrow \perp > p$ (contradiction-related bad)

However, it is difficult to make sense of a statement saying that something is better or worse than a tautology or a contradiction. If we wish to base our identification of the neutral elements on evaluative comparisons that we can actually make, then the solution must be sought elsewhere.

The most influential identification of neutral elements was proposed by Roderick Chisholm and Ernest Sosa [4]. They defined "good" as "better than something that is equal in value to its negation" and "bad" as "worse than something that is equal in value to its negation". For instance, let us assume that it is (morally) neither good nor bad for a person to read crime fiction. According to this definition, any action that is (morally) better than reading crime novels is a good action. Since Chisholm and Sosa used the term "indifferent" for "equal in value to its own negation", these can be called the "indifference-related" versions of "good" and "bad":

$Gp \leftrightarrow (\exists q)(p > q \sim \neg q)$ (indifference-related good)
$Bp \leftrightarrow (\exists q)(\neg q \sim q > p)$ (indifference-related bad)

Although this pair of definitions is conceptually related to Brogan's negation-related good and bad, the two pairs of definitions do not coincide unless rather strict demands are put on the structure of the preference relation [10]. The two pairs of definitions share the disadvantage of sometimes giving rise to predicates for "good" and "bad" that do not satisfy positivity respectively negativity. The indifference-related definitions also have the additional disadvantage of sometimes giving rise to predicates for "good" and "bad" that do not satisfy the even more elementary postulates mutual exclusiveness and non-duplicity [13, pp. 123–124].

The following definitions were introduced in order to obtain predicates for "good" and "bad" that satisfy these postulates for a wider category of preference relations [10]:

$Gp \leftrightarrow (\forall q)(q \succeq^{\star} p \rightarrow q > \neg q)$ (canonical good)
$Bp \leftrightarrow (\forall q)(p \succeq^{\star} q \rightarrow \neg q > q)$ (canonical bad)

Here, $\geq^{\star}$ stands for the ancestral of $\geq$. This means that $p \geq^{\star} q$ holds if and only if either $p \geq q$ or there is a series $r_1, \ldots, r_n$ of sentences such that $p \geq r_1 \geq \ldots \geq r_n \geq q$.

Whenever $\geq$ satisfies reflexivity, canonical good and bad satisfy the required postulates for a plausible interpretation of "good" and "bad" (including mutual exclusivity, closeness, non-duplicity of both predicates, positivity of "good", and negativity of "bad").[5] Furthermore, this pair of predicates is a generalization of negation-related good and bad in the following sense: If the preference relation is such that negation-related good satisfies positivity and negation-related bad satisfies negativity, then these negation-related predicates coincide with the canonical ones [13, p. 123].

In summary, we have well-functioning methods for defining the classificatory value terms "good" and "bad" from the comparative ones. We will now turn to the much less discussed issue of defining the comparative terms from the classificatory ones.

27.10 From Classificatory to Comparative Value

The philosophical significance of the above-mentioned definitions of classificatory values in terms of comparative ones has sometimes been put to question. To the extent that natural language can tell us anything about the structure of concepts, it points in the direction of treating classificatory rather than comparative notions as the primitive concepts from which others should be defined. There does not seem to be any natural language in which the classificatory terms are derived from the comparative ones. Instead, derivation in the opposite direction seems to be a universal pattern. As examples of this, the English "better" is believed to originate from a comparative form of a Proto-Indo-European adjective meaning "good", and the French "meilleur" from a comparative form of a Proto-Indo-European word meaning "strong" [25].

According to Henry Kyburg [17, p. 382] "[t]o apply a classificatory term is often to invoke an implicit comparison." But this only holds subject to two important

[5] An even weaker property than reflexivity, namely ancestral reflexivity ($p \geq^{\star} p$), is sufficient for this result.

provisos. First, classifying statements have to be available about more than one object. This is why only the second of the following two statements has comparative implications:

The fish is good in this restaurant.
The fish is good in this restaurant but the meat is not.

Secondly, the comparative implications may depend on the context. Consider again the example in Sect. 27.7 about the two chessplayers, Jennifer and Robert. In the context of the large tournament it is reasonable to say that neither of them is a good player. In the context of the local club we tend to describe Jennifer but not Robert as a good player. If we want to derive a comparison between two objects from classificatory statements about them, then we have to determine the context of these classificatory statements. As the chess-player example illustrates, a smaller context tends to yield more nuances than some of the larger contexts. This gives us a reason to choose the smallest possible context in which classificatory statements about both objects can be made, i.e. the context containing only these two objects.

Using this insight, Johan van Benthem [26] defined comparative concepts in terms of the corresponding classificatory ones as follows:

x is α-er than y if and only if: In the context $\{x, y\}$, x is α while y is not α [26, p. 195].

This is a general recipe that can be applied to concept pairs such as tall/taller, rich/richer etc. In preference logic we take it for granted that worseness is nothing else than converse betterness. Therefore, this recipe can be interpreted in two ways depending on whether we read $x > y$ as "x is better than y" or as "y is worse than x":

$x > y$ if and only if $G_{\{x,y\}}x$ & $\neg G_{\{x,y\}}y$ (goodness-based preference)
$x > y$ if and only if $B_{\{x,y\}}y$ & $\neg B_{\{x,y\}}x$ (badness-based preference)

It is easy to see that these two definitions are not equivalent and also that neither of them is plausible in all cases. Let x be good and not bad in the context $\{x, y\}$, and let y be neither good nor bad in the same context. Then $x > y$ holds according to first definition but $\neg(x > y)$ according to the second. This seems to be speak in favour of the first definition since we would expect $x > y$ to hold in this case. But next, let x be neither good nor bad in the context $\{x, y\}$, and let y be bad but not good in the same context. Then $\neg(x > y)$ holds according to the first definition but $x > y$ according to the second, which seems to speak in favour of the second definition.

To solve this problem we can replace the goodness- and badness-based definitions by the following one that takes both goodness and badness into account [16]:

$x > y$ if and only if either $G_{\{x,y\}}x$ & $\neg G_{\{x,y\}}y$ or $B_{\{x,y\}}y$ & $\neg B_{\{x,y\}}x$ (bivalently based preference)

Indifference and weak preference can be defined in the same vein:

$x \sim y$ if and only if $G_{\{x,y\}}x \leftrightarrow G_{\{x,y\}}y$ and $B_{\{x,y\}}x \leftrightarrow B_{\{x,y\}}y$

$x \geq y$ if and only if either: (i) $G_{\{x,y\}}x$, (ii) $B_{\{x,y\}}y$, or
(iii) $\neg G_{\{x,y\}}x \;\&\; \neg G_{\{x,y\}}y \;\&\; \neg B_{\{x,y\}}x \;\&\; \neg B_{\{x,y\}}y$

With these definitions we obtain the standard relationship between weak preference, strict preference, and indifference, i.e. $x \geq y \leftrightarrow x > y \vee x \sim y$. Furthermore, $\geq$ satisfies completeness. If G and B satisfy five of the conditions mentioned in Sect. 27.7, namely mutual exclusiveness, non-reversal of both good and bad, upward difference, and downward difference, then $\geq$ satisfies transitivity [16].

In summary, with this focus on the minimal comparison class it is possible to define comparative values in terms of classificatory ones. Since we have already seen that definitions in the opposite direction are available, this means that the two classes of value terms are definable in terms of each other. However, the two directions of these definitions form a rather disharmonious pair. When we go from the classificatory to the comparative terms we need to have context indices on the classificatory predicates that we start with, but the comparative predicates that we obtain do not come with such indices. When we go in the other direction, from comparative to classificatory predicates, no context indices are obtained for the latter. It remains an open question whether a framework can be constructed in which comparative and classificatory value terms are fully interdefinable.

27.11 Between Quantitative and Comparative Values

We will now turn to the right-hand side of the value triangle, namely that which connects comparative and quantitative value expressions. One direction, namely that from quantitative to comparative values, is easily obtained. For any value function v we can equate preference or betterness ($>$) with having higher value and indifference ($\sim$) with having the same value:

Exact numerical representation:
$x > y$ if and only if $v(x) > v(y)$
$x \sim y$ if and only if $v(x) = v(y)$

It follows directly that the preference relation defined in this way will be complete and transitive.

The derivation of quantitative from comparative value is a somewhat more intricate matter. Suppose that we have an (admittedly strange) preference relation $\geq$ such that $x > y$, $y > z$, and $z > x$. A value function corresponding to this relation would have to be such that $v(x) > v(y)$, $v(y) > v(z)$, and $v(z) > v(x)$, which is clearly impossible. It has in fact been shown that if the comparison class is countable, then a preference relation $\geq$ is reconstructible in terms of a numerical function if and only if it satisfies both completeness and transitivity [21].

One of the most discussed mechanisms for intransitivity is indiscernibility. Consider John who prefers coffee with as little sugar as possible. However, his ability to taste the difference between cups of coffee with different amounts of sugar is limited. He can only taste the difference if it is more than 0.15 grams. If we present him with the three cups x, with 2 grams of sugar, y with 2.1 grams and z with 2.2 grams, then he is able to taste the difference between x and z, but neither that between x and y nor that between y and z. This will yield a preference relation such that $x \sim y$, $y \sim z$, and $x > z$. Such a preference relation is not representable with a numerical function if we use the exact value representation introduced above. However, it can be represented if we include the limit of discrimination, in this case 0.15 grams, into the representation as follows:

Constant-threshold numerical representation:
$x > y$ if and only if $v(x) - v(y) > \delta$, where δ is a positive real number.

In other contexts, δ in this formula can be interpreted as a limit distinguishing those differences in value that are worthy of consideration from those that are negligible (even though they may be discernible). When interpreted as a discrimination limit, δ is often called a "just noticable difference" (JND). It has been shown that a preference relation over a finite comparison class can be numerically represented with a constant threshold if and only if it satisfies completeness and the following two properties [23]:

$x > y > z \rightarrow (x > w) \vee (w > z)$ (semi-transitivity)
$(x > y)$ & $(z > w) \rightarrow (x > w) \vee (z > y)$ (interval order property)

This construction can be generalized. The most general numerical structure that is still intuitively reasonable is arguably that in which the threshold is allowed to depend on both objects under comparison:

Doubly-variable-threshold numerical representation:
$x > y$ if and only if $v(x) - v(y) > \sigma(x, y)$, where σ is a function such that $\sigma(x, y) > 0$ for all x and y.

It has been shown that a preference relation over a finite comparison class can be numerically represented with a doubly variable threshold if and only if it satisfies acyclicity [1].

In summary, we can easily go from numerical to comparative values, and we can also go in the opposite direction, provided that the preference relation is acyclic and that we use the extra device of a threshold that can be interpreted as a limit of discrimination or of negligibility.

27.12 Between Quantitative and Classificatory Values

Let us now look at the bottom side of the value triangle, and begin with the right-to-left direction, i.e. the issue whether classificatory values can be defined in terms of quantitative ones. Such definitions can be modelled on the definitions in terms of comparative values that were discussed in Sect. 27.8. Thus, the negation-related definition of "good" and "bad" can be transferred to a quantitative framework as follows [18]:

$$Gp \leftrightarrow v(p) > v(\neg p)$$
$$Bp \leftrightarrow v(\neg p) > v(p)$$

These definitions have the advantage that G and B will always satisfy non-duplicity and mutual exclusivity. However, without restrictions on v they will not in general satisfy positivity respectively negativity.[6] A further problem with this pair of definitions, if unaided by restrictions on v, is that it allows violations of the postulate *bivalent sensitivity* (Gp & $Bq \rightarrow p > q$).[7]

The indifference-related definitions of "good" and "bad" can also be transferred to a quantitative framework. We can gain in lucidity (without losing in generality) by assuming that value assignments have been calibrated so that indifferent things have the value zero. Then "good" and "bad" can be defined as follows:

$$Gx \leftrightarrow v(x) > 0$$
$$Bx \leftrightarrow 0 > v(x)$$

This definition is in even greater need than the negation-related one of support from requirements on the structure of v. In particular, unless we disallow v from assigning positive values to both a statement and its negation, or negative values to them both, non-duplicity of G and B will not be satisfied.[8]

In the opposite direction, from classificatory to quantitative values, a simple construction is available provided that G and B satisfy mutual exclusiveness ($\neg(Gx$ & $Bx)$). We can then define the numerical function v as follows:

If Gx then $v(x) = 1$
If $\neg Gx$ & $\neg Bx$ then $v(x) = 0$
If Bx then $v(x) = -1$

The results of these deliberations are summarized in Fig. 27.3. We have found that definitions are obtainable in both directions along all three sides in the value triangle, although in some cases we needed "tricks" in the form of extensions of the formal apparatus such as context indices. In the rest of this chapter we will consider the connections between these value terms and two other categories of statements that they have often been associated with.

[6]To see that positivity of G does not follow, let $v(p) = v(q) = v(\neg q) = 0$ and $v(\neg p) = -1$. Then Gp and $q \geq p$ but $\neg Gq$.

[7]This can be seen from an example such that $v(\neg q) = 1$, $v(p) = v(q) = 0$, and $v(\neg p) = -1$.

[8]This can be seen from an example such that $v(p) = v(\neg p) = 1$ and $v(q) = v(\neg q) = -1$.

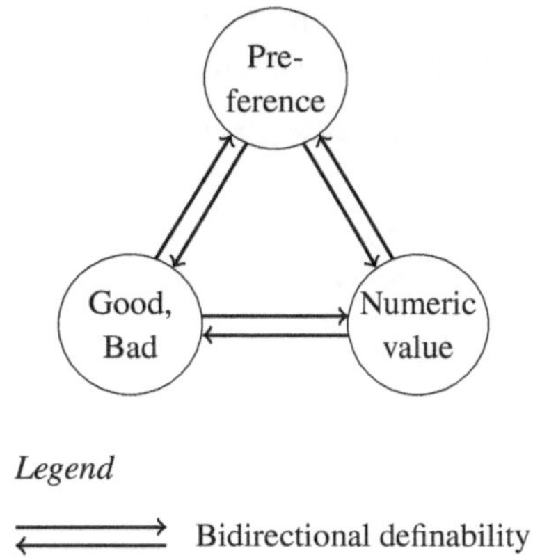

Fig. 27.3 Definability relations in the value triangle

27.13 Choices and Values

Statements about choices refer to actions, whereas statements about values (such as preferences) refer to states of mind. The difference comes out clearly if we consider states of affairs that we cannot choose. I prefer winning € 10,000 in a fair lottery to winning € 5,000 in the same fair lottery, but it is impossible for me to choose winning € 10,000 in this lottery, since if I could make such a choice then the lottery would not be fair.

Admittedly, if we adopt a behaviourist stance according to which states of mind do not exist other than as propensities to act, then preferences can be equated with hypothetical choices, and choices with actualized preferences. But this is a problematic metaphysical standpoint that should not be taken for granted in a formal analysis. Therefore it is advisable to treat choices as belonging to another category than values, from which follows that they are not interdefinable.

But lack of interdefinability does not mean lack of interconnections. We expect rational choices to be guided by preferences. There is something strange in choosing $\neg p$ while preferring p to $\neg p$. Of course there may be reasons to do so, for instance that the preferences in question do not include all the choice-relevant aspects of the alternatives. But some kind of justification is needed in cases like these, and it is certainly worth investigating what it means for choices to be guided by preferences. We should expect preference-guided choices to be restricted by the preferences, perhaps even derivable from them. We may ask what consequences it may have for the structure of choices that they have such connections with preferences, and conversely we may ask what structure preferences should have in order to ensure that their guidance gives rise to choices with desirable structural properties. All these are questions that we can (and should) ask without blurring the distinction between the different categories that choices and preferences belong to.

In order to perform such studies we need a formal representation of (hypothetical) choices.[9] The standard approach is to use choice functions for that purpose:

C is a *choice function* for a set A if and only if it is a function such that for all B:

(1) If $\emptyset \neq B \subseteq A$, then $\emptyset \neq C(B) \subseteq B$.
(2) Otherwise, $C(B)$ is undefined.

Various rationality principles for choice functions have been proposed, such as the following:

If $B_1 \subseteq B_2$ then $B_1 \cap C(B_2) \subseteq C(B_1)$ (Property α, the Chernoff property)

The most obvious way to construct a choice function out of a preference relation $\geq$ is to have the function always choose those elements that are at least as good as anything else that could have been chosen:

$$C(B) = \{x \in B \mid (\forall y \in B)(x \geq y)\}$$

Conversely, from a given choice function C we can construct a preference relation, based on choices from two-member sets:

$$x \geq y \text{ if and only if } x \in C(\{x, y\})$$

The interrelations between choices and preferences that can be obtained with these two definitions have been studied in considerable detail. It turns out that if a preference relation satisfies standard rationality criteria, then the choice function that it gives rise to will satisfy the major rationality criteria for choices (such as the above-mentioned Property α and others in the same style). Conversely, if a choice function satisfies these principles, then the preference relation that it gives rise to will in its turn satisfy the standard rationality criteria for preferences. Furthermore, retrievability holds in both directions: If we use these definitions to go from preferences to choices and then from choices back to preferences, then we regain the preference relations that we started with. Similarly, if we go from choices to preferences and then back to choices, then the original choice function will be regained. We therefore have pathways yielding full interdeterminability between preferences satisfying reasonable (but contestable) rationality criteria and choices guided by these preferences. But as already indicated, it is important to recognize that these are relations of interdeterminability, not interdefinability, since choices and preferences belong to different conceptual categories between which extensional but not intensional equivalence is possible.

[9]See Chap. 29 for additional information on choice functions and their connections with preference relations.

27.14 Norms and Values

In deontic logic, the logic of norms, it is generally recognized that there are three major groups of normative expressions in ordinary language, namely prescriptive, prohibitive, and permissive expressions.[10] In the formal language, they are represented by the corresponding three types of predicates. Here, prescriptive predicates such as "ought", "obligatory", and "morally required" will be denoted by O. Permissive predicates such as "permitted" and "allowed" will be denoted by P, and prohibitive ones such as "forbidden", "prohibited", and "morally wrong" by F. The three types of predicates are standardly and sensibly assumed to be interdefinable in the following way:

$Pp \leftrightarrow \neg O \neg p$ and $Pp \leftrightarrow \neg Fp$
$Fp \leftrightarrow O \neg p$ and $Fp \leftrightarrow \neg Pp$
$Op \leftrightarrow \neg P \neg p$ and $Op \leftrightarrow F \neg p$

The three categories of normative statements form a "norm triangle" with much more simple and direct definitions than those that we needed in our investigations of the value triangle. (See Fig. 27.4.)

As we saw in Sect. 27.2, normative and evaluative expressions belong to different categories in terms of their meanings, and therefore they cannot be interdefinable. But they are nevertheless strongly connected, and we expect them to cohere in some way or other. Therefore it is meaningful to search for possible relations of determinability between the two categories.

The most common proposal for a connection between predicates for norms and values is to identify what ought to be done with the best. This may be called the best-ought connection. It has strong support in the utilitarian camp. G.E. Moore, in a *locus classicus*, identified the assertion "I am morally bound to perform this

Fig. 27.4 The norm triangle with its interdefinabilities

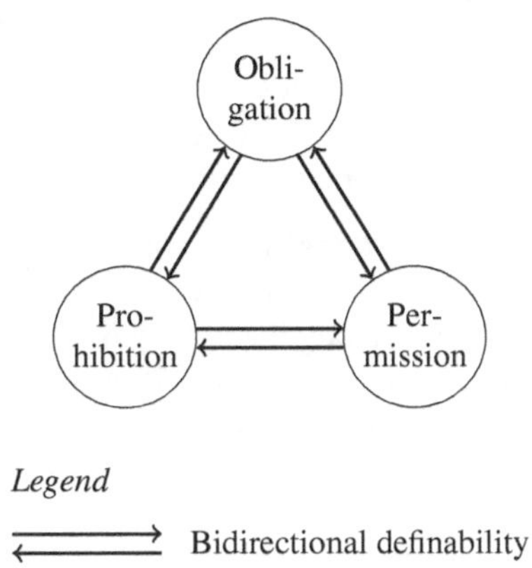

[10]See Chap. 32 for more information.

action" with the assertion "This action will produce the greatest possible amount of good in the Universe" [19, p. 147]. Another proposal, put forward by Gupta [8] and von Kutschera [27], equates "ought" with "good". This may be called the good-ought connection. Both these proposals equate a prescriptive predicate with a value predicate that satisfies positivity (i.e. a predicate H such that $Hq \,\&\, p \geq q \;\rightarrow\; Hp$). However, all proposals of this kind are highly problematic since they are threatened by counter-examples with the following structure [12]:

(1) p and q are mutually exclusive.
(2) $O(p \vee q)$
(3) $\neg Op$
(4) $\neg Oq$
(5) Either $p \geq (p \vee q)$ or $q \geq (p \vee q)$.

It follows straight-forwardly that if an example of this type can be found for a prescriptive predicate O, then that predicate cannot be equivalent with any positive value predicate. Such examples can indeed readily be found. One way to construct them is to let p and q represent two jointly exhaustive ways to satisfy the same moral requirement, and such that the difference between p and q is morally irrelevant. For instance, p may signify that I pay my debt to Adam by letting Simone bring my money to him, and q that I pay the debt in any other way.

Examples such as this, and others that can be constructed with the same structure, make a negative conclusion inevitable: No prescriptive predicate is extensionally equivalent with any value predicate that satisfies positivity. However, there are other ways to connect norm and value predicates to each other. Two other interesting options are to connect a permissive predicate P to a positive value predicate and to connect a prohibitive predicate F to a negative predicate (i.e. a predicate H such that $Hp \,\&\, p \geq q \;\rightarrow\; Hq$). Interestingly enough, these two options are equivalent. Let O, P, and F be three norm predicates that are interdefinable as explained above. It is easy to show that the following three properties of their connections with a preference relation $\geq$ are equivalent:

(1) P satisfies positivity ($Pq \,\&\, p \geq q \;\rightarrow\; Pp$),
(2) F satisfies negativity ($Fp \,\&\, p \geq q \;\rightarrow\; Fq$), and
(3) O satisfies contranegativity ($Op \,\&\, (\neg p \geq \neg q) \rightarrow Oq$).

One way to make this concrete is to connect a prohibitive term such as "forbidden" or "wrong" with the negative value term "bad" (the bad-wrong connection). Then an action is wrong if and only if it is bad. It ought to be performed if and only if it is bad not to perform it, and it is allowed if and only if it is not bad not to perform it. This definition assigns what seem to be suitable logical properties to the normative terms, but the norm–value interface that it provides is not flawless. A major reason for this is that the words "bad" and "wrong" do not necessarily have exactly the strengths necessary for exact interdeterminability. As was pointed out in another context by Chisholm and Sosa [5, p. 326], there are actions of "permissive ill-doing", i.e. "minor acts of discourtesy which most of us feel we have a right to perform (e.g. taking too long in the restaurant when others are known to be waiting)." Such acts are arguably morally bad but not morally forbidden. Therefore the bad-wrong connection should only be seen as a very rough approximation. It is, however, an interesting approximation since it provides us with full interdeterminability (though not interdefinability).

27.15 Conclusion

The major conclusions from these considerations are summarized in Fig. 27.5. Interdefinability holds internally among the three different kinds of normative predicates, and also – with some tailoring of the formal structure – among the three types of value terms. There are also connections of interdeterminability both between norms and values and between values and actions. Several of the connections indicated in the diagram, in particular those involving monadic value predicates, have been surprisingly little studied. Further investigations may possibly lead to the discovery of improved definability relations among the different formal representations of values and norms.

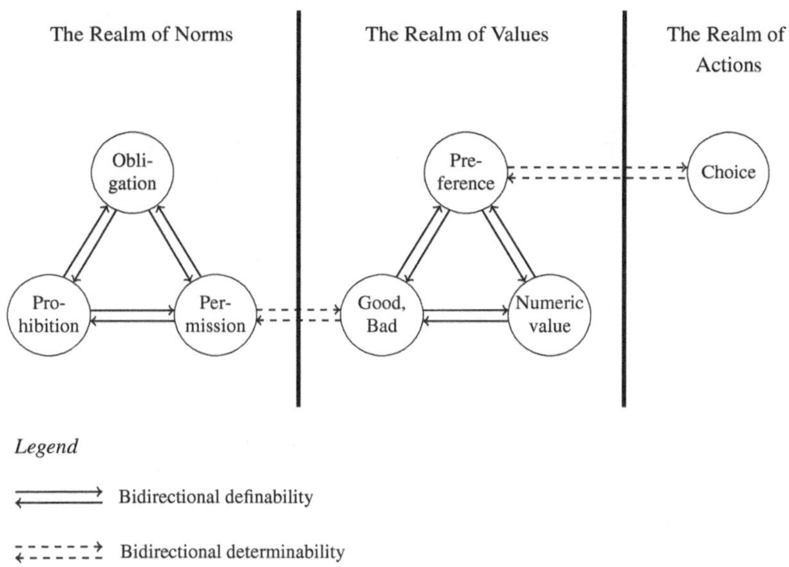

Fig. 27.5 Interdefinabilities and interdeterminabilities among statements belonging to the three realms of norms, values, and actions

References and Recommended Readings

1. Abbas, M. (1995). Any complete preference structure without circuit admits an interval representation. *Theory and Decision, 39*, 115–126.
2. Brogan, A. P. (1919). The fundamental value universal. *Journal of Philosophy, Psychology and Scientific Methods, 16*, 96–104.
3. Carnap, R. (1950). *The logical foundations of probability*. Chicago: University of Chicago Press.
4. Chisholm, R. M., & Sosa, E. (1966). On the logic of 'Intrinsically Better'. *American Philosophical Quarterly, 3*, 244–249.
5. Chisholm, R. M., & Sosa, E. (1966). Intrinsic preferability and the problem of supererogation. *Synthese, 16*, 321–331.
6. Danielsson, S. (1968). *Preference and obligation*. Uppsala, Sweden: Filosofiska Föreningen.
7. ∗ Gabbay, D., Horty, J., Parent, X., van der Meyden, R., & van der Torre, L. (Eds.). (2013) *Handbook of deontic logic and normative systems* (Vol. 1). London: College publications. [Covers most aspects of the logic of normative predicates.]
8. Gupta, R. K. (1959). Good, duty and imperatives. *Methodos, 11*, 161–167.
9. Halldén, S. (1957). *On the logic of 'Better'*. Lund: Gleerup.
10. Hansson, S. O. (1990). Defining 'good' and 'bad' in terms of 'better'. *Notre Dame Journal of Formal Logic, 31*, 136–149.
11. Hansson, S. O. (1991). Norms and values. *Crítica, 23*, 3–13.
12. Hansson, S. O. (1999). Preferences and alternatives. *Crítica, 31*, 53–66.
13. ∗Hansson, S. O. (2001). *The structure of values and norms*. Cambridge: Cambridge University Press. [Detailed discussion of the relationships among preferences, monadic values, and norms.]
14. Hansson, S. O. (2006). Category-specified value statements. *Synthese, 148*, 425–432.
15. Hansson, S. O. (2015). Representing supererogation. *Journal of Logic and Computation, 25*(2), 443–451.
16. Hansson, S. O., & Liu, F. (2014). From good to better. Using contextual shifts to define preference in terms of monadic value. In A. Baltag & S. Smets (Eds.), *Johan van Benthem on logical dynamics* (pp. 729–747). Cham: Springer.
17. Kyburg, H. E. (1997). Quantities, magnitudes, and numbers. *Philosophy of Science, 64*, 377–410.
18. Lenzen, W. (1983). On the representation of classificatory value structures. *Theory and Decision, 15*, 349–369.
19. Moore, G. E. ([1903] 1951). *Principia Ethica*. Cambridge: Cambridge University Press.
20. Moore, G. E. (1912). *Ethics*. London: Oxford University Press.
21. Roberts, F. S. (1979). *Measurement theory*. In G.-C. Rota (Ed.), *Encyclopedia of mathematics and its applications* (Vol. 7). Reading: Addison-Wesley.
22. ∗Rønnow-Rasmussen, T., & Zimmerman, M. J. (Eds.). (2005). *Recent work on intrinsic value*. Dordrecht: Springer. [Highly useful collection of papers on formal and informal value theory.]
23. Scott, D., & Suppes, P. (1958). Foundational aspects of theories of measurement. *Journal of Symbolic Logic, 23*, 113–128.
24. ∗Sen, A. (1970). *Collective choice and social welfare*. San Francisco: Holden-Day. [Unusually clarifying presentations of the relationships between choice and value.]
25. Ultan, R. (1972). Some features of basic comparative constructions. *Stanford Working Papers on Language Universals, 9*, 117–162.
26. van Benthem, J. (1982). Later than late: On the logical origin of the temporal order. *Pacific Philosophical Quarterly, 63*, 193–203.
27. von Kutschera, F. (1975). Semantic analyses of normative concepts. *Erkenntnis, 9*, 195–218.
28. von Wright, G. H. (1963). *The logic of preference*. Edinburgh: Edinburgh University Press.
29. von Wright, G. H. (1972). The logic of preference reconsidered. *Theory and Decision, 3*, 140–169.

Chapter 28
Value Theory (Axiology)

Erik Carlson

Abstract This chapter deals with an area of study sometimes called "formal value theory" or "formal axiology". Roughly characterized, this area investigates the structural and logical properties of value properties and value relations, such as goodness, badness, and betterness. There is a long-standing controversy about whether goodness and badness can, in principle, be measured on a cardinal scale, in a way similar to the measurement of well-understood quantitative concepts like length. Sect. 28.1 investigates this issue, mainly by comparing the properties of the relations "longer than" and "better than". In Sect. 28.2, some attempts to define goodness and badness in terms of the betterness relation are discussed, and a novel suggestion is made. Sect. 28.3, finally, contains an attempt to define the recently much discussed value relation "on a par with" in terms of the more familiar betterness relation.

This chapter deals with an area of study sometimes called "formal value theory" or "formal axiology". Roughly characterized, this area investigates the structural and logical properties of value properties and value relations, such as goodness, badness, and betterness. In contrast, "substantial value theory" or "substantial axiology" seeks to determine what is good and bad, and what is better than what. A third branch of value theory, usually called "meta-ethics", although "meta-axiology" would perhaps be a more appropriate term, discusses the ontological status of value properties, and the semantics of value terms and value judgements. Most philosophers would agree that the demarcations between the three areas are not sharp. Some of the issues to be discussed in this chapter arguably straddle the distinction between formal and substantial axiology, in particular.

The main focus of most axiological investigations is *intrinsic* or *final* value; i.e., the value a thing has "in itself", or "for its own sake". There is not space here for

E. Carlson (✉)
Uppsala University, Uppsala, Sweden
e-mail: Erik.Carlson@filosofi.uu.se

S. O. Hansson, V. F. Hendricks (eds.), *Introduction to Formal Philosophy*, Springer Undergraduate Texts in Philosophy, https://doi.org/10.1007/978-3-319-77434-3_28

trying to provide precise definitions of these concepts. [A helpful discussion and overview of the literature on intrinsic value can be found in the "Introduction" to Rønnow-Rasmussen and Zimmerman [29].]

28.1 "Longer than" and "Better than"

A useful way of introducing several central problems of formal axiology is to ask to what extent value (i.e., goodness and badness) is similar to a familiar and well-understood quantitative concept like length. Many philosophers have assumed that value can, at least in principle, be measured additively, analogously to the measurement of length. [An early example is Bentham [4]] Others have denied this. Often, however, this discussion is conducted without much recognition of what it takes for the additivity assumption to be true. Let us therefore state the conditions necessary for standard extensive or additive measurement, and then ask whether value can reasonably be thought to satisfy these conditions. Primarily, this will amount to comparing the properties of the relations "longer than" and "better than".

Letting $\succsim$ denote "at least as long as", we can define "longer than" , denoted $\succ$, and "equally long as", denoted $\sim$, as follows: $a \succ b$ iff (if and only if) $a \succsim b \wedge \neg(b \succsim a)$; $a \sim b$ iff $a \succsim b \wedge b \succsim a$. If $X = \{a, b, c \ldots\}$ is a set of items that have length, the relational structure $(X, \succsim)$ has the following properties:

Completeness. For any a and b in X, $a \succsim b \vee b \succsim a$.
Transitivity. If $a \succsim b \wedge b \succsim c$, then $a \succsim c$.
Concatenation. Any a and b in X can be put together, or "concatenated", into an item, denoted $a \circ b$, that also has length.
Monotonicity. $a \succsim b$ iff $a \circ c \succsim b \circ c$ iff $c \circ a \succsim c \circ b$.
Weak associativity. $a \circ (b \circ c) \sim (a \circ b) \circ c$.
Archimedeanness. For any a and b in X, there is a positive integer n, such that $na \succ b$, i.e., a concatenation of n copies of a is longer than b.

The last condition involves a certain amount of idealization, since there may not actually exist a sufficient number of copies of a.

These properties together imply that $(X, \succsim, \circ)$ is a "closed extensive structure". [This is a slight simplification. Actually, a somewhat more complicated Archimedean condition is needed. See [21], 73.] This, in turn, means that there is a function f with real numbers as values, such that (i) $f(a) \geq f(b)$ iff $a \succsim b$, and (ii) $f(a \circ b) = f(a) + f(b)$. Further, another function g satisfies properties (i) and (ii) iff g is a "similarity transformation" of f; i.e., iff there is a real number $x > 0$, such that, for all a in X, $g(a) = xf(a)$. This amounts to measurement on a ratio scale. Thus, if $f(a) = 4$ and $f(b) = 2$, it follows that a is twice as long as b.

Now, let X be a set of value bearers, and let $\succsim$, $\succ$, and $\sim$ denote "at least as good as", "better than", and "equally good as", respectively. Which of the above conditions can be expected to hold? It appears that they are all more or less controversial. Let us briefly consider each condition in turn.

Completeness Many substantial axiologies are pluralistic, recognizing value bearers of different kinds. Suppose, for example, that friendship and pleasure both have value. It then seems somewhat implausible that $a \succsim b \vee b \succsim a$ holds for every instance a of friendship and every instance b of pleasure. Such appeals to intuition, against completeness, are sometimes buttressed by the "small improvement argument". Let a be an instance of friendship and let b be an instance of pleasure, such that we are disinclined to claim either that $a \succ b$, or that $b \succ a$. Does it follow that $a \sim b$? If so, anything better than b must be better than a (given that $\sim$ is an equivalence relation). Consider an instance of pleasure b^+, which is just like b, only slightly more intense. Although $b^+ \succ b$, we will probably not judge that $b^+ \succ a$. Hence, the small improvement argument concludes, $\neg(a \sim b)$, implying that $\succsim$ is not complete.

An objection to the small improvement argument is that our unwillingness to judge that $b^+ \succ a$ only proves that we, in the first comparison, did not (rationally) judge that $a{\sim}b$. It does not prove that we judged that $\neg(a{\sim}b)$. To refrain from making a judgement of equality is not to make a judgement of nonequality. The small improvement argument presupposes, however, a judgement of the latter kind [27].

Transitivity The transitivity of $\succsim$ has been questioned by a number of philosophers. An interesting type of alleged counterexample is due to Stuart Rachels [26] and Larry Temkin [31]. Their examples can be seen as applications of the following general assumptions:

(1) For any painful experience, no matter what its intensity and duration, it would be better to have that experience than one that was only slightly less intense but twice as long.
(2) There is a continuum of painful experiences ranging in intensity from extreme forms of torture to the mild discomfort of, say, a hangnail.
(3) A mild discomfort for the duration of one's life would be preferable to two years of excruciating torture, no matter the length of one's life.

Rachels and Temkin argue from assumptions 1 to 3 to the conclusion that $\succ$ is not transitive. (If $\succ$ is not transitive, $\succsim$ cannot be transitive, either. For suppose that $\succsim$ is transitive, while $\succ$ is not. There is then a case such that $a \succ b \wedge b \succ c \wedge a \sim c$. This implies that $c \succsim a \wedge a \succsim b \wedge \neg(c \succsim b)$, contradicting the assumption that $\succsim$ is transitive.)

Other philosophers claim that we can know *a priori* that $\succ$ is always transitive, since transitivity is part of the meaning of comparatives like "better than". Thus, John Broome finds it "self-evident" that any comparative relation is necessarily transitive. Since this is a conceptual truth, "not much argument is available to support it directly" ([6], 51). A defense of the transitivity assumption must then consist mainly in responses to apparent counterexamples. Broome notes that many such examples involve large numbers, and argues that our intuitions about large numbers are unreliable. We may, for example, be unable to grasp what it would be like to have a hangnail for thousands of years. It is far from clear, though, that intuitively plausible counterexamples to transitivity must involve large numbers (see, e.g., [23]).

If Rachels and Temkin are right, $\succ$ is sometimes not only nontransitive, but *cyclical*. That is, there are value bearers $a, b, c, \ldots, y, z$, such that $a \succ b, b \succ c, \ldots, y \succ z$, and $z \succ a$. Let us call a structure of this kind a "betterness cycle". Whether or not there are betterness cycles, certain structural restrictions apply to any such case. For example, no betterness cycle can contain both good and bad options. The following propositions are surely universal truths:

(4) No option is both good and bad (all things considered).
(5) If a is good and $b \succ a$, then b is good.
(6) If a is bad and $a \succ b$, then b is bad.

Let a be an arbitrary option in a betterness cycle. If a is good, iterated applications of (5) entail that every option in the ordering is good. If a is bad, iterated applications of (6) entail that every option is bad. It thus follows from (4) to (6) that no betterness cycle contains both good and bad options. This is of some significance, since certain putative examples of betterness cycles contain intuitively good as well as intuitively bad options (see, e.g., [25]).

Concatenation Whether the concatenation condition is satisfied may depend on what kinds of entities are bearers of value. If the value bearers are taken to be propositional entities, concatenation is naturally identified with conjunction. This immediately leads to a problem, however, since conjunction is idempotent; i.e., $a \wedge a = a$. Given reflexivity of $\sim$, this means that $a \circ a \sim a$. If monotonicity and weak associativity hold, the assumption that $a \circ a \sim a$, for all a, implies that all value bearers are equally good.

A possible solution to this problem is to define $a \circ a$ as the conjunction of a with a numerically different but qualitatively identical propositional entity. For example, if a is the state of affairs that Alf is happy to degree 10, $a \circ a$ could be identified with the conjunction $a \wedge a^*$, where a^* is the state that Alf's counterpart in some other possible world is happy to degree 10. ($a^* \circ a^*$ then has to be identified with the conjunction of a^* and a third state a^{**}.)

If other kinds of entities, for example material objects, are among the value bearers, concatenation might be defined in terms of mereological fusion, rather than conjunction. Since also mereological fusion is usually understood as idempotent, the problem of how to understand self-concatenation remains. It should be noted, though, that self-concatenation must be defined by means of identical "copies" also in the context of length or mass measurement (See [21], 3f.)

Monotonicity The monotonicity condition is closely connected to G. E. Moore's famous principle of "organic unities". Moore claimed that, in some cases, "the intrinsic value of a whole is neither identical with nor proportional to the sum of the values of its parts" ([22], 184). One of Moore's examples of an organic unity is the state of being conscious of a beautiful object. Moore took such a state to be of great intrinsic value, containing as parts (in some sense) the object and the state of being conscious. But neither of these parts has, according to Moore, much intrinsic value considered in isolation.

However, Moore's way of formulating the principle of organic unities is unfortunate, since it is meaningful to add the values of two items only if value is measurable on an additive ratio scale. Measurability on such a scale implies, in turn, that the value of a whole *is* proportional to the sum of the values of its parts (see [12]). On a literal interpretation, therefore, Moore's claim does not make much sense. Arguably, what Moore really meant to assert was simply that the monotonicity condition does not hold. It is, indeed, easy to think of putative counterexamples to monotonicity. To borrow a case from Roderick Chisholm ([16], 306) suppose that a and b are two identical beautiful paintings, and that c is a beautiful piece of music. Suppose also that the value of contemplating a, b or c is the same. It nevertheless seems that the whole consisting in the contemplation of a and c is better than the whole consisting in the contemplation of a and b. Some philosophers have suggested, however, that a restricted version of the monotonicity assumption suffices for the purposes of value measurement [18, 32].

Weak Associativity If concatenation is identified with conjunction or mereological fusion, the weak associativity condition probably holds. As regards a concatenation operation involving physical interaction between objects, on the other hand, weak associativity may be questionable. As Fred Roberts notes, "combining a with b first and then bringing in c might create a different object from that obtained when b and c are combined first. To give an example, if a is a flame, b is some cloth, and c is a fire retardant, then combining a and b first and then combining with c is quite different from combining b and c first and then combining with a." ([28], 125) Clearly, this difference could be evaluatively relevant.

Archimedeanness The Archimedean condition has been denied by many philosophers. To cite just two examples, Franz Brentano judged it "quite possible for there to be a class of goods which could be increased *ad indefinitum* but without exceeding a given finite good" [5], while W. D. Ross believed that, although virtue and pleasure are both good, "*no* amount of pleasure is equal to any amount of virtue, [...] in fact virtue belongs to a higher order of value, beginning at a point higher on the scale of value than pleasure ever reaches [...]". ([30], 150)

Gustaf Arrhenius has argued that the existence of such "superior goods" has an implausible implication. Assuming a and b to be good, Arrhenius defines a as "weakly superior" to b iff there is a positive integer m, such that $ma \succ nb$, for every positive integer n. Now let $a_1, \ldots, a_k$ be a finite sequence of items, such that $a_1 \succ a_2 \succ \ldots \succ a_{k-1} \succ a_k$, and a_1 is weakly superior to a_k. Arrhenius shows that any such sequence must contain a pair a_i, a_{i+1}, such that a_i is weakly superior to a_{i+1}. However, he believes that for most or all types of goods, the sequence $a_1, \ldots, a_k$ can be chosen so that the difference in value between adjacent items is only marginal. Hence, the assumption that a_1 is weakly superior to a_k implies that a_i

is only marginally better than, although weakly superior to a_{i+1}. This, Arrhenius contends, is implausible ([1], 301).

The defender of superior goods can retort that since Arrhenius does not explain what a "marginal" value difference is, he provides no ground for denying that weak superiority is compatible with a merely marginal difference. As Arrhenius acknowledges, there need not be a pair a_j, a_{j+1} in the above sequence, such that a_j is *strongly* superior to a_{j+1}, in the sense that $a_j \succ na_{j+1}$, for all positive integers n. It might thus be claimed that strong, but not weak, superiority is incompatible with a merely marginal difference ([1], 301; [2], 138). Another response to Arrhenius' argument would be to deny the claim that any value difference can be spanned in a finite number of steps, such that each step involves only a marginal difference. If we share Ross' view that any amount of virtue is better than any amount of pleasure, why should we believe that a finite number of marginal worsenings could bridge the value gap between an instance of virtue and an instance of pleasure? Indeed, it could be argued that the claim of superiority essentially involves the denial of this contention. If so, Arrhenius' argument begs the question.

The most problematic of the conditions we have discussed are perhaps completeness, Archimedeanness, and monotonicity. It can be shown, however, that if value is represented by other mathematical entities than real numbers, measurement on a kind of generalized ratio scale is possible even if the former two conditions do not hold [7, 8, 10]. On the other hand, the truth of some version of the monotonicity condition appears essential for any form of extensive measurement, and, in fact, for measurement on any scale stronger than an ordinal scale.

28.2 Defining "Good" and "Bad" in Terms of "Better than"

On the face of it, there is another important difference between value, on the one hand, and quantities like length, on the other. There are *bad* things, i.e., things with *negative* value, but there are no things with negative length (speculative physics aside). Moreover, the "zero point", dividing the good from the bad things, appears to be absolute, rather than dependent on the scale of measurement. (This is in contrast to, e.g., temperature. Some temperatures are positive if measured on the Fahrenheit scale, but negative if measured on the Celsius scale.) Many philosophers have attempted to define goodness and badness in terms of betterness. Such definitions, if possible, would arguably be desirable for reasons of theoretical simplicity. In a very influential paper, Roderick Chisholm and Ernest Sosa [17] proposed the following definitions, with "I", "N", "G", and "B" standing for, respectively, "intrinsically indifferent", "intrinsically neutral", "intrinsically good", and "intrinsically bad":

D1. $a \sim b$ iff $\neg(a \succ b) \wedge \neg(b \succ a)$.
D2. Ia iff $\neg(a \succ \neg a) \wedge \neg(\neg a \succ a)$.
D3. Na iff $(\exists b)$ (I$b \wedge a \sim b$).
D4. Ga iff $(\exists b)$ (I$b \wedge a \succ b$).

D5. Ba iff $(\exists b)$ (I$b \wedge b \succ a$).

Chisholm's and Sosa's theory assumes equivalents to the following axioms:

A1. If $a \succ b$, then $\neg(b \succ a)$.
A2. If $a \succ c$, then $a \succ b \vee b \succ c$.
A3. If I$a \wedge$ Ib, then $a \sim b$.
A4. If Ga or B$\neg a$, then $a \succ \neg a$.

An important feature of Chisholm's and Sosa's theory is that, unlike many previous theories, it does not assume that a state of affairs is good if it is better than its negation, and bad if it is worse than its negation. The state that there are happy egrets is, they assume, intrinsically good, while the state that there are no happy egrets is intrinsically neutral. (The latter state is not intrinsically bad, since the mere absence of happiness does not "rate any possible universe a minus".) Analogously, the state that there are unhappy egrets is intrinsically bad, whereas the state that there are no unhappy egrets is neutral. (The latter state is not intrinsically good, since the mere absence of unhappiness does not "rate any possible universe a plus".)

D1 excludes the possibility that a and b are incomparable with respect to intrinsic value, in the sense that $\neg(a \succ b) \wedge \neg(b \succ a) \wedge \neg(a \sim b)$. (Cf. the discussion of completeness of $\succsim$, in Sect. 28.1) If incomparability is possible, D1 is not an appropriate definition of "is equal in intrinsic value to". Furthermore, b may be incomparable to each of a and c, although $a \succ c$. This would mean that A2 is violated.

Philip Quinn [24] has objected to Chisholm's and Sosa's logic on precisely the grounds that it illegitimately rules out the possibility of incomparability. Even if we assume a hedonistic axiology, Quinn remarks, it is far from obvious that the value of Smith's enjoying the taste of apples is comparable to the value of her enjoying the sound of Beethoven's Ninth Symphony. Quinn argues, nonetheless, that universal comparability can be shown to be true. He retains Chisholm's and Sosa's axioms A3 and A4, and proposes, in addition, the following axioms:

A5. *Transitivity*. If $a \succsim b \wedge b \succsim c$, then $a \succsim c$.
A6. $(a \succsim (a \vee b) \vee b \succsim (a \vee b)) \wedge ((a \vee b) \succsim a \vee (a \vee b) \succsim b)$.
A7. $(a \succsim (a \vee b) \vee (a \vee b) \succsim a)$ iff $(b \succsim (a \vee b) \vee (a \vee b) \succsim b)$.

Quinn shows that A5 to A7 entail that $a \succsim b$ or $b \succsim a$ holds for all a and b. In other words, universal comparability (completeness) is true.

A6 is hardly unassailable, though. To use Quinn's own example, what are the grounds for assuming that either a = *Smith enjoys apples*, or b = *Smith enjoys Beethoven*, is at least as good as $a \vee b$? If it is intuitively plausible to judge a and b incomparable, it appears equally plausible to judge each of these states incomparable to their disjunction.

Sven Ove Hansson [20] has suggested a more general definition of goodness and badness in terms of betterness. Hansson's proposal assumes neither universal comparability nor transitivity of betterness. However, his and, to the best of my knowledge, every other extant proposal presuppose that the value bearers are propositional entities, which can be negated. Since many value theorists believe that non-propositional entities, such as persons or material objects, can have intrinsic or final value, a definition format that does not rely on negation or indifference is desirable. Assuming that at least some value bearers can be concatenated, there is a fairly simple way to construct such a format [Carlson [13] contains further discussion of the proposal sketched below].

First, we define an item a as "universally null" iff, for all value bearers b, such that $a \circ b$ is a value bearer: $a \circ b \sim b$. Thus, a universally null item does not affect the intrinsic value of any whole of which it is a part. (In Chisholm's and Sosa's parlance, such an item rates any possible universe a zero.)

Let us assume the following four axioms:

A8. $\succsim$ is a quasi-order; i.e., reflexive and transitive.
A9. There is at least one universally null item that is a value bearer.
A10. For any universally null value bearers a and b, whose coexistence is logically possible, $a \circ b$ is a value bearer.
A11. For any complex value bearer $a \circ b$, it holds that $a \circ b \sim b \circ a$.

On the basis of these axioms, and letting "UN" abbreviate "universally null", we may propose the following definitions of "intrinsically good", "intrinsically bad", and "intrinsically neutral":

D6. Ga iff $(\exists b)$ (UN$b \wedge a \succ b$).
D7. Ba iff $(\exists b)$ (UN$b \wedge b \succ a$).
D8. Na iff $(\exists b)$ (UN$b \wedge a \sim b$).

If some value bearers are incomparable, there may be reason to assume the existence of a fourth value category, in addition to intrinsic goodness, badness, and neutrality (See [11]). This category, which may be labelled "intrinsic indeterminacy", is readily incorporated into our proposal. Defining "is incomparable in intrinsic value to", symbolized $||$, as $a \,||\, b$ iff $\neg(a \succsim b) \wedge \neg(b \succsim a)$, we may define "intrinsically indeterminate", abbreviated "IND", as follows:

D9. INDp iff $(\exists b)$ (UN$b \wedge a \,||\, b$).

The following twenty propositions can be derived from A8 to A11 and D6 to D9:

(i) If UN$a \wedge$ UNb, then $a \sim b$.
(ii) For any a, exactly one of the following is true: Ga, Ba, Na, or INDa.
(iii) If G$a \wedge b \succsim a$, then Gb.
(iv) If G$a \wedge a \,||\, b$, then G$b \vee$ INDb.
(v) If B$a \wedge a \succsim b$, then Bb.
(vi) If B$a \wedge a \,||\, b$, then B$b \vee$ INDb.
(vii) If N$a \wedge b \succ a$, then Gb.
(viii) If N$a \wedge a \succ b$, then Bb.
(ix) If N$a \wedge a \sim b$, then Nb.
(x) If N$a \wedge a \,||\, b$, then INDb.

(xi) If IND$a \wedge b \succ a$, then G$b \vee$ INDb.
(xii) If IND$a \wedge a \succ b$, then B$b \vee$ INDb.
(xiii) If IND$a \wedge a \sim b$, then INDb.
(xiv) If G$a \wedge$ Bb, then $a \succ b$.
(xv) If G$a \wedge$ Nb, then $a \succ b$.
(xvi) If G$a \wedge$ INDb, then $a \succ b \vee a \,||\, b$.
(xvii) If B$a \wedge$ Nb, then $b \succ a$.
(xviii) If B$a \wedge$ INDb, then $b \succ a \vee a \,||\, b$.
(xix) If N$a \wedge$ Nb, then $a \sim b$.
(xx) If N$a \wedge$ INDb, then $a \,||\, b$.

The proofs of these propositions are simple and will not be stated here.

Apart from being more generally applicable than earlier suggestions, D6 to D9 have the virtue of not prejudging questions in substantial axiology. Importantly, they permit organic unities of various sorts. For example, it is consistent with these definitions that a concatenation of intrinsically good items is intrinsically bad, or that a concatenation of intrinsically bad items is intrinsically good.

28.3 Comparability and Parity

It has usually been taken for granted, in line with our definition of incomparability in Sect. 28.2, that two value bearers, a and b, are comparable with respect to value iff $a \succ b$, or $b \succ a$, or $a \sim b$. If so, universal comparability just means that $\succsim$ is complete. In recent years, however, this view has been questioned. Ruth Chang has argued that there is a positive value relation, "on a par with", that is incompatible with the familiar relations. If two items are on a par, they are comparable with respect to value, although neither item is better than the other, and they are not equally good. As possible examples of parity Chang mentions the value relationships between two artists, such as Mozart and Michelangelo, or between two careers, such as one in accounting and one in skydiving, or between two Sunday enjoyments, such as an afternoon at the museum and one hiking in the woods [15].

Chang's characterization of the parity relation is sketchy and not very clear. She does not, for example, discuss the logical properties of the relation. Symmetry should surely hold. If a is on a par with b, then b is on a par with a. Further, since parity is assumed to be incompatible with value equality, and since equality is a reflexive relation, parity must be irreflexive. A symmetric and irreflexive relation cannot be transitive. Since *in*transitivity is out of the question, parity, at least as conceived by Chang, thus has to be nontransitive; i.e., neither transitive nor intransitive. (Chang has confirmed, in personal communication, that she understands the relation as symmetric, irreflexive, and nontransitive.)

Given that parity, as understood by Chang, has these logical properties, it can be defined in terms of the standard value relations. We retain our assumption that $\succsim$

is a quasi-order on the relevant set X of value bearers, and introduce the following definitions:

D10. An item $a \in X$ is an "upper semibound" of a set $S \subseteq X$ iff there is no $b \in S$, such that $b \succ a$.
D11. An item $a \in X$ is a "minimal upper semibound" of a set $S \subseteq X$ iff there is no upper semibound b of S, such that $a \succ b$.
D12. An item $a \in X$ is a "lower semibound" of a set $S \subseteq X$ iff there is no $b \in S$, such that $a \succ b$.
D13. An item $a \in X$ is a "maximal lower semibound" of a set $S \subseteq X$ iff there is no lower semibound b of S, such that $b \succ a$.

Next, we define a relation $\succcurlyeq$, which we may call "almost better than":

D14. $a \succcurlyeq b$ iff a is either (i) a minimal upper semibound of the set of $c \in X$, such that $\neg(c \succ b)$, or (ii) a maximal lower semibound of the set of $d \in X$, such that $\neg(b \succ d)$.

Let us say that a is "almost worse than" b iff $b \succcurlyeq a$. With the help of $\succcurlyeq$, we define "on a par with", denoted by $\asymp$:

D15. $a \asymp b$ iff (i) $\neg(a \succsim b) \wedge \neg(b \succsim a)$, and (ii) $(\exists c)$ $(c \succ a \wedge c \succcurlyeq b$, or $c \succ b \wedge c \succcurlyeq a$, or $a \succ c \wedge b \succcurlyeq c$, or $b \succ c \wedge a \succcurlyeq c)$.

Less formally put, two items are on a par just in case neither is at least as good as the other, but there is a third item that is either better than one of them and almost better than the other, or worse than one of them and almost worse than the other.

Given the following four assumptions, of which (9) is the axiologically most significant one, it can be shown that D15 yields a *necessary* condition for parity:

(7) If a and b are on a par, then $\neg(a \succsim b) \wedge \neg(b \succsim a)$.
(8) If $(\forall c)$ $(c \succ a$ iff $c \succ b$, and $a \succ c$ iff $b \succ c)$, then a and b are not on a par.
(9) If a and b are on a par, there is an item that is either (i) better than any item that is better than exactly one of a and b (call such an item "superior" to a and b), or (ii) worse than any item that is worse than exactly one of a and b (call such an item "inferior" to a and b).
(10) Every nonempty set $S \subseteq X$ with an upper (lower) semibound has a minimal upper (maximal lower) semibound.

Assumptions (7) to (10) imply that if a and b are on a par, then $a \asymp b$. Suppose that two items a and b are on a par. Hence, by (7), they are not standardly related. By (9), there is a superior or an inferior item, relative to a and b. Suppose that there is a superior item. By (8), there is an item that is better, or an item that is worse, than exactly one of a and b. Assume, first, that the former possibility obtains; e.g., that there is an item that is better than a, but not better than b. Let $S = \{c\colon c \succ a \wedge \neg(c \succ b)\}$. Since there is a superior item, S has an upper semibound. Hence, by (10), S has a minimal upper semibound, e. Let $S^* = \{d\colon \neg(d \succ b)\}$. We shall show, by *reductio*, that e is a minimal upper semibound of S^*, implying that $e \succcurlyeq b$ and $a \asymp b$.

If e is not an upper semibound of S^*, there is an $f \in S^*$, such that $f \succ e$. Since $e \succ a$, transitivity yields that $f \succ a$. But then $f \in S$, contradicting the assumption that e is an upper semibound of S. Hence, e is an upper semibound of S^*. If e is not a *minimal* upper semibound of S^*, there is an upper semibound g of S^*, such that $e \succ g$. But, since $S \subseteq S^*$, if g is an upper semibound of S^*, it is an upper semibound of S. This contradicts the assumption that e is a minimal upper semibound of S. Hence, e is a minimal upper semibound of S^*. By D14, therefore, $e \not\succcurlyeq b$. Since $e \succ a$, it follows that $a \asymp b$.

Now, suppose instead that there is no item that is better than exactly one of a and b. By (8), there is then an item that is worse than exactly one of the two items. Assume, thus, that $a \succ c \wedge \neg(b \succ c)$. Since there is a superior item, there is an item better than b. Further, since by assumption, any item that is better than b is better than a, and since $a \succ c$, it holds, for all d, that $d \succ b$ implies $d \succ c$. Hence, b is an upper semibound of the set $S = \{d: \neg(d \succ c)\}$. Moreover, since $b \in S$, b is a *minimal* upper semibound of S. It follows that $b \not\succcurlyeq c$. Since $a \succ c$, we conclude that $a \asymp b$.

We have thereby shown that if a and b are on a par, assumptions (7), (8) and (10) imply that $a \asymp b$, given the existence of an item superior to a and b. A similar argument shows that $a \asymp b$ follows from (7), (8) and (10), if there is an inferior item. Hence, (7) to (10) imply that if a and b are on a par, then $a \asymp b$. That is, D15 states a necessary condition for parity.

Does D15 also state a *sufficient* condition? In other words, is it always the case that if $a \asymp b$, then a and b are on a par? For this to hold, the following two claims must be true:

(11) If $a \asymp b$, then a and b are comparable.
(12) If $\neg(a \succsim b) \wedge \neg(b \succsim a)$, and a and b are comparable, then they are on a par.

Chang argues explicitly for (12), and (11) appears plausible given other assumptions she makes. Hence, D15 seems to be a satisfactory definition of Chang's notion of parity. [Carlson [9] contains a discussion of the plausibility of assumptions (7), (8), (9), (11) and (12), as well as of a different version of (10).]

References and Recommended Readings[1]

1. Arrhenius, G. (2005). Superiority in Value. *Philosophical Studies, 123*, 97–114.
2. Arrhenius, G., & Rabinowicz, W. (2005). Millian Superiorities. *Utilitas, 17*, 127–146.
3. Beardsley, M. (2005 [1965]). *Intrinsic value* (pp. 61–75). In Rønnow-Rasmussen and Zimmerman (2005).
4. Bentham, J. (1996 [1789]). *An introduction to the principles of morals and legislation*. Oxford: Clarendon Press.
5. Brentano, F. (1969 [1907]). Loving and hating. In R. M. Chisholm (Ed.), *The origin of our knowledge of right and wrong*. New York: Routledge and Kegan Paul.

[1] Asterisks (*) indicate recommended readings.

6. Broome, J. (2004). *Weighing lives*. Oxford: Oxford University Press.
7. Carlson, E. (2008). Extensive measurement with incomparability. *Journal of Mathematical Psychology, 52*, 250–259.
8. Carlson, E. (2010a). Generalized extensive measurement for Lexicographic orders. *Journal of Mathematical Psychology, 54*, 345–351.
9. Carlson, E. (2010b). Parity Demystified. *Theoria, 76*, 119–128.
10. Carlson, E. (2011a). Non-Archimedean extensive measurement with incomparability. *Mathematical Social Sciences, 62*, 71–76.
11. Carlson, E. (2011b). Defining goodness and badness in terms of betterness without negation. In E. Dzhafarov & L. Perry (Eds.), *Descriptive and normative approaches to human behavior* (pp. 51–66). Hackensack: World Scientific.
12. Carlson, E. (2015). Organic Unities. In I. Hirose & J. Olson (Eds.), *Oxford handbook of value theory* (pp. 285–299). Oxford: Oxford University Press.
13. Carlson, E. (2016). 'Good' in terms of 'Better'. *Noûs, 50*, 213–223.
14. Chang, R. (Ed.). (1997). *Incommensurability, incomparability, and practical reason*. Cambridge, MA/London: Harvard University Press.
15. * Chang, R. (2002). The possibility of parity. *Ethics, 112*, 659–688. [Seminal paper, arguing that two items may be "on a par", and hence evaluatively comparable, although neither is at least as good as the other.]
16. Chisholm, R. M. (2005 [1986]). Organic unities (pp. 305–318). In Rønnow-Rasmussen and Zimmerman (2005).
17. * Chisholm, R. M. & Sosa, E. (1966). On the logic of 'intrinsically better'. *American Philosophical Quarterly, 3*, 244–249. [Classic discussion of how to define goodness and badness in terms of betterness.]
18. Danielsson, S. (1997). Harman's equation and the additivity of intrinsic value. In L. Lindahl, P. Needham, & R. Sliwinski (Eds.), *For good measure: Philosophical essays dedicated to Jan Odelstad on the occasion of his fiftieth birthday* (pp. 23–34). Uppsala: Uppsala University.
19. Danielsson, S. (2005 [1998]). Harman's equation and non-basic intrinsic value (pp. 371–378). In Rønnow-Rasmussen and Zimmerman (2005).
20. Hansson, S. O. (1990). Defining 'good' and 'bad' in terms of 'better'. *Notre Dame Journal of Formal Logic, 31*, 136–149.
21. Krantz, D. H., Duncan Luce, R., Suppes, P., & Tversky, A. (1971). *Foundations of measurement, vol. 1: Additive and polynomial representations*. New York/London: Academic Press.
22. Moore, G. E. (1903). *Principia Ethica*. Cambridge: Cambridge University Press.
23. Packard, D. J. (1975). A note on Wittgenstein and cyclical comparatives. *Analysis, 36*, 37–40.
24. Quinn, P. L. (2005 [1977]). Improved foundations for a logic of intrinsic value (pp. 241–248). In Rønnow-Rasmussen and Zimmerman (2005).
25. Quinn, W. S. (1990). The puzzle of the self-torturer. *Philosophical Studies, 59*, 79–90.
26. Rachels, S. (1998). Counterexamples to the transitivity of *Better Than*. *Australasian Journal of Philosophy, 76*, 71–83.
27. Regan, D. (1997). Value, comparability, and choice (pp. 129–150). In Chang (1997).
28. * Roberts, F. S. (2009 [1979]). *Measurement theory: With applications to decisionmaking, utility, and the social sciences*. Cambridge: Cambridge University Press. [Comprehensive but relatively accessible introduction to measurement theory, with applications relevant to value theory.]
29. * Rønnow-Rasmussen, T. & Zimmerman, M. J. (eds.). (2005). *Recent work on intrinsic value*. Dordrecht: Springer. [Useful collection of papers on formal and substantial axiology.]
30. Ross, W. D. (1930). *The right and the good*. Oxford: Clarendon Press.
31. Temkin, L. S. (1996). A continuum argument for intransitivity. *Philosophy & Public Affairs, 25*, 175–210.
32. * Zimmerman, M. J. (2001). *The Nature of Intrinsic Value*. Lanham: Rowman & Littlefield. [In-depth treatment of the structural properties of intrinsic value.]

Chapter 29
Preference and Choice

Sven Ove Hansson

Abstract Preferences and choices have central roles in moral philosophy, economics, and the decision sciences in general. In a formal language we can express and explore the properties of preferences, choices, and their interrelations in a precise way, and uncover connections that are inaccessible without formal tools. In this chapter, the plausibility of different such properties is discussed, and it is shown how close attention to the logical details can help dissolve some apparent paradoxes in informal and semi-formal treatments.

29.1 Philosophical Problems of Preference

Comparative terms such as "better" and "equally good" have a prominent role both in everyday discussions and in specialized treatments of value in philosophy and economics. In spite of being understood by all of us, the meaning of these terms is in much need of clarification, as can be seen from the following three examples:

The Paradox of the Outvoted Democrat, Wollheim's Paradox (Wollheim [29]) Susan has worked hard in a campaign to save the regiment in her hometown from a close-down. When Parliament finally decides to close it down, she is very disappointed since she would very much prefer the regiment to be kept rather than being closed down. But she is also a strong supporter of democratic decision-making, and she would certainly not support the half-baked plans of some young officers to defy the decision and carry on as usual. Hence, she prefers that the

---

S. O. Hansson (✉)
Division of Philosophy, Royal Institute of Technology (KTH), Stockholm, Sweden
e-mail: soh@kth.se

S. O. Hansson, V. F. Hendricks (eds.), *Introduction to Formal Philosophy*, Springer Undergraduate Texts in Philosophy, https://doi.org/10.1007/978-3-319-77434-3_29

regiment be closed down rather than being kept. How can she at the same time prefer the closing down of the regiment to its continued existence and the other way around?

Preference Holism On the face of it, what we prefer and disprefer are small components of the world. I may prefer a cup of tea to a cup of coffee, or listening to Beethoven's second rather than his first symphony. But preferences referring to such small items presuppose that these items can be exchanged in isolation from the rest of the world, which may not always be the case. In a sense, such preferences are always conditional on what the rest of the world is like. I would not prefer having tea to having coffee if I had a medical condition that made tea poisonous to me. If, for some unknown reason, there will be more human suffering in the world if I listen to the second symphony rather than the first, then presumably I will prefer listening to the first. The only preferences that can hold unconditionally would be preferences that refer to the complete state of the world. But then, what can we mean by such holistic preferences and what is the relation between them and the more ordinary types of preferences that we express in our common lives?

Preferences and Choices

- Congratulations! You have won the lottery. You can now choose between a trip to London and one to Paris. Which do you choose?
- I choose to go to Paris.
- Why do you prefer a trip to Paris rather than one to London?'
- I don't.
- I'm sorry, I don't understand. I thought you said you have chosen Paris.
- Yes I did. I choose Paris but I prefer London.
- What do you mean? How can you choose one and prefer the other? Are you sure that Paris is your choice?
- Of course. What's the problem?

We expect preferences and choices to cohere. But since preferences and choices are quite different entities, it is not fully clear what it means for them to cohere. And if they do, what are the effects of their mutual coherence on the structure of our preferences, and on the structure of our choices?

With preference logic we can solve these and other problems of preference and choice. As a bonus, preference logic opens up new philosophical issues and insights that would not otherwise have been available to us.

29.2 The Basic Concepts of Preference Logic

Preference logic makes use of three comparative value concepts, namely "better" (strict preference), "equal in value to" (indifference), and "at least as good as" (weak preference). They are usually denoted by the symbols $>$, $\sim$, respectively $\geq$ (or by P, I, respectively R).

This formal language is idealized in several ways. In ordinary language we make a distinction between the subjective notion of preference and the presumably more objective notion of betterness. If you say that *Falstaff* is better than *Aida*, then you indicate a more objective or at least more generally applicable standpoint than if you say that you prefer *Falstaff* to *Aida*. In preference logic, no distinction is made between these two notions since they are assumed to have the same formal structure.

Furthermore, $A > B$ is taken to represent "B is worse than A" as well as "A is better than B" [3]. This is not in exact accordance with ordinary English. I consider the *Magic Flute* to be a better opera than *Idomeneo*, but it would be misleading to say that I consider *Idomeneo* to be worse than the *Magic Flute*. We tend to use "better" when focusing on the goodness of the higher-ranked of the two alternatives, and "worse" when emphasizing the badness of the lower-ranked one [7, p. 13]. This distinction is not made in preference logic.

Preference logic is devoted to the preferences of rational individuals. Therefore, if a proposed principle for preference logic does not correspond to how we actually think and react, then this may either be because the principle is wrong or because we are not fully rational in some of the cases it covers.

29.3 The Set of Alternatives

The objects of preference are represented by the *relata* (*alternatives*) of the preference relation, i.e. A and B in $A > B$. They can be taken to be primitive objects with no further structure. However, in economics they are usually vectors that represent bundles of goods. In philosophical logic, they are usually sentences (or propositions) representing states of affairs. Sentences appear to be the best general-purpose representation that we have for the objects of our preferences. If Xiuxiu prefers fish to meat, then that can be expressed as a preference for (the state of affairs expressed in) "Xiuxiu eats fish" over (that expressed in) "Xiuxiu eats meat". Sentences can also be combined to form composite relata. Hence, a preference for drinking white wine (w) rather than red whine (r) when eating fish (f) can be expressed with conjunctive relata: $(w \wedge f) > (r \wedge f)$.

In order to specify a system of preference logic it is necessary to state what its relata are, in other words to identify its *alternative set*. Many problems in informal discussions on preferences can only be clarified if a precisely defined set of alternatives is introduced. (See Sect. 29.7.)

29.4 Constitutive Logical Properties

The following properties of the three comparative relations are taken to be part of the very meaning of preference and indifference:

(1) $A > B \rightarrow \neg(B > A)$ (asymmetry of strict preference)
(2) $A \sim B \rightarrow B \sim A$ (symmetry of indifference)
(3) $A \sim A$ (reflexivity of indifference)
(4) $A > B \rightarrow \neg(A \sim B)$ (incompatibility of preference and indifference)
(5) $A \geq A$ (reflexivity of weak preference)
(6) $(A \geq B) \leftrightarrow (A > B) \vee (A \sim B)$
(7) $(A > B) \leftrightarrow (A \geq B) \wedge \neg(B \geq A)$
(8) $(A \sim B) \leftrightarrow (A \geq B) \wedge (B \geq A)$

Using these properties, we can simplify the formal structure in either of two ways. One option is to use $>$ and $\sim$ as primitive notions, i.e. notions not defined in terms of any other notions. We can then define $\geq$ according to (6). The other option is to use $\geq$ as a primitive notion and define $>$ and $\sim$ according to (7) and (8). Formally, this works out equally well in both directions. If we use $>$ and $\sim$ as primitives, assume that they satisfy (1)–(4) and define $\geq$ according to (6), then the remaining properties (5), (7) and (8) all hold. Conversely, if we use $\geq$ as the sole primitive, assume that it satisfies (5) and define $>$ and $\sim$ according to (7) and (8), then the remaining properties (1)–(4) and (6) can easily be shown to hold [23].

The choice between these two ways to simplify the logic is fairly inconsequential. Using $\geq$ as the sole connective is preferable from the viewpoint of formal simplicity, but the use of $>$ and $\sim$ seems more conducive to conceptual clarity.

When $>$ and $\sim$ are defined from $\geq$ via (7) and (8) they are called the *strict part*, respectively the *symmetric part*, of $\geq$.

29.5 Completeness

In most applications of preference logic, it is taken for granted that the following property is satisfied:

$(A \geq B) \vee (B \geq A)$ (completeness or connectedness)

Given (2) and (6) it is equivalent with:

$$(A > B) \vee (A \sim B) \vee (B > A)$$

The assumption of completeness is often convenient since it provides us with a formal structure that is easier to work with. However, as shown in Box 29.1, in many situations it seems perfectly rational to have incomplete preferences.

Box 29.1 Incomplete preferences

1. Lack of information
Filomena does not know anything about *Falstaff* or *Aida*. Therefore she does not consider one of them to be better than the other and neither does she consider them to be of equal value.

2. Insufficiently specified alternatives
When asked which he prefers, £500 or that his daughter gets a better grade in math, Ali has no answer to give. The reason is that the comparison is insufficiently specified. Does he have an offer to bribe the teacher? Or is he offered an extra course for his daughter that he only has to pay if she gets a better grade?

3. Costliness of acquiring preferences
There are about 90 brands of cheese in the local grocery store. In order to make her preference relation complete over all of these brands, Alice would have to buy samples of all of them and engage in extensive comparative testing. Since she is not very fond of cheese, making her preferences complete would not be worth the effort or the costs.

4. Morally questionable preferences
José has a nightmare in which a terrorist forces him to choose which of his three children will be killed. If he makes no choice, then all three of them will be killed. When he wakes up, José realizes that in such a situation, he would have to make a choice. However, he feels that he would be a worse person if he knew beforehand what his preferences would then be.

29.6 Transitivity

By far the most discussed logical property of preferences is the following:

$(A \geq B) \wedge (B \geq C) \rightarrow (A \geq C)$ (transitivity of weak preference)

It logically implies a whole herd of similar but logically weaker properties such as the following [23]:

$(A \sim B) \wedge (B \sim C) \rightarrow (A \sim C)$ (transitivity of indifference)
$(A > B) \wedge (B > C) \rightarrow (A > C)$ (transitivity of strict preference)
$(A \sim B) \wedge (B > C) \rightarrow (A > C)$ (IP-transitivity)
$(A > B) \wedge (B \sim C) \rightarrow (A > C)$ (PI-transitivity)
There is no series $A_1, \ldots, A_n$ of alternatives such that $A_1 > \ldots > A_n > A_1$ (acyclicity).

Transitivity is often taken to be an obvious requirement of rationality. If I consider A to be at least as good as B, and B at least as good as C, could there be any reason for me not to consider A to be at least as good as C? Well in fact there could. Box 29.2 exhibits the major types of examples that have been used as arguments against transitivity.[1]

Box 29.2 Intransitive preferences

1. Indistinguishable differences add up and become distinguishable
Aaron cannot taste the difference between wines A and B or between wines B and C, but he is able to taste the difference between A and C, and he likes A better [5, p. 34].

2. Negligible differences add up and become relevant
A self-torturer has a pain-inducing device implanted in her body. The device has 1001 settings, from 0 (off) to 1000. Each increase leads to a noticeable but negligible increase in pain. Each time that she advances the dial by one setting she receives £10,000, but there is no way for her to retreat. In the end the pain is so unendurable that she would gladly relinquish her fortune and return to 0 [17].

3. A trifle does not affect comparisons between disparate objects
A boy is indifferent between receiving a bicycle or a pony, and he is also indifferent between receiving a bicycle with a bell and a pony. However, he prefers receiving a bicycle with a bell to receiving just a bicycle [13].

4. Diverging preferences over several dimensions are reduced to one dimension
In an experiment performed in the 1950s, 62 college students were asked several questions about which of two potential marriage partners they preferred. The questions were so arranged that all three pairwise combinations of the following three persons were covered: A who was described as very intelligent, plain looking, and well off, B who was portrayed as intelligent, very good looking, and poor, and C who was reported to be fairly intelligent, good looking, and rich. 17 of the students exhibited the circular preference pattern $A > B > C > A$. This can be explained by these students always choosing the partner who was superior in two out of the three criteria [14]. This appears to be a mechanism at play in many situations where preferences over several dimensions have to be reduced to a single dimension [15].

[1]On preference transitivity, see also Chap. 31.

Sometimes preferences can be constructed from numerical values. Let u be a function that assigns a real number $u(A)$ to each element A of the alternative set. Then u is a *numerical representation* of $\geq$ if and only if it holds for all A and B that:

$$A \geq B \text{ if and only if } u(A) \geq u(B)$$

It has been shown that a preference relation has a numerical representation if and only if it is both complete and transitive. (This only holds under some rather technical conditions, but these conditions are satisfied whenever the alternative set is either finite or countably infinite. [6, pp. 27–29], [20, pp. 109–110])

29.7 The Outvoted Democrat

We are now equipped to deal with the paradox of the outvoted democrat that was mentioned above. Let r denote that the regiment in Susan's hometown stays in place and $\neg r$ that it does not. Susan wants to keep the regiment, so she prefers r to $\neg r$. But a decision to the contrary has been made, and since she wants democratic decisions to be implemented she also prefers $\neg r$ to r.

In order to make sense of this we must observe that r and $\neg r$ are insufficient to describe the alternatives. Susan's preferences also refer to the decision that has been made. Let Dr denote that a democratic decision has been made in favour of r, and similarly $D\neg r$ that a democratic decision has been made in the other direction. Then instead of merely r and $\neg r$ we have to consider the four alternatives $Dr \wedge r$, $Dr \wedge \neg r$, $D\neg r \wedge r$, and $D\neg r \wedge \neg r$. We can expect her preferences to be as in Fig. 29.1.

Our next task is to derive preferences for r and $\neg r$ from these preferences over composite states of affairs. It is reasonable to assume that these should be preferences *ceteris paribus* or all things being equal.

What does it mean to prefer r to $\neg r$ "everything else being equal"? Since $D\neg r$ holds in the actual world, one interpretation is "when $D\neg r$ is not changed". Susan prefers $D\neg r \wedge \neg r$ to $D\neg r \wedge r$, i.e. she prefers $\neg r$ to r when $D\neg r$ is kept constant. This accounts for her preference for $\neg r$ over r.

But there is also another interpretation. Alternatively, the background factor to be kept constant is whether or not the democratic decision is respected. Let us introduce the predicate R such that Rr means that the democratic decision with respect to r or $\neg r$ is respected. We can now rewrite the four alternatives. $Dr \wedge r$ is equivalent with $Rr \wedge r$, $D\neg r \wedge r$ with $\neg Rr \wedge r$, etc. This gives rise to the reformulation of the preference ordering that is shown in Fig. 29.2.

$$Dr \wedge r$$
$$\vee$$
$$D\neg r \wedge \neg r$$
$$\vee$$
$$D\neg r \wedge r$$
$$\vee$$
$$Dr \wedge \neg r$$

Fig. 29.1 The outvoted democrat's preferences, as expressed with the predicate D for "A democratic decision has been made to the effect that…". $\vee$ denotes transitive strict preference

$$Rr \wedge r$$
$$\vee$$
$$Rr \wedge \neg r$$
$$\vee$$
$$\neg Rr \wedge r$$
$$\vee$$
$$\neg Rr \wedge \neg r$$

Fig. 29.2 The same preferences, expressed with the predicate R for "The democratic decision concerning … is respected"

Since Susan prefers $Rr \wedge r$ to $Rr \wedge \neg r$, and she also prefers $\neg Rr \wedge r$ to $\neg Rr \wedge \neg r$, there can be no doubt that with this construction, she prefers r to $\neg r$. In this way, the ambiguity of the phrase "everything else being equal" makes it possible for the outvoted democrat to strictly prefer, at the same time, r to $\neg r$ and $\neg r$ to r, without being inconsistent [8].

29.8 Preference Holism

In the example of the outvoted democrat, we had preferences over complete alternatives (such as $Dr \wedge \neg r$) but also preferences over smaller units (such as r). More generally speaking, what is the relation between preferences on these two levels? In most philosophical discussions, the complete alternatives are much larger entities than in the voting example. Usually, they are taken to be possible worlds, i.e. sets of sentences that represent everything that can be said about the state of the world [18].

The most common approach is to assume that there is an underlying, holistic preference relation over the complete alternatives (possible worlds) from which preferences over smaller things are derivable as *ceteris paribus* preferences. This should of course not be seen as a faithful representation of actual deliberative or evaluative processes. Instead, the holistic preference relation should preferably be conceived as a reconstruction used to describe a coherence requirement on preferences. With this caveat, how can preferences over sentences be reconstructed as derivable from underlying preferences over possible worlds?

Georg Henrik von Wright, one of the pioneers of preference logic, attempted to explain the notion of *ceteris paribus* by means of counting differences in terms of logically independent atomic states of the world [28]. He assumed that there are n logically independent states of affairs $p_1 \ldots p_n$. It then holds for each world w and each atom p_k that w contains either p_k or $\neg p_k$. The similarity between worlds is measured by counting the number of atoms about which they agree.

Unfortunately, this simple construction does not work. Its major weakness is that the choice of atomic states is logically arbitrary. Consider the following two ten-atom worlds:

$$w_1 = \mathrm{Cn}(\{p_1, p_2, p_3, p_4, p_5, p_6, p_7, p_8, p_9, p_{10}\})$$

$$w_2 = \mathrm{Cn}(\{\neg p_1, p_2, p_3, p_4, p_5, p_6, p_7, p_8, p_9, p_{10}\})$$

where Cn is the consequence operator that takes us from any set of sentences to the set of all its logical consequences. w_1 and w_2 appear to be very similar, and it seems as if a comparison between them can be used for a *ceteris paribus* comparison between p_1 and $\neg p_1$. But now consider the sentences $r_2, \ldots r_{10}$, so defined that for each of them, $r_k \leftrightarrow (p_1 \leftrightarrow p_k)$. We can then rewrite w_1 and w_2 in the following alternative way:

$$w_1 = \mathrm{Cn}(\{p_1, r_2, r_3, r_4, r_5, r_6, r_7, r_8, r_9, r_{10}\})$$

$$w_2 = \mathrm{Cn}(\{\neg p_1, \neg r_2, \neg r_3, \neg r_4, \neg r_5, \neg r_6, \neg r_7, \neg r_8, \neg r_9, \neg r_{10}\})$$

Written in this way, w_1 and w_2 seem to be quite dissimilar. Since there are no objectively given logical atoms, logic cannot help us to choose between these two ways to compare the two sets. A measure of similarity that can be used to (re)construct *ceteris paribus* preferences will have to make use of more information than what is inherent in the logic.

29.9 Choice Functions

To investigate the relationship between preference and choice we need a formal representation also of the latter concept. The standard representation is a *choice function*. A choice function is defined over a set $\mathcal{A}$ of alternatives, and for each subset of that set it chooses, intuitively speaking, the most choiceworthy alternatives. Formally, C is a choice function for $\mathcal{A}$ if and only if it is a function such that for each subset $\mathcal{B}$ of $\mathcal{A}$, $C(\mathcal{B})$ is a subset of $\mathcal{B}$ that is non-empty if $\mathcal{B}$ is non-empty.

$C(\mathcal{B})$ can have more than one element. Since the alternatives are taken to be mutually exclusive, this does not mean that the agent chooses more than one alternative, only that there is more than one alternative that she is willing to choose. Which of these alternatives she ends up with is a matter of picking rather than choosing [26]. Hence, if a, b, and c are three marriage partners, then $C(\{a, b, c\}) = \{a, b\}$ does not indicate a bigamous proclivity but equal propensities to choose a or b.[2]

Among the rationality properties that have been proposed for choice functions, the following two are arguably the most important ones [22]:

Chernoff (property α) [4]
If $\mathcal{B}_1 \subseteq \mathcal{B}_2$ then $\mathcal{B}_1 \cap C(\mathcal{B}_2) \subseteq C(\mathcal{B}_1)$.

Property β
If $\mathcal{B}_1 \subseteq \mathcal{B}_2$ and $X, Y \in C(\mathcal{B}_1)$, then: $X \in C(\mathcal{B}_2)$ if and only if $Y \in C(\mathcal{B}_2)$

Suppose that we are choosing the best novelists from different categories. According to Chernoff, if one of those chosen in the category of European novelists is French, then (s)he is also one of those chosen in the category of French novelists. According to property β, if one of those chosen in the category of European novelists is French, then all those chosen in the category of French novelists must also be among those chosen in the category of European novelists.

Although these choice principles hold in many cases, it is not difficult to find examples in which they do not seem to hold, see Box 29.3.

---

[2]Obviously, choice functions can be defined so that picking is not needed: A *monoselective* choice function [11] is one that selects a single element out of any non-empty set to which it can be applied.

Box 29.3 Violations of the choice axioms

1. A choice can aim at another position than the top position
If the host offers Hao to take a fruit from a bowl with a big apple, a small apple, and an orange, then he will choose the big apple. However, if there is only a small and a big apple then he will (out of politeness) choose the smaller one [1] (This violates Chernoff.).

2. The alternative set carries information about the alternatives
An acquaintance whom Elena meets in the street offers her to come home to him for tea. In the choice between having tea at his house and going home she intends to opt for the former. But then he adds an additional option, namely to have some cocaine at his house. Among the three alternatives that she now has, she chooses to go home [25] (This violates Chernoff.).

In exactly the same situations, her friend Graciela would be indecisive in the first case (i.e. both having tea and going home are in her choice set), whereas she would choose to go home in the second case. (This violates property β.)

29.10 Preference-Based Choice

Preferences often have the function of guiding our choices. Sometimes it is even maintained that choice is nothing else than revealed preference [19, 21]. In order to clarify what it means for choices to be determined by preferences we can consider a choice function C that is derived from a preference relation $\geq$ as follows:

$$C(\mathcal{B}) = \{X \in \mathcal{B} \mid (\forall Y \in \mathcal{B})(X \geq Y)\}$$

A choice function C is *relational* if and only if it is based on some preference relation $\geq$ in this way. The formal connections are quite neat. It is possible to base a choice function on a given preference relation $\geq$ if and only if $\geq$ satisfies completeness and acyclicity. All such choice functions (i.e. all relational choice functions) satisfy Chernoff. A relational choice function satisfies property β if and only if the underlying preference relation is transitive [22].

29.11 Choice-Based Preference

Conversely, we can take choice as primary and define preferences in terms of a choice function. The obvious way to do this is to identify preference with "choice from two-member sets" [2], as follows:

$$p \geq q \text{ if and only if } p \in C(\{p, q\})$$

If this construction is applied to a choice function that has in its turn been derived from a preference relation in the way shown above, then the original preference relation will be recovered [23].

The definition of preference as (hypothetical) choice is popular among economists. This approach makes it possible to take an agnostic stance on mental processes, and treat preference relations merely as technical means to express well-organized propensities to choose.

From a philosophical point of view this interpretation of preferences is far from unproblematic. We often entertain preferences in matters in which we have no choice. Consider Vladimir who has bought a lottery ticket. He would prefer winning a luxury cruise to the Bahamas rather than winning a gift voucher worth £20,000 in his local grocery store. Since one cannot, by definition, choose to win, preferences such as these cannot be choice-guiding. Indeed, if he were given a choice between the cruise and the voucher, he would choose the voucher. The act of choosing something may have negative characteristics (such as shame at choosing something useless) that the event of winning it does not have [9, p. 22]. Furthermore, the first example in Box 29.3 shows that even in matters where we have a choice, the definition of preference as binary choice gives rise to difficulties in "interpreting preference thus defined as preference in the usual sense with the property that if a person prefers x to y then he must regard himself to be better off with x than with y" [24, p. 15].

29.12 A Central Dilemma in Preference Logic

There are two properties that we have a strong tendency to ascribe to preferences, and yet turn out to be difficult to reconcile [9, pp. 20–23]. One of these is *pairwiseness*: Whether a preference statement such as $A \geq$ B, $A > B$, or $A \sim B$ holds should depend exclusively on the properties of A and B, and not be influenced by other elements of the set of alternatives. It should therefore make no difference if we compare A and B when deliberating on the elements of the set $\{A, B\}$ or when deliberating on the elements of the set $\{A, B, C, D, E\}$. The other property is *choice-guidance*, which means that the logical properties of preferences should be compatible with their use as guides for choosing among the elements of the alternative set.

In combination, these two principles imply the further principle of *binary choice*, i.e. that comparisons of only two alternatives at a time are sufficient to determine a choice among all the alternatives. The examples in Box 29.3 show that this principle is not always plausible.

The tension between pairwiseness and choice-guidance is a central dilemma in the theory of preference. It is also an example of a philosophical insight that could only be gained with the help of formal representations of preferences.

Acknowledgements I would like to thank Karin Edvardsson Björnberg and Philippe Mongin for very useful comments on an earlier version of this text.

References and Proposed Readings

1. Anand, P. (1993). The philosophy of intransitive preference. *Economic Journal, 103*, 337–346.
2. Arrow, K. (1977). Extended sympathy and the possibility of social choice. *American Economic Review, 67*, 219–225.
3. Brogan, A. P. (1919). The fundamental value universal. *Journal of Philosophy, Psychology, and Scientific Methods, 16*, 96–104.
4. Chernoff, H. (1954). Rational selection of decision functions. *Econometrica, 22*, 422–443.
5. Dummett, M. (1984). *Voting procedures*. Oxford: Clarendon Press.
6. * Fishburn, P. C. (1970). *Utility theory for decision-making*. New York: Wiley. [Includes a thorough introduction to the logic of preferences.]
7. * Halldén, S. (1957). *On the logic of 'Better'*. Lund: C.W.K. Gleerup. [An early classic in preference logic.]
8. Hansson, S. O. (1993). A resolution of Wollheim's paradox. *Dialogue, 32*, 681–687.
9. *Hansson, S. O. (2001). *The structure of values and norms*. Cambridge: Cambridge University Press. [Includes detailed investigations of preference holism.]
10. * Hansson, S. O. (2001). Preference logic. In D. Gabbay & F. Guenthner (Eds.), *Handbook of philosophical logic* (Vol. 4, 2nd ed., pp. 319–393). Dordrecht: Kluwer. [Overview of preference logic, including proofs of many central results.]
11. Hansson, S. O. (2013). Maximal and perimaximal contraction. *Synthese, 190*, 3325–3348.
12. * Hausman, D. M. (2012). *Preference, value, choice, and welfare*. Cambridge: Cambridge University Press. [Overview of formal and informal issues relating to preferences, with a particular emphasis on economic applications.]
13. Lehrer, K., & Wagner, C. (1985). Intransitive indifference: The semiorder problem. *Synthese, 65*, 249–256.
14. May, K. O. (1954). Utility and the aggregation of preference patterns. *Econometrica, 22*, 1–13.
15. Mongin, P. (2000). Does optimization imply rationality? *Synthese, 124*, 73–111.
16. *Moulin, H. (1985). Choice functions over a finite set: A summary. *Social Choice and Welfare, 2*, 147–160. [An accessible presentation of the properties of choice functions.]
17. Quinn, W. S. (1990). The puzzle of the self-torturer. *Philosophical Studies, 59*, 79–90.
18. Rescher, N. (1967). Semantic foundations for the logic of preference. In N. Rescher (Ed.), *The logic of decision and action* (pp. 37–62). Pittsburgh: University of Pittsburgh Press.
19. Reynolds, J. F., & Paris, D. C. (1979). The concept of 'Choice' and Arrow's theorem. *Ethics, 89*, 354–371.
20. Roberts, F. S. (1979). *Measurement theory*. In G.-C. Rota (Ed.), *Encyclopedia of mathematics and its applications* (Vol. 7). Reading: Addison-Wesley.
21. Samuelson, P. (1938). A note on the pure theory of consumer behaviour. *Economica, 5*(17), 61–71.
22. Sen, A. (1969). Quasi-transitivity, rational choice and collective decisions. *Review of Economic Studies, 35*, 381–393.
23. *Sen, A. (1970). *Collective choice and social welfare*. San Francisco: Holden-Day. [Accessible treatment of the properties of choice functions and how they relate to preference relations.]
24. Sen, A. (1973). *Behaviour and the concept of preference*. London: London School of Economics, Inaugural Lecture.
25. Sen, A. (1993). Internal consistency of choice. *Econometrica, 61*, 495–521.
26. Ullmann-Margalit, E., & Morgenbesser, S. (1977). Picking and choosing. *Social Research, 44*, 757–785.

27. ∗ von Wright, G. H. (1963). *The logic of preference*. Edinburgh: Edinburgh University Press. [An early classic in preference logic.]
28. von Wright, G. H. (1972). The logic of preference reconsidered. *Theory and Decision, 3*, 140–169.
29. Wollheim, R. (1962). A paradox in the theory of democracy. In P. Laslett (Ed.), *Philosophy, politics and society* (Second Series, pp. 71–87). Oxford: Blackwell.

Chapter 30
Preference Change

Fenrong Liu

Abstract The notion of preference is important in philosophy, decision theory, and many other disciplines. It is the interplay of information and preferences that provides the driving force behind what we actually do. The chapter adds a new focus and argues that preference is not static, instead, it *changes dynamically* when triggered by various kinds of events. We show that how a wide variety of preference changes can be modeled in logic, thereby providing the formal philosopher with a natural extension of the scope of inquiry in the area of preference.

30.1 Introduction

The notion of preference is important in philosophy and many other disciplines. It is the interplay of information and preferences that provides the driving force behind what we actually do. Reasoning about preference to explain or predict behavior has therefore been a long-standing interest of logicians, starting from [25, 75]. Initially perhaps a marginal area, these are now a central part of studies of agency. More generally, notions of preference, choice and utility have long been indispensable in economics, decision theory, and game theory. And there are further philosophical motivations, too. Due to its close relations with normative notions like "good" and "bad", preference is an important topic in deontic reasoning (see e.g., [26, 71]).

Against this background, the present chapter adds a new focus, though still staying within the sphere of logic. The point is that preference is not static, given once and for all. Instead, it *changes dynamically* when triggered by various kinds of events, such as learning new facts or accepting new commands. What we really need to understand then, is not just reasoning about given preferences, but also about the various changes that these can undergo. Logical investigations of preference change started in the 1990s, and by now there is a small, but growing body of literature:

F. Liu (✉)
Department of Philosophy, Tsinghua University, Beijing, China
Tsinghua University – The University of Amsterdam JRC for Logic, Beijing, China
e-mail: fenrong@tsinghua.edu.cn

S. O. Hansson, V. F. Hendricks (eds.), *Introduction to Formal Philosophy*, Springer Undergraduate Texts in Philosophy, https://doi.org/10.1007/978-3-319-77434-3_30

see [24, 27, 43, 70]. In this chapter, our focus will be on preference dynamics, and how this can be captured by logical modeling. For reasons of unified exposition, we will use the dynamic-epistemic paradigm of [44], but we do provide references to other equally valid and attractive approaches. As usual, a formal logical analysis helps us describe actual reasoning scenarios, but it can also clarify the philosophical foundations of a field. Our aim is to show that logical models of preference change can help with both.

30.2 A Budget of Preference Changes

Many authors have pointed out the importance and ubiquity of preference dynamics, coming in several varieties. And if our preferences change, this also has implications for related notions such as *rationality* viewed as choosing one's 'sufficiently good' available actions. Changes in preference imply changes in rational actions. Or, in a deontic setting, changes in the 'betterness order' of situations imposed by a moral authority imply changes in our obligations: moral behavior is a dynamic process.

Intrinsic preference change While the general uses of preference are wide-ranging, we follow the literature in citing small domestic examples. For a start, some preference changes occur spontaneously, or at least, without a clear reason [27]:

Example 1 "I grow tired of my favorite brand of mustard A, and start to like brand B better."[1]

One might prefer a more lofty example here, such as suddenly repenting of one's criminal past and its evaluation of alternative situations. Preference changes of this primitive kind are ubiquitous, and they also occur in economics and cognitive science. But one can also think of deontic scenarios. Suppose that some moral authority states a new norm, and I decide to obey it – perhaps just one command "Pay your taxes". This changes my evaluation ordering of the relevant situations: earlier ties may now acquire a deontic preference order.[2]

Information driven preference change Other preference changes have more structure, with incoming information as a trigger. Van Benthem and Liu [69] give the example of taking a trip, where preferences may change because of new information. This example of [40] is in the same line:

Example 2 "Initially, I desire to eat sushi from this plate. Then I learn that this sushi has been made with old fish. Now I desire not to eat this sushi."

[1]One might argue that this preference change has a *cause*, if not a reason: 'getting bored' with what we have. But we will not pursue this line.

[2]One might also say that this preference change is triggered by information that a moral authority made the statement.

This is close to von Wright's example of preferring hock to claret, with a doctor telling one that claret is better for one's health.

Changes in belief and preference Clearly, then, receiving new information can lead to new preferences. Often, there is an intermediary propositional attitude here: a change in one's beliefs, [30]:

Example 3 "The belief that fluoride prevents dental cavities can lead a person to prefer fluoride toothpaste to others. If she comes to disbelieve this connection, she may well abandon this preference."

Belief changes are seldom spontaneous, they have triggers, [44]:

Example 4 Belief change through information change "Alice considers buying a house, based on low cost (C) and a good neighborhood (N), with the second criterion more important than the first. There is a choice between two houses d_1 and d_2. First Alice prefers d_2 over d_1 because she believes that, though both houses have equally good neighborhoods, d_2 is cheaper. But then Alice reads in a reliable newspaper that not $N(d_2)$, and accepting this information, she changes her beliefs – while her preference switches to house d_1.

Preference change involving time and world change But there are yet other triggers. When the world changes, our preference may change along with it:

Example 5 [30] "Consider a person who prefers one apple today to two apples tomorrow, yet prefers two apples in 51 days to one apple in 50 days."

Here the passage of time is correlated with preference change.[3] A more local example of the same phenomenon appears in [40]:

Example 6 "It is a nice afternoon, and I would like to take a walk. Then it starts to rain. I do not want to have a walk anymore."

Here the change in preferences is triggered by a change in the world. These scenarios can also involve reasons for one's preference:

Example 7 [43] "Alice is going to buy a house. Her criteria are: low cost, high quality, and good neighborhood, in that order. One day, Alice wins a lottery prize of ten million dollars. Now she considers quality most important, then neighborhood, then cost."

This reason-based kind of preference change can be modeled in terms of changing 'priorities' underpinning the preference – as we shall see later.

This survey is by no means complete, and the philosophical literature has many further interesting scenarios of preference change, such as the "Sour Grapes" of [33]. But our point will have been made: preference changes are natural. Moreover, beyond the simple one-step examples that we have shown, preference changes can

[3] This raises delicate issues of consistency of agents' preferences over time, that have been studied in philosophy and economics, see [24].

accumulate over time in the area of *games*, witness the discussion of rationalization strategies in Chap. 12 of [44]. A more empirical example is behavior of players in auctions, which often diverges from a priori equilibrium predictions since preference changes in actual play change the structure of the game [47].

In what follows, we will show how preference change is accessible to techniques from logical dynamics [66].Thus, reasoning about preference change can be added to the existing circle of logics describing given preferences at some moment in time. Next, we show how the varieties of preference change encountered in the above can be defined and sorted out in this logical perspective, by bringing in further features. First, we add reasons for preferences, in the form of a priority structure of relevant criteria that can be modified dynamically. Next we discuss the entanglement of preference with information, knowledge and belief, and analyze the resulting richer dynamics of information and evaluation change.

In each case, we will emphasize main ideas, rather than technical theorems. Also, we will take care to identify choice-points and open problems that become visible in the lens of logical analysis.

30.3 Logical Dynamics of Preference Change

To make the above issues more precise, we will use the logician's standard apparatus of formal models and formal languages. With that in place, as we shall see, we can then also define changes in preference, and investigate repertoires of what might be considered natural preference changes.

Preference models Formal models embody a choice-point in conceptualizing a given notion. In this chapter, we follow standard practice, and say that preferences arise from a comparison between given alternatives, that one can think of as 'worlds' or 'situations'. Thus, preference is typically associated with an ordering of worlds, indicating that one alternative is 'better' than another. This is standard in decision theory and game theory – where the ordering can of course depend on the agent. Preference logics then study the abstract properties of different comparative structures in suitable formal languages [28].

A natural starting point are *modal preference models*, being tuples $\mathfrak{M} = (W, \leq, V)$, where W is a set of possible worlds ('situations', 'states', 'outcomes'), $\leq$ is a reflexive and transitive relation,[4] and V is a valuation assigning truth values to proposition letters at worlds. We read the relation $s \leq t$ as 'world t is at least as

---

[4]In this chapter, we use pre-orders since we want to allow for cases where different worlds are incomparable. Total preference orders, the norm in areas like game theory, provide an interesting specialization for our analysis.

good as world s'. If $s \leq t$ but not $t \leq s$, we call t *strictly better* than s, written $s < t$. If $s \leq t$ and $t \leq s$, then worlds s and t are *indifferent*.[5]

These models support modal languages that can analyze a good deal of the usual reasoning about preference. We refer to the Chap. 29 by Sven Ove Hansson in this Handbook for more on this.

Preference change as model change Modal preference models as such are not yet dynamic. But now suppose an action or event takes place that affects the current preference order. Say, a 'public suggestion' in favor of a proposition φ might cancel any old preference for $\neg\varphi$-worlds over φ-worlds (cf. [69] on "open the door"). Here is how this can be done:

The event changes the current model to one with a new preference order.

Here is how this works more precisely. Given any modal preference model $(\mathfrak{M}, s)$, with actual world s, the public suggestion action $\sharp\varphi$ changes the current preference order $\leq$ as follows. The new preference relation becomes:

$$\leq^* \quad = \quad \leq -\{(s,t) \mid \mathfrak{M}, s \models \varphi \text{ and } \mathfrak{M}, t \models \neg\varphi\}.$$[6]

The philosophical reader may find this analysis a bit crude. A suggestion is a speech act coming from one agent, the preference change is a subsequent voluntary response by, presumably, another agent. Our analysis lumps these together – and what we have really described is an act of *taking a suggestion*. More refined views of speech acts are found in [58], while [77] gives a more structured analysis of deontic actions. However, we can also simplify our reading of the event $\sharp\varphi$. Perhaps, it was just a spontaneous act, as in our first example of a pure preference change in Sect. 30.2, with the agent suddenly acquiring an aversion to $\neg\varphi$-worlds.

Other forms of preference change Stronger preference changes occur as well. Consider a 'strong command' telling us to make sure that φ is better regardless of anything else. Incorporating this wish of some over-riding authority can be modeled in the same style as before with a relation transformer $\Uparrow \varphi$:

Given any modal preference model $(\mathfrak{M}, s)$, the new preference relation $\Uparrow \varphi$ is defined as: "make all φ-worlds become better than all $\neg\varphi$-worlds, whether or not they were better before – but within these two zones, retain the old ordering.[7]

[5]These models are often called 'betterness models' since one may want to reserve the term 'preference' for an induced relation between *propositions* viewed as types of situations. We will not discuss this propositional view of preference, though it is in harmony with our analysis.

[6]This very operation was proposed in the early source [70].

[7]An alternative notation for this and other preference transformations is in terms of 'program notation' for the new relation created our of the old relation R. For radical upgrade, this would be $\Uparrow \varphi(R) := (?\varphi; R; ?\varphi) \cup (?\neg\varphi; R; ?\neg\varphi) \cup (?\neg\varphi; \top; ?\varphi)$.

The space of options Suggestions and radical commands are extremes on a spectrum of preference changes. Many further options exist, often in analogy with relational transformations from the theory of belief revision [23, 59, 64]. In fact, our model transformation format allows for infinitely many varieties of preference change, including many that make no intuitive sense. Getting a better grasp of what are natural preference changes inside this class seems an open problem for philosophical analysis.

What we have described here is a format for single acts or events of preference change.[8] This can be modified to deal with further intuitive aspects of such changes. For instance, the force of changes may depend systematically on features ignored here, such as relative authority of the issuers of commands, or tendencies toward change of the agent whose preference is affected.

Dynamic logics of preference change We have now given a semantic update mechanism for modeling changes in preference. But is there also a systematic logic of such changes? We need a formalism that can do two things: (i) express the relevant changes, and (ii) describe their effects in terms of what agents prefer after the change has taken place. This can be achieved by a syntax in the style of dynamic-epistemic logics.[9] In addition to formulas, there are now *action expressions* $\sharp\varphi$ and $\Uparrow\varphi$ for any formula φ of the language, and also, there are *dynamic modalities* $[A]\psi$ for any action expression A. A typical example is the interpretation of the suggestion modality:

> $[\sharp\varphi]\psi$ is true in a model $\mathfrak{M}$ at world s iff, in the new model after φ has been publicly suggested, ψ holds at s.

The resulting logic is completely axiomatizable over the static base logic of preference [69]. In particular, one key principle is a 'recursion axiom' stating just when a preference modality holds for the new ordering after a suggestion:

- $\langle\sharp\varphi\rangle\langle\leq\rangle\psi \leftrightarrow (\neg\varphi \wedge \langle\leq\rangle\langle\sharp\varphi\rangle\psi) \vee (\langle\leq\rangle(\varphi \wedge \langle\sharp\varphi\rangle\psi))$.

Details are not relevant here. But it is important to see what such axioms do. They express the effect of a preference change on models in terms of what is true in the language of preference, the preferred medium of analysis in the philosophical literature. This dynamic recursion step may be compared with finding the key difference equations that describe the progression of a dynamical system. Once this recursive law is understood, it can also be used to analyze other changes, such as those in derived preferences between propositions (cf. [44]). There is no need for additional principles at this higher propositional level: the reasoning will follow automatically.

---

[8]Hansson [29] looked at the changes caused by multiple or several sentences and showed its relationship with single step change.

[9]We will not aim at full generality in our formulations. Readers can find technical details in [4, 66, 73].

30.4 Other Logical Approaches to Preference Change

The dynamic-epistemic approach is not the only method for dealing with preference change. Here is a quick review of other important approaches.

AGM-style preference revision An early in-depth analysis of preference change is given in [27] using AGM-style belief revision theory. Key operations are *preference revision*, *preference contraction*, but also 'preference addition' and 'preference subtraction', where preferences evolve as alternatives are added to, or removed from a current set of worlds. Postulates were proposed for each of these dynamic operators, including connections between revision and contraction. This has inspired many follow-up studies in philosophy and decision theory, cf. [40, 48, 61], and the recent collection [24].

Dynamic semantics The dynamic update semantics for conditionals proposed by [74] takes the meaning of default conditionals to be systematic changes in a language user's current plausibility ordering among worlds, without eliminating any alternatives, but changing their relative positions. Following this line, [72] study relation changes in a deontic setting after new information comes in – arguing that one knows the meaning of a normative sentence if ones knows the change it brings about in the deontic betterness relation of its recipient. The authors also propose a deontic logic for prescriptive obligations in update semantics.[10]

Agency in computer science and AI Another relevant strand comes from studies of agency. Boutilier and Goldszmidt [11] model conditionals in terms of changes of a current plausibility order that will make the conditional true. More generally, dynamic changes of desires, intentions, and their relations with preference have been studied in the context of planning and agent theory, cf. [14, 36].

These different strands are not in conflict, and they are often related from a technical viewpoint. For instance, the AGM-style approach to change and the dynamic-epistemic one touch at many points, [6, 59, 64] provide various comparisons and merges. Likewise, the first full-blown dynamic-epistemic treatment of belief revision in [2, 3] was inspired by Spohn's 'ranking models'. And there are many more instances of parallel developments and mutual influences.

We have shown how the dynamics of preference change can be modeled in a systematic logical fashion, while also pointing out that there are several legitimate and interconnected ways of doing this. We now turn to two further important aspects of preference, which pose a challenge to our analysis of change so far: reasons for preferences, and the entanglement of preference with information-driven epistemic attitudes like knowledge and belief.

[10] An early source for the idea that conditionals effect model changes is [60], which explains conditionals as changing current rankings among worlds.

30.5 Reasons and Priorities

Reasons for preferences So far, we have analyzed preferences without looking at further 'reasons'. But as [75] pointed out, in contrast to 'intrinsic preferences' which are just there, 'extrinsic' preferences often do have underlying considerations. Here is a simple decision scenario:

Example 8 [44] "Alice is going to buy a house. For her, there are several things to consider: the cost, the quality and the neighborhood, strictly in that order. All these criteria are clear-cut for her. For instance, the cost is good if it is inside her budget, otherwise it is bad. Her decision is then determined by the fact whether the alternatives have the desirable properties, and also by the given order of importance for the properties."

Underlying structures of criteria occur in many places in philosophy. In epistemology, reasons for beliefs or even knowledge are an important feature [55], and belief revision theory employs 'belief bases' for similar purposes (cf. [56], which also elaborates analogies with the economic and decision-theoretic literature). Likewise, deontic 'ideality orderings' of worlds often come with an underpinning in terms of structured moral criteria [68]: the reasons for our current moral evaluation of worlds. Gabbay [19], Grossi [22] also show how currently studied argumentation networks are structures of interdependent reasons with varying priorities.

Priority graphs Criteria affecting preference can be diverse, ranging from general principles to individual facts. Moreover, their ranking can have many patterns, from total orders to pre-orders allowing for incomparable or conflicting considerations. The following formal model allows for all of these.

We will extend the sparse modal preference models $\mathfrak{M} = (W, \leq, V)$ of Sect. 30.3 with their primitive preference order $\leq$ among worlds to richer structures that bring out reasons for this ordering. Our first important notion comes from the pioneering paper [1]:

A *priority graph* $\mathrm{G} = \langle \mathcal{P}, < \rangle$ is a strictly partially ordered set of nodes, labeled by propositions in some relevant language L.

The graph order represents the priorities among the given propositions. What their language will be depends on the application. Often, it is a propositional language describing simple properties of worlds (or objects) – but one may also have priorities among epistemic or other intensional propositions.

Preference among worlds is now induced by the priority structure[11]:

Let $\mathrm{G} = \langle \mathcal{P}, < \rangle$ be a priority graph, and $\mathfrak{M}$ a propositional model in which the language L defines properties of worlds in the domain W of $\mathfrak{M}$. The *induced preference relation* $\leq_{\mathrm{G}}$ is defined as follows:

$$y \leq_{\mathrm{G}} x := \forall P \in \mathcal{P}((Py \rightarrow Px) \vee \exists P' < P(P'x \wedge \neg P'y)).$$

[11] We will state only one way of deriving preferences, taken from [1].

Here is an informal explanation. In principle, the preference relation $y \leq_G x$ wants object x to have every property in the graph G that object y has. But there can be 'compensation'. In case y has P while x does not, this is still admissible, provided there is some property P' with higher priority in the graph where x does better: x has P' while y lacks it. In case the graph G is a total order, this reduces to the well-known *lexicographical ordering* of objects.

Priority graphs have turned out useful in many areas. Andréka et al. [1], Girard [20] and Liu [44] give further uses and technical theory. Here, we just note that, given a priority graph G, the induced preference model $\mathcal{M}_G = (W, \leq_G, V)$ is a natural representation of a reason-based extrinsic preference.

Matters of representation But now a question arises. Given that priority-based preference is a much richer structure than mere betterness order among worlds, have we perhaps lost generality? The answer is negative. De Jongh and Liu [15], Liu [44] prove *representation theorems* showing that every connected preference order on worlds can be represented as induced by a total priority graph, while every pre-order can be induced by some strictly partially-ordered graph. Thus, the usual modal logic of preference models still applies fully. Liu [45] points out that these representations undercut von Wright's distinction between intrinsic and extrinsic preferences, since we can construct a priority structure behind any intrinsic preference order – though its 'reasons' may be artificial. Indeed, the present perspective suggests richer *two-level models* of worlds with a preference order plus a priority graph inducing this order. This offers a much richer way of describing preference reasoning.[12]

Dynamics of priority and calculus of reasons Now let us return to the topic of change. Change in preference is often induced by change in priority structure: new criteria may come in, old ones lose relative force, or may even be deleted entirely. Recall the earlier Example 7, it shows vividly how changes in a priority graph can happen, due to events that have taken place. This priority dynamics can be modeled in terms of *graph change*. Given a priority graph G, there are obvious options for placing a new proposition A. One can make it the highest priority, or the lowest, or one may rank it just side by side with G. The background is a 'calculus of priority graphs' developed algebraically in [1], and in modal logic in [20]. Its main operations are two forms of composition. Given any two priority graphs G, G′,

- the *sequential composition* G; G′ adds the graph G on top of G′ in the order: all nodes in the first come before all those in the second,
- the *parallel composition* G $\|$ G′ is the disjoint union of the graphs G and G′, without any order links between them.

The main conceptual point here is this. Once we have priority graphs encoding reasons for preference, these structured reasons themselves become an explicit object of study. Thus, reasons are a natural complement to preference, their structure

[12] van Benthem and Grossi [67] suggest that classic scenarios from meta-ethics provide systematic cues for extracting, not just deontic inferences as is usually done, but also normative priority structure, as well as relevant changes in both.

deserves separate attention, and their dynamics of change is well within the reach of formal methods.

Connecting changes at two levels How are our two accounts of preference change, one at the level of betterness order, and one at the level of priority structure, related? For our basic examples, a perfect harmony reigns:

> Consider a preference model $\mathfrak{M}$ whose relation $\leq$ is induced by a priority graph G. Taking a suggestion A in $\mathfrak{M}$ gives a new model whose relation is induced by the priority graph G $\parallel$ A, where A is the one-point priority graph with just the proposition A. Next, consider a priority graph G $= (\mathcal{P}, <)$ inducing a preference relation $\leq$ on a model $\mathfrak{M}$. Prefixing a new proposition A to G induces the new preference relation $\Uparrow A(\leq)$.

But this harmony is not always present. Liu [45] finds that natural syntactic operations on priority graphs may lack matching betterness transformers.[13] Thus, in the end, priority graphs offering reasons for preference are the richer perspective from which to understand preference change.

Preference is at the heart of decision and rational choice theories. In recent work at the interface of preference logic, philosophy, and social science, themes such as reason-based preference have come to the fore, with further lines of their own. Dietrich and List [16] point out that, though existing decision theory gives a good account of how agents make choices given their preferences, issues of where these essential preferences come from and how they can change are rarely studied.[14] The authors propose a model in which agents' preferences are based on 'motivationally salient properties' of alternatives, consistent sets of which can be compared using a 'weighing relation'. Two intuitive axioms are identified in this setting that precisely characterize the property-based preference relations. Starting from similar motivations, [51] studies reason-based preference in more complex doxastic settings, drawing on ideas from similarity-based semantics for conditional logic. Essentially, preference results here from agents' comparing two worlds, one having some property and the other lacking it, close to their actual world, and comparing these based on relevant aspects of utility. The framework supports extensive analysis in modal logic, including illuminating results on frame correspondence and axiomatization. Osherson and Weinstein [52] gives an extension to preference in the presence of quantifiers, while [53] makes a link between these preference models and deontic logic.

30.6 Preference, Knowledge and Belief

Entanglement of information and evaluation We have now dealt with several aspects of our various examples of preference change in Sect. 30.2. But one important issue

---

[13] An example is deleting redundant double occurrences of the same proposition.

[14] These are the two main topics of [44].

remains: the intuitive entanglement of an evaluative notion like preference with information-driven epistemic notions like knowledge and belief. This entanglement may take various forms. For instance, I may prefer a certain object to another right now, because I do not yet *know* about some decisive flaw. Or I may prefer taking an umbrella despite the inconvenience of carrying it, if I *believe* that it is going to rain. This mixture of preference with knowledge and belief seems essential to agency. It also emerges in more conceptual questions such as this: is preference subject to epistemic *introspection*? If I have a preference, do I know that I have it? If the answer is positive, the notion of preference must have enough epistemic content to support this inference. In what follows, we will not resolve these issues, but merely show how the models of this chapter can be extended to study such forms of entanglement.

Preference and knowledge Models for preference in the presence of knowledge have been proposed by Fillion [18], Pacuit et al. [54] and other authors. A simple version just merges our basic preference models with standard epistemic ones (cf. [69]):

> Any modal preference model $\mathfrak{M} = (W, \leq, V)$ and epistemic model $\mathfrak{M} = (W, \sim, V)$ with $\sim$ an epistemic accessibility relation over worlds, yield an *epistemic preference model* $\mathfrak{M} = (W, \sim, \leq, V)$.

These models interpret a combined formal language with preference modalities and epistemic knowledge operators, both interpreted as usual. Now entangled statements become expressible mixing knowledge and preference. The following illustrations represent (a) epistemic introspection on 'preference', and (b) a possible tension between preference and knowledge:

(a) $\langle\leq\rangle\varphi \to K\langle\leq\rangle\varphi$: Positive betterness introspection
(b) $\langle\leq\rangle\varphi \wedge K\neg\varphi$: Regret about things that we know cannot be.[15]

Epistemic accessibility models are just one way of modeling knowledge, and alternatives exist, cf. [17, 32, 34, 49]. Our claim is that entanglement with preference makes sense throughout, while the methodology of preference change as model change in this chapter works across a wide variety of such models. The same points can be made about our next topic:

Preference and belief A similar approach works for belief. A convenient format uses *plausibility models*, where worlds in epistemic equivalence classes are ordered by some binary relation of greater plausibility (cf. [5]). This yields finer qualitative distinctions inside epistemic ranges, where belief is truth in all most plausible worlds only. One advantage of this setting is its easy treatment of conditional beliefs, a basic notion not reducible to absolute belief. This time, the simplest merged models are of the form $\mathfrak{M} = (W, \preceq, \leq, V)$, with W a set of worlds, $\preceq$ a doxastic relation 'at least as plausible as', and $\leq$ our earlier relation of 'at least as good as', with V again a valuation for proposition letters. Again, many entangled kinds of

[15] Such statements are crucial to analyzing off-equilibrium play in games.

statement can be expressed in the matching bi-modal language of preference and belief, cf. [44].

A similar model with two relations was also proposed in [10], reading plausibility as 'being as normal as'. As for entangled notions, Boutilier defines *conditional ideal goal* (IG) as saying that ϕ is an ideal goal with condition ψ if and only if the best of the most normal ψ worlds satisfy ϕ. Still deeper forms of entanglement of preference and belief and their formal properties are studied in [39], where preference between propositions refers only to their most plausible worlds.

Changes in knowledge and belief Let us first state briefly how these models for knowledge and belief support changes in both. We start with an action that changes the domain of the model, modeling incoming new 'hard information', which has been the inspiration for much of the dynamic-epistemic literature.

Public announcement logic (PAL) studies the logical rules of knowledge change under public announcements. Given an epistemic model $\mathfrak{M} = (W, \sim, V)$, public announcements $!\varphi$ of true propositions ϕ change the model into a submodel:

> Consider any model $\mathfrak{M}$ where formula ϕ is true at world s. The updated model $(\mathfrak{M}|\phi, s)$ ("$\mathfrak{M}$ *relativized to* ϕ *at* s") is the submodel of $\mathfrak{M}$ whose domain is the set $\{t \in \mathfrak{M} \mid \mathfrak{M}, t \models \phi\}$.

Complete logics for this kind of information change can be found in the literature (cf. [44, 73] on adding preference.) Here is the typical recursion axiom for knowledge after update:

$$\langle !\varphi \rangle \langle K \rangle \psi \leftrightarrow \varphi \wedge \langle K \rangle \langle !\varphi \rangle \psi .$$

However, if we want to model the realistic phenomenon of 'regret' about worlds that are no longer epistemic options, epistemic updates for $!\varphi$ should not remove the $\neg\varphi$-worlds, since we might still want to refer to them. One way of doing this is by changing public announcement $!\varphi$ to a milder relation-changing operation $\dagger\varphi$ of 'link-cutting'. The new model $\mathfrak{M}_{\dagger\varphi}$ is the original $\mathfrak{M}$ with its worlds and valuation unchanged, but with accessibility relations $\sim$ replaced by this subrelation:

> keep only those $\sim$-links that do not cross between the φ- and $\neg\varphi$-zones of $\mathfrak{M}$.

Again, recursion axioms and complete dynamic logics for this operation can be found in a number of places in the literature. Here is the axiom for new knowledge as an illustration:

$$\langle \dagger\varphi \rangle \langle K \rangle \psi \leftrightarrow (\varphi \wedge \langle K \rangle (\varphi \wedge \langle \dagger\varphi \rangle \psi)) \vee (\neg\varphi \wedge \langle K \rangle (\neg\varphi \wedge \langle \dagger\varphi \rangle \psi)).$$

Now let us turn from knowledge change to the dynamics of *belief*. As with preference change, the relevant events change a current order, viz. the plausibility relation. Van Benthem [64] studies two sorts of belief change, *radical* revision $\Uparrow \varphi$ and *conservative* revision, and gives complete dynamic-epistemic axiomatizations.

As before, the effect of these relation transformers on key modalities is stated in recursion axioms describing new belief after the event. As an illustration, here is the recursion axiom for new beliefs under radical upgrade from [64]:

$$[\Uparrow \varphi]B\psi \leftrightarrow (E\varphi \wedge B([\Uparrow \varphi]\psi|\varphi)) \vee B[\Uparrow \varphi]\psi.$$ [16]

Preference change and changes in knowledge and belief Now we are in a position to deal with the mixture of informational and preference dynamics noted in Sect. 30.2. If an entangled notion of preference has both betterness and epistemic modalities, its truth value may be affected by both information and betterness changes. To see how this works, we also need 'mixed recursion axioms', telling us how an information change affects a betterness modality, or vice versa. These tend to be simple commutations.[17] However, we can also merge preference with beliefs in more intimate ways, intersecting plausibility and betterness relations in the model. Liu [44] has a systematic study of entanglement phenomena and their dynamic logics.

With these things in place, the reader has all the equipment needed to analyze all scenarios given in Sect. 30.2. Even so, the above analysis is just a 'proof of concept' leaving many questions. We assumed that the basic relations for betterness, epistemic accessibility, and plausibility are independent. Things change when we impose philosophically motivated *mutual constraints* on these. One must then check that proposed model transformations for preference change respect these constraints – for which there is no general guarantee. Another conceptual issue behind the scenes in Sect. 30.2, and in much of the literature, is this. To which extent are direct changes in a given preference ordering of worlds *encodable* alternatively as results of pure information changes? In other words, do we really need a separate notion of preference change if we can already analyze information changes? We must leave these issues open here.

30.7 Conclusion and Further Directions

This chapter has shown how a wide variety of preference changes can be modeled in logic, thereby providing the formal philosopher with a natural extension of the scope of inquiry in the area of preference, while also providing new tools for dealing with these and related phenomena. Our main emphasis has been on the modeling of change by updating current models, thereby adding a 'dynamic dimension' to existing semantics in the area of philosophical logic. In doing so, we have not

[16] A complete dynamic logic of belief change also needs recursion axioms for conditional beliefs and for the existential modality.

[17] Here are two illustrations, with link-cutting operation as our knowledge update and 'suggestion' as our betterness upgrade: (i) $\langle \dagger\varphi\rangle\langle\leq\rangle\psi \leftrightarrow \langle\leq\rangle\langle\dagger\varphi\rangle\psi.$, (ii) $\langle\sharp\varphi\rangle\langle K\rangle\psi \leftrightarrow \langle K\rangle\langle\sharp\varphi\rangle\psi$.

exhausted the state of the art in the field, and we have skimmed over some challenges that lie ahead for this style of analysis. Here are a few that should be of interest beyond the narrower circle of logicians.

Social settings and groups This chapter has emphasized preference changes for single agents. But natural scenarios are often social, involving more than one agent. In particular, agents are connected to each other by various social relations. An agent often changes her preferences and beliefs because of peer pressure from friends or neighbours or yet other social sources. These phenomena have been studied extensively by logicians in recent years, cf. [7, 13, 21, 31, 41, 46, 76]. Beyond this multi-agent interaction, there is also formation of groups as actors in their own right with information and preferences, a key topic in social epistemology or social choice theory. A fundamental issue here is to understand the tension, or ideally: the cooperation, between individual and group attitudes. There are various proposals to this effect. For instance, voting is a way of creating group preferences attuned with individual preferences [63], deliberation is another, involving informational communication in the process [42], and recently, models of group belief have been proposed based on the evolutionary dynamics of trust and social influence, see [8].

Long-term temporal perspective Preference changes as discussed here were single steps. But 'one at a time' is not enough to understand the essential process nature of many topics important to philosophers, such as the structure of conversation, the functioning of a system of norms, or scientific inquiry. This long-term character is not just a matter of iterating single steps. As has been argued in Bovens and Ferreira [12], Hansson [29], Hoshi [35] 'procedural information' about possible trajectories of the current process may be essential, too. Thus the dynamic logics presented here need to interface with temporal logics of the sort studied in the philosophy of action, for instance those of [9, 62].

Other areas of philosophy We have noted several times that preference change also occurs in normative reasoning, of a moral nature or in terms of practical 'best action'. These are areas where our analysis makes good sense, but much remains to be done in connecting up. Dastani [68] have illustrations in deontic logic, [44, 65] in game theory.

Likewise, the ideas of this chapter make sense in the semantics of natural language, enriching standard accounts of 'dynamic meaning': cf. [38] for a case study of imperatives. A much richer set of examples can be found in [77, 78].

Utility and probability. Entanglement of information and preference is standard in quantitative disciplines using utility and probability, in particular, in decision theory and game theory (cf. for instance, [37, 50, 57]). Our analysis of preference change can be extended to these areas, but it will have to deal with more sophisticated entangled static notions like 'expected value', while the space of dynamic operations on quantitative models is also much vaster. Interfacing logic and probability is natural in many fields, and preference change is definitely one of them.

References

1. Andréka, H., Ryan, M., & Schobbens, P.-Y. (2002). Operators and laws for combining preferential relations. *Journal of Logic and Computation, 12*, 12–53.
2. Aucher, G. (2003). A combined system for update logic and belief revision. Master's thesis, MoL-2003-03. ILLC, University of Amsterdam.
3. Aucher, G. (2008). Perspectives on belief and change. Ph.D. thesis, Université de Toulouse.
4. Baltag, A., Moss, L., & Solecki, S. (1998). The logic of common knowledge, public announcements, and private suspicions. In I. Gilboa (Ed.), *Proceedings of the 7th Conference on Theoretical Aspects of Rationality and Knowledge (TARK 98)* (pp. 43–56).
5. Baltag, A., & Smets, S. (2006). Dynamic belief revision over multi-agent plausibility models. In *Proceedings of the 7th Conference on Logic and the Foundations of Game and Decision Theory (LOFT 06)*, Liverpool.
6. Baltag, A., & Smets, S. (2008). A qualitative theory of dynamic interactive belief revision. In M. W. G. Bonanno & W. van der Hoek (Eds.), *Logic and the foundations of game and decision theory* (Texts in logic and games, Vol. 3). Amsterdam: Amsterdam University Press.
7. Baltag, A., Christoff, Z., Hansen, J. U., & Smets, S. (2008). Logical models of informational cascades. In J. van Benthem & F. Liu (Eds.), *Logic across the university: Foundations and applications, Proceedings of the Tsinghua Logic Conference* (Studies in logic, Vol. 47, pp. 405–432). London: College Publications.
8. Baltag, A., Liu, F., Shi, C., & Smets, S. (2017). Belief aggregation via Markov chain. Ongoing work, ILLC, University of Amsterdam.
9. Belnap, N., Perloff, M., & Xu, M. (2001). *Facing the future*. Oxford: Oxford University Press.
10. Boutilier, C. (1994). Conditional logics of normality: A modal approach. *Artificial Intelligence, 68*, 87–154.
11. Boutilier, C., & Goldszmidt, M. (1993). Revision by conditional beliefs. In *Proceedings of AAAI'93* (pp. 594–599).
12. Bovens, L., & Ferreira, J. L. (2010). Monty Hall drives a wedge between Judy Benjamin and the Sleeping Beauty. A reply to Bovens. *Analysis, 70*(3), 473–481.
13. Christoff, Z. (2016). *Dynamic logics of networks. Information flow and the spread of opinion*. Ph.D. dissertation, ILLC, University of Amsterdam.
14. Dastani, M., Huang, Z., & van der Torre, L. (2000). Dynamic desires. *Journal of Applied Non-classical Logics, 12*, 200–202.
15. de Jongh, D., & Liu, F. (2009). Preference, priorities and belief. In T. Grune-Yanoff & S. Hansson (Eds.), *Preference change: Approaches from philosophy, economics and psychology* (Theory and decision library, pp. 85–108). Dordrecht: Springer.
16. Dietrich, F., & List, C. (2013). Where do preferences come from? *International Journal of Game Theory, 42*(3), 613–637.
17. Dretske, F. (1981). *Knowledge and the flow of information*. Oxford: Blackwell.
18. Fillion, N. (2007). Treating knowledge and preferences in game theory via modal logic. Technical report, The University of Western Ontario.
19. Gabbay, D. M. (2010). Sampling logic and argumentation networks: A manifesto. Manuscript, King's College London.
20. Girard, P. (2008). Modal logics for belief and preference change. Ph.D. thesis, Stanford University.
21. Girard, P., Liu, F., & Seligman, J. (2011). Logic in the community. In *Proceedings of the 4th Indian Conference on Logic and its Applications* (LNCS, Vol. 6521, pp. 178–188). Springer.
22. Grossi, D. (2011). An application of model checking games to abstract argumentation. In J. H. van Ditmarsch & S. Ju (Eds.), *Proceedings of the 3rd International Workshop on Logic, Rationality and Interaction (LORI 2009)*. (FoLLI-LNAI, Vol. 6953 pp. 74–86). Springer.
23. Grove, A. (1988). Two modellings for theory change. *Journal of Philosophical Logic, 17*, 157–170.

24. Grune-Yanoff, T., & Hansson, S. (Eds.) (2009). *Preference change: Approaches from philosophy, economics and psychology* (Theory and decision library). Springer.
25. Hallden, S. (1957). *On the logic of "better"*. Lund: C.W.K. Gleerup. [Pioneer work in preference logic.]
26. Hansson, S. (1990). Preference-based deontic logic. *Journal of Philosophical Logic, 19*, 75–93.
27. Hansson, S. (1995). Changes in preference. *Theory and Decision, 38*, 1–28. [The first work to study preference change using AGM framework.]
28. Hansson, S. (2001). Preference logic. In D. Gabbay & F. Guenthner (Eds.), *Handbook of philosophical logic* (Vol. 4, pp. 319–393). Dordrecht: Kluwer Academic.
29. Hansson, S. (2010). Multiple and iterated contraction reduced to single-step single-sentence contraction. *Synthese, 173*, 153–177.
30. Hansson, S., & Grüne-Yanoff, T. (2006). Preferences. In *Stanford encyclopedia of philosophy*. Stanford. http://plato.stanford.edu/entries/preferences/.
31. Hansen, P. G., & Hendricks, V. F. (2014). *Infostorms: How to take information punches and save democracy*. Cham: Copernicus Books/Springer.
32. Hendricks, V. (2003). Active agents. *Journal of Logic, Language and Information, 12*, 469–495.
33. Hill, B. (2009). Three analyses of source grapes. In T. Grune-Yanoff & S. Hansson (Eds.), *Preference change: Approaches from philosophy, economics and psychology* (Theory and decision library, pp. 27–56). Springer, Dordrecht.
34. Holliday, W. H. (2010). Epistemic logic and relevant alternatives. In M. Slavkovik (Ed.), *Proceedings of the 15th Student Session of the European Summer School in Logic, Language and Information* (pp. 4–16).
35. Hoshi, T. (2009). Epistemic dynamics and protocol information. Ph.D. thesis, Stanford University.
36. Icard, T., Pacuit, E., & Shoham, Y. (2010). Joint revision of beliefs and intentions. In *Proceedings of the Twelfth International Conference on the Principles of Knowledge Representation and Reasoning (KR 2010)* (pp. 572–574). AAAI Publications.
37. Jeffrey, R. (1965). *The logic of decision*. Chicago: University of Chicago Press.
38. Ju, F., & Liu, F. (2011). Update semantics for imperatives with priorities. In J. L. H. van Ditmarsch & S. Ju (Eds.), *Proceedings of the 3nd International Workshop on Logic, Rationality and Interaction (LORI 2011)* (FoLLI-LNAI, Vol. 6953, pp. 127–140). Springer.
39. Lang, J., van der Torre, L., & Weydert, E. (2003). Hidden uncertainty in the logical representation of desires. In *Proceedings of the 18th International Joint Conference on Artificial Intelligence (IJCAI'03)* (pp. 189–231).
40. Lang, J., & van der Torre, L. (2008). From belief change to preference change. In *Proceedings of the 18th European Conference on Artificial Intelligence (ECAI-2008)* (pp. 351–355). [Explore various kinds of preference change, and the connection between belief revision and preference change.]
41. Liang, Z., & Seligman, J. (2011). A logical model of the dynamics of peer pressure. *Electronic Notes in Theoretical Computer Science, 278*, 275–288.
42. List, C. & Pettit, P. (2002). Aggregating sets of judgments: An impossibility result. *Economics and Philosophy, 18*, 89–110.
43. Liu, F. (2008). Changing for the better: Preference dynamics and agent diversity. Ph.D. thesis, ILLC, University of Amsterdam.
44. Liu, F. (2011a). *Reasoning about preference dynamics* (Synthese library, Vol. 354). Springer, Dordrecht.
45. Liu, F. (2011b). A two-level perspective on preference. *Journal of Philosophical Logic, 40*, 421–439. [Provide a richer structure of reason-based preference and study the dynamics of reasons and preference.]
46. Liu, F., Seligman, J., & Girard, P. (2014). Logical dynamics of belief change in the community. *Synthese, 191*(11), 2403–2431.
47. McClure, S. (2011). Decision making. Lecture slides SS100. Stanford University.

48. Nayak, A., Nelson, P., & Polansky, H. (1996). Belief change as change in epistemic entrenchment. *Synthese, 109*(2), 143–174.
49. Nozick, R. (1981). *Philosophical explanations*. Cambridge: Harvard University Press.
50. Osborne, M., & Rubinstein, A. (1994). *A course in game theory*. Cambridge: The MIT Press.
51. Osherson, D., & Weinstein, S. (2012). Preference based on reasons. *The Review of Symbolic Logic, 5*(1), 122–147.
52. Osherson, D., & Weinstein, S. (2014). Quantified preference logic. arXiv:1208.2921
53. Osherson, D., & Weinstein, S. (2014). Deontic modality based on preference. arXiv:1409.0824
54. Pacuit, E., Parikh, R., & Cogan, E. (2006). The logic of knowledge based on obligation. *Synthese, 149*, 311–341.
55. Reisner, A., & Steglich-Petersen, A. (Eds.) (2011). *Reasons for belief*. Cambridge: Cambridge University Press.
56. Rott, H. (2001). *Change, choice and inference: A study of belief and revision and nonmonotonic reasoning*. Oxford: Oxford University Press.
57. Savage, L. (1954). *The foundations of statistics*. New York: Wiley.
58. Searle, J. R., & Veken, D. v. d. (1985). *Foundations of illocutionary logic*. Cambridge: Cambridge University Press.
59. Segerberg, K. (2001). The basic dynamic doxastic logic of AGM'. In M.-A. Williams & H. Rott (Eds.), *Frontiers in belief revision* (pp. 57–84). Dordrecht: Kluwer Academic.
60. Spohn, W. (1988). Ordinal conditional functions: A dynamic theory of epistemic states. In W. Harper & B. Skyrms (Eds.), *Causation in decision, belief change and statistics II* (pp. 105–134). Dordrecht: Kluwer Academic.
61. Spohn, W. (2009). Why the received models of considering preference change must fail. In T. Grune-Yanoff & S. Hansson (Eds.), *Preference change: Approaches from philosophy, economics and psychology* (Theory and decision library, pp. 109–121). Springer, Dordrecht.
62. Thomason, R., & Gupta, A. (1980). A theory of conditionals in the context of branching time. *Philosophical Review, 89*(1), 65–90.
63. Uckelman, J., & Endris, U. (2008). Preference modeling by weighted goals with max aggregation. In G. Brewka & J. Lang (Eds.), *Proceedings of the 11th International Conference on Principles of Knowledge Representation and Reasoning (KR-2008)* (pp. 579–587). Menlo Park: AAAI Press.
64. van Benthem, J. (2007). Dynamic logic for belief revision. *Journal of Applied Non-Classical Logic, 17*, 129–156.
65. van Benthem, J. (2011a). Exploring a theory of play. In K. R. Apt (Ed.), *Proceedings of the 13th Conference on Theoretical Aspects of Rationality and Knowledge (TARK-2011)* (pp. 12–16). ACM.
66. van Benthem, J. (2011b). *Logical dynamics of information and interaction*. Cambridge: Cambridge University Press. [Provide both conceptual and technical introduction to dynamic epistemic logic, as well as its applications.]
67. van Benthem, J., & Grossi, D. (2011). Normal forms for priority graphs. Technical Report PP-2011-02, ILLC, University of Amsterdam.
68. van Benthem, J., Grossi, D., & Liu, F. (2010). Deontics = betterness + priority. In G. Governatori & G. Sartor (Eds.), *Proceedings of the 10th International Conference on Deontic Logic in Computer Science, DEON 2010* (LNAI, Vol. 6181, pp. 50–65). Springer.
69. van Benthem, J., & Liu, F. (2007). Dynamic logic of preference upgrade. *Journal of Applied Non-Classical Logic, 17*, 157–182.
70. van Benthem, J., van Eijck, J., & Frolova, A. (1993). Changing preferences. Technical Report, CS-93-10, Centre for Mathematics and Computer Science, Amsterdam.
71. van der Torre, L. (1997) Reasoning about obligations: Defeasibility in preference-based deontic logic. Ph.D. thesis, Rotterdam.
72. van der Torre, L., & Tan, Y. (1999). An update semantics for deontic reasoning. In P. McNamara & H. Prakken (Eds.), *Norms, logics and information systems* (pp. 73–90). Amsterdam: IOS Press.

73. van Ditmarsch, H., van der Hoek, W., & Kooi, B. (2007). *Dynamic epistemic logic*. Berlin: Springer.
74. Veltman, F. (1996). Defaults in update semantics. *Journal of Philosophical Logic, 25*, 221–261.
75. von Wright, G. (1963). *The logic of preference*. Edinburgh: Edinburgh University Press. [Foundational work in preference logic]
76. Xue Y., & Parikh, R. (2015). Strategic belief updates through influence in a communit. *Studies in Logic, 8*,124–143.
77. Yamada, T. (2008). Logical dynamics of some speech acts that affect obligations and preferences. *Synthese, 165*(2), 295–315.
78. Yamada, T. (2010). Scorekeeping and dynamic logics of speech acts. Manuscript, Hokkaido University.

Chapter 31
Money-Pumps

Sven Ove Hansson

Abstract A money-pump is a thought experiment involving a person whose preferences form a circle. Repeatedly, she pays some money to go from one alternative to another that she likes better. When she has paid her way around the full circle she is back at the starting-point, but with less money. Money-pumps have been used to show that certain preference patterns are irrational since they make a person exploitable. Formal tools can be used to analyze money-pumps in a precise manner, distinguish between different types of money-pumps, and investigate decision strategies to avoid their pernicious effects. Although money-pumps are rather contrived constructions, these investigations have practical relevance since there seem to be decision situations in the real world with the same structure.

31.1 Introduction

Joan is a dedicated stamp-collector with a strong urge to own the stamps she likes the most. There are three stamps – we can call them a, b, and c – that she has quite determined but perhaps also somewhat unusual attitudes to: She prefers a to b, b to c, and c to a.

One day she enters a stamp shop with stamp a. The stamp-dealer offers her to trade in a for c, if she pays 1 euro. She willingly accepts the deal.

Next, the stamp-dealer takes out stamp b from a drawer, offering her to swap c for b, against another payment of 1 euro. She accepts. But when she is on her way out of the shop, the dealer calls her back and advises her that it only costs 1 euro to change back to a, the very stamp that she had in her pocket when she entered the shop. Since she prefers it to b, she pulls out a third euro coin. She walks out of the shop with the same stamp as when she

S. O. Hansson (✉)
Division of Philosophy, Royal Institute of Technology (KTH), Stockholm, Sweden
e-mail: soh@kth.se

S. O. Hansson, V. F. Hendricks (eds.), *Introduction to Formal Philosophy*, Springer Undergraduate Texts in Philosophy, https://doi.org/10.1007/978-3-319-77434-3_31

entered, 3 euros poorer and presumably content to have made three good deals in just a few minutes.

No explanation is needed why this is called a money-pump. Presumably, if Joan had stayed in the shop the dealer could have repeated the procedure indefinitely, pumping all the money she had away from her. Nor can there be any doubt that it is the unusual structure of her preferences (cyclic strict preferences) that made her susceptible to this form of exploitation. Money-pumps were invented by Frank P. Ramsey [20, p. 182] as a thought experiment intended to show the irrationality of intransitive preferences.[1] In a classical formulation:

> "Suppose an individual prefers y to x, z to y, and x to z. It is reasonable to assume that he is willing to pay a sum of money to replace x by y. Similarly, he should be willing to pay some amount of money to replace y by z and still a third amount to replace z by x. Thus, he ends up with the alternative he started with but with less money." [23, p. 45]

The standard use of money-pumps is to show that certain preference patterns are irrational since they can make you lose money. This is a pragmatic argument, i.e. an argument concerning some principle that "appeals to the desirable/undesirable consequences of [that principle's] satisfaction/violation" [19, p. 289].[2] Money-pumps are useful examples in discussions of the status of pragmatic arguments, but in this chapter the focus will instead be on how we can use formal language to investigate their structure.

31.2 The Major Types of Money-Pumps

Joan's transactions with the stamp dealer exemplify the classical form of a money-pump, also called a *money-pump of the first kind* [8]. Such money-pumps are based on cycles of strict preferences. Money-pumps of the first kind have often been invoked to show that it is imprudent to act upon cyclic strict preferences. The mechanism at hand is illustrated in Fig. 31.1. Obviously, it is not crucial that the cycle has three elements; a longer cycle could produce the same effect. (The same applies to cycles with one or two elements, but such cycles are much less credible than those with at least three elements.)

But there are also other types of money-pumps.

> Half an hour later, Joan's friend Kahil, who is also a philatelist, pays a visit to the shop. He is indifferent between stamps a and b, and also between stamps

[1] Let $\geq$ denote "at least at good as" (weak preference) and $>$ "better than". Then preferences are transitive if and only if it holds for all x, y, and z that if $x \geq y$ and $y \geq z$, then $x \geq z$. Transitive preferences can easily be shown not to have $>$-cycles, i.e. cycles of strict preference.

[2] On pragmatic arguments, see also [3, 10]. Another important example of pragmatic arguments are Dutch books, see Box 19.3 in Chapter 19.

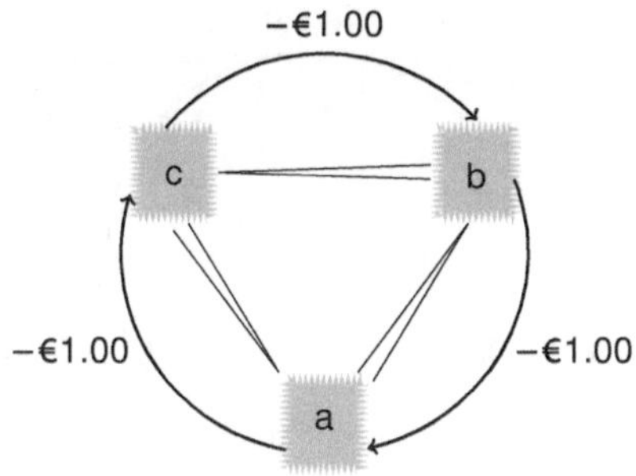

Fig. 31.1 Joan's deal. She prefers *a* to *b*, *c* to *a*, and *b* to *c*, and is willing to pay €1.00 for each move from a less to a more preferred option. This is a money-pump of the first kind

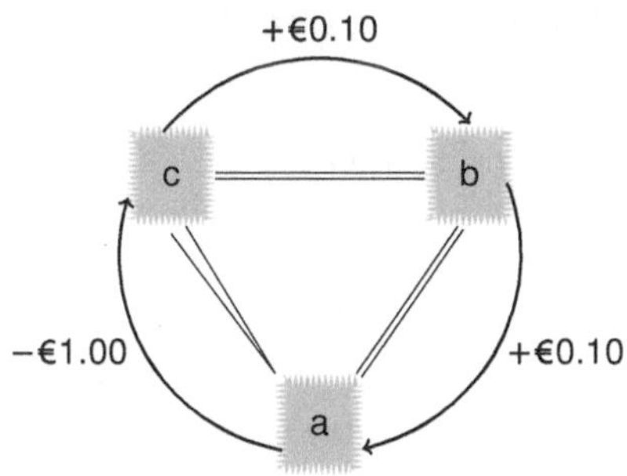

Fig. 31.2 Kahil's deal. He is indifferent between *a* and *b*, and also between *b* and *c*, but he prefers *c* to *a*. He is willing to pay € 1.00 for a move from a less to a more preferred option. Against a compensation of € 0.10 he will move between options that he is indifferent between. This is an accumulating money-pump of the second kind

> *b* and *c*, but he prefers *c* to *a*. Strangely enough, just like Joan he enters the shop carrying stamp *a*.

Can the stamp-dealer extract money from Kahil as well? In fact he can, but now he must apply a modified strategy. The first move is identical:

> The dealer offers Kahil to exchange stamp *a* for stamp *c* against a modest fee of € 1.00. Kahil accepts. Next, the dealer offers to pay him 10c (€ 0.10) if he is willing to take stamp *b* instead of stamp *c*. Since he is indifferent between *b* and *c*, he accepts the bid. After that the dealer proposes yet another deal: € 0.10 for changing from *b* to *a*. Since he is indifferent between these two stamps as well, Kahil accepts. A few minutes later he leaves the shop with *a*, the same stamp that he came with, € 0.80 euros less in his purse and the relish of having completed three favourable deals in just a few minutes.

Kahil's experience in the stamp shop exemplifies a money-pump of the second kind. It differs from the first kind in having not only preference steps but also indifference steps. The dealer has to use some of the money he gains in the preference steps in order to pass through the indifference steps. Since the pump collects money in each

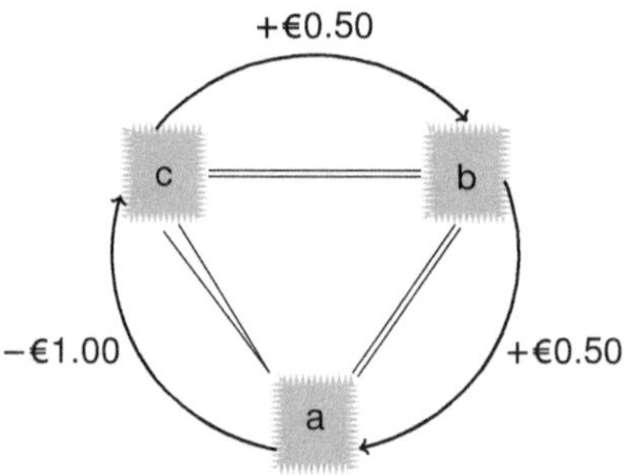

Fig. 31.3 Li Mei's deal. She is indifferent between a and b, and also between b and c, but she prefers c to a. She is willing to pay € 1.00 for a move from a less to a more preferred option. Against a compensation of € 0.50, she moves between options that she is different between. This is a non-accumulating money-pump of the second kind

round that the customer abides, it is an *accumulating money-pump of the second kind* (Fig. 31.2).

This is a busy day for the shopkeeper.

> About an hour after Kahil left, Li Mei enters the shop. Her preferences are just the same as Kahil's, i.e. she is indifferent between stamps a and b, and also between b and c, but prefers c to a. By a most remarkable coincidence she also has a copy of stamp a with her.
>
> She is an unusally charming lady, and the stamp-dealer takes a liking to her. He does not want to extract money from her but he likes having her in the shop. Being the kind of person he is, the only way to entertain her that he knows of is to buy and sell stamps with her. First he lets her exchange stamp a for stamp c against 1 euro. Then he pays her 50 cents to make her change to stamp b, and after that another 50 cents to change back to a. Then he lets her change back to c against 1 euro. She is quite amused, so he goes on to repeat the procedure quite a few times. Almost half an hour later she leaves the shop, with stamp a in her pocket and a friendly smile on her face that makes him hope intensely that she will come back soon to deal with him again.

This is also a money-pump of the second kind. It is illustrated in Fig. 31.3. Li Mei loses no money, but nevertheless her engagement in this procedure has a smack of irrationality. Since she leaves with the same possessions as when she entered (in terms of both stamp and money) this is a *non-accumulating money-pump of the second kind.*

As already mentioned, money-pumps of the first kind operate with cycles of strict preferences. Using P to denote strict preferences, such cycles are called PPP-cycles or P^3-cycles if they have three elements, $PPPP$-cycles or P^4-cycles if they have four elements, etc. Money-pumps of the second kind can be built on any cycle of indifference and strict preference that has at least one step of strict preference [2]. Using R to denote a step that is either indifference or strict preference, these are

Fig. 31.4 The first steps in the three money-pumps in the stamp shop

Joan *Money-pump of the first kind*	Kahil *Accumulating money-pump of the second kind*	Li Mei *Non-accumulating money-pump of the second kind*
$\langle a, 0\rangle$	$\langle a, 0\rangle$	$\langle a, 0\rangle$
$\langle c, -1.00\rangle$	$\langle c, -1.00\rangle$	$\langle c, -1.00\rangle$
$\langle b, -2.00\rangle$	$\langle b, -0.90\rangle$	$\langle b, -0.50\rangle$
$\langle a, -3.00\rangle$	$\langle a, -0.80\rangle$	$\langle a, 0\rangle$
...	...	...

called $R^n P$-cycles, where n is any number of steps that may be either indifference or strict preference.

But in addition to the objects that the cyclic preferences refer to, money-pumps also operate with money. All the deals that Joan, Kahil, and Li Mei made with the shop keeper involved both the exchange of stamps and the exchange of different sums of money. Therefore, a precise description of the preferences involved has to picture them as preferences over two-dimensional states, where the two dimensions are possessions of stamps and money. Strictly speaking, Joan, Kahil and Li Mei did not choose among the stamps a, b, and c. Instead, they chose among combinations of these stamps and various sums of money. To clarify this, a vectorized notation is appropriate [8]. Let $\langle x, v\rangle$ denote that the costumer owns stamp x and has a net outcome of v euros from her dealings with the shopkeeper. Figure 31.4 shows, in this notation, the stages that the three customers went through in the shop.

31.3 Money-Pumps in Disguise

Although money has an essential role in the money-pumps, it can be replaced by other media of compensation. Bearing this in mind, it is easy to see that quite a few of the examples put forward in the literature on preference rationality turn out to be money-pumps in disguise – the disguise consisting in something else taking the role that money has in the classical money-pumps. In particular, the non-accumulating money-pump of the second kind appears to have been repeatedly invented and reinvented.

Michael Dummett [6, p. 34] has provided us with an example in which a person cannot distinguish between wines a and b or between wines b and c, but is perfectly capable of distinguishing between a and c and likes c better. Thus her preferences over these wines can be summarized as $c > a \sim b \sim c$. The wine-vendor sells wine b only with a bottle of a new beer that he is trying to popularize, and he sells wine a only with two bottles of that same beer. The price of a bottle of wine c is the same as that of wine b plus one beer and also the same as that of wine a plus two beers. The customer puts more weight on the difference in taste between wines c and a than on the free beer. We should therefore expect her preferences over the

Fig. 31.5 The preferences involved in the lawn-crossing example

$$\begin{array}{c} \langle l_0, 0 \times t\rangle \\ \vee \\ \langle l_n, n \times t\rangle \\ \vee \\ \langle l_{n-1}, (n-1) \times t\rangle \\ \vee \\ \dots \\ \vee \\ \langle l_2, 2 \times t\rangle \\ \vee \\ \langle l_1, 1 \times t\rangle \\ \vee \\ \langle l_0, 0 \times t\rangle \end{array}$$

composite alternatives to conform with the following cyclic pattern:

$$\langle \text{wine } c, 0 \text{ beer}\rangle > \langle \text{wine } a, 2 \text{ beers}\rangle > \langle \text{wine } b, 1 \text{ beer}\rangle > \langle \text{wine } c, 0 \text{ beer}\rangle$$

This corresponds exactly to Li Mei's preference pattern from above: just replace the three stamps with the equally named wines and exchange the money for beer bottles (at € 0.50 per bottle). We can easily imagine Dummett's customer at the store counter with a bottle of wine *c*, replacing it by a bottle of wine *b* and a beer (since she tastes no difference between the wines), then by a bottle of wine *a* and two beers (again, she tastes no difference between the two wines), and then by a bottle of wine *c* and no beer (since this wine tastes better), and then (if she is irresolute enough) again switching to wine *b* and a beer, etc.

The "lawn-crossing example" has been much discussed in the literature on utilitarianism ([9, p. 107], [15]). Each time you cross a particular lawn, you make a perceptible time gain. No single crossing makes a (perceptible) difference in the condition of the lawn, but a large number of crossings may completely destroy it. Let t denote the gain in time from each crossing and l_n the condition of the lawn after n crossings. This gives rise to the multi-stage cyclic pattern of combined preferences shown in Fig. 31.5. This is a non-accumulating money-pump of the second kind, and here "time is money", i.e. time-gain takes the role of money in the money-pump.

In an ingenious thought experiment by Warren S. Quinn, a medical device has been implanted into the body of a person (the self-torturer). The device has 1001 settings, from 0 (off) to 1000. Each step upwards on the scale leads to a negligible increase in pain. Each week, the self-torturer "has only two options – to stay put or to advance the dial one setting. But he may advance only one step each week, and he may never retreat. At each advance he gets $10,000." In this way he may "eventually

Fig. 31.6 The preferences involved in Quinn's self-torturer example

$$\begin{array}{c} \langle s_0, \$0\rangle \\ \vee \\ \langle s_{1000}, \$10.000.000\rangle \\ \vee \\ \langle s_{999}, \$9.999.000\rangle \\ \vee \\ \dots \\ \vee \\ \langle s_2, \$20.000\rangle \\ \vee \\ \langle s_1, \$10.000\rangle \\ \vee \\ \langle s_0, \$0\rangle \end{array}$$

reach settings that will be so painful that he would then gladly relinquish his fortune and return to 0" [16, p. 79]. The cycle that this gives rise to is shown in Fig. 31.6. Just like the lawn-crossing example, this is a variant of the non-accumulating money-pump of the second kind.

A somewhat more complex type of money-pump has been proposed in order to show that if an agent has intransitive indifferences ($x \sim y$ and $y \sim z$ but not $x \sim z$) then that agent is also prone to have cyclic preferences ($x > y$, $y > z$ and $x > z$). This is of course interesting since the latter pattern is usually conceived as more severely irrational than the former. The first version of this argument is due to Yew-Kweng Ng [14] but we will use an elegant example put forward by George Schumm [22].

Schumm invites us to consider a Mr. Smith who chooses between three boxes of Christmas tree ornaments. Each box contains one red, one blue, and one green ball. The balls of box 1 are denoted r_1, b_1, and g_1, those of box 2 r_2, b_2, and g_2, and those of box 3 r_3, b_3, and g_3. Mr. Smith cannot see any difference between r_1 and r_3 or between r_3 and r_2, but he sees a difference between r_1 and r_2, and he prefers the former. Hence his evaluation of the red balls follows the following pattern:

$$r_1 > r_2 \sim r_3 \sim r_1$$

His preference patterns for the blue and green balls are as follows:

$$b_3 > b_1 \sim b_2 \sim b_3$$

$$g_2 > g_3 \sim g_1 \sim g_2$$

When comparing the three boxes, he prefers Box 1 to Box 2 since, to his eye, they contain equally attractive blue balls and green balls, while Box 1 contains the prettier red ball. Analogously, he prefers Box 2 to Box 3 since the only noticeable difference is the more beautiful green ball of the former, and Box 3 to Box 1 due to its superior blue ball. Thus his preferences over the boxes exhibit cyclic strict preferences (PPP) although the underlying preferences over balls were only subject to a less damaging form of cycle (IIP, where I denotes indifference). From this Schumm drew the following conclusion:

> "The case of Smith shows, I think, that one cannot plausibly abandon the transitivity of indifference without giving up that of preference as well. To be sure, the two principles are *logically* independent in the sense that neither one, when taken together with uncontroversial axioms, implies the other. But given any proposed counterexample to the transitivity of indifference... one can always construct, on the foregoing model, an equally compelling counterexample to the transitivity of strict preference" [22, p. 437].

31.4 The Solution: Should We Be Resolute or Sophisticated?

It is fairly obvious that money-pumps only work if the person to be pumped makes her decisions rather short-sightedly. If Joan came to the shop with a good plan for what to achieve there, or if she had fully understood what was going on, she would not have ended up with $\langle a, -3.00\rangle$. As Frederick Schick pointed out, if the agent sees the full picture, "he may well reject the offer and thus stop the pump.... He need not act as if he wore blinders." [21, pp. 117–118]

Currently there are two major, competing proposals for how best to behave in order to avoid being money-pumped. Edward McClennen [11, 12] argues that the right solution is to be a *resolute* decision-maker. This means that one makes long-term plans and sticks to them. If Joan were a resolute decision-maker, then she would have decided from the beginning what to buy in the shop, and she would have stuck to that plan. Such a plan would certainly not have had $\langle a, -3.00\rangle$ as its goal state. A resolute decision-maker does not have much reason to ponder how she may react and behave in various future situations. To the contrary: "Rather than regimenting present choice of a plan to projected future choice, the required alignment can be secured, in principle, in just the reverse manner, by regimenting future choice to the originally adopted plan" [12, p. 231].

The other proposal, championed by Wlodek Rabinowicz, is to be a *sophisticated* decision-maker. This means that you foresee how you would act in various future decision situations, and adjust your current choices according to that. If Joan were a sophisticated decision-maker, then she would have thought through beforehand the various offers that the stamp-dealer could make. She would have discovered the path leading to $\langle a, -3.00\rangle$, and knowing where it leads she would certainly have refrained from following it. In the common types of money-pumps, as presented above, both the resolute and the sophisticated strategy protect against exploitation. However, other decision problems have been constructed in which the sophisticated decision-maker can be exploited with a money-pump, whereas the resolute decision-maker is still protected [17, 18].

31.5 Money-Pumps in Real Life?

Although we seldom find money-pumps in the real world, they may be useful tools of thought for analyzing what rationality demands, not only in scholarly contexts but also on a more personal level. As noted by Peter Fishburn, "the mere possibility of a money pump could encourage people to reexamine expressed preferences. This would likely lead to transitive revisions in some cases." [7, p. 118] Such preference revisions have been shown to take place in experimental settings [4].

Money-pumps can also serve to show the relevance of agenda-setting and agenda-shifts in decision-making. The importance of agenda-setting has been confirmed by results from social choice theory showing how a clever and well-informed agenda-setter can "design an agenda to reach virtually any point in the alternative space" [13, p. 1087]. Arguably, agents in real life cannot plan how to act under all possible future circumstances; when subject to "surprise choices" we may all be vulnerable to money-pumping [5]. And even in the absence of surprises, we often make our decisions based on a smaller decision horizon than what would have been possible. A typical example would be a smoker's multiply repeated decision to smoke one more cigarette.

Chrisoula Andreou has argued that important environmental decisions have the structural properties of money-pumps. Environmental damage is often the accumulated effect of actions each of which has no noticeable detrimental effect on the environment. We may therefore be collectively in the same position as a smoker or as Quinn's self-torturer, whose problems are caused by the fact that "if he is going to quit, he is better off, relative to his concerns, quitting at the next setting rather than at the current setting" and who will therefore "end up in excruciating pain" [1, p. 102]. This may be one of the reasons why we so often fail to take the measures necessary to avoid environmental damage.

This is a fine illustration of how careful investigations of philosophical and logical issues with no apparent practical relevance may end up having direct implications for the conduct of human affairs. We need to break stalemates that impede progress in climate policies. Which is the best strategy? Should we be resolute or sophisticated?

Acknowledgements I would like to thank Richard Bradley and Wlodek Rabinowicz for very useful comments on an earlier version of this text.

References and Recommended Readings

1. Andreou, C. (2006). Environmental damage and the puzzle of the self-torturer. *Philosophy and Public Affairs, 34*, 95–108.
2. Bossert, W., & Suzumura, K. (2008). A characterization of consistent collective choice rules. *Journal of Economic Theory, 138*, 311–320.
3. Cantwell, J. (2003). On the foundations of pragmatic arguments. *Journal of Philosophy, 100*, 383–402.
4. Chu, Y., & Chu, R. (1990). The subsidence of preference reversals in simplified and marketlike experimental settings: A note. *American Economic Review, 80*, 902–911.
5. Cubitt, R. P., & Sugden, R. (2001). On money pumps. *Games and Economic Behavior, 37*, 121–160.
6. Dummett, M. (1984). *Voting procedures*. Oxford: Clarendon Press.
7. Fishburn, P. (1991). Nontransitive preferences in decision theory. *Journal of Risk and Uncertainty, 4*, 113–134.
8. Hansson, S. O. (1993). Money-pumps, self-torturers and the demons of real life. *Australasian Journal of Philosophy, 71*, 476–485.
9. Harrison, J. (1953). Utilitarianism, universalisation, and our duty to be just. *Proceedings of the Aristotelian Society, 53*, 105–134.
10. Jordan, J. (2010). Pragmatic arguments. In C. Taliaferro, P. Draper, & P. L. Quinn (Eds.), *A companion to philosophy of religion* (2nd ed., pp. 425–433). Malden: Wiley-Blackwell.
11. ∗ McClennen, E. F. (1990). *Rationality and dynamic choice. Foundational explorations*. Cambridge: Cambridge University Press. [The best source on resolute decision-making.]
12. McClennen, E. F. (1997). Pragmatic rationality and rules. *Philosophy and Public Affairs, 26*, 210–258.
13. McKelvey, R. D. (1979). General conditions for global intransitivities in formal voting models. *Econometrica, 47*, 1085–1112.
14. Ng, Y. (1977). Sub-semiorder: A model of multidimensional choice with preference intransitivity. *Journal of Mathematical Psychology, 16*, 51–59.
15. Österberg, J. (1989). One more turn on the lawn. In S. Lindström & W. Rabinowicz (Eds.), *In so many words. Philosophical essays dedicated to Sven Danielsson on the occasion of his fiftieth birthday* (pp. 125–133). Uppsala: Uppsala University, Department of Philosophy.
16. ∗Quinn, W. S. (1990). The puzzle of the self-torturer. *Philosophical Studies, 59*, 79–90. [One of the most elegant thought experiments with a money-pump structure.]
17. ∗Rabinowicz, W. (2000). Money pump with foresight. In M.J. Almeida (Ed.), *Imperceptible harms and benefits*, (pp. 123–154). Dordrecht: Kluwer. [Together with his 2001 paper the best source on sophisticated decision-making.]
18. ∗Rabinowicz, W. (2001). A centipede for intransitive preferrers. *Studia Logica, 67*, 167–178. [Complements the foregoing.]
19. Rabinowicz, W. (2006). Levi on money pumps and diachronic Dutch books. In E. J. Olsson (Ed.), *Knowledge and inquiry. Essays on the pragmatism of Isaac Levi* (pp. 289–312). Cambridge: Cambridge University Press.
20. Ramsey, F. P. ([1931] 1950). *The foundations of mathematics and other logical essays*. London: Routledge & Kegan.
21. Schick, F. (1986). Dutch bookies and money pumps. *Journal of Philosophy, 83*, 112–119.
22. Schumm, G. F. (1987). Transitivity, preference, and indifference. *Philosophical Studies, 52*, 435–437.
23. Tversky, A. (1969). Intransitivities of preferences. *Psychological Review, 76*, 31–48.

Chapter 32
Deontic Logic

Sven Ove Hansson

Abstract Deontic logic is the logic of normative concepts such as obligation, permission, and prohibition. This non-technical overview of the area has a strong emphasis on the connections between deontic logic and problems discussed in moral philosophy. Major issues treated are the distinction between ought-to-be and ought-to-do, the various meanings of permissive expressions, the logical relations among norms, the paradoxes of deontic logic, and the nature of moral conflicts and moral dilemmas. It is concluded that deontic logic has resources for precise treatment of important issues in moral philosophy, but in order to make full use of these resources, more co-operation between logicians and moral philosophers is needed.

32.1 Introduction

Example 1
HOST: Please, don't take the apple!
GUEST: But you told me I could have an apple or an orange. Have you changed your mind?
HOST: No, I haven't changed my mind. I gave you permission to take an apple or an orange. You are still permitted to do so. But I haven't allowed you to take an apple.
GUEST: Can I have an orange?
HOST: But of course!

Example 2
MR. WEISENHEIMER: You are not allowed to enter this garden.
MS. WRIGHT: But I promised the owner to mow her lawn when she is away, and I am obliged to fulfil my promise.

S. O. Hansson (✉)
Division of Philosophy, Royal Institute of Technology (KTH), Stockholm, Sweden
e-mail: soh@kth.se

S. O. Hansson, V. F. Hendricks (eds.), *Introduction to Formal Philosophy*, Springer Undergraduate Texts in Philosophy, https://doi.org/10.1007/978-3-319-77434-3_32

MR. WEISENHEIMER: So what?
MS. WRIGHT: In order to do so I have to enter her garden. Since I am under an obligation to mow her lawn I have a permission to enter her garden to do so.
MR. WEISENHEIMER: You are rushing to conclusions. I do not see how a permission can follow from an obligation. Permissions and obligations are very different things.

Example 3
ADULTERER: I have put myself in a terrible situation. I have promised Anne to get a divorce and then marry her. She has waited for me more than five years. Now she is pregnant with my child and she entreats me to take the decisive step. But I also still love my wife, and I have promised never to leave her. What should I do?
MORALIST: Since you can only be married to one person you should not have promised two persons to be married to them. That is what is wrong.
ADULTERER: I know that. But please tell me what I should do.
MORALIST: I have already told you.

In discussions on moral norms, we tend to assume that they have certain structural properties. Obligations should be consistent, and certain norms imply other norms. Deontic logic is the discipline that attempts to uncover the logical laws of our normative concepts and systematize their structural properties.

The logic of norms is complex and has been difficult to unveil. Furthermore, the contacts between deontic logic and informal moral philosophy have not been sufficiently close and sometimes not even sufficiently friendly. According to one moral philosopher, "deontic logic has so far created more problems than it has solved" [25]. But recent developments in deontic logic seem to bring the two disciplines closer to each other.

Some philosophers have claimed that deontic logic is an oxymoron since it applies (truth-valued) logic to subject-matter that does not refer to truth or falsehood. But this is criticism that can easily be answered, provided that we use the logical apparatus as a *model* of normative concepts. A model need not share all the properties of that which is modelled. An economist can use a model in which monetary value is infinitely divisible, although real money comes in discrete units. Similarly, we can use a model expressed in truth-valued logic for subject matter that is not truth-functional.

32.2 The Basic Parallel with Modal Logic

There are three major types of normative expressions: prescriptive, prohibitive, and permissive expressions. In the formal language, they are represented by logical expressions containing the predicates O (*o*bligation, ought), P (*p*ermission), and F (prohibition, *f*orbiddance).

There is a striking analogy between the logical relations in the following two triads:

necessary – impossible – possible
obligatory – forbidden – permitted

Something x is impossible if and only if not-x is necessary. Similarly, x is forbidden if and only if not-x is obligatory. Furthermore, it is possible that x if and only if it is not necessary that not-x. In the same way, x is permitted if and only if not-x is not obligatory, etc. These analogies were known already in the twelfth century [18]. Deontic logicians still adhere to them. But of course, there are other principles of modal logic that cannot be transferred to deontic logic. In particular, what is necessary is true, but what is morally obligatory is (unfortunately) often untrue.

32.3 Forming the Deontic Language

The sentences of a deontic language are formed recursively in three steps, as shown in Fig. 32.1. Atomic sentences devoid of normative content $(p, q, \dots)$ are the starting material. In the first step, truth-functional combinations such as $p \vee q$ and $r \rightarrow (q \,\&\, s)$ are formed from the atomic sentences. In the second step, deontic operators such as O, P, or F are prefixed to the outcomes of the first step. This gives rise to (atomic) deontic sentences such as Op, $P(q \vee r)$, and $F(p \rightarrow r)$. The third step consists in forming truth-functional combinations of the outcomes from the second step. This gives rise to expressions such as $Op \rightarrow P(p \vee q)$, $Fp \vee P(p \& \neg r)$, etc. The expressions obtained in this third step form the common core of deontic languages. It is important to observe that the elements formed in the second step (such as $O(p \vee q)$) are retained in the third step, whereas the expressions from the first step (such as $\neg p \rightarrow q \vee r$) have been lost.

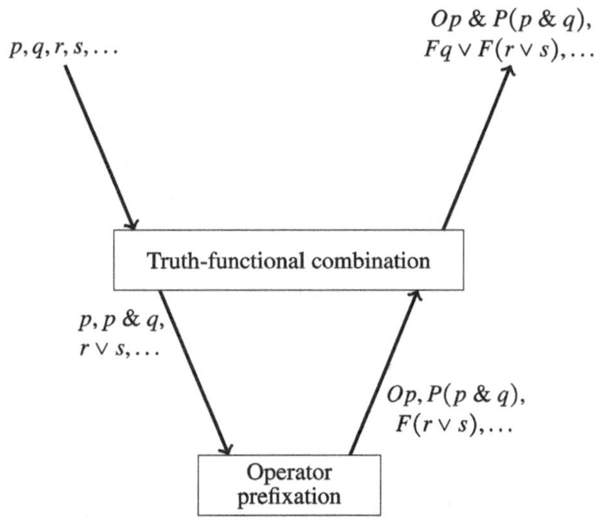

Fig. 32.1 The formation of a deontic language. Atomic sentences (upper left) go through a process of truth-functional combination. Then operators are affixed, and finally a new process of truth-functional combination takes place

Three types of extensions of this common core of deontic languages are common. First, formulas from the first step can be included directly in the third, giving rise to "mixed" formulas such as $p \,\&\, Op$ and $p \rightarrow Oq$. Secondly, the second and third step can be cyclically repeated, giving rise to sentences with nested deontic operators, i.e. deontic operators within the scope of other such operators: OOp, $FPFp$, $O(Pp \vee O\neg p)$, $O\neg(Op \,\&\, O\neg p)$, etc. Thirdly, we can include, at the second stage, operators representing other non-truthfunctional concepts than deontic ones, most commonly operators for necessity ($\Box$), possibility ($\Diamond$), and conditionality ($\Rightarrow$).

32.4 Interpreting the Deontic Language

Some of the normative terms in natural language are also used for non-normative purposes.[1] As one example of this, the word "must" can denote either obligation or necessity. ("You must help her." – "You must be wrong.") In a somewhat similar way, the word "ought" is ambiguous between two meanings:

You ought to help your destitute brother. (ought-to-do, Tunsollen)
There ought to be no injustice in the world. (ought-to-be, Seinsollen)

In the first of these sentences, "ought" is prescriptive, and a synonym of "obligatory", "duty", and "morally required". The second sentence is an example of ought-sentences that "are not prescriptive at all, either prudentially or morally, but express valuations. Such is 'Everybody ought to be happy'. This is not a prescription or command to anybody to act or to refrain." [27, p. 195]

Most other prescriptive predicates do not have the particular ambiguity that "ought" has. It would not make much sense to say that there is a duty for the world to contain no injustice or that it is obligatory that everyone be happy. Since deontic logic represents norms it should only be concerned with the prescriptive sense of "ought". The non-prescriptive meanings of that word have to be left out, just like the non-prescriptive meanings of "must".

The meaning of expressions for *permission* differs between deontic logic and natural language. When we say that an action is permitted we usually imply that its omission is also permitted. Therefore it would be strange to say to a convict serving life sentence: "You may stay in prison tomorrow." Generally speaking, "when saying that an action is permitted we mean that one is at liberty to perform it, that one may either perform the action or refrain from performing it". [26, p. 161] (bilateral permission). In deontic logic, however, "being permitted to perform an action is compatible with having to perform it". (ibid.) (unilateral permission) Hence Pp says that p is permitted but it does not tell us whether or not $\neg p$ is also permitted.

A major reason why unilateral permission is preferred in deontic logic is that it is related to obligation in the same way as possibility to necessity, so that the above-mentioned parallel with modal logic can be upheld. Another reason is that bilateral

[1]For more details, see Chap. 1.

permission can be straightforwardly defined in terms of unilateral permission (as Pq & $P\neg q$, "q is permitted and not-q is also permitted").

Prescriptions, permissions, and prohibitions all come in different *degrees of stringency* (strength). Every child understands the difference in stringency between the two instructions "do not speak with food in your mouth" and "do not erase the hard disk on mom's computer". Natural language contains resources for expressing such differences. "Must" is more stringent than "ought", and "ought" is more stringent than "should" [6]. In the formal language we can express the difference by including operators representing different strengths of moral requirement, but this has only seldom been done [2, 10, 22].

32.5 Standard Deontic Logic

Although he had many forerunners, Georg Henrik von Wright can rightly be said to have founded modern deontic logic. In his famous 1951 paper, he set forth a number of axioms that in his view characterize rational reasoning about norms. This was just before the advent of possible world semantics. The pioneers of that area were all aware of von Wright's work, and discussed deontic logic in their early writings. They also all spotted the crucial difference between modal and deontic possible worlds semantics, namely that the accessibility relation should be reflexive in the former but not in the latter type of semantics, so that $\Box p \rightarrow p$ holds but not $Op \rightarrow p$ [30]. Stig Kanger [17], Jaakko Hintikka [14, p. 12], and Saul Ktipke [19, p. 95] all realized what it takes to create a deontic semantics, but William Hanson [7] seems to have been the first to write out a full construction of possible world semantics for deontic logic.

William Hanson constructed deontic semantics so that it differs from modal semantics only in the respect already mentioned, namely that the accessiblity relation was not reflexive. This elegant construction yields the logical principles that von Wright had proposed for deontic logic, but with one exception: In his 1951 paper, von Wright had proclaimed a principle of "deontic contingency" for tautologies, namely: "A tautologous act is not necessarily obligatory", $\neg O(p \vee \neg p)$. Possible world semantics instead validated the opposite principle:

$$O(p \vee \neg p) \text{ (the axiom of the empty duty)}$$

In spite of its intuitive implausibility ([15, p. 191], [20, p. 31]) this postulate was rapidly accepted as a tribute to logical elegance.

Most deontic logicians were primarily interested in unnested deontic sentences, i.e. sentences in which no deontic operator is positioned within the scope of another such operator. (This is sensible, due to the difficulties in interpreting expressions such as OOp and $P(Op \vee q)$.) With this limitation, the accessibility relation of modal-style deontic logic can be replaced by a strikingly simple semantic construction:

Ideal Worlds Intersection (IWI)
There is a subset $\mathcal{I}$ of the set $\mathcal{W}$ of possible worlds, such that:
For all p, Op holds if and only if $p \in w$ for all $w \in \mathcal{I}$.

$\mathcal{I}$ is called the set of "ideal" worlds (or "deontically ideal" or "(deontically) perfect" worlds). It is easy to show that the sentences that are valid in this simple model coincide with those that are derivable from the following three axioms:

$Op \rightarrow \neg O \neg p$,
$Op \;\&\; Oq \leftrightarrow O(p \& q)$, and
$O(p \vee \neg p)$

The first two of these axioms were present in von Wright's 1951 paper, whereas the third is the axiom of the empty duty that had to replace its negation in order to fit into the semantics. The second axiom is equivalent to the combination of the following two:

$Op \;\&\; Oq \rightarrow O(p \& q)$ (agglomeration)
If Op, and p logically implies q, then Oq. (necessitation)

In 1969 Bengt Hansson introduced the term "standard deontic logic" (SDL) to denote the deontic logic that can be characterized either by these axioms or by the semantic principle of Ideal Worlds Intersection.

It was realized at an early stage that SDL in its original form lacks a credible account of conditional obligation. A sentence such as "If you insult her then you ought to apologize to her" can be semi-formalized in the format "If p then Oq", but what should a full formalization look like? The two obvious options in the SDL language, namely $O(p \rightarrow q)$ and $p \rightarrow Oq$, could easily be shown to give rise to absurd conclusions. To solve this problem Bengt Hansson [8] introduced a two-place predicate $O(q|p)$ ("If p then q is obligatory").[2] He also proposed a simple semantic principle for conditional obligation. Instead of just dividing the possible worlds into ideal and non-ideal worlds, we can order them in more than two grades. Immediately beneath the ideal worlds we have the second-best worlds, beneath them the third-best worlds, etc. In order to determine the conditional obligations relative to some statement p, we restrict our attention to worlds in which p is true. Then $O(q \mid p)$ ("If p then q is obligatory") holds if and only if q holds in all those worlds that are best among the worlds in which p is true.[3]

However, in spite of its formal elegance, SDL has been under constant attack due to the implausible results that have been derived from it. In the following two sections we are going to have a look at some of that criticism, before turning to more constructive developments.

[2]The dyadic predicate can replace the monadic one, since we can define Op as $O(p \mid \top)$, where $\top$ is a tautology.

[3]Important results on dyadic SDL can be found for instance in [23, 24, 29].

32.6 Free Choice Permission

Recently when a neighbour asked me if he could borrow a crowbar, I showed him my crowbars and said:

You may borrow either the big or the small crowbar.

In saying so I offered him a choice between the two tools. However, in another context the same sentence could have another meaning. Suppose that the tools belonged to someone else who had authorized me to lend one of them to the neighbour. However, I had forgotten which of the two he could borrow. Then I could have said:

You may borrow either the big or the small crowbar, but I do not know which.

The first case illustrates *free choice permission*. In natural language, this is by far the most common meaning of permissive expressions that refer to a disjunction. A formal permission operator that represents it should expectedly satisfy the postulate

$$P(a \vee b) \rightarrow Pa \ \&\ Pb.$$

However, this principle does not hold in SDL. This seems to have been first noted by von Wright [32, pp. 21–22]. His discovery gave rise to an extensive search for a plausible operator of free choice permission.

The most obvious solution would be to just add the axiom $P(a \vee b) \rightarrow Pa \ \&\ Pb$ to the SDL axioms. However, it was soon realized that the combination of this axiom with the original postulates would give rise to a whole series of implausible results. ([32, p. 21], [16, p. 61], [21, p. 140]) We can take David Makinson's derivation as an example:

$O(\neg a \& \neg b) \rightarrow O\neg a$ (Holds in SDL.)
$O\neg(a \vee b) \rightarrow O\neg a$ (Equivalent sentences are exchangeable in SDL.)
$\neg O\neg a \rightarrow \neg O\neg(a \vee b)$ (Due to sentential logic.)
$Pa \rightarrow P(a \vee b)$ (Definition of P.)
$Pa \rightarrow Pb$ (Since $P(a \vee b) \rightarrow Pb$, due to the added postulate.)

What this means is that if something is permitted (Pa) then so is everything else (Pb). This is obviously an intolerable result. Therefore we cannot solve the "free choice problem" by just turning the SDL permission operator into an operator of free choice permission. Having realized this, deontic logicians tried instead to add a second permission operator P_c to supplement rather than replace the SDL operator P. Arguably, this could most naturally be done with the definition

$$P_c(a \vee b) \leftrightarrow Pa \ \&\ Pb.$$

However, this definition has implausible consequences, as shown for instance in the following derivation:

$Pa \rightarrow (Pa \,\&\, P(a \vee b))$ (Due to $Pa \rightarrow P(a \vee b)$ that holds in SDL.)
$Pa \rightarrow P_c(a \vee (a \vee b))$ (The definition of P_c.)
$Pa \rightarrow P_c(a \vee b)$ (Exchangeability of logical equivalents in SDL.)

Hence, if you are permitted to borrow a book from the library (Pa), then you have a free choice to either borrow or steal the book ($P_c(a \vee b)$). (More morbid examples are not difficult to construct.)

Several other, more complex constructions of free choice operators have been tried out, but they have all been shown to have absurd consequences. [12] The underlying reason for this is that they all rely on the following assumption that has usually been taken for granted:

The single sentence assumption: Free choice between a and b can be represented as a property of a single sentence, namely $a \vee b$.

Provided that logically equivalent sentences are interchangeable, the single sentence assumption has the following implication:

If $a \vee b$ is equivalent with $c \vee d$, then there is a free choice permission between a and b if and only if there is a free choice permission between c and d.

It is not difficult to find examples showing that this leads to absurd conclusions:

The vegetarian's free lunch [12][4]
In this restaurant I may have a meal with meat or a meal without meat. Therefore I may either have a meal and pay for it or have a meal and not pay for it.
Proof: Let m denote that you have a meal with meat, v that you have a meal without meat, and p that you pay. $P(m \vee v)$ is equivalent with $P(((m \vee v)\&p) \vee ((m \vee v)\&\neg p))$.

To sum up, (free choice) permission is a permission to perform either the action represented by the sentence a or that represented by the sentence b. We have found that it is not a function of a single sentence $a \vee b$ but a function of the two sentences a and b. It must be represented as a function of two variables, not one. Similarly, (free choice) permission to perform either a, b, or c is a function of three variables, etc. Therefore, free choice permission should be represented as a property of a set of action-describing sentences ($\{a, b\}$ respectively $\{a, b, c\}$), rather than a property of the disjunction of these sentences ($a \vee b$, respectively $a \vee b \vee c$) [10, pp. 130–131].

[4]Also discussed in the Chap. 1.

32.7 The Deontic Paradoxes

The problem with free choice permission is that the following deontic principle does not hold in SDL, although it is intuitively plausible:

$P(p \vee q) \rightarrow Pp \;\&\; Pq$ ("If p-or-q is permitted, then so is p, and so is q.")

There are also many cases in which the difference goes the other way around: A property holds in SDL but it is not intuitively plausible. Such divergences are called "deontic paradoxes". The best-known of them is Ross's paradox. It is a counterexample to the following SDL theorem:

$Op \rightarrow O(p \vee q)$: (If p is obligatory then so is p-or-q.)

Ross [28] proposed the following counter-example:

> "If you ought to mail the letter, then you ought to either mail or burn it." [28, p. 62]

Several other deontic paradoxes have been proposed. One of the most ingenious is Åqvist's [1] knower paradox:

> "If the police officer ought to know that Smith robbed Jones, then Smith ought to rob Jones."

This is a counterexample to the SDL principle of necessitation. (If Op holds and p logically implies q, then Oq holds as well.) The paradox also makes use of the epistemic principle that only that which is true can be known.

As was pointed out by von Wright [33], all the major deontic paradoxes rely on necessitation. Necessitation in its turn follows from the basic construction of the possible worlds semantics for SDL, namely that a sentence is valid if and only if it holds in all elements of a certain set of possible worlds. The paradoxes put this construction in doubt.

This construction has also been criticized on more fundamental ethical grounds. Important types of moral obligations that are recognized and much discussed in moral philosophy are difficult to account for in SDL semantics. This applies for instance to obligations of compensation and reparation. Suppose that John sees a small child fall into the pool in front of him. It would be easy for him to save the child's life. Does he have an obligation to do so? According to SDL semantics we have to consider what his actions would have been in an ideal world. In an ideal world, the child would presumably not have fallen into the water. (This may apply even if we consider the ideal worlds to be ideal only in terms of obligation-fulfilment. If the child's parents had fulfilled their obligations, then the accident would not have happened.) Hence, in the ideal worlds the child would not have been in danger, and John could not have saved it. It follows that John is under no moral obligation to save the child. This example is due to Holly Goldman, according to whom SDL "ignores the fact that particular obligations flow from abstract principles *together with* contingent features of the world", and these features "do not appear in *all* the morally best worlds" [5, p. 244].

Preventive actions are almost as difficult as compensatory ones to account for in SDL semantics. In an ideal world there will be no acts of violence or racism, and consequently no one will act to prevent such misdeeds. Therefore, if our obligations in the actual world consist in doing what we would have done in the ideal worlds, then there can be no obligation in the actual world to act against violence or racism.

In summary, it does not seem appropriate to identify our obligations with how we would act in an ideal world. Such an identification would amount to a recommendation to act as if we lived in an ideal world. But that is bad advice. Acting as one would have done in an ideal world is the behaviour that is expected to follow from wishful thinking, not from well-considered moral deliberation [11].

32.8 Alternative Semantics

Instead of judging the obligatoriness of actions according to the value (ideality) of the worlds in which these actions take place, we can relate obligatoriness to the value of these actions themselves. This can be done by relating normative predicates to an underlying preference relation $\geq$ ("is better than or equal in value to"). The following definition will then have a central role:

> A predicate H is *positive* with respect to a preference relation $\geq$ if and only if it holds for all p and q that if Hp and $q \geq p$, then Hq (i.e., if p has the H-property and q is at least as good as p, then q has the H-property).

A plausible preference-based deontic logic can be founded on the simple principle that the permissive predicate P is positive with respect to some underlying preference relation $\geq$ [10]. In other words, if p is permitted, and q is at least as good as p, then q is also permitted. In contrast, the prescriptive predicate O cannot reasonably be assumed to satisfy positivity. To see this, suppose that you have an unexpected, hungry visitor. Let p denote that you give your hungry visitor something to eat and q that you serve her a gourmet meal. It is quite plausible to value q at least as highly as p. But even if we do so we can hold p to be morally required without also holding q to be morally required. In other words, we can have Op, $q \geq p$ and $\neg Oq$, which shows that the obligation predicate O does not satisfy positivity.[5]

In a framework based on the positivity of permission, the validity of postulates in deontic logic will depend on the properties of the underlying preference relation. Standard preference relations give rise to a deontic logic in which the necessitation postulate (the source of the deontic paradoxes) does not hold, but several other, more plausible postulates hold, such as the following: [10]

[5]However, if P satisfies positivity, and O is definable from P in the usual way ($Op \leftrightarrow \neg P \neg p$), then O satisfies contranegativity: If Op and $\neg p \geq \neg q$ then Oq.

$Op \,\&\, Oq \rightarrow O(p\&q)$
("If each of p and q is by itself obligatory, then so is p-and-q.")

$O(p\&q) \rightarrow Op \vee Oq$
("If p-and-q is obligatory, then so is either p or q, or both.")

$P(p\&q) \,\&\, P(p\&\neg q) \rightarrow Pp$
("If each of p-and-q and p-and-not-q is permitted, then so is p.")

32.9 Deontic Inconsistencies and Moral Conflicts

The structure of moral dilemmas is readily expressible in deontic logic. If both Op ("p is obligatory") and $O\neg p$ ("not-p is obligatory"), then the dictates of the O operator cannot be completely complied with, and no fully acceptable course of action is available. The view that moral dilemmas are impossible implies that such combinations of obligations should be excluded from the logic, and then $Op \,\&\, Oq$ cannot hold if $p\&q$ is logically false. The competing view that moral dilemmas are possible does not impose that restriction on deontic logic. It can be explicated in a formal framework that distinguishes among obligations of different strengths, and uses these distinctions to account for the resolution of moral dilemmas [9]. To see how this works, let us first consider the following example:

MORALIST: You have a large debt that is due today. You should pay it.
SPENDTHRIFT: It is impossible for me to do so. I don't have the money.
MORALIST: I know that.
SPENDTHRIFT: Yes, and I already know what my obligations are. Please, as a moralist, tell me instead what I should do.
MORALIST: I have already told you. You should pay your debt.

Our Moralist is unhelpful, since she refuses to accept the shift in perspective demanded by Spendthrift when asking what she should do. With this phrase, Spendthrift calls for action-guidance. The "should" of "You should pay your debt" is not suitable for action-guidance, since it requires something that Spendthrift cannot do. The shift in focus demanded by Spendthrift can be described as a shift from a morally adequate prescriptive predicate O_M to a predicate O_A that is suitable for action-guidance. Let p designate that she pays off her debts. Then $O_M p$ ("p is morally required") holds, but so does $\neg O_A p$ ("p is not required by proper moral action guidance"). Furthermore, let q denote that she pays her creditors at least as much as she can without losing her means of subsistence. Then both $O_M q$ and $O_A q$ hold, i.e. q is both morally required and required by proper moral action guidance.

We can now apply this distinction to a typical example of a moral dilemma. Suppose that you can either save A's life (p) or B's life (q), but not both. You are equally morally required to perform each of these incompatible actions, but your obligation to perform at least one of them ($p \vee q$) is still stronger. From a moral point of view, arguably the most adequate deontic operator will be one that requires both that you save A and that you save B. Hence $O_M p$ and $O_M q$ hold, and so does $O_M(p \vee q)$. Since p and q cannot both be realized, we then have a moral dilemma in terms of O_M. From the viewpoint of action-guidance, since p and q are incompatible, $O_A p$ and $O_A q$ cannot both hold, and since we have no reason to prefer one of them to the other, neither of them holds. However, $p \vee q$ is morally required to a higher degree than either p or q, and we therefore have good reasons to assume that $O_A(p \vee q)$ holds. Hence, from the action-guiding point of view, it is obligatory to perform at least one of the two actions, and permissible to perform either to the exclusion of the other.

The case of the Adulterer, as presented above in Sect. 32.1, illustrates the same point. In the dialogue our Moralist stuck to standards of moral requirement that were rather unhelpful since they made all courses of action impermissible. The Adulterer needs a weaker notion of moral requirement that does not require the impossible. This weaker notion must be such that staying with his wife (w) and marrying his mistress (m) are not both prescribed (but for our present purpose we can leave it unsettled whether $O_A w$ & $\neg O_A m$, $\neg O_A w$ & $O_A m$, or $\neg O_A w$ & $\neg O_A m$ holds for such a notion of moral requirement in this case).

Thus applied, the distinction between moral and action-guiding deontic predicates provides a formal account of how moral dilemmas can exist, yet be resolvable in terms of a weaker notion of moral requirement that is suitable for action-guidance. It should be observed that the moral "ought" is not eliminated, only supplemented by an obeyable "ought" that is suitable for action-guidance.

Hopefully, these examples can illustrate that deontic logic has resources for precise treatment of important moral issues. In order to develop this potential, cooperation between logicians and moral philosophers is needed.

References and Proposed Readings

1. Åqvist, L. (1967). Good samaritans, contrary-to-duty imperatives, and epistemic obligations. *Noûs, 1*, 361–379.
2. Dellunde, P., & Godo, L. (2008). Introducing grades in deontic logic. In R. van der Meyden & L. van der Torre (Eds.), *DEON 2008* (LNAI, Vol. 5076, pp. 248–262).
3. * Føllesdal, D., & Hilpinen, R. (1970). Deontic logic: An introduction. In R. Hilpinen (Ed.), *Deontic logic: Introductory and systematic readings* (pp. 1–35). Reidel: Dordrecht. [An old but still very readable introduction to deontic logic.]
4. *Gabbay, D., Horty, J., Parent, X., Meyden, R., & Torre, L. (Eds.) (2013). *Handbook of deontic logic and normative systems*. (Vol. 1). London: College publications. [Comprehensive coverage of most topics in deontic logic.]
5. Goldman, H. S. (1977). David Lewis's semantics for deontic logic. *Mind, 86*, 242–248.
6. Guendling, J. E. (1974). Modal verbs and the grading of obligations. *Modern Schoolman, 51*, 117–138.
7. Hanson, W. H. (1965). Semantics for deontic logic. *Logique et Analyse, 31*, 177–190.
8. * Hansson, B. (1969). An analysis of some deontic logics. *Noûs, 3*, 373–398. [A classic in dyadic deontic logic.]

9. Hansson, S. O. (1999). But what should I do? *Philosophia, 27*, 433–440.
10. ∗Hansson, S. O. (2001). *The structure of values and norms*. Cambridge: Cambridge University Press. [Overview with an emphasis on alternative semantics.]
11. Hansson, S. O. (2006). Ideal worlds – Wishful thinking in deontic logic. *Studia Logica, 82*, 329–336.
12. ∗Hansson, S. O. (2013). The varieties of permission. In Gabbay et al. (2013), pp. 195–240. [Detailed overview of the logical and semantical issues concerning permission.]
13. ∗Hilpinen, R., & McNamara, P. (2013). Deontic logic: A historical survey and introduction. In Gabbay et al. (2013), pp. 1–134. [An excellent history and overview of the whole area.]
14. Hintikka, J. (1957). Quantifiers in deontic logic. *Societas Scientiarum Fennica, Commentationes Humanarum Literarum, 23*(4).
15. Jackson, F. (1985). On the semantics and logic of obligation. *Mind, 94*, 177–195.
16. Kamp, H. (1973). Free choice permission. *Proceedings of the Aristotelian Society, 74*, 57–74.
17. Kanger, S. ([1957] 1971). New foundations for ethical theory. Reprinted in R. Hilpinen (Ed.), *Deontic logic: Introductory and systematic readings* (pp. 36–58). Dordrecht: Synthese Library.
18. Knuuttila, S. (1981). The emergence of deontic logic in the fourteenth century. In R. Hilpinen (Ed.), *New studies in deontic logic* (pp. 225–248) Dordrecht: Reidel.
19. Kripke, S. (1963). Semantical analysis of modal logic I. Normal modal propostional logic. *Zeitschrift für mathematische Logik und Grundlagen der Mathematik, 9*, 67–96.
20. Lenk, H. (1978). Varieties of commitment. *Theory and Decision, 9*, 17–37.
21. Makinson, D. (1984). Stenius' approach to disjunctive permission. *Theoria, 50*, 138–147.
22. McNamara, P. (1996). Must I do what I ought? (or will the least I can do?) In M. A. Brown and J. Carmo (Eds.), *Deontic logic, agency and normative systems* (DEON'96: Third International Workshop on Deontic Logic in Computer Science, 11–13 Jan 1996, pp. 154–173). Sesimbra: Springer.
23. Parent, X. (2008). On the strong completeness of Åqvists dyadic deontic logic G. In R. van der Meyden and L. van der Torre (Eds.) *DEON 2008* (LNAI, Vol. 5076, pp. 189–202).
24. Parent, X. (2010). A complete axiom set for Hansson's deontic logic DSDL2. *Logic Journal of IGPL, 18*, 422–429.
25. Purtill, R. L. (1980). Review of al-Hibri, deontic logic. *Southwestern Journal of Philosophy, 11*, 171–174.
26. Raz, J. (1975). Permissions and supererogation. *American Philosophical Quarterly, 12*, 161–168.
27. Robinson, R. (1971). Ought and ought not. *Philosophy, 46*, 193–202.
28. Ross, A. (1941). Imperatives and logic. *Theoria, 7*, 53–71.
29. Spohn, W. (1975). An analysis of Hansson's dyadic deontic logic. *Journal of Philosophical Logic, 4*, 237–252.
30. Wolenski, J. (1989). Deontic logic and possible worlds semantics: A historical sketch. *Studia Logica, 49*, 273–282.
31. ∗von Wright, G. H. (1951). Deontic logic. *Mind, 60*, 1–15. [*The* classic in deontic logic; still very readable.]
32. von Wright, G. H. (1968). An essay in deontic logic and the general theory of action. *Acta Philosophica Fennica, 21*, 1–110.
33. von Wright, G. H. (1981). On the logic of norms and actions. In R. Hilpinen (Ed.), *New studies in deontic logic* (pp. 3–35). Reidel, Dordrecht.
34. ∗von Wright, G. H. (1999). Deontic logic: A personal view. *Ratio Juris, 12*, 26–38. [A personal account of the history of deontic logic.]

Chapter 33
Action Theories

Andreas Herzig, Emiliano Lorini, and Nicolas Troquard

Abstract We present the main logical theories of action. We distinguish theories identifying an action with its result from theories studying actions in terms of both their results and the means that result is obtained. The first family includes most prominently the logic of seeing-to-it-that and the logic of bringing-it-about-that. The second includes propositional dynamic logic and its variants. For all these logics we overview their extensions by other modalities such as modal operators of knowledge, belief, and obligation.

33.1 Introduction

Actions such as raising one's arm, switching on a computer, jumping a traffic light, killing somebody, or waltzing are investigated in several areas of philosophy, among others in philosophy of action, philosophy of language and philosophy of law. Through the analogy between actions and programs the concept is also relevant in computer science, in particular in artificial intelligence, multi-agent systems and theoretical computer science. Several other concepts are intimately related to action. One that is directly related is that of the *ability* to act. Mental attitudes and norms also play an essential role in the study of action.

It has been attempted since Aristotle to systematise the analysis of action. Taking advantage of the mathematical advances in predicate logic, ontological perspectives on action were proposed in the form of first-order theories in the mid-twentieth century and have been very influential in philosophy. Concurrently, various research programs investigated the logic of action as such, trying to uncover the grand principles. These approaches are dominated by a modal view of action, and a first

A. Herzig (✉) · E. Lorini
Institut de Recherche en Informatique de Toulouse (IRIT), CNRS, France
e-mail: Andreas.Herzig@irit.fr; Emiliano.Lorini@irit.fr

N. Troquard
Free University of Bozen-Bolzano, Italy
e-mail: nicolas.troquard@unibz.it

S. O. Hansson, V. F. Hendricks (eds.), *Introduction to Formal Philosophy*, Springer Undergraduate Texts in Philosophy, https://doi.org/10.1007/978-3-319-77434-3_33

survey of this field is in a 1992 special issue of Studia Logica [44]. The present chapter overviews the resulting logics of action. We start by introducing the main issues at stake.

33.1.1 Actions as Events Brought About by Agents

It is generally considered that an action can be identified with an *event* that is brought about by an *agent* [14, 48], as exemplified by Belnap talking about "an agent as a wart on the skin of an action" [5]. The dedicated term in the literature is that an agent is *agentive* for an event. Examples of events are that an arm goes up, that a computer starts, that somebody dies, etc. So my action of switching the computer on is identified with me bringing about the event that the computer starts.

Essentially, there exist two different semantical accounts of events: the first account identifies an event with a *set of possible worlds*, also called a *proposition*; the second account identifies an event with a *binary relation between possible worlds*, also called a transition relation. In the first view, events are facts of the world, identified with propositions: subsets of the set of possible worlds where the event occurs. To these propositions the usual set-theoretic operations can be applied. We thus obtain a way to interpret complex events and actions that are built with the logical connectives of propositional logic, such as negation, conjunction, and material implication. In the second view, the transition relations of atomic events are a given, and the transition relation of a complex event is built up from them.

33.1.2 Action as Result vs. Action as 'Means+Result'

The two views on the semantics of events yielded two traditions of logics of action. The difference is reflected by two different logical forms of action sentences they consider: the first family is about sentences such as "I bring it about that the computer is on" and focuses on *the result of an action*; the second family is about sentences such as "I bring it about that the computer is on by pushing the power button" and focuses on both *the result and the means by which it is obtained.*

The first family are the so-called logics of agency. The logic of seeing-to-it-that (STIT) [4, 6] and the logic of bringing-it-about-that (BIAT) [18, 19, 40] are two sub-families. These logics are studied in philosophy of action and more recently in multi-agent systems. The second family contains variants and extensions of propositional dynamic logic (PDL). These latter logics were introduced and studied in theoretical computer science, but were also investigated by philosophers.

33.1.3 Potential Action

A notion that is often studied along with actual agency is the mere existence of a potential action. "He could have done otherwise"; "She can win this match"; "The Democrats have a strategy to undermine the influence of the Senate whatever the rest of the electorate does"; "I can switch on the light if you want"; "He can! But he would be lucky!" Loaded with many distinct but somewhat overlapping meanings, this notion has been called ability, capability, opportunity, power, etc. In this chapter we will simply use *ability* as an umbrella term for potential action.

Some meanings of the term *ability* have not yet been satisfyingly formalised in logic. One in particular is Kenny's sense of ability [28]: I am able to do an action if when I try to do that action under normal conditions then I usually succeed. Kenny's example is that of an expert dart player who is able to hit the bullseye while a layman is not. Although very close to our real world experience, one difficulty is to meaningfully capture in a formalism that ability is not a sufficient condition for actual agency and that actual agency is not evidence of ability *à la* Kenny.

Yet, possible action has been studied alongside actual action in some logical formalisms. All of the logics presented in this chapter that deal with both actual and potential agency subscribe the principle 'actual agency implies potential agency', for short: 'do implies can'.

- BIAT logic is about actual agency. Elgesem has added a notion of ability to bring about a proposition. In his logic an ability still can exist without actual agency: a lion in a zoo can catch a zebra. Both agency and ability are primitive concepts in his logic (although they are defined by means of the same semantic structure).
- STIT logic is primarily about actual agency and potential agency. It is equipped with quantification over possible unrolling of events. Potential agency for a proposition is then reduced to the existence of an unrolling of events where actual agency for that proposition is expressed.
- Coalition logic CL [37] and alternating-time temporal logic ATL [1] are about the ability of an agent to ensure something whatever the other agents do. There is no notion of actual agency, and the language does not explicitly refer to action terms.
- The standard version of PDL [21] enables to talk about the possibility of the occurrence of an event and about what is true afterwards. Linear versions of PDL also allows one to capture actual agency. Furthermore, there are variants of PDL which allow one to represent both actual action and potential action such as PDL with actual actions [31] and DLA [23].

Table 33.1 classifies the logics that we are going to overview in this chapter according to the distinctions 'potential and/or actual agency' and 'result vs. result+means'.

Table 33.1 Logical form and concepts of the logics of this chapter

	Result	Means + result
Potential only	CL, ATL	PDL
Actual only	BIAT	Linear PDL
Potential + actual	STIT, Elgesem's BIAT	PDL with actual actions, DLA

33.1.4 Actions and Mental Attitudes

Our actions are determined by our beliefs and desires: I switch my computer on because I want to know the weather forecast and believe I can find it on the Internet, or because I believe I got email and want to read it, or because I want to send an email and believe my Internet connection is not down.

According to an influential view due to Bratman, desires do not directly lead to actions, but it is rather the intermediate mental attitude of intention that triggers actions [7]. Cohen and Levesque designed a logic adding modal operators of belief and choice to PDL within which intention can be defined [17].

33.1.5 Actions and Deontic Concepts

What we do is not only influenced by our mental attitudes, but also by obligations and prohibitions. Indeed, there are cases where agents perform actions independently of their beliefs and desires merely because they are obliged to do so; think e.g. of soldiers blindly obeying their commander.

Meyer gave a logical account of obligation and action that is based on PDL [34], while Horty based his account on STIT [24, 25].

33.1.6 The Rest of This Chapter

We are now going to present the main logics of action and discuss their basic logical principles. In the next section we introduce the family of those logics allowing us to talk about actions in terms of their results: BIAT and STIT. Thereafter we present the family of logics allowing us to talk about actions in terms of results and means to achieve these results: PDL and its linear variants. For each family we discuss the interplay with ability, mental attitudes and norms.

Throughout this chapter $\phi, \psi, \ldots$ denote formulas and $i, j, \ldots$ denote agents (individuals) that populate the world.

33.2 Action as Result

According to Belnap and Perloff's 'stit-thesis' every agentive sentence can be transformed into a sentence of the form "i sees to it that ϕ", where i is an agent and ϕ is a proposition. In other words, an action is identified with the result it brings about. The sentence "agent i sees to it that ϕ" itself can then be viewed as a proposition. This allows for a purely logical analysis of agentive sentences.

Let us start by formulating several principles that all of the logics in this section satisfy.

First, if we view agentive sentences as propositions then it is natural to require that the set of worlds where ϕ is true contains the set of worlds where i is agentive for ϕ. This is a *principle of success*: the proposition "i sees to it that ϕ" should imply the proposition ϕ. In other words, it should be valid that if i sees to it that ϕ then ϕ is true. Note that it follows from this principle that an agent can never see to it that $\phi \wedge \neg\phi$.

Second, the different approaches agree about the *principle of aggregation*: "if i sees to it that ϕ and i sees to it that ψ then i sees to it that $\phi \wedge \psi$".

Third and as already discussed in the introduction, action implies ability. This is a *do implies can* principle: "if i sees to it that ϕ then i is able to achieve ϕ".

Fourth, a bringing about of a proposition is not sensitive to the syntactical formulation of that proposition. For example, if Zorro and Don Diego Vega are the same person and one considers that their being dead is the same proposition, then Sgt. Gonzales bringing about that Zorro is dead is equivalent to Sgt. Gonzales bringing about that Don Diego Vega is dead. This is the *principle of equivalents for actual agency*. A similar principle can be formulated for potential agency.

All variants of STIT and of BIAT satisfy the principles of success, of aggregation, 'do implies can', and equivalents for agency. Beyond these standard principles there are quite some differences that have been captured by quite different semantics. We therefore present the two families separately.

The main difference between BIAT logic and STIT logic is that the latter satisfies a principle of independence of agents while the former does not: in STIT it is assumed that each combination of the agents' individual actions can be *chosen jointly*, while this is not required in BIAT. It may be argued that while the principle of independence of agents is acceptable in the case of *choice* (or *trying*), it is less so in the case of *action*. Suppose two agents are standing in front of a room door and intend to enter the room. The door is too narrow to allow them to successfully enter at the same time, even though each agent can successfully enter if the other agent does nothing. While the two agents can simultaneously decide/try to enter the room, their attempts will fail to be performed successfully.

After the presentation of each family of logics we briefly mention extensions by concepts such as knowledge, belief, intention, and obligation.

33.2.1 The Logic of Bringing-it-About-That BIAT

BIAT logic, the logic of bringing-it-about-that, dates back to Kanger and Pörn [27, 40].[1] We here present Elgesem's semantics [19] whose validities were axiomatised by Governatori and Rotolo [20]. The semantics is in terms of selection function models $\langle W, \{f\}_i, V\rangle$ where W is some set of possible worlds, $V : \mathcal{P} \longrightarrow 2^W$ is a valuation function mapping propositional variables to subsets of W, and for every agent i, $f_i : W \times 2^W \rightarrow 2^W$ is a selection function associating a proposition to every possible world and proposition. The object $f_i(w, X)$ is the set of those worlds where i realises the ability he has in w to bring about his goal X. Therefore i is able to bring about X at w if $f_i(w, X)$ is nonempty; and i brings about X at w if w belongs to $f_i(w, X)$.

The functions f_i have to satisfy the following additional constraints:

- $f_i(w, X) \subseteq X$, for every $X \subseteq W$ and $w \in W$;
- $f_i(w, X_1) \cap f_i(w, X_2) \subseteq f_i(w, X_1 \cap X_2)$, for every $X_1, X_2 \subseteq W$ and $w \in W$;
- $f_i(w, W) = \emptyset$, for every $w \in W$.

The first two constraints correspond to the principle of success and to the principle of aggregation. The third constraint says that an agent cannot be agentive for a tautology.

The language of BIAT logic has modal operators of agency $\mathtt{Biat}_\mathtt{i}$ and modal operators of ability $\mathtt{Can}_i$, one of each for every agent i. The formula $\mathtt{Biat}_\mathtt{i}\phi$ reads "i brings it about that ϕ", and the formula $\mathtt{Can}_i\phi$ reads "i can achieve ϕ".[2]

The truth conditions are as follows:

$$
\begin{array}{lll}
M, w \models p & \text{iff} & w \in V(p); \\
M, w \models \mathtt{Biat}_\mathtt{i}\phi & \text{iff} & w \in f_i(w, ||\phi||_M); \\
M, w \models \mathtt{Can}_i\phi & \text{iff} & f_i(w, ||\phi||_M) \neq \emptyset.
\end{array}
$$

In the last two conditions the set $||\phi||_M$ is the extension of ϕ in M, i.e. the set of possible worlds where ϕ is true: $||\phi||_M \stackrel{\text{def}}{=} \{w \in W : M, w \models \phi\}$.

Alternative semantic characterisations of the operators $\mathtt{Biat}_\mathtt{i}$ exist in the literature: Pörn proposed to simulate it by combining two more elementary modal operators that are normal [40]; Carmo *et col.* have used neighborhood semantics [42]. However, there are no completeness results for these alternative semantics.

So, what are the axioms of BIAT, i.e., what are the formulas of the language that are true in every model? As announced above, the axioms of success, aggregation, and 'do implies can' are all valid in BIAT logic, and the rule of equivalents preserves BIAT validity:

---

[1]There is no well-established name in the literature, we therefore opted for the acronym BIAT, justs as the well-established STIT stands for 'seeing-to-it-that'.

[2]Instead of $\mathtt{Biat}_\mathtt{i}$ Jones and Pörn use $\mathtt{E}_i$ and Elgesem uses $\mathtt{Does}_i$. Instead of $\mathtt{Can}_i$ Elgesem uses $\mathtt{Ability}_i$.

$$\texttt{Biat}_\texttt{i}\phi \rightarrow \phi \tag{33.1}$$

$$(\texttt{Biat}_\texttt{i}\phi \wedge \texttt{Biat}_\texttt{i}\psi) \rightarrow \texttt{Biat}_\texttt{i}(\phi \wedge \psi) \tag{33.2}$$

$$\texttt{Biat}_\texttt{i}\phi \rightarrow \texttt{Can}_i\phi \tag{33.3}$$

$$\frac{\phi \leftrightarrow \psi}{\texttt{Biat}_\texttt{i}\phi \leftrightarrow \texttt{Biat}_\texttt{i}\psi} \tag{33.4}$$

$$\frac{\phi \leftrightarrow \psi}{\texttt{Can}_i\phi \leftrightarrow \texttt{Can}_i\psi} \tag{33.5}$$

A subject that has been a source of disagreement in the literature is whether an agent can bring about a logical tautology. Can John bring it about that $2 + 2 = 4$? BIAT rules it out:

$$\neg\texttt{Can}_i\top \tag{33.6}$$

is an axiom. Together with the 'do implies can' axiom of Eq. (33.3), it entails that $\neg\texttt{Biat}_\texttt{i}\top$ is valid. That is, no agent is agentive for a tautology.

The principle that is maybe most surprisingly absent is the axiom of monotony $\texttt{Biat}_\texttt{i}(\phi \wedge \psi) \rightarrow (\texttt{Biat}_\texttt{i}\phi \wedge \texttt{Biat}_\texttt{i}\psi)$: i may bring it about that $\phi \wedge \psi$ without necessarily bringing it about that ϕ. $\texttt{Biat}_\texttt{i}$ is therefore not a normal modal 'box' operator. The same is the case for the logic of the ability operators $\texttt{Can}_i$. Moreover, they do not satisfy the principle $\texttt{Can}_i(\phi \vee \psi) \rightarrow (\texttt{Can}_i\phi \vee \texttt{Can}_i\psi)$; to see this take $\psi = \neg\phi$. Therefore the latter cannot be modal 'diamond' operators either. Moreover they do not satisfy $\phi \rightarrow \texttt{Can}_i\phi$. Due to these last two properties Elgesem's ability operators satisfy what Brown calls Kenny's constraint [11].

In presence of several agents, these operators can be combined to express interesting properties of interaction. One can say for instance that an agent i makes (resp. can make) another agent j bring it about that ϕ, in formula: $\texttt{Biat}_\texttt{i}\texttt{Biat}_\texttt{j}\phi$ (resp. $\texttt{Can}_i\texttt{Biat}_\texttt{j}\phi$). Following the common law maxim "quid facit per alium facit per se", some authors consider that when i makes another agent bring about something then i himself brings about that something [13]. Others disagree [19]. Troquard [46], in a group extension of BIAT suggests a principle $\texttt{Biat}_\texttt{i}\texttt{Biat}_\texttt{j}\phi \rightarrow \texttt{Biat}_{\{\texttt{i},\texttt{j}\}}\phi$, where $\texttt{Biat}_{\{\texttt{i},\texttt{j}\}}\phi$ indicates that the group composed of i and j brings about ϕ together. Aiming at another kind of compromise, Santos *et al.* have proposed a logic with two kinds of agency operators: one of indirect agency (noted $\texttt{G}_i$) satisfying the above principle and another one of direct agency (noted $\texttt{E}_i$) which does not (and instead satisfies $\texttt{E}_i\texttt{E}_j\phi \rightarrow \neg\texttt{E}_i\phi$) [41, 42].

Our next family of logics will validate this principle, and much more.

33.2.2 The Logic of Seeing-to-it-That STIT

While the temporal aspects were kept abstract in BIAT logics, the semantics of STIT logics inherits the Ockhamist conception of time [50] where the truth of statements is evaluated with respect to a moment that is situated on a particular history through time (that is identified with a sequence of moments). This is one of the reasons why the models of STIT logics that we are going to present now [4, 24, 25] are more intricate. A systematic comparison between Belnap et al.'s semantics for STIT and other semantics for STIT such as the Kripke-style semantics by [30] and the bundled-tree semantics by [16] has been recently proposed by [15].

A STIT model is based on a *tree of moments* which are the possible states of the world. Every moment occurs at an *instant*, a mere time-stamp. A *history* is a maximal path in the tree. When a moment belongs to a history we say that the history passes through the moment. Time is therefore indeterministic, and indeterminism is due mainly to agents making choices where they could have chosen otherwise: at every moment m, each of the agents has a repertoire of *choices*, and each of these choices consists in selecting a subset of the histories passing through m. The future is understood to be on one of the selected histories. Then the future lies among the histories at the intersection of the choices taken by all agents. Whatever each of the agents chooses, the intersection of all the agents' choices must be non-empty. This is the *independence constraint*.

Formulas are evaluated in a STIT model M with respect to moment-history pairs (m, h) such that m is on h. A significant variety of modalities of agency have been studied within STIT logic, with sometimes only little differences. We are going to mainly talk about two of them that are rather different: the *achievement stit* and the *Chellas stit*. Both have in common with the BIAT modality the principles of Eqs. (33.1), (33.2), (33.3), (33.4), and (33.5) of the section "The Logic of Bringing-it-About-That BIAT". The achievement stit moreover satisfies the principle of Eq. (33.6), while the Chellas stit does not.

The theories are also equipped with an operator of historical possibility $\Diamond$. The formula $\Diamond\phi$ reads "there is a possible history passing through the current moment such that ϕ". Formally speaking, given a history h and a moment m passing through h (i.e., such that $m is on h$), the formula $\Diamond\phi$ is interpreted as follows:

$$M, h, m \models \Diamond\phi \quad \text{iff} \quad M, h', m \models \phi \text{ for some history } h' \text{ such that } m \text{ is on } h'.$$

We can define the dual modal operator $\Box$ by stipulating $\Box\phi \stackrel{\text{def}}{=} \neg\Diamond\neg\phi$ and thereby express the fact that "ϕ is settled true at the current moment".

The original stit modality proposed by Belnap and Perloff [6] is the *achievement stit*. Let us write $\texttt{AStit}_{\texttt{i}}$ for that modal operator. An agent i sees to it that ϕ if a previous choice of i made sure that ϕ is true at the current instant, and ϕ could have been false at this instant had i done otherwise.

$$M, h, m \models \texttt{AStit}_\texttt{i}\phi \quad \text{iff} \quad \begin{array}{l} \text{there is a moment } m_0 \text{ preceding } m \text{ on } h \text{ such that} \\ (1)\ M, h', m' \models \phi \text{ for every } h' \text{ and } m' \text{ such that} \\ \quad \text{(i) } h \text{ and } h' \text{ are in the same choice of } i \text{ at } m_0, \\ \quad \text{(ii) } m' \text{ is on } h' \text{ and at the same instant as } m; \\ (2) \text{ there is a history } h'' \text{ and a moment } m'' \text{ at the} \\ \quad \text{same instant as } m \text{ with } M, h'', m'' \not\models \phi. \end{array}$$

Just as in **BIAT** logic, the idea of achievement is conveyed by validity of the principle of success ($\texttt{AStit}_\texttt{i}\phi \to \phi$) and by the principle that no agent sees to a tautology ($\neg\texttt{AStit}_\texttt{i}\top$).

Now comes a rather fascinating insight from such a complex modality. If $\texttt{AStit}_\texttt{i}\phi$ is i doing ϕ, one can capture that agent i refrains from doing ϕ by the formula $\texttt{AStit}_\texttt{i}\neg\texttt{AStit}_\texttt{i}\phi$. What the logic tells us is that *doing* is equivalent to *refraining from refraining from doing*:

$$\texttt{AStit}_\texttt{i}\phi \leftrightarrow \texttt{AStit}_\texttt{i}\neg(\texttt{AStit}_\texttt{i}\neg\texttt{AStit}_\texttt{i}\phi).$$

(Precisely, this holds under the assumption that an agent does not perform an infinite number of non-vacuous choices during a finite interval of time.)

Horty and Belnap [25] simplified the achievement stit into the *deliberative stit* where the decisive choice of the action is at the current moment. The idea of deliberativeness resides in that an agent is currently seeing to something but could as well see to something else. The logic of the *Chellas stit* further simplifies the deliberative stit by removing the negative part from the truth condition. Let us write $\texttt{CStit}_\texttt{i}$ for Chellas's stit operator. Its semantics is as follows:

$$M, h, m \models \texttt{CStit}_\texttt{i}\phi \quad \text{iff} \quad \begin{array}{l} M, h', m \models \phi \text{ for every } h' \text{ such that } h \text{ and } h' \\ \text{are in the same choice of } i \text{ at } m. \end{array}$$

Hence the Chellas stit operator is a simple quantification over the histories that the current choice of the agent allows. A trained logician may observe that 'being in the same choice cell' is an equivalence relation and that every operator $\texttt{CStit}_\texttt{i}$ therefore obeys the principles of modal logic **S5**.

While the axiom of monotony is invalid in **BIAT** logic, the corresponding formula is valid for the Chellas stit:

$$\texttt{CStit}_\texttt{i}(\phi \wedge \psi) \to (\texttt{CStit}_\texttt{i}\phi \wedge \texttt{CStit}_\texttt{i}\psi). \tag{33.7}$$

The striking principle of the Chellas stit that earned it its name (because Chellas has been a strong advocate, see [44]) is:

$$\Box\phi \to \texttt{CStit}_\texttt{i}\phi. \tag{33.8}$$

In words, an agent cannot avoid what is settled; in particular he can and does bring about every tautology.

Just as the achievement stit, both the Chellas stit operator and the deliberative stit operator satisfy that refraining from refraining from doing is doing (even without the assumption that an agent does not perform an infinite number of non-vacuous choices during a finite interval of time).

A common feature of all STIT logics is that the agents' choices are constrained to be *independent*, while they are not necessarily so in BIAT logic. This can be nicely characterised in the logic of the Chellas stit by the principle

$$(\Diamond \mathtt{CStit}_{\mathtt{i}}\phi \wedge \Diamond \mathtt{CStit}_{\mathtt{j}}\psi) \to \Diamond(\mathtt{CStit}_{\mathtt{i}}\phi \wedge \mathtt{CStit}_{\mathtt{j}}\psi), \text{ for } i \neq j. \quad (33.9)$$

It follows that when i and j are different then $\Diamond \mathtt{CStit}_{\mathtt{i}}\phi \wedge \Diamond \mathtt{CStit}_{\mathtt{j}}\neg\phi$ is unsatisfiable (because $\mathtt{CStit}_{\mathtt{i}}\phi \to \phi$ is valid and because $\Diamond$ is a normal modal operator). This principle can straightforwardly be extended from two agents i and j to any finite number of agents and is central in Xu's axiomatisation of the Chellas stit ([4, Chap. 17]). In contrast, there is no BIAT formula corresponding to Eq. (33.9), simply because the right hand side of the implication cannot be expressed (due to the absence of an operator of historic possibility in the existing BIAT logics).

A somewhat surprising consequence of the independence of agents is the validity of the following 'make do implies settled' principle:

$$\mathtt{CStit}_{\mathtt{i}}\mathtt{CStit}_{\mathtt{j}}\phi \to \Box\phi, \text{ for } i \neq j. \quad (33.10)$$

In words, i can make j see to it that ϕ only if ϕ is settled. This highlights that unlike in BIAT, in STIT logics we cannot reason about the power of agents over others. While this principle may be felt to be unfortunate from the point of view of common sense, it accommodates well with social choice theory and game theory. In [3] it is shown that the schema of Eq. (33.10) is actually equivalent to the schema of Eq. (33.9) and that its generalisation to any finite number of agents can substitute Xu's axiom of independence in the axiomatisation of STIT.

We just mention that when combined with the operator of historical possibility, the Chellas stit operator can express the *deliberative stit operator* $\mathtt{DStit}_{\mathtt{i}}$ as follows:

$$\mathtt{DStit}_{\mathtt{i}}\phi \overset{\text{def}}{=} \mathtt{CStit}_{\mathtt{i}}\phi \wedge \Diamond\neg\phi.$$

The other way round, the Chellas stit operator can be expressed by $\mathtt{DStit}_{\mathtt{i}}$ as:

$$\mathtt{CStit}_{\mathtt{i}}\phi \overset{\text{def}}{=} \mathtt{DStit}_{\mathtt{i}}\phi \vee \Box\neg\phi.$$

The Chellas stit operator together with historical possibility also allows to express by $\Diamond \mathtt{CStit}_{\mathtt{i}}\phi$ that an agent has the *ability* to see to it that ϕ. The schema $\mathtt{CStit}_{\mathtt{i}}\phi \to \Diamond \mathtt{CStit}_{\mathtt{i}}\phi$ is valid and provides a 'do implies can' principle. While the aggregation principle is clearly invalid for that ability operator, it satisfies monotony and the principle $\Diamond \mathtt{CStit}_{\mathtt{i}}\top$. Hence every $\mathtt{CStit}_{\mathtt{i}}$ is a normal modal diamond operator (violating therefore Kenny's constraint for ability operators).

33.2.3 Extensions

33.2.3.1 Temporal Operators

Broersen et al. [10] have added the temporal operators of linear-time temporal logic LTL to the stit language. In that language they introduce another modality of ability different from the above as $\Diamond \texttt{CStit}_{\texttt{i}} \texttt{X}\phi$, where X is the temporal 'next' operator. They show that this definition of ability matches the ability operator of Pauly's coalition logic CL [37]. They also show that the further addition of the 'eventually' modality of LTL allows one to reduce alternating-time temporal logic ATL [1] to that temporal extension of STIT.

Lorini recently extended the stit language by future tense and past tense operators and provided a complete axiomatization for this temporal extension of stit [30]. The semantics for temporal stit used by Lorini is based on the concept of temporal Kripke stit model which extends Zanardo's concept of Ockhamist model [50] with a choice component.

Ciuni and Zanardo extended the stit language by (restricted) branching-time operators of computational tree logic CTL and proved a completeness result [16].

33.2.3.2 Mental Attitudes and Deontic Concepts

Starting with Kanger and Lindahl [29], many researchers working on logics of agency were interested in deontic concepts such as the obligation or the permission to act. Starting from the neighbourhood semantics for BIAT logic, Santos *et al.* added a modal operator of obligation Obl to the language [12, 41, 42]. Then the formula $\texttt{OblBiat}_{\texttt{i}}\phi$ expresses that agent i is obliged to bring it about that ϕ.

Horty proposed to integrate obligation into branching-time structures by means of a function idl which for every moment m selects the ideal histories among all the histories running through m: those where all the obligations are fulfilled [24].

$$M, h, m \models \texttt{Obl}\phi \text{ iff } M, h', m \models \phi \text{ for every } h' \text{ such that } h' \in idl(m).$$

Much less work was done on the integration of mental attitudes into logics of agency. For some first attempts see [9, 49]. More recently, some authors have worked on the combination of epistemic logic and STIT logic by enriching the STIT semantics with names for choices and action tokens [26, 33].

33.2.3.3 Resource-Sensitive Agency

In [38, 39], Porello and Troquard have proposed a variant of BIAT logic, where the modality of agency is used to formalise agents using, transforming, and producing consumable resources. Using Linear Logic in place of classical logic, one can write sentences like

$$(egg \otimes egg \otimes \texttt{Biat}_{\texttt{i}}(egg \otimes egg \multimap omelet)) \multimap omelet,$$

saying that if agent i transforms two eggs into one omelet, and two eggs are available, then one omelet can be produced. On the other hand, *omelet* does not follow from $egg \otimes \texttt{Biat}_{\texttt{i}}(egg \otimes egg \multimap omelet)$ as the resources are too few.

33.3 Action as 'means+result'

The preceding analysis of actions was merely in terms of their results. Another tradition studies not only the result, but also the means the agent employs to attain that result. The logical form of such sentences is "i brings it about that ϕ by doing α".

If we identify "i does α" as "i brings it about that ψ", for some appropriate proposition ψ, then we end up with an analysis of a dyadic agency operator, as studied by Segerberg [45].

We will not present that view in more detail here and just note that Segerberg's logic turns out to be an instance of the action theory that we are going to present now. Instead of identifying events and actions with propositions, that theory views them as 'things that happen', coming with some change in the world. It is then natural to interpret events and actions as *transitions* between possible worlds, just as computer programs running from an initial state to an end state. This view is taken by propositional dynamic logic PDL, which has *names* to identify these transitions. It is a view of action whose development has benefited from the synergies between philosophy and the formal science of computer programming.

The availability of names for actions allows us to build complex actions from atomic actions. The latter may then be identified with *basic actions*: actions that make up an agent's repertoire. In practice, the choice of granularity for the set of these actions depends on the application at hand. While raising an arm could be taken as a basic action when modeling a voting procedure, a choreographer might want to decompose the raising of an arm into more basic performances of bodily movements.

In the interpretation of actions, an edge between two possible worlds may stand for two different things, depending on how the events of the world will unroll: first, it might be an *actual transition* corresponding to the event actually taking place; second, it might be a *possible transition* that does not actually occur. The logic that we are going to present now mainly adopts the latter perspective.

33.3.1 Propositional Dynamic Logic PDL

Standard PDL has names for events. In this section we describe an *agentive version* of PDL as used in several places in the artificial intelligence literature (e.g., [22, 35]. In that version, atomic actions take the form $i{:}\alpha$ where i is an agent and α is an atomic event. Complex actions—alias programs—are then built recursively from these atomic actions by means of the PDL connectives ";" (sequential composition), "∪" (nondeterministic composition), "*" (iteration), and "?" (test). For instance, the complex event

$$\pi_1 = (\neg treeDown?; i{:}\mathbf{chop})^*; treeDown?$$

describes i's felling a tree by performing the atomic 'chop' action until the tree is down.

The language of PDL has modal operators $\texttt{Poss}_\pi$ where i is an agent and π is an action. The formula $\texttt{Poss}_\pi\phi$ reads "there is a possible execution of π after which ϕ is true".[3] Due to indeterminism, there might be several possible executions of π. While $\texttt{Poss}_\pi$ quantifies existentially over these executions, the dual modal operator $\texttt{After}_\pi$ quantifies universally. It is definable from the former by $\texttt{After}_\pi\phi \stackrel{\text{def}}{=} \neg\texttt{Poss}_\pi\neg\phi$.

While in the 'action-as-result' view of BIAT and STIT logics actions are interpreted as propositions, in PDL an atomic action $i{:}\alpha$ is interpreted as a set of edges of the transition relation: there is an edge from world w_1 to world w_2 that is labeled $i{:}\alpha$ if it is possible to execute $i{:}\alpha$ in w_1 and w_2 is a possible outcome world. The set of all these edges makes up the accessibility relation $R_{i:\alpha}$ associated to $i{:}\alpha$. Complex actions are then interpreted by operations such as relation composition in the case of sequential composition ";" or set union in the case of nondeterministic composition "∪". For instance, our example action π_1 is interpreted by the set of couples (w, w') such that one can go from w through finite **chop**-paths running through possible worlds satisfying $\neg$*treeDown* and whose last possible world w' satisfies *treeDown*.

The formula $\texttt{Poss}_\pi\phi$ is true at a world w if there is a couple (w, w') in R_π such that ϕ is true at world w':

$$M, w \models \texttt{Poss}_\pi\phi \ \text{ iff } M, w' \models \phi \text{ for some } w' \text{ such that } wR_\pi w'.$$

The formula $\texttt{Poss}_\pi\phi$ therefore expresses a weak notion of ability: the action π might occur and ϕ could be true afterwards. The modal operators $\texttt{Poss}_\pi$ are normal modal diamond operators. Hence the axiom $\texttt{Poss}_\pi(\phi \vee \psi) \rightarrow \texttt{Poss}_\pi\phi \vee \texttt{Poss}_\pi\psi$ is valid (violating therefore Kenny's principle for ability operators).

As we have announced above, Segerberg's dyadic agency operator can be viewed as an instantiation of PDL. His atomic events α take the form $\delta_i\psi$ where ψ is a proposition. In that framework he argues for principles such as transitivity: when i brings about ϕ_2 by bringing about ϕ_1 and i brings about ϕ_3 by bringing about ϕ_2, does i bring about ϕ_3 by bringing about ϕ_1? This can formally be written as $(\texttt{After}_{\delta_i\phi_1}\phi_1 \wedge \texttt{After}_{\delta_i\phi_2}\phi_3) \rightarrow \texttt{After}_{\delta_i\phi_1}\phi_3$.

[3]The standard notation is $\langle\pi\rangle\phi$; we here deviate in order to be able to distinguish actual action from potential action.

33.3.2 Linear-Time Propositional Dynamic Logic PDL

Probably Cohen and Levesque were the first to adapt PDL in order to model actual agency [17]. The modalities are interpreted in *linear-time* PDL models: every world w has a unique history running through it. We distinguish modal operators of actual action by writing them as $\texttt{Happ}_\pi \phi$, read "π is performed, and ϕ is true afterwards". Then the following principle for basic actions characterises linear PDL models:

$$(\texttt{Happ}_{i:\alpha}\top \wedge \texttt{Happ}_{j:\alpha'}\phi) \rightarrow \texttt{Happ}_{i:\alpha}\phi \tag{33.11}$$

Cohen and Levesque's linear PDL being only about actual action, Lorini and Demolombe [31] proposed a logic combining PDL operators of potential action $\texttt{Poss}_{i:\alpha}$ with linear PDL operators of actual action $\texttt{Happ}_{i:\alpha}$. In this logic, that we call here PDL with actual actions, the 'do implies can' principle takes the form of the valid schema for atomic actions:

$$\texttt{Happ}_{i:\alpha}\phi \rightarrow \texttt{Poss}_{i:\alpha}\phi. \tag{33.12}$$

Another logic which allows us to represent both actual action and potential action is the *Dynamic Logic of Agency* (DLA) [23]. That logic combines linear PDL operators of actual action $\texttt{Happ}_{i:\alpha}$ with the historical possibility operator of STIT logic: potential action is expressed by the formula $\Diamond\texttt{Happ}_{i:\alpha}\phi$ which has to be read "there is a possible history passing through the current moment such that agent i performs α, and ϕ is true afterwards".

An extension of DLA with program constructions of PDL, called Ockhamist PDL (OPDL), has been recently proposed in [2]. It is shown that both PDL and Full Computation Tree Logic CTL* can be polynomially embedded into OPDL.

33.3.3 Extensions

33.3.3.1 PDL Plus Knowledge and Belief

The first to add a modal operator of knowledge to a PDL-like logic was Moore [36]. This allowed him to formulate and study a principle of perfect recall (aka 'no forgetting') $\texttt{Know}_i\texttt{After}_\alpha\phi \rightarrow \texttt{After}_\alpha\texttt{Know}_i\phi$, as well as the converse principle of 'no miracles' (aka 'no learning'). Similar axioms for belief have also been studied in the literature, in particular under the 'denomination successor state axiom for knowledge' in artificial intelligence [43]. Principles of perfect recall and 'no miracles' play a central role in public announcement logic and more generally dynamic epistemic logics. These logics consider particular atomic events: announcements of (the truth of) formulas. Such events do not change the world, but only the agents' epistemic states. An overview of dynamic epistemic logics can be found in [47].

33.3.3.2 PDL Plus Obligations

Meyer's account extends PDL by a *violation constant* V that was first proposed by Anderson. Agent i's being forbidden to do basic action α is then reduced to all possible executions of α by i resulting in possible worlds where V is true; and i's permission to do α is reduced to some execution of α resulting in a possible world where V is false. In formulas:

$$\texttt{Perm}(i{:}\alpha) \stackrel{\text{def}}{=} \texttt{Poss}_{i:\alpha}\neg \texttt{V}$$
$$\texttt{Forb}(i{:}\alpha) \stackrel{\text{def}}{=} \neg\texttt{Perm}(i{:}\alpha) \stackrel{\text{def}}{=} \neg\texttt{Poss}_{i:\alpha}\neg\texttt{V} \stackrel{\text{def}}{=} \texttt{After}_{i:\alpha}\texttt{V}$$

One may account for the obligation to perform an action by stipulating that every non-performance of α by i results in a violation state. It is however subject to debate how the complement of an action should be defined (see e.g. the discussion in [8]).

33.3.3.3 Linear PDL Plus Belief and Intentions

Cohen and Levesque have analysed intention in linear PDL [17]. In their account intentions are defined in several steps from the concept of *strongly realistic preference*: among the worlds that are possible for an agent there is a subset the agent prefers. There is a modal operator $\texttt{Pref}_i$ for each agent i, and $\texttt{Pref}_i\phi$ reads "i chooses ϕ to be true".[4] Such a notion of preference is strongly realistic in the sense that belief logically implies preference. Furthermore, there are the temporal operators "eventually" (noted F), "henceforth" (noted G), and "until" (noted U) that are interpreted on histories of linear PDL models just as in linear-time temporal logic LTL.

The incremental construction is then as follows. (1) Agent i has the *goal* that ϕ if i prefers that ϕ is eventually true, formally $\texttt{Goal}_i\phi \stackrel{\text{def}}{=} \texttt{Pref}_i\texttt{F}\phi$. (2) i has the *achievement goal* that ϕ if i has the goal that ϕ and believes that ϕ is currently false, formally $\texttt{AGoal}_i\phi \stackrel{\text{def}}{=} \texttt{Goal}_i\phi \wedge \texttt{Bel}_i\neg\phi$. (3) i has the *persistent goal* that ϕ if i has the achievement goal that ϕ and will keep that goal until it is either fulfilled or believed to be out of reach, formally $\texttt{PGoal}_i\phi \stackrel{\text{def}}{=} \texttt{AGoal}_i\phi \wedge (\texttt{AGoal}_i\phi)\,\texttt{U}\,(\texttt{Bel}_i\phi \vee \texttt{Bel}_i\texttt{G}\neg\phi)$. (4) i has the *intention* that ϕ if i has the persistent goal that ϕ and believes he can achieve that goal by an action of his. The formal definition requires quantification over i's actions; we do not go in the details here.

Lorini and Herzig [32] complemented Cohen and Levesque's approach by integrating the concept of an *attempt* to perform an action. The central principle there is "can and attempts implies does": if i intends to (attempt to) perform α and α is feasible then α will indeed take place. This principle is a sort of converse to the 'do implies can' principle.

[4]The original notation is $\texttt{Choice}_i$ instead of $\texttt{Pref}_i$, but we preferred to avoid any confusion with the concept of choice in stit theory.

References and Recommended Readings

1. Alur, R., Henzinger, T. A., & Kupferman, O. (1997). Alternating-time temporal logic. In *Proceedings of the 38th IEEE Symposium on Foundations of Computer Science*.
2. Balbiani, P., Lorini, E. (2013). Ockhamist propositional dynamic logic: A natural link between PDL and CTL*. In *Proceedings of the 20th International Workshop on Logic, Language, Information, and Computation (WOLLIC 2013)* (Lecture notes in computer science, Vol. 8071, pp. 251–265). Springer
3. Balbiani, P., Herzig, A., & Troquard, N. (2008). Alternative axiomatics and complexity of deliberative STIT theories. *Journal of Philosophical Logic, 37*(4), 387–406.
4. * Belnap, N., Perloff, M., & Xu, M. (2001). *Facing the future: Agents and choices in our indeterminist world*. Oxford: Oxford University Press. [A compilation of a over a decade of work of the authors on agency in branching-time.].
5. Belnap, N. (1991). Backwards and forwards in the modal logic of agency. *Philosophy and Phenomenological Research, 51*(4), 777–807.
6. Belnap, N., & Perloff, M. (1988). Seeing to it that: A canonical form for Agentives. *Theoria, 54*(3), 175–199.
7. Bratman, M. E. (1987). *Intentions, plans, and practical reason*. Cambridge/London: Harvard University Press.
8. Broersen, J. (2003). *Modal action logics for reasoning about reactive systems*. PhD thesis. Amsterdam: Vrije Universiteit Amsterdam.
9. Broersen, J. (2011). Making a start with the STIT logic analysis of intentional action. *Journal of Philosophical Logic, 40*, 399–420.
10. Broersen, J., Herzig, A., & Troquard, N. (2006). Embedding alternating-time temporal logic in strategic STIT logic of agency. *Journal of Logic and Computation, 16*(5), 559–578.
11. Brown, M. A. (1992). Normal bimodal logics of ability and action. *Studia Logica, 52*, 519–532.
12. Carmo, J., & Pacheco, O. (2001). Deontic and action logics for organized collective agency, modeled through institutionalized agents and roles. *Fundamenta Informaticae, 48*, 129–163.
13. Chellas, B. F. (1969). *The logical form of imperatives*. Stanford: Perry Lane Press.
14. Chisholm, R. M. (1964). The descriptive element in the concept of action. *Journal of Philosophy, 61*, 613–624.
15. Ciuni, R., & Lorini, E. (2017). Comparing semantics for temporal STIT logic. *Logique et Analyse* (to appear).
16. Ciuni, R., & Zanardo, A. (2010). Completeness of a branching-time logic with possible choices. *Studia Logica, 96*(3), 393–420.
17. Cohen, P. R., & Levesque, H. J. (1990). Intention is choice with commitment. *Artificial Intelligence, 42*(2–3), 213–261.
18. Elgesem, D. (1993). *Action theory and modal logic*. Ph.D. thesis. Institut for filosofi, Det historiskfilosofiske fakultetet, Universitetet i Oslo.
19. Elgesem, D. (1997). The modal logic of agency. *Nordic Journal of Philosophical Logic, 2*(2), 1–46.
20. * Governatori, G., Rotolo, A. (2005). On the axiomatization of Elgesem's logic of agency and ability. *Journal of Philosophical Logic, 34*, 403–431. [A semantics for the logic of bringing-it-about-that in terms of neighbourhood frames.].
21. * Harel, D., Kozen, D., Tiuryn, J. (2000). *Dynamic logic*. Cambridge: MIT Press. [A standard textbook for dynamic logics.].
22. Herzig, A., Longin, D. (2004). C&L intention revisited. In D. Dubois, C. Welty, & M.-A. Williams (Eds.), *Proceeding of the 9th International Conference on Principles on Principles of Knowledge Representation and Reasoning (KR2004)* (pp. 527–535). AAAI Press.
23. Herzig, A., & Lorini, E. (2010). A dynamic logic of agency I: STIT, abilities and powers. *Journal of Logic, Language and Information, 19*, 89–121.
24. * Horty, J. F. (2001). *Agency and deontic logic*. New York: Oxford University Press. [A thorough analysis of obligations to do in the models of branching-time and choice of agents.].

25. Horty, J., & Belnap, N. (1995). The deliberative STIT: A study of action, omission, ability and obligation. *Journal of Philosophical Logic, 24*(6), 583–644.
26. Horty, J., & Pacuit, E. (2017). Action types in STIT semantics. *Review of Symbolic Logic, 10*, 617–637.
27. Kanger, S., Kanger, H. (1966). Rights and parliamentarism. *Theoria, 32*, 85–115.
28. Kenny, A. (1975). *Will, freedom, and power*. Oxford: Blackwell.
29. Lindahl, L. (1977). *Position and change: A study in law and logic*. Dordrecht: D. Reidel publishing Company.
30. Lorini, E. (2013). Temporal STIT logic and its application to normative reasoning. *Journal of Applied Non-Classical Logics, 23*(4), 372–399.
31. Lorini, E., & Demolombe, R. (2008). Trust and norms in the context of computer security: Toward a logical formalization. In R. Van der Meyden & L. Van der Torre (Eds.), *Proceedings of the International Workshop on Deontic Logic in Computer Science (DEON 2008)* (LNCS, Vol. 5076, pp. 50–64). Springer.
32. Lorini, E., & Herzig, A. (2008). A logic of intention and attempt. *Synthese KRA, 163*(1), 45–77.
33. Lorini, E., Longin, D., & Mayor, E. (2014). A logical analysis of responsibility attribution: Emotions, individuals and collectives. *Journal of Logic and Computation, 24*(6), 1313–1339.
34. Meyer, J.-J. Ch. (1988). A different approach to deontic logic: Deontic logic viewed as a variant of dynamic logic. *Notre Dame Journal of Formal Logic, 29*, 109–136.
35. Meyer, J.-J. Ch., van der Hoek, W., & van der Linder, B. (1999). A logical approach to the dynamics of commitments. *Artificial Intelligence, 113*(1–2), 1–40.
36. Moore, R. C. (1985). A formal theory of knowledge and action. In J. R. Hobbs & R. C. Moore (Eds.), *Formal theories of the commonsense world* (pp. 319–358). Norwood: Ablex.
37. * Pauly, M. (2002). A modal logic for coalitional power in games. *Journal of Logic and Computation, 12*(1), 149–166. [A now classic article on group abilities in game and social choice theory.].
38. Porello, D., & Troquard, N. (2014). A resource-sensitive logic of agency. In *ECAI 2014 – 21st European Conference on Artificial Intelligence* (pp. 723–728).
39. Porello, D., & Troquard, N. (2015). Non-normal modalities in variants of linear logic. *Journal of Applied Non-Classical Logics, 25*(3), 229–255.
40. Pörn, I. (1977). *Action theory and social science: Some formal models* (Synthese library, Vol. 120). Dordrecht: D. Reidel.
41. Santos, F., Jones, A. J. I., & Carmo, J. (1997). Action concepts for describing organised interaction. In R. H. Sprague (Ed.), *Proceeding of Thirtieth Annual Hawaii International Conference on System Sciences (HICSS-30)* (Vol. 5, pp. 373–382). IEEE Computer Society Press.
42. Santos, F., Jones, A. J. I., & Carmo, J. (1997). Responsibility for action in organisations: A formal model. In G. Holmström-Hintikka & R. Tuomela (Eds.), *Contemporary action theory* (Vol. 1, pp. 333–348). Kluwer Academic, Dordrecht.
43. Scherl, R., Levesque, H. J. (2003). The frame problem and knowledge producing actions. *Artificial Intelligence, 144*(1–2).
44. Segerberg, K. (Ed.). (1992). *"Logic of Action": Special issue of Studia Logica* (Vol. 51:3/4). Springer Heidelberg.
45. Segerberg, K. (1999). Two traditions in the logic of belief: Bringing them together. In H. J. Ohlbach & U. Reyle (Eds.), *Logic, language and reasoning: Essays in honour of Dov Gabbay* (Trends in logic, Vol. 5, pp. 135–147). Dordrecht: Kluwer Academic.
46. Troquard, N. (2014). Reasoning about coalitional agency and ability in the logics of "bringing-it-about". *Autonomous Agents and Multi-Agent Systems, 28*(3), 381–407.
47. van Ditmarsch, H. P., van der Hoek, W., & Kooi, B. (2007). *Dynamic epistemic logic*. Dordrecht/Boston: Kluwer Academic.
48. Von Wright, G. H. (1963). *Norm and action. A logical inquiry*. London: Routledge and Kegan Paul, .
49. Wansing, H., & Semmling, C. (2008). From BDI and STIT to BDI-STIT logic. *Logic and Logical Philosophy, 17*, 185–207.
50. Zanardo, A. (1996). Branching-time logic with quantification over branches: The point of view of modal logic. *Journal of Symbolic Logic, 61*, 1–39.

Part VII
Decision Theory and Social Philosophy

Chapter 34
Decision Theory: A Formal Philosophical Introduction

Richard Bradley

Abstract Decision theory is the study of how choices are and should be made.in a variety of different contexts. Here we look at the topic from a formal-philosophical point of view with a focus on normative and conceptual issues. After considering the question of how decision problems should be framed, we look at the both the standard theories of chance under conditions of certainty, risk and uncertainty and some of the current debates about how uncertainty should be measured and how agents should respond to it.

34.1 Introduction: Making Decisions

Decision problems abound. Consumers have to decide what products to buy, doctors what treatments to prescribe, hiring committees what candidates to appoint, juries whether to convict or acquit a defendant, aid organisations what projects to fund, and legislatures what laws to make. Descriptive decision theory aims to provide explanations for, and predictions of, the choices that are actually made by individuals and groups facing choices such as these. Normative decision theory, on the other hand, addresses the question of what decisions they should make and how they should make them: how they should evaluate the alternatives before them, what criteria they should employ, and what procedures they should follow. Our focus will be on the latter.

Decision problems arise for agents – entities with the resources to coherently represent, evaluate and change their environments in various possible ways – typically within the context of ongoing personal and institutional projects, activities or responsibilities. These projects together with the environment, both natural and social, provide the givens for the decision problems the agent faces: her resources for acting, her information and often her standards for evaluating outcomes, as well

R. Bradley (✉)
London School of Economics and Political Science, London, UK
e-mail: r.bradley@lse.ac.uk

S. O. Hansson, V. F. Hendricks (eds.), *Introduction to Formal Philosophy*, Springer Undergraduate Texts in Philosophy, https://doi.org/10.1007/978-3-319-77434-3_34

as the source of the problems she must respond to. Lastly, for agents to face a genuine decision problem they must have options: actions that they are capable of performing and equally of foregoing if they so choose. Some examples will illustrate the variety of forms such problems can take.

1. *Take a bus?* You have an appointment that you don't want to miss. If you walk you will arrive a little late. If you take the bus and the traffic is light, you should arrive a little ahead of time. On the other hand if the traffic is heavy then you will arrive very late, perhaps so late that the appointment will be lost. Is it worth risking it?
2. *Another slice of cake*. I have a weakness for chocolate cake which is contributing to a weight problem. My host offers me another slice of cake. Should I accept? I don't want to end up with diabetes or some other obesity related health problem, but one slice of cake will make very little difference and bring pleasure to both me and my host.
3. *Free condoms*. By supplying condoms free, rates of transmission of venereal disease can be considerably reduced. But there is the possibility that it will also encourage sexual activity thereby partially or even completely offsetting the benefits of a decreased transmission rate by virtue of the increase in the number of sexual liaisons.
4. *Road Building*. A new motorway linking two cities will reduce travelling time between the two of them and increase trade, with benefits for inhabitants of both cities. But those living close to the road will suffer from increased pollution and noise, as well as a fall in the value of their houses. Should it be built?

Many decision problems of the kind displayed in these examples can be described in the following way. A decision maker or decision making body has a number of options before them: the actions they can take or policies they can adopt. The exercise of each option is associated with a number of possible consequences, some of which are desirable from the perspective of the decision maker's goals, others are not. Which consequences will result from the exercise of an option depends on the prevailing features of the environment: whether traffic is light or heavy, how overweight I am, whether land prices are falling, and so on.

Let us call the set of environmental features relevant to the determination of the consequence of the exercise of any of the options, a state of the world. Then a decision problem can be represented by a matrix showing, for each available option, the consequence that follows from its exercise in each relevant state of the world. In our first example, for instance, taking the bus has the consequence of having to buy a ticket and arriving late in the event of heavy traffic and paying for a ticket and arriving early in the event of light traffic. This decision problem can be represented by a simple table such as the following:

More generally if A^1 through A^m are the m options open to the decision maker, s_1 through s_n are n possible states of the world (these must be mutually exclusive and exhaust all the possibilities), and C_1^1 through C_n^m are the $m \times n$ consequences that might follow from the choice, then a decision problem can be represented by a state-consequence matrix of the following kind:

	Heavy traffic	*Light traffic*
Take a bus	Arrive late Pay for a ticket	Arrive early Pay for a ticket
Walk	Arrive a little late No ticket needed	Arrive a little late No ticket needed

What choices should be made when facing a decision problem of this kind will depend on the circumstances the agent faces and in particular the amount of information she holds about its various features. Standard presentations distinguish between conditions of *certainty*, when the true state of the world, and hence the outcome of the action, is known; *risk* or *uncertainty*, when either the probabilities of the outcomes associated with an option are known (risk) or the agent can reach a judgement as to how probable they are on the basis of the information she holds (uncertainty); and *ignorance*, when nothing is known about the states. There are however many ways of butchering the beast and expositors draw the line between these conditions in different places (and indeed sometimes use these terms differently). Intermediate cases are important too, most notably when the decision maker is partially ignorant of the relevant probabilities – a situation commonly termed ambiguity.

When the decision maker knows the true state of the world, decision theory says that she should pick the option she considers best. When she is uncertain as to the actual state of the world, she must make a judgement as to how probable it is that each of the possible states is actually the case and pick the option whose expected benefit is greatest relative to these probability judgements. For instance suppose that I consider the probability of heavy traffic to be one-half and the benefit or desirability of the various possible consequences to be as below:

	Heavy traffic	*Light traffic*
Take a bus	−2	1
Walk	−1	−1

Then the expected benefit of taking the bus is a probability weighted average of the benefits of its possible consequences, i.e. $(-2 \times 0.5) + (1 \times 0.5) = -0.5$. On the other hand, walking has a certain benefit of -1. So in this case I should take the bus. But had the probability of heavy traffic been a lot greater, then walking would have been the better action.

More formally, let P be a probability measure on the states of the world and u a utility measure on consequences (we will say more about what these measures are and where they come from in due course). Then a state-consequence matrix,

Table 34.1
State-consequence matrix

	States of the world			
Options	s_1	s_2	...	s_n
A^1	C_1^1	C_2^1	...	C_n^1
...	...	...	...	...
A^m	C_1^m	C_2^m	...	C_n^m

such as that of Table 34.1, induces a probability-utility matrix in which options are represented as random variables that assign a utility value to each state of the world (intuitively, the utility of the consequence of exercising the option in question in that state).

So represented, each option has an expected value that is jointly determined by the functions u and P. The expected value of option A_1, denoted by $E(A_1)$ is, for instance, $u(C_{11}) \cdot P(s_1) + \ldots + u(C_{1n}) \cdot P(s_n)$. More generally, if the number of possible states of the world is finite[1]:

$$E(A^i) = \sum_{j=1}^{n} u(C_j^i) \cdot P(s_j)$$

Now what standard decision theory recommends is choosing the option with the highest expected value. This is known as the maximisation of expected utility hypothesis.

We will examine the maximisation hypothesis in greater detail later on. First, however, we look at a number of issues regarding the formulation and representation of decision problems. In the subsequent sections we look at the relation between preference and choice on the one hand and preference and utility on the other, setting aside complications arising from uncertainty. In the third section we return to decision making under uncertainty. In the final section we look at decision making under ignorance.

34.2 Framing Decision Problems

Decision theory makes a claim about what option(s) it is rational to choose, when the decision problem faced by the agent can be represented by a state-consequence matrix of the kind exemplified by Table 34.1. It is very important to stress that the theory does not say that you *must* frame decision problems in this way. Nor does it say that agents *will* always do so. It just says that *if* they are framed in this way, then only options which maximise expected benefit should be chosen. Nothing precludes the possibility that the same decision situation can or must be framed in different ways. This is true in more than one sense.

[1]The restriction to a finite number of states of the world is made for simplicity, but the expected value will still be well defined even if we drop it.

Firstly, it may be that the problem is not naturally represented by a state-consequence matrix. When I consider whether or not to have another slice of cake, for instance, it is not so much my uncertainty about the consequences of doing so that makes the choice difficult for me, but the contrast between desirability of the short term consequences (good) and the long-term ones (bad). So this problem should be given a different representation. We discuss this issue below in Sect. 34.2.

Secondly, the problem may not be representable by any kind of decision matrix at all because we are unable to identify the various elements of it: what our options are, what the relevant factors are that determine the outcome of each option, or what the consequences are of exercising one or another of the identified options when these factors are present. We discuss this problem in Sect. 34.4.

Thirdly, sometimes no structuring at all may be required; for instance, when certain actions are morally or legally obligatory or when habit dictates the course you take. These cases don't disprove the principle of maximising expected benefit. The point is rather that when the outcome of an action is certain, deliberation is redundant: the high probability of particular events or the great desirability (or otherwise) of particular consequences swamp the contribution that other factors might make.

34.2.1 Locations of Benefit

A two-dimensional decision matrix gives a two factor representation of a choice problem; in Table 34.1, for instance, these are just the states of the world and the consequences that follow from exercising an option in that world. But the state of the world in which a consequence is realised is not the only factor that matters to our assessment of its significance: this can also depend on who is affected by the actions and at what time and place. As John Broome [10] puts it, the good associated with an outcome of the exercise of an option has a number of different *locations*: people, places, times, qualities and states of the world. The desirability of being served cold beer, for instance, depends on the location of this service: it's good if the beer is served to me, in the evening, with a smile and when I have not had a few too many already; bad when it's for my children, or first thing in the morning, or during a philosophy lecture.

Locations of benefit are easily confused with *perspectives* on benefit, because many of the sorts of things that serve as the former, also serve as the latter. A perspective is a standpoint from which a judgement is made. You and I may reach different judgements because our standpoint differs: we might have different evidence and reasoning skills, perhaps different interests and biases, that lead us see things differently. Our standpoint also varies with time – as we get older, for instance, our aesthetic standards 'mature' – and sometimes with place and social role. But the way in which benefit varies with perspective need not be the same as the way it varies with location. I might now judge that it would be good if I were seen by a dentist next week. On the other hand, next week I might judge that the dentist is best avoided. Thus what I judge now to be the benefit next week of making an appointment now to see the dentist will be judged next week as anything but a benefit (even though the benefit, as judged from any temporal standpoint, does not depend when it obtains).

A consequence, in this more refined picture, is something that happens at a multi-dimensional location. Any one of these dimensions may be used to construct a two-factor matrix representation of a decision problem. For instance, when the problem is like the cake-eating one we can work with a time-consequence decision matrix like the following, in which the consequences of the relevant options (having or foregoing another slice of cake) at each relevant point in time are displayed.

	Times	
Actions	Now	Future
Another slice of cake	Pleasure from eating Host will be pleased	Risk of obesity and ill-health
Forego more cake	Forego pleasure Disappoint host	Likelihood of good health

A table of this kind makes it easy for the decision maker to focus on the question of how to weigh up present and future costs and benefits. Similar tables can be drawn up to assist reasoning with other locations, having different columns for the different people affected by the actions or the places at which the consequences occur, for instance. In the road building example for instance the salient locations are the people affected by the possible policy decisions. A person-consequence table helps the decision maker focus on the issue of the distribution of the benefits and costs to different people associated with each policy.

How decisions depend on the distribution of benefit across different dimensions of locations has been studied in different branches of decision theory: across states in the theory of decision making under uncertainty, across people in social choice theory, across time in intertemporal decision theory, across different qualities in multicriteria decision theory and so on. Moreover, the formal similarities between decision problems involving different locations has been a rich source of inspiration for decision theorists and has encouraged abstract examination of assumptions about the relationship between evaluations of consequences, locations and options. For the rest of this essay however I will focus on the decision problems in which uncertainty about the state of the world is the central feature. In fact the focus will be even more narrow than this, for I will say nothing about the very important case in which the events of which we are uncertain are the actions of other agents, leaving treatment of this to the chapter on game theory. Nonetheless many of the basic lessons drawn from the discussion here will apply in these other fields as well.

34.2.2 Choosing a Frame

It is typically possible to represent the decision problem one faces in more than one way: for instance, by location-consequence matrices that differ with respect to the locations they pick out or with regard to how finely the locations and consequences are individuated. In particular, they may be specified more or less finely or precisely, with the implication that a decision problem can always be refined (or coarsened) by adding detail to (or removing detail from) the description of the states and the consequence. This raises the question as to whether there are better and worse ways of representing a decision problem and if so, what these are. There are two claims that I want to make in this regard: firstly that not all representations of a decision problem are equally good and, secondly, that many representations are nonetheless permissible. This latter point is of some importance because it follows that an adequate decision theory must be 'tolerant' to some degree of the manner in which the problem is represented and that the solution it gives to a decision problem should be independent of how the problem is represented.

Let us start with the first claim, that some representations of a problem are better than others. A representation of a decision problem should help us arrive at a decision by highlighting certain features of the problem and in particular those upon which the decision depends. There are at least two considerations that need to be traded off when talking about the usefulness of a representation: the quality of the decisions likely to be obtained and the efficiency of obtaining them. To make a good decision, a decision maker must give appropriate weight to the factors upon which the decision depends. In deciding whether to take an umbrella or not, for instance, I need to identify both the features of the possible outcomes of doing so that matter to me (e.g. getting wet versus staying dry) and the features of the environment upon which these outcomes depend (e.g. the eventuality of rain). Furthermore I need to determine how significant these features are: how desirable staying dry is relative to getting wet, how probable it is that it will rain, and so on. If my representation of the decision problem is too sparse I risk omitting features that are relevant to the quality of the decision. If I omit possible weather states from my representation of the umbrella-taking decision, then I will fail to take into account factors – in particular the probability of rain – upon which the correctness of the decision depends. So, *ceteris paribus*, a representation that includes more relevant features will be better than one that does not.

One way of ensuring that no relevant features are omitted is simply to list *all* the features of possible outcomes and states of the world. But drawing up and making use of such a list is clearly beyond our human capabilities and those of any real agents. Reaching judgements costs in terms of time and effort. Representations that include too many features will result in inefficient decision making requiring more resources than is justified.[2] So, *ceteris paribus*, a simpler representation will be better than a more complicated one.

Achieving a good trade-off between accuracy and efficiency is not just a matter of getting the level of complexity right. It is also a matter of identifying the most useful features to represent explicitly. It is useful to represent a feature if it is (sufficiently) relevant to the decision and if we can determine what significance to attach to it. A feature of the state of the world or of a consequence is relevant to a decision problem if the choice of action is sensitive to values that we might reasonably assign to these features. For instance, whether it is desirable to take an umbrella with me or not will be sensitive to the probability of rain, but not sensitive at all to the probability of a dust storm on Mars.

The second aspect of usefulness is equally important. A representation should be appropriate to our informational resources and our cognitive capabilities in specifying features of the environment that we are capable of tracking and features of consequences that we are capable of evaluating. If the weather is relevant to my decision as to whether to take an umbrella or not, but I am incapable of reaching a judgement on the likelihood of rain or (perhaps I have no information relevant to the question or I don't understand the information I have been given) then there is little point in framing the decision problem in terms of weather contingencies. A good representation of a problem helps us to bring the judgements we are able to make to bear on the decision problem.

It follows of course that whether a framing is a useful one or not will depend on properties of the decision maker (and in more than one way). Firstly whether the features of the problem it represents are relevant depends on what matters to the decision maker and hence what sort of considerations her decisions will be sensitive to. And secondly whether a representation facilitates decision making will depend on the cognitive abilities and resources of the decision maker. Both of these will vary from decision maker to decision maker and from one time and context to another. It is clearly desirable therefore that a decision theory be 'representation tolerant' to as great a degree as possible, in the sense of being applicable to a decision problem irrespective of how it turns out to be useful for the decision maker to represent it.

34.3 Modelling Uncertainty

The modern theory of decision making under uncertainty has its roots in eighteenth century debates over the value of gambles, with Daniel Bernouilli (in [4]) giving the earliest precise statement of something akin to the principle of maximising expected utility. The first axiomatic derivation of an expected utility representation of preferences is due to Frank Ramsey [27] whose treatment in many way surpasses those of later authors. But modern decision theory descends from Savage, not

[2]What level of resources is justified will of course depend on what is at stake.

Ramsey, and it is in his book '*The Foundations of Statistics*' that we find the first rigorous simultaneous derivation of subjective probabilities and utilities from what are clearly rationality conditions on preference.

It is to Savage too that we owe the representation of the decision problem faced by agents under conditions of uncertainty that is now standard in decision theory. Savage distinguishes three types of object: states, consequences and actions. States of the world completely capture all the possible facts that might prevail in the decision situation that affect the outcome of acting. Consequences, on the other hand, are the features of the world that matter to the decision maker, such as that he is in good health or wins first prize in a beauty contest or is allowed to sleep late on a Sunday morning. Actions are the link between the two, the means by which different consequences are brought about in different states of the world. Formally, for Savage, they are just functions from states to consequences.

Although this tripartite distinction is natural and useful, Savage imposes some quite stringent conditions on these objects and the relationships between our attitudes to them. Firstly, states are required to be causally independent of the action the agent performs, while consequences are causally dependent on both the action and the state of the world. Secondly, the desirability of each consequence is required by Savage to be independent of the state of the world in which they are realised and of our beliefs about them, and vice versa (Binmore [5] calls this Aesop's principle). Both these conditions must hold if the representation of a decision problem by the kind of state-consequence matrix given in Table 34.1 can be transformed into a probability-utility matrix of the kind given by Table 34.2. The first ensures that the same probabilities can be applied to the states in comparing acts and the second that the utilities attached to consequences are state-independent.

The upshot is that Savage's theory is not partition independent in the way that I argued was desirable. Decision makers must represent the problems they face in a way which respects the conditions of probabilistic independence of the states from the acts and desirabilistic independence of the consequences from both the states and the acts. It is not always natural for us to do so. For instance in our earlier example of a decision as to whether to walk or take a bus we considered consequences such as paying for a ticket. But the desirability of such consequences are not state-independent. In particular they depend on all the possible contingencies that might arise, such as a medical emergency or an unexpected credit card bill, that

Table 34.2
Utility-probability matrix

	States of the world		
Options	$P(s_1)$	...	$P(s_n)$
A^1	$u(C_1^1)$	...	$u(C_n^1)$
...	...	...	...
A^m	$u(C_1^m)$	...	$u(C_n^m)$

require me to spend money. If too many of them arise a ticket would simply be unaffordable, if not many do it may be a trivial expense.[3]

34.3.1 State Uncertainty

A second feature of the representation of decision problems by a probability-utility matrices requires discussion. For Savage, an agent's uncertainty about what to do derives entirely from her uncertainty about what the state of the world is. This 'fundamental' uncertainty is captured by a probability function on the states of the world, measuring the degrees to which the agent judges or believes each state to be the actual one. There are two criticisms of this view of uncertainty that should be considered.

Firstly, the Savage model seems to ignore other forms of uncertainty and in particular the uncertainty that we might have regarding what value to attach to consequences and the uncertainty we might have regarding what actions are available. Both will be examined in more detail below

Secondly, there seems to be a significant difference between being unsure about when someone will arrive because one lacks precise information about their time of departure, traffic conditions, the route they have taken, and so on, and having absolutely no idea when they will arrive because you don't know when or whether they have left, whether they are walking or driving or indeed whether they even intend to come. In the former case, the information one holds is such as to make it possible to assign reasonable probabilities to the person arriving within various time intervals. In the latter, one has no basis at all for assigning probabilities, a situation of radical uncertainty that we previously termed *ignorance*. It may be rare for us to be totally ignorant, but situations of partial ignorance (or ambiguity), in which the decision maker is unable to assign determinate probabilities to all relevant contingencies, are both common and important.

More generally, according to some critics Savage's representation fails to distinguish between the different levels of confidence we might have, or have reason to have, in our probability judgements. Compare a situation in which we are presented with a coin about which we know nothing and one in which we are allowed to conduct lengthy trials with it. In both situations we might ascribe probability one-half to it landing heads on the next toss: in the first case for reasons of symmetry, in the second because the frequency of heads in the trials was roughly 50%. It seems reasonable however to say that our probability ascriptions are more reliable in the second case than the first and hence that we should feel more confident

[3]Savage was perfectly aware of this objection and drew an important distinction between small-world and grand-world decision problems. But he never produced a theory which, to his own and others satisfaction, explained how to convert grand-world problems into small-world ones satisfying the two requirements.

in them. To take this into account our state of uncertainty might be represented not by a probability function but by a set of reliability judgements over possible probabilities, or more formally, by a function $R : \Pi \rightarrow [0, 1]$, defined on a set $\Pi = \{p_i\}$ of probability functions on the set of events, and such that $\sum_i R(p_i) = 1$. These reliabilities could be incorporated into decision making in various ways, but the most natural perhaps is to prescribe choice that maximises reliability weighted expected utility. It is not difficult to see that such a rule is formally equivalent to maximising expected utility relative to the probability function $\bar{p} = \sum_i p_i.R(p_i)$. This is not an objection to introducing second-order probabilities, but merely to point out that use of reliabilities is more natural in the context of belief formation, than in decision making.

34.3.2 Evaluative Uncertainty

The distinctions between certainty, risk and uncertainty are standardly used only to characterise the agent's state of knowledge of the world. But it is equally important to distinguish cases in which consequences have known, or given, objective values and those in which these values are either unknown and the decision maker must rely on subjective evaluations of them, or do not exist and the decision maker must construct them. The possibility of evaluative uncertainty is typically ignored by decision theorists, because of their (often unconscious) attachment to the view that what makes a consequence valuable or otherwise (to the agent) is just that she desires it to some degree, or that she prefers it to a greater or lesser extent to other consequences. If this view were correct, talk of evaluative uncertainty would be misleading as one is not normally uncertain about what one's own judgement on something is (just what it should be).

There are however at least two ways in which one can be uncertain about the value to attach to a particular consequence or, more generally, whether one consequence is preferable to another. Firstly one may be uncertain about the factual properties of the consequence in question. If possession of the latest Porsche model is the prize in a lottery one is considering entering, one may be unsure as to how fast it goes, how safe it is, how comfortable and so on. This is uncertainty of the ordinary kind and, if one wishes, it can be 'transferred' (subject to some qualifications discussed in the next section) from the consequence to the state of the world by making the description of the consequence more detailed. For example, the outcome of the lottery may be regarded as having one of several possible consequences, each an instantiation of the schema 'Win a car with such and such speed, such and such safety features and of such and such comfort', with the actual consequence of winning depending on the uncertain state of the world.

Secondly one can be unsure as to the value of a consequence, not because of uncertainty about its factual properties, but because of uncertainty about whether these properties are valuable or as to how valuable they are. One may know all the specifications, technical or otherwise, of the latest Porsche and Ferrari models, so that they can be compared on every dimension, but be unsure whether speed matters more than safety or comfort. Once all factual uncertainty has been stripped from a consequence by detailed description of its features, one is left with pure value uncertainty of this kind.

When we assume that values are given, we take this uncertainty to have been resolved in some way. This could be because we assume that there is a fact of the matter as to how good a consequence is or as to whether one outcome is better than another, a fact that would be detailed by the true axiology. But it could also be because the description of the decision problem itself comes with values 'built-in'. For instance, in a problem involving a decision between two courses of medical treatment, it may be that a limited number of value considerations apply in the assessment of these treatments: number of patients saved, amount of discomfort caused, and so on. The decision theorist will be expected in such circumstances to apply only the relevant values to the assessment of the options, and to set aside any other considerations that he or she might 'subjectively' consider to be of importance. A large number of applications of expected utility theory take place in this sort of environment, when the issue of what values to apply have been settled by prior public policy debate.

In many situations, however, values are not given in any of these ways and the agent may be uncertain as to the value she should attach to the relevant prospects. In these circumstances the utility that the agent assigns to a consequence will reflect a subjective value judgement expressing her evaluative uncertainty. What kind of judgement this is a matter of considerable controversy, in particular regarding whether it expresses beliefs about factual properties of the consequences on which its desirability depends, beliefs about the objective normative properties instantiated by the consequences, or a judgement of a different kind to a belief. Formally, on the other hand, the only matter of concern is whether such judgements are adequately captured by utility ascriptions. If they are (as I believe), then considerations of evaluative uncertainty will have interpretative, but not formal, implications for expected utility theory. If not, new formal tools will need to be developed.

34.3.3 Option Uncertainty

In the state-consequence representation of a decision problem that we started with, actions were associated with definite consequences, one for each state of the world. But in real decision problems we are often unsure about the relationship between actions, worlds and consequences in essence because we do not know what consequence follows in each possible state of the world from a choice of action. For instance, we may be uncertain as to whether taking an umbrella will certainly have the consequence of keeping us dry in the event of rain. Perhaps the umbrella has holes, or the wind will blow it inside out or the rain will be blown in from the sides. We can put this difficulty in slightly different terms. If an action is *defined* as a particular mapping from states to consequences, then no uncertainty can arise about its consequences. But what we will then be unsure about is which actions are actually available to us i.e. which of the various hypothetical actions are real options. Whether we describe the problem as uncertainty about what options we have or as uncertainty about the consequences, in each state of the world, of exercising any of the options we know we have, is of little substance, and I shall use the same term – option uncertainty – to denote both.

Decision theorists tend to 'push' this uncertainty into the states of the world, by refining their description until all such contingencies are taken care of. They will regard a state of the world as insufficiently described by the absence or presence of rain, and argue that one needs to specify the wind speed and direction, the quality of the umbrella, etc. There are two reasons why this strategy will not work on all occasions. Firstly because, according to our best scientific theories, the world is not purely deterministic. When the conditions under which a coin is tossed do not determine whether a coin will land heads or tails, for instance, the act of tossing the coin does not have a predictable consequence in each state of the world. And secondly, even if we are in a purely deterministic set-up, it may be subjectively impossible for the decision maker to conceive of and then weigh up all the relevant contingencies or to provide descriptions of the states of the worlds that are sufficiently fine-grained as to ensure that a particular consequence is certain to follow, in each state, from the choice of any of the options open to them.

There are three strategies for handling this problem. One way is to use descriptions of the states of the world that identify the set of the conditions sufficient for the determination of the consequence, given the performance of the action, without actually enumerating the conditions. For instance, instead of defining actions in terms of states and consequences, we could take actions and consequences as our primitives and then define states of the world as consequence-valued functions ranging over actions. Similar strategies are advocated in the philosophical literature. Lewis [25], for instance, treats states as 'dependency hypotheses', which are just maximally specific propositions about how consequences depend causally on acts, while Stalnaker's [32] suggests that a state of the world be denoted by a conjunction of conditional sentences of the form 'If action A were performed then consequence C would follow; if action A' were performed then consequence C' would follow; if . . . '. By pursuit of any version of this strategy, option uncertainty is transformed into a particular kind of state uncertainty, namely uncertainty as to the true mapping from actions to consequences or as to the truth of the conjunction of conditionals that describes it.

A second strategy is to coarsen the description of the consequences to the degree necessary to ensure that we can be certain it will follow from the exercise of an option in a particular state. As Richard Jeffrey [18] points out, consequences may be identified by nothing more than act-state pairs, such as taking an umbrella in the rain and taking it in the snow. In his approach the outcomes of acts are taken to be *logical* consequences of act-state descriptions, but the coarsening of consequence-descriptions necessary to ensure option certainty need not be as radical as this.

Pursuit of this strategy converts option uncertainty, not into ordinary uncertainty about the state of the world, but into uncertainty about the desirability of the consequence as described – one part of what I previously called value uncertainty. We may be sure that the act of taking an umbrella will have the consequence in a rainy state of being able to protect ourselves against the rain by opening the umbrella. But whether this is a good thing or not depends on contingencies that by assumption we are unable to enumerate or identify. How bad it is to get soaked, for instance, depends on how cold the rainwater is. And rain temperature may be a variable about whose determinants we know very little. Whatever utility value we assign to the coarse-grained consequence of having an umbrella as rain-protection will embody this uncertainty and hence should be susceptible to revision.

The last strategy to consider, also originating in Richard Jeffrey's work, is the most radical and involves embracing option uncertainty rather than trying to reduce it to some other kind of uncertainty. This requires to think of actions not as functions from states to consequences, but as probability distributions over consequences. We will discuss this strategy in greater detail later on when presenting Jeffrey's theory.

34.4 Choice and Preference

Earlier we claimed that when a decision maker faces no uncertainty she should choose the option with the best consequences. There are two basic assumptions involved here. The first is Consequentialism: the idea that the only thing relevant to choice in these circumstances is the outcome or consequence of so choosing and not any feature of the means or process by which this outcome is achieved.[4] The second assumption is that there exists some value standard applicable to the outcomes which licenses talk of one or more of them being best. Jointly they entail that the decision maker ought to choose the action with the best consequence.

The value standard can have different interpretations, which in turn will imply different readings of the ought expressed by the choice principle. When the relevant standard is a subjective one, such as that based on the decision maker's preferences, the ought expresses a requirement of rationality, namely that she make a choice that is consistent with her subjective evaluation of its outcome. When the standard is an objective one, the prescription is to choose the action that has the outcome that is objectively best.

[4]It should be noted that the assumption of Consequentialism does not rule out a role for non-consequentialist considerations, in particular in determining the composition of the set of options. For instance if some actions are not permissable because they would violate someone's rights then they would be excluded from the option set. What it does assume is that such non-consequentialist considerations do not enter beyond this point.

34.4.1 Preference Relations

Let us try and make these claims more exact. First, some basic vocabulary. Let $X = \{\alpha, \beta, \ldots\}$ be a set of objects and let R be a binary relation on X. We say that R is:

1. *Transitive* iff for all $\alpha, \beta, \gamma \in X$, $\alpha R\beta$ and $\beta R\gamma$ implies that $\alpha R\gamma$ (and intransitive otherwise)
2. *Complete* iff for all $\alpha, \beta \in X$, $\alpha R\beta$ or $\beta R\alpha$ (and incomplete otherwise)
3. *Reflexive* iff for all $\alpha \in X$, $\alpha R\alpha$ (and irreflexive otherwise)
4. *Symmetric* iff for all $\alpha, \beta \in X$, $\alpha R\beta$ implies $\beta R\alpha$
5. *Antisymmetric* iff for all $\alpha, \beta \in X$, $\alpha R\beta$ and $\beta R\alpha$ implies that $\alpha = \beta$
6. *Acyclic* iff for all $\alpha_1, \alpha_2, \ldots, \alpha_n \in X$, $\alpha_1 R\alpha_2, \alpha_2 R\alpha_3, \ldots, \alpha_{n-1} R\alpha_n$ implies that not $\alpha_n R\alpha_1$.

In conditions of certainty, the assumption of Consequentialism implies that an option may be identified with the consequence of choosing to exercise it. So we can let the same set of alternatives represent both the options amongst which the agent must choose and the outcome of doing so. (A note of caution: to say that the consequence is certain is not to say that it is fully specified, so there may be disguised uncertainty.)

The decision maker's value standard is represented by a binary relation $\succeq$ on this set. Intuitively '$\alpha \succeq \beta$' means, on a subjective interpretation, that β is not preferred to α; on an objective one, that β is not better than α . In accordance with standard terminology I will call $\succeq$ a weak preference relation, without meaning thereby to impose a subjective interpretation. The strict preference relation $\succ$, indifference relation $\approx$, and comparability relation $\bowtie$, all on the set of alternatives X, are then defined by:

1. $\alpha \succ \beta$ iff $\alpha \succeq \beta$ and not $\beta \succeq \alpha$
2. $\alpha \approx \beta$ iff $\alpha \succeq \beta$ and $\beta \succeq \alpha$
3. $\alpha \bowtie \beta$ iff $\alpha \succeq \beta$ or $\beta \succeq \alpha$.

It will be assumed throughout that $\succeq$, $\approx$ and $\bowtie$ are all reflexive, that $\approx$ and $\bowtie$ are also symmetric, and that $\succ$ is a symmetric. It is common to assume that these relations are weak orders, i.e. that they are both transitive and complete. But the status of these two properties is very different. There are compelling grounds, on both subjective and objective interpretations, for assuming transitivity. Some authors have even argued that it belongs to the logic of comparative relations that they should respect it (e.g. Broome [10]). Completeness on the other hand cannot plausibly be said to be a requirement of rationality. Not only are we often unable to reach a judgement or don't need to, but on occasion it would be wrong of us to do so, e.g. when we expect to receive decisive information in the near future. Nor are there compelling grounds for supposing that objective betterness is complete: some goods may simply be incommensurable.

Why then do decision theorists so often assume completeness? One reason is that it makes the business of proving representation theorems for decision principles a lot easier mathematically speaking. A second reason lies in the influence of the Revealed Preference interpretation of decision theory. On this view having a preference for one alternative over another just is to be disposed to choose the former over the latter when both are available. Since agents are plausibly disposed one way or another in any choice situation (some choice is made after all), it follows that revealed preferences must be complete. But this interpretation has little to offer decision theory construed as either an explanatory or a normative theory. For if preferences are simply choice dispositions then citing someone's preferences cannot provide either an explanation or a justification of what they choose.[5]

The third argument, that completeness should be regarded as a requirement of coherent extendability, is the most cogent. The idea is this: although it is not a requirement of rationality that we should have reached a preference judgement regarding all prospects, it should nonetheless be possible to extend our current set of preferences to one which is both complete and consistent by reaching judgements about new prospects. If our current judgements are coherently extendible, then we can be sure that reaching new ones will not require a revision of our current preferences in order to retain consistency. Or to put it the other way round, if our preferences are not coherently extendible then as we reach judgements on prospects about which we formerly had no opinion, we run the risk of finding ourselves with an inconsistent set of preferences. Indeed we are sure to if we make enough new judgements. This does not give us a decisive reason to conform to the requirement of coherent extendability, as inconsistency can be avoided by revising some of our old judgements when we make new ones. But it does suggest that, *ceteris paribus*, it is pragmatically desirable to do so.

Suppose we accept the case for conformity with the requirement of coherent extendability. Then by studying the case of complete preferences we can derive a set of constraints on our beliefs and desires that must be fulfilled in order that they too be coherently extendible. For instance, if we can show that the rationality of a complete set of preferences implies that our beliefs must have some particular property P, then we can conclude that our (incomplete) beliefs must have the property of being extendible to a set of beliefs having P.

34.4.2 Choice

Let X be a finite set of alternatives and C be a choice function on $\wp(X)$: a mapping from subsets $A \subseteq X$ to subsets $\varnothing \subset C(A) \subseteq A$. The choice function C will be said to be *specific* iff its range is restricted to singleton sets. Intuitively $C(A)$

[5]To be clear, it is the ontological doctrine just described that should be rejected, not the associated epistemological doctrine according to which knowledge of preferences ultimately rests on observations of choice. The latter, in contrast to the former, has much to recommend it.

is the set of objects from the set A that could be chosen: could permissibly be so in normative interpretations, could factually be so in descriptive ones. When C is specific a further interpretation is possible, namely that $C(A)$ is the object observed to be chosen from the set A.

We are especially interested in the case when a choice function C can be said to be based on, or determined by, a weak preference relation $\succeq$. A natural condition for this being the case is that an object is chosen from a set if and only if no other object in the set is strictly preferred to it. Formally:

$$\textbf{(PBC)}\ \alpha \in C(A) \Leftrightarrow \neg\exists\beta \in A : \beta \succ \alpha$$

PBC is sometimes called the *Maximality* condition. With a qualification that will be made a little later on, PBC seems necessary for preference based choice. But is it sufficient? Sen [30] suggests to the contrary that it is not enough that nothing be (comparably) better than what is chosen, it must also be the case that what is chosen is (comparably) no worse than any alternative. More formally, preference-based choice should satisfy Strong PBC or as it is more commonly called:

$$\textbf{(Optimality)}\ \alpha \in C(A) \Leftrightarrow \forall\beta \in A, \alpha \succeq \beta$$

To examine these proposals let us use the weaker criterion of maximality to derive a set-valued function on $\wp(X)$ from the agent's preferences by defining, for all $A \in \wp(X)$:

$$C_{\succeq}(A) := \{\alpha \in A : \neg\exists\beta \in A,\ \beta \succ \alpha\}$$

Then:

Theorem 1

(a) $C_{\succeq}$ is a choice function iff $\succeq$ is acyclic
(b) Choice function $C_{\succeq}$ satisfies Optimality $\Leftrightarrow$ $\succeq$ is complete.
(c) Choice function $C_{\succeq}$ is specific $\Leftrightarrow$ $\succ$ is complete.

The proof of (a) and (b) can be found in Sen [30], (c) follows immediately.

Two comments. Firstly, Theorem 1(b) shows that to require satisfaction of Strong PBC is to make completeness of an agent's preferences a condition for their choices to be preference-based. But this seems unreasonable. As we have seen, completeness has little normative appeal as a preference condition and someone with incomplete preferences whose choices satisfy PBC can be said to be making these choices in the light of their preferences to the maximum extent possible. On the other hand, as Theorem 1(c) shows, neither satisfaction of PBC nor of Strong PBC is sufficient for preference to determine the choice of a specific alternative. For when two alternatives are incomparable or indifferent then both are permissible choices. The upshot is that we should regard satisfaction of PBC as the mark of preference-based choice, noting that only when an agent's *strict* preferences are complete will this condition suffice for preference to determine choice completely.

Secondly, there are various ways of giving substance to the notion of being preference-based. On an *explanatory* reading, it means that the decision maker's preferences explain the choices that she makes by providing the reasons for them. On the other hand, on a *normative* reading, it means that the decision maker's preferences rationalise or justify the choices that she makes. Revealed Preference theorists regard neither of these interpretations as warranted and advocate a third, purely *descriptive* reading, according to which 'preference-based' means no more than that a choice function can be represented by a preference relation. The first two interpretations give primacy to preferences, with PBC doing service as a principle by which we infer properties of choice from properties of preferences. The last interpretation, on the other hand, gives primacy to the properties of choices and to the problem of deriving properties of preferences from them.

The main condition of Revealed Preference theory is the Weak Axiom of Revealed Preference (WARP), which says that if α should be chosen from a set containing β, then whenever β should be chosen and α is available, α should also be chosen. Formally, we follow Sen [30] in breaking this down into two conditions:

Axiom 2

(WARP) Suppose $\alpha, \beta \in B \subseteq A$. *Then:*
(Condition Alpha) If $\alpha \in C(A)$, *then* $\alpha \in C(B)$
(Condition Beta) If $\alpha, \beta \in C(B)$ *and* $\beta \in C(A)$, *then* $\alpha \in C(A)$

Theorem 3 *Let* C *be a choice function on* $\wp(X)$. *Then:*

(a) *C satisfies Alpha if there exists a relation $\succeq$ on X such that $C = C_{\succeq}$*
(b) *C satisfies WARP iff there exists a weak order $\succeq$ on X such that $C = C_{\succeq}$.*

Proof (a) Suppose that there exists a relation $\succeq$ on X such that $C = C_{\succeq}$, that $\alpha \in B \subseteq A$ and that $\alpha \in C(A)$. By definition $\forall \beta \in A$, $\alpha \succeq \beta$ or $\beta \not\succeq \alpha$. Hence $\forall \beta \in B$, $\alpha \succeq \beta$ or $\beta \not\succeq \alpha$. Then by definition, $\alpha \in C(B)$. (Note that the converse is not true: it does not follow that if C satisfies Alpha that $C_{\succeq}$ is a choice function.) The proof of (b) can be found in Sen [30]. ■

Theorem 3(b) seems to give Revealed Preference theory what it needs, namely a characterisation of both the conditions under which the agent's preferences are 'revealed' by her choices and of the properties of these preferences. In particular if her choices respect WARP then a transitive and complete weak preference relation can be imputed to her which, together with PBC, determines these choices. But this observation is of very little normative significance in the absence of a reason for thinking that choices should satisfy the WARP axiom. The problem is that, unless $\succeq$ is complete, a preference-based choice function need not satisfy condition Beta. Suppose, for example, that the agent cannot compare α and β, but that no object in B is preferred to either. So both are permissible choices. Now suppose that $A = B \cup \{\gamma\}$ and that γ is preferred to α but not comparable with β. Then β is a permissible choice but not α. Since it is no requirement of rationality that preferences be complete, I take it that WARP is not normatively compelling. Hence preferences are not fully revealed by the choices they determine.

Condition Alpha is sometimes called the Independence of Irrelevant Alternatives condition in view of the fact that it implies that the framing of the choice set shouldn't influence an agent's preferences. The fact that it is implied by PBC, in the sense given by Theorem 3(a), is grounds for thinking it should be respected by choices. But as Sen has pointed out, the composition of the choice set itself can matter. When offered the choice between staying for another drink or leaving the party, I might choose to stay. But if offered the choice between leaving the party, staying for a drink or staying to participate in a satanic ritual I may well choose to leave.

It seems therefore that what preference-based choice requires is something more subtle than picking non-dominated alternatives relative to a given preference relation. It is this: that we should not choose any alternative from a set, when there is another in that set that it strictly preferred to it, *given* the set of alternatives on offer. Making this criterion for preference-based choice formal is tricky. Nonetheless, as we shall see later on, it has important conceptual implications.

34.5 Utility Representations

Preference relations that are weak orders can be represented numerically, thereby allowing for an alternative characterisation of rational choice. More exactly, let us call a function $U : X \to \mathbb{R}$, a *utility representation* of the weak order $\succeq$, iff for all $\alpha, \beta \in X$:

$$\alpha \succeq \beta \Leftrightarrow U(\alpha) \geq U(\beta)$$

Then:

Theorem 4 *Suppose that the preference relation $\succeq$ is a weak order on a countable set X. Then there exists a function U that is a utility representation of $\succeq$. Furthermore U' is another such a utility representation iff U' is a positive monotone transformation of U i.e. there exists a strictly increasing function $f : \mathbb{R} \to \mathbb{R}$ such that $U' = f \circ U$.*

See Kranz et al. [23, Section 2.1] for a proof of this theorem. In case X is not countable, numerical representability is not assured for weak orders unless X has a 'dense' subset – one containing elements lying (in the weak order) between any two prospects in X. When the preference order is lexicographic for instance this condition will be violated. In contrast, any continuous weak relation on a connected topological space *is* numerically representable by a continuous function (see Kreps [24] for details), where the continuity of a relation is defined as follows:

Continuity: For any subset $\{\alpha_i\}$ such that $\alpha_1 \models \alpha_2 \models \ldots \models \alpha_n$ and $\beta \succsim \alpha_n \succsim \gamma$, $\beta \succsim \alpha_i \succsim \gamma$, for all large i.

These representation results have an obvious weakness: the assumption that preferences are complete. But it is in fact simple enough to generalise the result to all transitive preference relations, complete or otherwise, by defining a utility representation of the transitive relation $\succeq$ to be a set $\mathcal{U}$ of utility functions such that for all $\alpha, \beta \in X$, $\alpha \succeq \beta$ iff for all $U \in \mathcal{U}$, $U(\alpha) \geq U(\beta)$. Such a set $\mathcal{U}$ may be constructed by placing in it, for each possible 'completion' of $\succeq$, a utility function that represents the resultant weak order. It follows that the set will inherit the uniqueness properties of its elements i.e. that $\mathcal{U}$ will be unique up to positive monotone transformation. More formally, let us say that a preference relation $\succeq$ on a set X is represented by a set of real-valued functions Φ just in case for all $\alpha, \beta \in X$,

$$\alpha \succsim \beta \Leftrightarrow F \in \forall F \in \Phi, F(\alpha) \geq F(\beta)$$

Then:

Theorem 5 (Evren and OK [14, p. 5]) *Let $\succsim$ be a weak order on a set X. Then there exists a set Φ of real-valued functions that represents $\succeq$.*

Theorem 4, together with the discussion in the previous section, implies that choices that are preference-based, in the sense of satisfying Optimality, are utility maximising. But one must be careful not to attach too much significance to this characterisation of utility maximisation. The mere existence of the function U that represents preferences does not in itself explain the agent's preferences, nor does it justify them. It merely describes them numerically. The contrast with belief is instructive. Under certain conditions (which will be described later on), a correspondence can be established between beliefs and preferences over specific alternatives, namely those whose desirability depends on the truth of the contents of the beliefs in question. In this case we are inclined to speak of the beliefs being the cause or reason for the preference. This is because we have independent scientific grounds for attributing causal powers to beliefs. Similarly for preferences. But we have no such grounds for attributing causal or justificatory powers to the utilities of alternatives distinct from the agent's preferences for them. We might speak, as I will do, of a utility judgement and the considerations upon which it is based. But this is no more than shorthand for talk of preferences, in which transitivity and completeness are taken for granted. Such talk has its dangers: in particular it can encourage one to read more into the numbers than is allowed by the representation. But it is also convenient; hence our interest in being clear about their content.

Theorem 4 is rather weak, as the uniqueness properties of the utility representation it establishes make manifest. With stronger assumptions about the set of alternatives and preferences over them, more interesting properties of the utility representation can be derived and its uniqueness increased. In the next couple of sections we will characterise the conditions on weak orders under which there exists an additive utility representation of it. Since our primary interest is in the normatively significant properties of preference relations and corresponding utility representations, and not with problem of numerical representation itself, I will be somewhat cavalier in my statement of the more technical conditions on preference orders. For example, to obtain a cardinal representation of a weak order it is typically necessary to assume an Archimedean condition; to ensure, in effect, that any two objects in the domain of the weak order are comparable, no matter how far apart they lie in that order. The exact condition required will depend on the nature of the set of alternatives being assumed, and for this reason I will not spell it out each time. For the details on these, the most comprehensive source is Krantz et al. [23].

34.5.1 Conjoint Additive Structures

For a first extension we return to our initial representation of a decision problem as a matrix of locations and consequences. The objects of choice here are ordered sets of outcomes, one for each possible state of the world or, more generally, location. Consequently the set of alternatives forms a product set of the form $X = X_1 \times X_2 \times \ldots \times X_n$, where each X_i is the set of possible outcomes at the ith location. A profile $(x_1, x_2, \ldots, x_n) \in X$ could be a set of attributes of a good, for instance, or a set of allocations to individuals, or a set of events at different times.

This structure allows for stronger assumptions about rational preference and a correspondingly richer utility representation of them. For any subset K of the set of possible locations, let $X_K := \prod_{j \in K} X_j$. For any partitions $\{K, L\}$ of the set of locations, let (a, c) be the member of X where $a \in X_K$ denotes the values of the locations in K and $c \in X_L$ denotes the values of the locations in L. Then consider:

Axiom 6 (Strong Separability) *For all partitions {K,L} of the set of locations and for all $a, b \in X_K$ and $c, d \in X_L$:*

$$(a, c) \succeq (b, c) \Leftrightarrow (a, d) \succeq (b, d)$$

The axiom of strong separability appears in different contexts under a wide variety of names, most notably Joint Independence [23] and the Sure-thing Principle [29] for the case where locations are states. It has a strong claim to be the most interesting and important of the conditions regularly invoked by decision theorists. On the one hand, it does not have the same normative scope as the transitivity condition. For instance, consider its application to allocations to different individuals. In this context Strong Separability rules out a direct sensitivity to inequality, such as might be manifested in a preference for (a, a) over (b, a) and for (b, b) over (a, b). Similarly in applications to decisions with outcomes at different temporal locations it rules out a preference for novelty over repetition, such as might be manifested in a preference for (a, b) over (b, b) and for (b, a) over (a, a). On the other hand, in many applications and when outcomes are carefully described, the axiom does seem normatively compelling. We return to this in the discussion of decision making under uncertainty.

Let us say that a location l is *essential* just in case there exist x_l and y_l such that $(x_1, .., x_l, \ldots, x_n) \succ (x_1, .., y_l, \ldots, x_n)$ for all $x_1, \ldots, x_n \in X_{N-l}$. And let us say that $\succeq$ is *solvable* on X just in case $(x_1, .., \bar{x}_l, \ldots, x_n) \succ (y_1, .., y_l, \ldots, y_n) \succ (x_1, .., \underline{x}_l, \ldots, x_n)$ implies that there exists an x_l such that $(x_1, .., x_l, \ldots, x_n) \approx (y_1, .., y_l, \ldots, y_n)$. If the preference relation $\succeq$ is Archimedean and solvable on X, then we call the pair $\langle X, \succeq \rangle$ an additive conjoint structure. Then:

Theorem 7 *Let $\langle X, \succeq \rangle$ be an additive conjoint structure with at least three essential locations. Assume that $\succeq$ satisfies Strong Separability. Then there exists a utility representation U of $\succeq$ on X such that for all $(x_1, x_2, \ldots, x_n) \in X$:*

$$U(x_1, x_2, \ldots, x_n) = \sum_{j=1}^{n} u_j(x_j)$$

for some family of functions $u_j : X_j \to \mathbb{R}$. Furthermore if U' and the u'_j are another such a family of utility representations, then there exists constants $a, b, a_j, b_j \in \mathbb{R}$ such that $U' = aU + b$ and $u'_j = a_j.u_j + b_j$.

The proof of this theorem involves three main steps (see Krantz et al. [23] for details). First, we observe that by application of Theorem 4, there exists a utility representation U of $\succeq$ on X, unique up to positive monotone transformation. The second step is to derive location-relative preference relations from $\succeq$, in which essential use is made of Strong Separability. In the light of Theorem 4 this implies the existence of location-relative utility functions – the u_j – also unique up to positive monotone transformation. The final step is to show that judicious choice of scales for the u_j permits U to be expressed as a sum of them.[6]

Theorem 7 has many applications. For a historically important example suppose that the X_j are different individuals and the x_j allocations that are made to them. Then Theorem 7 asserts the existence of an additive utility representation of any set of strongly separable preferences over allocations to individuals. This is typically called a utilitarian representation of social decisions.

34.5.2 Linear Utility

We now consider an even richer structure on the objects and a stronger restriction on preferences sufficient to ensure the existence of a linear representation of them.

[6]It is important to note that it's essential to the possibility of an additive representation that no cross-locational comparisons are possible. For such comparisons would constrain the co-scaling of the u_j and there would then be no guarantee that the permitted co-scaling allowed for an additive representation.

A set of objects X is said to be a *mixture set* iff for all $\alpha, \beta \in X$ and any $0 \leq k \leq 1$, there exists an element in X, denoted by $k\alpha + (1-k)\beta$, such that:

1. If $k = 1$ then $k\alpha + (1-k)\beta = \alpha$
2. $k\alpha + (1-k)\beta = (1-k)\beta + k\alpha$
3. For all $0 \leq l \leq 1$, $l(k\alpha + (1-k)\beta) + (1-l)\beta = lk\alpha + (1-lk)\beta$

Axiom 8 (Linearity) *For all $\alpha, \beta, \gamma \in X$ and any $0 \leq k \leq 1$:*

$$\alpha \approx \beta \Leftrightarrow k\alpha + (1-k)\gamma \approx k\beta + (1-k)\gamma$$

Axiom 9 (Archimedean) *For all $\alpha, \beta, \gamma \in X$, if $\alpha \succ \gamma \succ \beta$ then there exist k and l such that:*

$$k\alpha + (1-k)\beta \succ \gamma \succ l\alpha + (1-l)\beta$$

Theorem 10 (Herstein and Milnor [17]) *Assume that X is a mixture set and that $\succeq$ is an Archimedean weak order on X that satisfies the Linearity axiom. Then there exists a utility representation U of $\succeq$ on X such that for all $\alpha, \beta \in X$:*

$$U(k\alpha + (1-k)\beta) = kU(\alpha) + (1-k)U(\beta)$$

Furthermore U' is another such a utility representation iff U' is a positive linear transformation of U i.e. there exists constants $a, b \in \mathbb{R}$ such that $U' = aU + b$.

One very important application of the idea of a mixture space is to lotteries. Let Z be a (finite) set of outcomes or 'prizes' and let the set of lotteries $\Pi = \{p_i\}$ be a set of a probability mass functions on these outcomes i.e. each $p_i \in \Pi$ is a function from the $z \in Z$ to the interval [0, 1] such that $\sum_z p_i(z) = 1$. For any $p_i, p_j \in \Pi$, let $kp_i + (1-k)p_j$, called the k-compound of p_i and p_j, denote the member of Π defined by:

$$(kp_i + (1-k)p_j)(z) := kp_i(z) + (1-k)p_j(z)$$

It follows that Π is a mixture set of lotteries.

When applied to lotteries the Linearity axiom is typically called the Independence axiom: it says that if two lotteries p_i and p_j are equally preferred then a k-compound of p_i and p_k is equally preferred to a k-compound of p_j and p_k. The Archimedean condition amounts to saying that no matter how good p_i is (how bad p_j is) there is some compound lottery of p_i and p_j which gives p_i such small (large) weight that p_k is strictly preferred to it (it is strictly preferred to it p_k). Or more pithily, everything can be traded off if the probabilities are right.

Theorem 11 (Von Neumann and Morgenstern) *Let $\succeq$ be an Archimedean weak preference order on Π that satisfies the Independence (Linearity) axiom. Then there exists a utility representation U of $\succeq$ on Π and a function $u : Z \to \mathbb{R}$ such that for all $p_i \in \Pi$:*

$$U(p_i) = \sum_{z \in Z} p_i(z).u(z)$$

See Kreps [24] for an instructive proof of this result. Von Neumann and Morgenstern's theorem is usually considered to belong to the theory of decision making under uncertainty and its appearance here bears out my earlier claim that the distinction between certainty and risk is a matter of perspective. When making decisions under risk we *know* what situation we face, even if we don't know what the final outcome will be. This makes it a convenient bridgehead to the topic of uncertainty.

34.6 Decisions Under Uncertainty

It is now time to make more precise the claim that in situations of uncertainty, choices should maximise expected utility. Although this prescription is still consequentialist in spirit the explicit introduction of uncertainty requires a more nuanced expression of what Consequentialism entails in these circumstances. More specifically, in these circumstances, the choice-worthiness of an action depends not only on the consequences of the action but also on the relative likelihood of the possible states of the world in which the consequences might be realised. The prescription to maximise expected utility is made relative to a specification of the probabilities of states of the world and utilities of the consequences. There are thus two relations that need to be examined: the value relation that we discussed before and a possibility or probability relation on the states of the world expressing the decision maker's state of uncertainty. Both the properties of these relations and of the quantitative representations of them are relevant to the derivation of the expected utility principle.

Like the value ordering, the possibility ordering can be given both a subjective and objective interpretation, as can the numerical probabilities based on it. This means that in principle the prescription to maximise expected utility is amenable to four different readings with quite different normative implications. If both are construed objectively (as in, for instance, Broome [10]) then the principle prescribes action which maximises the objective expectation of goodness. If preferences are subjective but probabilities are objective (as they are in Von Neumann-Morgenstern decision theory [35]) then the principle prescribes maximisation of the objective expectation of subjective preference. If both are construed subjectively (as in Savage [29]) then the prescription is to maximise the subjective expectation of subjective preference and so on.

As the normative claims of these different interpretations of expected utility theory are rather different, one should not expect that one type of argument will serve to justify all of them. What we can do however is to build a common platform for such arguments by identifying the properties of the two ordering relations that are necessary and sufficient for the existence of an expected utility representation that justifies (either by rationalising or by normatively validating) the decision maker's choice. By an expected utility representation, I mean an assignment of utilities to consequences and probabilities to states of the world such that the agent's preferences over options cohere with their expected utility.

More formally, let Ω be a set of consequences, $S = \{s_1, s_2, \ldots\}$ be a set of states of the world and $\mathcal{F} = \{A, B, C, \ldots\}$ be the set of subsets of S, called events. Finally let $\Gamma = \{\alpha, \beta, \gamma, \ldots\}$ be the set of actions, where an act is function from S to Ω. In the light of an earlier remark that the difference between states and consequences is pragmatic rather than ontological, it makes sense to treat the latter as a type of event, rather than following Savage in treating them as logically distinct. Formally this means that $\Omega \subseteq \mathcal{F}$.

Let $\succeq$ be a preference relation on the set of actions. A function $V : \Gamma \rightarrow \mathbb{R}$ is called an *expected utility representation* of $\succeq$ iff V is a utility representation of $\succeq$ and there exists a real valued function $u : \Omega \rightarrow \mathbb{R}$ and a probability function $P : \mathcal{F} \rightarrow \mathbb{R}$ such that for all $\alpha \in \Gamma$:

$$V(\alpha) = \sum_{s_i \in S} P(\{s_i\}).u(\alpha(s_i))$$

Our examination will be conducted in two steps. In the first we apply the Von Neumann and Morgenstern theory to decision making under risk, i.e. to conditions in which probabilities are given. And in the second we present Savage's derivation of such a probability from the agent's preferences.

34.6.1 Expected Utility Theory

Suppose that our decision problem takes the form given by Table 34.1. We want to know under which conditions a preference relation over the available options has an expected utility representation. Consider first a situation in which the probabilities of the states of the world are known, a circumstance to which Von Neumann-Morgenstern utility theory is usually applied. It is important to note that to do so we must assume that the decision problem we face can be adequately represented as a choice between lotteries over outcomes. For this it is not enough that we know the probabilities of the states, we must also assume that the only feature of these states which matters for our choice of action is their probability. In particular, the fact that an outcome is realised in one state or another must not influence its desirability. This is known as the assumption of *state-independence*. It appears in an explicit form in the axiomatisations of expected utility theory given by Savage and by Anscombe and Aumann, but is merely implicit in the adoption of the Von Neumann-Morgenstern theory in situations of risk.

Let us call an act that satisfies these assumptions a lottery act. Then, on the basis of Theorem 11, we can make the following claim:

Proposition 12 *If preferences over lottery acts are linear and Archimedean then they have an expected utility representation.*

Normatively the implication is that, given a value relation on outcomes and a probability on states of the world, the only permissible actions are those that maximise the expectation of a utility measure of the value relation. Note that the utility representation is itself constrained by the assumption that preferences are linear because these imply that the manner in which outcomes are weighed against each other is sensitive in a particular way to their probabilities i.e. the assumption encodes a view about how value articulates with probability. This will be reflected, for instance, in the fact that if a preferred outcome has half the probability of a less preferred one, then its value (as measured by utility) must be twice that of the latter if the decision maker is to remain indifferent between the two.

The manner in which utility is cardinalised imposes significant constraints on how utility is interpreted. Suppose for instance that an agent is risk averse with respect to money in the sense that she prefers £50 for certain to a gamble yielding £100 with 50% probability and £0 with 50% probability. Then an expected utility representation of her preferences requires that the utility difference between receiving £50 and receiving £100 will be less than the utility difference between receiving nothing and receiving £50.

Both Arrow [3] and Sen [30] make the following objection. This way of cardinalising utility mixes up the intrinsic value to the agent of the money received with her attitude to risk taking. For it doesn't allow us to distinguish cases in which the agent prefers the £50 for certain to the gamble because of the diminishing marginal value of money from the case in which she does because she dislikes taking risks and is not willing to endanger the £50 for an even chance of doubling her money. Defendants retort that the notion of the intrinsic value being invoked in this argument lacks clear meaning. To give it a content we must be able to say how, at least in principle, we could separate the value component of preferences from the risk component that distorts it, leading to a decomposition of utility into a risk and a value component. There are several recent attempts to do so (see [11, 36] and [34]) and although it remains to be seen whether any are fully adequate, the basic conceptual point remains valid: there may be more than one type of factor contributing to an agent's preferences (apart from her beliefs).

A quite different line of criticism concerns not the interpretation of the expected utility representation, but the claims about rational preference upon which it sits. The main focus of attention in this regard has been the axiom of Independence and its violation in the so-called Allais' paradox. To illustrate the paradox, consider two pairs of lotteries yielding monetary outcomes with the probabilities given in the following table (Table 34.3).

Allais [1] hypothesised that many people, if presented with a choice between lotteries I and II would choose I, but if presented with a choice between III and IV, would choose IV. Such a pattern of choice is, on the face of it, in violation of the Independence axiom since the choice between each pair should be independent of the common consequences appearing in the third column of possible outcomes.

Table 34.3 Allais' paradox

Lottery	Probability	0.01	0.1	0.89
	I	$1000,000	$1000,000	$1000,000
	II	$0	$5000,000	$1000,000
	III	$1000,000	$1000,000	$0
	IV	$0	$5000,000	$0

Nonetheless Allais' conjecture has been confirmed in numerous choice experiments. Moreover many subjects are not inclined to revise their choices even after the conflict with the requirement of the Independence axiom is pointed out to them. So the 'refutation' seems to extend beyond the descriptive interpretation of the axiom to include its normative pretensions.

There are two lines of defense that are worth exploring. The first is to argue that the choice problem is under-described, especially with regard to the specification of the consequences. One common explanation for subjects' choices in these experiments is that they choose I over II because of the regret they would feel if they chose II and landed up with nothing (albeit quite unlikely), but IV over III because in this case the fact that it is quite likely that they will not win anything whatever they choose diminishes the force of regret. If this explanation is correct then we should modify the representation of the choice problem faced by agents so that it incorporates regret as one possible outcome of making a choice. The same would hold for any other explanation of the observed pattern of preferences that refers to additional non-monetary outcomes of choices.

The second line of defensive argument points to the gap between preference and choice. As we noted before, the specification of the choice set can influence the agent's attitudes. This is just such a case. In general the attitude we take to having or receiving a certain amount of money depends on our expectations. If we expect $100, for instance, then $10 is a disappointment. Now the expectation created by presenting the agent with two lotteries to choose from is quite different in the case where the choice is between lotteries I and II and the one in which the choice is between lotteries III and IV. In the first case they are being placed in a situation in which they can expect to gain a considerable amount of wealth, while in the second they are not. In the first they can think of themselves as being given $1000,000 and then having the opportunity to exchange it for lottery II. In the second case they can think of themselves as being handed some much lesser amount (say, whatever they would pay for lottery III) and then being given the opportunity to exchange it for lottery IV. Seen this way it is clear why landing up with nothing is far worse in the first case than in the second. It is because of what one has given up for it. In the first case landing up with nothing as a result of choosing II is equivalent to losing $1000,000 relative to one's expectations, whereas in the second case it is equivalent to losing some much smaller amount.

Both of these defences are unattractive from the point of view of constructing a testable descriptive theory of decision making under uncertainty. The first approach makes it very hard to tell what choice situation the agents face, since the description of the outcomes of the options may contain subjective elements. The second approach makes it difficult to use choices in one situation as a guide to those that will be made in another, since all preferences are in principle choice-set relative. But from a normative point of view they go some way to defending the claim that the Independence axiom is a genuine requirement of rationality.[7]

If we accept the normative validity of the Independence axiom, then we can draw the following conclusion. When the choices that we face can be represented by lotteries over a set of outcomes then rationality requires that we choose the options with maximum expected utility relative to the given probabilities of their outcomes and a given value/preference relation. What this leaves unanswered is why we should think that decision making under uncertainty can be so represented. To answer it we must return to Savage.

34.6.2 Savage' Theory

Savage [29] proves the existence of an expected utility representation of preference in two steps. First he postulates a set of axioms that are sufficient to establish the existence of a unique probability representation of the agent's beliefs. He then shows that probabilities can be used to construct a utility measure on consequences such that preferences amongst gambles cohere with their expected utilities, first on the assumption that the set of consequences is finite and then for the more general case of infinite consequences. Since the second step is essentially an application of Von Neumann and Morgenstern's theory, we will focus on the first and in particular on his derivation of a qualitative probability relation over events.

Savage takes the preference relation to be defined over a very rich set of acts, namely all functions from states to consequences. Because of its importance, I have 'promoted' the definition of the domain of the preference relation to being an additional postulate.

P0 (*Rectangular field assumption*[8]): $\Gamma = \Omega^S$
P1 (*Ordering*) $\succeq$ is (a) complete and (b) transitive.

For any events $F, G \in \mathcal{F}$, let acts $\bar{\alpha}$ and $\bar{\beta}$ be the corresponding constant acts such that for all states s, $\bar{\alpha}(s) = F$ and $\bar{\beta}(s) = G$. Given this definition it is straightforward to induce preferences over consequences from the preferences over acts by requiring that $F \succeq G$ iff $\bar{\alpha} \succeq \bar{\beta}$.

Savage's next step is to assume that the preference relation is separable across events i.e. that the desirability of a consequence of an act in one state of the world is independent of its consequences in other states. He does so by means of his famous Sure-thing principle. Consider the actions displayed in the table below.

[7]For arguments that it is not a requirement of rationality see [11] and [33].
[8]I take this term from Broome [10].

	Events	
Actions	E	E'
α	X	Y
β	X^*	Y

Then action α should be preferred to action β iff consequence X is preferred to consequence X^*. This is because α and β have the same consequence whenever E is not the case, and so should be evaluated solely in terms of their consequences when E is the case. Consequently any other actions α' and β' having the same consequence as α and β respectively whenever E is the case, and identical consequences when E is not, should be ranked in the same order as α and β. More formally:

P2 (*Sure-thing Principle*) Suppose that actions α, β, α' and β' are such that for all states $s \in E$, $\alpha(s) = \alpha'(s)$ and $\beta(s) = \beta'(s)$ while for all states $s \notin E$, $\alpha(s) = \beta(s)$ and $\alpha'(s) = \beta'(s)$. Then $\alpha \succeq \beta$ iff $\alpha' \succeq \beta'$

In view of P2 we can coherently define the conditional preference relation 'is not preferred to, given B', denoted $\succeq_B$, by $\alpha \succeq_B \beta$ iff $\alpha' \succeq \beta'$, where the acts α' and β' are as defined in P2. Given P2, it follows from this definition that the conditional preference relation is complete and transitive. This puts us into territory familiar from the discussion of conjoint additive structures. Given P0–P2, Theorem 7 implies that there exists an additive utility representation of preferences over acts that is unique up to positive affine transformation, i.e. such that the value of each act is the sum of the state-dependent utilities of its consequences.

This representation does not disentangle the contributions of the probabilities of states from the desirabilities of the consequences. To go further, assumptions that ensure the comparability of the state-dependent utilities are needed. Let us call an event $E \in \mathcal{F}$ a null event iff $\alpha \approx_E \beta$, for all $\alpha, \beta \in \Gamma$. Then Savage postulates:

P3 (*State Independence*) Let $B \in \mathcal{F}$ be non-null. Then if $\alpha(s) = F$ and $\alpha'(s) = G$ for every $s \in B$, then $\alpha \succeq_B \alpha' \Leftrightarrow F \succeq G$

The state independence assumption ensures the *ordinal* uniformity of preferences across states, but is not strong enough to ensure the *cardinal* comparability of the state-dependent utilities. The next step is the crucial one for ensuring this as well as for obtaining a probability representation of the agent's attitudes to events. First Savage defines a 'more probable than' relation, $\trianglerighteq$, on the set of events. Consider the following pair of actions:

	Events
Action	E E'
α	X Y

	Events
Action	F F'
β	X Y

Actions α and β have the same two possible consequences, but α has the preferred consequence whenever E is the case and β has it whenever F is the case. Now suppose that consequence X is preferred to consequence Y. Then α should be preferred to β iff E is more probable than F because the action which yields the better consequence with the higher probability should be preferred to one which yields it with lower probability. More formally:

Definition 13 (Qualitative probability) Suppose $E, F \in \mathcal{F}$. Then $E \trianglerighteq F$ iff $\alpha \succeq \beta$ for all actions α and β and consequences X and Y such that:

(i) $\alpha(s) = X$ for all $s \in E$, $\alpha(s) = Y$ for all $s \notin E$,
(ii) $\beta(s) = X$ for all $s \in F$, $\beta(s) = Y$ for all $s \notin F$,
(iii) $X \succeq Y$

In effect the circumstances postulated by this definition provides a 'test' for when one event is more probable than another. A further postulate is required to ensure that this test can be used to compare any two events in terms of their relative probability.

P4 (*Probability Principle*) $\trianglerighteq$ is complete

To apply Theorem 7, our earlier representation theorem for additive conjoint structures, we need to confirm that the derived 'more probable than' relation is not only complete, but transitive and quasi-additive. In fact this follows straightforwardly from P0 to P4 and the definition of the 'more probable than' relation. Two further structural axioms are required to ensure that the qualitative probability relation defined by P4 can be represented numerically.

P5 (*Non-Triviality*) There exists actions α and β such that $\alpha \succ \beta$.
P6 (*Non-Atomicity*) Suppose $\alpha \succ \beta$. Then for all $X \in \mathcal{F}$, there is a finite partition of S such that for all $s \in S$:

(i) ($\alpha'(s) = X$ for all $s \in A$, $\alpha'(s) = \alpha(s)$ for all $s \notin A$) implies $\alpha' \succ \beta$.
(ii) ($\beta'(s) = X$ for all $s \in B$, $\beta'(s) = \beta(s)$ for all $s \notin B$) implies $\alpha \succ \beta'$.

P6 is quite powerful and implies that there are no consequences which are so good or bad, that they swamp the improbability of any given event A. Nonetheless neither it nor P5 raises any pressing philosophical issues. And using them Savage proves:

Theorem 14 *There exists a unique probability function P on $\mathcal{F}$ such that for all $E, F \in \mathcal{F}$:*

$$P(E) \geq P(F) \Leftrightarrow E \trianglerighteq F$$

It is not difficult to see how in principle this theorem can serve as the basis for deriving an expected utility representation. In essence what needs to be established is a correspondence between each act α and a lottery which yields each possible consequence $C \in \Omega$ with probability, $P(\alpha^{-1}(C))$. Then since Savage's postulates for preferences over acts with a finite number of consequences imply that the induced preferences over the corresponding lotteries satisfy the Von Neumann and Morgenstern axioms, the utility of each such act can be equated with that of the expected utility of the corresponding lottery. The proof of this is far from trivial and we won't examine it here – see Savage [29] or Kreps [24] for details.[9]

34.6.3 The Status of Savage's Axioms

34.6.3.1 The Sure-Thing Principle

The most controversial of Savage's axioms is undoubtedly the Sure-thing Principle, Savage's version of the separability condition that appears with different names in different contexts. Although the Independence axiom of Von Neumann and Morgenstern's decision theory is not implied by the Sure-thing principle alone (P3 in particular is also required), the criticism based on Allais' paradox is clearly applicable here as well as are the lines of defence previously sketched. We will not repeat this discussion. But it is worth drawing attention to one further issue. As is evident from the informal presentation of the Sure-thing principle, it is essentially a principle of dominance. That is to say that its intuitive appeal rests on the thought that since only the consequences of an action matter to its evaluation, if the consequences of one act are as least as good as those of another, and are better in at least one event, then this act is better overall. But this application of Consequentialism is mistaken. For it matters not just what consequences an action has, but how probable these outcomes are and in particular how probable it makes them. Two actions can have identical consequences but if one of them brings about the better consequences with a higher probability than the other then it should be preferred.

The upshot of this is that the Sure-thing is not unconditionally valid as a principle of rationality. It is binding only if the states of the world are probabilistically independent of the acts being compared by reference to these states. But this presents Savage with a very significant problem. Amongst other things, his representation theorem is supposed to establish conditions under which a probability measure of belief can be attributed to the decision maker. But it now seems that the attribution

[9] Savage in fact introduces one further postulate necessary for the extension of the expected utility representation to infinite consequences sets. This final postulate is very much in the spirit of the Sure-thing principle and as it does not raise any additional conceptual issues, I will not state it here.

process depends on being able to establish what the decision maker's beliefs are in order to determine whether the Sure-thing principle is applicable. So Savage needs to assume precisely that which he hopes to deduce. Remarkably this fundamental difficulty has been all but ignored in the wide ranging decision-theoretic literature on belief identification.

State-Independence

The axiom of State-Independence requires that if constant act α is preferred to constant act β, given some non-null event E, then α is preferred to β, for any other non-null event F. It is not hard to produce counterexamples. Consider an act which has the constant consequence that I receive £100 and suppose I prefer it to an act with the constant consequence that I receive a case of wine. Would I prefer receiving the £100 to the case of wine given any event? Surely not: in the event of high inflation for instance, I would prefer the case of wine. One could retort that receiving £100 is not a genuine consequence since its description fails to specify features relevant to its evaluation. Perhaps 'receiving £100 when inflation is low' might be closer to the mark. But then the rectangular field assumption forces us to countenance actions which have this consequence, namely my receiving £100 when inflation is low, in states of the world in which inflation is high. Such acts seems nonsensical however and it is hard to see how anyone could express a reasonable preference regarding them.

An objection of this kind was famously made by Robert Aumann in a letter to Savage in 1971.[10] Savage's reply is interesting. He suggests that "a consequence is in the last analysis an experience" [12, p. 79], the implication being that experiences screen out the features of the world causing them and hence have state-independent desirabilities. This is unpersuasive. Even the desirability of experiences are contingent on the state of the world. On the whole I prefer that I be amused than saddened (or experience amusement to experiencing sadness), but I surely do not prefer it, given that a close friend has died. A second objection is more fundamental. To identify consequences with subjective experiences is to risk confusing the outcome of an action with one's evaluation of it. When I want to make a decision, say about whether to go for a swim, I need to know first what the outcome of this decision will be in the various possible states of the world. Then I try and evaluate these outcomes.

To the objection that his theory countenances nonsensical or impossible acts, Savage retorts that it is not necessary that the such acts "...serve something like construction lines in geometry" [12, p. 79], and that they need not be available in order for one to say whether they would be attractive or not. But he seems to under-appreciate the problem. Consider the decision whether or not to buy a life insurance policy that pays out some sum of money in the event of one's death. Now the pay-out

[10]Printed, along with Savage's letter in reply, in Drèze [12, pp. 76–81].

is not a state-independent consequence in Savage's sense, for I am not indifferent between being paid while alive and being paid while dead. However the natural refinement gives us the consequence of 'pay-out and dead' which patently cannot be achieved in any state of the world in which I am alive.

State-Dependent Utility

Although the assumption of state-independence is essential to Savage's representation theorem (and many others, including the widely used Anscombe-Aumann theory [2]), it is not intrinsic to the principle that rationality requires picking the option whose exercise has greatest expected benefit. Indeed Savage's theory can be generalised to a state-dependent version in the following way. For each state of the world s_j let v_j be a real-valued (utility) function on consequences measuring their desirability in that state of the world. Then the probability-utility matrix induced by the decision problem takes the form:

	States of the world		
Options	$P(\{s_1\})$	...	$P(\{s_n\})$
A^1	$v_1(C_1^1)$	...	$v_n(C_n^1)$
...	...	...	...
A^m	$v_1(C_1^m)$	...	$v_n(C_n^m)$

The expected benefit on any option is given, as before, by the expected value of the random variable which specifies its consequences. In this case this is defined by:

$$EU(A^i) = \sum_{j=1}^{n} v_j(C_j^i).P(s_j)$$

Proving the existence of such a representation is straightforward: as we observed earlier, given P0, P1 and P2 we can apply Theorem 7 to establish the existence of u_j such that $U(A^i) = \sum_{j=1}^{n} u_j(C_j^i)$ and then for any probability mass function P on the states of the worlds define $v_j := \frac{u_j}{P(s_j)}$. The problem is that the choice of probability function P here is arbitrary and there is no reason to think that it measures the decision maker's degrees of belief.[11]

[11]See, for instance, Drèze [12], Karni et al. [21] and Karni and Mongin [20] for a discussion of this issue.

34.7 Decision Theory: Evidential and Causal

This discussion of Savage's axioms reveals that three conditions must be satisfied for the maximisation of expected utility to be a rational basis for choice (assuming the absence of option uncertainty). Firstly, there must be no option uncertainty. Secondly, the desirability of each of the consequences must be independent both of the state of the world in which it obtains and the action that brings it about. And thirdly, the states of world must be probabilistically independent of the choice of action. Taken separately it is often possible to ensure that for all practical purposes these conditions are met by taking care about how the decision problem is framed. But ensuring that all three are satisfied at the same time is very difficult indeed since the demands they impose on the description of the decision problem pull in different directions. For instance option uncertainty can be tamed by coarsening the description of outcomes, but eliminating state-dependence requires refining them.

This problem provides strong grounds for turning our attention to a rival version of subjective expected utility theory that is due to Richard Jeffrey and Ethan Bolker. Jeffrey [18] makes two modifications to the Savage framework. First, instead of distinguishing between the objects of preference (actions), those of belief (events) and those of desire (consequences), Jeffrey takes the contents of all of the decision maker's attitudes to be propositions. And secondly, he restricts the set of actions to those propositions that the agent believes he can make true at will.

The first of these modifications we have already in effect endorsed by arguing that the difference between states and consequences is pragmatic rather than logical. Furthermore, if the contents of propositions are given by the set of worlds in which it is true, then Jeffrey's set of propositions will simply be Savage's set $\mathcal{F}$ of events, the only difference between the two being that there is no restriction of consequences to maximally specific propositions in Jeffrey's framework. This small modification has a very important implication however. Since states/events and consequences are logically interrelated in virtue of being the same kind of object, consequences are necessarily state-dependent. This means that Jeffrey's theory is not subject to the second of the restrictions required for Savage's theory.

The second modification that Jeffrey makes is more contentious and requires a bit of explanation. If he followed Savage in defining actions as arbitrary functions from partitions of events to consequences, the enrichment of the set of consequences would lead to an explosion in the size of the set of actions. But Jeffrey argues that many of the actions so defined would be inconsistent with the causal beliefs of the decision maker. Someone may think they have the option (which we previously named 'taking the car') of making it true that if the traffic is light they arrive on time, and if it's heavy they arrive late, but not believe that they have the option of making it true that if the traffic is light they arrive late, and if it's heavy they arrive on time. Yet Savage's rectangular field assumption requires that such options exist and that the agent takes an attitude to them. But if the agent doesn't believe that such options are causally possible, then any attitudes we elicit with regard to them may be purely artifactual.

We can look at this issue in a slightly different way. As we noted in the discussion of option uncertainty, an agent may be uncertain as to what consequences its performance yields in each state of the world. So they may not know what actions qua mappings from states to consequences are available to them. Jeffrey's solution is to conceive of an action, not as a mapping from states to consequences, but as a subjective probability distribution over consequences that measures how probable each consequence would be if the action were performed. This means that when evaluating the act of taking an umbrella for instance, instead of trying to enumerate the features of the state of the world that will ensure that I stay dry if I take an umbrella, I simply assess the probability that I will stay dry if I take the umbrella and the probability that I will get wet anyhow (even if I take it). I should then perform the act which has the greatest conditional expected utility given its performance

Two features of this treatment are noteworthy. Firstly, it is no longer required that the states of the world be probabilistically independent of the available actions. On the contrary, actions matter because they shape probabilities. This dispenses with the third constraint on the applicability of Savage's theory. Secondly, agents are not able to choose between arbitrary probability distributions over consequences but are restricted to those probability distributions that they consider themselves able to induce through their actions. To put it somewhat differently, we may think of both Savage's and Jeffrey's actions as inducing Von Neumann and Morgenstern lotteries over consequences. But Jeffrey only countenances preferences over lotteries which conform with the agent's beliefs. This solves the problem of option uncertainty by endogenising it. The agent is not option uncertain about an action because what action it is (what probability distribution it induces) is defined subjectively i.e. in terms of the agent's beliefs.

34.7.1 Desirability Representations

Let us now turn to the representation of preferences in the Bolker-Jeffrey theory. Recall that for Jeffrey the content of both beliefs and desires are propositions. To emphasise the contrast with Savage, let us model propositions as sets of possible worlds or states of the world. Then an agent's beliefs are measured, as in Savage, by a probability measure P on $\mathcal{F}$, the set of all propositions, while her degrees of desire are represented by a real valued (desirability) function V on $\mathcal{F} - \{\bot\}$, the set of non-contradictory propositions, that satisfies:

Axiom 15 (Desirability) *If* $X \cap Y = \varnothing$, *and* $P(X \cup Y) \neq 0$, *then:*

$$V(X \cup Y) = \frac{V(X).P(X) + V(Y).P(Y)}{P(X \cup Y)}$$

The notion of a desirability function on the set of propositions extends the quantitative representation of the agent's evaluative attitudes from just the maximally specific ones (that play the role of consequences in Savage's theory) to the full set of them. The basic intuition behind the extension encoded in the desirability axiom is the following. How desirable some coarse grained proposition is depends both on the various ways it could be realised and on the relative probability of each such realisation, given the truth of the proposition. For instance, how desirable a trip to beach is depends on how desirable the beach is in sunny weather and how desirable it is in rainy weather, as well as how likely it is to rain or to be sunny, given the trip.

What properties must preferences on prospects satisfy if preferences are to have a desirability representation i.e. such that $X \succeq Y \Leftrightarrow V(X) \succeq V(Y)$? There are two:

Axiom 16 (Averaging) *If $X \cap Y = \varnothing$, then:*

$$X \succeq Y \Leftrightarrow X \succeq X \cup Y \succeq Y$$

Axiom 17 (Impartiality) *If $X \cap Z = Y \cap Z = \varnothing$, $X \approx Y \not\approx Z$ and $X \cup Z \approx Y \cup Z$, then for all $Z' \in \Omega$ such that $X \cap Z' = Y \cap Z' = \varnothing$, it is the case that $X \cup Z' \approx Y \cup Z'$.*

The Averaging axiom say that if X is preferred to Y then X should be preferred to the prospect that either X or Y is the case, since the latter is consistent with Y being the case while the former is not. It has a somewhat similar motivation to the Sure-thing principle, but is much weaker. In particular it is not directly vulnerable to the Allais' paradox.

The impartiality axiom allows for a partial separation of beliefs and desires. The idea is as follows. Suppose propositions X and Y are equally preferred and that Z is some proposition disjoint from and preferred to both. Then the disjunction of X and Z will be equally preferred to the disjunction of Y and Z iff X and Y are equally probable. If X were more probable than Y then the probability of Z conditional on $X \cup Z$ would be less than the probability of Z conditional on $X \cup Y$. And so the prospect of $X \cup Z$ would be less desirable that than of $Y \cup Z$ since it would yield the more desirable prospect (Z) with lower probability.

Theorem 18 ((Bolker [6])) *Let $\mathcal{F}$ be an atomless Boolean algebra of propositions. Let $\succeq$ be a continuous weak preference order on $\mathcal{F}$. Then there exists a probability measure P and signed measure U on Ω, such that for all $X, Y \in \mathcal{F} - \{F\}$, such that $P(X) \neq 0 \neq P(Y)$:*

$$X \succeq Y \Leftrightarrow \frac{U(X)}{P(X)} \geq \frac{U(Y)}{P(Y)}$$

Furthermore P' and U' are another such pair of measures on Ω iff there exists real numbers a, b, c and d such that (i) $ad - bc > 0$, (ii) $cU(T) + d = 1$, (iii) $cU + dP > 0$, and:

$$P' = cU + dP$$
$$U' = aU + bP$$

It follows that the desirability function V defined by for all $X \in \mathcal{F}$ such that $P(X) \neq 0$, $V(X) = \frac{U(X)}{P(X)}$, represents the preference relation $\succeq$ but only up to fractional linear transformation. The uniqueness properties here are considerably weaker than in Savage's framework and this is perhaps the reason for the unpopularity of Jeffrey's theory amongst economists and applied decision theorists. It is not an insurmountable problem however for there are ways of strengthening Bolker's representation theorem, either by postulating direct probability comparisons (see Joyce [19]) or by enriching the set of propositions by conditionals (see Bradley [9] and [7]).

34.7.2 Causal Decision Theory

Jeffrey's decision theory recommends choosing the action with maximum desirability. Two closely related questions arise. Firstly, is this the same recommendation as given by Savage's theory? And secondly, if not, which is correct? The answer to the first question is less clear cut than might be hoped. On the face of it the prescriptions are different: Jeffrey requires maximisation of the conditional expectation of utility, given the performance of the action, while Savage requires maximisation of unconditional expectation of utility. But since they represent actions differently these two prescriptions are not directly in conflict. In fact, there is a way of making them perfectly compatible. The trick is to represent a Savage-type action within the Jeffrey framework by an indicative conditional of the form 'If the state of the world is s_1, then the consequence is C_1; if the state of the world is s_2, then the consequence is C_2; ...'. Then, given some reasonable assumptions about the logic of conditionals and rational preferences for their truth, the desirability of action-conditionals will just be the expected desirability of the consequence to which it refers, relative to the probability of the states with which the consequences are associated. (See Bradley [7] for details.)

On this interpretation, Savage and Jeffrey's theories are both special cases of a larger Bayesian decision theory. Most causal decision theorist reject this view and regard Savage's theory as a precursor to modern causal decision theory, which prescribes not maximisation of desirability but maximisation of causal expected utility. The distinction is brought out rather dramatically by the famous Newcomb's paradox, but since this example raises issues tangential to the main one, let us use a more banal example. Suppose that I am deliberating as to whether to eat out at Chez Posh next week. Chez Posh is very expensive, so not surprisingly the probability of being rich given that one eats there is high. I now apply Jeffrey's theory as follows. There are three prospects of interest: A: I have a good meal, B: I will be rich and C: I eat at Chez Posh. Then assuming that eating at Chez Posh guarantees a good meal:

$$V(C) = V(A \cap B \cap C).P(B|C) + V(A \cap \neg B \cap C).P(\neg B|C)$$

Since both $P(B|C)$ and $V(A \cap B \cap C)$ are high, Jeffrey's theory recommends going. But if I cannot afford to go, this will be very bad advice!

The problem is easy to spot. Although the probability of being rich given that one eats at Chez Posh is high, deciding to eat there won't make one rich. On the contrary it will considerably aggravate one's penury. So desirability is a poor guide to choice-worthiness in this case and indeed in any case when the performance of an action is evidence for a good (or bad) consequence but not a cause of it. Causal decision theory proposes therefore that actions be evaluated, not in terms of desirability, but in terms of the efficacy of actions in bringing about desired consequences.

More formally for each option A^i let P be a probability mass function on states of the world with p^i_j being the probability that s_j would be the state of the world were option A^i exercised. Let u^i_j be the utility of the consequence that results from the exercise of option A^i in state s_j. Then a decision problem can be represented by the following probability-utility matrix.

	States of the world		
Options	s_1	...	s_n
A^1	(p^1_1, u^1_1)	...	(p^1_n, u^1_n)
...	...	...	...
A^m	(p^m_1, u^m_1)	...	(p^m_n, u^m_n)

In this general case, the requirement of rationality to pick the option whose exercise has greatest expected benefit is made precise by causal decision theory, as the requirement to choose the option with maximum causal expected utility (CU), where this is defined as follows:

$$CU(A^i) = \sum_{j=1}^{n} u^i_j . p^i_j$$

In the special case when p^i_j equals $P(\{s_j\}|A^i)$ then the value of an option is given by its conditional expectation of utility, V, where this is defined by:

$$V(A^i) = \sum_{j=1}^{n} u^i_j . P(\{s_j\}|A^i)$$

This is just Jeffrey's desirability measure. So on this interpretation Jeffrey's theory is special case of causal decision theory, applicable in cases where the probability of a consequence on the supposition that an action is performed is just the conditional probability of the consequence given the action.

34.8 Ignorance and Ambiguity

The agents modelled in the decision theories described in the previous two sections are not only rational, but logically omniscient and maximally opinionated. Rational in that their attitudes – beliefs, desires and preferences – are consistent both in themselves and with respect to one another; logically omniscient because they believe all logical truths and endorse all the logical consequences of their attitudes; and opinionated because they have determinate belief, desire and preference attitudes to all prospects under consideration either because they possess full information or because they are willing and able to reach judgements on every possible contingency.

Relaxations of all of these assumptions have been studied both within empirical and normative decision theory. Firstly there is a growing literature on bounded rationality which looks at the decision making of agents who follow procedural rules or heuristics. Most of this work has descriptive intent, but some of it retains a normative element in that it seeks to show how bounded agents, with limited computational resources, should make decisions given their limitations.[12] Secondly, the problem of logical omniscience has been tackled in different ways by either modelling agents' reasoning syntactically and restricting their ability to perform inferences or by introducing possible states of the world that are subjectively possible, but objectively not.[13] Finally, there has been a long standing debate about how agents should make decisions when they lack the information necessary to arrive at precise probabilistic judgements, which we look at below. More recently this has been supplemented by a growing literature on the requirements on rationality in the absence of the completeness assumption. With some notable exceptions this literature is almost entirely focused on incomplete probabilistic information.

34.8.1 Decisions Under Ignorance

Let us consider the extreme case first when the decision maker knows what decision problem she faces but holds no information at all regarding the relative likelihood of the states of the world: a situation termed ignorance in the literature. There are four historically salient proposals as to how to make decisions under these circumstances which we can illustrate with reference to our earlier simple example of the decision as to whether to take a bus or walk to the appointment. Recall that the decision problem was represented by the following matrix.

---

[12] See for instance, Simon [31], Gigerenzer and Selten [15, 37] and Rubinstein [28].

[13] See for instance, Halpern [16] and Lipman [26].

	Heavy traffic	*Light traffic*
Take a bus	−2	1
Walk	−1	−1

1. *Maximin*: This rule recommends picking the option with the best worst outcome. For instance, taking the bus has a worst outcome of −2, while walking has a worst outcome of −1. So the rule recommends walking.
2. *MaxMean*: This rule recommends picking the option with the greatest average or mean utility. For instance, taking the bus has a mean utility of −0.5, while walking has a mean utility of −1. So on this rule taking the bus is better.
3. *Hurwicz Criterion*: Let Max_i and Min_i respectively be the maximum and minimum utilities of the possible outcomes of action α_i. The Hurwicz criterion recommends choosing the option which maximises the value of $h.Max_i + (1 - h)Min_i$ where $0 \leq h \leq 1$ is a constant that measures the decision maker's optimism. In our example, for instance, the rule recommends taking the bus for any values of h such that $h > \frac{1}{3}$: roughly as long as you are not too pessimistic.
4. *Minimax Regret*: Let the regret in a state of the world associated with an action α be the difference between the maximum utility that could be obtained in that state of the world, given the available actions, and the utility of the consequence of exercising α. The minimax regret rule recommends picking the action with the lowest maximum regret. For instance, in our example the regret associated with taking a bus is 1 if the traffic is heavy and 0 if its light, while that associated with walking is 0 if the traffic is heavy but 2 if it is light. So the rule recommends taking the bus.

Each of these criteria faces serious objections. Minimax Regret violates the aforementioned Independence of Irrelevant Alternatives condition and for that reason is widely regarded as normatively unacceptable (but note that we criticised this condition on the grounds that the composition of the choice set can be relevant). With the exception of Maximin, none of the rules give recommendations that are invariant under all positive monotone transformations of utilities. But in the absence of probabilistic information how are the utilities to be cardinalised? The Maximin rule seems unduly pessimistic. For instance, even if taking a bus has utility of 1000 in case of light traffic it recommends walking. The Hurwicz criterion seems more reasonable in this regard. But both it and Maximin face the objection that refinements of the decision problem produce no reassessment in situations in which it should. Consider for instance the following modified version of our decision matrix in which we have added both a new possible state of the world – medium traffic – and a new option.

	Heavy traffic	*Medium traffic*	*Light traffic*
Take a bus	−2	−2	1
Walk	−1	−1	−1
Car	−2	1	1

On both the Hurwicz criterion and the Maximin rule, taking a bus and driving a car are equally good, even though taking a car weakly dominates taking the bus. So they seem to give the wrong prescription. It is possible to modify these two rules so that they deal with this objection. For instance, the Leximin rule adds to Maximin the condition that if two options have equally bad worst outcomes then they should be compared on the basis of their second worst outcomes, and if these are equal on the basis of their third worst, and so on. A lexicographic version of the Hurwicz criterion is also conceivable. But the possibility of ties amongst worst and best outcomes pushes us in the direction of considering all outcomes. In which case we need to consider what weights to attach to them. The answer implicitly assumed by the Maxmean rule is that we should give equal weights in the absence of any information by which they can be distinguished (this is known as the Principle of Indifference). Unfortunately this procedure delivers different recommendations under different partitionings of the event space, so Maxmean too is not invariant in the face of refinements of the decision problem.

The fact that all these proposals face serious objections suggests that we are asking too much from a theory of decision making under ignorance. In such circumstances it is quite plausible that many choices will be permissible and indeed that rationality does not completely determine even a weak ordering of options in every decision problem. If this is right we should look for necessary, rather than sufficient, conditions for rational choice. I have already implicitly helped myself to an obvious candidate for such a condition – Weak Dominance – in arguing against the Hurwicz criterion. Weak Dominance says that we should not choose action α when there exists another action β such that in every state of the world s, $\beta(s) \succsim \alpha(s)$ and in at least one state of the world $\bar{s}$, $\beta(\bar{s}) \succ \alpha(\bar{s})$. But however reasonable Weak Dominance may look at first sight, it is only valid as a principle in circumstances in which states of the world are probabilistically independent of the acts. And by assumption in conditions of complete ignorance we have no idea whether this condition is met or not. It is not that we cannot therefore use dominance reasoning, but rather that it cannot be a requirement of rationality that we do. The same applies to every kind of dominance principle. And I know of no other plausible candidates for necessary conditions on rational evaluation of options other than transitivity. Since no consistent set of beliefs is ruled out in conditions of complete ignorance, it is possible that there are no further constraints on preference either.

34.8.2 Decisions Under Ambiguity

We use the term ambiguity to describe cases intermediate between uncertainty and ignorance i.e. in which the decision maker holds some information relevant to the assessment of the probabilities of the various possible contingencies but not enough to determine a unique one. Cases of this kind provided early counterexamples to expected utility theory. Consider the following example due to Daniel Ellsberg [13] in which subjects must choose between lotteries that yield monetary outcomes that depend on the colour of the ball drawn from an urn. The urn is known to contain 30 red balls and 60 balls that are either black or yellow, but in unknown proportions.

When asked to choose between lotteries I and II and between III and IV, many people pick I and IV. This pair of choices violates the Sure-thing principle, which requires choices between the pairs to be independent of the prizes consequent on the draw of a yellow ball. They are also inconsistent with the way in which Savage uses the Probability Principle to elicit subjective probabilities. For it follows from his definition of the qualitative probability relation that lottery I is preferred to lottery II iff Red is more probable than Black and that IV is preferred to III iff not-Red (i.e. Black or Yellow) is more probable than not-Black (i.e. Red or Yellow). But this is inconsistent with the laws of probability. It is this latter feature that distinguishes Ellsberg's paradox from Allais'.

One plausible explanation for the observed pattern of choices is an attitude that has since been termed 'ambiguity aversion'. The thought is that when choosing between I and II, people pick the former because this gives them \$100 with a known probability (namely one-third) while lottery II yields the \$100 with unknown probability or, more precisely, with a probability that could lie between zero and two-thirds. Similarly when asked to choose between III and IV they choose the one that yields the \$100 with a known probability, namely IV. They make these choices because they are unable to assign probabilities to Black or to Yellow and because, *ceteris paribus*, they prefer gambles in which they know what they can expect to get over gambles that are ambiguous in the sense that they don't know what they can expect from them.

So much is largely common ground amongst decision theorists. What is much less settled are the normative and explanatory implications of the paradox. There are three views one might adopt on this question. The first is that the Ellsberg paradox shows that expected utility theory is descriptively false but not normatively so and that the observed pattern of choices is simply irrational. The second view is that a preference for betting on known probabilities over unknown ones is perfectly reasonable and hence that the paradox shows that expected utility theory is both descriptively and normatively inadequate. The third is that it shows neither inadequacy because the decision problem has not been properly represented in Table 34.4. In particular if subjects care about the range of chances of winning

Table 34.4 Ellsberg's paradox

	Ball colour		
Lottery	Red	Black	Yellow
I	\$100	\$0	\$0
II	\$0	\$100	\$0
III	\$100	\$0	\$100
IV	\$0	\$100	\$100

yielded by their choice then this property of outcomes, and not just the final winnings, should be represented.[14]

The second of these views has inspired a number of different proposals for decision rules rivalling that of maximisation of expected utility, but this literature is growing so fast that any survey is likely to find itself out of date very quickly and so I will confine myself to mentioning a few of the more salient ones. Most proposals start from the observation that the information subjects hold in Ellsberg's problem constrains them to a family of admissible probability functions on states, each assigning $\frac{1}{3}$ to Red, some value p in the interval $[0, \frac{2}{3}]$ to Black and a corresponding value $1 - p$ to Yellow. This family of probabilities, together with a utility function on the monetary prizes, then induces a corresponding family of admissible expected utilities for the alternatives under consideration. What distinguishes the various proposals is how they see subjects as using this set of expected utilities as a basis for choice.

Perhaps the predominant proposal is the Maximin EU rule (or MEU), which requires choice of the alternative with the greatest minimum expected utility. For instance, while lottery I has expected utility $\frac{1}{3} \times U(\$100) + \frac{2}{3} \times U(\$0)$, lottery II has expected utility in the range $[U(\$0), \frac{2}{3} \times U(\$100) + \frac{1}{3} \times U(\$0)]$. The minimum value for the latter is $U(\$0)$ (assuming that utility is a positive function of money), so lottery I is better according to the MEU criterion. On the other hand lottery IV is better than lottery III since it has a guaranteed expected utility of $\frac{2}{3} \times U(\$100) + \frac{1}{3} \times U(\$0)$ while III has minimum expected utility of $\frac{1}{3} \times U(\$100) + \frac{2}{3} \times U(\$0)$. So MEU prescribes just the choices observed in the Ellsberg paradox. On the other hand it also prescribes indifference between lotteries I and III, since both have minimum expected utilities of $\frac{1}{3} \times U(\$100)$, whereas most people would strictly prefer lottery III.

MEU, like the Maximin rule we looked at in the previous section, seems too extreme and there are other popular rules that allow for caution in the face of uncertainty about the probabilities which are less so. For instance the α-MEU rule prescribes choice of the alternative that maximises the α-weighted average of its minimum and maximum expected utility, where $\alpha \in [0, 1]$ is interpreted as index of the agent's pessimism. A rather different proposal is the 'smooth' ambiguity model of Klibanoff, Marinacci and Mukerji [22] which values actions in accordance with

[14]See for instance [8].

a weighted average of a concave transformation of the expected utilities, where the weights are thought of as the agent's degrees of belief for the possible probability distributions over states and the concave transformation expresses their level of aversion to ambiguity. These models are more compelling than MEU and raise interesting philosophical questions about the parameters that they introduce but have yet to receive much discussion in the philosophical literature.

Ellsberg's paradox also raises an important methodological issue. Should we regard the choices the agent makes in situations of ambiguity as expressions of her preferences or as expressions of some other non-preference reason for choice? If we take the former view then we may regard her attitudes to ambiguity as a further psychological constituent of her preferences, but leave intact the standard theory of the relation between preference and choice with its implication that preferences are complete. If we take the latter view, then we can leave in place the standard view about the relation between preference, belief and desire and treat ambiguity attitudes as additional determinants of choice. The former view is the one taken by the majority of decision theorists working in the field, perhaps because of a deep commitment to revealed preference theory. But it seems philosophically more satisfactory to regard preferences themselves as potentially incomplete whenever beliefs are less than fully determinate. But which view it is best to take depends in part on what is discovered about ambiguity attitudes: how stable they are, how responsive are they to information, as so on. So it is premature to draw any strong conclusions.

References

1. Allais, M. (1953). Le comportement de l'homme rationnel devant le risque: critique des postulats et axiomes de l'école Américaine. *Econometrica, 21*, 503–546.
2. Anscombe, F. J., & Aumann, R. J. (1963). A definition of subjective probability. *Annals of Mathematical Statistics, 34*, 199–205.
3. Arrow, K. J. (1963). *Social choice and individual values* (2nd ed.). New York: Wiley.
4. Bernouilli, D. (1954/1738). Exposition of a new theory on the measurement of risk (L. Sommer, Trans.). *Econometrica, 22*, 23–26.
5. Binmore, K. (2009). *Rational decisions*. Princeton: Princeton University Press.
6. Bolker, E. (1966). Functions resembling quotients of measures. *Transactions of the American Mathematical Society, 124*, 292–312.
7. Bradley, R. (2007). A unified Bayesian decision theory. *Theory and Decision, 63*, 233–263.
8. Bradley, R. (2016). Ellsberg's paradox and the value of chances. *Economics and Philosophy, 32*(2), 231–248.
9. * Bradley, R. (2017). *Decision theory with a human face*. Cambridge: Cambridge University Press [Presents a theory of rationality for bounded agents under conditions of severe uncertainty].
10. Broome, J. (1991). *Weighing goods*. Cambridge, MA: Basil Blackwell.
11. Buchak, L. (2013). *Risk and rationality*. Oxford: Oxford University Press.
12. Dreze, J. (1987). *Essays on economic decisions under uncertainty*. Cambridge: Cambridge University Press.

13. * Ellsberg, D. (1961). Risk, ambiguity, and the savage axioms. *Quarterly Journal of Economics, 75*, 643–669 [A very influential early critical assesment of Savage's theory].
14. Evren, O., & OK, E. (2011). On the multi-utility representation of preference relations. *Journal of Mathematical Economics, 47*, 554–563.
15. Gigerenzer, G., & Selten, R. (2002).*Bounded rationality*. Cambridge: MIT Press
16. Halpern, J. Y. (2001). Alternative semantics for unawareness. *Games and Economic Behavior, 37*, 321–339.
17. Herstein, I. N., & Milnor, J. (1953). An axiomatic approach to measurable utility. *Econometrica, 21*(2), 291–297.
18. * Jeffrey, R. C. (1983). *The logic of decision* (2nd ed.). Chicago: University of Chicago Press [An influential formulation of decision theory in terms of propositional attitudes].
19. * Joyce, J. M. (1999). *The foundations of causal decision theory*. Cambridge: Cambridge University Press [The most complete formulation and defence of causal decision theory].
20. Karni, E., & Mongin, P. (2000). On the determination of subjective probability by choice. *Management Science, 46*, 233–248.
21. Karni, E., Schmeidler, D., & Vind, K. (1983). On state-dependent preferences and subjective probabilities. *Econometrica, 51*, 1021–1031.
22. Klibanoff, P., Marinacci. M., & Mukerji, S. (2005) A smooth model of decision making under ambiguity. *Econometrica, 73*, 1849–1892.
23. Krantz, D., Luce, R. D., Suppes, P., & Tversky, A. (1971). *The foundations of measurement. Volume 1. Additive and polynomial representations*. New York: Academic.
24. * Kreps, D. M. (1988). *Notes on the theory of choice*. Boulder/London: Westview Press [Excellent advanced introduction to decision theory].
25. Lewis, D. (1981). Causal decision theory. *Australasian Journal of Philosophy, 59*, 5–30.
26. Lipman, B. (1998). Decision theory without logical omniscience: Toward an axiomatic framework for bounded rationality. *Review of Economic Studies*, 339–361.
27. * Ramsey, F. P. (1926). Truth and probability. In D. H. Mellor (Ed.), *Philosophical papers*. Cambridge: Cambridge University Press, 2008 [Classic essay on the foundations of subjective probability and expected utility].
28. Rubinstein, A. (1998). *Modeling bounded rationality*. Cambridge: MIT Press
29. * Savage, L. J. (1954/1972). *The foundations of statistics* (2nd ed.). New York: Dover [The most influential formulation of subjective expected utility theory].
30. * Sen, A. K. (1970). *Collective choice and social welfare*. San Francisco: Holden-Day [Classic textbook on choice functions and social choice].
31. Simon, H. (1955). A behavioral model of rational choice. *Quarterly Journal of Economics, 69*, 99–118.
32. Stalnaker, R. ([1972]/1981b). Letter to David Lewis. In W. Harper, R. Stalnaker, & G. Pearce (Eds.), *Ifs: Conditionals, belief, decision, chance, and time* (pp. 151–152). Dordrecht: Reidel.
33. Stefánsson, H. O., & Bradley, R. (2015). How valuable are chances? *Philosophy of Science, 82*(4), 602–625.
34. Stefánsson, H. O., & Bradley, R. (Forthcoming 2017). What is risk aversion? *British Journal of Philosophy of Science*.
35. * von Neumann, J., & Morgenstern, O. (1944). *Theory of games and economic behaviour*. Princeton: Princeton University Press [Classic formulation of expected utility theory].
36. Wakker, P. (2010). *Prospect theory: For risk and uncertainty*. Cambridge: Cambridge University Press.
37. Weirich, P. (2004). *Realistic decision theory: Rules for nonideal agents in nonideal circumstances*. New York: Oxford University Press

Chapter 35
Dynamic Decision Theory

Katie Steele

Abstract This chapter considers the controversial relationship between *dynamic* choice models, which depict a series of choices over time, and the more familiar *static* choice models, which depict a single 'one-shot-only' decision. An initial issue concerns how to reconcile the normative advice of these two models: Should an agent take account of the broader dynamic context when making a decision, and if so, in a *sophisticated* manner (the orthodox backwards induction approach), or rather in a *resolute* manner (which takes the past as well as the future to be significant)? Further controversies concern what the dynamic implications of an agent's preferences reveal about the (ir)rationality of these preferences.

35.1 'Dynamic' Versus 'Static' Decision Theory

On paper, at least, *dynamic* (otherwise known as *sequential*) and *static* decision models look very different. The static model has familiar tabular or normal form, with each row representing an available act/option, and columns representing the different possible states of the world, yielding different outcomes for each act. Such models apparently depict a single 'one shot only' decision. Dynamic models, on the other hand, have tree or extensive form—they depict a series of anticipated choice points, the later choices often following the resolution of some uncertainty.

These basic differences between the two types of models raise a number of questions about how, in fact, they relate to each other:

- Do dynamic and static decision models depict the same kind of decision problem?
- If so, what is the static counterpart of a dynamic decision model? Ultimately: How should one initiate a sequence of choices?

K. Steele (✉)
School of Philosophy, Australian National University, Canberra, ACT, Australia
e-mail: katie.steele@anu.edu.au

S. O. Hansson, V. F. Hendricks (eds.), *Introduction to Formal Philosophy*, Springer Undergraduate Texts in Philosophy, https://doi.org/10.1007/978-3-319-77434-3_35

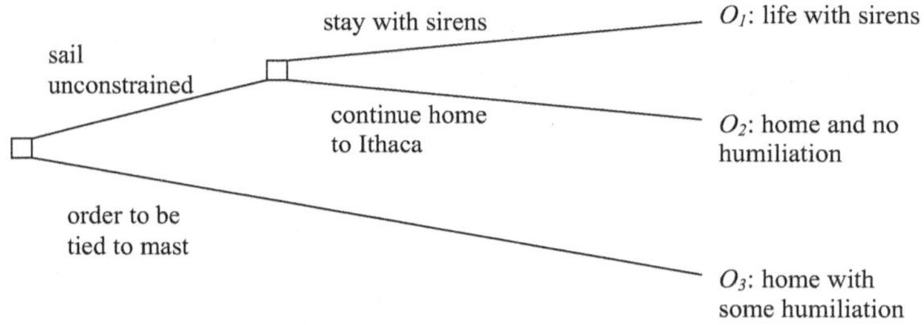

Fig. 35.1 Ulysses' dynamic decision problem

- Do dynamic decision models shed light on old normative questions concerning learning and choice rules?

These issues turn out to be rather controversial; this chapter will consider them in turn. Firstly, however, it is helpful to set the scene with a couple of examples.

A well-known dynamic decision problem is the one facing Ulysses on his journey home to Ithaca in Homer's great tale from antiquity. Ulysses must make a choice about the manner in which he will sail past an island inhabited by sweet-singing sirens, knowing that once he reaches the island, he must then choose whether to stop there indefinitely or to keep sailing. Ulysses' initial choice concerns whether to order the crew to tie him to the mast when nearing the island. If he makes the order, he will later have no further choices and the ship will sail onwards to Ithaca. If he does not make the order, he will later have the choice mentioned above. The final outcome depends on what sequence of choices Ulysses makes. The problem can be depicted in extensive form, as per Fig. 35.1. The square nodes represent the two choice points.

The second problem will be described only in abstract form: It is given in Fig. 35.2.[1] The model illustrates a dynamic decision involving some uncertainty. (This particular problem will also be useful for our discussion in later sections.) Circle nodes indicate points of uncertainty, where all of the branches have some probability of occurring, as per the beliefs of the agent in question. Square nodes represent choice points, as before, where the branches are the options the agent perceives as available at that choice point. In this particular decision problem, the first uncertainty to be resolved concerns whether some event E, or else its complement $\sim E$ turns out to be the case. The later uncertainty concerns whether event F or $\sim F$ is true. O_1, O_2, and O_3 refer to possible outcomes of the agent's sequence of choices, and delta is some small positive amount such that, say, $O_3 - \delta$ is slightly less preferable than O_3.

[1]This decision problem is from Rabinowicz [21, 599].

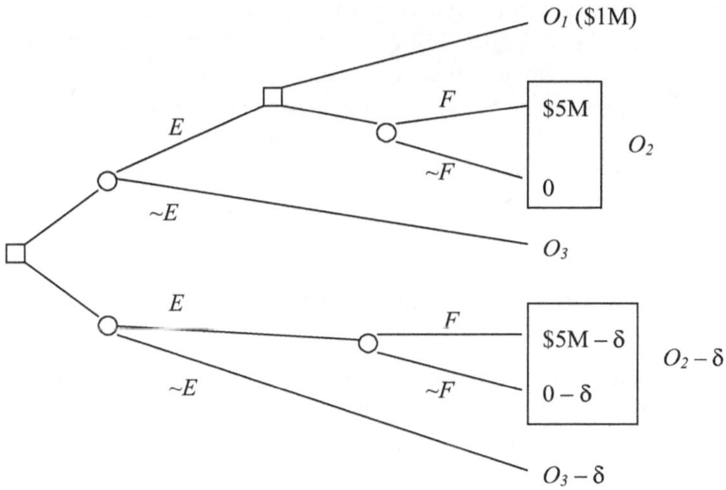

Fig. 35.2 Dynamic decision problem with uncertainty

35.2 Was Ulysses Rational?

It has been argued (e.g., [20], 202) that Ulysses' plight, while interesting from a dynamic perspective, is not in fact the appropriate topic for dynamic decision theory, because Ulysses is a flawed agent: he expects a change in attitude towards the sirens, for no good reason. According to Homer's story, Ulysses suffers a kind of weakness of will when he hears the sirens singing, and spontaneously changes his preferences, despite there being no new information available. This is an imperfect agent with problematic belief and desire changes, and not the appropriate subject for dynamic decision theory, or so the argument might go.

There is something to this line, but it introduces an uncomfortable rift: Dynamic and static decision theory must then deal with different subject matter. Ulysses is deemed irrational from the dynamic perspective, and thus not worthy of attention. On the other hand, surely decision theory has something to offer agents facing predicaments like Ulysses'. This is presumably the job of static decision theory.

One may certainly approach the relationship between dynamic and static decision theory in this way, namely that the former is the yardstick for assessing an agent like Ulysses' rationality over some period of time, perhaps his whole lifetime, while static decision theory assesses his rationality only at the 'present' time. In this way, Ulysses may be rational in the static sense when he orders the crew to tie him to the mast, while falling short of rationality in the more demanding, dynamic sense.

The worry is that the latter demanding notion of rationality—one that concerns an agent over an extended period of time—is of *mere* intellectual interest. If the aim is to offer pertinent practical guidance to an agent, it seems more fruitful to regard dynamic and static decision theory as concerned with the same decision problem, namely: What constitutes rational choice at the 'present' *live* point of decision? The dynamic model can be understood simply as a nice way of visualising the temporal series of choices and learning events that an agent predicts she will confront. On this reading, the key question, then, is: How should an agent choose her initial move in light of her predicted decision tree? To answer this, we need to determine how the future nodes of the decision tree should bear on the initial choice. How should the agent conceive, in static terms, of her choice problem?

35.3 How Should One Initiate a Sequence of Choices?

The questions left dangling in the previous section have generated a surprising amount of controversy. Three major approaches to dynamic choice have appeared in the literature. These are the *naive* or *myopic* approach, the *resolute* approach and the *sophisticated* approach. I join a number of others (e.g. [16, 18, 23]) in defending the latter, but there are also steadfast supporters of resolute choice (notably [20] and [17]). Myopic choice is best conceived as a useful contrast for the other two approaches.

Let us begin with the contrast case. A naive agent assumes that any path through the decision tree is possible, and so sets off in pursuit of whichever path they calculate to be optimal, given their present attitudes. For instance, a naive Ulysses would simply presume that he has three overall strategies to choose from: Either ordering the crew to tie him to the mast, or issuing no such order and later stopping at the sirens' island, or issuing no such order and later sticking his course. Ulysses prefers the outcome associated with the latter combination, and so he initiates this strategy by not ordering the crew to restrain him. Table 35.1 presents naive Ulysses' static decision problem. In effect, this decision model does not take into account Ulysses' present predictions regarding his future preferences.

There is no need to labour the point that the naive approach to dynamic choice is aptly named. Ulysses chooses to 'sail unconstrained and then go home to Ithaca', but, by his own lights, this combination of choices would not be realised; initiating the act would inevitably lead Ulysses to stay on the island of the sirens. The hallmark of the sophisticated approach, by contrast, is its emphasis on backwards planning: The sophisticated chooser does not assume that all paths through the decision tree, or in other words, all possible combinations of choices at the various choice nodes, are genuine options. The agent considers, rather, what they would be inclined to choose at later choice nodes if they were to arrive at the node in question. Indeed, the agent starts with the final choice nodes in the tree, and considers what would be

Table 35.1 Naive Ulysses

Act	Outcome
Sail unconstrained then stay with sirens	Life with sirens
Sail unconstrained then home to Ithaca	Reach home, no humiliation
Order tying to mast	Reach home, some humiliation

Table 35.2 Sophisticated Ulysses I

Act	Outcome
Sail unconstrained then stay with sirens	Life with sirens
Order tying to mast	Reach home, some humiliation

Table 35.3 Sophisticated Ulysses II

Act	Later prefer sirens ($p = 1$)	Later prefer Ithaca ($p = 0$)
Sail unconstrained	Life with sirens	Home, no humiliation
Order tying to mast	Home, some humiliation	Home, some humiliation

chosen at these nodes, given their predicted preferences at each of these positions. These predicted choices would presumably affect what the agent would choose at the second-last choice nodes. Once the second-last choices have been determined, the agent moves to the third-last choice nodes, and so on back to the initial choice node. The result is that only certain paths through the decision tree would ever be realised, if initiated.

Sophisticated Ulysses would take note of the fact that, if he reaches the island of the sirens unrestrained, he will want to stop there indefinitely, due to the transformative effect of the sirens' song on his preferences. He acknowledges the implications of this prediction and so deems the choice combination of 'not issuing an order for his crew to restrain him and then sticking to his course' to be an impossibility. Table 35.2 presents sophisticated Ulysses' static representation of his decision problem in terms of the combinations of choices that he predicts would arise. Note that there are only two feasible combinations of choices here. Table 35.3 gives an alternative static representation, whereby Ulysses' future preferences/choices are part of the state space; this static representation is, in a sense, more general, because it can easily be modified to incorporate probabilistic, as opposed to certain, predictions about future preferences/choices, (In Table 35.3, the later preference for staying with the sirens is certain; we see that the probability for this state, p, is equal to one.) Moreover, in Table 35.3, the 'acts' are limited to those things the agent can initiate at the moment of decision. This is in keeping with, for instance, Joyce's [11, 57–61] interpretation of acts in a static decision model.

Defenders of the resolute approach to dynamic choice dismiss problems like Ulysses', claiming that Ulysses cannot serve as a model agent for dynamic rationality, for the reason given in Sect. 35.2. Indeed, the resolute approach is particularly unconvincing in the context of Ulysses' decision problem. The key point of difference between the sophisticated and resolute approaches concerns how a *rational* agent may be expected to choose at future nodes of a decision tree. While the sophisticated approach assumes that an agent always chooses in accordance with their preferences at the time, the resolute approach holds that an agent may sometimes defer to their previous preferences or strategy—they may honour a previous commitment, despite present misgivings. The reason this recommendation is not very convincing in Ulysses' case is that there is no apparent reason why Ulysses, upon reaching the island of the sirens and not restrained by his crew, would ignore his current preferences and instead sail straight on. Indeed, depending how one understands the relationship between preference and choice, it is arguably contradictory to depict an agent choosing according to preferences other than their own preferences at the time.

Agents like Ulysses cannot appeal to the resolute approach to dynamic choice. Defenders of the resolute approach rather appeal to decision problems like the one in Fig. 35.2. Their aim is to defend both the resolute approach to dynamic choice and preferences that violate the *independence* axiom.[2] The agent's preferences, with respect to Fig. 35.2, are thus stipulated as follows (so as to violate independence): at all times she prefers O_1 to O_2, but she prefers the lottery that gives O_2 if E and O_3 if $\neg E$ to the lottery that gives O_1 if E and O_3 if $\neg E$. We can refer to the former lottery as L_2 and the latter as L_1. There is another lottery in Fig. 35.2, $L_2 - \delta$, where the value of δ is selected such that $L_2 > L_2 - \delta > L_1$.

Some examination reveals that a sophisticated agent with preferences as specified above (or more accurately, who *predicts* that her preferences will be as specified above) initially chooses 'down' in the problem in Fig. 35.2, which effectively amounts to the lottery $L_2 - \delta$. Note that this strategy is dominated by the one that amounts to L_2: 'up' at the initial choice node, and then 'down' at the second. According to McClennen [20], this is precisely the kind of situation in which the resolute, as opposed to the sophisticated, approach to the problem is more apt. At all times the agent prefers L_2 to $L_2 - \delta$, so it is in her best all-round interests to pursue the L_2 lottery and stick firmly to this plan, despite the fact that O_1 will look better than O_2 at the second choice node (i.e., 'up' rather than 'down'), should the agent reach this position.

We need not get sidetracked here by questions about the (ir)rationality of preferences that violate independence. The agent's preferences can simply be taken as given. The question is: Can such an agent reasonably expect to be a resolute chooser? That is: Would an agent with preferences as stipulated reasonably choose 'down' rather than 'up', were she to reach the second choice node in Fig. 35.2? Defenders of resolute choice say that the rational agent would indeed vindicate her previous decision to pursue the lottery L_2. In this author's opinion, that proposal defies the very meaning of preference. Of course, an agent may place considerable importance on honouring previous commitments. Any such integrity concerns, however, should be reflected in the description of final outcomes and thus in the agent's actual preferences at the time in question. Conceiving an agent's preferences as concerning more complex outcomes than initially supposed, is quite different from conceiving an agent's preferences to be out of step with her supposed choices at the time in question, which is what the resolute approach to sequential choice is committed to.

[2]Joyce [11, 86] gives the following informal statement of the *independence* axiom: 'a rational agent's preferences between (acts) A and $A*$ should not depend on circumstances where the two yield identical outcomes.'

What this discussion highlights is that controversies surrounding dynamic/sequential choice are essentially controversies about how an agent's decision at a time should be informed by her predicted future preferences/choices (in addition to predictions about future states of affairs). The naive approach recommends that an agent simply ignore predictions about her future attitudes. This is clearly problematic as it amounts to not taking into account all the available evidence when making a decision. The dispute between the resolute and sophisticated approaches is more fine-grained; it concerns how predicted future preferences inform corresponding future choices, in the context of the greater dynamic decision problem at hand.

While not orthodox, the most general translation of a dynamic decision problem to static form is to include future preferences/choices in the state space (see [29], Sect. 35.2). Table 35.3 employs this representation. The available acts are simply the options at the initial choice node. The outcomes of these acts depend not only on how things turn out in the external world, but also on what decisions the agent confronts later and the strategies she would then choose. These may all be aspects of the future that the agent is unsure about, and thus assigns probabilities between zero and one. (Note that the examples in this chapter, as per much of the discussion of sequential choice, are special cases where future preferences are known for sure, or in other words, as in Table 35.3, are assigned probability one.)

35.4 Normative Questions: Can Dynamic Decision Models Help?

We have seen how dynamic decision trees can help an agent like Ulysses take stock of his static decision problem, so that he can work out what to do 'now'. The literature on dynamic decision theory has more ambitious aims than this, however. Much discussion is devoted to more general normative questions: Must rational preferences conform to expected utility theory? What constitutes rational belief and preference change? Indeed, the work of a number of authors, including Hammond [6–10], Seidenfeld [23–27], McClennen [19, 20], Machina [17], Rabinowicz [21, 22], Skyrms [28], Steele [29], Bradley and Steele [3], and Buchak [4, 5] demonstrates that dynamic decision models provide a rich setting for investigating these familiar normative questions. As one might guess, the findings are controversial.

Our earlier discussion of the resolute approach to dynamic choice gave a glimpse of how dynamic-choice problems shed light on normative issues. Refer back to the problem in Fig. 35.2. This problem is useful for evaluating preferences that violate independence (as per cumulative prospect theory (see [17]) and the associated risk-weighted expected utility theory defended by Buchak [5]). The preferences specified in Sect. 35.3 violate independence, by design:

$O_1 > O_2$
$L_2 : (O_2 \text{ if } E; O_3 \text{ if } \neg E) > L_2 - \delta > L_1 : (O_1 \text{ if } E; O_3 \text{ if } \neg E)$

The dynamic choice problem in Fig. 35.2 can serve as a controlled experiment, so to speak, to test preferences of this sort. The experiment is 'controlled' because the agent has stable or constant preferences; she does not predict any rogue changes in belief or desire, as per Ulysses. Indeed, the agent predicts her preferences will change only due to learning new information that leads to a belief update in accordance with Bayes' rule. The question is whether these preferences are shown to be in some sense self-defeating, suggesting they are irrational.

Recall that the sophisticated agent with preferences as specified above will choose 'down' in Fig. 35.2, effectively selecting a strategy that amounts to $L_2 - \delta$. The embarrassment here is that there is another strategy in the dynamic tree—'up' initially and then 'down'—that effectively amounts to L_2, which clearly dominates $L_2 - \delta$. Of course, our agent is not guilty of choosing an option that is dominated by another *available* option. The problem, however, is that it is the agent's own preferences that make the dominating L_2 strategy unavailable to her. And for this reason, we might judge the preferences to be self-defeating or irrational.

A number of the authors mentioned above discuss this potential criterion for rational preferences, namely that dominating strategies in the dynamic setting should not be unavailable to an agent on account of her own (stable) preferences. Refer to this as the 'dominating-strategies' criterion. McClennen [20] upholds the criterion, and Rabinowicz [21] and Steele [29] express some support for it. Hammond [6, 7, 9, 10] defends an even stronger criterion that effectively requires all strategies in an extensive-form model, i.e., all combinations of choices, to be available to an agent. In other words, the agent's own preferences should not prevent her from pursuing what she 'now', i.e., at the outset, considers to be the best strategy. Hammond refers to this criterion as *consequentialism*, but this label is rather misleading.

Assuming sophisticated (rather than resolute) choice, the 'dominating-strategies' criterion rules out preferences that violate independence.[3] It is worth noting that this same criterion is the cornerstone of the well-known 'diachronic Dutch book argument', or at least the version in Skyrms [28]. In this case, the 'controlled experiment' takes a slightly different form: The agent at all times has preferences that conform to expected utility theory, and her basic desires are stable. It is the agent's learning or belief-update rule that is under scrutiny. Skyrms shows that a sophisticated agent whose belief-update rule is something other than *Bayesian conditioning*, may, in some cases, choose a dominated option because the dominating option is unavailable to her. Conversely, this is never the case for an agent who plans to update by Bayes' rule. On this basis, we are supposed to conclude that Bayesian conditioning is the uniquely rational belief-update rule.

[3]We thus see why McClennan [20] defends the resolute approach to dynamic choice. For similar reasons, Buchak [5] is also sympathetic to resolute choice.

Seidenfeld notably rejects both Hammond's consequentialism and the 'dominating-strategies' criterion just outlined. Seidenfeld seeks to defend decision theories that violate ordering (and only secondarily, independence); an example of such a theory is Levi's [15] *E-admissability* choice rule.[4] Like cumulative prospect theory (for instance), Levi's theory, in its general form, does not satisfy the 'dominating strategies' criterion (see [29]). Unlike cumulative prospect theory, however, Levi's theory does satisfy an alternative criterion for rational preference that concerns future choices between 'indifferents', or options the agent is indifferent between; Seidenfeld articulates and defends this criterion in a series of interchanges with other authors (i.e., in [23, 24, 26, 27]); the debate is discussed in detail in Steele [29].

We might label Seidenfeld's criterion the 'future indifferents' criterion. It holds that, in the controlled setting where preferences remain stable, if the agent will be indifferent between options at a later choice node, then she should be indifferent now between strategies that terminate in these options, and are otherwise identical prior to the choice node in question. If this criterion is not satisfied, as per theories that violate independence like cumulative prospect theory [23], there is no 'natural' way to evaluate the aforementioned strategies. The strategies have differing utilities, but either one may eventuate if the agent makes the appropriate initial choices. The obvious move here is to acknowledge only one strategy, with a final tie-breaking step. Steele [29] pursues this line of argument against Seidenfeld's criterion. There remain problems, however, if the evaluation of the single strategy depends on which tie-breaker is selected, as per preferences that violate independence.

Whether or not one affirms Seidenfeld's 'future indifferents' criterion for rational preference, the issues it raises demand consideration. Choice in the face of indifference has always been puzzling, but the problems take on new significance in the dynamic setting—in this setting there is a need to explicitly model predicted future choices, including choices between indifferents, in order to evaluate current options.

35.5 Concluding Remarks

The previous section gave a tour of the prominent dynamic-choice arguments concerning rational preference (and learning) that have been discussed in the literature. As indicated, there is persistent disagreement. Some of the disagreement concerns the sophisticated/resolute distinction discussed in Sect. 35.3. This is best considered a dispute about the meaning of the terms in a dynamic-choice model, in

[4]This is a lexical choice rule that can handle indeterminate belief and/or desire, represented by a set of probability-utility function pairs. The 'E-admissible' options are those that have maximal expected utility for at least one probability-utility representation in the set; these are the options that a rational agent *may permissibly* choose. The agent discriminates between 'E-admissible' options on the basis of her 'security' attitudes.

particular, future preference and its relationship to future choice. Beyond that, there is disagreement about what are reasonable features of choice in the dynamic setting, as discussed in Sect. 35.4.

A further topic of debate in the more recent literature concerns what rival decision theories say about the value of 'information' or evidence retrieval. This implicates dynamic choice as it concerns whether to choose one of a given set of options 'now' or rather wait to collect evidence that may be pertinent to the choice in question. A candidate rationality criterion is that one should always wait to retrieve further evidence if the evidence is 'free' and may influence one's choice; call this the 'free evidence' criterion. The criterion is discussed in Kadane et al. [12], Buchak [4], and Bradley and Steele [3] in relation to varying generalisations of expected utility theory. These authors reject the standard version of the 'free evidence' criterion, but arguably, the issues are not fully settled. Another topic deserving of further investigation is the possibilities for rational preference change; Bradley [2], for instance, considers cases beyond preference change in response to new information. The relationship between present uncertainty about future preferences and a 'preference for flexibility' with respect to available options in the future is a related issue that also deserves further investigation; for early works on this topic, see Koopmans [13], Kreps and Porteus [14], and Arrow [1].

References

Asterisks ($*$) indicate recommended readings.

1. Arrow, K. (1995). A note on freedom and flexibility. In K. Basu, P. Pattanaik, & K. Suzumura (Eds.), *Choice, welfare and development* (pp. 7–15). Oxford: Oxford University Press.
2. Bradley, R. (2009). Becker's thesis and three models of preference change. *Politics, Philosophy and Economics, 8*(2), 223–242.
3. Bradley, S., & Steele, K. S. (2016). Can free evidence be bad? Value of information for the imprecise probabilist. *Philosophy of Science, 83*(1), 1–28.
4. Buchak, L. (2010). Instrumental rationality, epistemic rationality, and evidence-gathering. *Philosophical Perspectives, 24*, 85–120.
5. Buchak, L. (2013). *Risk and rationality* Oxford: Oxford University Press.
6. Hammond, P. J. (1976). Changing tastes and coherent dynamic choice. *The Review of Economic Studies, 43*(1), 159–173.
7. Hammond, P. J. (1977). Dynamic restrictions on metastatic choice. *Economica, 44*(176), 337–350.
8. Hammond, P. J. (1988). Orderly decision theory: A comment on Professor Seidenfeld. *Economics and Philosophy, 4*, 292–297.
9. Hammond, P. J. (1988). Consequentialism and the independence axiom. In B. R. Munier (Ed.), *Risk, decision and rationality*. Dordrecht: D. Reidel.
10. Hammond, P. J. (1988). Consequentialist foundations for expected utility theory. *Theory and Decision, 25*, 25–78.
11. Joyce, J. M. (1999). *The foundations of causal decision theory*. Cambridge: Cambridge University Press.
12. Kadane, J. B., Schervish, M., & Seidenfeld, T. (2008). Is ignorance bliss? *Journal of Philosophy, 105*(1), 5–36.

13. Koopmans, T. C. (1962). *On flexibility of future preference. Cowles foundation for research in economics* (Cowles foundation discussion papers 150), Yale University.
14. Kreps, D. M., & Porteus, E. L. (1978). Temporal resolution of uncertainty and dynamic choice theory. *Econometrica, 46*(1), 185–200.
15. Levi, I. (1986). *Hard choices: Decision making under unresolved conflict*. Cambridge: Cambridge University Press.
16. Levi, I. (1991). Consequentialism and sequential choice. In M. Bacharach & S. Hurley (Eds.), *Foundations of decision theory*. Oxford: Basil Blackwell.
17. Machina, M. J. (1989). Dynamic consistency and non-expected utility models of choice under uncertainty. *Journal of Economic Literature, 27*, 1622–1668.
18. Maher, P. (1992). Diachronic rationality. *Philosophy of Science, 59*(1), 120–141.
19. McClennen, E. F. (1988). Ordering and independence: A comment on professor Seidenfeld. *Economics and Philosophy, 4*, 298–308.
20. * McClennen, E. F. (1990). *Rationality and dynamic choice: Foundational explorations*. Cambridge: Cambridge University Press.
21. *Rabinowicz. W. (1995). To have one's cake and eat it too: Sequential choice and expected-utility violations. *Journal of Philosophy, 92*(11), 586–620.
22. Rabinowicz, W. (2000). Preference stability and substitution of indifferents: A rejoinder to Seidenfeld. *Theory and Decision, 48*, 311–318.
23. Seidenfeld, T. (1988). Decision theory without "independence" or without "ordering". *Economics and Philosophy, 4*, 309–315.
24. Seidenfeld, T. (1988). Rejoinder [to Hammond and McClennen]. *Economics and Philosophy, 4*, 309–315.
25. Seidenfeld, T. (1994). When normal and extensive form decisions differ. *Logic, Methodology and Philosophy of Science, IX*, 451–463.
26. Seidenfeld, T. (2000). Substitution of indifferent options at choice nodes and admissibility: A reply to Rabinowicz. *Theory and Decision, 48*, 305–310.
27. Seidenfeld, T. (2000). The independence postulate, hypothetical and called-off acts: A further reply to Rabinowicz. *Theory and Decision, 48*, 319–322.
28. Skyrms, B. (1993). A mistake in dynamic coherence arguments? *Philosophy of Science, 60*(2), 320–328.
29. *Steele, K. S. (2010). What are the minimal requirements of rational choice? Arguments from the sequential-decision setting. *Theory and Decision, 68*, 463–487.

Chapter 36
Causal Decision Theory

Brad Armendt

Abstract Causal decision theory (*CDT*) is a general theory of rational decision, appropriate for simple or complex decision problems. It is an expected utility theory distinguished by its explicit attention to causal features of decision problems, and by the significance it attaches to those features. When the causal structure of a decision problem is uncomplicated, the recommendations of *CDT* and other theories generally agree. In more complex cases, however, *CDT* identifies rational decisions where other theories do not. Several varieties of *CDT* have been offered; they differ in their ways of representing beliefs about causal influence, but as decision theories they are very similar. Each of them was developed as a *subjective* expected utility theory, and that approach will be assumed here.

Causal decision theory (*CDT*) is a general theory of rational decision, appropriate for simple or complex decision problems. It is an expected utility theory distinguished by its explicit attention to causal features of decision problems, and by the significance it attaches to those features. When the causal structure of a decision problem is uncomplicated, the recommendations of *CDT* and other theories generally agree. In more complex cases, however, *CDT* identifies rational decisions where other theories do not. Several varieties of *CDT* have been offered; they differ in their ways of representing beliefs about causal influence, but as decision theories they are very similar. Each of them was developed as a *subjective* expected utility theory, and that approach will be assumed here.

B. Armendt (✉)
SHPRS, Arizona State University, Tempe, AZ, USA
e-mail: armendt@asu.edu

S. O. Hansson, V. F. Hendricks (eds.), *Introduction to Formal Philosophy*, Springer Undergraduate Texts in Philosophy, https://doi.org/10.1007/978-3-319-77434-3_36

36.1 A Basic Idea

The point of a deliberate action is to intervene in the world so as to *bring about* desirable results and/or to *prevent* undesirable results. Our motives for choosing among our available actions are rooted in a comparison of their expected results. Such results may be direct and obvious, or indirect and difficult to ascertain (think of effects on one's reputation, or on one's future habits). This consequentialist idea is that the values of actions lie in the values of their causal consequences. The idea is compatible with the fact that we often exercise our agency in ways that differ from deliberate, reasoned choice. When we improvise, explore, or do what seems fitting, at some level we may not foresee the causal consequences of our actions. But it is arguable that when we judge the value of what we do in such settings, we attend to effects of what we do.

Since we have limited information about what causes what in the world, our ability to foresee an action's causal consequences has limits. The decision-making value of an action is an *expected value*, where the expectation reflects the values of its various possible causal consequences, weighted by our assessments of how likely those consequences are to result from the action. (This assumes that the possible consequences, as we distinguish them, are incompatible with each other, and that the full list of possible consequences is a *partition*.)

Causal consequences abound. For any action *A*, if we care to, we can find a myriad of possible causal consequences traceable to *A* (for example, sunbathing on the beach might rearrange the grains of sand in many possible ways). A decision-maker will consider and be motivated by relatively few of these. Let us say, then, that our *Basic Idea* is:

BI: The values of actions are expectations of the values of their causal consequences that matter to the decision-maker.

This idea underlies causal decision theory (*CDT*).

What if acts have no causal consequences? The theory is about what the decision-maker takes to be causal influences, even if she is wrong about them. In a situation where a decision-maker believes that her alternatives have no causal consequences, *BI* has little to offer. *CDT* will then say that all such alternatives get no value from their (nonexistent) consequences, and they are equally choiceworthy. In a given utility assignment their utilities may be nonzero, but they will be equal.

Causal influences often occur in complex combinations; might there be situations to which *BI* is not easily applied? Yes. Can there be situations in which *BI* and *CDT* recommend actions that are less valuable than one would like? Again, yes. But a deficiency that arises when one's best option is disappointing is not a failure of *CDT*, if the theory recommends the best of the options that are available.

36.2 Historical Background

The decision theory developed by L.J. Savage [27] provided an elegant, rigorous, and influential account of personalist (subjective) probability and decision-making utility. The theory treats actions as functions from states of the world to consequences, and states as independent of actions. If the state of the world is known, the value of an action is the value of the consequence to which, in that state, the action leads. More generally, when the actual state is not certain, the value of an action is the weighted average, or expectation, of the values of the consequence it has in each possible state, where the weights are the unconditional probabilities of each state:

$$U(A) = \sum_j pr\left(S_j\right) U\left(C_A \;\&\; S_j\right) \tag{S1}$$

Since each act-state combination determines a consequence, and the states form a partition, we can also express Savage's utility rule in terms of the conjunctions of the action with the various states:

$$U(A) = \sum_j pr\left(S_j\right) U\left(A \;\&\; S_j\right) \tag{S2}$$

(Here and throughout this article, we use finite forms of such rules.) Savage's theory was a remarkable achievement, but for some purposes and applications that interest philosophers, its formal structure seems too rigid. In the framework of the theory, there is no doubt, given a state of the world, about what the precise outcome of a particular action will be; the decision-maker is supposed to deliberate about states and actions that are so finely specified as to make that so. But, as Savage knew, sometimes we find ourselves in situations that do not directly fit this structure, and the formal theory appears to require that we regiment our decision problems to an extent that sometimes exceeds what we can manage. We consider adding an egg to five others to make an omelet (Savage's example). It might be a good egg or a bad one (different possible states). Will adding one bad egg to five good ones yield a spoiled omelet? If we are not sure, then we do not know which consequence will result from the state of five good eggs and one bad one. Savage suggested that we handle this by using an expanded set of finer-grained states, each specifying the condition of the egg together with facts about adding a bad egg to good ones. But sometimes we may have little idea which sets of fine-grained facts will determine a given action's outcomes.

In the 1960s an appealing alternative was developed by Richard Jeffrey, relying on a theorem due to Ethan Bolker (Jeffrey [12]), and it came to be known as *evidential decision theory* (*EDT*). *EDT* allows the decision-maker to deliberate using states in which an action's outcome is not certain, as in the example of the omelet. Like the theories of Savage, Ramsey, and de Finetti, Jeffrey's *EDT* is a theory of subjective probability; probabilities measure the degrees of belief of the (rational) decision-maker. The theory has many virtues and was attractively packaged with other important ideas, and it became widely used among philosophers. Causal considerations are absent from the structure and principles of *EDT*.[1] That might be, and at times was, seen as one of its virtues, should *EDT* provide a satisfactory account of rational decision without them. But doubts arose that it could do so, and the doubts led to proposals for *CDT*, in several versions.

The expected utility (or desirability *des*) rule in Jeffrey's *EDT* applies to any proposition p and partition $\{q_j\}$ for which the $pr(q_j/p)$s are well defined. When applied to action A and a partition of possible states that matter to the decision-maker, it is

$$des(A) = \sum_j pr\left(S_j/A\right) des\left(A \,\&\, S_j\right). \tag{J1}$$

Note the role of the conditional probabilities $pr(S_j/A)$ of the state, given the act. States may be probabilistically independent of the act, but they need not be. When they are not, the probabilistic association influences how the values of the various states are weighted in the overall value of the action A. Jeffrey suggested that $des(A)$ be interpreted as the value of the *news that A is true*.[2] Jeffrey's suggestion applies to any element of the decision-maker's preference ranking; here it is applied to the action A.

In Jeffrey's theory, there can be more than one possible consequence that a given combination of state and action might yield, and if the set of all the possible consequences of A is a partition $\{C_i\}$, we have

$$des(A) = \sum_j pr\left(S_j/A\right) \sum_j pr\left(C_i/A \,\&\, S_j\right) des\left(C_i \,\&\, A \,\&\, S_j\right). \tag{J2}$$

Again we see that probabilistic associations between actions and states will affect the way that values of the various states and consequences are weighted in the overall value of the action A. If such associations arise because actions (are believed to) causally influence the states, this is compatible with *BI*. But if a probabilistic dependence of states on acts arises in some other way, there is room for conflict between *EDT* and *BI*.

---

[1]Though perhaps not always from its intended interpretation (of actions, *e.g.*), about which Jeffrey's views shifted over time. The noncausal character of the theory was pointed out by Jeffrey from its inception (Jeffrey [12], chapter 10), but see Jeffrey [13] and Joyce [15] for Jeffrey's later view.

[2]Think of how the policy of conditionalization recommends that your conditional probabilities $pr(-/p)$ guide your new $pr_n(-)$ probabilities, after p is learned for sure—that is, after you get the news that p.

36.3 Difficult Problems for Evidential Theory

When do probabilistic dependence of states on acts and direct causal dependence come apart? Suppose you believe that a common cause exerts influence over (a) your action, and also over (b) possible states of the world that matter to the values of the possible outcomes of your action. In such a context, probabilistic dependence and direct causal dependence diverge; one striking case is when your actions are probabilistic indicators of the actual state, but your actions exert *no* direct causal influence on the state. In other words, $pr(S_i/A) \neq pr(S_i)$, yet A neither promotes nor prevents S_i. Notice that you need not know what the common cause is; the situation could arise as long as you regard a relevant state S_i as probabilistically dependent on your action, yet at the same time reject the idea that your action causally influences the state. Jeffrey's expected utility (*des*) rule applies to any partition, so if the states $\{S_j\}$ form a partition, *EDT* endorses an evaluation of A using $\{S_j\}$, as in (J1) or (J2). The evaluation will then use probability weights that are at odds with what you believe A's causal influences to be. As we will see, *CDT* is designed to avoid that.

Call decisions that fit the preceding sketch, where members of a partition of significant states are probabilistically dependent on the available acts, yet no direct causal influence from acts to the states is believed to be present, (causally) *confounded decisions (CDs)*.[3] *EDT* uses the conditional probabilities $pr(S_i/A)$ as weights in its utility rule, and sometimes *EDT* does so in *CD*s where causal influence from action to state is absent. The difficulty is that this sometimes leads to incorrect recommendations. The best known example of a *CD* is *Newcomb's Problem*, introduced to philosophers by Robert Nozick [21].

> A being in whose power to predict your choices correctly you have great confidence is going to predict your choice in the following situation. There are two boxes, *B1* and *B2*. Box *B1* contains \$1,000; box *B2* contains either \$1,000,000 (\$M) or nothing. You have a choice between two actions: (1) taking what is in both boxes; (2) taking only what is in the second box. Furthermore, you know, and the being knows you know, and so on, that if the being predicts you will take what is in both boxes, he does not put the \$M in the second box; if the being predicts you will take only what is in the second box he does put the \$M in the second box. First the being makes his prediction; then he puts the \$M in the second box or not, according to his prediction; then you make your choice [22].

Some presentations specify that box *B1* is transparent, so you can see that it contains \$1000. The basis of the prediction is unspecified, but it is assumed to be something that obtains prior to or at the time the prediction is made; the predictor does not somehow observe your future act. (If you believe that he did, it is a different problem that need not be trouble for *EDT*.) A natural gloss on the story is that the

[3]More generally, in *CD*s there are significant state-partitions such that $pr(S_i/A)$ does not reflect the extent to which A is believed to have direct causal influence on S_i. *CDT* makes use of state-partitions for which direct causal influence is believed entirely absent.

basis of the prediction is a common causal influence on your choice and, through the predictor's actions, on the contents of box *B2*. Statements of the problem often specify that you believe the predictor is very reliable, but the challenge for *EDT* theory can arise even if you believe that the predictor's success rate is only slightly better than 50% (or a little better than that, if you give declining marginal utility to dollars).

Nozick [21] presented Newcomb's Problem as an illustration of a clash between two principles of choice, maximizing expected utility vs. dominance reasoning. Since the expected utility principle he used agrees with the treatment of an act's expected utility in Jeffrey's *EDT*, the clash is between *EDT* and dominance reasoning.

Evidential reasoning:

Suppose, to pick a number, you believe that the predictions are 90% reliable in the sense that the probability is .9 that the action chosen was correctly predicted. Also suppose that the possibility that the predictor misfills *B2* is negligible. Then we can regard '*\$M in B2*' and '*\$0 in B2*' as states, '*take B2*' and '*take B1 & B2*' as acts, with

$$pr\ (\$M\ in\ B2/take\ B2) = .9 \quad pr\ (\$0\ in\ B2/take\ B1\ \&\ B2) = .9$$
$$pr\ (\$0\ in\ B2/take\ B2) = .1 \quad pr\ (\$M\ in\ B2/take\ B1\ \&\ B2) = .1$$

Supposing that \$ measure the *des* of an outcome, *EDT* says

$$\begin{aligned} des\ (take\ B2) &= pr\ (\$M\ in\ B2/take\ B2)\, des\ (\$M\ in\ B2\ \&\ take\ B2) \\ &\quad + pr\ (\$0\ in\ B2/take\ B2)\, des\ (\$0\ in\ B2\ \&\ take\ B2) \\ &= .9\,(\$M) + .1\ \ (\$0) = \$900,000, \text{ while} \end{aligned}$$

$$\begin{aligned} &des\ (take\ B1\&B2) \\ &\quad = pr\ (\$M\ in\ B2/take\ B1\ \&\ B2)\, des\ (\$M\ in\ B2\ \&\ take\ B1\ \&\ B2) \\ &\qquad + pr\ (\$0\ in\ B2/take\ B1\ \&\ B2)\, des\ (\$0\ in\ B2\ \&\ take\ B1\ \&\ B2) \\ &\quad = .1\ (\$\mathrm{M} + \$\mathrm{T}) + .9\,(\$\mathrm{T}) = \$101,000 \end{aligned}$$

So if you maximize expected utility using *EDT*, you choose to take only what is in *B2*.

Dominance reasoning:

At the moment of decision, the contents of *B2* are already fixed. Either it contains \$M or it doesn't. Suppose it contains \$M; then taking both *B1* and *B2* is a better choice than taking *B2* alone and thereby leaving \$T on the table. Suppose it does not contain \$M; again, taking the contents of both boxes is the better choice. So either way, taking both boxes is better, you should take both boxes, and the recommendation by *EDT* is mistaken.

Now, as Jeffrey [12]) pointed out, it is easy to see that dominance reasoning is not generally reliable in the context of *EDT* when states are not probabilistically independent of acts. But in the Newcomb case, where you believe that the state is fixed before you act, and is not causally influenced by the act, dominance reasoning for the two-box answer is hard to dismiss. To follow *EDT*'s guidance would be to choose to generate evidence for a desired outcome while believing that the choice in no way promotes that outcome, and doing so at the cost of $T. Developers of causal decision theory took the two-box answer to be correct, and they proposed revisions of the expected utility calculation that a) agree with *EDT* in the many cases where probabilistic dependence is (believed) an accurate guide to causal influence, while b) correcting *EDT* in cases where probabilistic dependence and causal dependence (are believed to) come apart.

A second *CD* that appeared in the earliest discussions of *CDT* is the *Smoking Gene example*. The story given by Stalnaker [33] includes a hypothesis that was once considered by R.A. Fisher:

> Imagine a man deliberating about whether or not to smoke. There are two, equally likely hypotheses (according to his beliefs) for explaining the statistical correlation between smoking and cancer: (1) a genetic disposition to cancer is correlated with a genetic tendency to the sort of nervous disposition which often inclines one to smoke. (2) Smoking, more or less, causes cancer in some cases. If hypothesis (1) is true, he has no independent way to find out whether or not he has the right sort of nervous disposition.

An important point that Stalnaker makes with this example is that we may be uncertain about which causal structure is true, and that a good decision theory should be able to guide us in such cases. We will return to that point soon. Focus for the moment on the smoking gene hypothesis (1). If the hypothesis is correct, cancer is probabilistically dependent on smoking, because of the common influence of the disposition. *EDT* picks up on that, and drastically reduces the value given to smoking, even when the smoking gene hypothesis says (let us suppose) that smoking has no influence on one's genes, or on getting cancer. So if, contrary to Stalnaker's version of the example, the man knows that the smoking gene hypothesis (1) is correct, his decision problem resembles a Newcomb problem in which you know about the predictor's role in determining the contents of the opaque box. *EDT* will then recommend that the man refrain from smoking, even though its only relevant causal consequence is (let us suppose) a pleasure having positive value.

36.4 Causal Conditionals and the Gibbard-Harper Theory

A consequentialist expected utility theory expresses the utility of my action A in terms of the utilities of its possible consequences C_i, weighted by their probabilities. It is a natural idea to think of those weights as probabilities of conditionals *if I were to do A, then the result would be* C_i. As we have seen, Jeffrey's evidential theory does not do this, instead using conditional probabilities $pr(C_i/A)$ as the weights.

Do the conditional probabilities and the probabilities of the conditionals agree? In 1972, David Lewis presented a trivialization result undermining the idea that the probability of a conditional *if p then q* is always the same as the conditional probability of q given p [20]. With the Newcomb problem in mind, Robert Stalnaker [33] suggested that the generally correct way to evaluate the expected utility of A is to weight the value of each consequence C_i by the probability of the subjunctive causal conditional, $A \mathrel{\Box\!\!\rightarrow} C_i$, whose probability need not agree with $pr(C_i/A)$.

Allan Gibbard and William Harper developed Stalnaker's idea into a *CDT*; it was presented in 1975 and published in their [9]. They characterized two accounts of expected utility: one is essentially Jeffrey's evidential theory, calculated using conditional probabilities and denoted $\mathcal{V}$; the other, denoted $\mathcal{U}$, is calculated using probabilities of conditionals. What makes $\mathcal{U}$ a form of *CDT* is the interpretation Gibbard and Harper give to the conditionals. The conditionals are understood as causal and nonbacktracking; $A \mathrel{\Box\!\!\rightarrow} S_j$ expresses the idea that S_j would causally result from doing A (the idea that either S_j is inevitable, or that A would bring it about). When A in no way promotes or prevents S_j, the probability of the conditional, $pr(A \mathrel{\Box\!\!\rightarrow} S_j)$ is no different from the unconditional probability $pr(S_j)$. They explicitly make the working assumption that the conditionals satisfy a Conditional Excluded Middle principle so that, in the presence of other less controversial principles, partitions of possible consequences $\{C_i\}$, or of possible states $\{S_j\}$ can be counted on to generate partitions of conditionals $\{A \mathrel{\Box\!\!\rightarrow} C_i\}$, or $\{A \mathrel{\Box\!\!\rightarrow} S_j\}$.

So, in a *CD* where acts have no direct causal influence over states S_j, $\mathcal{U}$ weights the value of $C_i = (S_j \ \& \ A)$ by the unconditional probability $pr(S_j)$, and does so for each of the available actions, while $\mathcal{V}$ uses the different weights $pr(S_j/A)$. In such cases, though dominance reasoning may conflict with $\mathcal{V}$-theory, it is endorsed by $\mathcal{U}$-theory:

> *Newcomb's problem*: Since neither choice of *B2* alone nor choice of *B1* & *B2* influences the contents of the boxes, both $pr(take\ B2 \mathrel{\Box\!\!\rightarrow} S_j)$ and $pr(take\ B1\ \&\ B2 \mathrel{\Box\!\!\rightarrow} S_j)$ are equal to $pr(S_j)$, where the S_j are the possible arrangements of money in the boxes. So
>
> $$\begin{aligned}\mathcal{U}\,(take\ B2) &= pr\,(take\ B2 \mathrel{\Box\!\!\rightarrow} \$M\ in\ B2)\,\mathcal{U}\,(\$M)\\ &\quad + pr\,(take\ B2 \mathrel{\Box\!\!\rightarrow} \$0\ in\ B2)\,\mathcal{U}\,(\$0)\\ &= pr\,(\$M\ in\ B2)\,\mathcal{U}\,(\$M) + pr\,(\$0\ in\ B2)\,\mathcal{U}\,(\$0)\end{aligned}$$
>
> $$\begin{aligned}\mathcal{U}\,(take\ B1\ \&\ B2) &= pr\,(take\ B1\ \&\ B2 \mathrel{\Box\!\!\rightarrow} \$M\ in\ B2)\,\mathcal{U}\,(\$M + T)\\ &\quad + pr\,(take\ B1\ \&\ B2 \mathrel{\Box\!\!\rightarrow} \$0\ in\ B2)\,\mathcal{U}\,(\$T)\\ &= pr\,(\$M\ in\ B2)\,\mathcal{U}\,(\$M + T) + pr\,(\$0\ in\ B2)\,\mathcal{U}\,(\$T)\end{aligned}$$
>
> and whatever the probabilities are, the $\mathcal{U}$ of taking both boxes exceeds the $\mathcal{U}$ of taking only *B2*, by the *$T* in *B1*.

In ordinary situations, actions do not carry information about which causal conditional is true, but in *CD*s that is not so: $pr(take\ B2 \mathrel{\Box\!\!\rightarrow} \$M\ in\ B2\ /\ take\ B2)$ is greater than $pr(take\ B2 \mathrel{\Box\!\!\rightarrow} \$M\ in\ B2\ /\ take\ B1\ \&\ B2)$, since taking only *B2* is correlated with the predictor's filling *B2*. But we may still expect that actions do not exert causal influence over which conditionals are true; that is, over which of the possible causal structures that underpin the conditionals is the actual one.

In Gibbard and Harper's treatment, a single causal conditional $A \square\!\!\rightarrow C$ expresses the deterministic causation of *C* by *A*. A fuller representation of the relevant causal structure is given by a conjunction that specifies what each of the available actions would bring about: $(A_1 \square\!\!\rightarrow C_j)$ & $(A_2 \square\!\!\rightarrow C_k)$ & ... & $(A_n \square\!\!\rightarrow C_m)$, for all of the possible actions A_j, where the consequences are members of $\{C_i\}$. Following Lewis [19], call such a conjunction a *dependency hypothesis*, *DH*. Just as the decision-maker will typically be unsure which individual causal conditional holds, he will be unsure about which of the many possible dependency hypotheses in $\{DH_k\}$ is true. In a *CD*, individual conditionals may be probabilistically correlated with actions yet causally independent of them, and the same is so for dependency hypotheses. In the Gibbard-Harper theory, a conjunction *DH* & *A* of a dependency hypothesis and an action specifies a definite outcome C_i, and there is agreement between calculations of (*A*) in terms of a partition of individual conditionals:

$$\mathcal{U}(A) = \sum\nolimits_i pr\,(A \square\!\!\rightarrow C_i)\,\mathcal{U}\,(C_i) \tag{GH1}$$

and calculations in terms of unconditional probabilities of a partition of dependency hypotheses:

$$\mathcal{U}(A) = \sum\nolimits_j pr\,\left(j^{th} conjunction\ of\ conditionals\right) \times$$

$$\mathcal{U}\left(consequence\ of\ A\ according\ to\ DH_j\right)$$

$$\mathcal{U}(A) = \sum\nolimits_j pr\,\left(DH_j\right)\mathcal{U}\left(A\ \&\ DH_j\right). \tag{GH2}$$

Recall that Stalnaker's presentation of the smoking gene hypothesis envisions a man who is unsure whether the world fits that hypothesis, or a more usual story about the causes of cancer. In the Gibbard-Harper treatment, such uncertainty will be reflected in the probabilities he gives to dependency hypotheses that fit one possibility or the other. His choice will depend, as it should, on the relative weights he gives to the hypotheses.

36.5 K-Expectation and Other Theories

Other versions of *CDT* were developed soon after Gibbard and Harper's account. Brian Skyrms [29], David Lewis [19], and Howard Sobel [32] each replaced Gibbard and Harper's deterministic causal conditionals with ways of representing probabilistic causal influence. Detailed comparisons of the versions can be found in Lewis [19], Skyrms [28], and Joyce [14].

Lewis' theory, like Gibbard and Harper's, treats dependency hypotheses as conjunctions of causal conditionals, one conjunct for each available act. But Lewis' conditionals have chancy consequents, taking the general form $A \mathbin{\Box\!\!\rightarrow} [\mathrm{P}(S_j) = p]$, where $[\mathrm{P}(S_j) = p]$ asserts that the objective chances of the states S_j are given by the probability distribution p. Lewis rejected Conditional Excluded Middle for deterministic causal conditionals, mainly for the reason that worlds may be indeterministic, but he was willing to assume it for his dependency hypotheses. Sobel's theory, on the other hand, used what he called *practical chance conditionals* $A \diamond_x\!\!\rightarrow C$, which are understood to say 'if it were the case that A, then it might—with a chance of x—be the case that C.'

In contrast to the other theories, Skyrms' *K-expectation* theory has roots in accounts of probabilistic causation, and it omits conditionals entirely. Its dependency hypotheses are members of certain partitions of states, *K-partitions*, that satisfy conditions of richness and independence. The idea is that the decision-maker assessing the value of A entertains various hypotheses about ways the world might be that are a) not something he can influence by his action, and b) significant for the possible outcomes of A that he cares about. In a standard Newcomb problem, for example, {*B2 contains $M*, *B2 contains $0*} is a K-partition. A is evaluated this way: for each of the hypotheses, determine the value of A if that hypothesis is true; the overall value of A is the weighted average of those values, where the weights are the *unconditional* probabilities of the hypotheses. For this method to be reliable, the hypotheses must be sufficiently fine-grained; Skyrms [28] requires that the K_js be 'maximally specific specifications of factors outside our influence at the time of decision which are causally relevant to the outcome of our actions.' So the K-expected value of A is

$$U(A) = \sum\nolimits_j pr\left(K_j\right) U\left(A \,\&\, K_j\right), \tag{S1}$$

which agrees with (GH2) except for the different characterizations of the dependency hypotheses. What are the values $U(A \,\&\, K_j)$? Since each K_j settles which states outside the decision-maker's influence obtain, an evidential calculation of the value of $A \,\&\, K_j$ will not go astray due to confounding states. So if the members of partition $\{C_i\}$ describe the possible causal consequences of actions in sufficient detail to capture what the decision-maker cares about,

$$U\left(A \,\&\, K_j\right) = \sum\nolimits_i pr\left(C_i / A \,\&\, K_j\right) U\left(C_i \,\&\, A \,\&\, K_j\right).$$

By substitution into (S1), then, we get the utility rule for *K-expectation CDT*:

$$U(A) = \sum\nolimits_j pr\left(K_j\right) \sum\nolimits_i pr\left(C_i / A \,\&\, K_j\right) U\left(C_i \,\&\, A \,\&\, K_j\right). \tag{S2}$$

The Smoking Gene example: Suppose I am convinced of the smoking gene hypothesis. Let K_g be 'I have the gene,' and let $K_{\sim g}$ be 'I do not have the gene;' they form a K-partition. Let A_s and A_r be the actions of smoking and refraining, respectively. Let C be getting cancer, and P be enjoying the pleasure of smoking; then CP, $\sim CP$, $C{\sim}P$, $\sim C{\sim}P$ describe possible outcomes of A_s and A_r. Suppose their values are −999, 1, −1000, and 0, respectively. Suppose also that I enjoy the pleasure iff I smoke, and that $pr(K_g) = x$. Finally, suppose that the probability of C is .6 if I have the gene, whether or not I smoke, and that the probability of C is .1 if I do not have the gene, whether or not I smoke.

$$\begin{aligned} U(A_s) &= \textstyle\sum_j pr(K_j) \sum_i pr(C_i/A_s \,\&\, K_j)\, U(C_i \,\&\, A_s \,\&\, K_j) \\ &= x\left[pr(CP/A_s \,\&\, K_g)(-999) + pr(\sim CP/A_s \,\&\, K_g)(1)\right] + \\ &\quad (1-x)\left[pr(CP/A_s \,\&\, K_{\sim g})(-999) + pr(\sim CP/A_s \,\&\, K_{\sim g})(1)\right] \\ &= x\,[.6\,(-999) + .4(1)] + (1-x)\,[.1\,(-999) + .9(1)] = -500x - 99 \end{aligned}$$

$$\begin{aligned} U(A_r) &= x\left[pr(C{\sim}P/A_r \,\&\, K_g)(-1000) + pr(\sim C{\sim}P/A_r \,\&\, K_g)(0)\right] + \\ &\quad (1-x)\left[pr(C{\sim}P/A_r \,\&\, K_{\sim g})(-1000) + pr(\sim C{\sim}P/A_r \,\&\, K_{\sim g})(0)\right] \\ &= x\,[.6\,(-1000) + .4(0)] + (1-x)\,[.1\,(-1000) + .9(0)] = -500x - 100 \end{aligned}$$

So whatever the value of x, $U(A_s)$ exceeds $U(A_r)$ by 1, the value of the pleasure of smoking.

The states K_j are outside of the influence of the actions. Notice that in ordinary decision problems when they are also probabilistically independent of the actions, $pr(K_j) = pr(K_j/A)$, and substitution of the conditional probabilities into (S2) brings it into agreement with *EDT*, as expressed by (J2). So K-expectation theory and *EDT* agree in situations that are not *CD*s.

What if, as in Stalnaker's original presentation, I am unsure whether smoking causes cancer (H_2), or the smoking gene hypothesis (H_1) is true? Expand the K-partition. Assuming that whether or not I smoke in no way brings about one causal structure or the other, the elements of $\{H_1 \,\&\, K_g,\ H_1 \,\&\, K_{\sim g},\ H_2\}$ are outside my influence and sufficiently specific to form an adequate K-partition for this case.

36.6 If You're So Smart, Why Ain't You Rich?

A strong and persistent source of doubts about *CDT* is this fact: *CDT* leads to a recommendation to take both boxes in the Newcomb Problem, but it is entirely reasonable to expect that the average payout to decision-makers who take one box will exceed the average payout to those who take two boxes. Indeed, a good way to drive home the attraction of evidential reasoning is to imagine a line of people, each with an opportunity to play the Newcomb Problem once. Given a very reliable predictor, the vast majority of those who take one box receive \$*M*, while the vast majority of those who take two boxes receive \$*T*. How, we wonder, can taking two boxes be the rational act? The question was raised and answered by Gibbard and Harper [9], and the gist of their answer is this: The vast majority of one-boxers faced a significantly different situation than did the vast majority of two-boxers. Recall that the content of the opaque box is settled before the choice is made; some players are offered more fortunate alternatives than others. If and when you are given an opportunity to play, the outcomes of your alternatives are already defined. They may be the more favorable ones, or the less, but what you do won't change them. The question to ask about those who preceded you is, why are the one-boxers less rich than they could have been?

Correct as Gibbard and Harper's answer may be, the objection recurs in many later discussions of the Newcomb Problem and *CDT*. Defenders of *CDT* point out that resurrected versions of the objection usually fail to distinguish between two different decision problems: a) the problem you face in the midst of the Newcomb Problem, after the prediction is made, and b) a different problem that you might have faced, if you had anticipated an opportunity to play the Newcomb Problem, and you could have sought, before a prediction was made, to influence what it would be. But (b) is not the Newcomb Problem, and its options are no longer available when you are in the midst of the Newcomb Problem, after the prediction has been made. What *CDT* advises in (b) depends on a variety of factors (including, among others, the permanence of the state that would lead to a one-box prediction, and your expectations about other future decision situations that might arise while you are in that state). *CDT* might well recommend that, were you in (b), you should act so as to provide a basis for the predictor's making a one-box prediction, while it continues to recommend that you take two boxes in the Newcomb Problem.

The *why-ain't-you-rich?* objection need not be confined to the Newcomb Problem; in the Smoking Gene problem, more smokers contract cancer than do non-smokers, suggesting a *why-ain't-you-well?* objection. Perhaps that response is rarely heard because it is less tempting to think that choosing your genes is an available option in the midst of the Smoking Gene problem, than it is to think that influencing the past prediction is an available option in the Newcomb Problem.

36.7 Early Responses to *CDT*; Ratifiability and Deliberation Dynamics

Some early defenders of *EDT* responded to *CDT* by agreeing that the choices *CDT* recommends are correct, and by striving to show that users of *EDT* can arrive at those choices, too. Early efforts to rescue *EDT* as a general account of rational decision-making did not completely succeed, but important ideas grew out of them.

*CD*s occur when you believe that a confounding state exerts causal influence on your action, as well as on its outcome. Its path of influence might be something you notice. If some experience, some *tickle* will reliably indicate the influence of the smoking gene, then you can use the presence or absence of the tickle to ascertain whether the gene is present, and which dependency hypothesis *DH* is true. Further, if you know the correct *DH*, your problem is not a *CD* after all, and *EDT* will yield the correct recommendation. So goes the *tickle defense* of *EDT*. But why expect that such a convenient and reliable tickle will always be available? So the tickle defense as just stated does not establish that *EDT* will always make correct recommendations in *CD*s. A sophisticated variant of the defense was developed by Ellery Eells [7]; it involves not a tickle experience, but your beliefs and desires, which arguably *are* always present in rational deliberation, and your self-awareness of them. Eells argued that, when you are certain about what your beliefs and desires are, and certain that your beliefs and desires fully determine your action, it follows that your decision in the light of such knowledge is not a confounded one after all, and that *EDT* will guide you to the correct action.

Elements of Eells' treatment of *CD*s inspired another response to *CDT*, developed by Richard Jeffrey [12]. A decision-maker who is just at the point of carrying out her decision to do A can reflect on that fact, and she can incorporate the self-observation in an updated set of beliefs. Normally we do not expect the new beliefs to affect her evaluations of her options, but in *CD*s, they often will. A forward-looking version of this idea is to *suppose* that A is your final choice, and under that hypothesis, to reevaluate A and your other options. A is *ratifiable* iff under your hypothesis that A is your final choice, no other option has a value that exceeds A's. Let $V_{chA}(x)$ be the utility of x under the supposition that you choose A. Then A is ratifiable iff for every alternative choice B, $V_{chA}(A) \geq V_{chA}(B)$. So, in the Newcomb problem, under the hypothesis that both boxes are chosen, it is likely that box *B2* is empty, but choosing both boxes has greater value than choosing *B2* alone. So taking both boxes is ratifiable. On the other hand, under the hypothesis that only *B2* is chosen, it is likely that it contains *$M*. But under that hypothesis, choosing both boxes has greater value than choosing *B2* alone. So taking *B2* alone is not a ratifiable choice.

Jeffrey advocated *ratificationism*, which recommends choosing acts that are ratifiable; he regarded this as an addition to *EDT*. As Jeffrey himself pointed out, decisions may have one, more than one, or no ratifiable choices, so he did not claim that it was a completely general solution to decision-making. But *EDT* plus ratificationism yields better treatments of many *CD*s than does *EDT* alone.

Jeffrey also pointed out that in order to judge whether, under the hypothesis that A is the final choice, the desirability of doing A is greater than that of doing B, the decision-maker must find it conceivable that she chooses A, yet does B.[4] The gap between choice and performance is what makes this conceivable; choices are not always perfectly executed, slips 'twixt cup and lip' may occur. It turns out that this

[4]Ratifiability evaluations are based on beliefs and probabilities conditional on choosing A, yet doing B; such beliefs do not face the difficulties that would be faced by beliefs conditional on the contradiction that you do both of the incompatible acts A and B.

gap is what ultimately limits the success of efforts to rescue the general adequacy of *EDT* in *CD*s by Eells' appeal to strong self-knowledge of inputs to deliberation, or by ratificationism. Jeffrey himself presented Bas van Fraassen's example, in which a causally independent state is correlated with the direction in which the decision-maker may slip in the execution of her choice. Neither Eells' 'metatickle defense' nor ratificationism can guarantee that *EDT* will make the correct recommendation in such decision problems [4].

While ratificationism did not entirely save *EDT*'s claim to general adequacy, ratifiability is an important idea for rational decision-making in its own right [31]. Jeffrey's development of the idea used *EDT* to evaluate the choices ($V_{chA}(A)$ vs. $V_{chA}(B)$), but nothing bars us from using *CDT* to do so ($U_{chA}(A)$ vs. $U_{chA}(B)$).

Discussions of the merits of *EDT* and *CDT* also stimulated important philosophical work on *deliberation dynamics*, involving the idea, also present in some analyses of game-theoretic interactions, that your beliefs can evolve through self-awareness and learning during the course of your deliberation [29, 30]; also [5, 16]. More about this below.

36.8 Foundations for *CDT*

For any formal decision theory that offers quantitative evaluations of choiceworthy actions, it is appropriate to ask how the relevant quantities—utilities, probabilities, and others, if any—are fixed. A standard response is to provide an account of systems of rational preferences, and a demonstration that the quantities are associated with a decision-maker's system of preferences in a non-arbitrary way. *EDT* was given a beautiful foundation by Jeffrey and Bolker. *CDT*, like any decision theory, wanted a foundation too. Since *CDT* and *EDT* are in agreement when no *CD*s are involved, a good foundation for *CDT* should support the use of *EDT* in cases where it works, and should display those cases in an enlightening way. Skyrms [29] discussed the issue: for a given partition $\{K_j\}$, he suggested using Jeffrey-Bolker theory to construct utilities and probabilities for preferences conditional on each K_j, and then sketched a way of combining those conditional functions into an overall utility scale and a single unconditional probability distribution. Gibbard (1984, reported in [14]) took a different approach that provided conditions, relating states and causal counterfactuals, under which Savage's utility theory can accurately assess causal expected utility.

Brad Armendt [2, 3] provided a foundation for K-expectation *CDT* that does not rely on a prior specification of the K-partition. States that form appropriate K-partitions are identified by their behavior in the decision-maker's set of conditional preferences. The idea is that, in a *CD*, the unconditional ranking of action *A* will not agree with the ranking of *A* under the hypothesis that *A* is performed. (This is reminiscent of what is involved in judging whether *A* is ratifiable, but here the hypothesis is about the performance of *A*, rather than the choice of *A*.) But an unconditional ranking of *A* & K_j agrees with its conditional ranking under the hypothesis that A is done. The conditional preferences actually range over mixtures of propositions, *i.e.* lotteries, and the formal foundation relies on utility theorems by Fishburn and Herstein-Milnor. The mixing coefficients are a commonly employed device in formal utility theories, but they differ from the structure of the Jeffrey-Bolker foundation for *EDT*, which does not use them.

Jim Joyce [14] developed a foundation for *CDT* in the context of a general setting for theories of conditional preference and conditional belief. The conditioning involved may be either subjunctive supposition that captures causal relationships, or indicative supposition that captures evidential relationships *via* standard conditional probabilities. We have seen that, conditional on a dependency hypothesis K, *EDT*'s evidential values V_K and *CDT*'s causal values U_K agree. Joyce emphasized the point: he asserted that all value is news value, and that the difference between *EDT* and *CDT* lies in the epistemic perspective of the decision-maker. In deliberation about A, the right perspective comes from subjunctively supposing the performance of A; beliefs under that supposition guide the assessment of A.

Joyce's foundation is built on Jeffrey-Bolker axioms governing rational conditional preference, supplemented with axioms governing measures of rational conditional belief, that is, axioms for comparative conditional probability. No special partition is assumed; the set of conditions $\{C\}$ is taken to include at least the set of actions available to the decision-maker. The axioms guarantee, for each condition C, a utility function for preferences conditional on C, and a conditional *supposition function* $P(-||C)$. Further axioms unite the conditional functions into comprehensive utility and probability pair. The nature of the probabilities captured by the supposition function depends upon which additional principle(s) are assumed to govern suppositions; the foundation captures *CDT*, as intended, when a probability conditional on A is arrived at by subjunctively supposing that A.

36.9 Decision Instability and Deliberation Dynamics

At the end of their 1978 paper, Gibbard and Harper considered the issue of *stability* in rational choice, illustrated by the example of the man who met death in Damascus:

> Consider the story of the man who met death in Damascus. Death looked surprised, but then recovered his ghastly composure and said, 'I am coming for you tomorrow'. The terrified man that night bought a camel and rode to Aleppo. The next day, death knocked on the door of the room where he was hiding and said, 'I have come for you'.
>
> 'But I thought you would be looking for me in Damascus,' said the man.
>
> 'Not at all,' said death "that is why I was surprised to see you yesterday. I knew that today I was to find you in Aleppo."
>
> Now suppose the man knows the following. Death works from an appointment book which states the time and place; a person dies if and only if the book correctly states in what city he will be at the stated time. The book is

made up weeks in advance on the basis of highly reliable predictions. An appointment on the next day has been inscribed for him. Suppose, on this basis, the man would take his being in Damascus the next day as strong evidence that his appointment with death is in Damascus, and would take his being in Aleppo the next day as strong evidence that his appointment is in Aleppo.

Let K_D and K_A be 'Damascus is inscribed' and 'Aleppo is inscribed'; they form a K-partition. If, for example, $pr(K_D) > ½$ then *CDT* recommends going to Aleppo. But that choice seems unstable: when the man believes he is about to make it, he has new information that influences his beliefs, including his beliefs about K_D and K_A. The anticipation that he will choose Aleppo leads to a new belief $pr_n(K_A) > ½$, and $pr_n(K_D) < ½$, which makes Damascus the better option. But when the man believes he is about to choose Damascus, that new information again influences his beliefs, leading to $pr_{nn}(K_D) > ½$, ... and so on. Unlike the *CD*s we previously considered, where dominance reasoning worked and shifting $pr(K_j)$'s would not change which act is recommended, here a decision-maker with self-awareness will have difficulty arriving at a decision.

What should the man do? One plausible answer: toss a coin, or adopt some internal method of randomizing his choice, thereby pursuing a mixed strategy [10]. Neither pure act (remain in Damascus, go to Aleppo) is ratifiable, but a 50-50 mixture of those acts is. A good idea, but suppose that mixed strategies are ruled out as viable options—if the man were to use one, Death would know, and would interrupt his appointment-keeping to find the man wherever he is [34]. Fanciful examples aside, notice that problems that forbid or penalize mixed acts thereby impose a fairly restrictive exogenous constraint on the decision-maker's options. Having said that, however, we should be careful about 'solving' a decision problem by altering it with additional options, and offering a solution to the revised problem.

The sense of instability in problems like Death in Damascus arises from the possibility of reevaluating your options in light of information arising from your deliberations. This is a good setting for the theory of deliberation dynamics, mentioned earlier, where iterated or continuous updates of your beliefs inform your continuing deliberations, which in turn provide reasons for further belief updating. The theory applies to deliberation in general; other interesting settings include noncooperative games among Bayesian players [29, 30]. Under deliberation dynamics, your changing beliefs and evaluations follow trajectories that can behave in various ways, depending upon the problem that confronts you. When *CDT* is used to evaluate the options in a Newcomb Problem, deliberation yields straightforward convergence to high confidence that you will take both boxes, and that box *B2* will be empty, since an increasing confidence that you will take both boxes does not lead you to think it would be better to do otherwise.

The precise features of the trajectories of your beliefs will depend on the details of the dynamics: how much does the recognition that one action *A* looks better than another *B*, that $U_{t1}(A) > U_{t1}(B)$, lead you to increase your belief that you will do

A, $pr_{t2}(A)$? Various dynamics can be considered; for present purposes, the key idea is that they *seek the good*, raising the probabilities of actions exactly when you currently see those actions as better than others, or more precisely, as better than the *status quo*, which is your current expectation of the outcome of the problem you are deliberating about.

Returning to Death in Damascus, then, take seriously the idea that, during his deliberation, the man is attentive to his evaluations of his options, and that what he learns about them informs his beliefs about what he will soon do, and about what is inscribed in Death's appointment book. If at some time t_1 during his deliberation, he regards going to Aleppo to be the better action, so that $U_{t1}(A) > U_{t1}(D)$, and he realizes that he does, then he raises his belief that he will go to Aleppo, $pr_{t1}(A) > pr_{t1}(D)$, and also that Death is more likely to be there, $pr_{t1}(K_A) > pr_{t1}(K_D)$. Then, when he reevaluates his options at t_2 with those new beliefs, he sees Damascus as the better action, $U_{t2}(D) > U_{t2}(A)$, which gives him reason to revise his beliefs again. Under plausible dynamics for a rational agent, the oscillations in beliefs will dampen over time, and the man's beliefs $pr_{tn}(K_A)$ and $pr_{tn}(K_D)$ will converge to a stable equilibrium, where neither act is seen as better than the other. At that point, his tied evaluations give him no reason to further adjust the beliefs that underlie them. In the original Death in Damascus problem, at the equilibrium state, $pr_{eq}(K_A)$ and $pr_{eq}(K_D)$ are both ½, as are $pr_{eq}(A)$ and $pr_{eq}(D)$. His choice will be the outcome of some way of breaking the tie [5, 16]. A general feature of equilibrium states to which deliberation leads is that you see your available options as equally choiceworthy, as having equal expected utility. It may also happen that you then believe that you are as likely to do one act as the other, but that need not be so in problems that lack the symmetry of Death in Damascus.

Why should the man embark on this deliberative journey? There is at least this reason: a rational choice should be based on all of your relevant beliefs at the time you make it. So, *if* you believe at time t that Death is more likely to go to Aleppo than to Damascus, $pr_t(K_A) > pr_t(K_D)$, your evaluations at t of your options, $U_t(A)$ and $U_t(D)$, must use those beliefs. Or, to put it another way, a rational decision theory such as *CDT* is properly used only when those evaluations do so. Is it incumbent upon you to possess such beliefs in the midst of deliberation? We will return to that question.

The original version of Death in Damascus is a symmetric problem, but asymmetric versions are easily constructed; just add an incentive against travel that makes the outcomes of staying in Damascus a little better than the corresponding outcomes of traveling to Aleppo [26]. Or imagine that Death's appointment book more reliably predicts the traveler's presence when he is in one city than when he is in the other.

Discussions of unstable decision problems have become prominent in recent work on *CDT*. One reason is that they widen the scope of the theory beyond the problems where causal dominance reasoning applies. Another is that problems displaying instability have been offered as *counterexamples* to *CDT*.

36.10 Recent Debates

Andy Egan [8] challenged *CDT* with a set of examples that he judged to be counterexamples to the theory, and his challenge has received wide attention. One of the examples is the *Murder Lesion* problem:

> Mary is debating whether to shoot her rival, Alfred. If she shoots and hits [S & H], things will be very good for her. If she shoots and misses [S & M], things will be very bad. (Alfred always finds out about unsuccessful assassination attempts, and he is sensitive about such things.) If she doesn't shoot [$\sim S$], things will go on in the usual, okay-but-not-great kind of way. Though Mary is fairly confident that she will not actually shoot ... she thinks that it is very likely that if she were to shoot, then she would hit [$S \Box\!\!\rightarrow H$]. So far, so good. But Mary also knows that there is a certain sort of brain lesion [L] that tends to cause both murder attempts and bad aim at the critical moment. If she has this lesion, all of her training will do her no good—her hand is almost certain to shake as she squeezes the trigger. Happily for most of us, but not so happily for Mary, most shooters have this lesion, and so most shooters miss. Should Mary shoot?[5]

Following Egan, let $U(S \,\&\, H) = 10$, $U(S \,\&\, M) = -10$, and $U(\sim S) = 0$ throughout.[6] Mary's initial beliefs are that she is unlikely to shoot, $pr_i(S) < .5$. She also thinks that if she did, she would hit, $pr_i(S \Box\!\!\rightarrow H) > .5$. Her belief in that causal conditional is dependent on whether or not she shoots, since shooting is correlated with having the lesion; so $pr_i(S \Box\!\!\rightarrow H/\, S) < .5$. However, her initial unconditional belief in that conditional is high, as just specified, since she initially thinks S is unlikely.

With those initial beliefs, a *CDT* calculation will yield $U_i(S) > U_i(\sim S) = 0$, since the better outcome of S, namely S & H, is weighted by the high probability $pr_i(S \Box\!\!\rightarrow H)$, while the worse outcome S & M is weighted by the low probability $pr_i(S \Box\!\!\rightarrow M)$. So *CDT* recommends that Mary shoot. Egan regards that as a flawed recommendation:

> It's irrational for Mary to shoot. ... In general, when you are faced with a choice of two options, it's irrational to choose the one that you confidently expect will cause the worse outcome. Causal decision theory endorses shooting ... In general, causal decision theory endorses, ..., an irrational policy of performing the action that one confidently expects will cause the worse outcome. The correct theory of rational decision will not endorse irrational actions or policies. So causal decision theory is not the correct theory of rational decision.

---

[5] All quotes are from [8].

[6] There is symmetry in these payoffs, but perhaps not in the beliefs: *most* shooters have the lesion, but whether the same proportion of non-shooters lack it is unsaid; it's *nearly certain* that those with the lesion miss; in light of her training, Mary thinks it's *very likely* that she would hit. Nothing in what follows depends upon the problem being as symmetric as Death in Damascus is.

The act of shooting is intuitively irrational, Egan says, and widely judged to be so.

> . . . we have (or at least my informants and I have) clear intuitions that it's irrational to shoot or to press, and rational to refrain in *The Murder Lesion* . . .

There is a response to Egan's view of the example. Egan's case for the irrationality of *CDT*'s recommendation (that Mary shoot) is the intuitive irrationality of shooting. No basis for the intuition is offered, but it is not hard to feel, nor hard to explain. What is happening? Mary begins with the beliefs that she lacks the lesion, and that shooting would be effective; based on those beliefs *CDT* recommends she shoot.[7] That's the right recommendation given her beliefs and values at that time. But in preferring shooting, she probably has the lesion and will very likely miss. So Mary comes to confidently expect, and we who contemplate her problem come to confidently expect, that shooting will cause the worse outcome. That's what the first step in the deliberative process tells her. What is Egan's intuition, if not the result of taking that step? At that point, however, when Mary has *that* belief, *CDT* recommends that Mary *refrain*; that's what her current utilities will tell her. Egan applies the recommendation that *CDT* makes at one time (shoot) to a decision at a later time, after Mary's beliefs have changed, and he sees a flaw where none exists. The error is in the supposition that *CDT* is forever committed to its recommendation under Mary's initial beliefs.

> The resulting theory enjoins us to *do whatever has the best expected outcome, holding fixed our initial views about the likely causal structure of the world.* The following examples show that these two principles come apart, and that where they do, causal decision theory endorses irrational courses of action. (emphasis Egan)

The correct idea is that *CDT* enjoins us to do what has the best expected outcome, given our *current* views about the likely causal structure of the world.[8] This applies to us in the first person, as deliberators (use our current beliefs), and in the third person, as judges of what *CDT* recommends to others (use their current beliefs). So Egan's argument suffers from a mistake about what *CDT* recommends.

A second issue is that an intuition that refraining is uniquely rational has doubtful reliability. We are invited to deliberate a little bit, but not very far, about what to do, and to stop the deliberation at an arbitrary point, with no motivation given for stopping there. If Mary correctly assesses her options at that point, refraining is

---

[7] It's worth remarking that users of *CDT* are no more prone to murder, premature death, disease, or psycho-killing than anyone else. The window-dressing of commonly used examples should be more varied; outcomes might be small prizes or fines, rather than death. Many *games* exhibit instability: Battle-of-the-Sexes-with-a-twin, for example. The theoretical issues apply to small stakes as well as large, a point worth remembering when deliberation carries a cost.

[8] Egan actually states the requirement correctly, later in his paper in a different context (p.102), but it is clear that he relies on the incorrect version throughout the paper. Without it, what is the purported counterexample?

better, which provides her reason to think that, as a refrainer, she likely lacks the lesion, making shooting the better option after all (according to her) . . . , and so on. Even if the mistake about *CDT*'s recommendations were absent, the example would at best indicate a problem with joining *CDT* to a special unmotivated assumption about when deliberation must end. It establishes no problem for *CDT*, which is consistently a good guide to rational action. So says this line of response; let us call it the *Simple* response.

What, then, does *CDT* recommend in the *Murder Lesion* problem? If you have Mary's initial beliefs and deliberate no farther, it recommends shooting. If you have a different initial belief, that you probably have the lesion, it recommends refraining. If you have access to and appreciation of your deliberative states, *CDT* makes a succession of recommendations at each stage of your self-reflecting dynamical deliberation. The recommendations may change, but each one is correct for your beliefs and values at the moment it is made. Your dynamic deliberation may end in a variety of ways: you may get tired, you may have other things to do, the world may interrupt, or you may reach equilibrium. If your deliberation ends in action, the rational action to perform is the one recommended by your current beliefs, when deliberation ended. Recall that if you reach equilibrium, you see your options as equally worthy, and you shoot or refrain by breaking the tie. There is no single action, for every deliberator, that use of *CDT* insists upon in this, or in other unstable decision problems.

Arntzenius [5] and Joyce [16] develop a further, stronger response to Egan's Murder Lesion problem. Both argue that the uniquely rational outcome of your dynamical deliberation is to arrive at the equilibrium state to which it leads. You then regard shooting and refraining as equally good acts, and doing either one, through some way of breaking the tie, is rational. Arntzenius focuses more on the beliefs at equilibrium than on the selection of the act. Joyce gives more details: what *CDT* recommends all along are its evaluations at equilibrium, which rank each option equally. Why is that so? Consider the deliberations of a rational agent who has easy access to her beliefs and preferences during deliberation. Joyce argues that when information relevant to your choice is cost-free, it is rational to obtain and use it before making your decision final. The rational decision-maker should base her evaluations, we might say, on all of the *current evidence in her reach*. So the proper use of *CDT* incorporates all of the cost-free new information that deliberation provides, and in unstable problems, new information is always available until you reach equilibrium. *CDT*'s recommendation in an unstable decision problem like those we have considered is to 'pick' (choose via a tie-break) one of the acts that remains viable in equilibrium. In decision problems with more available actions, it may happen that some acts are excluded at equilibrium as sub-optimal, compared to others that are tied with maximum expected utility.

There is no great tension between the Arntzenius-Joyce response and the Simple response that preceded it above.[9] It is a good idea to make use of cost-free information. Both responses can agree that the deliberational equilibrium is a good state in which to make your decision. It may be debated whether you misapply *CDT* if you act before reaching it, but we do not pursue that here. It is surely an idealization that continued deliberation is cost-free (think of opportunity costs), and when it is not, it would seem that truncated deliberation can be rational. But in this context, it is worth keeping in mind that subjective rational decision theory interests us from both third-person and first-person points of view: as a theory that explains the choice-worthiness of actions for a decision-maker in light of her beliefs and desires, and as a tool that can help us ascertain what to do, as we reflect on our own relevant beliefs and desires. When we think of deliberation as a dynamic process in which utility assessments produce new evidence, and so on, the first-person perspective is in the foreground. But from the third-person perspective, it is an informative idea that a rational action is one that is endorsed in the equilibrium state.[10] Indeed, recall that the first-person use of heuristics can often help us see what, from the third-person perspective, is rational for us. Once we understand the path to equilibrium in an unstable problem, and realize that the recommendation at equilibrium is to 'pick,' we seem to have an excellent method for deciding the problem: forego the extended deliberation and pick. An illustration of the value of studying rational decision theory!

36.11 Further Topics and Conclusion

Challenges to *CDT*, and responses to challenges, continue to appear. Ahmed [1] presents sustained and interesting arguments that *CDT* is flawed, and that *EDT* is after all correct. Responses to Ahmed are found in Joyce [17, 18]. Unfortunately, we cannot adequately address these and other exchanges here.

Our *Basic Idea* points to the importance of understanding causal influence in deliberating about action. We might suspect, however, that our understanding of causation is partly rooted in our deliberative practices, and in an understanding

[9]In one respect, there is a difference. When you eventually do act A in an unstable decision problem, you will have grounds for regretting you did so. Arntzenius counts foresight, if you have it, of such regret against the rationality of the act; such foresight is avoided in the dynamical equilibrium state. If the foresight is possible in a deliberation truncated at t when *CDT* is correctly applied, the Simple response is more sanguine about the foreseen regret, and regards correct maximization of U_t as the criterion of rational action.

[10]Arntzenius [5] suggests that this is the way to think about the dynamical story: 'So, as long as we are idealizing, let us simply say that a rational person must always be in a state of deliberational equilibrium.' Whether that means that one must hold the *specific* beliefs that make the equilibrium is unclear. Whether it does or not, it amounts to an additional constraint on the use of *CDT*, if the person is not driven by rationality to get there, as he is by Joyce's epistemic norm that one must acquire cost-free information during deliberation.

of our agency. This is a thought with a long history; one person who explored it was Frank Ramsey [25]. More recently, discussions that explicitly focus on *CDT* and consider connections between our understanding of causation on one hand, and our decision-making and agency on the other, were given by Nancy Cartwright [6], Christopher Hitchcock [11], and Huw Price [23, 24]. Cartwright and Hitchcock generally favor *CDT*; Price argues for an understanding of causation that undermines the usual accounts of what the decision-maker rationally believes in the Newcomb Problem and other *CD*s, and so undermines *CDT*'s recommendations. One focal point in recent discussions is a question about what a deliberator can reasonably think, while deliberating, about the dependence of her impending act on states in the past such as the presence of the gene, or the basis of the Predictor's forecast. A view that is often attributed to Ramsey is that a deliberator must see her impending act as beyond the influence of past states other than her own, and as a bearer of no information about such past states. There is disagreement about its implications for *CDT* [1, 16, 24]. Here we just offer two inconclusive ideas. One is that causal influence is a matter of degree; a decision-maker who thinks she has any influence over her act, however small, has reason to deliberate about how to exercise the influence she wields. That thought by itself does not settle how she represents her influence in her deliberations. Another concerns the plausible idea that our understanding of causation is partly based on our experience of agency. If that is so, confounded decisions and their complexities are unlikely to have played a large role in such experiences, and it need not be entirely straightforward how we use the idea of causation, or how immediately our intuitive judgments fit, in confounded decisions.

The development of *CDT* has led to ideas and topics that extend far beyond the *CD* problems that provoked the theory. More than 40 years after its beginnings, *CDT* is now the preeminent rational decision theory that captures the essential role of causal beliefs in deliberation and decision. *CDT* is now both a starting point and a target for those engaged in further work.

References[11]

1. Ahmed, A. (2014). *Evidence, decision and causality*. Cambridge: Cambridge University Press.
2. Armendt, B. (1986). A foundation for causal decision theory. *Topoi, 5*, 3–19.
3. Armendt, B. (1988a). Conditional preference and causal expected utility. In W. Harper & B. Skyrms (Eds.), *Causation in decision, belief change, and statistics* (pp. 3–24). Dordrecht: Reidel.
4. Armendt, B. (1988b). Impartiality and causal decision theory. *In PSA, 1988*(*1*), 326–336. Philosophy of Science Association.
5. Arntzenius, F. (2008). No regrets, or: Edith Piaf revamps decision theory. *Erkenntnis, 68*, 277–297.

[11] Key references are marked with *.

6. Cartwright, N. (1979). Causal laws and effective strategies. *Nous, 13*(4), 419–437. Reprinted with additions in Cartwright, *How the Laws of Physics Lie*, 21–43, Oxford: Clarendon Press.
7. Eells, E. (1982). *Rational decision and causality*. Cambridge: Cambridge University Press.
8. Egan, A. (2007). Some counterexamples to causal decision theory. *Philosophical Review, 116*, 93–114.
9. *Gibbard, A. & Harper, W. (1978). Counterfactuals and two kinds of expected utility. In C. Hooker et al (Eds.), *Foundations and applications of decision theory*. Dordrecht: Reidel. Reprinted in W. Harper, *et al.* (eds.), *Ifs*, 153–190, Dordrecht: Reidel.
10. Harper, W. (1986). Mixed strategies and ratifiability in causal decision theory. *Erkenntnis, 24*, 25–36.
11. Hitchcock, C. (1996). Causal decision theory and decision-theoretic causation. *Noûs, 30*(4), 508–526.
12. *Jeffrey, R. (1983/[1965]). *The logic of decision* (2nd ed.). Chicago: University of Chicago Press.
13. Jeffrey, R. (2004). *Subjective probability: The real thing*. Cambridge: Cambridge University Press.
14. *Joyce, J. M. (1999). *The foundations of causal decision theory*. Cambridge: Cambridge University Press.
15. Joyce, J. M. (2007). Are Newcomb problems really decisions? *Synthese*, 156, 537–562.
16. *Joyce, J. M. (2012). Regret and instability in causal decision theory. *Synthese, 187*, 123–145.
17. Joyce, J. M. (2016). Review of Ahmed (2014). *Journal of Philosophy, 113*, 224–232.
18. Joyce, J. M. (forthcoming). Deliberation and stability in Newcomb problems and pseudo-Newcomb problems. In A. Ahmed (Ed.), *Newcomb's problem.*
19. *Lewis, D. (1981). Causal decision theory. *Australasian Journal of Philosophy, 59*, 5–30.
20. Lewis, D. (1976). Probabilities of conditionals and conditional probabilities. *Philosophical Review, 85*, 297–315.
21. Nozick, R. (1969). Newcomb's problem and two principles of choice. In N. Rescher et al. (Eds.), *Essays in honor of Carl S. Hempel* (pp. 114–146). Dordrecht: Reidel.
22. Nozick, R. (1993). *The nature of rationality*. Princeton: Princeton University Press.
23. Price, H. (1992). The direction of causation: Ramsey's ultimate contingency. *Philosophy of Science Association, 2*, 253–267.
24. Price, H. (2012). Causation, chance, and the rational significance of supernatural evidence. *Philosophical Review, 121*, 483–538.
25. Ramsey, F. (1929). General propositions and causality. In D. H. Mellor (Ed.), *F.P. Ramsey: Philosophical papers* (pp. 145–164). Cambridge: Cambridge University Press.
26. Richter, R. (1984). Rationality revisited. *Australasian Journal of Philosophy, 62*, 392–403.
27. Savage, L. J. (1972/[1954]). *The foundations of statistics* (2nd ed.). New York: Dover.
28. * Skyrms, B. (1980). *Causal necessity*. New Haven: Yale University Press.
29. *Skyrms, B. (1982). Causal decision theory. *Journal of Philosophy, 79*, 695–711.
30. *Skyrms, B. (1990a). *The dynamics of rational deliberation*. Cambridge: Harvard University Press.
31. Skyrms, B. (1990b). Ratifiability and the logic of decision. In P. French et al. (Eds.), *Midwest studies in philosophy XV* (pp. 44–56). Notre Dame: University of Notre Dame Press.
32. Sobel, J. H. (1986). Notes on decision theory: Old wine in new bottles. *Australasian Journal of Philosophy, 64*, 407–437.
33. Stalnaker, R. (1972). Letter to David Lewis. In W. Harper et al. (Eds.), *Ifs* (pp. 151–152). Dordrecht: Reidel.
34. Weirich, P. (1985). Decision instability. *Australasian Journal of Philosophy, 63*, 465–472.

Chapter 37
Social Choice and Voting

Prasanta K. Pattanaik

Abstract When individuals in a society have different preferences over the options available to the society, how should social decisions be taken so as to achieve a reasonable compromise? What are the principles that one should use in one's ethical evaluation of different states of the society? These ethical issues are at the centre of the theory of social choice and welfare. While they have been discussed and debated for centuries, what the modern theory of social choice and welfare has done is to bring to bear formal reasoning in exploring them. The literature that has developed in this area over the last 70 years or so is vast and it is not possible to give in this short review even a list of the major developments. What I seek to do here is to focus on a few of the most conspicuous landmarks in this literature.

37.1 Introduction

When individuals in a society have different preferences over the options available to the society, how should social decisions be taken so as to achieve a reasonable compromise? What are the principles that one should use in one's ethical evaluation of different states of the society? These ethical issues are at the centre of the theory of social choice and welfare. While they have been discussed and debated for centuries, what the modern theory of social choice and welfare has done is to bring to bear formal reasoning in exploring them. The literature that has developed in this area over the last 70 years or so is vast and it is not possible to give in this short review even a list of the major developments. What I seek to do here is to focus on a few of the most conspicuous landmarks in this literature.

P. K. Pattanaik (✉)
Department of Economics, University of California, Riverside, CA, USA
e-mail: prasanta.pattanaik@ucr.edu

S. O. Hansson, V. F. Hendricks (eds.), *Introduction to Formal Philosophy*, Springer Undergraduate Texts in Philosophy, https://doi.org/10.1007/978-3-319-77434-3_37

37.2 Two Aspects of Social Choice: Aggregation of Individual Preferences to Achieve Compromise vs. Social Welfare Judgments of an Ethical Observer

Consider the following exchange between two persons, *i* and *j* (where *j* is an Indian):

i: "Should India abolish death penalty altogether?"
j: "No, since 90 % of Indians believe that death penalty should be retained for very serious crimes such as premeditated murder."
i: "But what do *you* think India should do? Should India abolish death penalty?
j: "Yes, definitely. In my opinion, death penalty has no place in any civilized society because ..."

This exchange illustrates the two very different senses in which one can interpret the ethical problem of social choice or evaluation of social options. Implicit in *j*'s answer to *i*'s first question is the interpretation of the problem of social choice or evaluation of social options as a problem of aggregating the (possibly conflicting) opinions or preferences of the individuals in a given society so as to arrive at a reasonable compromise. Under this interpretation, typically the individuals' opinions or preferences are taken as given and the problem is simply one of aggregating these given opinions. In this context, an appeal to the fact that 90% of the population shares a particular preference can be a convincing reason for the society to do or not to do something. The second interpretation implicit in *i*'s second question and *j*'s answer to it is the interpretation of the social evaluation of options or the prescription for social choice as reflecting an individual's own ethical beliefs. An appeal to the shared opinion of an overwhelming majority of the society does not seem to be particularly relevant here; the individual evaluating the social options needs to provide independent justifications for her ethical beliefs.

The modern theory of social choice and welfare explores problems of social choice and social evaluation in both the senses mentioned above. In assessing the intuitive significance of many of the contributions to this theory, however, it is important to keep in mind the distinction, introduced by Little [8], Bergson [2], and Sen [15], between the two interpretations. In this essay, I concentrate on the literature that is concerned primarily with the first intuitive problem mentioned above, namely, the problem of arriving at a compromise in the presence of conflicting individual preferences or opinions.

37.3 Some Basic Notation and Definitions

Let $N = \{1, 2, \ldots, n\}$ denote a society. 1, 2, ..., and n ($\infty > n > 1$) denote the individuals in the society. I use the society in the usual sense of the term, though, for many purposes, the society can be interpreted in a flexible fashion to indicate any group of individuals (e.g., a committee). Let X denote the set of all conceivable social alternatives or options. X can be interpreted in different ways depending on the context. In welfare economics, the elements of X are often taken to be alternative complete description of the affairs in the society, though there is sometimes ambiguity about what exactly constitutes such a complete description. The elements of X are denoted by x, y, z, etc. Let $\mathcal{T}$ be the set of all binary weak preference relations ("at least as good as") R^* defined over X, such that R^* satisfies reflexivity over X (i.e., for all $x \in X$, xR^*x). Given $x, y \in X$ and $R^* \in \mathcal{T}$, xR^*y denotes that x is at least as good as y in terms of the binary weak preference relation R^*. For all $R^* \in \mathcal{T}$ and all $x, y \in X$, [xP^*y iff (xR^*y and not yR^*x)] and [xI^*y iff (xR^*y and yR^*x)]. P^* and I^* are to be interpreted, respectively, as the strict preference relation ("preferred to") and indifference relation ("indifferent to") corresponding to R^*.

Let $\mathcal{R}$ be the set of all $R^* \in \mathcal{T}$ such that R^* is an ordering over X, i.e., R^* satisfies the following three properties: (i) *reflexivity* over X; (ii) *connectedness* over X (for all distinct $x, y \in X$, xR^*y or yR^*x; and (iii) *transitivity* over X (for all $x, y, z \in X$, if xR^*y and yR^*z, then xR^*z). Let L be the set of all R^* in $\mathcal{R}$, such that R^* is linear, i.e., for all distinct $x, y \in X$, not [xR^*y and yR^*x]. Thus, L is the set of all preference orderings which do not permit indifference between distinct options.

Much in the theory of social choice can be formulated either in terms of a social ranking of options or in terms of the society's choices from different possible sets of feasible social options. I use the former formulation here.

Definition 37.1 A *social ranking rule* is a function $f\colon \mathcal{S}^n \longrightarrow \mathcal{T}$, where $\varnothing \neq \mathcal{S} \subseteq \mathcal{R}$.

$\mathcal{S}$ is to be interpreted as the set of all binary weak preference relations that an individual may have. The elements of $\mathcal{S}^n$ will be denoted by $(R_1, \ldots, R_n)$, $(R'_1, \ldots, R'_n)$, etc., and will be interpreted as profiles of individual weak preference relations. R_i, R'_i, etc., denote weak preference relations of individual i ($i \in N$). Thus, a social ranking rule $f\colon \mathcal{S}^n \longrightarrow \mathcal{T}$ is a function, which, for every profile of individual preferences in $\mathcal{S}^n$, specifies exactly one binary weak preference relation R in $\mathcal{T}$, R being interpreted as a social weak preference relation, or ranking, over X. xRy denotes that x is at least as good as y for the society. Typically, it is assumed that $\mathcal{S}=\mathcal{R}$, i.e., the set of all admissible preferences for an individual is the set of all orderings over X.

A social ranking R over X, has intuitive, though not logical, implications for social choice. For example, it will be intuitively rather odd to say that x is better for the society than y, but, given the choice between x and y, the society should choose y and reject x. Given the social ranking R and given a non-empty subset A of X, we say that $C(A, R) \equiv \{x \in A : xRy \text{ for all } y \in A\}$ is the *choice set* generated by R for A. Intuitively, $C(A, R)$ is the set of best alternatives in A, "best" being defined in terms of R. If A is the set of all feasible options before the society, then the society can choose any option in $C(A, R)$. It is possible to have an empty $C(A, R)$. For example, if xPy and yPz and zPx, then $C(\{x, y, z\}, R)$ is empty. In this case, R does not give much guidance about what the society should choose from $\{x, y, z\}$.

37.4 Arrow's Impossibility Theorem for Social Ranking Rules

What restrictions should one postulate for social ranking rules? This is the issue that Arrow ([1], 1963) addressed. He introduced four such restrictions.

Definition 37.2 Let $f : \mathcal{S}^n \longrightarrow \mathcal{T}$ be a social ranking rule. f satisfies:

(i) *Collective Rationality* iff $\mathcal{S}=\mathcal{R}$ and, for every $(R_1, \ldots, R_n)$ in $\mathcal{S}^n$, $R = f(R_1, \ldots, R_n)$ is an ordering;
(ii) *Weak Pareto Principle* iff, for every $(R_1, \ldots, R_n)$ in $\mathcal{S}^n$ and for all $x, y \in X$, if xP_iy for all $i \in N$, then xPy.
(iii) *Independence of Irrelevant Alternatives* iff, for all $(R_1, \ldots, R_n)$, $\left(R'_1, \ldots, R'_n\right) \in \mathcal{S}^n$, and for all $x, y \in X$, if for all $i \in N$, [xR_iy iff xR'_iy] and [yR_ix iff yR'_ix], then [xRy iff $xR'y$] and [yRx iff $yR'x$].
(iv) *Non-dictatorship* iff there does not exist $i \in N$, such that, for all $x, y \in X$ and all $(R_1, \ldots, R_n)$ in $\mathcal{S}^n$, if xP_iy, then xPy.

Collective rationality requires that, for every possible profile of individual preference orderings, the social ranking rule should specify an ordering as the social binary weak preference relation. Collective rationality can have two distinct types of justification. First, if the social weak preference relation, R, is to be used as the basis for social choice from a given set, A, of feasible social options, then R should generate a non-empty choice set for A. The restriction that R be an ordering is sufficient, though not necessary, to ensure that $C(A, R)$ will be non-empty for every finite non-empty subset A of X. A second justification for collective rationality can be that social choices from different possible sets of feasible options should be "rational", rational choices being conceived as choices that could be induced by an ordering (this is the conception of rational choice that economists typically use). The Weak Pareto Principle, embodying respect for unanimity, has been almost universally accepted in welfare economics. Independence of Irrelevant Alternatives requires that if the profile of individual orderings changes but every individual's ranking of two options, x and y, remains the same before and after the change, then the society's ranking of x and y must remain the same. This is sometimes justified by the pragmatic consideration that it leads to an economy of information needed for the social ranking over pairs of options: in the absence of this property, for the society to rank two alternatives, not only will it need information about how all individuals rank those two options, but it may also need information about the individuals' rankings with respect to other ("irrelevant") options. Another pragmatic justification for the condition is that violation of Independence of Irrelevant Alternatives gives individuals the opportunity to "misreveal" their preferences so as to change the social decision to their advantage (see Plott [12]). Finally, Non-dictatorship seems to be a reasonable condition: it simply requires that the society should not have a dictator, i.e., an individual such that whenever she strictly prefers any option x to any other option y, the society must rank x strictly above y irrespective of other individuals' preferences.

The following result due to Arrow constitutes one of the foundational results in the theory of social choice.

Theorem 37.1 (Arrow [1]): *If* $\#X \geq 3$, *then there does not exist any social ranking rule which simultaneously satisfies Collective Rationality, the Weak Pareto Principle, Independence of Irrelevant Alternatives, and Non-dictatorship.*

Given the apparent plausibility of the four conditions, the impossibility of satisfying all of them simultaneously (given the mild restriction that $\#X \geq 3$) has the flavor of a paradox. It is not, therefore, surprising that a significant part of the literature on the theory of social choice has been devoted to finding ways of escape from the dilemma posed by Arrow's result.

I would like to make two comments relating to the interpretation of Arrow's [1] theorem. First, since Arrow's framework makes the social ranking exclusively dependent on the profile of individual preference orderings, the question arises about the intuitive content of these preference orderings. One response to this question may be to say that an individual's preference ordering reflects all that the individual considers to be relevant in assessing the options, including, possibly, her ethical values (e.g., " a social state that involves excessive social and economic inequality is abhorrent" and "tigers have a right to survive and policies which will lead to their extinction are ethically unacceptable") as well as her self interest ("I shall be better off in x as compared to y"). This answer is adequate if the social choice problem is one of arriving at a compromise in the face of conflicting preferences, assumed to be given. It is not, however, adequate when one interprets the social ranking of options as reflecting an individual's judgments about social welfare. If I am giving *my* ethical assessment of alternative social options, then it is reasonable to expect that I should take into account all the individuals' personal well-being (it is possible that the survival of tigers and the extent of social inequality directly affects an individual's personal well-being), but it is not at all clear why I should take into account *their* ethical views about social and economic inequality or the survival of tigers in making *my* ethical assessment of social options (see Broome [3], p.12).

What happens if we interpret Arrow's theorem as a theorem about arriving at social welfare judgments on the basis of the different individuals' well-being corresponding to different social options? In this case, the individual orderings need to be interpreted as the orderings of social options in terms of the individuals' respective well-being. But note that, in this case, the very definition of a social ranking rule will make the social ranking of options dependent exclusively on the individual well-being *orderings* and will not allow us to take into account any cardinal information about individual well-being (e.g., information that the switch from x to y increases i's well-being more than the switch from z to w). Even if we relax the definition of a social ranking rule to permit cardinal information about the well-being of individuals, Independence of Irrelevant Alternatives with its focus on the individuals' rankings over pairs of options will have the effect of making all such cardinal information irrelevant for the social ranking. When the problem is one of aggregating the judgments or opinions of individuals, there may be some plausibility in ignoring how intensely an individual feels about one option being better than another, but ignoring cardinal information about the individuals' well-being would seem to be ethically unacceptable when the problem is one of discussing social welfare judgments. The framework of Arrow would seem to be more suitable for discussing how the society should arrive at a compromise given differing individual preferences than for discussing social welfare judgments.

37.5 The Impossibility of Paretian Liberalism

The literature inspired by Arrow [1] has given us numerous results demonstrating that a social ranking rule cannot satisfy certain apparently plausible conditions. I now take up one of these results, namely, the famous paradox of the Paretian liberal due to Sen [13, 14], which has had far-reaching influence on the theory of social choice and welfare.

Definition 37.3 Let $f : \mathcal{S}^n \longrightarrow \mathcal{T}$ be a social ranking rule. f satisfies:

(i) *Weak Collective Rationality* iff $\mathcal{S} = \mathcal{R}$ and, for every $(R_1, \ldots, R_n)$ in $\mathcal{S}^n$, R is reflexive and connected and P is acyclic, i.e., there do not exist $x_1, x_2, \ldots, x_m \in X$, such that $[x_1 P x_2$ and $x_2 P x_3$ and $\ldots$ and $x_{m-1} P x_m$ and $x_m P x_1]$;
(ii) *Minimal Liberalism* iff there exist distinct $i, j \in N$ and $x, y, z, w \in X$, such that $(x \neq y$ and $z \neq w)$, and

(5.1) for every $(R_1, \ldots, R_n)$ in $\mathcal{S}^n$, (if xP_iy, then xPy) and (if yP_ix, then yPx),

and

(5.2) for every $(R_1, \ldots, R_n)$ in $\mathcal{S}^n$, (if zP_jw, then zPw) and (if wP_jz, then wPz).

Acyclicity of P is much weaker than transitivity of R, and, hence Weak Collective Rationality is much weaker than Collective Rationality. Reflexivity and connectedness of R and acyclicity of P, together, are necessary and sufficient to ensure that $C(A, R)$ will be non-empty for every non-empty and finite subset A of X (see Sen [14], p.16). Minimal Liberalism was originally interpreted in terms of what might be called an individual's right to liberty in her "private" affairs. Under this interpretation, x and y figuring in the statement of Minimal Liberalism are visualized as two social states which are identical in all respects except for something (e.g., i's religion or the color of his shirt) that is considered to be in the personal or private sphere of individual i, and, similarly for z and w in the case of individual j. Thus, the condition stipulates that there are at least two distinct individuals in the society, each of whom enjoys decisiveness (or the "right" to decide) over some pair of distinct alternatives differing only with respect to her private life.

Theorem 37.2 (Sen [13, 14]): *There does not exist any social ranking rule which satisfies Weak Collective Rationality, Weak Pareto Principle, and Minimal Liberalism simultaneously.*

The condition of Minimal Liberalism constituted the first major departure from the dominant tradition of welfare economics, which considered information about people's preferences (or their utility) to be the only information relevant for the evaluation of social options.[1] If a social ranking rule satisfies Minimal Liberalism, then, not only do individual preferences matter for the social ranking of those two alternatives, but it also matters which two alternatives are under consideration: taking the interpretation of Minimal Liberalism in terms of an individual's decisiveness in matters relating to her private life, to invoke Minimal Liberalism one needs to know, besides the individual preferences, whether the options differ only with respect to somebody's private life.

If one accepts the interpretation of the condition in terms of individuals' rights to liberty in their private affairs, then Theorem 37.2 can be thought of as revealing a deep tension between such individual rights and the Weak Pareto Principle, which has been traditionally regarded as sacrosanct in economics. Many scholars (see, among others, Nozick [11], Gärdenfors [7], Sugden [16], and Gaertner et al. [6]) have argued that the interpretation of Minimal Liberalism in terms of individual rights is not quite compatible with our intuition about rights. Most of these scholars, however, acknowledge that Sen's intuitive insight into the tension between individual rights and the Weak Pareto Principle survives even under other formulations of individual rights suggested in the literature.

37.6 Two Voting Rules

In addition to exploring the implications of axioms regarding social choice/ social evaluation, which have a priori ethical appeal, the literature on the formal theory of social choice has also analyzed the structure of a large number of voting rules, which are basically different methods of reaching a compromise in the presence of differing preferences of individuals and many of which are often used in practice. Two of these voting rules, which have been studied over more than two centuries, stand out. The first is the majority voting rule, the formal structure of which was analyzed in detail by M. de Condorcet [5]. The second is Borda's rule advocated by J.-C. de Borda [4].

[1] This ethical position has been called "welfarism", which may not be an entirely felicitous term. Note that one can define welfarism more formally, but it is not necessary for my purpose here.

The Majority Ranking Rule

The first voting rule that I consider is the well-known majority ranking rule.

Definition 37.4 The *majority ranking rule* (MRR) is the social ranking rule f with domain $\mathcal{R}^n$, such that, for all $x, y \in X$ and all $(R_1, \ldots, R_n)$ in $\mathcal{R}^n$, xRy if and only if $\#\{i \in N : xP_iy\} \geq \# \{i \in N : yP_ix\}$.

It is easy to see that the MRR satisfies the Weak Pareto Principle, Independence of Irrelevant Alternatives, and Non-dictatorship, and that, under it, the social weak preference relation R is reflexive and connected for all $(R_1, \ldots, R_n)$ in $\mathcal{R}^n$. It is, however, well-known that, not only can the social weak preference relation R yielded by the MRR violate transitivity for some profiles of individual preference orderings, but even P can violate acyclicity under the MRR so that the choice set generated by R can be empty for some finite set of options and some profile of individual orderings. An example of this is the well-known voting paradox, where we have $N = \{1, 2, 3\}$ and $(R_1, R_2, R_3) \in \mathcal{R}^n$ is such that xP_1yP_1z, yP_2zP_2x, and zP_3xP_3y, so that the MRR yields xPy and yPz and zPx and $C(\{x, y, z\}, R)$ is empty. This is a major problem with the MRR. But how appealing is it to say that, if, at all, a majority winner exists in a set options, then the society should choose it from that set of options? To see this, it is helpful to see the properties of the MRR. One of the earliest studies of the properties of the MRR in the modern literature on social choice is to be found in May [9], who provided a characterization of the MRR in terms of a set of quite appealing properties.

While May's theorem clarifies the structure of the MRR and, in the process, demonstrates its several highly attractive properties, a very different justification for the MRR came from Condorcet [5] himself (for a lucid exposition of this perspective, see Young [19]). Suppose the number of individuals in the society is odd, we have a profile of linear individual orderings, and we have exactly two options, x and y, which have to be socially ranked and the society's ranking has to be either xPy or yPx. Further, suppose one of these two strict rankings is the "true" or "correct" ranking but it is not known which of them is the correct ranking and, a priori, the two rankings are equally likely to be correct. It does seem a little strange to characterize the ranking of options arrived at by aggregating individual preferences as "correct" or "incorrect". In some situations, however, it makes sense to talk about the correct ranking of x and y for the group. Consider the case of a trial by a jury, which Condorcet [5] discussed. A person is accused of a particular crime and the jury has to decide whether to convict him or not to convict him. All members of the jury share the same objective, namely, that the person should be convicted if and only if he is guilty. Let x denote that the person is convicted and let y denote that the person is not convicted. Given that all members of the jury have the shared objective of convicting the person if and only if he is guilty, and given that the person is either guilty or not guilty, in a very plausible sense exactly one of the two alternative strict rankings, xPy and yPx, is the correct ranking for the group, but it is not known which of them is correct. One can think of many other examples, where the two options are alternative policies for achieving a shared objective, and it seems plausible to talk about the "correct" group ranking of the two policies though it may not be known what exactly the correct ranking may be. Assume that n is odd and each individual's strict ranking of x and y has the same probability, q $\left(1 > q > \frac{1}{2}\right)$, of being the correct ranking. Condorcet [5] showed that, given the assumptions stated above, the probability that the social ranking of x and y under the MRR will be correct is

$$p = \sum\nolimits_{k=\frac{n+1}{2}}^{n} q^k (1-q)^{n-k} \left[\frac{n!}{k!\,(n-k)!}\right]$$

and that p approaches 1 as n becomes indefinitely large. Condorcet's remarkable result provides a strong justification for the MRR when there are exactly two alternatives. For an extension of Condorcet's probabilistic reasoning to the case of more than two alternatives, the reader may refer to Young [18, 19].

Borda's Ranking Rule

Our second voting rule is due to Borda [4], who was a contemporary of Condorcet and his intellectual rival.

Let R^* be a linear ordering over X. For all $x \in X$, let $s(x, R^*)$ denote $\#\{a \in X : xP^*a\} + 1$. Thus, if $X = \{x, y, z, w)$ and we have $xP^*yP^*zP^*w$, then $s(x, R^*) = 4$, $s(w, R^*) = 1$, and so on.

Definition 37.5 *Borda's ranking rule* (BRR) *is* the social ranking rule with domain L^n, such that, for all $(R_1, \ldots, R_n) \in L^n$ and all $x, y \in X$, xRy if and only if $\sum_{i \in N} s(x, R_i) \geq \sum_{i \in N} s(y, R_i)$.

Note that, to avoid some details not important for our purpose, I have defined Borda's ranking rule only for the case where the individual orderings are constrained to be linear. Given a profile of linear individual orderings, BRR proceeds as follows. For each option x and each individual ordering R_i, it specifies for x its "R_i – based score" denoted by $s(x, R_i)$. If x occupies the first position in the ordering R_i over X, then the R_i-based score of x is $\#X$; if x occupies the second position in the ordering, then its R_i-based score is $\#X$-1, and so on. Next, for every option in X, it sums up the R_i-based scores for x over all individuals i to get the "total score" of x. Finally, it ranks all the options on the basis of their respective total scores.

Several points may be noted here. First, for every profile of linear individual orderings, BRR yields a social ordering, and BRR satisfies the Weak Pareto Principle and non-dictatorship. But it can be easily shown that it violates Independence of Irrelevant Alternatives. Second, if X has exactly two alternatives, then it is clear that, for every profile of linear individual orderings[2] over X, BRR will yield the same social ranking as the MRR. Third, it is possible that, for some profile of linear orderings and some non-empty subset A of X, the social ranking yielded by MRR can define a unique best alternative in A, which is different from the

[2]Recall that in defining BRR, we have assumed that only linear individual orderings are permissible.

unique best alternative in A defined by the social ranking under BRR. To see this, let $N = \{1, 2, \ldots, 9\}$ and $X = \{x, y, z, w\}$, and let $(R_1, \ldots, R_9)$ be as follows (the options in a column are in a descending order of preference)

R_1, R_2, R_3, R_4	R_5, R_6, R_7	R_8	R_9
x	y	w	x
y	z	x	y
z	w	y	w
w	x	z	z

It can be checked that, given this profile, the choice set defined for X by the social ranking under MRR is $\{x\}$ while the choice set specified for X by the social ranking under BRR is $\{y\}$.

Like MRR, BRR has also been characterized in terms of highly plausible properties (see Nitzan and Rubinstein [10][3]; see also Young's [17] characterization of Borda's rule formulated in terms of social choice rather than in terms of a social ranking). Also, for Borda's rule formulated in terms of social choice (rather than in terms of a social ranking), Young [18, 19] provides a striking justification based on probabilistic reasoning analogous to, but different from, the probabilistic reasoning that Condorcet [5] used to justify the MRR.

37.7 Concluding Remarks

This essay has considered only a few contributions to the formal theory of social choice and welfare, which has emerged as an exceptionally rich and diverse area of study. These contributions, however, illustrate how the application of formal reasoning has yielded fresh insights into some very old issues in political and social philosophy.

References[4]

1. * Arrow, K. J. (1951, 1963). *Social choice and individual values* (1st ed., 1951; 2nd ed., 1963). New York: Wiley.
2. Bergson, A. (1954). On the concept of social welfare. *Quarterly Journal of Economics, 68*, 233–252.
3. Broome, J. (2009). Why Economics Needs Ethical Theory. In K. Basu & R. Kanbur (Eds.), *Arguments for a Better World* (pp. 7–14). Oxford: Oxford University Press.

[3]Nitzan and Rubinstein [10], however, allow individual preferences to be non-transitive.

[4]Recommended readings are indicated by asterisks before the names of the authors.

4. de Borda, J.-C. (1781). "Mémoire sur les Élections au Scrutin", *Histoire de l'Académie Royale des Sciences*, translated by Alfred de Grazia as "Mathematical derivation of an election system", *Isis* 44, Parts 1 &2, 1953 (pp. 42–51).
5. de Condorcet, M. (1785) *Essai sur l'application de l'analyse a la probabilite des decisions rendues a la pluralite des voix*. Paris.
6. Gaertner, W., P. K. Pattanaik, and K. Suzumura (1992), "Individual rights revisited", *Economica* 59, 1152–69.
7. Gärdenfors, P. (1981). Rights, games and social choice. *Noûs, 15*, 341–356.
8. * Little, I.M.D. (1952), "Social choice and individual values", *Journal of Political Economy 60*, 422–432.
9. * May, K. O. (1952). A set of independent necessary and sufficient conditions for simple majority decision. *Econometrica, 20*, 680–684.
10. Nitzan, S., & Rubinstein, A. (1981). A further characterization of Borda ranking method. *Public Choice, 36*, 153–158.
11. Nozick, R. (1974). *Anarchy, state and Utopia*. Oxford: Blackwell.
12. Plott, C. R. (1976). Axiomatic social choice theory. *American Journal of Political Science, 20*, 511–596.
13. * Sen, A. K. (1970). The impossibility of a Paretian Liberal. *Journal of Political Economy, 78*, 152–157.
14. Sen, A. K. (1970). *Collective choice and social welfare*. San Francisco: Holden Day.
15. Sen, A. K. (1977). Social choice theory: A re-examination. *Econometrica, 45*, 53–89.
16. Sugden, R. (1985). Liberty, preference, and choice. *Economics and Philosophy, 1*, 213–229.
17. * Young, H. P. (1974). An axiomatization of Borda's rule. *Journal of Economic Theory, 9*, 43–52.
18. * Young, H. P. (1988). Condorcet's theory of voting. *American Political Science Review, 82*, 1231–1244.
19. Young, H. P. (1997) Group choice and individual judgments. In D. C. Mueller (Ed.), *Perspectives on public choice* (pp. 181–200). Cambridge: Cambridge University Press.

Chapter 38
Judgment Aggregation

Philippe Mongin

Abstract Judgment aggregation theory generalizes social choice theory by having the aggregation rule bear on judgments of all kinds instead of barely judgments of preference. The theory derives from Kornhauser and Sager's doctrinal paradox and Pettit's discursive dilemma, which List and Pettit turned into an impossibility theorem – the first of a long list to come. After mentioning this formative stage, the paper restates what is now regarded as the "canonical theorem" of judgment aggregation theory (in three versions due to Nehring and Puppe, Dokow and Holzman, and Dietrich and Mongin, respectively). The last part of paper discusses how judgment aggregation theory connects with social choice theory and can contribute to it; it singles out two representative applications, one to Arrow's impossibility theorem and the other to the group identification problem.

38.1 A New Brand of Aggregation Theory

It is a commonplace idea that collegial institutions generally make better decisions than those in which a single individual is in charge. This optimistic view can be traced back to Enlighment theorists, such as Rousseau and Condorcet, and it permeates today's western judiciary organization, which is heir to this philosophical tradition. The more important a legal case, the more likely it is to be entrusted to a collegial court; appeal courts are typically collegial, and at the top of the legal organization, constitutional courts always are. However, the following, by now classic example from legal theory challenges the Enlightment view.

The author gratefully acknowledges Franz Dietrich's and Ashley Piggins's comments.

P. Mongin (✉)
CNRS & HEC Paris, France
e-mail: mongin@greg-hec.com

S. O. Hansson, V. F. Hendricks (eds.), *Introduction to Formal Philosophy*, Springer Undergraduate Texts in Philosophy, https://doi.org/10.1007/978-3-319-77434-3_38

A plaintiff has brought a civil suit against a defendant, alleging a breach of contract between them. The court is composed of three judges A, B and C, who will determine whether or not the defendant must pay damages to the plaintiff (d or $\neg d$). The case brings up two issues, i.e., whether the contract was valid or not (v or $\neg v$), and whether the defendant was or was not in breach of it (b and $\neg b$). Contract law stipulates that the defendant must pay damages if, and only if, the contract was valid and he was in breach of it. Suppose that the judges have the following views of the two issues, and accordingly of the case:

A	v	$\neg b$	$\neg d$
B	$\neg v$	b	$\neg d$
C	v	b	d

In order to rule on the case, the court can either directly collect the judges' recommendations on it, or collect the judges' views of the issues and then solve the case by applying contract law to these data. If the court uses majority voting, the former, *case-based* method delivers $\neg d$, whereas the latter, *issue-based* method returns first v and b, and then d. This elegant example is due to legal theorists Kornhauser and Sager [21]. They describe as a *doctrinal paradox* any similar occurrence in which the two methods give conflicting answers. What makes the discrepancy paradoxical is that each method is commendable on some ground, i.e., the former respects the judges' final views, while the latter provides the court with a rationale, so one would wish them always to be compatible. The legal literature has not come up with a clear-cut solution (see Nash [32]). This persisting difficulty casts doubt on the belief that collegial courts would be wiser than individual ones. Clearly, with a single judge, the two methods coincide unproblematically.

An entire body of work, now referred to as *judgment aggregation theory,* has grown out of Kornhauser and Sager's doctrinal paradox. As an intermediary step, their problem was rephrased by political philosopher Pettit [39], who wanted to make it both more widely applicable and more analytically tractable. What he calls the *discursive dilemma* is, first of all, the generalized version of the doctrinal paradox in which a group, whatever it is, can base its decision on either the *conclusion-based* or the *premiss-based* method, whatever the substance of conclusions and premisses may be. What holds of the court equally holds of a political assembly, an expert committee, and many other deliberating groups; as one of the promoters of the concept of deliberative democracy, Pettit would speculatively add political society as a whole. Second, and more importantly for our purposes, the discursive dilemma shifts the stress away from the conflict of methods to *the logical contradiction within the total set of propositions that the group accepts.* In the previous example, with $d \longleftrightarrow v \wedge b$ representing contract law, the contradictory set is

$$\{v, b, d \longleftrightarrow v \wedge b, \neg d\}.$$

Trivial as this shift seems, it has far-reaching consequences, because all propositions are now being treated alike; indeed, the very distinction between premisses and conclusions vanishes. This may be a questionable simplification to make in the legal context, but if one is concerned with developing a general theory, the move has clear analytical advantages. It may be tricky to classify the propositions into two groups, and it is definitely simpler to pay attention to whole sets of accepted propositions – briefly *judgment sets* – and inquire when and why the collective ones turn out to be inconsistent, given that the individual ones are taken to be consistent. This is already the problem of judgment aggregation.

In a further step, List and Pettit [24] introduce an aggregation mapping F, which takes profiles of individual judgment sets $(A_1, \ldots, A_n)$ to collective judgment sets A, and subject F to axiomatic conditions which they demonstrate are logically incompatible. Both the proposed formalism and impossibility conclusion are in the vein of social choice theory, but they are directed at the discursive dilemma, which the latter theory cannot explain in terms of its usual preference apparatus. At this stage, the new theory exists in full, having defined its object of study – the F mapping, or *collective judgment function* – as well as its method of analysis - it consists in axiomatizing F and investigating subsets of axioms to decide which result in an impossibility and which, to the contrary, support well-behaved rules (such as majority voting).

List and Pettit's impossibility theorem was shortly succeeded, and actually superseded, by others of growing sophistication, due to Pauly and van Hees [38], Dietrich [3], Dietrich and List [6], Mongin [29], Nehring and Puppe [35, 36], [10–12], and Dietrich and Mongin [9]. This lengthy, but still incomplete list, should be complemented by two papers that contributed differently to the progress of the field. Elaborating on earlier work in social choice theory by Wilson [45] and Rubinstein and Fishburn [42], and in a formalism that still belongs to that theory, Nehring and Puppe [33] inquired about which *agendas* of propositions turn the axiomatic conditions into a logical impossibility. Agendas are the rough analogue of preference domains in social choice theory. This concept raised to prominence in mature judgment aggregation theory, and Nehring and Puppe's characterization of impossibility agendas was eventually generalized by Dokow and Holzman [11], whose formulation has become the received one. On a different score, Dietrich [4] showed that the whole formalism of the theory could be deployed without making reference to any specific logical calculus. Only a few elementary properties of the formal language and the logic need assuming for the theorems to carry through. The so-called *general logic* states these requisites (see Dietrich and Mongin [9], for an up-to-date version). The first papers relied on propositional calculi, which turns out to be unnecessary. This major generalization underlies the theory as it is presented here, as well as in the more extensive overviews by Mongin and Dietrich [31] or Mongin [30]. (These two papers actually use the tag "logical aggregation theory" instead of the standard one "judgment aggregation theory" to emphasize the particular angle they adopt.)

The next section "A Logical Framework for Judgment Aggregation Theory" provides a syntactical, framework for the F function, using the general logic as a background. It states the axiomatic conditions on F that have attracted most attention, i.e., systematicity, independence, monotonicity and unanimity preservation. The issue of agendas arises in the ensuing Sect. 38.3, which presents an impossibility theorem in three variant forms, due to Nehring and Puppe, Dokow and Holzman, and Dietrich and Mongin, respectively. This is the central achievement of the theory by common consent – hence the label "canonical theorem" adopted here – but many other results are well deserving attention. For them, the reader is referred to the two reviews just mentioned, or at a more introductory level, those of List and Puppe [25] and Grossi and Pigozzi [17]. The final Sect. 38.4 sketches a comparison with social choice theory and discusses how judgment aggregation theory relates and contributes to the latter.

Several topics are omitted here. One is *probability aggregation*, which gave rise to a specialized literature already long ago (see Genest and Zidekh's [15] survey of the main results). Both commonsense and traditional philosophy classify judgments into certain and uncertain ones, so probability aggregation theoretically belongs to the topic of this chapter. However, we will comply here with the current practice of taking judgments in the restricted sense of judgments passed under conditions of certainty. Another, no doubt more questionable omission concerns those logics which the general logic excludes despite its flexibility; prominent among which are the *multi-valued* logics investigated by Pauly and van Hees [38], van Hees [43], and Duddy and Piggins [14], and the *non-monotonic* logics investigated by Wen [44]. Finally, we have omitted the topic of *belief merging*, or *fusion*, which emerged in theoretical computer science independently of judgment aggregation theory, but is now often associated with it. Although they represent the information stored in databases rather then by human agents, the computer scientists' "belief sets" or "knowledge bases" are analogous to judgment sets, and the problem of "merging" or "fusing" these items is analogous to the problem of defining a collective judgment function. Pigozzi [40] was one of the first to make this connection, and the reader can consult one of her up-to-date surveys (e.g., Pigozzi [41]). The computer scientists' solutions are particular cases of *distance-based judgment aggregation*, i.e., they depend on defining what it means for a judgment set to be closer to one judgment set than another. Miller and Osherson [28] thoroughly explore the abstract properties of distance metrics, while Lang et al. [22] provide a classification.

38.2 A Logical Framework for Judgment Aggregation Theory

By definition, a *language* $\mathcal{L}$ for judgment aggregation theory is any set of formulas $\varphi, \psi, \chi, \ldots$ that is constructed from a set of logical symbols $\mathcal{S}$ containing $\neg$, the Boolean negation symbol, and that is closed for this symbol (i.e., if $\varphi \in \mathcal{L}$, then $\neg\varphi \in \mathcal{L}$). In case $\mathcal{S}$ contains other elements, such as symbols for the remaining Boolean connectives or modal operators, they satisfy the appropriate closure properties. A *logic* for judgment aggregation theory is any set of axioms and rules that regulates the inference relation $\vdash$ on $\mathcal{L}$ and associated technical notions – logical truth and contradiction, consistent and inconsistent sets – while satisfying the general logic. Informally, the main requisites are that $\vdash$ be monotonic and compact, and that any consistent set of formulas can be extended to a complete consistent set. ($S \subset \mathcal{L}$ is *complete* if, for all $\varphi \in \mathcal{L}$, either $\varphi \in S$ or $\neg\varphi \in S$.) Monotonicity means that inductive logics are excluded from consideration, and compactness (which is needed only in specific proofs) that some deductive logics are. The last requisite is the standard Lindenbaum extendability property.

Among the many calculi that enter this framework, propositional examples stand out. They need not be classical, i.e., $\mathcal{S}$ may contain modal operators, like those of deontic, epistemic and conditional logics, each of them leading to a potentially relevant application. Each of these extensions should be double-checked, because some fail compactness. Although this may not be so obvious, first-order calculi are also permitted. When it comes to them, $\mathcal{L}$ is the set of *closed* formulas – those without free variables – and the only question is whether $\vdash$ on $\mathcal{L}$ complies with the general logic.

In $\mathcal{L}$, a subset X is fixed to represent the propositions that are in question for the group; this is the *agenda*, one of the novel concepts of the theory and one of its main focuses of attention. In all generality, X needs only to be non-empty, with at least one contingent formula, and to be closed for negation. The discursive dilemma reconstruction of the court example leads to the agenda

$$\overline{X} = \{v, b, d, d \leftrightarrow v \wedge b, \neg v, \neg b, \neg d, \neg(d \leftrightarrow v \wedge b)\}.$$

The theory represents judgments in terms of subsets $B \subset X$, which are initially unrestricted. These *judgment sets* – another notion specific to the theory – will be denoted by $A_i, A'_i, \ldots$ when they belong to the individuals $i = 1, \ldots n$, and by $A, A', \ldots$ when they belong to the group as such. A formula φ from one of these sets represents a proposition, in the ordinary sense of a semantic object endowed with a truth value. If φ is used also to represent a judgment, in the sense of a cognitive operation, this is in virtue of the natural interpretive rule:

(R) i judges that φ iff $\varphi \in A_i$, and the group judges that φ iff $\varphi \in A$.

Standard logical properties may be applied to judgment sets. For simplicity, we only consider two cases represented by two sets of judgments sets:

- the unrestricted set 2^X;
- the set D of *consistent* and *complete* judgment sets (consistency is defined by the logic and completeness is as above, but relative to X).

Thus far, the theory has been able to relax completeness, but not consistency (see, e.g., Dietrich and List [8]).

The last specific concept is the *collective judgment function* F, which associates a collective judgment set to each profile of judgment sets for the n individuals:

$$A = F(A_1, \ldots, A_n).$$

The domain and range of F can be defined variously, but we restrict attention to $F : D^n \to 2^X$, our baseline case, and $F : D^n \to D$, our target case, in which the collective sets obey the same stringent logical constraints as the individual ones. The present framework captures the simple voting rule of the court example, as well as less familiar examples. Formally, define *formula-wise majority voting* as the collective judgment function $F_{maj} : D^n \to 2^X$ such that, for every profile $(A_1, \ldots, A_n) \in D^n$,

$$F_{maj}(A_1, \ldots, A_n) = \{\varphi \in X : |\{i : \varphi \in A_i\}| \geq q\},$$

$$\text{with } q = \frac{n+1}{2} \text{ if } n \text{ is odd and } q = \frac{n}{2} + 1 \text{ if } n \text{ is even.}$$

Here, the range is not D because there can be unbroken ties, and so incomplete collective judgment sets, when n is even. More strikingly, for many agendas, the range is not D even when n is odd, because there are inconsistent collective judgment sets, as the court example neatly shows. By varying the value of q between 1 and n in the definition, one gets specific quota rules F^q_{maj}. One would expect inconsistency to occur with low q, and incompleteness with large q. Nehring and Puppe [33, 35] and Dietrich and List [7] investigate the F^q_{maj} in detail.

Having defined and exemplified F functions, we introduce some axiomatic properties they may satisfy.

Systematicity. For all formulas $\varphi, \psi \in X$ and all profiles $(A_1, \ldots, A_n)$, $(A'_1, \ldots, A'_n)$, if $\varphi \in A_i \Leftrightarrow \psi \in A'_i$ for every $i = 1, \ldots, n$, then

$$\varphi \in F(A_1, \ldots, A_n) \Leftrightarrow \psi \in F(A'_1, \ldots, A'_n).$$

Independence. For every formula $\varphi \in X$ and all profiles $(A_1, \ldots, A_n)$, $(A'_1, \ldots, A'_n)$, if $\varphi \in A_i \Leftrightarrow \varphi \in A'_i$ for every $i = 1, \ldots, n$, then

$$\varphi \in F(A_1, \ldots, A_n) \Leftrightarrow \varphi \in F(A'_1, \ldots, A'_n).$$

Monotonicity. For every formula $\varphi \in X$ and all profiles $(A_1, \ldots, A_n)$, $(A'_1, \ldots, A'_n)$, if $\varphi \in A_i \Rightarrow \varphi \in A'_i$ for every $i = 1, \ldots, n$, with $\varphi \notin A_j$ and $\varphi \in A'_j$ for at least one j, then

$$\varphi \in F(A_1, \ldots, A_n) \Rightarrow \varphi \in F(A'_1, \ldots, A'_n).$$

Unanimity preservation. For every formula $\varphi \in X$ and every profile $(A_1, \ldots, A_n)$, if $\varphi \in A_i$ for every $i = 1, \ldots, n$, then $\varphi \in F(A_1, \ldots, A_n)$.

By definition, F is a *dictatorship* if there is a j such that, for every profile $(A_1, \ldots, A_n)$,

$$F(A_1, \ldots, A_n) = A_j.$$

Given the unrestricted domain, there can only be one such j, to be called the *dictator*. The last property is

Non-dictatorship. F is not a dictatorship

It is routine to check that F_{maj} satisfies all the list. Systematicity means that the group, when faced with a profile of individual judgment sets, gives the same answer concerning a formula as it would give concerning a *possibly different* formula, when faced with a *possibly different* profile, supposing that the individual judgments concerning the first formula in the first profile are the same as those concerning the second formula in the second profile. Independence amounts to restricting this requirement to $\varphi = \psi$. Thus, it eliminates one claim made by Systematicity – i.e., that the identity of the formula does not matter – while preserving another – i.e., that the collective judgment of φ depends only on individual judgments of φ. That is, by Independence, the collective set A is defined *formula-wise* from the individual sets $A_1, \ldots, A_n$. By contrast, for any concept of distance envisaged in the distance-based literature (e.g., [22, 28]), if F is defined by minimizing the total distance of A to $A_1, \ldots, A_n$, F violates Independence. The collective judgment sets in this class of solutions are constructed from the individual sets *taken as a whole* and not formula-wise.

Systematicity was the condition List and Pettit's [24] impossibility theorem, but henceforth, the focus of attention shifted to Independence. The former has little to say for itself except that many voting rules satisfy it, but the latter can be defended as a *non-manipulability* condition. If someone is in charge of defining the agenda X, Independence will prevent this agent to upset the collective judgment on a formula by adding or withdrawing other formulas in X; this argument appears in Dietrich [3]. However, Independence does not block all and every form of manipulability, as Cariani, Pauly and Snyder [2] illustrate; they show that a suitable choice of the language $\mathcal{L}$ can influence the collective judgment.

Some writers take Monotonicity to be a natural addition to Independence. This condition requires that, when a collective result favours a subgroup's judgment, the same holds if more individuals join the subgroup. It can be defended in terms of democratic responsiveness, though perhaps not so obviously as the last two conditions, i.e., Unanimity preservation and Non-dictatorship.

The problem that has gradually raised to the fore is to characterize – in the sense of necessary and sufficient conditions – the agendas X such that no $F : D^n \to D$ satisfies Non-dictatorship, Independence, and Unanimity preservation. There is a variation of this problem with Monotonicity as a further axiomatic condition. The next section provides the answers.

38.3 The Canonical Theorem in Three Forms

The promised answers depend on further technical notions. First, a set of formulas $\mathcal{S} \subset \mathcal{L}$ is called *minimally inconsistent* if it is inconsistent and all its proper subsets are consistent. With a classical propositional calculus, this is the case for

$$\{v, b, d \leftrightarrow v \wedge b, \neg d\},$$

but not for

$$\{\neg v, \neg b, d \leftrightarrow v \wedge b, d\}.$$

Second, for $\varphi, \psi \in X$, it is said that φ *conditionally entails* ψ – denoted by $\varphi \vdash^* \psi$ – if $\varphi \neq \neg\psi$ and there is some minimally inconsistent $Y \subset X$ with $\varphi, \neg\psi \in Y$. This is trivially equivalent to requiring that $\{\varphi\} \cup Y' \vdash \psi$ holds for some minimal auxiliary set of premisses Y' that is contradictory neither with φ, nor with $\neg\psi$.

Now, an agenda X is said to be *path-connected* (another common expression is *totally blocked*) if, for every pair of formulas $\varphi, \psi \in X$, there are formulas $\varphi_1, \ldots, \varphi_k \in X$ such that

$$\varphi = \varphi_1 \vdash^* \varphi_2 \vdash^* \ldots \vdash^* \varphi_k = \psi.$$

Loosely speaking, agendas with this property have many, possibly roundabout logical connections. Finite agendas can be represented by directed graphs: the formulas φ, ψ are the nodes and there is an arrow pointing from φ to ψ for each conditional entailment $\varphi \vdash^* \psi$. The court agenda $\overline{X}$ is path-connected, as the picture below of conditional entailments illustrates (it does not represent all existing conditional entailments, but sufficiently many for the reader to check the claim) (Fig. 38.1).

(Here and in the next figures, an arrow pointing from one formula to another means that the former conditionally entails the latter, and the small print formulas near the head of the arrow are a choice of auxiliary premisses; $d \leftrightarrow v \wedge b$ is abridged as q.)

Now, we are in a position to state a version of the canonical theorem (see Dokow and Holzman [11], and Nehring and Puppe [36]; it originates in Nehring and Puppe [33]). From now on, we assume that $n \geq 2$.

Theorem (first form) *If X is path-connected, then no $F : D^n \to D$ satisfies Non-dictatorship, Unanimity preservation, Monotonicity and Independence. The agenda condition is also necessary for this conclusion.*

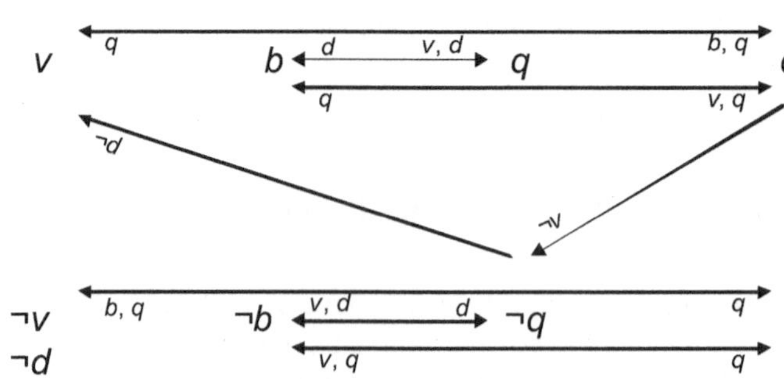

Fig. 38.1 The court agenda in the discursive dilemma version

To illustrate the sufficiency part, let us take $\overline{X}$ and F_{maj}, assuming that n is odd, so that F_{maj} has range D if and only if $F_{maj}(A_1, \dots, A_n)$ is consistent for all profiles $(A_1, \dots, A_n)$. The court example in the discursive dilemma version exhibits a profile contradicting consistency, and this shows that D is not the range of F_{maj}. The theorem leads to the same conclusion by a more general reasoning: since F_{maj} satisfies the four axioms and $\overline{X}$ is path-connected, D cannot be the range. By a converse to this entailment, when the agenda is *not* path-connected, there is no collective inconsistency even if the axiomatic conditions hold. This important addition is the necessity part of the theorem, which we do not illustrate here.

As it turns out, Monotonicity can be dropped from the list of axioms if the agenda is required to satisfy a further condition. Let us say that X is *even-number negatable* if there is a minimally inconsistent set of formulas $Y \subseteq X$ and there are distinct $\varphi, \psi \in Y$ such that $Y_{\neg\{\varphi,\psi\}}$ is consistent, where the set $Y_{\neg\{\varphi,\psi\}}$is obtained from Y by replacing φ, ψ by $\neg\varphi, \neg\psi$ and keeping the other formulas unchanged. This seems to be an unpalatable condition, but it is not demanding, as $\overline{X}$ illustrates: take

$$Y = \{v, b, d, \neg(d \leftrightarrow v \wedge b)\} \text{ and } \varphi = v, \psi = b,$$

and there are alternative choices of Y. The next result was proved by Dokow and Holzman [11] as well as, for the sufficiency part, by Dietrich and List [6].

Theorem (second form) *If X is path-connected and even-number negatable, then no $F : D^n \to D$ satisfies Non-dictatorship, Unanimity preservation, and Independence. If $n \geq 3$, the agenda conditions are also necessary for this conclusion.*

A further step of generalization is available. Unlike the work reviewed so far, it is motivated *not by the discursive dilemma, but by the doctrinal paradox*, and it is specially devised to clarify the premiss-based method, which is often proposed as a solution to this paradox (see Pettit [39], and some of the legal theorists reviewed by Nash [32]). Formally, we define the *set of premisses* to be a subset $P \subseteq X$, requiring only that it be non-empty and closed for negation, and reconsider the framework to account for the difference between P and its complement $X \setminus P$. Adapting the axioms, we define

Independence on premisses: same statement as for Independence, but holding only for every $p \in P$.

Non-dictatorship on premisses: there is no $j \in \{1, \dots, n\}$ such that $F(A_1, \dots, A_n) \cap P = A_j \cap P$ for every $(A_1, \dots, A_n) \in D^n$.

Now revising the agenda conditions, we say that X is *path-connected in P* if, for every pair $p, p' \in P$, there are $p_1, \dots, p_k \in P$ such that

$$p = p_1 \vdash^* p_2 \vdash^* \dots \vdash^* p_k = p'.$$

Note that formulas in $X \setminus P$ may enter this condition via the definition of conditional entailment $\vdash^*$. We also say that X is *even-number negatable in P* if there are $Y \subseteq$

X and $\varphi, \psi \in Y$ as in the above definition for being even-number negatable, except that "$\varphi, \psi \in Y \cap P$" replaces "$\varphi, \psi \in Y$" (i.e., the negatable pair consists of premisses). The two conditions can be illustrated by court agendas in the doctrinal paradox style.

If we stick to the agenda $\overline{X}$, the subset

$$\overline{P} = \{v, b, d \leftrightarrow v \wedge b, \neg v, \neg b, \neg(d \leftrightarrow v \wedge b)\}$$

best captures the judges' sense of what premisses are. However, the following construal may be more to the point. Suppose that judges do not vote on the law, but rather take it for granted and apply it – a realistic case from legal theory (see [21]). We model this, first by reducing the agenda to

$$\overline{\overline{X}} = \{v, b, d, \neg v, \neg b, \neg d\},$$

and second by including the formula $d \leftrightarrow v \wedge b$ into the inference relation, now defined by

$$S \vdash_{d \leftrightarrow v \wedge b} \psi \text{ if and only if } S \cup \{\varphi, d \leftrightarrow v \wedge b\} \vdash \psi.$$

In this alternative model, the set of premisses reads as

$$\overline{\overline{P}} = \{v, b, \neg v, \neg b\}.$$

Technically, the two construals are wide apart: $\overline{X}$ is both path-connected and even-number negatable in $\overline{P}$, whereas $\overline{\overline{X}}$ is even-number negatable but not path-connected in $\overline{\overline{P}}$, thus failing the more important agenda condition. The next two pictures – the first for $\overline{P}$ and the second for $\overline{\overline{P}}$ – illustrate the stark contrast (Fig. 38.2).

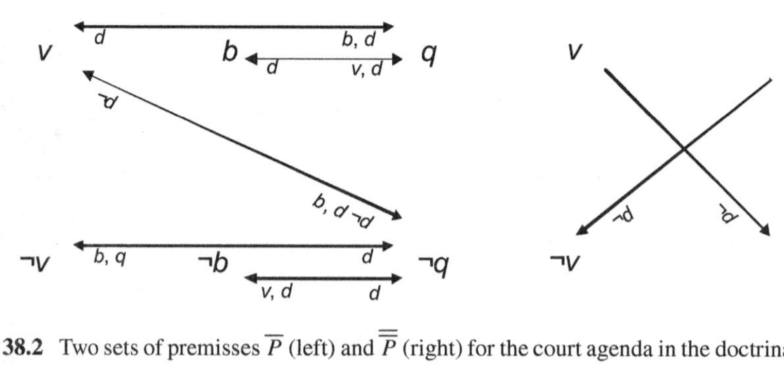

Fig. 38.2 Two sets of premisses $\overline{P}$ (left) and $\overline{\overline{P}}$ (right) for the court agenda in the doctrinal paradox version

(The first picture represents sufficiently many conditional entailments in $\overline{P}$ for the conclusion that $\overline{X}$ is path-connected in $\overline{P}$, and the second represents all conditional entailments in $\overline{\overline{P}}$, which are too few for $\overline{\overline{X}}$ to be path-connected in $\overline{\overline{P}}$.)

Having illustrated the new definitions, we state the result by Dietrich and Mongin [9] that puts them to use.

Theorem (third form) *If X is path-connected and even-number negatable in P, there is no $F : D^n \rightarrow D$ that satisfies Non-dictatorship on premisses, Independence on premisses and Unanimity preservation. If $n \geq 3$, the agenda conditions are also necessary for this conclusion.*

Note carefully that Unanimity preservation retains its initial form, unlike the other two conditions. If it were also restricted to premisses, one would check that no impossibility follows. Thus, the statement is best interpreted as an impossibility theorem for the premiss-based method, granting the normatively defensible constraint that unanimity should be preserved on all formulas. Anyone who accepts this addition – in effect, a whiff of the conclusion-based method – is committed to the unpleasant result that the premiss-based method is, like its rival, fraught with difficulties. As with the previous forms of the canonical theorem, solutions can be sought on the agenda's side by relaxing the even-number negatibility or – more relevantly – the path-connectedness condition. The $\overline{\overline{X}}$, $\overline{\overline{P}}$ reconstruction of the doctrinal paradox illustrates this way out; observe that F_{maj} is well-behaved in this case.

Legal interpretations aside, the third form of the theorem is more assertive than the second one. This is seen by considering $P = X$, a permitted limiting case. Having explored the canonical theorem in full generality, we move to the comparative topic of this paper.

38.4 A Comparison with Social Choice Theory

Judgment aggregation theory has clearly been inspired by social choice theory, and two legitimate questions are, how it formally relates, and what it eventually adds, to its predecessor. The F mapping resembles the *collective preference function* G, which takes profiles of individual preference relations to preference relations for the group. (Incidentally, the official terminology for G, i.e., the "social welfare function", is misleading since it jumbles up the concepts of preference and welfare.) The normative properties posited on judgment sets are evocative of those, like transitivity and completeness, which one encounters with preference relations, and the axiomatic conditions on F are most clearly related to those usually put on G. Systematicity corresponds to neutrality, Independence to independence of irrelevant alternatives, Monotonicity to positive responsiveness, and Unanimity preservation to the Pareto principle, not to mention the similar requisite of Non-dictatorship.

Conceptually, a major difference lies in the objects of the two aggregative processes. A *judgment*, as the acceptance or rejection of a proposition, is more general than a *preference* between two things. According to a plausible account, an agent, whether individual or collective, prefers x to y if and only if it judges that x is preferable to y, i.e., accepts the proposition that x is preferable to y. This clarifies the claim that one concept is more general than the other, but how does this claim translate into the respective formalisms?

We answer this question by following Dietrich and List's [6] footsteps. They derive a version of Arrow's [1] impossibility theorem in which the individuals and the group express *strict* preferences on the set of alternatives Z, and these preferences are assumed to be not only transitive, but also *complete*. Although these assumptions are restrictive from the viewpoint of social choice theory, the logical derivation elegantly shows how judgment aggregation theory can be linked to that theory. The first step is to turn the G mappings defined on the domain of preferences into particular cases of F. To do so, one takes a first-order language $\mathcal{L}$ whose elementary formulas xPy express "x is strictly preferable to y", for all $x, y \in Z$, and defines a logic for $\mathcal{L}$ by enriching the inference relation $\vdash$ of first-order logics with the axioms expressing the asymmetry, transitivity and completeness of P. The conditions for general logic hold. Now, if one takes X to be the set of elementary formulas of $\mathcal{L}$ and defines the set of judgment sets D from this agenda, it is possible to associate with each given G an $F : D^n \to D$ having the same informal content. The next step is to make good the results of judgment aggregation theory. Dietrich and List show that X satisfies the agenda conditions of the canonical theorem (second form). To finish the proof that Arrow's axiomatic conditions on G are incompatible, it is enough to check that they translate into those put on F in the theorem, so that the sufficiency part of the theorem applies.

Although social choice theory is primarily concerned with aggregating preferences, or related individual characteristics such as utility functions, it also extends in other directions, as illustrated by the work on *group identification*. Kasher and Rubinstein [19] consider a finite population N, each member of which is requested to partition N into two categories, conventionally labelled J and not-J. The question is to associate a collective partition with this process, and Kasher and Rubinstein answer it along social-choice-theoretic lines, i.e., by introducing a mapping H from profiles of individual partitions to collective partitions and submitting H to axiomatic conditions. Among other results, they show that if H determines the collective classification of i as J or not-J only from the individual classifications of i as J or not-J, and if H respects unanimous individual classifications of i as J or not-J, then H is a dictatorship. As List [23] suggests, this impossibility theorem can easily be derived from judgment aggregation theory by taking $\mathcal{L}$ to be a propositional language whose elementary formulas express "i is a J", for all $i \in N$. Then a reasoning paralleling that made for the Arrovian case leads to the dictatorship conclusion. The canonical theorem (second form) is again put to use, and just as in the earlier case, an important step is to check that the agenda conditions for this theorem hold. What turns out to be crucial in this respect is the assumption made by Kasher and Rubinstein that both individual and collective partitions are non-trivial (i.e., each partition classifies at least one individual as J and at least one individual as non-J).

There are other derivations of social choice results, most of them based on the canonical theorem or variants of it. For instance, Dokow and Holzman [12] recover a theorem by Gibbard [16] on quasi-transitive social preferences and oligarchies, and Herzberg and Eckert [18] explain how the infinite population variants of Arrow's theorem, as in Kirman and Sondermann [20], relate to infinite population extensions of the canonical theorem. Moreover, some of the derived social choice results are novel. Thus, Dokow and Holzman [12] obtain unnoticed variants of Gibbard's theorem. More strikingly, Dokow and Holzman's [13] analysis of collective judgment aggregation functions in the non-binary case delivers entirely new results concerning *assignment problems* (such as the problem of assigning a given number of jobs to a given number of candidates), and these problems arguably belong to social choice theory, although taken broadly.

Against this reassuring evidence, two reservations are in order. For one thing, the derivations from judgment aggregation theorems are often complex, which may discourage social choice theorists to use them despite the powerful generality of these theorems. A basic example is Arrow's theorem, which the canonical theorem permits recovering only in the version singled out by Dietrich and List [6]. To obtain the theorem in full, i.e., with weak preferences instead of strict ones, one way, due to Dokow and Holzman [12], is to derive first Gibbard's oligarchy theorem and then reinforce the assumptions, and another way, due to Dietrich [5], requires one to move to a richer judgment aggregation framework in which "relevance" constraints are put on the formulas of the agenda X. Either way is subtle, but perhaps disappointingly roundabout for social choice theorists. Another, less standard example concerns the generalization of Kasher and Rubinstein's [19] impossibility theorem to more than two categories. This turns out to be a non-trivial problem, and it can be solved using Dokow and Holzman's [13] apparatus of non-binary evaluations (see Maniquet and Mongin [26]). However, this resolution may seem to be exceedingly complex, given that a direct proof can be offered using standard tools in social choice theory (compare with Maniquet and Mongin [27]).

For another thing, the canonical theorem is an impossibility theorem, and so are other results we did not review here, like the theorems on oligarchy that generalize the canonical theorem. Admittedly, all these results fully *characterize* the agenda conditions for impossibility, so that they should not be interpreted only negatively; any failure of the necessary conditions corresponds to a possibility. However, there is no way to infer the precise form of the possibility in question, so these results should clearly be complemented with others, which will directly axiomatize judgment aggregation rules that are neither dictatorial nor oligarchical. The theory has actively followed this direct approach for voting rules, especially majority voting and its refinements [7, 34, 35, 37], but it should be applied more systematically elsewhere. The literature on belief merging may provide heuristic keys, although not every procedure for fusing databases delivers a plausible way of aggregating human judgments, and help resolve the legal and political issues that are at the core of judgment aggregation theory.

References and Recommended Readings

(* means "recommended as introductory material", and ** "recommended as advanced material")

1. Arrow, K. J. (1951). *Social choice and individual values*. New York: Cowles Foundation and Wiley; 2nd ed. (1963).
2. Cariani, F., Pauly, M., & Snyder, J. (2008). Decision framing in judgment aggregation. *Synthese, 163*, 1–24.
3. Dietrich, F. (2006). Judgment aggregation: (Im)possibility theorems. *Journal of Economic Theory, 126*, 286–298.
4. Dietrich, F. (2007). A generalized model of judgment aggregation. *Social Choice and Welfare, 28*, 529–565.
5. Dietrich, F. (2016). Aggregation theory and the relevance of some issues to others. *Journal of Economic Theory, 160*, 463–493.
6. *Dietrich, F., & List, C. (2007). Arrow's theorem in judgment aggregation. *Social Choice and Welfare, 29*, 19–33.
   (A paradigmatic and transparent application of judgment aggregation theory to social choice theory.)
7. Dietrich, F., & List, C. (2007). Judgment aggregation by quota rules: Majority voting generalized. *Journal of Theoretical Politics, 19*, 391–424.
8. Dietrich, F., & List, C. (2008). Judgment aggregation without full rationality. *Social Choice and Welfare, 31*, 15–39.
9. **Dietrich, F., & Mongin, P. (2010). The premiss-based approach to judgment aggregation. *Journal of Economic Theory, 145*, 562–582.
   (The source for the canonical theorem in the third version.)
10. Dokow, E., & Holzman, R. (2009). Aggregation of binary evaluations for truth-functional agendas. *Social Choice and Welfare, 32*, 221–241.
11. **Dokow, E., & Holzman, R. (2010). Aggregation of binary evaluations. *Journal of Economic Theory, 145*, 495–511.
    (The source for the canonical theorem in the second version, as well as an authoritative source for judgment aggregation results in general.)
12. Dokow, E., & Holzman, R. (2010). Aggregation of binary evaluations with abstentions. *Journal of Economic Theory, 145*, 544–561.
13. **Dokow, E., & Holzman, R. (2010). Aggregation of non-binary evaluations. *Advances in Applied Mathematics, 45*, 487–504.
    (An advanced mathematical treatment of judgment aggregation theory, with original applications to assignment problems.)
14. Duddy, C., & Piggins, A. (2013). Many-valued judgment aggregation: Characterizing the possibility/impossibility boundary. *Journal of Economic Theory, 148*, 793–805.
15. *Genest, C., & Zidek, J. V. (1986). Combining probability distributions: A critique and an annotated bibliography. *Statistical Science, 1*, 114–148.
    (A classic introduction to probability aggregation.)
16. Gibbard, A. (1969). Social choice and the Arrow conditions. WP, Department of Philosophy, University of Michigan. Published in *Economics and Philosophy, 30*, 2014, 269–284.

17. *Grossi, D., & Piggozzi, G. (2014). *Judgment aggregation: A primer*, E-book. San Rafael: Morgan & Claypool.
(A reader-friendly introduction to the field.)
18. Herzberg, F. S., & Eckert, D. (2012). The model-theoretic approach to aggregation: Impossibility results for finite and infinite electorates. *Mathematical Social Sciences, 64*, 41–47.
19. *Kasher, A., & Rubinstein, A. (1997). On the question "Who is a *J*?", *Logique et Analyse, 40*, 385–395.
(A reader-friendly work that started the group identification literature.)
20. Kirman, A., & Sondermann, D. (1972). Arrow's theorem, many agents, and invisible dictators. *Journal of Economic Theory, 5*, 267–277.
21. Kornhauser, L. A.,& Sager, L. G. (1993). The one and the many: Adjudication in collegial courts. *California Law Review, 81*, 1–59.
22. Lang, J., Pigozzi, G., Slavkovik, M., & van der Torre, L. (2011). Judgment aggregation rules based on minimization. In *Proceedings of the 13th Conference on the Theoretical Aspects of Rationality and Knowledge* (TARK XIII), ACM, 238–246. Extended version available at http://www.lamsade.dauphine.fr/\char126\relaxlang/papers/tark-extended.pdf.
23. List, C. (2008). Which worlds are possible? A judgment aggregation problem. *Journal of Philosophical Logic, 37*, 57–65.
24. List, C., & Pettit, P. (2002). Aggregating sets of judgments: An impossibility result. *Economics and Philosophy, 18*, 89–110.
25. List, C., & Puppe, C. (2009). Judgment aggregation: A survey. In P. Anand, C. Puppe, & P. Pattanaik (Eds.), *Oxford handbook of rational and social choice*. Oxford, Oxford University Press.
26. Maniquet, F., & Mongin, P. (2014). Judgment aggregation theory can entail new social choice results. HEC Paris Research Paper No. ECO/SCD-2014–1063.
27. Maniquet, F., & Mongin, P. (2016). A theorem on aggregating classifications. *Mathematical Social Sciences, 79*, 6–10.
28. Miller, M. K., & Osherson, D. (2009). Methods for distance-based judgment aggregation. *Social Choice and Welfare, 32*, 575–601.
29. Mongin, P. (2008). Factoring out the impossibility of logical aggregation. *Journal of Economic Theory, 141*, 100–113.
30. *Mongin, P. (2012). The doctrinal paradox, the discursive dilemma, and logical aggregation theory. *Theory and Decision, 73*, 315–345.
(A revised English version of Mongin and Dietrich, 2010. The most detailed review of judgment aggregation theory up to 2010, with the same commitment to logic as in the present chapter.)
31. Mongin, P., & Dietrich, F. (2010). Un bilan interprétatif de la théorie de l'agrégation logique. *Revue d'économie politique, 120*, 929–972.
32. *Nash, J. R. (2003). A context-sensitive voting protocol paradigm for multimembers courts. *Stanford Law Review, 56*, 75–159.
(A useful review of the legal literature on the doctrinal paradox, with an attempt at arbitrating between the case-based and issue-based methods of adjudication.)
33. Nehring, K., & Puppe, C. (2002). Strategy-proof social choice on single-peaked domains: Possibility, impossibility and the space between. WP of the Department of Economics, University of California, Davies.
34. Nehring, K., & Puppe, C. (2007). The structure of strategy-proof social choice. Part I: General characterization and possibility results on median spaces. *Journal of Economic Theory, 135*, 269–305.
35. Nehring, K., & Puppe, C. (2008). Consistent judgement aggregation: The truth-functional case. *Social Choice and Welfare, 31*, 41–57.
36. **Nehring, K., & Puppe, C. (2010). Abstract Arrowian aggregation. *Journal of Economic Theory, 145*, 467–494.
(The source for the canonical theorem in the first version and a convenient entry to Nehring and Puppe's earlier work.)

37. Nehring, K., Pivato, M., & Puppe, C. (2014). The Condorcet set: Majority voting over interconnected propositions. *Journal of Economic Theory, 151*, 268–303.
38. Pauly, M., & van Hees, M. (2006). Logical constraints on judgment aggregation. *Journal of Philosophical Logic, 35*, 569–585.
39. Pettit, P. (2001). Deliberative democracy and the discursive dilemma. *Philosophical Issues, 11*, 268–299.
40. Pigozzi, G. (2006). Belief merging and the discursive dilemma: An argument-based account to paradoxes of judgment aggregation. *Synthese, 152*, 285–298.
41. *Pigozzi, G. (2016). Belief merging and judgment aggregation. Stanford Encyclopedia of Philosophy, Winter 2016 edition. https://plato.stanford.edu/entries/belief-merging/.
(A concise and recent survey of belief merging.)
42. Rubinstein, A., & Fishburn, P. C. (1986). Algebraic aggregation theory. *Journal of Econom Theory, 38*, 63–77.
43. van Hees, M. (2007). The limits of epistemic democracy. *Social Choice and Welfare, 28*, 649–666.
44. Wen, X. (forthcoming). Judgment aggregation in nonmonotonic logic. *Synthese*. https://doi.org/10.1007/s11229-017-1391-2.
45. Wilson, R. (1975). On the theory of aggregation. *Journal of Economic Theory, 10*, 89–99.

Chapter 39
Logical Approaches to Law

John Woods

Abstract On the face of it, we might think that logic and the law were made for each other. Their intellectual identities are grounded in a shared stock of concepts: *argument, proof, evidence, inference, probability, relevance, presumption, precedent* or *analogy, plausibility, reasonability* and *explanation.* Provided that we understand logic broadly enough to include not only mathematical theories of deduction and induction, but also more recent attempts by computer scientists to investigate defeasible and default reasoning, there is not an item on this list that escapes the attention of logicians. If we also take note of brisk developments in dialogue logic and formal argumentation theory, the list of shared concepts enlarges accordingly, including among others, *leading questions* and *cross-examination.*

39.1 Conceptual Commonalities?

On the face of it, we might think that logic and the law were made for each other. Their intellectual identities are grounded in a shared stock of concepts: *argument, proof, evidence, inference, probability, relevance, presumption, precedent* or *analogy, plausibility, reasonability* and *explanation.* Provided that we understand logic broadly enough to include not only mathematical theories of deduction and induction, but also more recent attempts by computer scientists to investigate defeasible and default reasoning, there is not an item on this list that escapes the attention of logicians. If we also take note of brisk developments in dialogue logic and formal argumentation theory,[1] the list of shared concepts enlarges accordingly, including among others, *leading questions* and *cross-examination.*

[1] See, for example, Barth and Krabbe [1].

J. Woods (✉)
University of British Columbia, Vancouver, BC, Canada
e-mail: john.woods@ubc.ca

S. O. Hansson, V. F. Hendricks (eds.), *Introduction to Formal Philosophy*, Springer Undergraduate Texts in Philosophy, https://doi.org/10.1007/978-3-319-77434-3_39

It would not be wrong to say that we have had a golden age in logic, concerning whose beginnings it is an acceptable convenience to mention Frege's *Begriffschrift* of 1879. Logic has had a formal character ever since Aristotle. Golden age logic is formal, but in ways never envisaged by Aristotle. Golden age logic is also mathematical. There are at least three different but compossible ways in which this can be so – one having to do with *motivation,* another having to do with *methods*, and the third having to do with *matter*. Concerning motivation, sometimes a logic is purpose-built to accommodate a philosophical thesis about the foundations of mathematics. Thus classical logic – think here of *Principia Mathematica* – was built to accommodate logicism, that is, the thesis that arithmetic can be re-expressed without relevant loss in a purely logical notation. Intuitionist logic was likewise motivated. It was put together to accommodate mathematical constructivism, according to which only constructive proofs can be allowed.[2]

The second way in which a logic is mathematical is when its characteristic methods are mathematical. For example, the semantics of classical and modal logic, and virtually all the others in the deductive orbit, are set-theoretic through and through, and their meta-theories are structured in ways that permit proofs by mathematical induction. Inductive logics, in turn, are dominantly probabilistic, and the probabilities involved are those studied by the applied mathematics of games of chance.

Logic is mathematical in the third sense by virtue of its matter – that is to say, its subject-matter. Although some logicians draw a firm distinction between logic and mathematics, the majority view – certainly majority practice – is otherwise inclined, especially as relates to set theory. Accordingly, logic itself has come to be regarded as part of pure mathematics, as evidenced by virtually any page of the leading logic journals of the day.

There are those who think that the mathematicization of logic puts the commonality of concepts thesis in serious doubt. Everyone accepts that the intellectually foundational concepts of logic and the law are denoted by a shared lexicon of *names* – "argument", "proof", "evidence", "inference", and so on. But, say the commonality sceptics, if these names did in fact have the same referents, then the intellectually defining concepts of law would be open to mathematical articulation, which many critics think is too much for serious belief. On the contrary, they say, the concepts denoted by these common names are, in their legal contexts, *sui generis.* These reservations receive further support from methodological and procedural differences between the two disciplines. Logic expresses much of its formal character in the precision of its language, and the certainty, rigour and explicitness of its proofs, abetted by such linguistic artificialities and stipulations as may be needed to achieve those ends. It is conspicuously otherwise with the law. The medium in which legal reasoning is transacted is natural, not formal, language. Its proofs never rise to the standards of rigour demanded of logic, and the

[2]Roughly speaking, a constructive proof is one whose purported objects are expressly specifiable.

epistemological character of the law is largely one of implicitness.[3] For example, we will look in vain to the writings of theoretical jurisprudence for a definition that completes the biconditional schema, "A prosecutor's case constitutes proof of guilt beyond a reasonable doubt if and only if . . . .".

Broadly speaking, the *commonality thesis* asserts that the identically named foundational concepts of logic and law are the same concepts, whereas the – as I shall say – *two solitudes thesis* is that they are different concepts with shared names. This leaves the legally-minded logician with a number of options, in which we find varying answers to the question, "Assuming there were one, what would a logic of legal concept K look like?"

1. Accept the commonality thesis, and press on with the application of a received logic to concept K in legal contexts.
2. Accept the two solitudes thesis, and cease and desist. That is, take the position that there is no logic for K, that K is not a logic-worthy concept.
3. Accept the two solitudes thesis, but press on with *adaptations* of an existing logic to the peculiarities of K in legal contexts.
4. Accept the two solitudes thesis and *originate* a formalization better-suited to peculiarities of K in legal contexts.

Here are some examples to consider: Application of the probability calculus to the concept of probability in legal reasoning is sometimes thought to exemplify option one. Of course, for logicians, number two is the null option. Adjusting a plausibility logic[4] to the particular features of legal plausibility could be taken to exercise option three, and producing a built-from-the-ground-up logic for the concept(s) of legal presumption[5] would be an instance of option four.[6]

Excluding the null option, the present three give varying characterizations of the influence of legal concepts on the orthodoxies of formal philosophy. Option one presupposes the availability of an orthodox solution. Option three is a qualified challenge of the orthodox, and option four a repudiation of it.

We should note that these same options are also available to legal theorists, in the reverse direction, so to speak. As we have it now, it would appear that the dominant position of legal theorists is the two-solitudes scepticism of option two, whereas most of the activity of legally-minded logicians hovers in the vicinity of two-solitudes adaptations of option three.

Options three and four offer the would-be formalizer greatest creative potential. Faced with concepts *sui generis* to law, option three allows for the retrofitting of a technical apparatus already in service. Option four goes further. It envisages the prospect of a new logic for the concepts that resist accommodation in even a

---

[3] In this respect, among others, there are notable differences between common law and legal code traditions – between, for example, criminal law in England and France. In the interests of space, I shall confine my remarks to the common law tradition. A highly readable effort to bring to the fore the epistemological orientation of criminal procedures at common law is Laudan [9].

[4] In the manner, say, of Rescher [15].

[5] Although light on the formal side, see for example, Walton [17].

[6] For another variation of option four, see Horty [6]

reconfiguration of a standing formalism. It reflects the idea that concepts peculiar to the law, even when denoted by terms that also name concepts central to orthodox logical theory, require a *sui generis* logic rather than the adaptation of some existing system. This we might call the *sui generis-sui generis* thesis. One of the more interesting open questions of present-day logical theory is whether the logical requirements of legal reasoning are best served by accepting

The sui generis-sui generis thesis: *Sui generis* concepts require *sui generis* logics.

In what follows I shall focus on cases which suggest an affirmative answer to this question.

Throughout logic's golden age, option two has been by far the dominant position. It is said that the first logicians were Greek lawyers. Leibniz (1646–1716) was a lawyer, and Łukasiewicz (1878–1956) too, and Mill (1806–1873) might as well have been one. But the fact remains that, for well over a century, logic and the law have plied their respective trades unemcumbered by the slightest notice of one another. However, in the latter two decades of the twentieth century and into the present one, there have been stirrings of the other two options. In 1980, Chaim Perelman argued, in the manner of option three, for a restricted role for logic in the analysis of legal reasoning.[7] In this same spirit, Horn clause logics of logic programming were applied to the analysis of the British Nationality Act.[8] Important contributions from computer scientists also include Trevor Bench-Capon's [2] survey for an encyclopedia of computer science and technology[9] and Henry Prakken's legal modelling paper of 1997.[10] Also valuable is a recent collection of papers edited by Marilyn MacCrimmon and Peter Tellers, covering a number of approaches – e.g. fuzzy logics, and logics of uncertainty and probability,[11] and a 2003 paper by Prakken and his colleagues in which argument schemes are applied to the notion of legal evidence.[12] A recent volume of note is a book edited by Dov Gabbay and others on legal rationality.[13] Not to be overlooked is the much earlier engagement of legal issues by inductive logicians and probability theorists. An important and contentious contribution was Jonathan Cohen's [3] book,[14] which urged a distinction between two concepts of probability, only one of which Cohen thought was germane to the analysis of probabilistic and evidence-weighing reasoning in legal contexts. This latter he called inductive (or Baconian) probability, and the former aleatory (or Pascalian) probability. (The English word "aleatory" comes from the Latin word for dice-games.) Another earlier influence was deontic logic, part of

[7]Perelman [12].

[8]Sergot et al. [16].

[9]Bench-Capon [2].

[10]Prakken [13].

[11]MacCrimmon and Tellers [11].

[12]Prakken et al. [14].

[13]Gabbay et al. [5].

[14]Cohen [3].

whose motivation was an interest in deontological concepts of legal procedure – obligation, permission, and so on.[15]

When formal methods are applied to a concept, let us say that it constitutes a formalization of it. When speaking of a concept's meaning in pre-formalized linguistic practice, let us say that we are speaking of its intuitive meaning or, equivalently, of the *intuitive concept*. Consider now the question, "What is achieved by the formalization of a concept; for example, what do we learn about proof from a formalization of the concept knowledge?" There are four different answers to this *concept-engagement* question.

- *Analysis* An analysis of an intuitive concept K it makes its intuitive meaning *explicit.*[16]
- *Explication* An explication of an intuitive concept K preserves its pre-formalization meaning but does so in ways that make the intuitive meaning *precise.*
- *Rational reconstruction* A rationalization of a concept K involves the ascription to K of features not present in pre-formalized linguistic usage, but in a way that retains enough of the intuitive concept to make it intelligible to say that the rational construction at hand is a formalization of *it.* Rational reconstructions are semantic *make-overs.*
- *Stipulation* Here the formalization provides a nominal definition of a concept-lacking a prior presence in pre-formalized linguistic practice, while retaining meanings of the name of the original intuitive concept. Stipulations *make up* meanings.

The distinction between analysis and stipulation is roughly Kant's contrast between analysis and synthesis. Analysis, says Kant, is the business of making concepts clear, and synthesis the business of making clear concepts. Analysis is the purview of philosophy and synthesis the province of mathematics. In our schema, explication and rational reconstruction are hybrids, with explication more analysis-like and rational reconstruction trending rather more towards stipulation.[17] It is also important to note that the fourfold concept-engagement space is orthogonal to the fourfold-option space.

There are, of course, grey areas at each of these borders, but here are some quick examples. Some probability theorists think that to the extent that the intuitive meaning of probability resides in how prior probabilities are compounded, that aspect of its meaning is captured analytically by the axioms of the probability calculus. Some mathematicians take the view that the axioms of number theory

[15]The Deon Conferences are a good source. See, for example, http://www.doc.ic.ac.uk.deon02

[16]For lack of space, I pass over two important variations of the definitional notion of formalization, namely, implicit definitions and contextually eliminating definitions. For the first, think of the definition of the concept of number afforded by the axioms of number theory. For the second, think of the reduction of number theory to logic and set theory by contextual elimination.

[17]Kant [7, 8].

offer an explication of the intuitive concept of number. The notion of rational reconstruction we associate with Carnap. Its presence may be felt in Carnap's attempt to formalize physical objects as logical constructions of sense-data. Stipulation is the stock and trade of mathematics, as Kant noted. But it is also solidly at work in all of model-based science. For example, in population biology it is stipulated that populations are infinitely large; in neoclassical economics it is stipulated that utilities are infinitely divisible; in classical belief-change theories it is stipulated that belief is closed under logical consequence; and in rational decision theory it is stipulated that deciders have perfect information. Space doesn't permit further discussion of this fourfold distinction, except to say again that its partitions are not strict. As Quine famously quipped, one person's explication is another's stipulation.

39.2 Rationality

Model-based theories harbour a philosophically crucial distinction. It is the distinction between *descriptively adequate* theories and *normatively binding* theories. It is widely supposed that descriptively adequate theories successfully negotiate the relevant observational checkpoints, whereas normatively binding theories typically do not and need not. Consider, for example, mainstream formal theories of belief-change, in which it is stipulated that an agent proceeds rationally to the extent, among other things, that she closes her beliefs under consequence and, therefore, believes all logical truths. Since no human agent comes in any finite degree anywhere close to meeting these conditions, the theories that embody them are descriptive failures. Even so, it is commonly said that what such theories lack descriptively they more than make up for normatively; for they lay down conditions for the exercise of human reasoning *at its ideal best.*

If this were actually so, it would matter for both the moral and intellectual integrity of the law. Jurors have a duty to perform their functions rationally, including their own transitions from a required state of agnosticism about the accused's guilt to a state of belief capable of sustaining a verdict about it. Any theory of belief-change which, on pain of irrationality, mandates conditions infinitely beyond a juror's reach, triggers a further pair of options. One is that the law is an irrational disgrace. The other is that the orthodox approaches are wrong for the law, which, in turn, puts some (not much discussed) pressure on their normative presumptions.

The analysis of a concept presupposes its intuitive presence in preanalytic practice. An analysis may be said to be *conceptually faithful* to the degree that it preserves the presence of its intuitive analysandum. At the other end of the spectrum, synthesis, or the stipulation of new concepts, places a premium on the clarity that can be got from an inventive *mathematical virtuosity.* Concerning the option-space noted above, I have already mentioned the tendency of logicians to favour something like item three, in which the logical treatment of legal concepts is by way of existing logics reconfigured to accommodate the law's contextual peculiarities. As regards the present concept-engagement possibilities, it would appear that the formalizations effected by retrofitted logics fall oftener than not in the ambit of partial reconceptualizations. It is well to emphasize, however, that option four – the most creative of our four options – will in principle tolerate each of the four grades of conceptual engagement. But here, too, I am inclined to think that when the preferred approach to a legal concept is by way of a new logic, it is advisable to try for a conceptual explication, rather than a rational reconstruction. If the concept is *sui generis* to law, and its meaning is implicit in established legal usage, the first order of business should be as much clarification as the *intuitive* concept will bear. That is, the logic of the law should be careful not to give mathematical virtuosity too much sway over conceptual fidelity. It serves neither the lawyer nor the logician to produce formalizations of concepts of central importance to law that no lawyer could recognize as such without a crash course in a department of mathematics or computer science or technical philosophy. Similarly, in reaching their verdicts, the law presupposes (approvingly) that jurors have untutored access to these concepts, hence apply them in their intuitive senses.

39.3 Probability in the Law

It is widely agreed that a verdict of guilty in a criminal proceeding at common law is both unjust and epistemically untenable if, on the evidence presented at trial, the probability of guilt is insufficiently high. Thus high probability of guilt on the evidence heard at trial is, on this view, a necessary though not sufficient condition of a correctly arrived at decision to convict. One of the quite standard ways in which logicians seek to interact with the law is to apply the calculus of probability to this notion of probability. Seen this way, when a juror reaches his decision he calculates the probability of the accused's guilt based on the trial's evidence, and in so doing conforms his reasoning, albeit tacitly, to the conditions mandated by Bayes' Theorem, a standard formulation of which is:

$$P\,(G/E) = \frac{P\,(E/G) \times P\,(E/G)}{[P(G) \times P\,(E/G) + (\,P\,(\sim G) \times P\,(E/\sim G)\,]}$$

Here '*G*' denotes guilt, '*E*' evidence, and '*P*' probability. Accordingly, we may read the present instance of the theorem as asserting that the probability of guilt on the evidence presented (*P*(*G*/*E*)) is the prior probability that the accused is guilty (*P*(*G*)) *times* the likelihood accorded to the evidence by the hypothesis of guilt (*P*(*E*/*G*)) DIVIDED BY the prior probability of guilt (*P*(*G*)) *times* the guilt of the accused

(*P*(*E*/*G*)) *plus* the prior possibility of innocence (P(G)) *times* the likelihood of the evidence given the innocence of the accused (*P*(*E*/ *G*)).[18]

Upon reflection, it appears that implementation of Bayes' theorem is precluded by a juror's other duties, of which none is more important than the presumption of innocence. What this means in concrete terms is that in determining whether to convict the accused, at no time in this process can the hypothesis of guilt be given any probative consideration. In particular, in trying to make up their minds about a given piece of testimony, jurors may not use the hypothesis of guilt to determine or recompute the *credibility* of the evidence they have heard. Accordingly, the presumption of innocence precludes jurors from binding their reasoning to the *P*(*E*/*G*)-clause of the theorem. This leaves us with the following *upshot-space*:

- What legal procedure requires is a violation of Bayesian rectitude, and yet jurors on the whole manage to do what the law requires. In which case, jurors are probabilistic misfits.
- What legal procedure requires is a violation of Bayesian rectitude, and (if only tacitly) jurors manage to honour the Bayesian requirements. In which case, jurors are legal misfits.
- What legal procedure requires has nothing to do with what Bayes' theorem requires. In legal contexts, determining *P*(*G*/*E*) is not a Bayesian enterprise.

The present example lays down a valuable restriction on how to proceed. To the extent that *any* concept of probability is implicated in the determination of guilt or innocence, do not give it a Bayesian formalization. This is not a trivial conclusion. The probability calculus is a triumph of mathematical virtuosity. To date no mature and settled formalization of a non-Bayesian notion of legal probability has taken hold. Repairing this omission offers to logicians of the law the prospect of gainful employment.

39.4 Proof in the Law

At the criminal bar, conviction requires evidence that proves guilt beyond a reasonable doubt. Evidence presented at trial is entirely by way of testimony, of what witnesses say under oath. A witness can testify to a matter of fact or, if sworn as an expert, to a matter of opinion. Triers of fact (typically a jury) and triers of law (always a judge) are held to a strict duty of agnosticism both prior to the trial and at each of its phases, until the cessation of testimony, the presentation of closing arguments of counsel and the instruction of the jury by the judge. Neither do the triers have any independent means of confirming or disconfirming the propositions avowed by witnesses. Still less is the general epistemological question of the probativity of information gathered in this way an arguable matter in criminal

[18]It would be well to note that Bayes' theorem is not a definition of conditional probability. If it were, it would be viciously circular. It is in fact a rule for calculating a large class of conditional probabilities.

proceedings. The law's implicit epistemology makes it a non-negotiable assumption that sayso can be a reliable generator of proof beyond a reasonable doubt.

This creates an obvious problem for the criminal proof standard, especially when witnesses give conflicting testimony and the evidence is circumstantial in any significant degree. Is there a formal epistemologist who would allow as a *general* proposition that contradictory circumstantial testimony meets any standard of proof that he would antecendently have recognized? If the answer is No, the logician now has two more options to consider. One is to reject the law's concept of proof as epistemologically untenable. Another is to concede its epistemological legitimacy, and seek for the adaptation of an existing logic for its formalization or a purpose-built logic for it.

It is typical of cases in which defendents enter a plea of not guilty that witnesses will give conflicting testimony, thus confronting the juror with a critical pair of evidential duties to perform (or so it would appear). He must try to find a maximal consistent subset of the evidence that is most worthy of his belief. He must also determine whether that subset meets the requirement of proof of guilt beyond a reasonable doubt. An utterly natural question for a logician is, "What are the criteria for the correct performance of these tasks?" It is striking that jurisprudence does not answer this question. Indeed, it hardly even formulates it. This is not to say that the law gives no instruction on how to perform these duties. But what it doesn't do is specify those *criteria* in whose fulfillment dutiful compliance consists. Although there are occasional exceptions, the standard instruction to jurors from the bench runs along the following lines (again simplified):

- *If you believe witness W in regard to matter M, you must convict. If you believe witness W′ with respect to matter M, you must acquit. In regard to the rest of the testimony, if you find that what you believe of it merits conviction, then you must convict. If not, you must acquit.*

Concerning how to go about determining whether to believe a witness, the instruction is:

- *Pay attention. Try to understand the witness. Do not prejudice the issue or rush to judgement. Use your common sense.*

The last incorporates the venerable paradigm of the reasonable man (sic):

- *Form your judgement in the manner of the reasonable man, that is, by reasoning in the way of ordinary persons about ordinary things.*

It has long been recognized that jury deliberation is an exercise in practical, not theoretical, reasoning. The doctrine of the reasonable person carries an important suggestion about the logic of practical reasoning:

- *If there are criteria for the goodness of practical reasoning, their fulfillment in actual practice is inadvertent.*

All of this is crucial for a correct understanding of the epistemic status of the proof standard.

- *The criminal proof standard is not particularly high, and is attainable without tutelage by any reasonable layperson reasoning in the ordinary way of things.*

It would not be going too far to say that here is a concept of proof-determination that cuts sharply against the grain of orthodox epistemologies. In so doing, it raises obvious questions: Are there existing formal orthodoxies that can do the job for this notion of proof and of the decision consequent upon its positive application? It is notable that rational choice theory won't do. Its concept of the rational decider is one who seeks to maximize personal advantage. But the law's concept of the rational person is the intuitive concept: a rational agent is thoughtful and clear-headed, and by no means always selfish. Any theory of human performance that ties rationality to the pursuit of personal utilities are broadly utilitarian in character. But a juror's world is a deontological world in which preference is suppressed in favour of duty. So it is natural to wonder whether there is a mathematically virtuosic formal epistemology that could be contrived for this concept of proof. We may think that, to date, the most promising candidates are to be found in the proliferating literatures of defeasible, default and abductive reasoning. But even here, it is early days for definitive application to the law.[19]

As we begin to see, the criminal proof standard is less a matter of proof than of intellectually conscientious belief. If we used Woods [19] as our guide, a competent judge would instruct the jurors as follows:

- "With due regard for the instructions I have given you so far, and mindful of your duty to pay close attention to everything you've seen and heard at trial open-mindedly and without bias, if you find that you cannot in all intellectual conscientiousness convict the accused on that basis, then you must acquit. Equally, if you find that in all intellectual conscientiousness you cannot acquit the accused on that basis, then you must convict. Period."

It bears repeating that the criminal law's concept of proof is *sui generis.* It is not proof in the mathematical sense, and it is not proof in the information-theoretic sense. In a recent interview with Athanasios Christacopulous John Corcoran offers wise counsel to anyone eager to dive into the choppy waters of criminal proof:

> Before wrapping up my answer, I would like to remind myself and your readers of the points I have made several times. First, belief can be an obstacle to a proof because one of the marks of proof is its ability to resolve doubt. Second, we don't usually try to prove propositions we don't believe or at least suspect to be true. Third, the attempt to find a proof of something leads to doubts we never would have had. If you have a treasured belief you would hate to be without, do not try to prove it.[20]

Mind you, Corcoran is not speaking here of proof of criminal guilt, but his remarks serve to reinforce its *sui generis* conceptual character.

---

[19]See Woods [18] and [19], chapter 21. For a more thorough-going discussion, see also Woods [20], especially parts E and F.

[20]Corcoran and Christacopulous [4].

39.5 Inconsistency

It is typical that accusations of serious crimes are tried by juries, normally twelve in number. It is also typical that decisions to acquit or convict be unanimous, achieved by a bi-modal vote.[21] A jury's verdict is the product of twelve individual findings. Provided that they remain attentive and wide-awake, all twelve hear the same testimony. In what can only be regarded as a lexical misfortune, testimony is also referred to as evidence. In inductive settings, it is natural to suppose that propositions are evidential only if they are true, and there is no shortage of formal acknowledgement of this connection. But it is not a tenable connection for testimony, hence not a connection for evidence in the law's testimonial sense. This vitiates a suggestion of the previous section, namely, that the evidence on which a juror must base his finding is a maximal consistent subset of the total evidence heard. Since such subsets can contain elements which a juror disbelieves, the condition must be reformulated as calling for maximal consistent subsets of *believed* testimony.

The requirement that jurors not act on inconsistent evidence may strike one as a reasonable ideal for individual triers of fact, but it is not a condition that stands any realistic chance of realization in the aggregate. It is in principle possible that each juror bases his finding on different subsets of the total evidence, not all of them pairwise compatible. This is deeply consequential. It shows that the verdict in a criminal trial is not a unanimous finding on some aggregation of the total evidence, but rather is the sum of individual findings, each predicated on its own readings of the evidence.

Philosophical intuitions tug here in different directions. On the one hand, it is awkward that an accused can be opportuned by a verdict made by people who collectively give an inconsistent reading of the evidence. On the other hand, it might strike us as epistemically and morally promising that different, though non-compossible, readings of the evidence should lead twelve times out of twelve to the same finding across the board. As we have it now, there is no settled formal theory of inconsistency-management or collective agency that confronts this issue directly. Notwithstanding, it is possible to see in outline assumptions that a purpose-built logic should try to preserve and clarify. For simplicity, consider the extreme case in which each juror's consistent subset of the total evidence is incompatible with every other. We may suppose that the finding of an individual *juror* is both rational and legally permissible if and only if, consistent with the judge's instructions,

- *It is arrived at from a reading of the total testimony which a reasonable person, proceeding in the ordinary way about ordinary things, might reasonably have given;*

and

[21] In Scottish jurisprudence, a third option is allowed – "not proven".

- *The reasoning from his reading to his finding is that of a reasonable person reasoning in the ordinary way about ordinary things.*

We may now put it that a *jury's* verdict is both rational and legally permissible to the extent that the individual jurors' findings, each rational and legally permissible, in the sense at hand, sum to 12. For these conditions to be met, it is neither required nor desirable that individual readings – and individual findings – be aggregated. Equally it is neither required nor desirable that the jury's verdict be represented as the negotiated settlement of the jurors' competing theories of the case. A jury's verdict is not, therefore, the solution of a conflict resolution exercise.[22] In so saying, a number of theoretical paradigms – ranging from game theory in all its principal forms to voting preference theory – seem wrong for jury deliberation.

39.6 A Concluding Remark

As noted at the beginning, the law brims with concepts of considerable interest to logic. This presents the logician with a quite general methodological question. Is it best to press existing methods of formal representation into service – with or without some tweaking as needs be – or would these epistemologically laden concepts be better handled by formalizations purpose-built for them? There are advantages and tensions either way. Staying with the tried-and-true has the attractions of the familiar dab-hand, possibly at the expense of conceptual distortion. Purpose-built logics can be expected to do better on the score of conceptual fidelity, but for the most part they are not yet well-developed and are certainly not tried-and-true. So that is a cost. It is significant that these approaches do not exclude one another. It is perfectly possible for the logic of law to operate on each of these tracks, in the spirit of "let's see what works best." The law is a waiting feast for logicians. We should greet it with an experimental openness appropriate to its promise.

References and Recommended Readings[23]

1. Barth, E. M., & Krabbe, E. W. C. (1985). *From axiom to dialogue*. Berlin/New York: De Gruyter.
2. Bench-Capon, T. (1995). Knowledge based systems in the legal domain. In A. Kent & J. G. Williams (Eds.), *Encycopedia of computer science and technology* (pp. 163–186). New York: Dekker.

[22]Of course, in actual practice, jurors may argue with one another over their differing readings of the testimony, and may at times effect some reduction of the difference. But this is not intrinsic to the juror's function. Their task is to determine whether their findings agree, not necessarily their readings.

[23]Asterisks (*) indicate recommended readings.

3. * Cohen, J. (1977). *The provable and the probable*. Oxford: Clarendon Press. [A classic protest against over-use of the probability calculus in the analysis of legal reasoning and other contexts.]
4. Corcoran, J. & Christacopulous, A.. (2017). Interview with John Corcoran. *Academia.edu*, posted 10/03/2017.
5. Gabbay, D. M., Canivez, P., Rahman, S., & Thiercelin, A. (Eds.). (2010). *Approaches to legal rationality*. Dordrecht: Springer.
6. Horty, J. F. (2016). Norm change in the common law. In S. O. Hansson (Ed.), *David Makinson: Classical methods for nonclassical problems* (pp. 335–355). Berlin: Springer.
7. Kant, I. (1974a). *Inquiry concerning the distinctness of principles of natural theology and morality*. Indianapolis: Bobbs-Merrill. First published in 1764.
8. Kant, I. (1974b). *Logic*. Indianapolis: Bobbs-Merrill. First published in 1800.
9. * Laudan, L. (2006). *Truth, error and criminal law: An essay in legal epistemology*. Cambridge: Cambridge University Press. [Perhaps the earliest treatment of the epistemology implicit in the procedures of common law].
10. * Laudan, L.(2016). *The law's flaws: Rethinking trials and errors*, Volume 3 of the Law and Society series. London: College Publications. [A highly recommended follow-up to Laudan (2006).]
11. MacCrimmon, M., & Tellers, P. (2002). *The dynamics of judicial proof*. Heidelberg: Physica-Verlag.
12. * Perelman, C. (1980). *Justice, law and argument*. Dordrecht: Reidel. [A golden oldie.]
13. * Prakken, H. (1997). *Logical tools for modelling legal argument*. Dordrecht: Kluwer. [A good example of how computer models and dialogue-games and other such devices can try to shed light on the structure of legal thinking].
14. Prakken, H., Reed, C., & Walton, D. (2003). Argumentation schemes and generalization in reasoning about evidence. *Proceedings of ICAIL-03*.
15. * Rescher, N. (1976). *Plausible Reasoning*. Assen: Van Gorcum. [Essential reading, but problematic]
16. Sergot, M. J., Cory, T., Hammond, P., Kowalski, R., Kriwarczek, F., & Sadri, F. (1986). *British Nationality Act* (Vol. *29*, pp. 370–386). *Communications of the ACM*.
17. Walton, D. (1996). *Argumentation schemes for presumptive reasoning*. Mahwah: Lawrence Erlbaum.
18. Woods, J. (2010). Abduction and proof: A criminal paradox. In Gabbay et al. (Eds.), *Approaches to Legal Rationality* (pp. 217–238). Dordrecht, Springer.
19. * Woods, J. (2015). *Is legal reasoning irrational? An introduction to the epistemology of law*, volume 2 of the Law and Society series. London: College Publications. London: College Publications, Second edition in 2018 [The first university-level textbook in English on this subject.]
20. Woods, J. (2018). What strategicians might learn from the common law. *The IfCoLoG Journal of Logics and Their Applications*, forthcoming.